T0189899

Lecture Notes in Computer Science 13914

The series Lecture Notes in Computer Science (LNCS), including its subseries Lecture Notes in Artificial Intelligence (LNAI) and Lecture Notes in Bioinformatics (LNBI), has established itself as a medium for the publication of new developments in computer science and information technology research, teaching, and education.

LNCS enjoys close cooperation with the computer science R & D community, the series counts many renowned academics among its volume editors and paper authors, and collaborates with prestigious societies. Its mission is to serve this international community by providing an invaluable service, mainly focused on the publication of conference and workshop proceedings and postproceedings. LNCS commenced publication in 1973.

Shlomi Dolev · Ehud Gudes · Pascal Paillier
Editors

Cyber Security, Cryptology, and Machine Learning

7th International Symposium, CSCML 2023
Be'er Sheva, Israel, June 29–30, 2023
Proceedings

Editors
Shlomi Dolev
Ben-Gurion University of the Negev
Be'er Sheva, Israel

Ehud Gudes
Ben-Gurion University of the Negev
Be'er Sheva, Israel

Pascal Paillier
Zama
Meythet, France

ISSN 0302-9743 ISSN 1611-3349 (electronic)
Lecture Notes in Computer Science
ISBN 978-3-031-34670-5 ISBN 978-3-031-34671-2 (eBook)
https://doi.org/10.1007/978-3-031-34671-2

This Springer imprint is published by the registered company Springer Nature Switzerland AG
The registered company address is: Gewerbestrasse 11, 6330 Cham, Switzerland

Preface

CSCML, the International Symposium on Cyber Security, Cryptology and Machine Learning, is an international forum for researchers, entrepreneurs, and practitioners in the theory, design, analysis, implementation, or application of cyber security, cryptology, and machine learning systems and networks and, in particular, for conceptually innovative topics in these research areas. Information technology has become crucial to our everyday lives, an indispensable infrastructure of our society, and therefore a target for attacks by malicious parties. Cyber security is one of the most important fields of research these days because of these developments. Two of the (sometimes competing) fields of research, cryptography and machine learning, are the most important building blocks of cyber security.

Topics of interest for CSCML include: cyber security design; secure software development methodologies; formal methods, semantics, and verification of secure systems; fault tolerance, reliability, and availability of distributed secure systems; game-theoretic approaches to secure computing; automatic recovery self-stabilizing and self-organizing systems; communication, authentication, and identification security; cyber security for mobile systems and the Internet of Things; cyber security of corporations; security and privacy for cloud, edge, and fog computing; cryptocurrency; blockchain; cryptography; cryptographic implementation analysis and construction; secure multi-party computation; zero knowledge proofs; privacy-enhancing technologies and anonymity; post-quantum cryptology and security; machine learning and big data; anomaly detection and malware identification; business intelligence and security; digital forensics; digital rights management; trust management and reputation systems; and information retrieval, risk analysis, and DoS.

The 7th CSCML took place during June 29–30, 2023, Online, hosted by Ben Gurion University of the Negev, Beer-Sheva, Israel. The keynote speakers were Dorit Dor, Chief Technology Officer at Check Point Software Technologies; Aharon Aharon, Former CEO Israel Innovation Authority, Apple VP of Hardware Technologies, VP and GM of Apple Israel; David Simchi-Levi, Professor of Engineering Systems at MIT and head of the MIT Data Science Lab; Shlomit Wagman, Harvard University, Former Chair of Israel Anti-Money Laundering Authority; and Eli Ben-Sasson, Co-Founder and President of StarkWare, and Chairman of its Board of Directors.

The conference was organized in cooperation with the International Association for Cryptologic Research (IACR), and selected papers will appear in a dedicated special issue of the Journal of Cryptography and Communications.

This volume contains 21 contributions selected by the Program Committee from 70 submissions, and also includes 15 short papers, of at most 11 pages each. All submitted papers were read and evaluated in an open peer-review process in which, on average, submissions received three reviews by members of the Program Committee assisted by external reviewers. We thank the members of the Program Committee for all their hard work.

We gratefully acknowledge the support of IBM and Ben-Gurion University of the Negev (BGU), in particular BGU-NHSA, the BGU Lynne and William Frankel Center for Computer Science, and the BGU Department of Computer Science.

June 2023

Shlomi Dolev
Ehud Gudes
Pascal Paillier

Organization

Founding Steering Committee

Orna Berry	Google Cloud, Israel
Shlomi Dolev (Chair)	Ben-Gurion University of the Negev, Israel
Yuval Elovici	Ben-Gurion University of the Negev, Israel
Bezalel Gavish	Southern Methodist University, USA
Ehud Gudes	Ben-Gurion University of the Negev, Israel
Jonathan Katz	University of Maryland, USA
Rafail Ostrovsky	University of California, Los Angeles, USA
Jeffrey D. Ullman	Stanford University, USA
Kalyan Veeramachaneni	Massachusetts Institute of Technology, USA
Yaron Wolfsthal	IBM, Israel
Moti Yung	Columbia University and Google, USA

Organizing Committee

General Chair

Shlomi Dolev	Ben-Gurion University of the Negev, Israel

Program Chairs

Ehud Gudes	Ben-Gurion University of the Negev, Israel
Pascal Paillier	Zama, France

Organization Chair

Rosemary Franklin	Ben-Gurion University of the Negev, Israel

Program Committee

Vijay Atluri	Rutgers University, USA
Shivam Bhasin	Nanyang Technological University, Singapore
Chiranjib Bhattacharyya	Indian Institute of Science, Bengaluru, India

Carlo Blundo	Universitá degli Studi di Salerno, Italy
Florian Bourse	Independent Researcher, France
Anat Bremler-Barr	IDC, Israel
Yang Cao	Kyoto University, Japan
Claude Carlet	University of Paris 8, France and University of Bergen, Norway
Jean-Sebastien Coron	University of Luxembourg, Luxembourg
Jintai Ding	University of Cincinnati, USA
Nir Drucker	IBM Research, Israel
Cécile Dumas	CEA-Leti, Université Grenoble Alpes, France
Orr Dunkelman	University of Haifa, Israel
Pierre-Alain Fouque	Rennes University, France
Nurit Gal-Oz	Sapir Academic College, Israel
Louis Goubin	University of Versailles, France
Shay Gueron	University of Haifa, Israel and Amazon, USA
Helena Handschuh	Rambus Inc., USA
James Joshi	NSF, USA
Marc Joye	Zama, France
Sokratis Katsikas	NTNU, Norway
Ram Krishnan	University of Texas at San Antonio, USA
Kwok Yan Lam	Nanyang Technological University, Singapore
Mark Last	Ben-Gurion Unviersity of the Negev, Israel
Shujun Li	University of Kent, UK
Wenjuan Li	Hong Kong Polytechnic University, China
Subhamoy Maitra	Indian Statistical Institute, India
Maryam Majedi	University of Southern California, USA
Brad Malin	Vanderbilt University, USA
Oded Margalit	Ben-Gurion University and Citi Lab, Israel
Weizhi Meng	Technical University of Denmark, Denmark
Debdeep Mukhopadhyay	Indian Institute of Technology, Kharagpur, India
Martin Olivier	University of Pretoria, South Africa
Thomas Peyrin	Nanyang Technological University, Singapore
Rami Puzis	Ben-Gurion University, Israel
Vincent Rijmen	KU Leuven, Belgium and University of Bergen, Norway
Palash Sarkar	Indian Statistical Institute, India
Berry Schoenmakers	Eindhoven University of Technology, The Netherlands
Gil Segev	Hebrew University of Jerusalem, Israel
Sandeep Shukla	Indian Institute of Technology Kanpur, India
Renaud Sirdey	Commissariat à l'Energie Atomique, France
Paul Spirakis	University of Liverpool, UK

External Reviewers

Sponsors

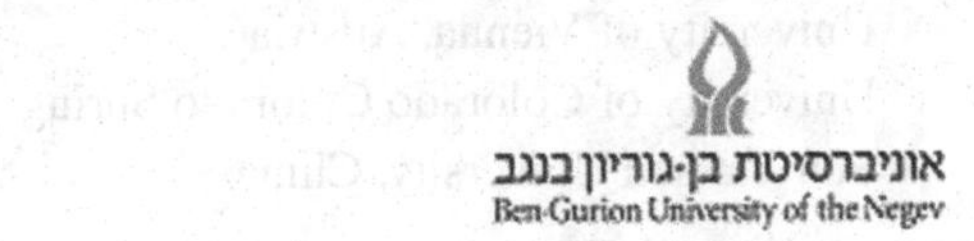

In cooperation with

Contents

Localhost Detour from Public to Private Networks 1
Yehuda Afek, Anat Bremler-Barr, Dor Israeli, and Alon Noy

Pseudo-random Walk on Ideals: Practical Speed-Up in Relation Collection for Class Group Computation 18
Madhurima Mukhopadhyay and Palash Sarkar

Efficient Extended GCD and Class Groups from Secure Integer Arithmetic 32
Berry Schoenmakers and Toon Segers

On Distributed Randomness Generation in Blockchains 49
Ilan Komargodski and Yoav Tamir

Efficient Skip Connections Realization for Secure Inference on Encrypted Data 65
Nir Drucker and Itamar Zimerman

Single Instance Self–masking via Permutations (Preliminary Version) 74
Asaf Cohen, Paweł Cyprys, and Shlomi Dolev

A Fusion-Based Framework for Unsupervised Single Image Super-Resolution 85
Divya Mishra, Itai Dror, Ofer Hadar, Daniel Choukroun, Shimrit Maman, and Dan G. Blumberg

Generating One-Hot Maps Under Encryption 96
Ehud Aharoni, Nir Drucker, Eyal Kushnir, Ramy Masalha, and Hayim Shaul

Building Blocks for LSTM Homomorphic Evaluation with TFHE 117
Daphné Trama, Pierre-Emmanuel Clet, Aymen Boudguiga, and Renaud Sirdey

CANdito: Improving Payload-Based Detection of Attacks on Controller Area Networks 135
Stefano Longari, Carlo Alberto Pozzoli, Alessandro Nichelini, Michele Carminati, and Stefano Zanero

Using Machine Learning Models for Earthquake Magnitude Prediction in California, Japan, and Israel 151
Deborah Novick and Mark Last

A Bag of Tokens Neural Network to Predict Webpage Age 170
Klaas Meinke, Tamis Achilles van der Laan, Tiberiu Iancu, and Ceyhun Cakir

Correlations Between (Nonlinear) Combiners of Input and Output of Random Functions and Permutations 181
Subhabrata Samajder and Palash Sarkar

PPAuth: A Privacy-Preserving Framework for Authentication of Digital Image 188
Riyanka Jena, Priyanka Singh, and Manoranjan Mohanty

Robust Group Testing-Based Multiple-Access Protocol for Massive MIMO 200
George Vershinin, Asaf Cohen, and Omer Gurewitz

The Use of Performance-Counters to Perform Side-Channel Attacks 216
Ron Segev and Avi Mendelson

HAMLET: A Transformer Based Approach for Money Laundering Detection 234
Maria Paola Tatulli, Tommaso Paladini, Mario D'Onghia, Michele Carminati, and Stefano Zanero

Hollow-Pass: A Dual-View Pattern Password Against Shoulder-Surfing Attacks 251
Jiayi Tan and Dipti Kapoor Sarmah

Practical Improvements on BKZ Algorithm 273
Ziyu Zhao and Jintai Ding

Enhancing Ransomware Classification with Multi-stage Feature Selection and Data Imbalance Correction 285
Faithful Chiagoziem Onwuegbuche, Anca Delia Jurcut, and Liliana Pasquale

A Desynchronization-Based Countermeasure Against Side-Channel Analysis of Neural Networks 296
Jakub Breier, Dirmanto Jap, Xiaolu Hou, and Shivam Bhasin

New Approach for Sine and Cosine in Secure Fixed-Point Arithmetic 307
Stan Korzilius and Berry Schoenmakers

How Hardened is Your Hardware? Guiding ChatGPT to Generate Secure Hardware Resistant to CWEs 320
Madhav Nair, Rajat Sadhukhan, and Debdeep Mukhopadhyay

Evaluating the Robustness of Automotive Intrusion Detection Systems Against Evasion Attacks 337
Stefano Longari, Francesco Noseda, Michele Carminati, and Stefano Zanero

On Adaptively Secure Prefix Encryption Under LWE 353
Giorgos Zirdelis

SigML: Supervised Log Anomaly with Fully Homomorphic Encryption 372
Devharsh Trivedi, Aymen Boudguiga, and Nikos Triandopoulos

HBSS: (Simple) Hash-Based Stateless Signatures – Hash All the Way to the Rescue! (Preliminary Version) 389
Shlomi Dolev, Avraam Yagudaev, and Moti Yung

Improving Performance in Space-Hard Algorithms 398
Hatice Kübra Güner, Ceyda Mangır, and Oğuz Yayla

A Survey of Security Challenges in Automatic Identification System (AIS) Protocol 411
Silvie Levy, Ehud Gudes, and Danny Hendler

A New Interpretation for the GHASH Authenticator of AES-GCM 424
Shay Gueron

Fast ORAM with Server-Aided Preprocessing and Pragmatic Privacy-Efficiency Trade-Off 439
Vladimir Kolesnikov, Stanislav Peceny, Ni Trieu, and Xiao Wang

Improving Physical Layer Security of Ground Stations Against GEO Satellite Spoofing Attacks 458
Rajnish Kumar and Shlomi Arnon

Midgame Attacks and Defense Against Them 471
Donghoon Chang and Moti Yung

Deep Neural Networks for Encrypted Inference with TFHE 493
Andrei Stoian, Jordan Frery, Roman Bredehoft, Luis Montero, Celia Kherfallah, and Benoit Chevallier-Mames

On the Existence of Highly Organized Communities in Networks of Locally Interacting Agents 501
Vasiliki Liagkou, Panagiotis E. Nastou, Paul Spirakis, and Yannis C. Stamatiou

Patch or Exploit? NVD Assisted Classification of Vulnerability-Related GitHub Pages 511
Lucas Miranda, Cainã Figueiredo, Daniel Sadoc Menasché, and Anton Kocheturov

Author Index 523

Localhost Detour from Public to Private Networks

Yehuda Afek[1], Anat Bremler-Barr[1,2(✉)], Dor Israeli[1], and Alon Noy[1]

[1] Tel Aviv University, Tel Aviv-Yafo, Israel
afek@post.tau.ac.il, bremler@idc.ac.il, dorisraeli@mail.tau.ac.il
[2] Reichman University, Herzliya, Israel

Abstract. This paper presents a new localhost browser based vulnerability and corresponding attack that opens the door to new attacks on private networks and local devices. We show that this new vulnerability may put hundreds of millions of internet users and their IoT devices at risk. Following the attack presentation, we suggest three new protection mechanisms to mitigate this vulnerability. This new attack bypasses recently suggested protection mechanisms designed to stop browser-based attacks on private devices and local applications [18,20].

Keywords: Browser Based Attack · Private Network · Localhost · IoT

1 Introduction

Internet web pages work by instructing the user's browser to issue requests and execute code coming from many different sources that are not necessarily the original source (origin) of the page being rendered. This behavior is vital in order to allow rich and modular web pages. For example, a news website needs to embed images from a photos website, and a store website advertises using Google Ads and exchanges data with Google's ads servers. Attackers have been abusing the above capability to launch *browser-based attacks* on internet users by embedding malicious requests in a website the victim is tricked to visit. Several mitigation mechanisms were introduced to mitigate different variants of this type of abuse, e.g., Same Origin Policy (SOP) [32], Cross-Origin Resource Sharing (CORS) [31], Same Site Cookies [33] and the recently proposed Private Network Access [20]. In this paper we first present a new and sophisticated browser-based attack scheme that circumvents the Private Network Access mitigation mechanism, and secondly, suggest three new mitigation mechanisms that protect against the new attack.

In the early days of the Internet, an attacker could, for example, steal a victim's bank password by simply tricking the victim to visit the attacker's malicious website. In the malicious website, the attacker embeds a Javascript code that makes the user's browser requesting the bank's web page with the user credentials. The browser then renders the bank page, and the attacker code

S. Dolev et al. (Eds.): CSCML 2023, LNCS 13914, pp. 1–17, 2023.
https://doi.org/10.1007/978-3-031-34671-2_1

can simply read the content of the rendered web page and retrieve the victim's credentials and secretive information.

To counter attacks like the above, most browsers have integrated the Same Origin Policy (SOP) [32] mitigation mechanism. SOP blocks the attacker's website from reading content arriving from an origin (e.g., http://bank.com) that is different from the initiating origin (e.g., http://attacker.com).

However, SOP does not mitigate all browser based attacks. It addresses attacks from one origin against a different origin, but does not completely prevent malicious requests on a target server, as it prevents a different origin *reading* the responses of such requests, unless specifically permitted by the destination origin's server (using a relaxation mechanism called Cross Origin Resource Sharing [31]). However SOP does not prevent the request from reaching the target, and variants of browser based attacks still persist, like DNS Rebinding attacks [22].

Newer browser based attacks on local applications and local devices have been discovered and noticed by the security community [1,17,18,20], enabling the penetration and attack of many IoT devices [21,23] and PCs [27,28]. An example attack against a local application is a vulnerability in Trend Micro's software [1]. The software installed a local web server on users' PCs, and it contained a vulnerability in which if the user visits a specific link (for example `http://localhost:4321/api?cmd`) on her browser, her PC is compromised (executing the `cmd` command on her PC, where `cmd` is any batch command, and 4321 is the Trend Micro local server port number). Therefore, attackers could embed such a link in their malicious website in order to compromise and penetrate Trend Micro's users' PCs. Again, SOP does not prevent this attack even though the origins differ, because the attacker does not need to read any information from localhost in order to compromise the user's PC.

Several mitigations have been recently proposed to prevent these attacks on private end points, i.e., "Private Network Access" (PNA) [20] by Mike West and Titouan Rigoudy of Chromium, and "Internal Network Policy" (INP) [18]. These mitigations extend the Cross Origin Resource Sharing (CORS) [31] mechanism in order to block by default such requests that are initiated from public origin web servers and are destined to private origins (as defined by IANA [24,25], e.g., http://10.0.0.6), and only allow the request if the destination server opts in. West et. al. even propose treating IoT devices more suspiciously, and block by default requests from private origins to local origins (e.g., to http://127.0.0.1:8888). We note that PNA successfully raises the bar against basic attacks in which the malicious website requests private and local resources directly. We have disclosed our finding with the PNA team [19] that have adopted one of the suggested mitigations to future versions of PNA.

In this paper we discover a three step attack against private networks, bypassing the recent mitigations, and demonstrating that the localhost may pose a greater threat to Internet users than previously thought. The key of the presented attacks is using a localhost web server as a stepping stone into the private network. Thus showing that localhost web servers are a special case of private network servers, and as suggested by [20] should be treated differently than IoT devices or other private network assets, due to several unique attributes of localhost servers that we point out and demonstrate in here. We analyzed several popular applications that install a local web server and found out that the total number of users with such applications may reach a billion (e.g., Dropbox, Steam, etc.). The severity, ease of exploitation, and wide availability of such local softwares, make this new attack a serious threat.

The basic attack has several variants. An elaborated variant is depicted in Fig. 1 and 2 which is a triple step detour reaching into a private network device using a localhost as a stepping stone. Simpler variants maliciously penetrate into the user PC to steal resources or make any other damage. The basic requirement is a localhost web server (e.g., http://localhost:1234) listening on the victim's PC with a corresponding public web site (e.g., https://www.allowed.com) that following [20], is opted-in and is allowed to make cross site requests to the local web server. In the first step of any of the variants, the victim is tricked to visit the attacker's malicious website (e.g., https://evil.com) which embeds an iframe to the allowed server which in turn is tricked the user browser to make a request to the corresponding local web server within the iframe. The last request is approved by the preflight exchange because the local web server permits accesses from https://allowed.com origin (the opt-in part). This special request triggers a chosen vulnerability in the local web server, bypassing the Private Network Access mitigation. The attack continues causing the local web server to issue a request from the user browser to an IoT device in the private network. We demonstrate such an attack using a vulnerability we found in a local web server and disclosed privately to its product security team, which has patched the vulnerabilities.

In the second part of the paper we propose three independent simple mitigation mechanisms against each step of the elaborated attack:

1. Mitigate lateral attacks against the private network as shown in our attack, by always requiring CORS [31] Preflights when requesting private or local resources, even if the origin is local or private. This mitigates the final step of the above attack, and protects the private network from an exploited local web server.
2. Protect local web servers from abuse using a default restricted rendering policy, utilizing existing mechanisms e.g., CSP [30]. This way we disable the attacker from harming the local application and also help mitigate some variants of the triple step attack.
3. Limit even further the requests from public origins to private network origins (and from private to local). Currently, Private Network Access [20] enables the private servers to allow-list their public counterpart defined by the public origin (using CORS). Our attack abuses this by fooling the allowed public website to be our allowed mediator to the local web server, and exploiting a vulnerability in a specific local page. The new mechanism that we propose here, enables the public allowed-server to specify to the browser which of its pages can initiate a request and to what private resource. This way, the attack

is limited and security developers can focus on these special pages, notice however that this mitigation by itself does not block all possible attacks in the first step of the triple step attack. The mechanism can be implemented by a new dedicated HTTP server header, or by updating the Content Security Policy mitigation mechanism.

In Section 2 we describe West et. al's recent Private Network Access mitigation mechanism. Then, in Section 3, we show that localhost web servers are special and more problematic then thought before. The Localhost Detour attack that bypasses the Private Network Access mitigation and allows attacking the private network is described in Section 4. Section 5 reports on public websites with corresponding local web servers, and an example vulnerability we disclosed. Then in Section 6 three mitigation mechanisms against localhost attacks are proposed. In our technical report [35] we briefly review relevant browser based vulnerabilities and corresponding mitigation mechanisms.

In this paper we review Private Network Access' security model where it's authors assume there is some security hierarchy (public space is less secure than private space, and private space is less secure than local space). Thus PNA initiates a CORS preflight only when an origin of less secure network space initiates a request to an origin of a more secure network space (e.g., public to local). In reality there is no such security hierarchy, and all network spaces are both dangerous and in danger. In our scenario, a local origin initiates a request to the private network. Because PNA assumes local origins are more secure than private origins, the mitigation does not initiate a CORS preflight. As we show, this means that local devices put the private space in danger. The solution is to relax the hierarchy assumption, and enforce preflights between all origins.

Ethical Disclosure. The Localhost Detour attack scheme (Fig. 2) has been disclosed to the Chromium Private Network Access repository [19] and is acknowledged by a team member that has confirmed the issue. They referenced to another post with a different scenario where an attacker could abuse the mitigation (e.g., between LAN devices). They also conclude that the problem arises because preflight is not enforced in these cases. Note that in their scenario only private-to-private preflights are required. In their response, and in a more recent post, they rightfully explain that their first priority is to roll out the PNA mitigation as is, and in a later version to upgrade the mitigation. The next version will supposedly include preflights whenever the recipient is private or local as also suggested here.

Moreover, in this paper we demonstrate the Localhost Detour Attack using a vulnerability we discovered in the Folding At Home software. As we note in the sequel, we made a responsible disclosure to F@H's security team and helped fix the problem. They were responsive and the fix was adopted into the recent version of the F@H client software.

2 Private Network Access

West and Rigoudy of the Chromium's security research team recently proposed a set of browser mitigation schemes, called Private Network Access (PNA) [20]. The goal of the PNA scheme (previously named CORS1918), is to protect local IoT devices, routers, and localhost servers from browser based attacks in the context of different, public or local origins. It significantly raises the bar for attackers and protects against most attacks in which a malicious website requests private or local resource.

This browser mitigation schemes essentially utilizes the already-adopted CORS [31] preflight protocol [29] to require the destination servers to opt-in to the special requests coming from their associated partners.

PNA employs the widely adopted IANA IPv4 [24] and IPv6 [25] address space conventions, to partition the IP address space into three categories:

1. *Local addresses* are those unique for the current device, usually referred as `localhost` (e.g., `127.0.0.0/8`).
2. *Private addresses* are those unique to the private network (e.g., `10.0.0.0/8`).
3. *Public addresses* are all other addresses, that are global to all devices (e.g., `8.8.8.8`).

They proceed to define *private network request* as all requests that are destined to either a local address and initiated by either a private address or a public address, or destined to a private address and initiated by a public address (i.e., $3 \rightarrow 2$, $3 \rightarrow 1$, and $2 \rightarrow 1$ in Table 1).

Table 1. Chromium's proposed Private Network Access mitigation logic as to when to always enforce secure context and CORS preflight. Empty boxes mean that PNA will not initiate preflights upon these cases. This logic does not cancel other mitigations such as SOP.

Initiator \ Destination	① Local	② Private	③ Public
① Local			
② Private	HTTPS + Preflight		
③ Public	HTTPS + Preflight	HTTPS + Preflight	

It seems however that the scope of the current design of the PNA mechanism excludes the security level of the initiator and destination. For example, an IoT device should protect itself against CSRF attacks (see full technical report [35]). PNA thus focuses on the crux of the problem, and makes an assumption regarding the security level of the initiator and the destination.

With PNA the browser identifies private network requests and validates that each is of secure context as defined in the WHATWG standard [34], i.e., using HTTPS scheme for the request. Next, the browser initiates a CORS preflight to the request's destination (even for simple CORS requests), with a newly defined header, as shown in Table 1. The new header and value is simply

```
Access-Control-Request-Private-Network: true
```

The browser then issues the original request only if the preflight response allows it according to the CORS header and value, e.g.:

```
Access-Control-Allow-Private-Network: true
```

The mitigation secures the destination web servers by default, as old servers or non supporting devices will not respond with the expected header, and thus will be protected by the browser. This secure-by-default approach means that PNA is not backwards compatible, because old non-public servers will not allow preflight requests initiated due to private network requests.

3 Localhost, the Achilles Heel

The critical step on which all of the attacks presented in this paper are based is the request from an allowed public server to a corresponding local web server. The public website could be for example http://www.allowed.com and the local web server is connected to localhost:1234, see Table 2 for several examples.

On today's PCs there are likely to be several local web servers installed by different products, e.g., Steam gaming client, Dropbox local client, JetBrains programming IDEs, NVIDIA Web Helper, Folding@Home client, Arduino Create agent, and until recently also Spotify. These applications create a local web server in order to execute code outside of the browser. For example, Folding At Home [7] has a program that allows the user to utilize their computer for distributed computing, something impossible within the browser. Thus these local servers usually execute with the installing user privileges and open the door to several different vulnerabilities, such as XSS (see full technical report [35]).

3.1 Localhost is Special

While some IoT devices also provide a web server listening in the private network at some private IP address (e.g., 10.0.0.126) or domain, there are at least four reasons to treat the localhost servers differently:

1. Local web servers are often an internal feature, and thus remain undocumented and run in the background unbeknown to the PC user. This is unlike a physical IoT device the user usually sees and usually had to buy and install. This implies that users might not be aware of a security issue in one of the local servers installed with or without their knowledge.
2. Local web servers are accessible through a known loopback IP address and port (e.g., http://127.0.0.1:12345), while IoT devices are usually dynamically assigned network-specific IP addresses by the router.

3. Local web servers run at a more privileged device, the user's PC, and at a higher privilege than regular browsers. E.g., Dropbox's local server can access and manipulate the file system, Steam's local server can run processes.
4. Local web servers interact with a public website, which even under PNA [20] can be approved to make requests on them.

3.2 Localhost: A Gateway to the Local and Private Network

As reviewed in Table 1 in Sect. 2, a public domain can request local or private resources only if the public origin is approved by a preflight. As said above, localhost servers usually interact with their corresponding public website, e.g., the public website requests information from the localhost software. With the introduction of PNA, the local servers would need to approve the new preflight request from their corresponding public websites. But then the approved web server might behave somewhat like a Trojan horse, letting in any XSS or other browser vulnerabilities found on the local server. Then, as can also be seen in Table 1, the local origin can request any origin without Private Network Access preflights. This means that a localhost web server can initiate requests to other localhost or IoT web servers in the private network.

4 Localhost Detour Attack

The Localhost Detour Attack presented here, manipulates a public web site that passes the PNA checks since it is authorized to make requests to its corresponding local web server. This could be achieved using an XSS vulnerability in the public website, and sometimes even without a vulnerability demonstrated in the sequel, in a full attack flow. Then the attack exploits a vulnerability in the local web server that was authorized by the public request, to maliciously access a device in the private network. In the notation of Table 1 the attack goes $3 \rightarrow 1$ and then $1 \rightarrow 2$ (or $1 \rightarrow 1$ for other attack options).

The attack is out of the PNA mitigation scope, as PNA assumes the initiator and destination are secure. We circumvent the mitigation by abusing common vulnerabilities and/or behaviours in the initiator and the destination. Thus, although the PNA mitigation does raises the bar and makes penetration much harder, ways around it are shown here.

4.1 Attack Threat Model

We assume the victim has some localhost HTTP web server S running on some local port P, usually installed with some popular software. We further assume that due to PNA, the local web server S allows some public domain D_A to initiate requests to it.

As for the attackers, we assume they can trick the victim to visit a malicious website D_E (i.e., phishing), and that they know that server S is running on a local port P on the victim's computer. This assumption is valid because the programs that run such local web servers are popular and use hard-coded ports which are the same for all computers, as we show in the next chapter.

Finally, we assume the attackers find a vulnerability in the local web server S and some way to trigger it from the allowed public website D_A. The former assumption is plausible as vulnerabilities exist and we back it up with CVE examples in the next section. The latter assumption is even weaker than the former, as it does not require a full vulnerability but just a way to request the local endpoint (i.e., redirect).

4.2 Attack Flow Example

Activating the vulnerability from the public website is almost always possible with an XSS vulnerability, but even posting a link on the allowed public website that points to the local address and phishing the victim to click it is sufficient, as the request's initiator's origin is allowed. The basic attack flow begins by the attacker exploiting an XSS found in the allowed public website to request the local web server. Sometimes the attacker does not even need a vulnerability in the allowed public website. A simple example is a public website that has a page that opens an iframe to the localhost web server (as we demonstrate in Sect. 5). A more complex example could be posting a link on the allowed public website that points to the local address and phishing the victim to click it is enough, as the request's initiator's origin is allowed.

The attack is illustrated in diagram 1, and its sequence of steps is:

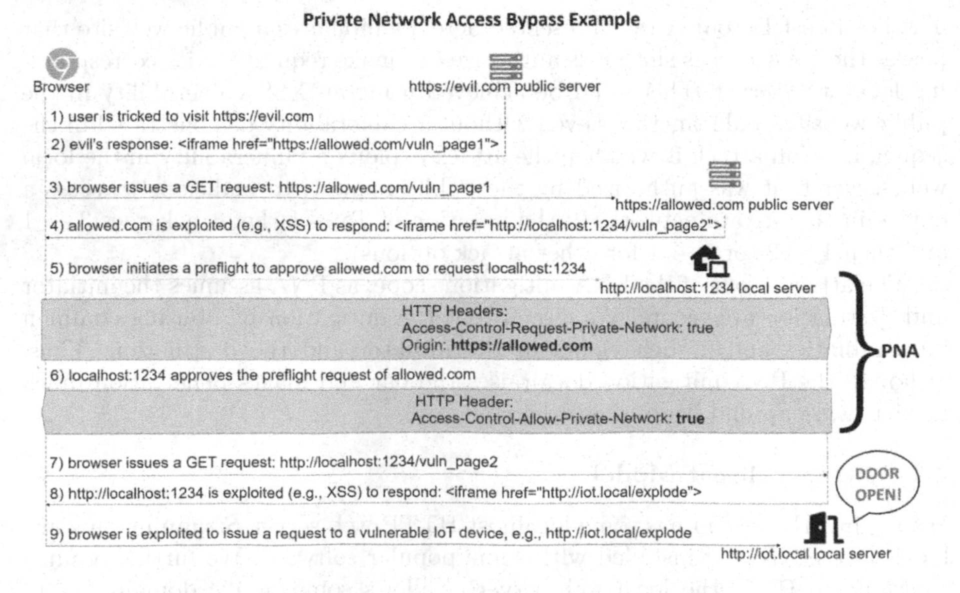

Fig. 1. Request-response illustration of the "Localhost Detour" Private Network Access mitigation bypass; The victim visits a malicious website evil.com, e.g. by phishing, and then the attacker starts the attack. The attack makes use of a page in allowed.com that activates a vulnerability in the appropriate localhost server, and from there the attacker can request private resource on the LAN and even the PNA mitigation scheme does not stop it.

1. User (victim) is tricked to visit https://evil.com/.
2. The server at evil.com responds with an HTML iframe pointing to: https://allowed.com/vuln_page1, where *vuln_page1* is a vulnerable page (e.g., an XSS) that is found on allowed.com's origin.
3. The browser renders the iframe and issues a request to the vulnerable page at:
https://allowed.com/vuln_page1.
4. The public server of https://allowed.com/vuln_page1 is exploited to return an attacker controlled response (e.g., due to an XSS). The response is an HTML iframe pointing to allowed.com's corresponding local web server listening on http://localhost:1234/vuln_page2. We assume *vuln_page2* is a vulnerable page on the local server's origin (XSS).
5. The browser renders the response from allowed.com and because allowed.com is a public origin that requests a private origin (localhost:1234), the browser initiates a preflight (due to PNA).
6. The local web server approves the preflight request.
7. The browser receives the approving preflight response and sends the request to the local server at address http://localhost:1234/vuln_page2.
8. The local web server handles the request and responds with an attacker controlled HTML. The localhosts' response is an HTML iframe pointing to http://iot.local/open_door, a vulnerable IoT device's private web server. Here we assume that *open_door* is vulnerable page on the IoT origin.
9. The victim's browser then renders the malicious response and issues a request to the vulnerable IoT endpoint, e.g., http://iot.local/explode. Private Network Access does not issue a preflight when the initiator's origin is local.

Although Private Network Access' RFC states that the devices' developers are responsible for their own security measures, we think this is not a realistic assumption to be made. The conclusion we take from our attack is that a vulnerable local server allows a resourceful attacker to circumvent the Private Network Access mitigation, and thus we must not leave the devices security unaccounted for.

4.3 General Attack Cookbook

Now that we have reviewed the attack example, we provide the attack cookbook recipe, as also illustrated in Fig. 2:

1. Find an application software that installs a local web server with a public counterpart (See Table 2 for a few examples). This could be a specific, prechosen software (e.g. a company wants to harden their software), or any such application. These programs can be found by scanning PCs for local ports it listens on, or by crawling the Internet for public websites that request localhost and investigating their software.
2. Find a vulnerability in that software's local web server. The vulnerability type could be anything, and this will define the capabilities of the attack. A different vulnerability like a redirect page, could allow an attacker to send a CSRF request to a neighbouring IoT device.

3. Understand the capabilities of the previous vulnerability to devise the final step of the attack. This could be for example to execute bash code on the victim's machine, to use known CSRF attacks against IoT devices, to scan the network etc.
4. Find a page in the public website counterpart that can request the local resource. This could be done in two ways, either by the page the developer intended to request the local web server, or by finding a minimal vulnerability in the website. For example, an XSS in the website could possibly allow an attacker to request any such local resource, but even a weaker vulnerability like open-redirect is sufficient.
5. Combine ingredients to a complete attack. This could usually be done by registering a domain and phishing the victim to visit it. The malicious website will include an iframe to the allowed website, which in turn will request the local resource. Additional methods exist, for example if the software's website allows users to post messages with links (like Amazon Stores), an attacker could post a malicious link to local resource, and phish victims to visit the link ("Click here for more deals!"), actually initiating a request to the local resource from the counterpart website's origin.

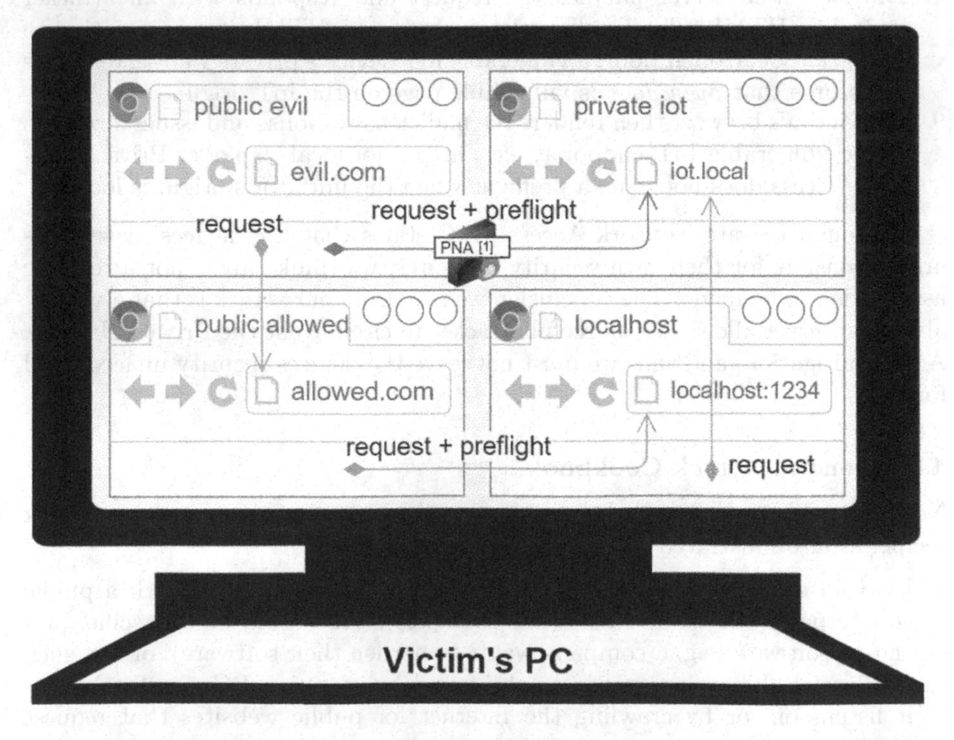

Fig. 2. PNA (red arrow) prevents public websites like evil.com from requesting private or local resources like iot.local. The Localhost Detour attack (green arrows) circumvents the mitigation by abusing a localhost webserver that allows some public origin like allowed.com. Thus evil.com requests iot.local indirectly by requesting allowed.com, which in turn requests localhost:1234, passing the PNA preflight check. Finally localhost requests iot.local without needing a preflight. (Color figure online)

5 Public-Local Matching Web Server Pairs

We noticed three main HTTP request methods by which a public web server requests the local one:

1. Iframe: some public domains (e.g., *Foldingathome.org* [7]) embed their local web server in an HTML iframe, making the local web server's interface part of the public website.
2. AJAX API: some public websites (e.g., *Dropbox.com* [5]) use the local web server as an API endpoint to retrieve information regarding the user (e.g., automate website login) or to instruct the product to act in some manner (e.g., open a file in a local editing application). The public websites usually use Fetch or Websocket requests.
3. HTML Links: some public websites (e.g., *Kubernetes.io* [14]) have HTML links pointing to the localhost server, as part of a usage guide, blog posts, or an interactive way to instruct the user to visit the application.

Table 2. List of products that create a local web server and the public origin we suspect will be allowed by the local server after Private Network Access mitigation mechanism is implemented. Each product's local HTTP port is specified, together with the reason the public origin requests it (e.g., Steam.com requests its local product for authentication API, and Kubernetes.io has HTML links to its localhost endpoint). We provide estimates to number of users and example CVEs in the local application to demonstrate the wide spread and availability of this critical issue.

Product Name and Origin	Port	Usage	#Users	examp. CVEs
Dropbox [5]	17600	API	700M	2019-12171 2018-20819
Steam Client [16]	27060	API	120M	2019-14743, 2019-15315
Jetbrains InetlliJ [6]	63342	Links	8M	2019-15037 2019-15848
Folding@Home [12]	7396	iframe	>4M	–
Spring.io [15]	8080	Links	>1M	2018-1270 2018-1271
Logitech Media Server [9]	9000	Links	–	2017-16568 2017-15687
Kubernetes CTL [14]	8001	Links	–	2019-1002100
Swagger [3]	3200	Links	–	2016-5682
Arduino Create Agent [4]	8992	API	–	–
Jupyter Notebook [11]	8888	Links	–	2015-6938 2018-8768
HedgeDoc [8]	9000	Links	–	2021-21259
MinIO [13]	9000	Links	–	2020-11012

Table 2 lists several examples of a public and local web servers, where the first makes HTTP requests to the second. Most usually these servers belong to the same product, like Steam or Dropbox, etc.

5.1 Three Examples

We provide three examples of products with such public-local web server pairs, one for each of the above three request types: iframe, AJAX API, and HTML links. Each of these examples would have to be approved by the CORS-preflight exchange if and when PNA is implemented in the browsers, in order not to break the corresponding product functionality. We further point out for each example a vulnerability an attacker may utilize in order to attack the localhost server.

Note that this type of vulnerabilities is common and we give several examples. We start with a vulnerability we have recently discovered, and verified in our lab (i.e., *folding at home*), and then continue to illustrate possible scenarios in known products (i.e., *Dropbox* and *InteliJ*). Notice we have not yet searched or know of open specific vulnerabilities in the later products, but only claim that these products are potentially open to the risk of being abused as part of an attack.

Folding at Home. Folding@Home [7] is a distributed computing software designed to help scientists find new therapeutics. The software raises a local web server on port 7396, and it servers as a dashboard (e.g., to view current research process). Users can access the dashboard through the browser directly (i.e., http://localhost:7396) or by using the software's official website at https://client.foldingathome.org/. We have successfully implemented and tested the following attack in our lab, and responsibly disclosed it to the company. Folding At Home acknowledged the issues and fixed the vulnerability in the official release.

The public website works by automatically opening an HTML iframe pointing to the local web server's index page. Thus, if an attacker had a vulnerability stored in the local server, they would not need a vulnerability in the public website, since the public website triggers it by itself.

In the Folding At Home website, researchers can create research teams, and add team information (e.g., name, image and website's URL). These details can be modified by anyone with the team's password, through the public website. The modification form does not validate that the input is legal (e.g., that the new team website URL does not contain illegal URL characters), i.e., the team name could be a javascript string. Right away as a user opens the local website index page, the information regarding their account is shown including their research team's information and team's website URL. The code that displays the team URL is shown below:

```
if (typeof data.team_url != 'undefined')
    team_name = '<a target="_blank" href="'
        + data.team_url + '">' + team_name + '</a>';
$('#box-points-team').html(team_name);
```

This code is vulnerable to an XSS (see full technical report [35]) attack because the team URL is not sanitized or validated. An attacker with the victim's team's password, can change the team URL to be:

```
"><script>MALICIOUS CODE</script><"
```

Then, when the user opens the local index page, the malicious javascript code is executed in their browser.

An attacker then could abuse the fact that the public website opens an iframe to the private local server, and utilize the above XSS vulnerability to attack the user private network. A malicious code attacking an IoT device with a CSRF (see full report [35]) vulnerability might look roughly as follows:

```
"><script>
    fetch("http://iot.local/explode")
</script><"
```

We disclosed this vulnerability report privately to Folding At Home's security team, and they have quickly responded. They fixed the issues, both the XSS vulnerability in the local web server and the validation issue in the public website.

Dropbox. Dropbox [5] is a popular cloud storage company with over 700 million users. The users can store files in the cloud by installing a software on their PC, and they can then manage their data through the public website (https://dropbox.com).

The software starts a local web server that listens on port 17600, and the public web site sends Websocket requests to it (e.g., to open a file locally).

An attacker could utilize an open-redirect vulnerability in the public website (which she yet has to find), e.g., [26], and embed it in an iframe in their malicious website, redirecting to the localhost web server. Thus the attacker can trigger a vulnerability in the local web server (again yet needs to be found), bypassing Private Network Access.

Jetbrains IntelliJ. Jetbrains' IntelliJ IDEs [10] are popular programming editors. Their software raises a local web server at port 63342. This local port serves several purposes, e.g., to open a file X in a local editor when a user visits a link of a specific form, (e.g., http://localhost:63342/open?file=X).

Jetbrains has an official public website used for bug tracking (at https://youtrack.jetbrains.com), where users and employees can describe their bugs and issues. Many times the issues regard the local web server and HTML links are posted with the intention to be clickable (as an interactive evidence). For this reason this origin will probably be trusted by the software.

Thus, attackers could utilize a vulnerability in the public domain, e.g., [2], and successfully request the local server. After that, the attacker can exploit vulnerabilities in the local server, e.g., [28], bypassing the Private Network Access.

6 Proposed Mitigations

In this section we describe three mitigation mechanisms that work independently to mitigate this family of attacks. The full chain of mitigation mechanisms is depicted in Fig. 3.

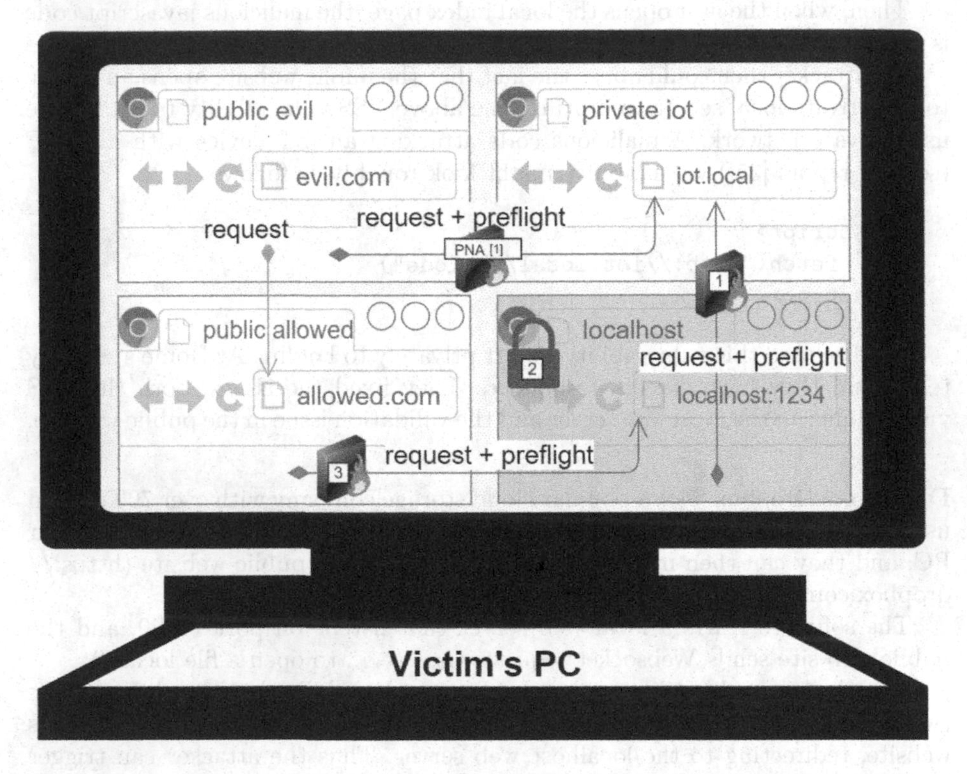

Fig. 3. Illustrative summery of our proposed mitigations against the Localhost Detour attack: (1) Always issue preflights to private and local origins in order to mitigate lateral movement (see also Table 3), (2) Default restrictive policies to protect local web servers from exploitation, and (3) Explicit server side destination declaration in order to minimize attack surface to specific pages.

6.1 Lateral Movement Mitigation

The final and third step of the attack is an access from the local server to a honest private server (e.g., on an IoT device). This could be mitigated by setting an equal footing between all sub IP spaces, as can be shown in Fig. 3 at mitigation number (1).

This mitigation can be enforced by simply extending the PNA mechanism to always issue preflight request/response to non-public origins, as already done in PNA from public origins to private and local origins.

The main goal of this mitigation is to block by default unintended requests, and this is achieved by initiating a preflight. Thus a requested private or local origin will have to knowingly inspect and allow the request. Unfortunately there is currently no good solution to certificates inside a private network, and this is why it is currently unrealistic in our opinion to require secure context. Note this is not perfect as it means the destination should validate the initiator's integrity on its own.

The proposed extension to PNA logic is depicted in Table 3.

Table 3. Summery of our proposed mitigation mechanisms against lateral movement from the exploited localhost to the private network. The suggestion modifies Private Network Access (PNA) (see Table 1) to always initiate a preflight authorization when the destination is local or private.

Initiator \ Destination	① Local	② Private	③ Public
① Local	Preflight_{new}	Preflight_{new}	
② Private	HTTPS + Preflight	Preflight_{new}	
③ Public	HTTPS + Preflight	HTTPS + Preflight	

6.2 Private Servers Protection

The second mitigation mechanism we propose is to impose restrictive default policy in the browser when it issues requests to the private network. This mitigation is designed to protect private and local server from cross site scripting and misuse (label (2) in Fig. 3), and is secure by default. If a product needs to disable the mitigation in order to implement some functionality, it can relax the mitigation.

6.3 Attack Surface Minimization

Finally, we propose to restrict the attack surface by what we call "explicit private destinations", as illustrated in Fig. 3 with indicator (3). The goal is to specify the public pages that are allowed to issue requests to the private network. This mitigation minimizes the public pages that an attacker can abuse, and the private pages the attacker can exploit. For more details see technical report [35].

7 Conclusions

This paper highlights a weak (perhaps the weakest) link in the chain of browser mitigation mechanisms, a localhost web server, and suggests new methods to patch this link. We reviewed browser-based attacks, and build upon Chromium's recent Private Network Access [20] mitigation mechanism. The Localhost Detour attack that is presented here, bypasses this mitigation and therefore points out a dangerous vulnerability. The attack abuses a vulnerability in the local web server and its corresponding public website in order to attack the private network. Moreover, developers of localhost web servers should consider this security aspect in their design.

Acknowledgment. We thank Dr. Amit Klein for very helpful discussions and advice.

References

1. Trendmicro node.js http server listening on localhost can execute commands (2016). https://bugs.chromium.org/p/project-zero/issues/detail?id=693&redir=1
2. Cve-2020-7913: Xss vulnerability on jetbrains youtrack (2020). https://cvedata.com/cve/CVE-2020-7913/
3. API documentation & design tools for teams | swagger (2021). https://swagger.io
4. Arduino digital store (2021). https://create.arduino.cc
5. Dropbox is building the world's first smart workspace. (2021). https://www.dropbox.com/home
6. Everything - jetbrains youtrack (2021). https://youtrack.jetbrains.com
7. Folding@home (2021). https://foldingathome.org/about/
8. Hedgedoc (2021). https://docs.hedgedoc.org/
9. Home - welcome to mysqueezebox.com! (2021). https://mysqueezebox.com
10. Jetbrains (2021). https://www.jetbrains.com
11. Jupyter blog (2021). https://blog.jupyter.org
12. Local folding@home web control (2021). https://client.foldingathome.org
13. Minio | the minio quickstart guide (2021). https://docs.min.io/
14. Production-grade container orchestration (2021). https://kubernetes.io/
15. Spring | home (2021). https://spring.io
16. Welcome to steam (2021). https://store.steampowered.com
17. Acar, G., Huang, D.Y., Li, F., Narayanan, A., Feamster, N.: Web-based attacks to discover and control local IoT devices. In: Proceedings of the 2018 Workshop on IoT Security and Privacy, pp. 29–35. ACM (2018)
18. Afek, Y., Bremler-Barr, A., Noy, A.: Eradicating attacks on the internal network with internal network policy. CoRR abs/1910.00975 (2019). http://arxiv.org/abs/1910.00975
19. Chromium: Chromium's response to an issue (2021). https://github.com/WICG/private-network-access/issues/38#issuecomment-766720523
20. Chromium, Mike West and Titouan Rigoudy: Private Network Access (2021). https://wicg.github.io/private-network-access/
21. Barda, D., Zaikin, R., Shriki, Y.: Keeping the gate locked on your IoT devices: vulnerabilities found on Amazon's Alexa (2020). https://research.checkpoint.com/2020/amazons-alexa-hacked/
22. Dorsey, B.: Attacking private networks from the internet with DNS rebinding (2018). https://medium.com/@brannondorsey/attacking-private-networks-from-the-internet-with-dns-rebinding-ea7098a2d325
23. Hemmings, J.: EE 4GEE mobile WiFi router - multiple security vulnerabilities writeup (2017). https://blog.jameshemmings.co.uk/2017/08/24/ee-4gee-mobile-wifi-router-multiple-security-vulnerabilities-writeup/#more-276

24. IANA: IPv4 Special-Purpose Address Registry (2021). https://www.iana.org/assignments/iana-ipv4-special-registry
25. IANA: Ipv6 special-purpose address registry (2021). https://www.iana.org/assignments/iana-ipv6-special-registry
26. kumar: Dropbox open redirect - misconfigured regex (2017). https://www.kumar.ninja/2017/07/dropbox-open-redirect-misconfigured.html
27. Leitschuh, J.: Zoom zero day: 4+ million webcams & maybe an RCE? Just get them to visit your website! (2019). https://infosecwriteups.com/zoom-zero-day-4-million-webcams-maybe-an-rce-just-get-them-to-visit-your-website-ac75c83f4ef5
28. Milne, J.: JetBrains IDE remote code execution and local file disclosure (2016). http://blog.saynotolinux.com/blog/2016/08/15/jetbrains-ide-remote-code-execution-and-local-file-disclosure-vulnerability-analysis/
29. Mozilla: Preflight request (2019). https://developer.mozilla.org/en-US/docs/Glossary/Preflight_request
30. Mozilla: Content Security Policy (2019). https://developer.mozilla.org/en-US/docs/Web/HTTP/CSP
31. Mozilla: Cross-Origin Resource Sharing (CORS) (2019). https://developer.mozilla.org/en-US/docs/Web/HTTP/CORS
32. Mozilla: Same-origin policy (2019). https://developer.mozilla.org/en-US/docs/Web/Security/Same-origin_policy
33. Rowan Merewood, C.: SameSite cookies explained (2019). https://web.dev/samesite-cookies-explained/
34. WHATWG: HTML: Secure Context (2021). https://html.spec.whatwg.org/multipage/webappapis.html
35. Afek, Y., Bremler-Barr, A., Israeli, D., Noy, A.: Technical report - Localhost detour from public to private networks. https://deepness-lab.org/publications/localhost-detour-from-public-to-privatenetworks/

Pseudo-random Walk on Ideals: Practical Speed-Up in Relation Collection for Class Group Computation

Madhurima Mukhopadhyay[1(✉)] and Palash Sarkar[2]

[1] Indian Institute of Technology, Kanpur, Kanpur, India
mukhopadhyaymadhurima@gmail.com
[2] Indian Statistical Institute, Kolkata, India
palash@isical.ac.in

Abstract. We introduce a technique to obtain practical speed up for relation collection in class group computations. The idea is to perform a pseudo-random walk over the ideals. The ideals visited by the walk are used in the manner exactly as in the previous algorithm due to Gélin (2018). Under the heuristic assumption that the ideals visited by the walk behave as the ideals randomly generated in Gélin's algorithm, the asymptotic complexity of the new algorithm remains the same as that of Gélin's algorithm. The main advantage of the new method over Gélin's method is that the pseudo-random walk requires a single ideal multiplication to generate the next ideal in the walk, whereas Gélin's algorithm requires a number of ideal multiplications to generate each ideal to be tested. We have made Magma implementations of both the new algorithm and Gélin's algorithm. Timing results confirm that there is indeed a substantial practical speed-up in relation collection by the new algorithm over Gélin's algorithm.

Keywords: Class group · pseudo-random walk

MSC (2010): 11Y40 · 94A60

1 Introduction

A basic problem in computational algebraic number theory is to compute the class group of the ring of integers of a number field and more generally, the class group of an order of the ring of integers. The complexity of the computational problem is expressed in terms of the discriminant $\Delta_{\mathcal{K}}$ of the number field $\mathcal{K}$. Presently, there is no polynomial time (in $\log|\Delta_{\mathcal{K}}|$) algorithm for computing the class group. Shanks [15,16] had proposed the first algorithm for computing the class group in time $O(|\Delta_{\mathcal{K}}|^{1/5})$ assuming that the generalised Riemann hypothesis (GRH) holds.

S. Dolev et al. (Eds.): CSCML 2023, LNCS 13914, pp. 18–31, 2023.
https://doi.org/10.1007/978-3-031-34671-2_2

Later improvements have provided sub-exponential algorithms for the problem, where for $N > 0$, $\alpha \in [0,1]$ and $c \geq 0$, the sub-exponential expression $L_N(\alpha, c)$ is defined as follows.

$$L_N(\alpha, c) = \exp\left((c + o(1))(\log N)^{\alpha}(\log\log N)^{1-\alpha}\right). \tag{1}$$

Hafner and McCurley [13] proposed the first sub-exponential algorithm for computing the class group of imaginary quadratic number fields also assuming GRH. The runtime of their algorithm is $L_{|\Delta_{\mathcal{K}}|}(1/2, \sqrt{2})$. For general number fields, Buchmann [5] extended the Hafner-McCurly algorithm to obtain a sub-exponential algorithm when the extension degree is fixed. Several ideas which improve the implementation of Buchmann's algorithm were introduced by Cohen, Diaz Y Diaz and Olivier [8]. Biasse and Fieker [3] described an algorithm to compute the class group of any number field in time $L_{|\Delta_{\mathcal{K}}|}(2/3 + \epsilon, c)$, for arbitrarily small ϵ and $c > 0$ without any restriction on the extension degree. This algorithm was simplified by Gélin [10] and the asymptotic complexity was improved. It is to be noted that the complexity analyses of all the algorithms are heuristic; one of the main heuristic assumptions being that smoothness results for general ideals are assumed to apply to the restricted set of ideals considered by the algorithms.

Sub-exponential class group computation algorithms are index calculus algorithms having two dominant steps, namely the relation collection step and the linear algebra step. The linear algebra step essentially consists of computing the Smith Normal Form of a matrix which is the output of the relation collection step. Progress in the class group computation algorithms has mainly been in the relation collection step.

In the present work, we provide a new heuristic algorithm to improve the practical runtime of the relation collection step in Gélin's algorithm [10]. The core idea of previous ideal reduction algorithms is to randomly generate ideals on which lattice techniques are used to obtain principal ideals of bounded norms which can then be tested for smoothness over a predefined factor basis. Presently, the most simplified form of this algorithm is due to Gélin [10]. The random generation of the ideals in Gélin's algorithm requires exponentiating some randomly chosen prime ideals in the factor basis and then multiplying the resulting ideals together. This requires performing a number of ideal multiplications.

We propose to perform a pseudo-random walk on ideals. Once the initial ideal has been computed, each step of the walk requires a single ideal multiplication. The ideals generated in each step can be processed using lattice techniques as in Gélin's algorithm. The parameters which determine the asymptotic complexity of Gélin's algorithm [10] are not changed. So, under the heuristic assumption that the ideals in the pseudo-random walk behave like random ideals, the asymptotic complexity of the new algorithm remains unchanged from that of Gélin's algorithm.

We have made a Magma implementation of both the new algorithm and Gélin's algorithm [10]. Since our goal is to compare the new algorithm with Gélin's algorithm [10], we did not try to compute the class group for one particular field. Instead, we chose four number fields and conducted experiments with both the algorithms to collect a number of relations. The timing results obtained from these experiments indeed confirm the speed improvement of the new algorithm over Gélin's algorithm [10]. In particular, for the four fields with which we experimented, the ratio of the times required by Gélin's algorithm to the new algorithm is about 3.

Related Work

A related line of work considers algorithms with improved complexity when the coefficients of the generating polynomial of the number field are small. Biasse [2] and Biasse and Fieker [3] showed an $L_{|\Delta_{\mathcal{K}}|}(a, \cdot)$ algorithm for certain fields where a can be as low as 1/3. Gélin and Joux [12] extended the result of Biasse and Fieker [3] to a larger class of fields by describing an algorithm for obtaining polynomials with small coefficients for certain fields. The most recent work in this line is the sieving based relation collection algorithm due to Gélin [11].

The set of relations obtained for computing the class group are also used for computing the regulator[1] and a set of generators of the unit group. In fact, the regulator and the class number (i.e., the order of the class group) are required to be computed together. The techniques for decomposing an ideal over a factor basis are also used to obtain algorithms for the principal ideal problem. We refer to [7] for details.

Cryptographic Applications

As discussed above, there is no polynomial time algorithm to compute the class group and the best known algorithms have sub-exponential complexity. So, for a number field with a sufficiently large discriminant, the order of the class group can be considered to be unknown. Such a hidden order group forms the basis for several cryptographic primitives [4,6,9]. Concrete suggestions for instantiating these primitives have been made using class groups of imaginary quadratic fields. In principal, though, the class group of a general number field can also be used. The security versus efficiency question of using class groups of general number fields versus those of imaginary quadratic fields remains to be studied. Progress in algorithms for computing class groups influences the choice of number fields over which the relevant cryptographic primitives can be securely implemented.

2 Preliminaries

Let $\mathbb{Q}$, $\mathbb{R}$ and $\mathbb{C}$ respectively denote the fields of rational, real and complex numbers. The ring of integers will be denoted by $\mathbb{Z}$. In the following we provide a brief overview of some relevant facts regarding the problem of computing the class group. Further details can be found in [7].

[1] The definition of the regulator of a number field is somewhat complicated. We do not need this notion in the present work and so we do not provide the definition. The reader may refer to [7] for the definition.

An algebraic number field is a finite extension of $\mathbb{Q}$. More concretely, $\mathcal{K}$ is an algebraic number field defined by $T(X)$ if $\mathcal{K} \simeq \mathbb{Q}[X]/(T(X))$ for some irreducible $T(X) \in \mathbb{Z}[X]$. The extension degree of $\mathcal{K}$ over $\mathbb{Q}$ is equal to the degree of $T(X)$.

Suppose $T(X)$ is of degree n. Then $T(X)$ has n roots in $\mathbb{C}$. Since the complex roots occurs in pairs, let r_1 be the number of real roots of $T(X)$ and $2r_2$ be the number of complex roots of $T(X)$, where $r_1 + 2r_2 = n$. Suppose $\xi_1, \ldots, \xi_n$ are the roots of $T(X)$. For $i = 1, \ldots, n$, the embedding σ_i of $\mathcal{K}$ into $\mathbb{C}$ is defined in the following manner. For $a_0, \ldots, a_{n-1} \in \mathbb{Q}$,

$$\sigma_i(a_0 + a_1X + \cdots + a_{n-1}X^{n-1}) \mapsto a_0 + a_1\xi_i + \cdots + a_{n-1}\xi_i^{n-1}.$$

So $\sigma_1, \ldots, \sigma_n$ are n embeddings of $\mathcal{K}$ into $\mathbb{C}$. The embedding σ_i is called real or complex according to whether ξ_i is real or complex.

The norm and trace functions $\mathcal{N}_{\mathcal{K}/\mathbb{Q}}$ and $\mathcal{T}_{\mathcal{K}/\mathbb{Q}}$ respectively from $\mathcal{K}$ to $\mathbb{Q}$ are defined as follows. For $\alpha \in \mathcal{K}$

$$\mathcal{N}_{\mathcal{K}/\mathbb{Q}}(\alpha) = \prod_{i=1}^{n} \sigma_i(\alpha), \; \mathcal{T}_{\mathcal{K}/\mathbb{Q}}(\alpha) = \sum_{i=1}^{n} \sigma_i(\alpha).$$

The norm and trace functions are respectively multiplicative and additive on $\mathcal{K}$. For computational purposes, it is important to be able to compute the norm of an element. Any element $\alpha \in \mathcal{K}$ can be written as $\alpha = A(X)/d$ where $d \in \mathbb{Z}$ and $A(X) \in \mathbb{Z}[X]$. From Proposition 4.3.4 of [7], we have

$$\mathcal{N}_{\mathcal{K}/\mathbb{Q}}(\alpha) = \frac{1}{d^n} \text{Res}\,(T(X), A(X)) \tag{2}$$

where Res denotes the resultant.

An element $\alpha \in \mathcal{K}$ is said to be an integral element of $\mathcal{K}$, if it satisfies a monic polynomial with integer coefficients. The set of all integral elements of $\mathcal{K}$ is denoted by $\mathcal{O}_{\mathcal{K}}$. The set $\mathcal{O}_{\mathcal{K}}$ is a ring and is also an n-dimensional $\mathbb{Z}$-module (i.e., a lattice). A $\mathbb{Z}$-basis of $\mathcal{O}_{\mathcal{K}}$ is called an integral basis of $\mathcal{K}$. The field of fractions of $\mathcal{O}_{\mathcal{K}}$ is $\mathcal{K}$. For any $\alpha \in \mathcal{K}$, there is an integer a, such that $a\alpha \in \mathcal{O}_{\mathcal{K}}$.

Let $b_1, \ldots, b_n$ be an integral basis of $\mathcal{K}$. The discriminant $\Delta_{\mathcal{K}}$ of the number field $\mathcal{K}$ is defined to be

$$\Delta_{\mathcal{K}} = (\det(\sigma_i(b_j)))^2 = \det\left(\mathcal{T}_{\mathcal{K}/\mathbb{Q}}(b_ib_j)\right).$$

A fractional ideal $\mathfrak{a}$ in $\mathcal{K}$ is an $\mathcal{O}_{\mathcal{K}}$-submodule of $\mathcal{K}$ for which there is a $d \in \mathcal{O}_{\mathcal{K}}$ such that $d\mathfrak{a} \subseteq \mathcal{O}_{\mathcal{K}}$. The element d is said to be a denominator of $\mathfrak{a}$. It is easy to argue that $d\mathfrak{a}$ is an ideal of $\mathcal{O}_{\mathcal{K}}$. An (ordinary) ideal of $\mathcal{O}_{\mathcal{K}}$ is a fractional ideal (taking $d = 1$) and is called an integral ideal. If $\mathfrak{a}$ and $\mathfrak{b}$ are fractional ideals, then the product $\mathfrak{ab}$ is defined in exactly the same manner as for integral ideals. Any fractional ideal $\mathfrak{a}$ of $\mathcal{O}_{\mathcal{K}}$ is invertible, i.e., there is a fractional ideal $\mathfrak{b}$, such that $\mathfrak{ab} = \mathcal{O}_{\mathcal{K}}$. The inverse of $\mathfrak{a}$ is $\mathfrak{a}^{-1} = \{\alpha \in \mathcal{K} : \alpha\mathfrak{a} \subseteq \mathcal{O}_{\mathcal{K}}\}$. Any fractional idea $\mathfrak{a}$ can be written as $\mathfrak{b}/\mathfrak{c}$, where $\mathfrak{b}$ and $\mathfrak{c}$ are integral ideals; in fact, one may choose $\mathfrak{c} = d\mathcal{O}_{\mathcal{K}}$ where d is a denominator of $\mathfrak{a}$.

Let $\mathcal{I}_{\mathcal{O}_\mathcal{K}}$ be the set of all fractional ideals of $\mathcal{O}_\mathcal{K}$. Then $\mathcal{I}_{\mathcal{O}_\mathcal{K}}$ is a commutative group under multiplication with $\mathcal{O}_\mathcal{K}$ as the identity element. For $\alpha \in \mathcal{K}$, the set $(\alpha) = \alpha\mathcal{O}_\mathcal{K}$ is said to be a principal fractional ideal of $\mathcal{O}_\mathcal{K}$. Let $\mathcal{P}_{\mathcal{O}_\mathcal{K}}$ be the set of all principal fractional ideals of $\mathcal{O}_\mathcal{K}$. Then $\mathcal{P}_{\mathcal{O}_\mathcal{K}}$ is a subgroup of $\mathcal{I}_{\mathcal{O}_\mathcal{K}}$. The quotient group $\mathrm{Cl}_{\mathcal{O}_\mathcal{K}} = \mathcal{I}_{\mathcal{O}_\mathcal{K}}/\mathcal{P}_{\mathcal{O}_\mathcal{K}}$ is said to be the ideal class group of $\mathcal{O}_\mathcal{K}$. It is known that $\mathrm{Cl}_{\mathcal{O}_\mathcal{K}}$ is finite and the order of $\mathrm{Cl}_{\mathcal{O}_\mathcal{K}}$ is said to be the class number of $\mathcal{K}$.

Let $\mathfrak{a}$ be an integral ideal of $\mathcal{O}_\mathcal{K}$. It can be proved that the quotient group $\mathcal{O}_\mathcal{K}/\mathfrak{a}$ is finite. The norm of $\mathfrak{a}$, written as $\mathcal{N}(\mathfrak{a})$ is defined to be the cardinality of $\mathcal{O}_\mathcal{K}/\mathfrak{a}$. If $\mathfrak{a}$ is a principal ideal $\alpha\mathcal{O}_\mathcal{K}$, then it can be proved that $\mathcal{N}(\mathfrak{a}) = |\mathcal{N}(\alpha)|$. The notion of norm can be extended to fractional ideals. Suppose $\mathfrak{a}$ is a fractional ideal which is written as $\mathfrak{a} = \mathfrak{b}/\mathfrak{c}$, where $\mathfrak{b}$ and $\mathfrak{c}$ are integral ideals. Then $\mathcal{N}(\mathfrak{a}) = \mathcal{N}(\mathfrak{b})/\mathcal{N}(\mathfrak{c})$. A fractional ideal $\mathfrak{a}$ can be written as the ratio of two integral ideals $\mathfrak{b}$ and $\mathfrak{c}$ in more than one way. The value of the norm of $\mathfrak{a}$ does not depend upon the choice of $\mathfrak{b}$ and $\mathfrak{c}$.

Unique factorisation holds for the fractional ideals of $\mathcal{O}_\mathcal{K}$. If $\mathfrak{a}$ is a fractional ideal, then there are unique prime ideals $\mathfrak{p}_1, \ldots, \mathfrak{p}_k$ and non-zero integers $e_1, \ldots, e_k$ such that $\mathfrak{a} = \mathfrak{p}_1^{e_1} \cdots \mathfrak{p}_k^{e_k}$.

A basic computational problem in algebraic number theory is to compute the class group. Let K be a number field and $\mathcal{B}$ be a set of ideals of $\mathcal{O}_\mathcal{K}$ such that any $[\mathfrak{a}] \in \mathrm{Cl}_{\mathcal{O}_\mathcal{K}}$ can be written as $[\mathfrak{a}] = [\mathfrak{p}_1]^{e_1} \cdots [\mathfrak{p}_k]^{e_k}$ for some $\mathfrak{p}_1, \ldots, \mathfrak{p}_k \in \mathcal{B}$ and integers $e_1, \ldots, e_k$. Let $N = \#\mathcal{B}$ and $\mathcal{B} = \{\mathfrak{p}_1, \ldots, \mathfrak{p}_N\}$. Then there is a surjective homomorphism φ from $\mathbb{Z}^N$ to $\mathrm{Cl}_{\mathcal{O}_\mathcal{K}}$ which takes $(e_1, \ldots, e_N)$ to $[\mathfrak{p}_1]^{e_1} \cdots [\mathfrak{p}_N]^{e_N}$. By the first isomorphism theorem, $\mathrm{Cl}_{\mathcal{O}_\mathcal{K}}$ is isomorphic to $\mathbb{Z}^N/\mathrm{Ker}(\varphi)$. It is easy to see that $\mathrm{Ker}(\varphi)$ is a sub-lattice of $\mathbb{Z}^N$.

In view of the above, computing $\mathrm{Cl}_{\mathcal{O}_\mathcal{K}}$ boils down to determining $\mathcal{B}$ and computing $\mathrm{Ker}(\varphi)$. For the moment suppose that $\mathcal{B}$ is given. A relation is an equation of the form

$$\mathfrak{p}_1^{e_1} \cdots \mathfrak{p}_N^{e_N} = \langle x \rangle \tag{3}$$

where x is some element of $\mathcal{K}$. Suppose A is an $m \times N$ matrix whose rows are vectors $(e_1, \ldots, e_N) \in \mathbb{Z}^N$ corresponding to relations of the form given by (3). The structure of $\mathrm{Cl}_{\mathcal{O}_\mathcal{K}}$ is given by the Smith normal form (SNF) of A.

Obtaining relations of the form (3) forms the main computational task of the algorithm for computing the structure of the class group. The linear algebra part consists of computing the SNF of the relation matrix A. There is a further verification step that is required. The SNF of A possibly provides a factor of the class number. One computes an approximation of the regulator from the obtained relations and verifies that the product of the class number and the regulator is approximately one. If this does not hold, then the SNF of A provides a proper factor of the class number. In that case, more relations need to be obtained and the linear algebra and verification steps repeated. For more details of these two steps we refer to [7,10].

The set $\mathcal{B}$ is defined to be the set of all prime ideals whose norms are at most a constant B. Then $\#\mathcal{B} \approx B/\log B$. Assuming the Extended Riemann

Hypothesis, Bach [1] showed that choosing B equal to $12(\log|\Delta_\mathcal{K}|)^2$ is sufficient to ensure that the classes of the ideals in $\mathcal{B}$ generate the entire class group. By Bach's bound we mean the quantity $12(\log|\Delta_\mathcal{K}|)^2$ and denote it by C. Choosing $B = C$ is not sufficient to ensure that sufficiently many relations can be found. The value of B is chosen to be higher than Bach's bound. Following Gélin [10], B is set to be equal to $L_{|\Delta_\mathcal{K}|}(\delta, c_b)$. For appropriate choices of δ and c_b the overall runtime of the algorithm is sub-exponential. We refer to [10] for details.

A crucial issue in the relation collection has been pointed out by Gélin [10]. The relation collection phase is completed when the number of relations is larger than $\#\mathcal{B}$ and when all ideals of norms below Bach's bound are involved in at least one relation.

2.1 Generating Relations

Suppose as above that $\mathcal{B} = \{\mathfrak{p}_1, \ldots, \mathfrak{p}_N\}$ is the set of all prime ideals having norms less than B where B itself is at least as large as $12(\log|\Delta_\mathcal{K}|)^2$. For the asymptotic analysis, B is chosen to be $B = L_{|\Delta_\mathcal{K}|}(\delta, c_b)$ for suitable values of δ and c_b. The set $\mathcal{B}$ is said to be the factor basis. The main task of the class group computation algorithm is to generate relations of the form (3).

The basic idea for obtaining relations is to generate random principal ideals and then check whether they are smooth over $\mathcal{B}$. In practice, given a principal ideal $\langle x\rangle$, one checks whether the positive integer $\mathcal{N}(\langle x\rangle) = |\mathcal{N}(x)|$ is B-smooth. If this holds, then $\langle x\rangle$ is factored.

Given a random element of x of $\mathcal{O}_\mathcal{K}$, the probability of $|\mathcal{N}(x)|$ being B-smooth is very small. So, simply trying out random elements of $\mathcal{O}_\mathcal{K}$ will not lead to an efficient algorithm. To ensure that the probability of $|\mathcal{N}(x)|$ being B-smooth is reasonable, it is required to choose x in a manner which ensures that $|\mathcal{N}(x)|$ satisfies some pre-determined upper bound. The literature contains various methods of choosing x such that $|\mathcal{N}(x)|$ is below a desired bound.

Given an element $\alpha \in \mathcal{O}_\mathcal{K}$, its canonical embedding $\sigma(\alpha)$ into $\mathbb{C}^n$ is $\sigma(\alpha) = (\sigma_1(\alpha), \ldots, \sigma_n(\alpha))$. Since the complex embeddings occur in pairs, the vector $(\sigma_1(\alpha), \ldots, \sigma_n(\alpha))$ can be considered to be represented by a vector from $\mathbb{R}^n$. So, henceforth, we will consider $\sigma(\alpha)$ be a vector in $\mathbb{R}^n$. Consider the lattice $\sigma(\mathfrak{a})$ obtained by the embedding of an ideal $\mathfrak{a}$ of $\mathcal{O}_\mathcal{K}$. Suppose v is a short vector in $\sigma(\mathfrak{a})$ and let x_v be the corresponding element in $\mathfrak{a}$ such that $\sigma(x_v) = v$. Then $|\mathcal{N}(x_v)|$ is small and so the principal ideal $\langle x_v\rangle$ also has small norm.

Algorithm 1 describes Gélin's method of generating relations using ideal reduction. The parameters of the algorithm are k, A, β and B. The parameter k determines the number of ideals to be considered in each iteration, A determines the maximum value of the exponent to which the ideals are raised, the parameter β determines the block size of the block Korkine-Zolotarev reduction (BKZ-reduction), and the parameter B is the bound on the norms of the ideals in the factor basis. Gélin specifies k and A to be poly$(\log|\Delta_\mathcal{K}|)$. Further, the value of β is to be chosen such that the overall complexity is subexponential in $|\Delta_\mathcal{K}|$.

In Step 5 of Algorithm 1, a single vector from the BKZ reduction is returned. It is, however, possible that several vectors from the BKZ reduction have sufficiently small norms and a linear combination of these vectors can be tried. This saves the number of BKZ reductions and leads to practical speed-ups. Such improvement does not change the asymptotic complexity. We refer to [10] for details of this improvement and other implementation notes.

Algorithm 1: Gélin's method for generating relation using ideal reduction [10].

Input: The factor base $\mathcal{B} = \{\mathfrak{p}_1, \ldots, \mathfrak{p}_N\}$.
Output: The set of generated relations.

1 **while** *sufficiently many relations have not been obtained* **do**
2   Choose k random prime ideals $\mathfrak{p}_{j1}, \mathfrak{p}_{j2}, \ldots, \mathfrak{p}_{j_k}$ from $\mathcal{B}$
3   Choose k random exponents $e_{j1}, e_{j2}, \ldots, e_{j_k}$ from $\{1, 2, \ldots, A\}$
4   Set $\mathfrak{a} = \prod_{i=1}^{N} \mathfrak{p}_i{}^{e_i}$ with $e_i = 0$ if $i \notin \{j_1, \ldots, j_k\}$
5   Compute the smallest element v of the BKZ$_\beta$ reduced basis of the lattice $\sigma(\mathfrak{a})$
6   Obtain the algebraic integer x_v corresponding to v
7   Set $\mathfrak{b}$ as the unique ideal such that $\langle x_v \rangle = \mathfrak{a}\mathfrak{b}$
8   **if** $|\mathcal{N}(\mathfrak{b})|$ *is B-smooth* **then**
9   Obtain the factorization of $\mathfrak{b}$ such that $\mathfrak{b} = \prod_{i=1}^{N} \mathfrak{p}_i{}^{e'_i}$
10   Store the relation $\langle x_v \rangle = \prod \mathfrak{p}_i{}^{e_i + e'_i}$

Step 8 of Algorithm 1 checks whether $\mathcal{N}(\mathfrak{b})$ is B-smooth. The actual requirement is that the ideal $\mathfrak{b}$ is $\mathcal{B}$-smooth. If $\mathcal{N}(\mathfrak{b})$ is B-smooth, then it follows that $\mathfrak{b}$ is $\mathcal{B}$-smooth and a factorisation of $\mathfrak{b}$ can be obtained from a factorisation of $\mathcal{N}(\mathfrak{b})$. We refer to [10] for details.

Notation: Before proceeding further, we fix some notation.

B upper bound on the size of the ideals in the factor basis;
$\mathcal{B}$ the factor basis;
N number of prime ideals in the factor basis, i.e., $N = \#\mathcal{B}$;
C Bach's bound, i.e., $C = 12(\log |\Delta_\mathcal{K}|)^2$;
$\mathcal{C}$ the set of prime ideals whose norms are at most C;
N_b number of prime ideals whose norms are at most C, i.e., $N_b = \#\mathcal{C}$;
β block size of BKZ reduction.

3 Generating Relations from a Pseudo-random Walk on Ideals

Consider Algorithm 1. The asymptotic complexity of the algorithm is determined by B and β. The parameter β determines the upper bound on the norm of the element x_v; B and β determine the smoothness probability of the ideal $\langle x_v \rangle$.

We do not propose to change these parameters. Instead, we focus on the concrete complexity of an individual iteration. This consists of computing the ideal $\mathfrak{a}$, computing the BKZ-reduction, and the smoothness check of the norm of $\langle x_v \rangle$. The portions on BKZ-reduction and the smoothness check are also not modified. Our focus is on reducing the cost of computing the ideal $\mathfrak{a}$.

In Step 4 of Algorithm 1, the ideal $\langle \mathfrak{a} \rangle$ is computed as

$$\mathfrak{a} = \prod_{i=1}^{k} \mathfrak{p}_{j_i}^{e_{j_i}}. \tag{4}$$

Since each e_{j_i} is chosen randomly from $\{1, 2, \ldots, A\}$, computing $\mathfrak{p}_{j_i}^{e_{j_i}}$ requires about $\log_2 A$ ideal multiplications. So, the total cost of computing $\mathfrak{a}$ in (4) is about $(k-1)\log_2 A$ ideal multiplications. Below we describe a new algorithm for generating $\langle \mathfrak{a} \rangle$ which requires a single ideal multiplication.

Our idea is to perform a pseudo-random walk on a sufficiently large set of ideals. Each step of the walk will generate an ideal $\mathfrak{a}$ to which the BKZ-reduction and the rest of Algorithm 1 can be applied. The cost of each step will consist of a single ideal multiplication. The pseudo-random walk that we construct is inspired by Pollard's rho algorithm.

Suppose $[\mathfrak{t}_1, \ldots, \mathfrak{t}_m]$ is a list of m pre-computed ideals. Let $\mathcal{H}$ be a hash function which maps an ideal to the set $\{1, \ldots, m\}$. The requirement on $\mathcal{H}$ is that the output should be more or less uniformly distributed. For $i \geq 0$, the pseudo-random walk proceeds as follows. An ideal $\mathfrak{a}_0$ is chosen and for $i \geq 1$, $\mathfrak{a}_i$ is obtained as follows: let $m_i = \mathcal{H}(\mathfrak{a}_{i-1})$ and set $\mathfrak{a}_i = \mathfrak{a}_{i-1} \cdot \mathfrak{t}_{m_i}$.

The list $[\mathfrak{t}_1, \ldots, \mathfrak{t}_m]$ and some additional information are stored in a table $\mathfrak{T}$. We explain how the ideals $\mathfrak{t}_1, \ldots, \mathfrak{t}_m$ are constructed and the entries of the table $\mathfrak{T}$.

Let $\mathcal{C} = \{\mathfrak{q}_1, \ldots, \mathfrak{q}_{N_b}\}$ be the prime ideals whose norms are below Bach's bound. Let κ be a parameter and set $q = \lfloor N_b/\kappa \rfloor$ and $r = N_b - q\kappa$ so that we have $N_b = q\kappa + r = r(\kappa+1) + (q-r)\kappa$. The set $\mathcal{C}$ is randomly partitioned into q groups where the first r groups each have $\kappa + 1$ ideals and the last $q - r$ groups each have κ ideals. The ideals in each of the groups are multiplied together and the products are stored in $\mathfrak{T}$. Along with the product ideal, the information identifying the ideals that have been multiplied to obtain the product is also stored in $\mathfrak{T}$. The random partitioning of the ideals in $\mathcal{C}$ into groups, multiplying together the ideals in each group and storing them in table $\mathfrak{T}$ is carried out a total of R times. So, the number of entries in $\mathfrak{T}$ is qR, i.e., $m = qR$. Further, each ideal in $\mathcal{C}$ is represented a total of R times in the table $\mathfrak{T}$. The method for constructing $\mathfrak{T}$ is shown in Algorithm 2.

The entries of $\mathfrak{T}$ are pairs. For $0 \leq i \leq m-1$, $\mathfrak{T}[i]$ is an entry of the form $(\mathfrak{b}, (j_1, \ldots, j_s))$, where $\mathfrak{b} = \mathfrak{q}_{j_1} \cdots \mathfrak{q}_{j_s}$. By $\mathfrak{T}[i]$.ideal we will denote the ideal $\mathfrak{b}$ and by $\mathfrak{T}[i]$.index we will denote the tuple $(j_1, \ldots, j_s)$.

The table $\mathfrak{T}$ is constructed in a pre-computation phase. This pre-computation consists of about $Rq\kappa \approx RN_b$ ideal multplications. When R is a constant, the number of ideal multiplications required to construct $\mathfrak{T}$ is negligible with respect to the number of ideal multiplications required in all the iterations for relation collection.

Algorithm 2: Construction of the pre-computed table $\mathfrak{T}$.

```
Input: The set of prime ideals C = {q_1, ..., q_{N_b}} whose norms are at most
       Bach's bound.
Output: The table T.
1  q ← ⌊N_b/κ⌋, r ← N_b − qκ
2  T ← ∅
3  J ← {1, 2, ..., N_b}
4  for i_1 ← 1 to R do
5      I ← J
6      for i_2 ← 1 to q do
7          if i_2 ≤ r then
8              s ← (κ + 1)
9          else
10             s ← κ
11         {j_1, j_2, ..., j_s} ← random set of s distinct integers chosen from I
12         I ← I \ {j_1, j_2, ..., j_s}
13         b ← q_{j_1} ··· q_{j_s}
14         Append (b, (j_1, ..., j_s)) to T
```

How long should the pseudo-random walk proceed? There are several aspects to this question.

1. As the walk progresses, both the number and the exponents of the prime ideals occurring as factors in the ideals visited by the walk increases. So, a long walk can result in ideals with large norms.
2. From a practical point of view, in our Magma implementation we have observed that as the norms of the ideals increase, it becomes difficult to construct the lattices corresponding to the ideals.

In view of the above two points, long walks are not feasible, at least for Magma implementation. One way is to continue the walk as long as it is feasible to construct the associated lattice. Alternatively, one may put an *a priori* bound on the length of an individual walk. Determining the bound requires performing some initial experiments to obtain an idea of the number of steps that the walk can proceed without encountering the failure of lattice construction.

The algorithm for relation collection based on the pseudo-random walk is shown in Algorithm 3. The parameters β and B are the same as those in Algorithm 1. It is assumed that the table $\mathfrak{T}$ has been constructed prior to the execution of Algorithm 3. Recall that the ideals stored in $\mathfrak{T}$ are products of either κ or $\kappa + 1$ prime ideals whose norms are below N_b. Algorithm 3 uses an additional parameter κ_0 which determines the number of prime ideals to be multiplied together to obtain the starting ideal of a walk. The variable wlen records the current length of the walk, while the parameter λ represents the maximum length of any walk.

As above, let $\mathcal{C} = \{\mathfrak{q}_1, \ldots, \mathfrak{q}_{N_b}\}$ be the set of all prime ideals whose norms are below Bach's bound. The actual factor basis $\mathcal{B} = \{\mathfrak{p}_1, \ldots, \mathfrak{p}_N\}$ consists of all prime ideals whose norms are below the bound B. We assume that for $i = 1, \ldots, N_b$, $\mathfrak{p}_i = \mathfrak{q}_i$, i.e., the first N_b prime ideals in $\mathcal{B}$ have norms below Bach's bound.

Algorithm 3: Relation collection using pseudo-random walk.

Input: The factor base $\mathcal{B} = \{\mathfrak{p}_1, \ldots, \mathfrak{p}_N\}$.
Output: The set of generated relations.

```
1  while sufficiently many relations have not been found do
2      exp[i] = 0 for i = 1,...,N
3      Choose s randomly from {1,...,κ0}
4      Choose i_1,...,i_s randomly from {1,...,N_b}
5      Set a = p_{i_1} ··· p_{i_s}
6      exp[i] ← 1 for i = i_1,...,i_s;
7      wlen ← 1;
8      while wlen ≤ λ do
9          Compute the smallest element v of the BKZ_β reduced basis of the
           lattice σ(a)
10         Obtain the algebraic integer x_v corresponding to v
11         if |N(x_v)/N(a)| is B-smooth then
12             Set b as the unique ideal such that ⟨x_v⟩ = ab
13             Obtain the factorization of b such that b = ∏_{i=1}^{N} p_i^{e'_i}
14             for i ← 1 to N do
15                 exp[i] ← exp[i] + e'_i
16             Store the relation ⟨x_v⟩ = ∏_{i=1}^{N} p_i^{exp[i]} (i.e. ⟨x_v⟩ = ab)
17         ℓ ← H(a)
18         a ← a · T[ℓ].ideal
19         exp ← exp + T[ℓ].index
20         wlen ← wlen + 1;
```

In Algorithm 3, the ideals $\mathfrak{a}$ which are visited by the pseudo-random walk are products of prime ideals whose norms are below C. This is because a walk starts with such an ideal and for each step, the present ideal is multiplied with an ideal from the pre-computed table. Since the ideals in the pre-computed table are also of the same type, the property of being products of prime ideals whose norms are below C holds for all the ideals $\mathfrak{a}$ visited by the walk. On the other hand, the smoothness check of $\mathcal{N}(\mathfrak{b})$ is with respect to B. As a result, the algorithm tries to ensure that $\mathfrak{b}$ is smooth over the factor basis $\mathcal{B}$ rather than $\mathcal{C}$. Trying to ensure the smoothness of $\mathfrak{b}$ over $\mathcal{C}$ will result in the theoretical smoothness probability being too low. In practice, however, it may be possible to work with $\mathcal{C}$ as the factor basis as we found during our experiments. Recall that one of the stopping criterion for relation collection is that each element of $\mathcal{C}$ is involved in at least one relation. Since the ideals $\mathfrak{a}$ are products of ideals from $\mathcal{C}$, this criterion becomes somewhat easier to ensure.

Remarks

1. The list exp in Algorithm 3 is an array of N integers. This list is likely to be very sparse. So, it would be more efficient to represent exp as a list of pairs $[(i_1, e_1), \ldots, (i_s, e_s)]$, such that $\mathsf{exp}[i] = e_j$ if $i = i_j$, $j = 1, \ldots, s$ and $\mathsf{exp}[i] = 0$, otherwise. Another possibility is to represent exp as a list of integers where i_1 is repeated e_1 times, i_2 is repeated e_2 times, and so on. Since the integers $e_1, e_2, \ldots$ are quite likely to be equal to 1, this representation would be even more compact than storing the pairs $(i_1, e_1), (i_2, e_2), \ldots$. The operations on exp are to be suitably modified to be used with a compact representation.
2. The computation of the ideal $\mathfrak{b}$ in Step 7 of Algorithm 1 requires computing $\mathfrak{a}^{-1}$. Since $\mathcal{N}(\mathfrak{b})$ may not turn out to be B-smooth in Step 8, the computation of the ideal $\mathfrak{b}$ may not be required. Slightly altering the sequence of instructions, it is possible to avoid the computation of $\mathfrak{a}^{-1}$ in the cases where $\mathcal{N}(\mathfrak{b})$ is not B-smooth. Since $\mathcal{N}(\mathfrak{b}) = \mathcal{N}(x_v)/\mathcal{N}(\mathfrak{a})$, Algorithm 3 checks for the B-smoothness of $|\mathcal{N}(x_v)/\mathcal{N}(\mathfrak{a})|$ and computes $\mathfrak{b}$ only if the check is successful.
3. Note that the norms of the ideals in $\mathfrak{T}$ are known. The table $\mathfrak{T}$ can be expanded to store the norms of the ideals. This speeds up the computation of the norm of $\mathfrak{a}$ required in Step 11 of Algorithm 3. Along with the current ideal $\mathfrak{a}$ of the walk, its norm is also stored. The norm of the next ideal in the walk is obtained by multiplying the norm of the current ideal $\mathfrak{a}$ and the norm of the ideal in the location $\mathfrak{T}[\mathcal{H}(\mathfrak{a})]$.
4. The pseudo-random walk is a sequential procedure. Parallelism can be incorporated by starting independent pseudo-random walks. The table $\mathfrak{T}$ remains the same for all the walks. The starting ideals are chosen independently. This allows the separate walks to proceed independently.
5. Algorithm 3 has been described keeping in mind that the precision is fixed. Ideals in the initial steps of the walk have smaller norms compared to ideals in the later stages of the walk. So, an alternative approach would be to start with a smaller precision and increase the precision as the walk progresses. Since precision determines the efficiency of the various computational steps, working with a lower precision in the initial stages would lead to improved efficiency. The idea of increasing precision as the walk progresses can work up to a certain point, since as the precision becomes too large, the computation slows down considerably. At that point, it is better to switch to a new initial point and start a new walk.

We compare the number of ideal multiplications required by Algorithms 1 and 3. For the comparison, we ignore the number of ideal multiplications required to prepare the table $\mathfrak{T}$. This is a one-time activity whose cost has been mentioned above; amortised over the iterations required for generating all the relations, this cost is negligible. A pseudo-random walk in Algorithm 3 proceeds for at most λ steps. These λ steps require at most $\kappa_0 + \lambda - 2$ ideal multiplications. In comparison, Algorithm 1 requires about $\lambda(k-1)\log_2 A$ ideal multiplications for generating λ ideals $\mathfrak{a}$ in λ iterations. The other costs of both Algorithms 1 and 3, namely BKZ-reduction, smoothness checking and ideal factorisation, remain the same, though there is the issue of the norms of the ideals $\mathfrak{a}$ to be considered. We discuss this point below.

Though in principle, the reduction in the number of ideal multiplications should lead to improvement in time, there is a practical aspect that needs to be kept in mind. In practice, the time to multiply two ideals depends on the norms of the ideals. Even if one of the ideals has a relatively large norm, there is a noticeable increase in the time to compute the product. While determining the walk length, this aspect needs to be kept in the mind. As the walk length increases, so does the norms of the ideals visited by the walk. This means that even though each step of the walk requires a single multiplication, the time for this multiplication increases with the length of the walk. So, if a walk is too long, then it may turn out that computing the next ideal of the walk takes a very long time. In effect, this means that for Algorithm 3 to be competitive with Algorithm 1, the walk length should not be too long. In our experiments, we have fixed the walk length so that the norms of the ideals visited by the walk in Algorithm 3 are less than the norms of the ideals generated by Algorithm 1.

We further consider the issue of increase in the norms of the ideals visited by the walk. The first ideal in the walk has norm at most $\kappa_0 C$. At each step, the ideal in the walk is multiplied by κ or $\kappa+1$ prime ideals having norms at most C. So, the ideal obtained after i steps of the walk has norm at most $(\kappa_0 + i(\kappa+1))C$ which is at most $(\kappa_0 + \lambda(\kappa+1))C$ since $i \leq \lambda$. In comparison, the ideals $\mathfrak{a}$ generated in each iteration of Algorithm 1 have norms to be at most about $kAB/2$. So, the norms of the ideals in the walk in Algorithm 3 are at most the norms of ideals the ideals $\mathfrak{a}$ in Algorithm 1 if the condition $(\kappa_0 + \lambda(\kappa+1))C \leq kAB/2$ holds. The parameters may be chosen to satisfy this condition. Note that in Algorithm 1, the norms of all the ideals are about $kAB/2$, while for Algorithm 3, the norms of the ideals in the initial steps of the walk are lower.

A pseudo-random walk chooses the first ideal randomly while the subsequent ideals are chosen deterministically. A crucial requirement in the asymptotic analysis of the algorithm is the smoothness probability of random ideals. Since the ideals appearing in a walk do not have independent randomness, one has to heuristically assume that the result on the smoothness probability of random ideals applies to the ideals occurring in the walk. Our experiments show that there is no substantial effect on the smoothness probability.

Let Π denote the probability that an ideal $\mathfrak{a}$ in the walk leads to an ideal $\mathfrak{b}$ computed in Step 12 of Algorithm 3 which is smooth over $\mathcal{B}$ and hence leads to a relation. About $1/\Pi$ ideals need to be considered to obtain a single relation. A total of about N relations are required. Consequently, about N/Π ideals need to be considered to obtain all the relations. Each walk visits λ ideals. The total number of walks required to consider N/Π ideals is $N/(\Pi\lambda)$. The total number of starting points is $\sum_{s=1}^{\kappa_0}\binom{N_b}{s}$. So, to ensure that $N/(\Pi\lambda)$ walks are possible, we must have

$$\sum_{s=1}^{\kappa_0}\binom{N_b}{\kappa_0} \geq \frac{N}{\Pi\lambda}. \tag{5}$$

The parameter κ_0 has to be chosen to satisfy (5).

As proved in [10], the norm of $\mathfrak{b}$ considered in Step 12 of Algorithm 3 satisfies the bound $\mathcal{N}(\mathfrak{b}) \leq \beta^{n(n-1)/(2(\beta-1))}\sqrt{|\Delta_{\mathcal{K}}|}$. Further, the B-smoothness of $\mathcal{N}(\mathfrak{b})$ depends on the value of B. We do not suggest any change to either the value of β or to the value of B. The difference between Algorithms 1 and 3 is in the generation of the ideal $\mathfrak{a}$. As discussed above, the norm of $\mathfrak{a}$ considered by Algorithm 3 is never more than the norm of $\mathfrak{a}$ considered by Algorithm 1. The net effect of all these considerations is that the asymptotic result obtained in [10] for Algorithm 1 also holds for Algorithm 3. The advantage of Algorithm 3 over Algorithm 1 is in improved practical efficiency.

4 Implementation

We have implemented Algorithms 1 and 3 using Magma, version V2.22-3. We did not perform the entire class group computation. Rather, we performed two kinds of experiments. The first set of experiments compares the times required for relation collection by Algorithms 1 and 3. These experiments show that in general Algorithm 3 performs better than Algorithm 1. The second set of experiments performs the entire relation collection step for two fields having discriminants of sizes about 157 and 256 bits. By entire relation collection we mean that the number of relations collected is a little more than the number of prime ideals whose norms are below the Bach bound. The ability to collect such a large set of relations demonstrates that Algorithm 3 can actually work in practice.

Various issues arise while implementing Algorithm 1. For guidance in determining parameters, we considered the asymptotic analysis. This, however, fails to provide sufficient information and in some cases lead to substantially less efficient parameter choices. Various implementation issues and timing results are available in the full version of this paper [14]. The Magma codes for Gélin's algorithm and the algorithm described in this paper are available at the following link.

https://github.com/Madhurima11/Class-Group-Computation

5 Conclusion

In this paper, we have introduced a technique to perform a pseudo-random walk over ideals. After the first step, each step of the walk requires a single ideal multiplication. The ideals visited by the walk are used for relation collection in exactly the same manner as used in Gélin's algorithm [10]. The practical advantage over Gélin's algorithm is the reduction in the number of ideal multiplications required to generate the next ideal to be tested. Our Magma implementations of both the new algorithm and Gélin's algorithm confirm that there is indeed a practical speed-up.

Acknowledgement. We would like to thank Alexandre Gélin for various discussions regarding class group and its computational aspects. We also appreciate the help provided by Allan Steel, Claus Fieker, Geoffrey Bailey, Jean-François Biasse, John Cannon for computations with lattices using Magma. Thanks to the reviewers for their comments.

References

1. Bach, E.: Explicit bounds for primality testing and related problems. Math. Comput. **55**(191), 355–380 (1990)
2. Biasse, J.-F.: An $L(1/3)$ algorithm for ideal class group and regulator computation in certain number fields. Math. Comput. **83**(288), 2005–2031 (2014)
3. Biasse, J.-F., Fieker, C.: Subexponential class group and unit group computation in large degree number fields. LMS J. Comput. Math. **17**(A), 385–403 (2014)
4. Boneh, D., Bünz, B., Fisch, B.: Batching techniques for accumulators with applications to IOPs and stateless blockchains. In: Boldyreva, A., Micciancio, D. (eds.) CRYPTO 2019. LNCS, vol. 11692, pp. 561–586. Springer, Cham (2019). https://doi.org/10.1007/978-3-030-26948-7_20
5. Buchmann, J.: A subexponential algorithm for the determination of class groups and regulators of algebraic number fields. Séminaire de théorie des nombres, Paris **1989**(1990), 27–41 (1988)
6. Bünz, B., Fisch, B., Szepieniec, A.: Transparent SNARKs from DARK compilers. In: Canteaut, A., Ishai, Y. (eds.) EUROCRYPT 2020. LNCS, vol. 12105, pp. 677–706. Springer, Cham (2020). https://doi.org/10.1007/978-3-030-45721-1_24
7. Cohen, H.: A Course in Computational Algebraic Number Theory. Graduate Texts in Mathematics, vol. 138, p. 88. Springer, Heidelberg (1993). https://doi.org/10.1007/978-3-662-02945-9
8. Cohen, H., Diaz, F.D.Y., Olivier, M.: Subexponential algorithms for class group and unit computations. J. Symb. Comput. **24**(3–4), 433–441 (1997)
9. Dobson, S., Galbraith, S.D., Smith, B.: Trustless unknown-order groups. Cryptology ePrint Archive, Report 2020/196 (2020). https://ia.cr/2020/196
10. Gélin, A.: On the complexity of class group computations for large degree number fields. arXiv preprint arXiv:1810.11396 (2018)
11. Gélin, A.: Reducing the complexity for class group computations using small defining polynomials. arXiv preprint arXiv:1810.12010 (2018)
12. Gélin, A., Joux, A.: Reducing number field defining polynomials: an application to class group computations. LMS J. Comput. Math. **19**(A), 315–331 (2016)
13. Hafner, J.L., McCurley, K.S.: A rigorous subexponential algorithm for computation of class groups. J. Am. Math. Soc. **2**(4), 839–850 (1989)
14. Mukhopadhyay, M., Sarkar, P.: Pseudo-random walk on ideals: practical speed-up in relation collection for class group computation. Cryptology ePrint Archive, Paper 2021/792. https://eprint.iacr.org/2021/792
15. Shanks, D.: Class number, a theory of factorization, and genera. In: Proceedings of Symposia in Pure Mathematics, vol. 20, pp. 415–440 (1969)
16. Shanks, D.: The infrastructure of a real quadratic field and its applications. In: Proceedings of the 1972 Number Theory Conference, pp. 217–224 (1972)

Efficient Extended GCD and Class Groups from Secure Integer Arithmetic

Berry Schoenmakers and Toon Segers(✉)

Department Math & CS, TU Eindhoven, Eindhoven, The Netherlands
{l.a.m.schoenmakers,a.j.m.segers}@tue.nl

Abstract. In this paper we first present an efficient protocol for the secure computation of the extended greatest common divisor, assuming basic secure integer arithmetic common to many MPC frameworks. The protocol is based on Bernstein and Yang's constant-time 2-adic algorithm, which we adapt to work purely over the integers. This yields a much better approach for the MPC setting, but raises a new concern about the growth of the Bézout coefficients. By a careful analysis we are able to prove that the Bézout coefficients in our protocol will never exceed $3\max(a, b)$ in absolute value for inputs a and b. Next, we present efficient protocols for implementing class groups of imaginary quadratic number fields in the MPC setting. We start from Shanks' original algorithms for the efficient composition of binary quadratic forms and combine these with our particular adaptation of a forms reduction algorithm due to Agarwal and Frandsen. We will formulate this result in terms of secure groups, which are introduced as oblivious data structures implementing finite groups in a privacy-preserving manner. Our results show how class group operations can be run efficiently between multiple parties operating jointly on secret-shared group elements. We have integrated secure class groups in MPyC along with other instances of secure groups such as Schnorr groups and elliptic curves.

1 Introduction

Recall that a finite group consists of a finite set $\mathbb{G}$ of group elements (including the identity element) together with an associative group operation $*$ and the corresponding inverse operation. In this paper we will formulate our protocols implementing class groups in an MPC setting in terms of secure groups. A secure group is introduced as a cryptographic scheme for a group $(\mathbb{G}, *)$ supporting a secret-shared representation of group elements from $\mathbb{G}$ and an oblivious implementation of the group operation $*$. We note that Bar-Ilan and Beaver [BIB89, Lemma 6] already considered a generic protocol for secure inversion of group elements, as a natural generalization of secure matrix multiplication. A secure group will cover further tasks such as protocols for group inversion, equality tests of group elements, and random sampling, all operating on secret-shared values.

Our main goal is to support secure class groups. We focus on ideal class groups of imaginary quadratic fields, which have been studied extensively for

S. Dolev et al. (Eds.): CSCML 2023, LNCS 13914, pp. 32–48, 2023.
https://doi.org/10.1007/978-3-031-34671-2_3

use in cryptography by Buchmann et al. (see [BV07] and references therein) and recently got renewed interest (see, e.g., [Wes19, BHR+21, DGS22]). For the implementation of the class group operation, we present efficient protocols for the composition and reduction of binary quadratic forms.

A key step in our protocol for the composition of forms is the secure computation of the extended greatest common divisor. We start from the constant-time algorithms proposed by Bernstein and Yang [BY19]. Their approach relies on the use of 2-adic arithmetic but we will work purely over the integers for an efficient solution in the MPC setting. The core of the protocol centers around Bernstein and Yang's divsteps2 algorithm which takes a fixed number of roughly 3ℓ iterations when run on numbers of bit length at most ℓ. The work for each iteration is dominated by secure comparisons with numbers of bit length at most $\log_2 \ell$, requiring $O(\log \ell)$ secure multiplications each. We develop a novel analysis showing that the Bézout coefficients are bounded above by $3\max(a, b)$ in absolute value when computing the extended greatest common divisor of numbers a and b. The number of $O(\ell \log \ell)$ secure multiplications compares favorably to, e.g., a basic binary gcd algorithm [MvOV96, Algorithm 14.61] which requires $O(\ell^2)$ secure multiplications. Other binary gcd algorithms such as Bojanczyk and Brent [BB87] (and related algorithms as considered by [BY19]) as well as the algorithm attributed to Penk [Knu69, Exercise 4.5.2.39] also lead to $O(\ell^2)$ secure multiplications due to the use of either full-size ℓ-bit secure comparisons or inner loops requiring $O(\ell)$ secure multiplications in the worst case.

To obtain an efficient protocol for the reduction of forms we start from the algorithm proposed by Agarwal and Frandsen [AF06]. A key property of their approach is that the use of full integer divisions is avoided in the main loop of the form reduction algorithm. We show how to adapt their algorithm to the MPC setting such that the work for each iteration of the main loop is limited to a small constant number of secure comparisons and operations of similar complexity requiring $O(\ell)$ secure multiplications each. For discriminant $\Delta < 0$ of bit length ℓ and forms with coefficients all of bit length at most ℓ, we show that $\ell/2$ iterations suffice for the reduction, which leads to $O(\ell^2)$ secure multiplications overall.

We conclude the paper with a brief discussion of the support for secure groups in MPyC and applications in threshold cryptography and verifiable MPC. Apart from the implementation of class groups, MPyC also supports various other secure groups such as elliptic curves and quadratic residues in a multiparty setting. We use these secure groups to implement noninteractive zero-knowledge proofs where the role of the prover is distributed between multiple parties.

Roadmap. The paper is organized as follows. Section 2 introduces preliminaries for secure multiparty computation. Section 3 presents the protocol for the extended greatest common divisor. Section 4 introduces a cryptographic scheme for secure groups. Section 5 presents generic protocols related to secure groups for the following tasks: conditional (if-else), random sampling, inversion and exponentiation. Section 6 presents specific protocols for secure class groups. Section 7 concludes with remarks about implementation and applications.

2 MPC Setting

We consider an MPC setting with m parties tolerating a dishonest minority of up to t passively corrupt parties, $0 \leq t < m/2$. The basic protocols for secure addition and multiplication over a finite field rely on Shamir secret sharing [BGW88, GRR98]. We write $[\![a]\!]$ (or sometimes $[\![a]\!]_p$) for a Shamir secret sharing of a finite field element $a \in \mathbb{F}_p$. For the scope of this paper, it suffices to consider finite fields of prime order p, where we assume $p > m$. For secure integer arithmetic, we use the common representation for signed integers a in a bounded range, e.g., with $\text{len}(a) \leq \ell \ll p$, where $\text{len}(a) = \lceil \log_2(|a| + 1) \rceil$. The amount $\text{len}(p) - \ell$ of excess bits is often referred to as "headroom".

To operate on secret-shared integer values we use common protocols for secure integer arithmetic as building blocks. E.g., we let $[\![r]\!] \leftarrow \mathsf{random}(\mathbb{F}_p)$ denote the generation of a uniformly random $r \in_R \mathbb{F}_p$ and $[\![a_0]\!], \ldots, [\![a_{\ell-1}]\!] \leftarrow \mathsf{bits}([\![a]\!])$ denote the bit decomposition of a. Also $[\![d]\!] \leftarrow \mathsf{if\_else}([\![c]\!], [\![a]\!], [\![b]\!])$ denotes the secure conditional, where $d = c(a-b)+b$. A protocol that generates a vector of j secret-shared bits in $\mathbb{F}_p \cap \{0, 1\}$ is denoted by $\mathsf{random\ bits}(\mathbb{F}_p, j)$. See [Hoo12, Tof07, DFK+06] for these and other common MPC protocols, which we will use as black boxes.

We also use secure integer division $([\![q]\!], [\![r]\!]) \leftarrow \mathsf{divmod}([\![a]\!], [\![b]\!])$ with $a = bq + r$ and $0 \leq r < b$ as a primitive. Our implementation is based on the Newton-Raphson method [ACS02, CS10]. Finally, we assume an efficient protocol for securely determining the bit length $[\![\text{len}(a)]\!]$. See Sect. 7.

3 Secure Extended GCD

This section presents a new protocol for computing the extended greatest common divisor (xgcd) of secret-shared integers. In the next section we use this protocol for the group operation for class groups of imaginary quadratic fields in MPC.

We first present Algorithm 1, which can be viewed as an alternative to the divsteps2 algorithm by Bernstein and Yang [BY19], on which it is based, but without the use of any 2-adic arithmetic. Algorithm 1 works purely over the integers. To compute, for instance, a modular inverse we do not need any (potentially costly) pre- or postprocessing (in contrast to Bernstein and Yang's algorithm recip2), which would complicate the conversion to an MPC protocol. Hence, our xgcd algorithm may also be of independent interest for other settings in which the use of 2-adic arithmetic is not straightforward. (Using precomputation for modular inverses can be beneficial, particularly in applications that reuse the modulus. In settings where inputs are not reusable, the precomputation cannot be reused. This is the case in our application of the xgcd to compute class group operations.)

As starting point we take the constant-time extended gcd algorithm divsteps2 by Bernstein and Yang [BY19, Figure 10.1]. Our protocol xgcd, see Protocol 1, retains the algorithmic flow of their divsteps2 algorithm, which is controlled by the following step function [BY19, Section 8]:

Algorithm 1. $\mathsf{xgcd}(n, a, b)$ $\qquad n, a, b \in \mathbb{N}$, a odd

1: $\delta, f, g, u, v, q, r \leftarrow 1, a, b, 1, 0, 0, 1$ $\qquad \triangleright\ f = ua + vb,\ g = qa + rb$
2: **for** $i = 1$ **to** n **do**
3: $\quad g_0 \leftarrow g \bmod 2$
4: $\quad$ **if** $\delta > 0$ and $g_0 = 1$ **then**
5: $\quad\quad \delta, f, g, u, v, q, r \leftarrow -\delta, g, -f, q, r, -u, -v$
6: $\quad$ **if** $g_0 = 1$ **then**
7: $\quad\quad g, q, r \leftarrow g + f, q + u, r + v$
8: $\quad$ **if** $r \bmod 2 = 1$ **then**
9: $\quad\quad q, r \leftarrow q - b, r + a$
10: $\quad \delta, g, q, r \leftarrow \delta + 1, g/2, q/2, r/2$
11: **if** $f < 0$ **then**
12: $\quad f, u, v \leftarrow -f, -u, -v$
13: **return** f, u, v

$$\mathsf{divstep}(\delta, f, g) = \begin{cases} (1 - \delta, g, (g - f)/2), & \text{if } \delta > 0 \text{ and } g \text{ is odd}, \\ (1 + \delta, f, (g + (g \bmod 2)f)/2), & \text{otherwise.} \end{cases} \tag{1}$$

Throughout, variable f is ensured to be an odd integer, and therefore the divisions by 2 are without remainder in both cases of the step function. Bernstein and Yang argue that this step function compares favorably with alternatives from the literature, e.g., the Brent–Kung step function [BK85] and the Stehlé–Zimmermann step function [SZ04]. The computational overhead of function $\mathsf{divstep}$ is small, and the required number of iterations n as a function of the bit lengths of the inputs a and b compares favorably with the alternatives. Concretely, with $(\delta_n, f_n, g_n) = \mathsf{divstep}^n(1, a, b)$ for odd a, Bernstein and Yang prove that $f_n = \gcd(a, b)$ and $g_n = 0$ holds for $n = \mathsf{iterations}(\ell)$, where $\ell = \max(\mathrm{len}(a), \mathrm{len}(b))$ and

$$\mathsf{iterations}(d) = \begin{cases} \lfloor (49d + 80)/17 \rfloor, & \text{if } d < 46, \\ \lfloor (49d + 57)/17 \rfloor, & \text{otherwise.} \end{cases} \tag{2}$$

Hence, $\mathsf{iterations}(\ell) \approx 3\ell$.

The first major change compared to Bernstein and Yang's $\mathsf{divsteps2}$ algorithm is that we entirely drop the use of truncation for f and g. In our MPC setting, f and g are secret-shared values (over a prime field of large order) and therefore limiting the sizes of f and g is not useful. The second major change is that we will avoid the use of 2-adic arithmetic entirely, by ensuring that the Bézout coefficients will remain integral throughout all iterations. Concretely, this means that we will make sure that coefficients q and r are even before the division by 2 at the end of each iteration: if q and r are odd, we use $q - b$ and $r + a$ instead, which will then be even, because a is odd by assumption. We thus obtain Algorithm 1 as an alternative to Bernstein-Yang's constant-time algorithm.

Protocol 1. $\mathsf{xgcd}(n, [\![a]\!], [\![b]\!])$ $\quad n, a, b \in \mathbb{N}$, a odd

1: $[\![\delta]\!], [\![f]\!], [\![g]\!], [\![v]\!], [\![r]\!] \leftarrow 1, [\![a]\!], [\![b]\!], 0, 1$
2: **for** $i = 1$ **to** n **do**
3: $\quad [\![g_0]\!] \leftarrow [\![g \bmod 2]\!]$
4: $\quad$ **if** $[\![\delta]\!] > 0$ and $[\![g_0]\!] = 1$ **then**
5: $\quad\quad [\![\delta]\!], [\![f]\!], [\![g]\!], [\![v]\!], [\![r]\!] \leftarrow -[\![\delta]\!], [\![g]\!], -[\![f]\!], [\![r]\!], -[\![v]\!]$
6: $\quad$ **if** $[\![g_0]\!] = 1$ **then**
7: $\quad\quad [\![g]\!], [\![r]\!] \leftarrow [\![g]\!] + [\![f]\!], [\![r]\!] + [\![v]\!]$
8: $\quad$ **if** $[\![r \bmod 2]\!] = 1$ **then**
9: $\quad\quad [\![r]\!] \leftarrow [\![r]\!] + [\![a]\!]$
10: $\quad [\![\delta]\!], [\![g]\!], [\![r]\!] \leftarrow [\![\delta]\!] + 1, [\![g]\!]/2, [\![r]\!]/2$
11: **if** $[\![f]\!] < 0$ **then**
12: $\quad [\![f]\!], [\![v]\!] \leftarrow -[\![f]\!], -[\![v]\!]$
13: $[\![u]\!] \leftarrow ([\![f]\!] - [\![v]\!] * [\![b]\!])/[\![a]\!]$
14: **return** $[\![f]\!], [\![u]\!], [\![v]\!]$ $\quad \triangleright f = ua + vb = \gcd(a, b)$

We turn Algorithm 1 into a secure protocol operating on secret-shared values as follows. We drop variables q and u from the main loop, and instead set $u = (f - vb)/a$ at the end. Overall, this saves $3n$ secure multiplications for q and n secure multiplications for u, that are otherwise needed to implement the if-then statements obliviously. See Protocol 1 for the result.

All operations on the remaining variables δ, g, f, v, r are done securely. The secure computations of $[\![g \bmod 2]\!]$ and $[\![r \bmod 2]\!]$ are done by using a secure protocol for computing the least significant bit. The secure computation of $[\![\delta > 0]\!]$ is the most expensive part of each iteration. However, by taking into account that δ is bounded above by $i + 1$, we can further reduce the cost of this secure comparison.

The result is a secure protocol with a single loop of $n \approx 3\ell$ iterations, where ℓ denotes the maximum bit length for a and b. Per iteration, the work is proportional to $O(\log \ell)$ secure multiplications, as the comparison for $\log \ell$ bit numbers determines the complexity of the adapted divstep. The resulting $O(\ell \log \ell)$ compares favorably to, e.g., the binary extended gcd from [MvOV96, Algorithm 14.61], which would require $O(\ell^2)$ secure multiplications.

For the general case, where a is not known to be odd, we use an auxiliary protocol to securely count the number of trailing zeros of a given secret-shared integer. We have designed and implemented a novel approach requiring $O(\ell)$ secure multiplications in $O(1)$ rounds for this task. Moreover, we have designed and implemented protocols for the secure modular inverse of a modulo b (provided $\gcd(a, b) = 1$), and for computing $\gcd(a, b)$ and $\mathrm{lcm}(a, b)$ securely. See Sect. 7.

Bounding the Bezout Coefficients

We conclude this section with a result on the bounds for the Bézout coefficients computed by our xgcd algorithm (and protocol). The novelty of this result is due to the fact that the bounds are explicit (avoiding big-O notation), while the xgcd algorithm avoids full-sized comparisons to control the size of intermediate variables. To this end, we analyse Algorithm 1 and show in Theorem 1 below that the Bézout coefficients are bounded by $3a$.

For the analysis, we partition an execution of Algorithm 1 into k consecutive runs as follows. A new run starts with each assignment to variable v: the first run starts with the initialization $v \leftarrow 0$ in the first line and each next run starts with an update $v \leftarrow r$ in line 5.

By v_j we denote the value assigned to v at the start of the jth run, $1 \leq j \leq k$. Hence, $v_1 = 0$ and v_k will be the final value of v. We define $v_0 = -1$.

Let e_j denote the length of the jth run, which is defined as the number of times δ is incremented (in line 10) during the run. Hence $\sum_{j=1}^{k} e_j = n$. Note that $e_1 = 0$ is possible, but $e_j \geq 1$ for $2 \leq j \leq k$.

Let δ_j be the value of δ at the end of the jth run. We define $\delta_0 = -1$. Then $\delta_j \geq 1$ for $1 \leq j \leq k-1$. Note that $\delta_k \leq 0$ is possible, but this will be irrelevant for the analysis. Also, $e_j = \delta_j + \delta_{j-1}$ for $1 \leq j \leq k$.

We want to show that all v_j are bounded by induction on j.

At the start of the jth run, r is set to $-v_{j-1}$. At the end of each loop iteration the value of r, which is ensured to be even at that point, is halved. This will limit the growth of $|r|$. On the other hand, $|r|$ may grow because of the additions of v and/or a. Since $\delta > 0$ precisely during the last $\delta_j - 1$ iterations of the jth run, however, it follows that g must then be even, and therefore $|r|$ can only grow because of the addition of a.

To analyze the extreme values for r at the end of each run, we introduce the following three auxiliary functions:

$$\begin{aligned}
\phi(r, v, N) &= (r + v(2^N - 1))/2^N, \\
\psi(r, v, \delta, \delta') &= \phi(\phi(r, \min(v, 0), \delta + 1), 0, \delta' - 1), \\
\Psi(r, v, \delta, \delta') &= \phi(\phi(r, \max(v, 0) + a, \delta + 1), a, \delta' - 1).
\end{aligned}$$

The relevant monotonicity properties of these functions are stated as follows.

Lemma 1. *Functions $\phi(r, v, N)$, $\psi(r, v, \delta, \delta')$, and $\Psi(r, v, \delta, \delta')$ are increasing in r and nondecreasing in v.*

The next two lemmas establish the basic bounds for v_j, which follow almost directly from the way we have defined ψ and Ψ.

Lemma 2. *We have $v_0 = -1$, $v_1 = 0$, and for $2 \leq j \leq k$:*

$$\psi(-v_{j-2}, v_{j-1}, \delta_{j-1}, \delta_j) \leq v_j \leq \Psi(-v_{j-2}, v_{j-1}, \delta_{j-1}, \delta_j).$$

Lemma 3. *$v_j \leq \Psi(-v_{j-2}, \Psi(-v_{j-3}, v_{j-2}, \delta_{j-2}, \delta_{j-1}), \delta_{j-1}, \delta_j)$ for $j \geq 3$.*

Proof. From Lemma 2, we know that $v_j \leq \Psi(-v_{j-2}, v_{j-1}, \delta_{j-1}, \delta_j)$. Since function $\Psi(r, v, \delta, \delta')$ is nondecreasing in v, we get the claimed bound for v_j because we also have $v_{j-1} \leq \Psi(-v_{j-3}, v_{j-2}, \delta_{j-2}, \delta_{j-1})$ (using that $j - 1 \geq 2$). □

Finally, we prove our main result, establishing a nontrivial bound for the Bézout coefficients v_j.

Theorem 1. *$|v_j| \leq 3a$ for $0 \leq j \leq k$.*

Proof. The proof is by induction on j. Since $v_0 = -1$ and $v_1 = 0$, the bound clearly holds for $j \leq 1$ as $a \geq 1$. For v_2, we have the following bounds from Lemma 2:

$$\psi(1, 0, \delta_1, \delta_2) \leq v_2 \leq \Psi(1, 0, \delta_1, \delta_2).$$

Since $\psi(1, 0, \delta_1, \delta_2) \geq 0$ and $\Psi(1, 0, \delta_1, \delta_2) \leq a$, the bound also holds for $j = 2$.

For $j \geq 3$, we first prove the lower bound $v_j \geq -3a$. From Lemma 2, we have

$$v_j \geq \psi(-v_{j-2}, v_{j-1}, \delta_{j-1}, \delta_j).$$

Since $\psi(r, v, \delta, \delta')$ is increasing in r and nondecreasing in v, we then get

$$v_j \geq \psi(-3a, -3a, \delta_{j-1}, \delta_j),$$

as $-v_{j-2} \geq -3a$ and $v_{j-1} \geq -3a$ follow from the induction hypothesis.

Hence, the lower bound for v_j holds, as

$$\psi(-3a, -3a, \delta_{j-1}, \delta_j) = \phi(-3a, 0, \delta_j - 1) \geq -3a.$$

Next we prove the upper bound $v_j \leq 3a$. From Lemma 3, we have

$$v_j \leq \Psi(-v_{j-2}, \Psi(-v_{j-3}, v_{j-2}, \delta_{j-2}, \delta_{j-1}), \delta_{j-1}, \delta_j).$$

Since $\Psi(r, v, \delta, \delta')$ is increasing in r and nondecreasing in v, we get

$$v_j \leq \Psi(-v_{j-2}, \Psi(3a, v_{j-2}, \delta_{j-2}, \delta_{j-1}), \delta_{j-1}, \delta_j),$$

as $-v_{j-3} \leq 3a$ on account of the induction hypothesis.

To complete the proof we distinguish the cases $v_{j-2} \leq 0$ and $v_{j-2} > 0$.

If $v_{j-2} \leq 0$, we see that

$$\Psi(\alpha a, v_{j-2}, \delta_{j-2}, \delta_{j-1}) = \phi(\phi(3a, a, \delta_{j-2} + 1), a, \delta_{j-1} - 1),$$

which is independent of v_{j-2} and is bounded above by $\phi(3a, a, 2) = 3a/2$.

Since $\Psi(r, v, \delta, \delta')$ is nondecreasing in v, we therefore have

$$v_j \leq \Psi(-v_{j-2}, 3a/2, \delta_{j-1}, \delta_j).$$

And as $\Psi(r, v, \delta, \delta')$ is increasing in r, we conclude

$$v_j \leq \Psi(3a, 3a/2, \delta_{j-1}, \delta_j),$$

as $-v_{j-2} \leq 3a$ on account of the induction hypothesis. This leads to

$$v_j \leq \phi(3a, 5a/2, 2) = (3a + 15a/2)/4 = 21a/8 < 3a.$$

If $v_{j-2} > 0$, we see that

$$\Psi(3a, v_{j-2}, \delta_{j-2}, \delta_{j-1}) = \phi(\phi(3a, v_{j-2} + a, \delta_{j-2} + 1), a, \delta_{j-1} - 1)$$

still depends on v_{j-2}. To prove the upper bound for v_j in this case, we therefore introduce:

$$f(v, \delta, \delta', \delta'') = \Psi(-v, \Psi(3a, v, \delta, \delta'), \delta', \delta''),$$

for $0 < v \leq 3a$, hence $\Psi(r, v, \delta, \delta') = \phi(\phi(r, v + a, \delta + 1), a, \delta' - 1)$.

For the derivative of f with respect to v we get, for $\delta, \delta', \delta'' \geq 1$:

$$\frac{\partial f}{\partial v} = \left(3 \cdot 2^{\delta+\delta'} - 2^{\delta'+1} - 2^{\delta+1} + 1\right) 2^{-\delta-2\delta'-\delta''} > 0,$$

hence we set $v = 3a$ at its maximal value.

Finally, we have that $g(\delta, \delta', \delta'') = f(3a, \delta, \delta', \delta'')$ is increasing in δ and decreasing both in δ' and in δ'', for $\delta, \delta', \delta'' \geq 1$:

$$\begin{aligned}
\frac{\partial g}{\partial \delta} &= a\left(2^{\delta'+1} - 1\right) 2^{-\delta-2\delta'-\delta''} \log 2 > 0 \\
\frac{\partial g}{\partial \delta'} &= -a\left(7 \cdot 2^{\delta+\delta'} - 3 \cdot 2^{\delta+2} - 2^{\delta'+1} + 2\right) 2^{-\delta-2\delta'-\delta''} \log 2 < 0 \\
\frac{\partial g}{\partial \delta''} &= -a\left(7 \cdot 2^{\delta+\delta'} + 2^{\delta+2\delta'+1} - 3 \cdot 2^{\delta+1} - 2^{\delta'+1} + 1\right) 2^{-\delta-2\delta'-\delta''} \log 2 < 0.
\end{aligned}$$

Therefore, the maximum value of f is

$$\lim_{\delta \to \infty} f(3a, \delta, 1, 1) = 3a,$$

which is the desired upper bound for v_j. □

4 Definition of Secure Groups

Let $(\mathbb{G}, *)$ denote an arbitrary finite group, written multiplicatively. To define secure group schemes we will use $[\![a]\!]_{\mathbb{G}}$ to denote a secure representation of a group element $a \in \mathbb{G}$. In general, such a secure representation $[\![a]\!]_{\mathbb{G}}$ will be constructed from one or more secret-shared finite field elements. Also, we may compose secure representations, e.g., we may put $[\![a]\!]_{\mathbb{G}} = ([\![a']\!]_{\mathbb{G}'}, [\![a'']\!]_{\mathbb{G}''})$ as secure representation of $a = (a', a'') \in \mathbb{G}$ for a direct product group $\mathbb{G} = \mathbb{G}' \times \mathbb{G}''$.

A secure group scheme should allow us to apply the group operation $*$ to given $[\![a]\!]_{\mathbb{G}}$ and $[\![b]\!]_{\mathbb{G}}$, and obtain $[\![a * b]\!]_{\mathbb{G}}$ as a result. We will refer to this as a secure application of the group operation, or as a secure group operation, for short. Similarly, a secure group scheme may allow us to perform a secure inversion, which lets us obtain $[\![a^{-1}]\!]_{\mathbb{G}}$ for a given $[\![a]\!]_{\mathbb{G}}$. Another common task is to generate a random sample $[\![a]\!]_{\mathbb{G}}$ with $a \in_R \mathbb{G}$, hence to obtain a secure representation of a group element a drawn from the uniform distribution on $\mathbb{G}$.

To cover a representative set of tasks, we define secure groups as follows.

Definition 1. *A* **secure group scheme** *for* $(\mathbb{G}, *)$ *comprises protocols for the following tasks, where* $a, b \in \mathbb{G}$.

Group operation. *Given* $[\![a]\!]_\mathbb{G}$ *and* $[\![b]\!]_\mathbb{G}$, *compute* $[\![a * b]\!]_\mathbb{G}$.
Inversion. *Given* $[\![a]\!]_\mathbb{G}$, *compute* $[\![a^{-1}]\!]_\mathbb{G}$.
Equality test. *Given* $[\![a]\!]_\mathbb{G}$ *and* $[\![b]\!]_\mathbb{G}$, *compute* $[\![a = b]\!]$.
Conditional. *Given* $[\![a]\!]_\mathbb{G}, [\![b]\!]_\mathbb{G}$ *and* $[\![x]\!]$ *with* $x \in \{0, 1\}$, *compute* $[\![a^x b^{1-x}]\!]_\mathbb{G}$.
Exponentiation. *Given* $[\![a]\!]_\mathbb{G}$ *and* $[\![x]\!]$ *with* $x \in \mathbb{Z}$, *compute* $[\![a^x]\!]_\mathbb{G}$.
Random sampling. *Compute* $[\![a]\!]_\mathbb{G}$ *with* $a \in_R \mathbb{G}$ *(or, close to uniform).*
En/decoding. *For a set* S *and an injective map* $\sigma : S \to \mathbb{G}$:
- Encoding. *Given* $[\![s]\!]$, *compute* $[\![\sigma(s)]\!]_\mathbb{G}$.
- Decoding. *Given* $[\![a]\!]_\mathbb{G}$ *with* $a \in \sigma(S)$, *compute* $[\![\sigma^{-1}(a)]\!]$.

By default, all inputs and outputs to these protocols are secret-shared. In addition, the scheme also comprises variants of default protocols where some of the inputs and outputs are public and/or private.

By definition, a secure group scheme thus includes an ordinary group scheme where all protocols operate on public values. Also note that there may be multiple encoding/decodings for a group $\mathbb{G}$, each defined on a specific set S.

5 Generic Protocols for Secure Groups

The implementation of basic tasks such as the group operation and en/decoding strongly depend on the representation of the secure group. For other tasks, however, we may look for generic implementations. This section presents generic protocols for the following four tasks from Definition 1: conditional (if-else), random sampling, inverse, and exponentiation.

Secure Conditional. The secure conditional can be evaluated in terms of secure exponentiation in either of these two generic ways, if applicable: as $[\![a]\!]_\mathbb{G}^{[\![x]\!]} * [\![b]\!]_\mathbb{G}^{1-[\![x]\!]}$, or as $([\![a]\!]_\mathbb{G}/[\![b]\!]_\mathbb{G})^{[\![x]\!]} * [\![b]\!]_\mathbb{G}$, where $x \in \{0, 1\}$. This way it suffices to implement the basic operation $[\![a]\!]_\mathbb{G}^{[\![x]\!]}$ with $x \in \{0, 1\}$. And if base a is publicly known, it even suffices to implement $a^{[\![x]\!]}$ (with Protocol 7 as a good option to implement this operation).

In pseudocode, the secure conditional is denoted as if-else($[\![x]\!], [\![a]\!]_\mathbb{G}, [\![b]\!]_\mathbb{G}$), with the obvious variations if a and/or b are publicly known. The condition $[\![x]\!]$ should always be secret.

Secure Random Sampling. The task of secure random sampling from a group $\mathbb{G}$ corresponds to generating $[\![a]\!]_\mathbb{G}$ with $a \in_R \mathbb{G}$ (or, close to uniform), see Definition 1. In this section we present several methods for secure random sampling.

A generic protocol is obtained by letting party $\mathcal{P}_i$ generate a random group element $a_i \in_R \mathbb{G}$ privately, and then use t (threshold) secure group operations to form $[\![a]\!]_\mathbb{G} = \prod_i [\![a_i]\!]_\mathbb{G}$ for a subset of $t + 1$ parties. A potential drawback of this generic method is the cost of t secure group operations, which may be avoided if we use more direct methods, e.g., through efficient encodings for the group or direct use of the underlying representation of the group. For instance, to sample a point (x, y) on an elliptic curve, we may first generate $[\![x]\!]$ at random, and then solve for $\pm[\![y]\!]$, rejecting x if no solutions exist. Another potential drawback is the lack of public verifiability if parties generate random group elements privately. A common way to achieve verifiability is to start from publicly verifiable random bits, and use these bits to deterministically generate random samples.

Sampling in the Black Box Group Model. Given the structure of $\mathbb{G}$, we may reduce generating $[\![a]\!]_{\mathbb{G}}$ with $a \in_R \mathbb{G}$ to generating a random $[\![x]\!]$ with $x \in_R \mathbb{Z}_n$, in case $\mathbb{G} = \langle g \rangle$ is a cyclic group of order n, say. This extends to arbitrary abelian groups if we are given a generating set.

For arbitrary groups for which we do not know the structure of the group, however, we adopt a different approach referred to as sampling in the black box group model. An important property of algorithms such as Dixon's is that the problem of secure random sampling is reduced to the secure generation of random bits. If these bits are generated in a publicly verifiable random way, it follows that the entire protocol for random sampling in a group can be made verifiable.

For nonabelian groups, the algorithm of Dixon [Dix08, Theorem 1] is the state-of-the-art for provable complexity bounds. Its applicability suffers from an unknown constant. However, [Dix08, Theorem 3, Lemma 13(b)] offers a procedure that addresses this. In the second half of this section we argue why this technique compares favorably to other common heuristics.

Dixon's Algorithm. For our purpose, we apply Dixon's algorithm in the clear to construct a public array of group elements. Given $0 \leq \varepsilon < 1$, the algorithm is then used to securely generate ε-uniformly distributed random elements, meaning that each group element has probability $(1 \pm \varepsilon)/|\mathbb{G}|$ to appear as output. The number of group operations to construct a random element generator is proportional to $\log^2 |\mathbb{G}|$, see [Dix08, Remark 2].

For $a_1, \dots, a_j \in \mathbb{G}$, let a *random cube of length* j, *Cube*$(a_1, \dots, a_j)$, be defined by the probability distribution $\{a_1^{\epsilon_1} \cdots a_j^{\epsilon_j} : (\epsilon_1, \dots, \epsilon_j) \in_R \{0,1\}^j\}$. Given a generating set $\mathcal{G}$ of $\mathbb{G}$, the following theorem shows that we can construct a random cube of length proportional to $\log |\mathbb{G}|$ that is 1/4-uniform with a given probability.

Theorem 2. ([Dix08, **Theorem** 1]). *Let $\mathcal{G} = \{g_1, \dots, g_d\}$ be a generating set of $\mathbb{G}$. Let $W_j = Cube(g_j^{-1}, \dots, g_1^{-1}, g_1, \dots, g_j)$ be a sequence of cubes, where for $j > d$, g_j is chosen at random from W_{j-1}. Then for each $\delta > 0$, there is a constant K_δ, independent of d or $\mathbb{G}$, such that with probability at least $1 - \delta$, distribution W_j is 1/4-uniform when $j \geq d + K_\delta \log |\mathbb{G}|$.*

Practical Procedure. Constant K_δ in Theorem 2 may still make the implementation impractical. The following theorem states that we can avoid the constant K_δ and reduce the cube length if we start from a distribution W that is close to

Protocol 2. random($\mathbb{G}, x$) random cube $Z_j(\boldsymbol{g})$ for $\mathbb{G}$

1: $[\![r_0]\!], \ldots, [\![r_{j-1}]\!] \leftarrow$ random-bits(j)
2: $[\![g_i^{xr_i}]\!]_{\mathbb{G}} \leftarrow$ if-else($[\![r_i]\!], g_i^x, 1_{\mathbb{G}}$), for $i = 0, \ldots, j-1$
3: $[\![a^x]\!]_{\mathbb{G}} \leftarrow \prod_i [\![g_i^{xr_i}]\!]_{\mathbb{G}}$
4: **return** $[\![a^x]\!]_{\mathbb{G}}$ ▷ random $\mathbb{G}$-element to the power x, $x \in \mathbb{Z}$

Protocol 3. inverse($[\![a]\!]_{\mathbb{G}}$)

1: $[\![r]\!]_{\mathbb{G}} \leftarrow$ random($\mathbb{G}$)
2: $a * r \leftarrow [\![a]\!]_{\mathbb{G}} * [\![r]\!]_{\mathbb{G}}$
3: $[\![a^{-1}]\!]_{\mathbb{G}} \leftarrow [\![r]\!]_{\mathbb{G}} * (a * r)^{-1}$
4: **return** $[\![a^{-1}]\!]_{\mathbb{G}}$

the uniform distribution U on $\mathbb{G}$ w.r.t. to the statistical (or, variational) distance $\Delta[W; U] = \frac{1}{2}\sum_{a\in\mathbb{G}} |W(a) - U(a)| = \max_{A\subseteq\mathbb{G}} |W(A) - U(A)|$. [Dix08, Lemma 13(b), page 11] describes the procedure in detail.

Theorem 3. **([Dix08, Theorem 3(c)]).** *Let U be the uniform distribution on $\mathbb{G}$ and suppose W is a distribution with $\Delta[W; U] \leq \varepsilon < 1$. Let $a_0, \ldots, a_{j-1} \in \mathbb{G}$ be chosen independently according to distribution W. If $Z_j = Cube(a_0, \ldots, a_{j-1})$, then with probability at least $1 - 2^{-h}$, Z_j is 2^{-k}-uniform when*

$$j \geq \beta(2\log_2 |\mathbb{G}| + h + 2k),\ \textit{where}\ \beta := \log_2(2/(1+\varepsilon)). \tag{3}$$

Given a random cube Z_j of length j that is sufficiently close to uniform random, Protocol 2 is a general protocol for secure random sampling from a group $\mathbb{G}$. The protocol requires j secure random bits and $j-1$ secure group operations. The result is raised to the power x, where $x = 1$ is intended as the default case, and this can be extended to several powers.

Secure Inverse. The sampling protocols from Sect. 5 allow us to invert secure group elements for groups with known or unknown order. Protocol 3 implements this functionality following the classical approach by [BIB89].

Secure Exponentiation. We start this section with a generally applicable protocol for secure exponentiation $[\![a^x]\!]_{\mathbb{G}}$ based on the binary representation of $[\![x]\!]$, see Protocol 4. The computational complexity is dominated by about 2ℓ group operations and ℓ calls to lsb. For the round complexity we see that the ℓ iterations of the loop are done sequentially. The protocol can be optimized in lots of ways. For instance, it is more efficient to compute the bits of $[\![x]\!]$ all at once, including the sign bit, using a standard solution.

For abelian groups the linear dependency on ℓ for the round complexity can be removed, as we show in Protocol 5. The improvement is that the ℓ iterations of the loop can now be done in parallel. Using standard techniques for constant rounds protocols (see, e.g., [BIB89, DFK+06]) the overall round complexity can be made independent of ℓ.

Protocol 4. exponentiation($[\![a]\!]_{\mathbb{G}}, [\![x]\!]$)

1: $[\![b]\!]_{\mathbb{G}}, [\![y]\!] \leftarrow$ if-else($[\![x < 0]\!]$, (inverse($[\![a]\!]_{\mathbb{G}}$), $-[\![x]\!]$), ($[\![a]\!]_{\mathbb{G}}, [\![x]\!]$)) ▷ skip if $x \geq 0$ ensured
2: $[\![c]\!]_{\mathbb{G}} \leftarrow [\![1]\!]_{\mathbb{G}}$
3: **for** $i = 0$ **to** $\ell - 1$ **do** ▷ $\text{len}(x) \leq \ell$ assumed
4: $\quad [\![e_i]\!] \leftarrow$ lsb($[\![y]\!]$)
5: $\quad [\![c]\!]_{\mathbb{G}} \leftarrow$ if-else($[\![e_i]\!], [\![b]\!]_{\mathbb{G}} * [\![c]\!]_{\mathbb{G}}, [\![c]\!]_{\mathbb{G}}$)
6: $\quad [\![y]\!] \leftarrow ([\![y]\!] - [\![e_i]\!])/2$
7: $\quad [\![b]\!]_{\mathbb{G}} \leftarrow [\![b]\!]_{\mathbb{G}}^2$
8: **return** $[\![c]\!]_{\mathbb{G}}$

Protocol 5. exponentiation($[\![a]\!]_{\mathbb{G}}, [\![x]\!]$) $\mathbb{G}$ abelian

1: $[\![x_0]\!], \ldots, [\![x_{\ell-1}]\!] \leftarrow$ bits($[\![x]\!]$) ▷ $\text{len}(x) + 1 \leq \ell$ assumed for two's complement
2: $[\![r]\!]_{\mathbb{G}}, [\![r^{-1}]\!]_{\mathbb{G}}, \ldots, [\![r^{-2^{\ell-1}}]\!]_{\mathbb{G}} \leftarrow$ random($\mathbb{G}, 1, -1, \ldots, -2^{\ell-1}$)
3: $b \leftarrow [\![a]\!]_{\mathbb{G}} * [\![r]\!]_{\mathbb{G}}$ ▷ $b = a * r$
4: **for** $i = 0$ **to** $\ell - 1$ **do**
5: $\quad [\![c_i]\!]_{\mathbb{G}} \leftarrow$ if-else($[\![x_i]\!], b^{2^i} * [\![r^{-2^i}]\!]_{\mathbb{G}}, 1_{\mathbb{G}}$) ▷ in parallel
6: **return** $[\![c_0]\!]_{\mathbb{G}} * \cdots * [\![c_{\ell-2}]\!]_{\mathbb{G}} * [\![c_{\ell-1}]\!]_{\mathbb{G}}^{-1}$

These protocols can be optimized if either a or x is public. For instance, if a is public, the list of powers $a, a^2, a^4, \ldots$ can be computed locally, and if x is public, one can use techniques such as addition chains.

However, if one of the two inputs a or x is public, we can obtain efficient protocols by securely randomizing the other input. Protocol 6 solves the case that x is public, assuming that the group is abelian. The protocol uses secure random sampling from $\mathbb{G}$ to obtain both a random group element and its $(-x)$th power.

Protocol 7 solves the case that a is public. The protocol random-pair(a) outputs a random exponent $[\![r]\!]$ together with $[\![a^{-r}]\!]_{\mathbb{G}}$ for public input $a \in \mathbb{G}$. For public output a^x, we can also use public a^{-r} in the first step of the protocol. Finally, in the special case that a is an element of large order, parties may directly use their Shamir secret shares, and perform Lagrange interpolation in the exponent to raise a to the power $[\![x]\!]$.

6 Secure Class Groups

We focus on ideal class groups of imaginary quadratic fields, using $\text{Cl}(\Delta)$ to denote the class group with discriminant $\Delta < 0$. We also use $\text{Cl}(\Delta)$ to denote the isomorphic form class group of integral binary quadratic forms $f(x, y) = ax^2 + bxy + cy^2$ with discriminant $\Delta = b^2 - 4ac < 0$. These forms are written as $f = (a, b, c)$ with $a, b, c \in \mathbb{Z}$, where it is implied that f is primitive, that is, $\gcd(a, b, c) = 1$, and f is positive definite, that is $a > 0$.

For the composition of two forms $f_1, f_2 \in \text{Cl}(\Delta)$ we will use the algorithm due to Shanks [Sha71], as presented by Cohen [Coh93, Algorithm 5.4.7]. We slightly adapt the algorithm, skipping some case distinctions that were introduced for

Protocol 6. exponentiation($[\![a]\!]_{\mathbb{G}}, x$) public exponent x, $\mathbb{G}$ abelian

1: $[\![r]\!]_{\mathbb{G}}, [\![r^{-x}]\!]_{\mathbb{G}} \leftarrow$ random($\mathbb{G}, 1, -x$)
2: $a * r \leftarrow [\![a]\!]_{\mathbb{G}} * [\![r]\!]_{\mathbb{G}}$
3: $[\![a^x]\!]_{\mathbb{G}} \leftarrow (a * r)^x * [\![r^{-x}]\!]_{\mathbb{G}}$
4: **return** $[\![a^x]\!]_{\mathbb{G}}$

Protocol 7. exponentiation($a, [\![x]\!]$) public base a

1: $[\![r]\!], [\![a^{-r}]\!]_{\mathbb{G}} \leftarrow$ random-pair(a)
2: $x + r \leftarrow [\![x]\!] + [\![r]\!]$ ▷ $x + r$ should be sufficiently random
3: $[\![a^x]\!]_{\mathbb{G}} \leftarrow a^{x+r}[\![a^{-r}]\!]_{\mathbb{G}}$
4: **return** $[\![a^x]\!]_{\mathbb{G}}$

efficiency, see Algorithm 2. Apart from the computationally nontrivial reduction performed in the final step of the algorithm, the resulting algorithm only requires two xgcd's and one integer division. This compares favorably with alternatives such as the classical composition algorithm by Dirichlet [LD63] (see also [Cox11, Lemma 3.2] and [Lon19, Algorithm 6.1.1]).

As secure representation of a form $f = (a, b, c) \in \mathbb{G}$ we define $[\![f]\!]_{\mathbb{G}} = ([\![a]\!]_p, [\![b]\!]_p, [\![c]\!]_p)$, for a sufficiently large prime p. The prime p should be sufficiently large such that the intermediate forms computed by Algorithm 2 do not cause any overflow modulo p (also accounting for the "headroom" needed for secure comparison and secure integer division). Using Protocol 1 for secure xgcd and a protocol for secure integer division, Algorithm 2 is then easily transformed into a protocol for the secure composition of forms.

To reduce a form $f = (a, b, c)$, the classical reduction algorithm by Lagrange (see [BV07, Algorithm 5.3]) runs in at most $2 + \lceil \log_2(a/\sqrt{\Delta}) \rceil$ steps [BV07, Theorem 5.5.4], each step requiring an integer division. Our goal is to avoid the expensive (secure) integer divisions where possible.

To this end, we take the binary reduction algorithm by [AF06, Algorithm 3] as our starting point. (Recently, [DEH22] arrived at an equivalent algorithm to design a quantum circuit to reduce binary quadratic forms.) We minimize the number of comparisons by exploiting the invariant $a > 0$ and $c > 0$. The algorithm reduces forms by the following transformations in $SL_2(\mathbb{Z})$:

$$S = \begin{pmatrix} 0 & -1 \\ 1 & 0 \end{pmatrix} \quad T_m = \begin{pmatrix} 1 & m \\ 0 & 1 \end{pmatrix}.$$

The total number of iterations of the main loop required to achieve $|b| \leq 2a$ is at most len(b), if $f = (a, b, c)$ is the given form. This follows from the fact that if $|b| \leq 2a$ does not hold yet, an iteration of the main loop will reduce len(b) by at least 1. We will ensure that len(b) $\leq$ len(Δ), such that it suffices to run the main loop for len(Δ) iterations, independent of the input. Note that we need to test $a > c$ in each iteration as well.

As an important optimization, we limit the number of iterations to len(Δ)/2 by noting that it suffices to reduce b until len(b) $<$ len(Δ)/2. This ensures that

Algorithm 2. $\mathsf{compose}(f_1, f_2)$ $\quad$ f_1, f_2 primitive, positive definite, reduced

Input: $f_1 = (a_1, b_1, c_1)$ and $f_2 = (a_2, b_2, c_2)$
1: $s \leftarrow (b_1 + b_2)/2$
2: $d', x_1, y_1 \leftarrow \mathsf{xgcd}(a_1, a_2)$ $\quad$ ▷ $x_1a_1 + y_1a_2 = d' = \mathsf{gcd}(a_1, a_2)$
3: $d, x_2, y_2 \leftarrow \mathsf{xgcd}(s, d')$ $\quad$ ▷ $x_2s + y_2d' = \mathsf{gcd}(s, d')$
4: $v_1, v_2 \leftarrow a_1/d, a_2/d$
5: $r \leftarrow (y_1y_2(s - b_2) - x_2c_2) \bmod v_1$ $\quad$ ▷ integer division
6: $a_3 \leftarrow v_1v_2$
7: $b_3 \leftarrow b_2 + 2v_2r$
8: $c_3 \leftarrow (b_3^2 - \Delta)/(4a_3)$
9: $f_3 = (a_3, b_3, c_3)$
10: **return** $\mathsf{reduce}(f_3)$

Algorithm 3. $\mathsf{reduce}(f)$ $\quad$ $f \in \mathrm{Cl}(\Delta)$ primitive, positive definite

Input: $f = (a, b, c)$
1: **for** $i = 1$ **to** $\lceil \mathrm{len}(\Delta)/2 \rceil$ **do** $\quad$ ▷ invariant $a > 0$ and $c > 0$
2: **if** $b > 0$ **then** $s_b \leftarrow 1$ **else** $s_b \leftarrow -1$ $\quad$ ▷ using $\mathrm{len}(b) < \mathrm{len}(\Delta) - i$
3: $j \leftarrow \mathrm{len}(b) - \mathrm{len}(a) - 1$
4: $m \leftarrow -s_b 2^j$ $\quad$ ▷ compute 2^j together with $\mathrm{len}(a)$ and $\mathrm{len}(b)$ at no extra cost
5: **if** $s_b b > 2a$ **then** $\quad$ ▷ $|b| > 2a$
6: $f \leftarrow fT_m$ $\quad$ ▷ $fT_m = (a, b + 2ma, m^2a + mb + c)$
7: **if** $a > c$ **then**
8: $f \leftarrow fS$ $\quad$ ▷ $fS = (c, -b, a)$
9: $m \leftarrow \lfloor \frac{a-b}{2a} \rfloor$ $\quad$ ▷ integer division
10: $f \leftarrow fT_m$
11: **if** $a > c$ **then**
12: $f \leftarrow fS$
13: **if** $(a + b)(a - c) = 0$ and $b < 0$ **then** $\quad$ ▷ ensure $b \geq 0$ if $a = -b$ or $a = c$
14: $b \leftarrow -b$ $\quad$ ▷ $f \leftarrow fS = fT_1$
15: **return** f

$|b| \leq \sqrt{|\Delta|}$ after the main loop. If $|b| \leq 2a$ does not hold yet, then it follows that $a < |b|/2 \leq \sqrt{|\Delta|/4}$, and we only need one "normalization" step to ensure $|b| \leq a$.

The result is presented as Algorithm 3. This algorithm avoids the integer division and multiple comparisons in the main loop of Lagrange's reduction algorithm, at the cost of three secure comparisons and two secure bit length computations in the main loop. To compute the bit length securely we use a novel protocol avoiding the use of a full bit decomposition.

For secure encoding to class groups, a given integer s will be mapped to a form (a, b, c) by computing a as a simple function of s and setting b as the square root of Δ modulo $4a$, see Algorithm 4. We note that this encoding improves upon the encoding proposed by [Sch03] (compare to [Sch99] as well), which relies on using prime numbers for a. Our algorithm avoids the need for primality tests, which are dominating the computational cost for the encoding algorithm.

Algorithm 4. encode$(s, \mathrm{Cl}(\Delta))$ $\quad$ s sufficiently small w.r.t. $|\Delta|$

1: $n = s \cdot gap_\ell$
2: $a \leftarrow n - 1 - (n \bmod 4)$
3: **repeat**
4: $\quad a \leftarrow a + 4$ $\quad \triangleright\ a \equiv 3 \pmod 4$
5: $\quad b \leftarrow \Delta^{\frac{a+1}{4}} \bmod a$
6: **until** $b^2 \equiv \Delta \pmod a$ and $\gcd(a, b) = 1$
7: **if** $\Delta \not\equiv b \pmod 2$ **then**
8: $\quad b \leftarrow a - b$
9: $f \leftarrow (a, b, \frac{b^2 - \Delta}{4a})$ $\quad \triangleright\ \Delta \equiv b^2 \pmod 4$
10: $d \leftarrow n - a$
11: **return** f, d

7 Concluding Remarks

We have presented protocols for the extended gcd and the class group operations in an MPC setting. Python implementations of these protocols have been integrated in the MPyC package [Sch18]. This package covers protocols for secure integer arithmetic involving basic operations like $+, *, <$. More advanced operations such as secure integer division and secure computation of the bit length are also included in the MPyC package (details for these protocols will be covered in other work).

The implementation in MPyC also extends to other secure groups, including Schnorr groups and elliptic curves. Many results pertaining to threshold cryptosystems and threshold signatures from the literature can thus be stated in terms of secure groups. For example, secure groups make it straightforward to conduct the prover side of a Σ-protocol, a compressed Σ-protocol [AC20] or succinct noninteractive argument of knowledge (SNARK) in MPC. This allows for convenient constructions of *verifiable MPC* protocols [SVdV16], i.e., proof systems that enable MPC parties to create a publicly verifiable proof of correctness of the MPC computation, even in the extreme case that all MPC parties are corrupt. A demonstration of verifiable MPC is available in a separate repository [SS22].

Acknowledgements. We thank Alessandro Danelon, Mark Abspoel, Niek Bouman, Thomas Attema, and the anonymous reviewers for their valuable comments. This work has received funding from the European Union's Horizon 2020 research and innovation program under grant agreement No 780477 (PRIViLEDGE).

References

[AC20] Attema, T., Cramer, R.: Compressed *Sigma*-protocol theory and practical application to Plug & play secure algorithmics. In: Micciancio, D., Ristenpart, T. (eds.) CRYPTO 2020. LNCS, vol. 12172, pp. 513–543. Springer, Cham (2020). https://doi.org/10.1007/978-3-030-56877-1_18

[ACS02] Algesheimer, J., Camenisch, J., Shoup, V.: Efficient computation modulo a shared secret with application to the generation of shared safe-prime products, 417–432 (2002)

[AF06] Agarwal, S., Frandsen, G.S.: A new GCD algorithm for quadratic number rings with unique factorization. In: Correa, J.R., Hevia, A., Kiwi, M. (eds.) LATIN 2006. LNCS, vol. 3887, pp. 30–42. Springer, Heidelberg (2006). https://doi.org/10.1007/11682462_8

[BB87] Bojanczyk, A.W., Brent, R.P.: A systolic algorithm for extended GCD computation. Comput. Math. Appl. **14**(4), 233–238 (1987)

[BGW88] Ben-Or, M., Goldwasser, S., Wigderson, A.: Completeness theorems for non-cryptographic fault-tolerant distributed computation (extended abstract). In: Proceedings of Symposium on Theory of Computing (STOC '88), pp. 1–10. ACM (1988)

[BHR+21] Block, A.R., Holmgren, J., Rosen, A., Rothblum, R.D., Soni, P.: Time- and space-efficient arguments from groups of unknown order. In: Malkin, T., Peikert, C. (eds.) CRYPTO 2021. LNCS, vol. 12828, pp. 123–152. Springer, Cham (2021). https://doi.org/10.1007/978-3-030-84259-8_5

[BIB89] Bar-Ilan, J., Beaver, D.: Non-cryptographic fault-tolerant computing in constant number of rounds of interaction, 201–209 (1989)

[BK85] Brent, R.P., Kung, H.T.: A systolic algorithm for integer GCD computation. In 1985 IEEE 7th Symposium on Computer Arithmetic (ARITH), pages 118–125 (1985)

[BV07] Buchmann, J.A., Vollmer, U.: Binary quadratic forms - an algorithmic approach, volume 20 of Algorithms and computation in mathematics. Springer (2007)

[BY19] Bernstein, D.J., Yang, B.-Y.: Fast constant-time GCD computation and modular inversion. IACR Trans. Cryptogr. Hardw. Embed. Syst. **2019**(3), 340–398 (2019)

[Coh93] Cohen, H.: A course in computational algebraic number theory, volume 138 of Graduate texts in mathematics. Springer (1993)

[Cox11] Cox, D.A.: Primes of the form $x^2 + ny^2$: Fermat, class field theory, and complex multiplication, volume 34. John Wiley & Sons (2011)

[CS10] Catrina, O., Saxena, A.: Secure computation with fixed-point numbers, 35–50 (2010)

[DEH22] David, N., Espitau, T., Hosoyamada, A.: Quantum binary quadratic form reduction. IACR Cryptol. ePrint Arch. p. 466 (2022)

[DFK+06] Damgård, I., Fitzi, M., Kiltz, E., Nielsen, J.B., Toft, T.: Unconditionally secure constant-rounds multi-party computation for equality, comparison, bits and exponentiation, pp. 285–304 (2006)

[DGS22] Dobson, S., Galbraith, S., Smith, B.: Trustless unknown-order groups. Math. Cryptol. **1**(2), 25–39 (2022)

[Dix08] Dixon, J.D.: Generating random elements in finite groups. Electron. J. Comb., 15(1) (2008)

[GRR98] Gennaro, R., Rabin, M.O., Rabin, T.: Simplified VSS and fast-track multiparty computations with applications to threshold cryptography. In: Proceedings of Principles of Distributed Computing, PODC '98, pp. 101–111. ACM (1998)

[Hoo12] Hoogh, de, S.J.A.: Design of large scale applications of secure multiparty computation : secure linear programming. PhD thesis, Technische Universiteit Eindhoven, Department of Mathematics and Computer Science (2012)

[Knu69] Knuth, D.E.: The art of computer programming, volume 2: Seminumerical algorithms (1969)

[LD63] Dirichlet, P.L.: Vorlesungen über Zahlentheorie (1863)

[Lon19] Long, L.: Binary quadratic forms. GitHub https://github.com/Chia-Network/vdf-competition/blob/master/classgroups.pdf (2019). Accessed 23 Jan 2020

[MvOV96] Menezes, A., van Oorschot, P.C., Vanstone, S.A.: Handbook of Applied Cryptography. CRC Press (1996)

[Sch99] Schaub, J.: Implementierung von Public-Key-Kryptosystemen über imaginär-quadratischen Ordnungen (Master's thesis). Technische Universität Darmstadt, Fachbereich Informatik (1999)

[Sch03] Schielzeth, D.: Realisierung der elgamal-verschlüsselung in quadratischen zählkorpern (Master's thesis). Technische Universität Berlin. http://www.math.tu-berlin.de/kant/publications.html (2003)

[Sch18] Schoenmakers, B.: MPyC secure multiparty computation in Python. GitHub https://github.com/lschoe/mpyc (2018)

[Sha71] Shanks, D.: Class number, a theory of factorization, and genera. In: Proceedings of the Symp. Math. Soc., 1971, volume 20, pages 41–440, 1971

[SS22] Schoenmakers, B., Segers, T.: Verifiable MPC. GitHub. https://github.com/toonsegers/verifiable_mpc (2022)

[SVdV16] Schoenmakers, B., Veeningen, M., de Vreede, N.: Trinocchio: privacy-preserving outsourcing by distributed verifiable computation. In: Manulis, M., Sadeghi, A.-R., Schneider, S. (eds.) ACNS 2016. LNCS, vol. 9696, pp. 346–366. Springer, Cham (2016). https://doi.org/10.1007/978-3-319-39555-5_19

[SZ04] Stehlé, D., Zimmermann, P.: A binary recursive GCD algorithm. In: Buell, D., (edt.), Algorithmic Number Theory, pages 411–425, Berlin, Heidelberg. Springer, Berlin Heidelberg (2004)

[Tof07] Toft, T.: Primitives and applications for multi-party computation. PhD thesis, Aarhus University (2007)

[Wes19] Wesolowski, B.: Efficient verifiable delay functions. In: Ishai, Y., Rijmen, V. (eds.) EUROCRYPT 2019. LNCS, vol. 11478, pp. 379–407. Springer, Cham (2019). https://doi.org/10.1007/978-3-030-17659-4_13

On Distributed Randomness Generation in Blockchains

Ilan Komargodski[1,2](✉) and Yoav Tamir[1]

[1] Hebrew University of Jerusalem, Jerusalem, Israel
ilank@cs.huji.ac.il
[2] NTT Research, Sunnyvale, USA

Abstract. We present a smart contract for generating unbiased randomness on a blockchain/ledger-style system. Our smart contract can be stored on a blockchain and be executed publicly whenever needed. In particular, our protocol is suitable for leader-election-style applications and for mitigating front-running attacks.

We prove correctness and security of our smart contract within a formal model of a distributed ledger extended by financial incentives and smart contracts. To the best of our knowledge, our formalization is the first to capture the concept of "time" in the context of smart contracts. Furthermore, we show that our protocol is incentive compatible, namely, under some reasonable financial assumption, following the honest execution, even for an all but one corruptions, gives the most financial benefit.

Technically, our smart contract utilizes recently-introduced non-malleable time-lock puzzles. At a very high level, these are cryptographic commitments that can be "force opened" after a predefined delay. Using those, we implement a commit-and-reveal-style protocol but where we consider the particulars of the blockchain setting.

1 Introduction

Generating randomness in a distributed manner has been one of the most foundational tasks in cryptography and distributed computing. Protocols for generating randomness should satisfy the natural property that at the end of an (honest) execution, the common output of all parties is uniformly distributed. Moreover, even if some of the parties collude and deviate from the prescribed protocol, they should not be able to significantly bias the honest parties' common output. The security (a.k.a. fairness) of such protocols is defined as the bias of the honest parties' common output, i.e., the distance from the uniform distribution.

This problem has been studied since the 80's and a host of results are known. In the honest majority setting, where more than half of the parties are honest and guaranteed to follow the prescribed protocol, perfectly fair protocols are known. Indeed, the randomness generation functionality is a special case of secure general multiparty computation [GMW87,BGW88,CCD88]. However, in the dishonest majority setting, the situation is more complex as explained by Cleve's result concerning coin flipping protocols, i.e., protocols for generating a single bit of

S. Dolev et al. (Eds.): CSCML 2023, LNCS 13914, pp. 49–64, 2023.
https://doi.org/10.1007/978-3-031-34671-2_4

randomness. Cleve's seminal work [Cle86] showed that every R-round coin flipping protocol (in the dishonest majority setting) can be biased by a $1/R$ factor. Cleve's lower bound applies in many settings. Namely, it applies even if cryptographic assumptions are used and even if strong trusted setup assumption are made.

Applications. Of course, joint randomness generation in the dishonest majority setting is extremely useful, especially in modern applications where it is infeasible to guarantee an honest majority (such as many permission-less ledger-style applications). We give two applications that are of interest. First, in modern Proof-of-Stake (PoS) blockchains, the leader of the next round is chosen at random from a "special" subset of parties (i.e., stake holders). For fairness it is necessary that this choice will be unbiased/random. Second, it is known that if the "leader" of some round can control the order of transactions within a block, it can exploit public knowledge together with information from pending transactions network, to its own favor. This problem is typically referred to as "front-running" and the natural solution for this problem is to let the leader only submit a set of transactions to include in a block and then randomly shuffle it using a freshly chosen random permutation. The last step, clearly, requires fresh unbiased randomness.

Existing Approaches. There are several existing approaches, each of which has its own advantages and drawbacks. We survey some of them next, see Appendix A for a wider survey. Some suggest to use primitives like threshold signatures or specific instances of (standard) secure multiparty computation protocols. These, however, typically require an honest majority and are therefore insufficient in many settings. Another approach is to directly use an unfair protocol, like a hash-and-reveal style protocol. In such protocols, all participants first commit to their votes (using, say, a cryptographic hash function) and then after everyone committed, they all reveal their votes. As mentioned, these protocols are unfair as an attacker (that only aborts before revealing their vote) can significantly bias the output. But, the point is that such biasing could be detected by the honest parties that can later punish the cheating party. The protocol itself needs to be repeated if a cheater is detected. This abort-retry-style attack causes either an inefficiency or forces to lock significant deposits. This is perhaps the most common approach, adopted (for instance) by the Ethereum based Randao protocol [Ran17].

The last approach that we mention is one based on cryptographic timing assumptions, including time-lock puzzles [RSW96], timed-commitments [BN00], and verifiable delay functions [Bon+18, Pie19, Wes20, Feo+19]. The latter primitive, referred to as a VDF, is a function that is slow to compute, even with considerable parallel computing, but a solution is easily verifiable. As Boneh et al. [Bon+18] suggested in their original work, VDFs directly yield a method for generating randomness, but the problem is that all existing constructions require some form of trusted setup which we would like to avoid in fully decentralized systems.

Time-lock puzzles [RSW96] and timed commitments [BN00], at a very high level, are a special type of commitments that can be "force opened" after a delay. Timed commitments also have the property that any puzzle has a solution. Both of these objects can be used to prevent abort and guarantee the reveal of the underlying value after some predefined period of time. Apparently, slightly changing the model by assuming a bound on the network delay, Cleve's lower bound does not hold. Indeed, Boneh and Naor [BN00] constructed timed commitments and suggested to use them to get a fair two-party coin flipping protocol.

In the context of mitigating front-running attacks, it was recently mentioned by Baum et al. [Bau+21a] that one can use time-lock puzzles to get a multi-party fair coin flipping protocol. In their suggestion, each party uses a time-lock puzzle to "commit" on a random string and after some predetermined time they all open their puzzles. Puzzles that are not willingly opened are force-opened. While it may seem okay at a first glance, it is not hard to get convinced that this protocol is completely broken due to mauling/man-in-the-middle attacks.

This attack can be solved using a refined variant of time-lock puzzles called *non-malleable* time-lock puzzled, recently introduced and constructed by Freitag et al. [Fre+20]. This is a variant of time-lock puzzles where it is provably impossible to carry out meaningful man-in-the-middle attacks. Freitag et al. [Fre+20] showed that their new primitive can indeed by used to obtain a multi-party fair coin flipping protocol. However, their protocol is somewhat theoretical and it is not clear whether and how it can be used in the setting of blockchains or more generally in decentralized finance applications.

In this work, based on the approach of Freitag et al. [Fre+20], we describe a new practical blockchain-based protocol along with a smart contract for distributedly generating fair unbiased randomness. At a superficial level, our protocol is similar to the Randao protocol [Ran17], but as we show our protocol prevents the abort-retry-style attack described above, which significantly decreases fees and deposits required. Our analysis and formalization are within a new formal model of a Blockchain with a Smart-Contract functionality, suited to the "timing" setting. This may be of independent interest.

2 Model

Our model is an extension of the Blockchain-Hybrid Model as introduced in Choudhuri, Goyal, and Jain [CGJ19]. The Blockchain is modeled as a global ledger – all parties can write and read data from it using a global oracle. The data is always written along with a signature containing the ID of the writing party. Our model is also synchronous with a global clock, meaning that all parties have a shared time counter. We extend the Blockchain-Hybrid Model in two ways. We add financial incentives by giving each party a constant initial amount of funds, modeled as a public counter. We later show that our protocol is efficient under the simple assumptions that all adversaries wish to maximize their collective sum of funds, and that they do not profit by delaying the protocol. We also add the functionality of Smart-Contracts to the Blockchain.

Smart-Contracts. Smart-Contracts (SC) were introduced in the original Ethereum white paper [But15], and are nowadays implemented in most cryptocurrencies. An SC is a set of instructions (code) that is committed to the blockchain to be executed by the network for a fee depending on the length and running time of the program. The SC can have an internal memory, and like standard parties, has a balance on the Blockchain, can make and receive transactions, and can access the blockchain ledger. It can also expose "API" functions to the network to be activated by other parties (which requires them also to pay the execution fee) to be run publicly by the SC.

We formally model this functionality as follows; Each party can create a Smart-Contract by sending a set of API functions $(F_1, F_2, ..F_k)$ to the Blockchain oracle. Each function F is a set of instructions describing a function taking parameters (C, ID, P), C is an amount of funds transacted to the SC by the caller, ID the ID of the caller, and P is any additional data the function requires. The functions can access a shared memory area in which they can save state between calls, the shared memory can be viewed publicly by all parties.

To call (execute on the network) a function of the smart contract, the caller sends to the oracle the ID of the SC function to be executed, along with (T, C, P), where T is a time delay in the global clock units. The callers fund balance is reduced by C funds. If The caller not have enough funds at the call time, the call is rejected by the oracle and it not committed. All T, C and P are written to the Blockchain by the oracle to be viewed publicly, signed by the callers ID. The oracle then checks if the execution of $F(T, C, P)$ on a single tape TM would finish successfully in less than T time. If so, after a delay of T time, it publishes the result of the computation to the Blockchain, along with any transactions made by it. Otherwise it does nothing. The transactions made by the SC are signed by its own ID. In a case where multiple calls to the SC are done in parallel, the oracle executes them one by one by their arrival order. Calls that were done exactly in the same time are ordered arbitrarily. We price the execution time for SCs such that no party has enough funds to execute a SC function taking super polynomial time.

The global clock functionality is also modeled by a global oracle. As all parties are essentially Turing Machines, every instruction they execute is a single time unit. Each party can call the oracle to receive the current time.

3 Incentivesed TLP-Based Fair Coin-Flipping

We propose a simple commit and reveal scheme based on the random coin flipping protocol introduced in [Fre+20]. The approach is similar to Randao but where the commitment is done with TLPs instead of a hash. This way parties can only delay the result by not revealing their commitment and cannot bias it. This enables the protocol to be fair even in an all but one dishonest parties model, and specifically resistant to the aborting/last reveler attacks. We extend the protocol by adding financial incentives enforced by a Smart Contract. Specifically, to participate in the protocol each party must commit funds, which will be returned to them only if they reveal their committed value on time. This leads to incentive compatibility: there is no reason, apart from merely delaying the reveal of the out, to deviate from the honest behaviour. Therefore, we show that a minimal deposit is required to incentivize all parties to reveal their values in time.

Technically, we use *non-malleable time-lock puzzles* (nmTLPs) as proposed in [Fre+20]. Time-lock puzzles are a mechanism for sending messages "to the future", by allowing a sender to quickly generate a puzzle with an underlying message that remains hidden until a receiver spends a moderately large amount of time solving it. Further, non-malleability, roughly, guarantees that it is impossible to "maul" a puzzle into one for a related message without solving it. We give the details of the construction in Sect. 3.2.

Remark 1 (Alternatives to nmTLPs). There are several possible alternative to the non-malleable time-lock puzzles of [Fre+20] that we mention next. First, consider the non-malleable time-commitment construction of [Thy+21] (actually it satisfies a slightly stronger notion of security called CCA). Their construction is not a time-lock puzzle but rather a time-commitment—this is due to the fact that the setup of their puzzle takes as much time as solving it (i.e., there is no trapdoor for efficient generation). Additionally, they rely on class groups, a somewhat newer cryptographic class of assumptions than classical RSA-based repeated squaring. On the positive side, they manage to get a homomorphic property which allows them to get a more scalable evaluation phase (because they need to solve only one puzzle while we need to solve all puzzles that were not opened). The second construction we mention is the UC-secure time-lock puzzle of [Bau+21b]. This is yet another stronger notion of security than merely non-malleability and so also sufficient for our purpose. However, since this notion of security is very strong, it cannot be realized from "standard" cryptographic assumption, i.e., they use programmable random oracles (and show that it is necessary). Their construction, similarly to ours, does not have homomorphic properties.

3.1 The Protocol

The protocol relies on nmTLPs as a main ingredient. Furthermore, the protocol uses a random oracle H and consists of five stages:

1. **Start:** Start the protocol by initializing a Smart Contract (SC).
2. **Commit:** All parties wishing to participate in the protocol commit to a random value using a nmTLP and pays the participation fee to the SC.
3. **Reveal:** Parties reveal their values and recover their participation fee.
4. **Bounty Force Reveal:** Puzzles that were not revealed in the previous stage can be solved by any party to claim half of the participation fee of the non-revealing party, the other half is distributed evenly between the participating parties.
5. **Result:** After all the puzzles have been opened, all committed values are summed appropriately to output the final value.

We now give some intuition for the construction of the nmTLPs.

3.2 Publicly-Verifiable Non-malleable Time-Lock Puzzles

The construction follows [Fre+20]. We refer the reader there for the theoretical background and some of the proofs (exact references are given below). We also refer to [Pie19] for further explanations and additional possible implementation accelerations.

Basic Construction. The basis of the TLP we use comes from [RSW96]:

- Their TLP used a puzzle of the form (x, N, T) where $N = p \cdot q$ for p,q primes, T is a number representing the time delay parameter, and x is a random element of the group Z_N^*.
- The solution for such a puzzle is $y = x^{2^T}$.

To encrypt a value s, one generates p and q, and, using the size of the group Z_N^* (which is $\phi(N) = (p-1)(q-1)$), calculates y by $y = x^{2^T} = x^{2^T \bmod |Z_N^*|} \bmod N$. Finally, one can XOR y with s, denote as z, and release it along with (x, N, T). That is, the construction takes as input (s, r) where s is a secret to encrypt and r is the randomness used, and outputs $TLP(s, r) = (x, N, T, z)$.

It is conjectured that, if one does not know the order of the group, at least T sequential steps of computation are required to calculate y. In other words, the encrypted secret would remain hidden until time T. This conjecture is considered standard by now as it was first made by Rivest, Shamir, and Wagner [RSW96] more than 25 years ago and it still stands.

Public-Verifiability and Non-malleability. In order to prevent the need of all involved parties to each execute the lengthy process of solving a puzzle in case it is not opened by the original encryptor, we wish that a solved puzzle would be easily verifiable. Syntactically, this adds a "Verify" functionality to a TLP which verifies that a solution and a proof are valid.

Additionally, the TLP must satisfy another property called non-malleability to make the commit-and-reveal random coin flipping scheme secure. Generally speaking, it means that an adversary cannot use a given puzzle to create a different puzzle with a correlated value.

The construction of the TLP as defined above is now extended in a way that gives both public verifiability and non-malleability. Given (r, s), instead of generating $TLP(s, r)$, a new randomness is generated as $r' = H(s, r)$ for H being a random oracle, and a puzzle is created for the secret (s, r) using randomness r' as $TLP((s, r), r')$. This construction enables an easy verification of a valid puzzle: simply verify that the given puzzle is generated by (s, r), which are given as the revealed secret of the puzzle. Additionally, we also get the required security property of non malleability. Intuitively, any valid puzzle must be generated by the secret it contains (s, r), and by the randomness $r' = H(s, r)$. Thus, assuming the security of the TLP from [RSW96], one cannot create any valid puzzle in polynomial-time without knowing (s, r) generate it. We refer to [Fre+20] for the full proofs.

Note that the verification described above is insufficient, as it only works for proving the solutions of valid puzzles, and there is no proof that a puzzle is invalid. For this cause, a proof of repeated squaring (due to [Pie19]) is used to prove that the solution for a puzzle is (s, r) but (s, r) do not generate the given puzzle (assuming the puzzle was verified to be in the right form), and so is invalid. The proof showing that $y = x^{2^t}$ uses the Fiat-Shamir heuristic applied on an interactive proof described as follows (following [Pie19]): If $t = 1$, the verifier directly checks that $x^2 = y$. Otherwise, the prover sends to the verifier the value $\mu = x^{2^{t/2}}$ and the verifier replies with $r \leftarrow [2\lambda]$, which is the next challenge. The prover and verifier recursively engage in a protocol to prove that $(x^r \mu)^{2^{t/2}} = \mu^r y$. The proof system uses the group QR_N^+ of signed quadratic residues, when N is a product of two $(\lambda + 1)$ safe primes. This group has some properties important for the proof system: it is of size 1/4 of the size of Z_N^*, enabling it to be large enough. Also, it has no subgroups smaller than 2^λ which is required for the proof of (statistical) security, and it has a fast membership test.

Construction. We now show the complete construction of the protocol. We use λ as the security parameter and global constants, T, R, and C, defined below. Time is measured in the units of the global clock.

- T is length both of the stage where parties send their commitments, and the stage where they reveal them.
- R is the time delay parameter used in the TLPs. Practically, it should be chosen so that the standard party could solve the puzzle in time $100T$, providing security against adversaries with 100 times faster sequential computing power.
- C is the amount of funds a party must commit to participate in the protocol. Any party could decide not to open its puzzle and lose its participation fees (or almost half of them if it claims the bounty on its own puzzle in the last moment), and cause a delay of $100T$ in the protocol. Thus, to make such a delay unprofitable, the fee should be calibrated to be larger than the expected profit an adversary could gain by a delay $100T$ in the protocol.

We define two helper functions:

- $Mul_N(a, b)$ - multiplication of a and b in the group of signed quadratic residues QR_N^+:
  1. Let $x = a \cdot b \mod (N)$.
  2. Output $min(x, N - x)$

  Note that all multiplications in the protocol are done using Mul_N.

- $Check\_Membership(x, N)$ - checks whether x is a member of the group of signed quadratic residues QR_N^+. We use $Jacobi(\cdot,\cdot)$ which is a function calculating the Jacobi symbol, a calculation that can be done in polynomial time, see implementation at [Wik].
  1. Output $0 \leq x \leq (N-1)/2 \,\&\, Jacobi(x, N) = 1$

The TLP is implemented as follows:

- $Gen(1^\lambda, t, s; r)$: Creates a TLP for the value s using randomness r and security parameter λ:
  1. Let $r_h = H(s, r)$, $p = 0$, $q = 0$ and $g = 0$.
  2. *# We now find two safe primes which act as a secret key of the TLP*
     For $i \in [1, \lambda]$:
     (a) Let x = random prime from $[2^{\lambda+1}, 2^{\lambda+2})$.
     (b) If $(x-1)/2$ is not also prime: continue.
     (c) If $p == 0$: let $p = x$ and continue.
     (d) If $q == 0$: let $q = x$ and exit the loop.
        # p and q are found with high probability by a conjecture on the distribution of safe primes, presented in [Sho05]
  3. If p or q are 0: abort.
  4. *# We need to also find a random element of the group* $QR_{p\cdot q}^+$.
     For $i \in [1, \lambda]$:
     (a) Let g be a random number from $[1, p \cdot q)$.
     (b) If $Check\_Membership(x, p \cdot q)$: exit the loop.
        # g is found with high probability as the size of QR_N^+ *is* $\phi(N)/4$.
  5. If $g = 0$: abort.
  6. Let $y = g^{2^t \bmod ((p-1)(q-1))}$.
  7. Output $(x, p \cdot q, s \oplus y)$.
- $Solve(1^\lambda, mcrs, t, x, N, c)$: force opens a TLP by executing the sequential computation, providing the secret and an easily verifiable proof:
  1. If $Check\_Membership(x, N)$ is not True: output $(\perp, \perp)$.
  2. *# We calculate the secret s and a proof* $(\pi_1, ..., \pi_{\log(t)})$.
     Let $\pi_1 = x^{2^{t/2}}$.
  3. Let y $= \pi_1^2$.
  4. Let $x_0 = x, y_0 = y$
  5. For $i \in [1, \log t - 1]$:
     (a) Let $r_i = H(1^\lambda, mcrs, t, x, N, y, c, \pi_1, ..., \pi_i)$.
     (b) Let $x_i = x_{i-1}^{r_i} \cdot \pi_i$.
     (c) Let $y_i = \pi_i^{r_i} \cdot y_{i-1}$.
     (d) Let $\pi_{i+1} = x_{i-1}^{2^{t/2^i}}$.
  6. Let s$' = y \oplus c$ and parse s' as s, r.
  7. If $z = Gen(1^\lambda, t, s; r)$: $output(s, r)$.
  8. Otherwise, let $\pi = (\pi_1, ..., \pi_{\log(t)})$ and output $(\perp, (s, r, \pi))$.
- $Verify(1^\lambda, mcrs, t, z, s_z, \pi)$: Verifies a solution for a TLP, noting that the verified solution may also be $\perp$ indicating the puzzle is invalid:

1. If $s_z \neq \perp$: parse $\pi = r$. Output True if and only if $z = Gen(1^\lambda, t, s_z; r)$.
   # As $s_z = \perp$, we now only need verify the puzzle is indeed invalid:
2. Parse z as (x, N, c). If failed, output True if and only if $\pi = \perp$.
3. In every parsing step following, if we fail to parse we output False.
4. Parse $\pi = (s, r, (\pi_1, ..., \pi_{\log(t)}))$.
5. If $z = Gen(1^\lambda, t, s; r)$: output False.
6. If $Check\_Membership(x, N)$ is not True, output False.
7. For $i \in [1, \log t - 1]$:
   (a) Let $r_i = H(1^\lambda, mcrs, t, x, N, y, c, \pi_1, ..., \pi_i)$.
   (b) Let $x_i = x_{i-1}^{r_i} \cdot \pi_i$.
   (c) Let $y_i = \pi_i^{r_i} \cdot y_{i-1}$.
   (d) Let $\pi_{i+1} = x_{i-1}^{2^{t/2^i}}$.
8. Output True if and only if $x_{\log t}^2 = y_{\log t}$.

3.3 The Smart Contract

The protocol uses an SC defined with the following API functions. At any stage, if a verification or parsing fails, the function simply stops and does not commit any changes to the ledger or memory.

- START(C′, ID, P): starts a random coin flipping protocol, can be called by any party, but only a single protocol can be run at a time:
  1. Verify that the field START_TIME does not exists in the SC's memory.
  2. Create the following fields in memory:
     (a) START_TIME initialized to the current time.
     (b) MCRS initialized to the empty string.
     (c) RESULT initialized to 0.
     (d) PUZZLES initialized to an empty array
     (e) PARTICIPANTS initialized to an empty array
- COMMIT(C′, ID, P): commits a party's puzzle to the current protocol. The calling party must deposit C funds to participate.
  1. Verify that C′ = C, and that the START_TIME field exists in memory and contains a value larger than the current time minus T.
  2. Parse P as (z, crs).
  3. Verify that z does not already exist in any (ID′, z) in PUZZLES.
  4. Append (ID, z) to PUZZLES, and ID to PARTICIPANTS.
  5. Append crs to MCRS in the memory.
- REVEAL(C, ID, P): reveals the value of a puzzle. If done before $2T$ time has passed since the protocol start, the entire deposit is returned to the party that created the puzzle (the hardness of the TLP implies that no party other than the creating party is able to solve the puzzle in less than 2T times). Otherwise, the solving party gets half the deposit, a quarter is divided between all protocol participants, and another quarter is disposed of.
  1. Verify that the START_TIME field exists and that START_TIME $+ T \leq$ *current_time*.
  2. Parse P as (z, s, π).

3. Verify that (ID′, z) exists in memory for some ID.
4. Verify that $Verify(1^\lambda, MCRS, R, z, s, \pi)$ outputs True.
5. If $s \neq \perp$: update RESULT to $RESULT \oplus s$.
6. Remove (ID, z) from PUZZLES.
7. If START_TIME + $2T \leq current\_time$: return C funds to the party ID.
8. Otherwise: transact $C/2$ funds to ID, dispose $C/4$ funds, and divide the rest $C/4$ funds equally between all IDs in PARTICIPANTS.
9. If there are no more (ID, z) in memory, the protocol is over - publish (RESULT, START_TIME, PARTICIPANTS) to the public ledger, and delete all fields from memory.

The implementation of the protocol for each party is as follows:

1. **Start:** the protocol starts by one of the parties creating the Smart Contract described above (if not already existing) and calling the START function. Other parties that wish to participate verify that the code of the created SC is as described.
2. **Commit:** This stage starts with the call to the START function, and ends after T time. Each party that wishes to participate in the protocol does as following during this stage:
   (a) Generates random numbers s, crs and r large enough for the *Gen* function.
   (b) Calculate z=$Gen(1^\lambda, 100T, s; r)$.
   (c) Call the SC COMMIT function with (C, λ, (z, crs)).
3. **Reveal:** This stage starts after T time has passed and ends after 2T time. Each party that committed a number in the previous stage should now reveal it:
   (a) Call the SC REVEAL function with (0, (z, s, r)).
4. **Bounty Force Reveal** Starts after 100T time until all puzzles are opened. At this stage parties try to solve all unopened puzzles to gain the bounty and end the protocol:
   (a) Get MCRS from the memory of the SC.
   (b) While there is another unsolved puzzle Z in the PUZZLES memory section:
      i. Parse z as (x, N, c), if failed call the SC REVEAL function with $(0, (\lambda)^2, \perp, \perp)$ and continue to the next puzzle.
      ii. Calculate $(s, \pi) = Solve(1^\lambda, MCRS, 100T, x, N, c)$.
      iii. Call the SC REVEAL function with $(0, (z, s, \pi))$.
5. **Result:** After all puzzles were opened, the protocol result would be on the public ledger under the SC's ID coupled with START_TIME - (RESULT, START TIME, PARTICIPANTS).

4 Analysis

Our protocol satisfies fairness, optimistic efficiency and public verifiability, as defined in [Fre+20] and incentivesed honesty which we now define.

1. **Fairness:** No malicious adversary (controlling all but one parties) can bias the output of the protocol, even by aborting early. Namely, as long as there is at least one honest participating party, the output will be a uniformly random value. More formally, the fairness for the protocol is defined as follows:

Lemma 6.2 *(From* [Fre+20]*). For any distinguisher D, there exists a negligible function negl such that for all $\lambda \in N$, the following holds. Suppose that at most $n(\lambda) \in poly(\lambda)$ parties participate for the commit phase, and at least one honest party runs the commit phase. Let s be the output of the open phase at the end of the protocol for security parameter λ, and let $r \leftarrow \{0,1\}^{L(\lambda)}$. It holds that $|Pr[D(s) = 1] - Pr[D(r) = 1]| \leq negl(\lambda)$.*

2. **Optimistic Efficiency:** If all participating parties are honest, then the protocol terminates within two message rounds (in our case at time less than $2T$), and all parties can efficiently verify the output of the protocol without force opening any puzzle by the lengthy sequential computation required for it.
3. **Public Verifiability:** In the case that any participating party is dishonest and does not publish their solution, any party can break the puzzle in a moderate amount of time (in our case $100T$) and provide a publicly verifiable proof of the solution. We even require that an honest party can prove that a published puzzle has no valid solution.
4. **Incentive Compatible:** Any all-but-one coalition of parties has a financial incentive to act honestly and reveal their puzzle in time. We show this while assuming that delaying the protocol's result could give a financial benefit only lesser than $1/4$ of the participation fee.

Remark 2 (Game-based vs. simulation-based fairness). We use a game-based definition of fairness, as it is simpler and sufficient for our settings. It is possible to obtain a simulation-style fairness, following [Fre+20, Appendix B], with a minor change to the protocol of feeding the previous output to a random oracle and outputting the result.

4.1 Fairness, Optimistic Efficiency, and Public Verifiability

We will now show fairness, optimistic efficiency, and public verifiability by arguing that the proof for those properties given in [Fre+20] for their multi-party coin flipping protocol holds for our protocol as well, due to the similarity between the protocols.

The model in [Fre+20] is quite similar to ours: They use a public bulletin model, in which any party may "publish" a message that all other parties will see within some fixed time. Our model essentially extends this model by adding SCs, financial incentives, and signing every message published with the publisher's ID, instead of anonymously as in the bulletin model.

We now compare the functionality differences between the protocols. The financial incentives and funds usage functionality are a major addition made for our protocol, but, as it only gives incentives to act honestly and an adversary may ignore them, it does not affect the actual security properties of the protocol. Other than this, the protocols have essentially has the same functionality and stages, but with two main differences:

1. In our protocol, messages are published to the SC and verified by it before they are committed and published, differing from [Fre+20] where all messages are immediately published to the public bulletin. Note that an honest party's execution is not affected by this as it always sends valid messages (and puzzles).
2. The published messages are not signed with the publisher's ID in [Fre+20], and so parties can publish anonymous messages. This functionality is not specifically used by honest parties in the honest flow of the protocol, thus the honest flow of the protocol in [Fre+20] would execute the same even if all messages were signed.

Thus, the main difference regarding the above properties is that a malicious adversary in [Fre+20] has more "power", as it can theoretically abuse the functionality of publishing anonymous or invalid messages. As the fairness proof for Lemma 6.2 is proved in [Fre+20] for their protocol, it follows that it holds against stronger adversaries than required for our protocol, and thus holds for our protocol as well.

The properties of optimistic efficiency, and public verifiability are related only to honest parties execution, and so the proofs in [Fre+20] also transfers to our protocol. We rely on the following corollary proven for the protocol in [Fre+20]:

Corollary 6.2 *(From* [Fre+20]*). Let $B, L : N \longrightarrow N$ where $B(\lambda) = 23L(\lambda)$. Assuming the B-repeated squaring assumption for RSWGen, there exists a multiparty coin flipping protocol that outputs $L(\lambda)$ bits and satisfies optimistic efficiency, fairness, and public verifiability. The protocol supports an unbounded number of participants and requires no adversary-independent trusted setup. Security is proven in the auxiliary-input random oracle model.*

4.2 Incentive Compatible

We now show that even an all-but-one coalition of dishonest parties has no incentive to deviate from the honest flow. Let C be the participation fee, γ be the financial incentive to delay the protocol in $100T$ time, Φ the bounty reward for solving an unsolved puzzle, and κ the amount the is divided between participants in the case of a force reveal. Assume there is an all but one coalition of size $n - 1$. If a party (or a few) in the coalition does not reveal its puzzle, the coalition will lose at least $C - (\Phi + \frac{\kappa}{n})$ of the reward, even if it manages to claim the force open bounty. For simplicity, we assume that the honest party has enough parallel computing power to compute all unsolved puzzles at about $100T$ time. We assume this as practically, parties that did not participate in the protocol but are members of the blockchain would also be able to help solve puzzles and claim rewards, so there should be many parties willing to help solve puzzles and finish the protocol at time $100T$, even if most of the participating parties would be malicious. In addition, we assume that the financial cost of running the computation of solving a puzzle is negligible compared to the force-open bounty. In such case, after $100T$ time the honest party would finish solving the unsolved puzzles and the protocol would finish. It follows that as long as $\gamma > C - (\Phi + \frac{\kappa}{n})$, there is no incentive to deviate and not reveal a puzzle, and the protocol is incentive compatible. We chose $\Phi = T/2$ and $\kappa = T/4$, but any amounts that adhere the above would make the protocol Incentive Compatible.

Finally, we note that, unlike in the Randao protocol, the reward for biasing or aborting the protocol can be arbitrarily large since the fairness property of the protocol says that the protocol cannot be aborted or biased.

4.3 Performance

Practically, due to the incentive compatibility property, it is reasonable to assume that most of the protocol's executions would go by the honest flow and terminate almost immediately. If this is not the case, the participation fees are probably not really lower than the incentive to delay and should be increased. Now, by the Optimistic Efficiency property, the protocol would end in time $2T$. T could be a few seconds, only long enough to give enough time for each participant to broadcast its commitment.

In dishonest flow, where some number of parties decide not to reveal their values, the run-time will lengthen at least to $102T$, and may be larger depending on how many parties would help to solve the unopened puzzles. Practically we assume that there would always be enough honest parties in the network (ones that are not participating in the protocol in its current round), that would be incentivesed to solve puzzles for the bounty, so the run-time will not exceed much from $102T$.

Acknowledgements. Ilan Komargodski is the incumbent of the Harry & Abe Sherman Senior Lectureship at the School of Computer Science and Engineering at the Hebrew University, supported in part by an Alon Young Faculty Fellowship, by a grant from the Israel Science Foundation (ISF Grant No. 1774/20), and by a grant from the US-Israel Binational Science Foundation and the US National Science Foundation (BSF-NSF Grant No. 2020643).

A Overview of Existing Solutions

There are some more existing solutions to the problem of generating fair randomness. In particular, each Proof-of-Stake-based blockchain uses a tailored solution for the specific randomness required for its next block's leader selection process. There are solutions based on verifiable secret sharing (VSS), such as Ouroboros [Kia+17], which is resilient up to an honest majority but has a disadvantage of quadratic communication requirements. Other works, such as

Algorand [Gil+17] or Ouroboros Praos [Dav+18], have used a Verifiable Random Function-based (VRF) approach to generate the randomness. They generally use a method where some parties plug the previous round's randomness in their private key-based VRF to generate new randomness. That is actually deterministic but heuristically unpredictable and unbiasable. Another approach is using Verifiable Delay Functions (VDFs), as will be done in Ethereum 2.0. The main disadvantage of VDF based protocols is that they currently rely on some form of trusted setup. Also, as far as we know, every known "timing-based" solution can create randomness only with a significant time delay, as they must have a large enough time buffer to mitigate faster adversaries. In addition, a faster computing adversary would be able to know the generated randomness slightly before the honest parties finish their computation. Our solution suffers from a similar "delay" problem (but we do not need a trusted setup!). And yet, the VDF-based solution was chosen by the Ethereum foundation, mainly because of its simplicity. For generating the trusted setup they design an independent solution in the form of a "ceremony" where few stake-holders execute a special multiparty computation protocol to generate the setup. Ethereum foundation currently chose to use a 102-minute VDF, and also funds an open source project to develop a fast specific ASIC for their VDF. In order to avoid waiting 102 min for each block they cleverly start the randomness generation process enough time in advance. (We can do the same).

The Randao protocol [Ran17] was previously proposed as a general solution for randomness generation. Its main advantages are its simplicity, speed, and (apparent) resilience to all but one dishonest party setting. It uses a simple commit and reveal scheme, with deposits to avoid Last-Revealer attacks: In the first stage, each party chooses a private random number r and commits to it publicly by sending $\mathsf{hash}(r, \text{seed})$. Each party also deposits a certain amount of funds in a smart contract. In the second stage, every party reveals its r and seed, which are verified using the hash from round one, and the final randomness produced is the (modulus) sum of the randomness sent by every party. The deposit is returned to each party that revealed their votes. If a party did not reveal its vote, its deposit is taken and distributed equally between those who did reveal at the current round. In such a case, the round is aborted and the protocol restarts. Practically, the incentive for parties to participate in the randomness generation protocol, and risk losing the deposit, is given by a fee taken from those who require randomness, which is distributed among the participants upon a successful randomness generation. As mentioned, the protocol is not secure/fair as it is vulnerable to Last-Revealer attacks. An adversary can wait until all other parties reveal their votes, and then calculate what the randomness would result in if he also reveals. He would then be able to decide not to reveal if the result is not satisfactory to him, which will then result in the protocol restarting and new randomness generated. The incentive of not losing the deposit makes it less attractive to defect in this way, but if the resulting randomness will decide some highly profitable event, such as a lottery, the deposit must be extremely large in order to keep the protocol safe by making it unprofitable to defect. Note that in the Randao setting, the randomness requester decides by itself how much deposit to put, and the participation fee is relative to it. Ethereum 2.0 uses the Randao scheme concatenated with a VDF to avoid the last revealer attack (at the cost of generating a trusted setup).

References

[Cle86] Cleve, R.: Limits on the security of coin flips when half the processors are faulty, pp. 364–369 (1986)

[GMW87] Goldreich, O., Micali, S., Wigderson, A.: How to play any mental game or a completeness theorem for protocols with honest majority. In: STOC, pp. 218–229. ACM (1987)

[BGW88] Ben-Or, M., Goldwasser, S., Wigderson, A.: Completeness theorems for non-cryptographic fault-tolerant distributed computation (extended abstract). In: STOC, pp. 1–10 (1988)

[CCD88] Chaum, D., Crépeau, C., Damgård, I.: Multiparty unconditionally secure protocols (extended abstract). In: STOC, pp. 11–19 (1988)

[RSW96] Rivest, R.L., Shamir, A., Wagner, D.A.: Time-lock puzzles and timed-release crypto. Technical report, USA (1996)

[BN00] Boneh, D., Naor, M.: Timed commitments. In: Bellare, M. (ed.) CRYPTO 2000. LNCS, vol. 1880, pp. 236–254. Springer, Heidelberg (2000). https://doi.org/10.1007/3-540-44598-6_15

[Sho05] Shoup, V.: A Computational Introduction to Number Theory and Algebra. Cambridge University Press, USA (2005). ISBN 0521851548

[But15] Buterin, V.: A next generation smart contract & decentralized application platform (2015)

[Gil+17] Gilad, Y., et al.: Algorand: scaling byzantine agreements for cryptocurrencies. In: Proceedings of the 26th Symposium on Operating Systems Principles, SOSP 2017, pp. 51–68. Association for Computing Machinery, Shanghai (2017). https://doi.org/10.1145/3132747.3132757. ISBN 9781450350853

[Kia+17] Kiayias, A., Russell, A., David, B., Oliynykov, R.: Ouroboros: a provably secure proof-of-stake blockchain protocol. In: Katz, J., Shacham, H. (eds.) CRYPTO 2017. LNCS, vol. 10401, pp. 357–388. Springer, Cham (2017). https://doi.org/10.1007/978-3-319-63688-7_12. ISBN 978-3-319-63688-7

[Ran17] Randao.org: Randao: Verifiable Random Number Generation (2017). https://www.randao.org/whitepaper/Randao_v0.85_en.pdf

[Bon+18] Boneh, D., Bonneau, J., Bünz, B., Fisch, B.: Verifiable delay functions. In: Shacham, H., Boldyreva, A. (eds.) CRYPTO 2018. LNCS, vol. 10991, pp. 757–788. Springer, Cham (2018). https://doi.org/10.1007/978-3-319-96884-1_25

[Dav+18] David, B., Gaži, P., Kiayias, A., Russell, A.: Ouroboros praos: an adaptively-secure, semi-synchronous proof-of-stake blockchain. In: Nielsen, J.B., Rijmen, V. (eds.) EUROCRYPT 2018. LNCS, vol. 10821, pp. 66–98. Springer, Cham (2018). https://doi.org/10.1007/978-3-319-78375-8_3. ISBN 978-3-319-78374-1

[CGJ19] Choudhuri, A.R., Goyal, V., Jain, A.: Founding secure computation on blockchains. In: Ishai, Y., Rijmen, V. (eds.) EUROCRYPT 2019. LNCS, vol. 11477, pp. 351–380. Springer, Cham (2019). https://doi.org/10.1007/978-3-030-17656-3_13

[Feo+19] De Feo, L., Masson, S., Petit, C., Sanso, A.: Verifiable delay functions from supersingular isogenies and pairings. In: Galbraith, S.D., Moriai, S. (eds.) ASIACRYPT 2019. LNCS, vol. 11921, pp. 248–277. Springer, Cham (2019). https://doi.org/10.1007/978-3-030-34578-5_10

[Pie19] Pietrzak, K.: Simple verifiable delay functions. In: Innovations in Theoretical Computer Science Conference, ITCS, pp. 60:1–60:15 (2019)

[Fre+20] Freitag, C., et al.: Non-malleable time-lock puzzles and applications. Cryptology ePrint Archive, Report 2020/779 (2020). https://ia.cr/2020/779. 2020

[Wes20] Wesolowski, B.: Efficient verifiable delay functions. J. Cryptol. **33**(4), 2113–2147 (2020). https://doi.org/10.1007/s00145-020-09364-x

[Bau+21a] Baum, C., et al.: SoK: mitigation of front-running in decentralized finance. In: IACR Cryptol. ePrint Arch. 162 (2021)

[Bau+21b] Baum, C., David, B., Dowsley, R., Nielsen, J.B., Oechsner, S.: TARDIS: a foundation of time-lock puzzles in UC. In: Canteaut, A., Standaert, F.-X. (eds.) EUROCRYPT 2021. LNCS, vol. 12698, pp. 429–459. Springer, Cham (2021). https://doi.org/10.1007/978-3-030-77883-5_15

[Thy+21] Thyagarajan, S.A.K., et al.: Efficient CCA timed commitments in class groups. In: CCS, pp. 2663–2684. ACM (2021)

[Wik] Wikipedia. Jacobi symbol entry. https://en.wikipedia.org/wiki/Jacobi_symbol. Accessed 16 Oct 2022

Efficient Skip Connections Realization for Secure Inference on Encrypted Data

Nir Drucker(✉) and Itamar Zimerman

IBM Research, Haifa, Israel

drucker.nir@gmail.com

Abstract. Homomorphic Encryption (HE) is a cryptographic tool that allows performing computation under encryption, which is used by many privacy-preserving machine learning solutions, for example, to perform secure classification. Modern deep learning applications yield good performance for example in image processing tasks benchmarks by including many skip connections. The latter appears to be very costly when attempting to execute model inference under HE. In this paper, we show that by replacing (mid-term) skip connections with (short-term) Dirac parameterization and (long-term) shared-source skip connection we were able to reduce the skip connections burden for HE-based solutions, achieving ×1.3 computing power improvement for the same accuracy.

Keywords: shared-source skip connections · Dirac networks · Dirac parameterization · homomorphic encryption · privacy preserving machine learning · PPML · encrypted neural networks · deep neural networks

1 Introduction

The use of Homomorphic Encryption (HE) to construct Privacy-Preserving Machine Learning (PPML) solutions, e.g., secure Deep Neural Network (DNN) inference on the cloud, becomes more and more realistic. For example, Gartner [11] predicted that in 2025, 50% of large enterprises will adopt HE-based solutions. In addition, we see many companies and academic institutes collaborate in global activities such as HEBench [23] and the HE standardization efforts [2]. The main reason is, of course, that HE allows finance and health organizations to comply with regulations such as GDPR [10] and HIPAA [5] when uploading sensitive data to the cloud.

One principle scenario of HE-based PPML solutions involves two entities: a user and a semi-honest cloud server that performs Machine Learning (ML) computation on HE-encrypted data. Specifically, the cloud offers an ML as a Service (MLaaS) solution, where it first trains a model in the clear, e.g., a DNN, and then, uses it for inference operations on the clients' data. On the other side, the client first generates its own HE keys, stores the secret key, and uploads the

S. Dolev et al. (Eds.): CSCML 2023, LNCS 13914, pp. 65–73, 2023.
https://doi.org/10.1007/978-3-031-34671-2_5

public and evaluation keys to the cloud. Subsequently, upon demand, it encrypts secret samples and submits them to the cloud that uses the client's public and evaluation keys to perform the model inference operation. The final encrypted results are sent back to the client who decrypts them using its private key.

The clients' data is kept confidential from the server during the entire protocol due to HE, while the cloud model is never sent to the client, which allows the cloud to monetize its MLaaS service. In this paper, we focus on this scenario but stress that our study can be used almost without changes in many other threat models.

One downside of HE-based solutions is their latency. While there are many software and hardware improvement that make HE-based solutions practical such as [1,16], there is still a gap between computing on encrypted data and computing on cleartexts, where our goal is to reduce that gap. Our starting point is a recent study [4] that pointed out on skip connections as a major contributor to the overall latency of secure inference solutions that use DNNs. The authors of [4] suggested removing the skip connection at the cost of some accuracy degradation or replacing the skip connections using several heuristics. We continue this line of work by suggesting using modern techniques that allow training DNNs while maintaining good accuracy. Specifically, we replace mid-term skip connections in DNNs with short-term (Dirac parameterization) [25] and long-term (Shared source skip connection) [22].

Our Contribution. We used ResNet50, a state-of-the-art network in terms of size that can run efficiently under HE [4,16], as our baseline. We modified it to be HE-friendly, a term that we explain later, and apply the above techniques. Our experiments show that using this approach we were able to reduce the number of HE bootstrap operations by $\times 1.36 - 1.75$ and thus the overall CPU time by $\times 1.3$.

Organization. The document is organized as follows. In Sect. 2 we provide some background about HE. We describe skip connections and their variants in Sect. 3. Our experiments and results are presented in Sect. 4 and we conclude in Sect. 5.

2 Homomorphic Encryption (HE)

HE is an encryption scheme that encrypts input plaintext from a ring $\mathcal{R}_1(+, *)$ into ciphertexts in another ring $\mathcal{R}_2(\oplus, \odot)$, i.e., it contains the encryption function $\text{Enc} : \mathcal{R}_1 \rightarrow \mathcal{R}_2$ and decryption function $\text{Dec} : \mathcal{R}_2 \rightarrow \mathcal{R}_1$, where a scheme is correct if $\text{Dec}(\text{Enc}(x)) = x$. In addition, HE schemes include homomorphic addition and multiplication operations such that $\text{Dec}(\text{Enc}(x) \oplus \text{Enc}(y)) = x + y$ and $\text{Dec}(\text{Enc}(x) \odot \text{Enc}(y)) = x * y$ see survey in [13]. In our experiment, we use CKKS [6,7] an approximately correct scheme, i.e., for some small $\epsilon > 0$ that is determined by the key, it follows that $|x - \text{Dec}(\text{Enc}(x))| \leq \epsilon$ and the same modification applies to the other equations.

Chain Index and Bootstrapping. HE ciphertexts and particularly CKKS ciphertexts have a limit on the number of multiplications they can be involved with before a costly bootstrap operation is required. To this end, every ciphertext includes an extra metadata parameter called the "multiplication chain index" (a.k.a. modulus chain index) or CIdx. Ciphertexts start with a CIdx of 0 and after every multiplication of two ciphertexts with CIdx of x and y, the result has a CIdx of $\max(x, y) + 1$, where at least a ReScale operation is required. This process continues until the ciphertext reaches the predefined limit, which was originally set by the client to achieve the desired level of security and performance. To enable further computation on a ciphertext, a Bootstrap operation is performed to reduce its CIdx, or even reset it back to 0. In general, many HE-based applications attempt to minimize the number of Bootstrap invocations and this is also our goal in this paper.

There are two options for adding or multiplying two ciphertexts c_1, c_2 with CIdx $= x, y$, respectively, where w.l.o.g $x > y$: a) adjust c_1 to have a CIdx $= y$ by invoking ReScale(Bootstrap(c_1), y); or b) invoke ReScale(c_2, x). This first option is costlier because it invokes both ReScale and Bootstrap in advance while the other approach leaves the bootstrap handling to future operations. However, this approach is preferred when c_1 is expected to be added to multiple ciphertexts with lower chain indices. In that case, we perform only one Bootstrap operation on c_1 instead of many on the other operations' results. An automatic bootstrapping placement mechanism is expected to consider the above.

HE Packing. Some HE schemes, such as CKKS [6], operate on ciphertexts in a homomorphic Single Instruction Multiple Data (SIMD) fashion. This means that a single ciphertext encrypts a fixed-size vector, and the homomorphic operations on the ciphertext are performed slot-wise on the elements of the plaintext vector. To utilize the SIMD feature, we need to pack and encrypt more than one input element in every ciphertext. The packing method can significantly impact both bandwidth, latency, and memory requirements. In this paper we decided to rely on IBM HELayers, which provides efficient packing capabilities for DNNs through the use of a new data structure called tile tensors [1]. We stress that adding or multiplying two ciphertexts that represent different tile tensor shape is problematic and an extra transformation is needed. While automatically handled by HELayers, one of goals is to also save these transformations.

3 Skip Connections

Skip connections, a.k.a, residual connections [14], are crucial components in modern network architectures. Given several layers f(x), applying skip-connection to the layer means wrapping f(x) with a function $S_f(x) = f(x) + x$. For real-world applications, networks without skip connections are hard to train, especially very deep networks. Skip connection solves optimization issues such as (i) vanishing gradients (ii) exploding gradients [20], or (iii) shattering gradients [3]. In practice, modern architectures heavily rely on skip connections e.g., Transforms [24] ViT [9], LLMs, CNNs [17], WaveNet [18], GPT [21], and Residual Net (ResNet) who has became one of the most cited DNN of the 21st century. When considering cleartext networks, skip-connections require only a simple addition, and thus provide an efficient solution that enables easier optimization of DNNs. Moreover, they also play a fundamental role in modern Deep Learning (DL) solutions. While skip-less networks exist, and new variants appear from time to time e.g., [19,25], they are rarely used in real-world applications, as they tend to perform poorly in complex scenarios and noisy data.

3.1 Handling Skip Connections in HE

Observation 3.1 of [4] explains the relation between skip connections and bootstrapping operations.

Observation 1 (observation 3.1 [4]**).** *Given a skip connection layer* $S_f(x) = x + f(x)$*, where* f *is a combination of some layers. When running under HE,*

1. $CIdx(S_f(x)) \in \{CIdx(x), CIdx(f(x))\}$.
2. *When* $CIdx(x) \neq CIdx(f(x))$ *the skip connection implementation invokes either a ReScale or a Bootstrap operation and may increase the overall multiplication depth of the network by* $|CIdx(x) - CIdx(f(x))|$.

In addition, the authors of [4] explained that the cost of $S_f(x)$ can be even higher because the input x or $f(x)$ may need to go through some transformations before adding them together, which is the case with the HELayers SDK. Given the latency costs associated with implementing skip connections under HE, [4] proposed either removing skip connections by first training a network and then gradually removing the connections, which resulted in some accuracy degradation, or that they suggested some heuristics to reroute these connections, which offers a tradeoff between latency and accuracy. Here, we suggest another heuristic that brings knowledge from the AI domain into the HE domain. Particularly, we replace DNNs skip connections with Dirac parameterization [25] and shared-source skip connection [22]. Informally speaking, shared-source skip connections connect the output of the initial layer or input with the output of different locations in the network. The reason this reduces the number of bootstraps is that after the initial layer, the chain index is very low. In addition, we can aim to add these connections only to layer outputs that share the same tile tensor shape and thus save reshape operations. Dirac parameterization is explained as follows: let $\sigma(x)$ be a function that combines non-linearity and batch normalization, then a standard convolutional layer in ResNet is of an explicit form $y = x + \sigma(W \odot x)$, where a Dirac parameterization is of the form $y = \sigma(diag(a)I + W) \odot x = \sigma(diag(a)x + W \odot x)$. This addition, helps training and does not affect the latency of the secure inference.

Figure 1 shows a standard ResNet network with skip connection in their original places. In contrast, Fig. 2 shows our modification. Because HE only supports polynomial operations, we first replace non-polynomial layers with polynomial layers (red font layers) to achieve a HE-friendly network. Specifically, we replace MaxPool with AVGPooling and ReLU activations with polynomial activations similar to [12]. Subsequently, we removed all mid-term skip connections and added long-term shared-source connections (green arrows) from the output of

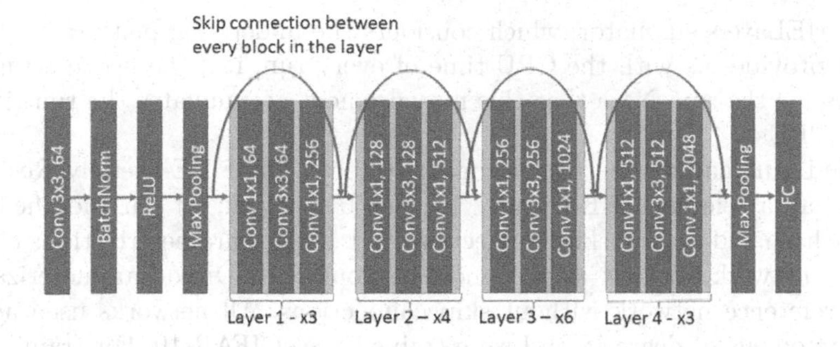

Fig. 1. An illustration of ResNet50 every layer contains several blocks and there is a skip connection between every block.

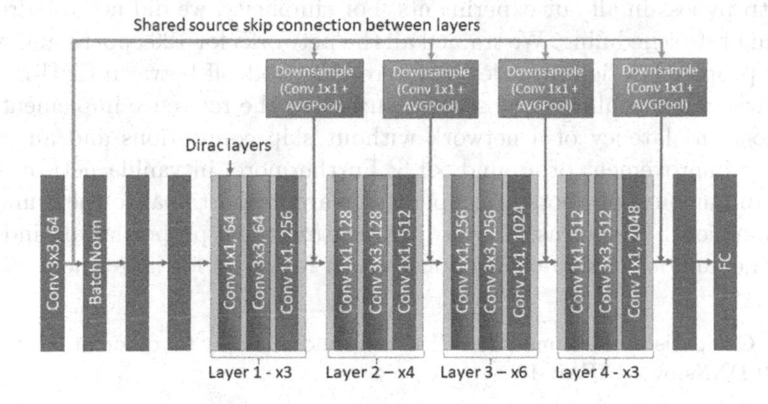

Fig. 2. An illustration of our modified HE-friendly ResNet50. Every layer contains several blocks and there is a shared-source skip connection from the first layer output to the output of the four other layers. Red layers were modified to make the network HE-friendly as in [12] (Color figure online).

the first convolution layer to every one of the four layers' outputs. To ensure that the dimensions match we added 1×1 convolutional and average pooling layers to these connections. Note that these layers can be performed on the server side but also by the client if we consider a split network, where the first layer is performed on the client side. This offers a tradeoff between latency and bandwidth. Finally, we added low-term Dirac parameterization to the first two convolution layers of every block, where the stride is 1 (orange blocks).

4 Experiments

In our experiments, we used a single NVIDIA A100-SXM4-40GB GPU with 40 GB of memory for training and for secure inference an Intel® Xeon® CPU E5-2699 v4 @ 2.20 GHz machine with 44 cores (88 threads) and 750 GB memory. In addition, for inference we used HELayers [15] version 1.5.2, where we set the underlying HE library to HEaaN [8] targeting 128 bits security. Specifically, we used HELayers simulator, which considers the underlying platform capabilities and provides us with the CPU-time of every run, i.e., the needed compute resources for the run. Note that this measurement accumulates the run time of all used CPUs.

Table 1 summarizes the test accuracy results of four HE-friendly ResNet50 variants: a) a reference HE-friendly ResNet50 network; b) our modified network with shared-source skip connection and Dirac parameterization; c) The reference network without skip connections but with Dirac parameterization; d) The reference network without skip connections. All networks used activation polynomials of degree 8 and were trained over CIFAR-10. For training, we used PyTorch as our library, AdamW as the optimizer (with all default hyper-parameters, and learning rate of $1e{-}3$), a batch size of 50, and the standard cross-entropy loss in all our experiments. For simplicity, we did not use dropout or learning rate scheduling. We trained all the networks for 120 epochs and we see that our proposed design provides an interesting tradeoff between CPU-runtime and accuracy. It has almost the same accuracy as the reference implementation but almost the latency of a network without skip connections and an overall CPU-time improvement of around $\times 1.3$. Furthermore, in vanilla networks, the latency and amount of bootstraps operations are proportional to the number of skip-connections. In contrast, in our architecture, these properties are independent of the number of skip-connections, which is crucial for larger networks.

Table 1. Comparison of accumulated CPU-time and accuracy for different HE-friendly ResNet50 DNNs over CIFAR-10.

Network architecture	CPU-time (h)	# bootstraps	Test accuracy Non HE-friendly	Test accuracy HE-friendly
Reference	18.06	2,568	91.67	91.46
W/o skip connections	12.9	1,888	88.74	88.68
W/ Dirac params	12.9	1,888	90.87	90.74
Our variant	13.4	1,888	91.25	91.08

Figure 3 compares the training status of the three HE-friendly ResNet50 variants. For these, it reports the test accuracy (x-axis) per training epoch (y-axis). The reference HE-friendly network is represented by the blue line, our modified network by the red line, and a network where the skip-connections were completely removed by a green line.

Table 2 extends Table 1 and shows the accumulated CPU-time improvement when using our modified HE-friendly ResNet50 as reported by HELayers simulator [1] with activation functions of different polynomial degrees. We observe that compared to the reference network we got an improvement of $\times 1.18$, $\times 1.3$, and $\times 1.34$ in the consumed compute resources when using polynomial activations of degrees 2, 4, and 8, respectively. On the other hand, our network consumed slightly more compute resources compared to a network completely without skip

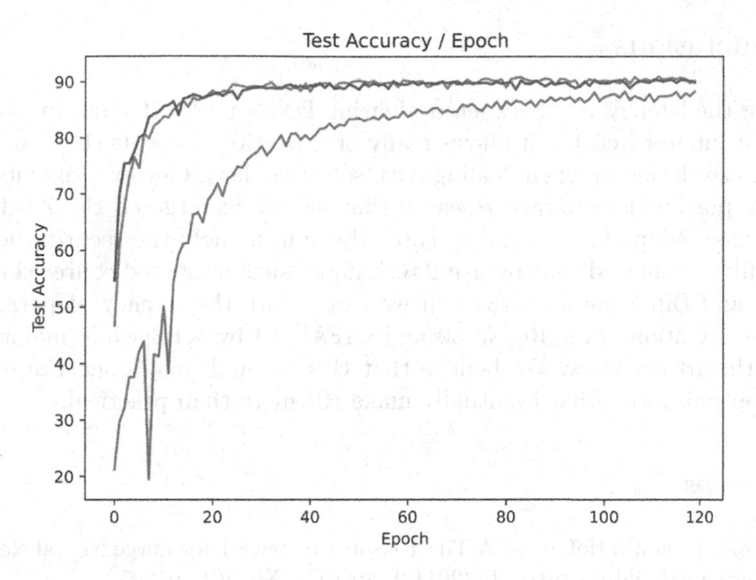

Fig. 3. Test accuracy per training epoch of 3 HE-friendly ResNet50 network variants: Reference (blue line), Our modified network (red line), No skip-connection network (green line). (Color figure online)

Table 2. Performance comparison and number of bootstraps used by HELayers to run different HE-friendly ResNet50 DNNs over CIFAR-10 with activations approximation of different polynomial degrees.

Activation poly. deg.	Network architecture	CPU-time (h)	# bootstraps
2	Reference	12.33	1,664
	W/o skip connections	9.13	1,304
	Our variant	10.40	1,480
4	Reference	11.36	1,376
	W/o skip connections	7.68	784
	Our variant	8.69	784
8	Reference	18.06	2,568
	W/o skip connections	12.92	1,888
	Our variant	13.38	1,888

connections, specifically, $\times 0.88$, $\times 0.87$, and 0.97, for networks with polynomial activations of degrees 2,4, and 8, respectively. Nevertheless, as shown in Fig. 3 it achieved the accuracy of the reference implementation. We note that some of the compute resources reduction was due to the reduction in the number of performed bootstraps. However, another reduction comes from the more effective utilization of tile tensors by avoiding several reshaping operations when using shared-source connections.

5 Conclusions

Reducing the latency gap between evaluating DNNs under HE and in cleartext is of great interest because it allows many organizations to scale their computation and use the cloud when dealing with sensitive data. Our experiments show that it is possible to combine research that originally targeted the AI domain to achieve speedups in a second research domain, namely the security domain. Specifically, we showed that by using techniques such as shared-source skip connections and Dirac parameterization we can reduce the latency of performing inference operations over ResNet50 and CIFAR-10 by $\times 1.3$ while maintaining state-of-the-art accuracy. We believe that this research may open the door to further optimizations that eventually make HE more than practical.

References

1. Aharoni, E., et al.: HeLayers: A Tile Tensors Framework for Large Neural Networks on Encrypted Data. CoRR abs/2011.0 (2020). arXiv:2011.01805
2. Albrecht, M., et al.: Homomorphic encryption security standard. Technical report, HomomorphicEncryption.org, Toronto, Canada (2018). https://HomomorphicEncryption.org
3. Balduzzi, D., Frean, M., Leary, L., Lewis, J.P., Ma, K.W.D., McWilliams, B.: The shattered gradients problem: If resnets are the answer, then what is the question? In: Precup, D., Teh, Y.W. (eds.) Proceedings of the 34th International Conference on Machine Learning. Proceedings of Machine Learning Research, vol. 70, pp. 342–350. PMLR (2017). https://proceedings.mlr.press/v70/balduzzi17b.html
4. Baruch, M., et al.: Sensitive Tuning of Large Scale CNNs for E2E Secure Prediction using HE (2023)
5. Centers for Medicare & Medicaid Services: The Health Insurance Portability and Accountability Act of 1996 (HIPAA) (1996). https://www.hhs.gov/hipaa/
6. Cheon, J.H., Han, K., Kim, A., Kim, M., Song, Y.: A full RNS variant of approximate homomorphic encryption. In: Cid, C., Jacobson Jr., M.J. (eds.) Selected Areas in Cryptography - SAC 2018, SAC 2018. Lecture Notes in Computer Science, vol. 11349, pp. 347–368. Springer, Cham (2019). https://doi.org/10.1007/978-3-030-10970-7_16
7. Cheon, J.H., Kim, A., Kim, M., Song, Y.: Homomorphic encryption for arithmetic of approximate numbers. In: Takagi, T., Peyrin, T. (eds.) ASIACRYPT 2017. LNCS, vol. 10624, pp. 409–437. Springer, Cham (2017). https://doi.org/10.1007/978-3-319-70694-8_15
8. CryptoLab: HEaaN: Homomorphic Encryption for Arithmetic of Approximate Numbers (2022). https://www.cryptolab.co.kr/eng/product/heaan.php
9. Dosovitskiy, A., et al.: An image is worth 16x16 words: Transformers for image recognition at scale. arXiv preprint arXiv:2010.11929 (2020)
10. EU General Data Protection Regulation: Regulation (EU) 2016/679 of the European Parliament and of the Council of 27 April 2016 on the protection of natural persons with regard to the processing of personal data and on the free movement of such data, and repealing Directive 95/46/EC (General Data Protection Regulation). Official Journal of the European Union 119 (2016). http://data.europa.eu/eli/reg/2016/679/oj

11. Gartner: Gartner identifies top security and risk management trends for 2021. Technical report (2021). https://www.gartner.com/en/newsroom/press-releases/2021-03-23-gartner-identifies-top-security-and-risk-management-t
12. Gilad Bachrach, R., Dowlin, N., Laine, K., Lauter, K., Naehrig, M., Wernsing, J.: Cryptonets: applying neural networks to encrypted data with high throughput and accuracy. In: International Conference on Machine Learning, pp. 201–210 (2016). http://proceedings.mlr.press/v48/gilad-bachrach16.pdf
13. Halevi, S.: Homomorphic Encryption. In: Lindell, Y. (ed.) Tutorials on the Foundations of Cryptography: Dedicated to Oded Goldreich, pp. 219–276. Springer, Cham (2017). https://doi.org/10.1007/978-3-319-57048-8_5
14. He, K., Zhang, X., Ren, S., Sun, J.: Deep residual learning for image recognition. In: Proceedings of the IEEE Conference on Computer Vision and Pattern Recognition (CVPR) (2016). https://openaccess.thecvf.com/content_cvpr_2016/html/He_Deep_Residual_Learning_CVPR_2016_paper.html
15. IBM: HELayers SDK with a Python API for x86 (2021). https://hub.docker.com/r/ibmcom/helayers-pylab
16. Lee, E., et al.: Low-complexity deep convolutional neural networks on fully homomorphic encryption using multiplexed parallel convolutions. In: Chaudhuri, K., Jegelka, S., Song, L., Szepesvari, C., Niu, G., Sabato, S. (eds.) Proceedings of the 39th International Conference on Machine Learning, vol. 162, pp. 12403–12422. PMLR (2022). https://proceedings.mlr.press/v162/lee22e.html
17. Liu, Z., Mao, H., Wu, C.Y., Feichtenhofer, C., Darrell, T., Xie, S.: A convnet for the 2020s. In: Proceedings of the IEEE/CVF Conference on Computer Vision and Pattern Recognition, pp. 11976–11986 (2022)
18. van den Oord, A., et al.: Wavenet: a generative model for raw audio. In: 9th ISCA Speech Synthesis Workshop, pp. 125–125 (2016). https://www.isca-speech.org/archive_v0/SSW_2016/abstracts/ssw9_DS-4_van_den_Oord.html
19. Oyedotun, O.K., Shabayek, A.E.R., Aouada, D., Ottersten, B.: Going deeper with neural networks without skip connections. In: 2020 IEEE International Conference on Image Processing (ICIP), pp. 1756–1760 (2020). https://doi.org/10.1109/ICIP40778.2020.9191356
20. Pascanu, R., Mikolov, T., Bengio, Y.: On the difficulty of training recurrent neural networks. In: Dasgupta, S., McAllester, D. (eds.) Proceedings of the 30th International Conference on Machine Learning. Proceedings of Machine Learning Research, vol. 28, pp. 1310–1318. PMLR (2013). https://proceedings.mlr.press/v28/pascanu13.html
21. Radford, A., Narasimhan, K., Salimans, T., Sutskever, I., et al.: Improving language understanding by generative pre-training (2018). https://www.cs.ubc.ca/amuham01/LING530/papers/radford2018improving.pdf
22. Tai, Y., Yang, J., Liu, X.: Image super-resolution via deep recursive residual network. In: 2017 IEEE Conference on Computer Vision and Pattern Recognition (CVPR), pp. 2790–2798 (2017). https://doi.org/10.1109/CVPR.2017.298
23. The HEBench Organization: HEBench (2022). https://hebench.github.io/
24. Vaswani, A., et al.: Attention is all you need. In: Advances in Neural Information Processing Systems, vol. 30 (2017). https://proceedings.neurips.cc/paper/2017/file/3f5ee243547dee91fbd053c1c4a845aa-Paper.pdf
25. Zagoruyko, S., Komodakis, N.: Diracnets: Training very deep neural networks without skip-connections (2017). arXiv:1706.00388

Single Instance Self–masking via Permutations (Preliminary Version)

Asaf Cohen, Paweł Cyprys(✉), and Shlomi Dolev

Ben-Gurion University of the Negev, Beer Sheva, Israel
cyprysp@gmail.com

Abstract. Self–masking allows the masking of success criteria, part of a problem instance (such as the sum in a subset-sum instance) that restricts the number of solutions. Self–masking is used to prevent the leakage of helpful information to attackers; while keeping the original solution valid and, at the same time, not increasing the number of unplanned solutions.

Self–masking can be achieved by xoring the sums of two (or more) independent subset sum instances [4,5], and by doing so, eliminate all known attacks that use the value of the sum of the subset to find the subset fast, namely, in a polynomial time; much faster than the naive exponential exhaustive search.

We demonstrate that the concept of self–masking can be applied to a single instance of the subset sum and a single instance of the permuted secret-sharing polynomials.

We further introduce the benefit of permuting the bits of the success criteria, avoiding leakage of information on the value of the i'th bit of the success criteria, in the case of a single instance, or the parity of the i'th bit of the success criteria in the case of several instances.

In the case of several instances, we permute the success criteria bits of each instance prior to xoring them with each other. One basic permutation and its nesting versions (e.g., π^i) are used, keeping the solution space small and at the same time, attempting to create an "all or nothing" effect, where the result of a wrong π trials does not imply much.

1 Introduction

There is a need for an efficient candidate to serve as a one-way function (OWF). Say, a function that is more efficient than SHA, that is, a function that is easy to compute yet has no known attack for easily finding preimage and/or collisions, is important, as such functions can influence the daily computation invested in commitments and signatures. Commitments based on candidates for[1] one-way functions are used in many scenarios, including in obtaining Zero Knowledge

[1] Note that a provable one-way function implies $P \neq NP$.

Partially supported by the Rita Altura Trust Chair in Computer Science and the Israeli Science Foundation (Grant No. 465/22).

S. Dolev et al. (Eds.): CSCML 2023, LNCS 13914, pp. 74–84, 2023.
https://doi.org/10.1007/978-3-031-34671-2_6

Proofs (ZKP) [1,7], whereas OWF is used as a commitment primitive. For a signature example, consider signatures in the style of Lamport's signature [10]. Lamport's signature is an example of using OWF to facilitate a (one-time) digital signature.

There are known candidates for one-way functions for which an algorithm that can be used in practice has successfully inverted them. See the example in [13] for instances of the subset-sum. Indeed, throughout history, one of the main goals of crypto-analysts is to find such breaks, and in many cases, they were successful, e.g., the recent result [11].

Background. Permutations are often used in the design of one-way functions as they provide a source for non-linearity, see, e.g., [14].

We suggest using the permutation of items in vectors, either the items to choose the sub-set from or the y-values used in reconstructing (secret–sharing) polynomials. Specifically, in the case of the polynomials, we construct a function from the set of all permutations on n elements to a finite field element, the free coefficient of a (secret–sharing) polynomial. Computing the free coefficient requires only an inner product, yet when the permutation is unknown, it is harder to inverse. One piece of evidence for such difficulty is shown in the sequel, proving that random permutations result in an approximately uniform probability on the output space.

To further enhance the challenge, we incorporate self–masking. Roughly speaking, the goal of self–masking is similar to code obfuscation, where the instance is given and defines the solution/functionality but adds a level of pseudo-randomness. The practicality of the masking technique [4,5] depends heavily on the hardness of reconstructing the self–masked parts and the number of additional solutions the masking introduces. The success criteria for the reconstruction are encoded to prevent an easy reveal of the self–masked parts.

To make the presentation self–contained and still short, we present only definitions that are explicitly used in our analysis.

Self-masking is a technique to conceal part of a computation task instance(s), for example, the required sum in the subset-sum instance(s), by using a function, possibly bitwise xor, with another analogous part of an (independently chosen) other instance(s). Self-masking may preserve correlation among the solutions, possibly by having sums implied by the same indices of items of the participating instances, and in this way (further) restrict the number of possible unplanned solutions for the combined instances [4,5].

Here we further extend the self-masking techniques to the case of a single instance, using the representation of the original instance to conceal a part (e.g., the desired sum of the subset) by applying a function on the part, using the randomness used to produce the instance, for example, the bits of the vector of items (participating in the subset sum). For a more comprehensive background on one-way functions self–masking and related applications, see, e.g., [4,5,8,9].

In [5], it is suggested to produce a random sorted array by randomly choosing n values, choosing the values by which each previous item in the array is incremented (for simplicity, assume integers with no bounds) then randomly permute the items. Both these operations can be performed in $O(n)$ steps. However, sorting the array (i.e., finding the reverse random permutation) in a comparison-based sort requires $\Omega(n \log n)$ steps. It is further suggested in [5], to allow other orders beyond the sorted order to be the successful order and to define the success using polynomial P, where $P(i)$, $1 \leq i \leq n$ is the value of a (randomly selected) entry in an array. Then, $P(0)$ and permuted values of the array entries are given, requiring finding the reverse permutation. Here, we significantly extend the basic suggestion of [5], e.g., we suggest shuffling the bits of $P(0)$ (without increasing the number of undesired solutions) and proving the uniform distribution of the solving permutations.

Note that using polynomials over items in a finite field and Lagrange interpolation is similar to the secret sharing technique suggested in [12], where ignoring even one item yields a uniform distribution of $P(0)$.

Paper Organization. The rest of the paper is organized as follows. Section 2 gives a very short presentation of self–masking ideas in the current literature and exemplifies its use in the subset–sum problem. Section 3 gives the main results for an OWF based on one instance of the secret–sharing polynomial problem. Section 4 depicts an algorithm that incorporates the self–masking ideas with the OWF instance presented herein. Section 5 concludes the paper. Many details and proofs are omitted from this short version and can be found in [2].

2 Subset Sum with Permutations

We start with a few definitions and settings to make the presentation as self–contained as possible.

Subset Sum. Given a subset-sum instance $A_{n,l} = (a_1, a_2..a_n)$ and b, such that each a_i and b are of ℓ bits. Find a subset of the elements summed up to b (mod 2^{l+1}).

MsbLsb. Is the (long) sequence of $n\ell$ bits, starting from the most significant bit (MSB) of the first number/item, continuing to the first MSB of the second number/item, until all n MSBs bits are used, then turning to use the next to the MSB row collecting in a similar fashion, additional n bits, and so on, until at last, the LSB bits join the created sequence to form a sequence of $n\ell$ bits. We call the obtained sequence *MsbLsb* sequence. Note that other (more sophisticated) constructions for harvesting many randomly chosen bits from the randomly chosen items can be suggested.

Permutations. To harden the reconstruction of the critical parts, say the sum b of the required subset, the self–masking technique presented in [4] can be extended by applying permutation defined by the *MsbLsb* sequence as a permutation index and constructing the actual permutation by using the mapping defined in e.g., [6]. Given an (integer) i index of a permutation in the lexicographical order of the permutations. Unique permutation hashing [6] output the i'th permutation, where i is the index of the permutation in the lexicographic order of permutations, as follows, outputs the first index of the permutation to be the index of the bucket (just as done in bucket sort where buckets are indexes are 1 to n, and each bucket size is $n!/n$ values) in which i is mapped to, say this bucket is j. Next, the scope is the mapping to a bucket in the j'th bucket, partitioned to $n-1$ buckets each of size $(n-1)!/(n-1)$ values, and eliminating the j index from consideration, continuing this way to define the entire permutation explicitly.

We suggest the following particular masking (as an easy example from many possible options). The first ℓ bits in *MsbLsb* are xored with b. Let *mb* (masked b) denote the xor result, concatenate *mb* with the next m bits of *MsbLsb* to form *emb* (extended masked b) a sequence of $\ell + m$ bits.

Use the rest of the bits in *MsbLsb* to choose (almost) uniformly a permutation index in the range 1 to $(\ell + m)!$ (use mod$(\ell + m)!$ as needed) permute *emb* accordingly. And randomly permute $a_1, a_2 ... a_n$. The function's output is the permuted *emb* and the (randomly) permuted $a_1, a_2, ... a_n$.

To reverse the function, one has to produce *emb* that fits the $\ell + m$ bits of a permutation of $a_1, a_2 ... a_n$ (essentially returning to the initial unimportant order of the subset sum order), the main indicator for correct reverting is the m extended bits. There are even more restrictions related to the existing sum in the spirit of [4].

Note that the m bits of *MsbLsb* that extend the *mb* to form *emb*, serve as a success criteria combination, yielding exponentially smaller probability for a collision as m grows. Namely, the longer m is, the smaller the probability of finding more than one permutation that yields a fitting m value.

3 Permuting Secret Sharing Polynomials

In this section, we turn to the scope of polynomials (rather than subset-sum) extending the idea sketched in [5], which uses (secret sharing) polynomials over a finite field (introduced in [12]) to create a function that is hard to invert yet easy to verify. The construction is motivated by the following pictorial "story".

3.1 The Combination Lock

Consider an ordered number set $Y = (y_1, y_2, \ldots, y_n)$, that when permuted in a particular combination(s), can open a safe, and y_0 as a challenge associated with Y, which is publicly known and can serve to easily prove that a certain party has the right combination, without opening the safe in practice.

If n is big enough, then a useful lock can be established, together with an easy-to-implement proof of having a key to the safe. In the worst case, one needs to try $n!$ possibilities to open the safe while proving that one has the key that can be linear in n.

The safe lock is opened when the order of the elements in the vector Y corresponds to the particular number, y_0, in the following way: y_0 is the free coefficient in a polynomial of degree $n-1$ defined over a finite field (just like the safe locker is defined over a finite number of possible digits) by the sequence of points $(1, Y_\pi(1)), (2, Y_\pi(2)), \ldots, (n, Y_\pi(n))$, where Y_π is the permuted Y. Note that the first index in each of the above points is regarded as the x coordinate of the point. In the sequel, we prefer to choose the x values of these points to be random rather than the simplest vector $X = (1, 2, \ldots, n)$ of the x values (yet, we make sure while randomly selecting X, that there are no repetitions of x values in X). Note that these x values are exposed and known to all parties.

When we provide the opening criteria, a naive burglar will try all $n!$ possibilities by using Lagrange interpolation until a polynomial with identical y_0 is found, which in turn may take too much time, for the limited time the burglar may afford. However, it is possible that an ingenious burglar can use the free coefficient y_0 to reconstruct the permutation of Y in a much more efficient way. First, in a key part of this paper, we wish to show that while there are less than $n!$ permutations to test, choosing the values of X and Y vectors appropriately results in a system that is hard to invert, in a sense to be rigorously defined later. Moreover, we may incorporate the two instances of self-masking idea [4,5], where two combination lockers are installed, and their free coefficients are not given as before, but, instead, the bitwise xor of the free coefficients is given. Now, the burglar can open the safe when she reconstructs the given xor value of the free coefficients.

The safe can be designed to restrict the permutation of one of the lockers to be the opposite permutation of the other. Thus, when the burglar tries one combination in a locker, she has to set the reverse permutation in the second, compute Lagrange for each permuted lock, and xor the y_0s.

To make the challenge harder, we hide the value of each of the y's, hopefully restricting the burglar from finding them in a "blind" fashion[2] so the burglar is caught before opening the safe. We may shuffle/permute the bits of y_0 of the first polynomial, using the (secret) permutation for the numbers of the other polynomial elements, and vice versa, before xoring their bits.

3.2 Presentation of the Problem

In this paper, we introduce self-masking for one instance of a problem, i.e., where we have one vector Y, and one instance of y_0. The choice of n, yielding $n!$ possible permutations (though, as mentioned, the actual number may be smaller), should be coordinated with the choice of a field F so that the probability of guessing a permutation that opens the safe among the permutations that yield the y_0 in the single instance problem, or the xor of the shuffled free coefficients (the two-instances problem; in our story, possibly permutations used by other bank managers) is negligible. Thus, we would prefer values of $n!$ coordinated with the value of $\ell = \lg|F|$; this relation can be tuned according to the security parameters (for restricting the number of collisions) required.

The *polynomial permuting* problem of dimensions n and ℓ, to be denoted $PP(n, \ell)$, is defined as follows. Let $F = GF(2^\ell)$ be the finite field of 2^ℓ elements,

[2] We try to enforce exhaustive search as much as we can; obviously, a success to do so in a provable way is the long-standing problem of $P \neq NP$.

and assume some lexicographic order on the elements of F. Let further $X = (x_1, \ldots, x_n)$ be n distinct nonzero elements of F.

Definition 1. *For any X, Y, define by $P_{X,Y}$ the unique polynomial of degree $n-1$ over F, satisfying $P_{X,Y}(x_i) = Y(i), i = 1, \ldots n$.*

Since $P_{X,Y}$ above is unique, there is a unique y_0 in F such that $P_{X,Y}(0) = y_0$. Define $[n] = \{1, 2, \ldots, n\}$ and let Π denote the set of all possible permutations on $[n]$. Further denote by Y_π, $\pi \in \Pi$, the permuted Y according to the permutation π. We can now define $PP(n, \ell)$.

Definition 2. *An input to $PP(n, \ell)$ is a tuple (X, Y, y_0) of length $2n+1$ over $F = GF(2^\ell)$. It is required to decide if there is a permutation $\pi \in \Pi$ s.t. the unique polynomial $P_{X,Y_\pi}(\cdot)$ of degree $n-1$ over F, satisfying $P_{X,Y_\pi}(x_i) = Y_\pi(i), i = 1, \ldots n$, also satisfies $P_{X,Y_\pi}(0) = y_0$.*

That is, $PP(n, \ell)$ requires to determine if under some permutation π, the polynomial P_{X,Y_π} in Definition 1 has y_0 as its free coefficient. Intuitively, we think of X as fixed, and any permutation of Y has some y_0 that satisfies these conditions. In other words, given Y and *only y_0 which satisfies the conditions under some permutation of Y, Y_π, but without giving π itself,* implicitly encodes the permutation π. Our goal is to show that revealing this permutation explicitly is hard.[3] Specifically, define a *function ensemble*, parameterized by X and Y as follows.

Definition 3. *Fix X and Y, both in F^n. Denote by $f_{X,Y} : \Pi \to F$ the function $f_{X,Y}(\pi) = P_{X,Y_\pi}(0)$.*

We wish to show that under some choices of X and Y, and some relationships between l and n, the function $f_{X,Y}(\pi)$ is hard to invert, in the sense that choosing a random $\pi \in \Pi$ results in a uniformly distributed y_0.

3.3 Results

Theorem 1. *Assume X and Y are chosen independently at random, both as uniform i.i.d. vectors of length n over $F = GF(2^\ell)$. Fix a permutation σ. Then randomly choosing a permutation $\pi \in \Pi$ to invert $f_{X,Y}(\sigma)$, that is, to have $f_{X,Y}(\pi) = f_{X,Y}(\sigma)$, has a success probability*

$$Pr\left\{f_{X,Y}(\pi) = f_{X,Y}(\sigma)\right\} \leq (1 - \epsilon_{\ell,n}) \left(\frac{1}{2^\ell} + \frac{2}{e(n-1)!}\right) + \epsilon_{\ell,n}, \tag{1}$$

with $\epsilon_{\ell,n} \to 0$ as $\ell \to \infty$ for any fixed $n > 0$.

The proof is omitted from this version and can be found in [2].

[3] We do not know if PP is solvable in polynomial time; however, the applicability of the masking technique is independent of this question.

Theorem 1 asserts that the probability of a particular secret appearing when the y elements are uniformly chosen and then uniformly permuted is approximately uniform across all possible secrets for proper choices of n and l. Thus, there is no a-priori benefit in preferring one secret over another or one permutation over another. Moreover, given a certain instance with a vector Y, the number of permutations that are mapped (collisions in terms of OWF) to any particular secret is approximately uniform across all possible secrets too.

3.4 Numerical Results

We present the visualised result of an experimental investigation, which sought to examine the impact of the order of the finite field and the number of points n on the distribution of the $P(0)$ values of the polynomial in multiple experiments. Figure 1 provide a visual representation of the distribution of the $P(0)$ values of the polynomial defined over a finite field with order 2^{ℓ} and n number of points, respectively (averaged over several experiments).

As evident from Fig. 1, the average distribution of the polynomial's $P(0)$ value exhibits a negligible degree of variability for the finite field's order and the number of points used. It is indeed very close to uniform in all experiments whose results can be found at [3].

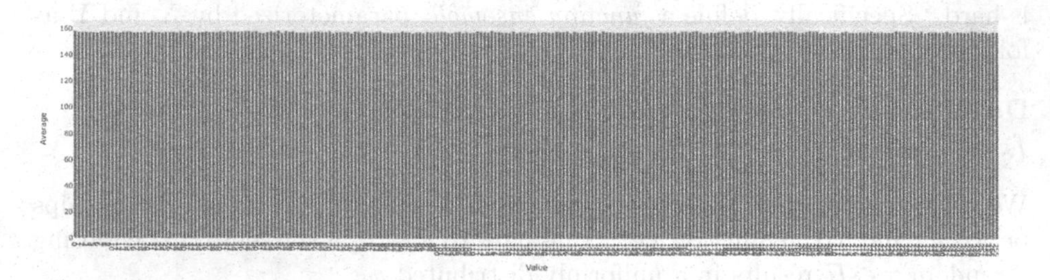

Fig. 1. Averaged distribution of $P(0)$ value over a Finite Field with order 256 and 8 points.

To further understand the results, we include the table below. As mentioned, a primary objective of the study was to examine the impact of selecting the order of the finite field F and the number of points n on the observed outcomes. $W = 1000$ denotes the number of experiments conducted. We further explain the table entries.

- F: the order of the finite field. F is used in the proposed algorithm to perform all the arithmetic operations.
- n: the number of points used in the algorithm, n is also related to the order of the finite field in this specific example, $|F| = 2^n$.
- μ_{min} and μ_{max}: the minimal and maximal values of the average counts, respectively. Counts are averaged over the W experiments, each count refers to one permutation leading to the specific secret (a value in F). The difference between μ_{min} and μ_{max} is denoted by Δ_{μ}.

- σ and σ^2: the standard deviation and variance, respectively, to provide information about the spread of the computed averages.
- $P(A) = P(0)$: the probability that any element from the finite field is a secret (assuming uniform distribution).
- $P(\mu_{min}), P(\mu_{max})$: represents the success probability when the adversary would choose a minimal average value or maximal average value, which is defined as the ratio of μ_{min} to the order of the finite field or μ_{max} respectively.
- $P(\mu_{max})$: the success probability when the adversary would choose a value with a maximal average, which is defined as the ratio of μ_{min} to the order of the finite field.
- $\mathcal{A}_{adv}$: this parameter is defined as a percentage of the maximal potential advantage that an adversary can get by exploiting knowledge of the distribution of $P(0)$, specifically by selecting the average value with the highest number of repetitions. It is calculated as $\mathcal{A}_{adv} = (P(\mu_{max}) - P(\mu_{min})) \cdot 100\%$

Table 1. Results of experiments with varying finite field $\mathbb{F}_{|}$ and n defined as number of used points.

$\mathbb{F}_{\vert}$	n	μ_{min}	μ_{max}	Δ_{μ}	σ	σ^2	P(A)=P(0)	$P(\mu_{min})$	$P(\mu_{max})$	$\mathcal{A}_{adv}$
16	4	1.447	1:5460	0.0990	0.0007	0.0272	0.0625	0.0904	0.0966	0.6188
32	5	3.643	3.9440	0.3010	0.0048	0.0693	0.0313	0.1138	0.1233	0.9406
64	6	11.009	11.4870	0.4780	0.0115	0.1074	0.0156	0.1720	0.1795	0.7469
128	7	38.955	39.7780	0.8230	0.0344	0.0185	0.0078	0.3043	0.3108	0.6430
256	8	156.324	158.457	2.1330	0.1836	0.4285	0.0039	0.6106	0.6190	0.8332

As evident from Table 1, the observed outcome of the experiment is similar to the choice of finite field and the number of points used. Results of the proposed algorithm were obtained using the SAGE software[4]. Implementation of the interactive charts with the results of our experiments are available for the reader online, along with the numerical values in [3].

4 Algorithm – Techniques Integration Sample

In this section, we present an implementation based on the proposed ideas. The algorithm demonstrates the use of permutations in the scope of self–masking for two (secret sharing) polynomials. Thus, enhancing the single self-masking effect on self–masking a single (secret sharing) polynomial, presented above. The design uses a single secret permutation π that encodes (also in its nested forms) the masking in several masking permutations. The permutations of the

[4] SageMath version 9.7, Release Date: 2022-09-19.

elements of Y_1 (according to π) and the elements of Y_2 (according to π^{-1}). Then, the permutation of the bit of y_{01}, and y_{02} (according to different random bits defined by more different nested versions of π). Only then is the mutual xor used. Thus, a single π serves as proof for the commitment but is used in many forms, roughly speaking, similar to the use of a seed.

In order to provide a clear and concise representation of the implementation, we present the pseudo-code below. Note, for the sake of simplicity we omit the possibility of choosing random x values for the randomly chosen Y_1 and Y_2 coordinates and present the restricted version in which $x = 1, 2, \ldots, n$.

Algorithm 1: Polynomial based self masking algorithm

```
Input: Points_Number = n, Field_Size = GF(2^ℓ)
Result: xor_original, Y1, Y2

1  Function Generate_Points(Points_Number, Field_Size):
2      F = FiniteField(Field_Size)
3      while x < len(Points_Number) do
4          Y1 = sort(F.get_random_elements_without_repetitions())
5          Y2 = sort(F.get_random_elements_without_repetitions())
6      return Y1, Y2
7
8  Function Generate_Input():
9      Y1, Y2 = Generate_Points(n, 2^ℓ)
10     π = Generate_permutation
11     y01 = free_coefficient (lagrange_interpolation(Y1, π))
12     y02 = free_coefficient (lagrange_interpolation(Y2, π^-1))
13     y01_shuffled = shuffle(y01, π^-2, π^-3)
14     y02_shuffled = shuffle(y02, π^2, π^3)
15     xor_original = y01_shuffled ⊕ y02_shuffled
16     return xor_original, Y1, Y2
```

Function *Generate_Points* (line 1), uses two arguments *Points Number* and *Field_Size.* At the beginning of this function, the F object is created using the $FiniteField$ class (lines 4-5). The constructor of this class takes an integer value as an argument, which determines the size of the field. Then, in the loop, list Y_1 (and independently later list Y_2) is created from *Points Number* distinct values randomly selected from F and then sorted. Lastly, the generated sorted lists are returned.

The function *Generate_Input* starts with the invocation of the *Generate_Points* function with the number of points n and the field 2^ℓ. As a result, two sorted lists of numbers Y_1 and Y_2 are returned, each consisting of n distinct numbers in the field. Then, a permutation π is randomly selected (line 10) and applied to Y_1. The permuted numbers in Y_1 are regarded as y coordinates of n points with the n smallest distinct x coordinates. The y coordinates are paired to the x coordinates according to the order of the y's in the permuted Y_1 and the growing order of the x coordinates. Then, Lagrange interpolation is applied to the n points, finding the free coefficient of the polynomial of degree $n-1$ that they uniquely represent (line 11). The same is done for Y_2, but this time the y coordinates are ordered according to π^{-1}. Note that we prefer to correlate the operation on the two arrays based on knowing the solution (the permutation π) and use different correlations based on the solution, hence the choice of π and π^{-1}.

Using the sequence of the bits defined by (the permutation indexes of) π^2 and π^3 (possibly even π^4..., depending on the number of bits needed to encode a permutation of ℓ bits), we shuffle (permute the bits) of the free coefficient y_{01} (line 13) and similarly, using π^{-2} and π^{-3}, we shuffle the bits of y_{02} (line 14), and bitwise *xor* the resulting y_{01}*_shuffled* with y_{02}*_shuffled* to form *xor_original* (line 15). Lastly, we return the xor results and the two sorted vectors, each consisting of n distinct numbers in the field.

We note that our schemes work when the numbers are not necessarily distinct, and we choose to restrict the use of distinct numbers as an optimization, avoiding equivalent permutations. Also, note that one can tune the field and number of $y's$ to support independent (rather than correlated) permutations for Y_1 and Y_2.

The basic self masking technique for PP is as follows: the input for a masked function $[f]$ is a triple (Y_1, Y_2, π), where Y_1, Y_2 are two independent n-subsets of $GF(2^\ell)$ and $\pi \in SYM(n)$ is a permutation. Note that a symmetric group consists of all possible permutations of a finite set of distinct elements. Let $y_{01} = P_{Y_1,\pi}(0)$ and $y_{02} = P_{Y_2,\pi}(0)$. Then:

$$[f_{n,\ell}]\,(Y_1, Y_2, \pi) = (y_{01} \oplus y_{02}, Y_1, Y_2). \tag{2}$$

That is, the values of y_{01} and y_{02}, corresponding to inputs (Y_1, π) and (Y_2, π), mask each other by $y_{01} \oplus y_{02}$.

Let $N(Y, y)$ denote the number of “collisions” corresponding to $y_0 \in F$ when the input set is Y. i.e.

$$N(Y, y) = |\{\pi : P_{Y,\pi}(0) = y\}|. \tag{3}$$

Ideally, we would like that $N(Y, y)$ is almost the same for all $y \in F$ (i.e., it is either $\lfloor \frac{n!}{2^\ell} \rfloor$ or $\lceil \frac{n!}{2^\ell} \rceil$). Unfortunately, this is not the case for the PP case: for a given set Y of n elements, the distribution of $N(Y, y)$ is not uniform (and is dependent on the set Y).

5 Concluding Remarks

Permuting an array in $O(n)$ steps while providing success criteria for reversing the permutation, a permutation selected from the $n!$ possible permutations, is advocated in [5]. We extend the xor only approach suggested in [4] to using permutation for a single (and multiple) instance case of the subset–sum problem. We also presented an algorithm building on the concepts of the new one-way function and its self–masking. In [4], the parity of the i'th bit in b_1 and b_2 has been exposed; here, the permutation usage masks the parity of the corresponding bits.

We also analyze the feasibility of such permuting approach in the scope of creating a one-way function from one instance of the permuted (secret–sharing) polynomial problem. The idea can be extended to multiple instances and self–masking. One can xor permuted success criteria of more than two instances (possibly enlarging ℓ as needed) to enhance the effect of mutual permuted one-time-pad.

Acknowledgment. We thank Shlomo Moran for his input throughout the research.

References

1. Babai, L., Moran, S.: Arthur-merlin games: a randomized proof system, and a hierarchy of complexity classes. J. Comput. Syst. Sci. **36**(2), 254–276 (1988). https://doi.org/10.1016/0022-0000(88)90028-1
2. Cohen, A., Cyprys, P., Dolev, S.: Single instance self-masking via permutations. Cryptology ePrint Archive, Paper 2023/416 (2023). https://eprint.iacr.org/2023/416
3. Cohen, A., Cyprys, P., Dolev, S.: Repository with results of the described experiments (2023). http://ptinyurl.com/5wb2t6cf
4. Cyprys, P., Dolev, S., Moran, S.: Self masking for hardering inversions. IACR Cryptol. ePrint Arch. 1274 (2022). https://eprint.iacr.org/2022/1274
5. Dolev, H., Dolev, S.: Toward provable one way functions. IACR Cryptol. ePrint Arch. 1358 (2020). https://eprint.iacr.org/2020/1358
6. Dolev, S., Lahiani, L., Haviv, Y.: Unique permutation hashing. Theor. Comput. Sci. **475**, 59–65 (2013). https://doi.org/10.1016/j.tcs.2012.12.047
7. Goldwasser, S., Micali, S., Rackoff, C.: The knowledge complexity of interactive proof-systems (extended abstract). In: Sedgewick, R. (ed.) Proceedings of the 17th Annual ACM Symposium on Theory of Computing, 6–8 May 1985, Providence, Rhode Island, USA, pp. 291–304. ACM (1985). https://doi.org/10.1145/22145.22178
8. Håstad, J., Impagliazzo, R., Levin, L.A., Luby, M.: A pseudorandom generator from any one-way function. SIAM J. Comput. **28**, 12–24 (1999)
9. Impagliazzo, R., Naor, M.: Efficient cryptographic schemes provably as secure as subset sum. J. Cryptol. **9**(4), 199–216 (1996)
10. Lamport, L.: Constructing digital signatures from a one way function. Technical report CSL-98 (1979). This paper was published by IEEE in the Proceedings of HICSS-43 in January, 2010
11. Perlner, R., Kelsey, J., Cooper, D.: Breaking category five sphincs+ with sha-256. Cryptology ePrint Archive, Paper 2022/1061 (2022). https://eprint.iacr.org/2022/1061
12. Shamir, A.: How to share a secret. Commun. ACM **22**(11), 612–613 (1979)
13. Shamir, A.: A polynomial time algorithm for breaking the basic Merkle-Hellman cryptosystem. In: Chaum, D., Rivest, R.L., Sherman, A.T. (eds.) CRYPTO, pp. 279–288. Plenum Press, New York (1982)
14. Sharma, R., Mishra, P., Kumar, Y., Gupta, N.: Differential δ-uniformity and non-linearity of permutations over Zn. Theor. Comput. Sci. **936**, 1–12 (2022)

A Fusion-Based Framework for Unsupervised Single Image Super-Resolution

Divya Mishra[1(✉)], Itai Dror[1], Ofer Hadar[1], Daniel Choukroun[2], Shimrit Maman[3], and Dan G. Blumberg[4]

[1] School of Electrical and Computer Engineering, Beer-Sheva, Israel
divya@post.bgu.ac.il
[2] Department of Mechanical Engineering, Beer-Sheva, Israel
[3] Homeland Security Institute, Beer-Sheva, Israel
[4] Department of Geography and Environmental Development, Ben Gurion University of the Negev, 84105 Beer-Sheva, Israel

Abstract. Image super-resolution has been a continuously demanding topic in the computer-vision community in recent decades and has witnessed impressive applications in increasing spatial resolution in every field like medicine, agriculture, remote sensing, defense security, and many more applications. Further, deep learning-based image super resolution methods have shown tremendous improvement in reconstruction performance. However, most of the recent state-of-the-art deep learning-based methods for image super-resolution assume an ideal degradation by the bicubic kernel on standard dataset approaches and perform poorly on real-world satellite images in practice, as real degradations are far away and more complex in nature than pre-defined assumed kernels. Motivated by this real-time challenge, our idea is to enhance the 600 m spatial-resolution image, which is extremely low, and implicitly defines image-specific features in an iterative way without defining any fixed explicit degradation for image super-resolution. Besides, we also did a comparative study based on a No-Reference Image Quality Assessment. The evaluation is done both qualitatively (vision based) and quantitatively without recurring to a reference image for quality assessment. The proposed framework outperforms by incorporating domain knowledge from recently implemented unsupervised single-image blind super-resolution techniques.

Keywords: Super-resolution · Feature estimation · Data fusion · Unsupervised image super-resolution

1 Introduction

BGUSAT, the first Israeli research CubeSat is a nanosatellite joint venture between the Ben-Gurion University of the Negev, IAI (Israel Aerospace Indus-

Supported by a grant (Grant No. 3-17380) from the Ministry of Science and Technology, Israel.

S. Dolev et al. (Eds.): CSCML 2023, LNCS 13914, pp. 85–95, 2023.
https://doi.org/10.1007/978-3-031-34671-2_7

tries), and ISA (Israeli Space Agency). It is a Low Earth Orbit (LEO) 3U CubeSat imaging the Earth in the Short Wave Infra Red (SWIR) spectrumsingle-imageite has been fully operational since Feb. 15th, 2017, and has already collected many images of the Earth, claiming mission success for the technology part. Image enhancement is highly required in order to get more intelligence from the acquired images that are already at 600 m spatial resolution. Our idea is to extend the work, particularly for images from BGUSat-a nano-satellite. Further, deep learning-based image super-resolution methods have shown tremendous improvement in reconstruction performance. However, most of the recent state-of-the-art deep learning-based methods for image super-resolution assume an ideal degradation kernel (like bicubic down-sampling) on standard datasets. These approaches perform poorly on real-world satellite images in practice since real degradations are far away and more complex in nature than pre-defined assumed kernels. With this drawback in mind, various state-of-the-art kernel estimation-based methods have evolved via iterative approaches like Iterative Kernel Correction (IKC) [5], InternalGAN (InGaN) [1] and Correction filter [6] for blind super-resolution. However, iterative kernel estimation-based approaches are not only time-consuming but also require complex objective functions along with regularization. Motivated by this real-time challenge, this paper introduces a forward approach based on fusion as shown in Fig. 1 that implicitly defines image-specific features in a way that suits perceptually according to the human visual system, without defining any fixed explicit degradation for image super-resolution like bicubic interpolation assumed to be as ideal degradation in most of the deep learning based scenarios.

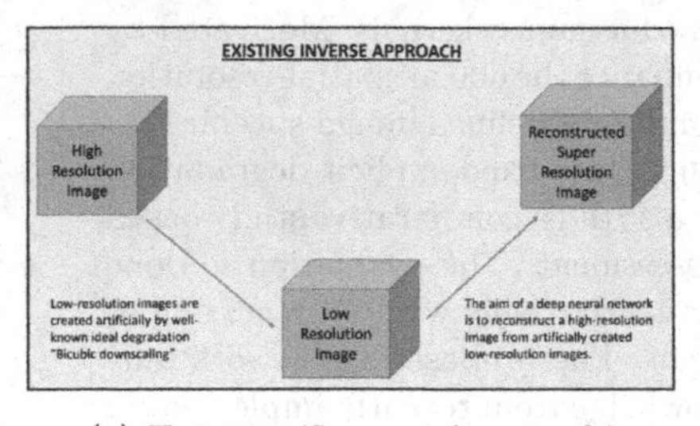

(a) Existing 'Inverse Approach'

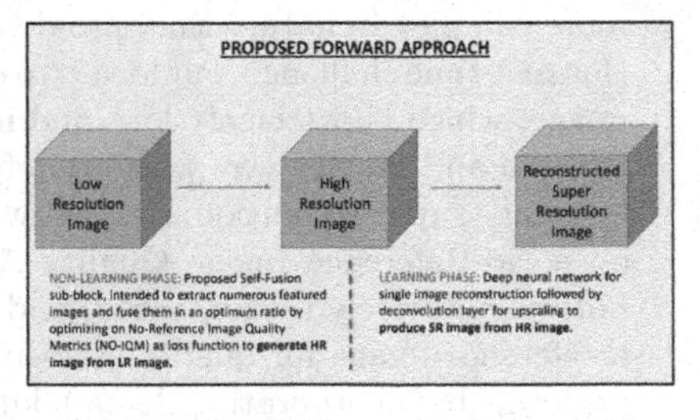

(b) Proposed 'Forward Approach'

Fig. 1. A difference between (a) existing 'Inverse Approach' and (b) proposed 'Forward Approach'.

Several methods are used to extract feature information from these existing better perceptual quality images. In order to generate fake LR images, a common norm is followed by using traditional bicubic interpolation. This is an ideal degradation which is assumed to be in most of the learning-based scenarios. However, this is not always true for real-world scenes. Real-world remote-sensing images have distortions from multiple sources, and learning-based scenarios effects like clouds, haze, and many more. And, hence real-world world LR images are not able to perform practically on those learned ideal scenarios. Motivating by this challenge, particularly for real-world image real-worldly low resolution, we proposed a general architecture based on the scenarios of different featured images generated from different frequency extracting filters.

1.1 Challenges in Unsupervised Learning Methods

Following are the recent challenges in real world single-image super-resolution:

- It is hard to understand the grouping of images since one doesn't have a standard ground truth HR image of the respective LR.
- The recent state-of-the-art unsupervised deep learning-based models rely on large unpaired datasets, which are sometimes difficult to arrange in case of real-world scenarios.
- The unsupervised learning models are more complex than supervised models.
- The models rely on huge computational resources like GPUs and long training time.

1.2 Our Contributions

Following are the major achievements of our proposed Self-Fusion architecture for image super-resolution:

- The model is entirely unsupervised in the way, it needs only one LR image to process. Here, no ground-truth data is required, either paired or unpaired from other standard satellites.
- The existing fusion-based models applied in the field of remote sensing images require two or more input images as input to process. But, we need only single images to initiate the process.
- One do not require geo-registered images since all featured images are the same copy in terms of location.
- The proposed architecture is relatively simple and, compared to other relevant architectures, requires less time and space computational resources.
- The architecture does not depend on particular image distribution and works for many different types of image distribution.

2 Related Work

Unsupervised image super-resolution is always a challenging task as compared to supervised learning scenarios, due to the presence of GT high-resolution images. Several methods are used to extract feature information from these existing better perceptual quality images. In order to generate fake LR images, a common norm is followed by using traditional bicubic interpolation. This is an ideal degradation which is assumed to be in most of the learning-based scenarios. The example models in this series are SRCNN [2], the first CNN-based model for image super-resolution, and FSRCNN [3], which improves the processing speed of existing SRCNN. Later residual networks ResNet [23] are proposed to leverage the benefit of residual features left out in the long process of training as the network elongates from backward to forward. The skip connections helped here to support the left-out features from the previous layer. Further, in [11], the author proved how bilateral blur outstands than a gaussian blur for making better LR pairs in CNN-based architecture for SR-based applications. Further, in [13] by utilizing the bilateral filter, particularly for SRCNN, it has been proved that the convergence time is less with similar image quality at less number of iterations.

SRGAN [9] is the first deep learning model for image super-resolution tasks. Above all, these techniques need fake LR images that are not even come closer to real-world degradations. Real-world remote-sensing images have distortions from multiple sources like sensor noise, and environmental effects like clouds, haze, and many more. And, hence these real-world LR images is not able to perform practically on those learned ideal scenarios. Zero-Shot Super-Resolution ZSSR [18] is the first CNN-based unsupervised learning method on an absolutely single image. The idea is to assume the original LR image is a better-quality version and down-sample it to fake the LR image further. In this way, this idea of self-supervision by leveraging the benefit of paired images is used. The problem with this is, it makes extremely poor-resolution images more challenging to reconstruct at higher scales. Also, it is slower in processing. Image-specific iterative networks are also proposed like Iterative Kernel Correction (IKC) [5], Internal-GAN (InGaN) [1] and Correction filter [6]. The idea of these Iterative methods is to train a neural network for image-specific kernels and priors. These methods are time-consuming and do not work for higher scaling factors. Also, they are not optimum for real-world images. In order to mitigate this challenge, we proposed a forward approach in spite of the traditional inverse approach to first create an HR image from existing extremely poor-quality images and then reconstruct followed by up-sample them with the extra benefit of deep learning. Recently, deep learning-based fusion models are also employed for image super-resolution tasks. In 2019, IFCNN [24] is the first fusion-based deep learning architecture proposed. They suggested a general framework for fusion from images under different lighting conditions. In 2020, U2Fusion [21], is proposed to enhance the image quality suffered from different lightening conditions. In 2021, UMEF [16] is another fusion network proposed for particularly multiple scene images from different exposure scenes. Also, in [4], the BGUSat dataset, which is also utilized in our case, is used for image super-resolution using multiple sequences at different time frames.

3 Proposed Framework

We proposed a forward model to super-resolve, particularly the real-world low-quality images. The idea is first to remove the noise and extract different frequency details from potential filters that are already embedded in LR image. The complete fusion-based proposed framework is shown in Fig. 2. By different frequency, we mean:

- Low-frequency information corresponds to image smoothing features.
- Mid-band frequency details correspond to texture and contour features.
- High-frequency information that relates to edge features a of an image.

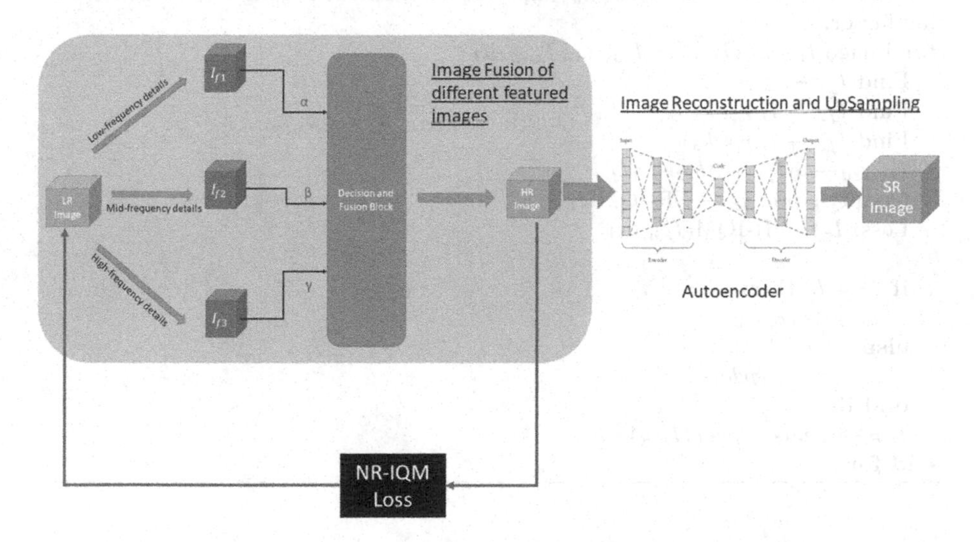

Fig. 2. Proposed fusion-based framework for Unsupervised Single-Image Super-Resolution.

The different features are extracted from widely available filters in image processing. One can try optimizing different combinations of filters designed for particular applications. For example: for extracting low-frequency features (for image smoothening), gaussian, bilateral [tasks median filters can be used, for mid-frequency feature extraction (like texture and lighting details), wiener filter, Gabor filter, anisotropic gaussian filter conditions, Laplace filter, and granulometric analysis and for high-frequency information like sharp edges and curves, gradient-based techniques, Sobel operators, canny edge extractor.

The fusion weights α, β and γ are decided and updated iteratively based on the no-reference image quality metric (NR-IMQ) as a loss function. This loss function is tweaking between the original LR BGUSat image and produced HR image of the same size. One can use the possible NR-IMQ (Blind or Reference-less Image Spatial QUality Evaluator: BRISQUE [14], Natural Image Quality Evaluator: NIQE [15] and Perception-based Image QUality Evaluator: PIQUE [19]) metrics that are based on natural scene statistical model (NSS) [17] of an image and hence maintains the perceptual quality too. The produced HR image looks more artificial in nature. In order to save the perceptual detail, autoencoders are employed to leverage the great reconstruction property through convolutional neural networks. One can play with different parameter settings and variations of autoencoder architectures. To get SR image. the convolutional

Algorithm 1. Algorithm to Produce Super-Resolution Image: I_{SR}

Input: Low-resolution image I_{LR}
Output: Super-resolution image I_{SR}
Paraemeters: α, β and γ (Fusion weights), k_1: low-f quency extraction filter; k_2: mid-frequency extraction filter, k_3: high-frequency extraction filter and s is the scaling factor.
for Image I_{LR} in $(I_{x1}, I_{x2}, I_{x3}, ..., I_{xn})$ **do**
  Find $I_{f1} \leftarrow I_{LR} * k_1$
  Find $I_{f2} \leftarrow I_{LR} * k_2$
  Find $I_{f3} \leftarrow I_{LR} * k_3$
  $I_{Fused} \leftarrow \alpha I_{f1} + \beta I_{f2} + \gamma I_{f3}$
  Quality score: $Q \leftarrow$ NR-IQM(I_{LR})
  Loss: $L \leftarrow$ NR-IQM(I_{Fused})

  if $Q > L$ **then**
    $I_{HR} \leftarrow I_{Fused}$
  else
    $\alpha, \beta, \gamma \leftarrow Update$
  end if
  $I_{SR} \leftarrow Autoencoder(I_{HR}) * s$
end for

upsampling2D layers are introduced after image reconstruction in order to produce a super-resolved image at desired scaling factor. We proposed a general architectural design to fuse different featured images from various frequency components. Algorithm 1. is designed in an iterative fashion by keeping no-reference image quality metrics on the check to decide the best-fused HR image. The parameters are (1) k_1: low-frequency extraction filter; (2) k_2: mid-frequency extraction filter; (3) k_3: high-frequency extraction filter (4) Q: is the NR-IQM score for reference LR image.

4 Results and Discussion

The model experiments on two modalities of test images:

- **Single band BGUSat images:** More than 2000 BGUSat images are currently available, but we tested on two images for demonstration purposes: one for the Israel Dead-sea region and another for the China border image. Further, For a scaling factor of 2, we have compared our results with popular unsupervised image super-resolution based on image-specific CNN called ZSSR [18], for both images, as shown in Fig. 3 and Fig. 4, the texture and edges are sharper and visually convincing. These enhanced images are of better visual quality.
- **RGB Colored remote sensing images of UC-Merced dataset** [22]: We tested the performance of the proposed model on two RGB-colored remote sensing images: one from the class harbor and another from the class building

whose visual results at scaling factor of 2 are shown as in Fig. 5 and Fig. 6 respectively. The baseline comparison has been done with existing unsupervised methods like EDSR [10], ZSSR [18], LapSRN [8] and MIP [20]. This proves that the process does not depend on a particular image distribution and works for many different types of image distributions and hence has good generalization.

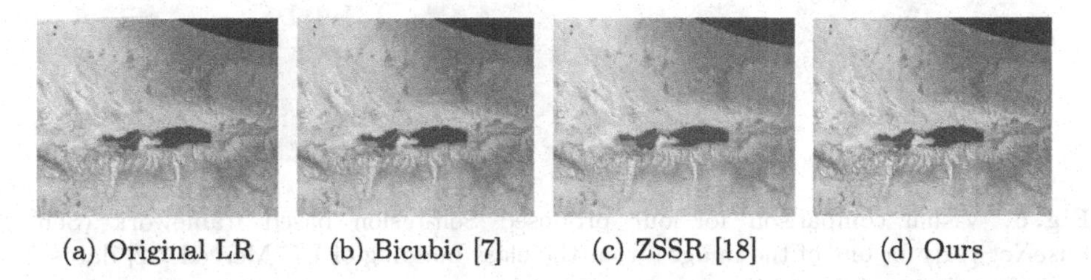

(a) Original LR (b) Bicubic [7] (c) ZSSR [18] (d) Ours

Fig. 3. Visual comparison for our proposed Self-fusion based framework (Self-FuseNet [12]) for Israel dead sea image at scaling factor of 2.

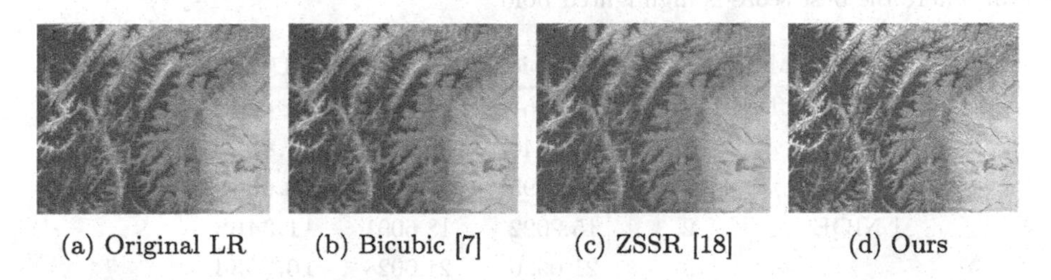

(a) Original LR (b) Bicubic [7] (c) ZSSR [18] (d) Ours

Fig. 4. Visual comparison for our proposed Self-fusion based framework (Self-FuseNet [12]) for China border image at scaling factor of 2.

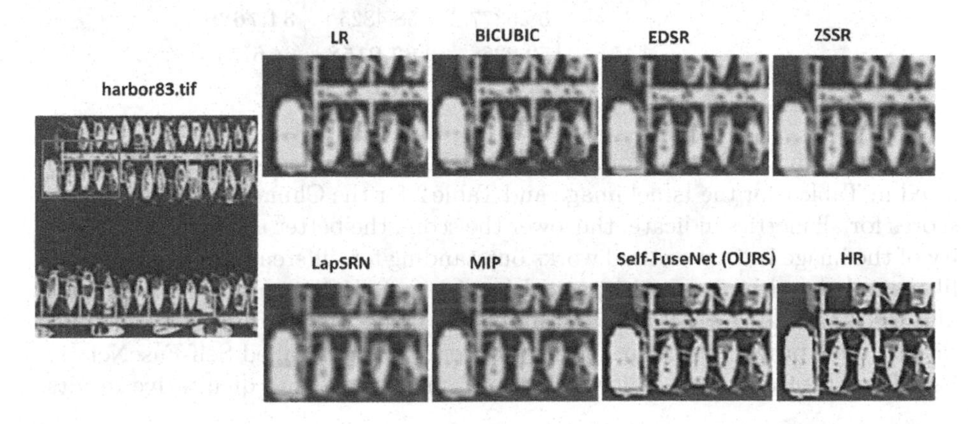

Fig. 5. Visual comparison for our proposed Self-fusion based framework (Self-FuseNet [12]) for one of the images from the class harbor of UC-Merced [22] dataset at scaling factor of 2.

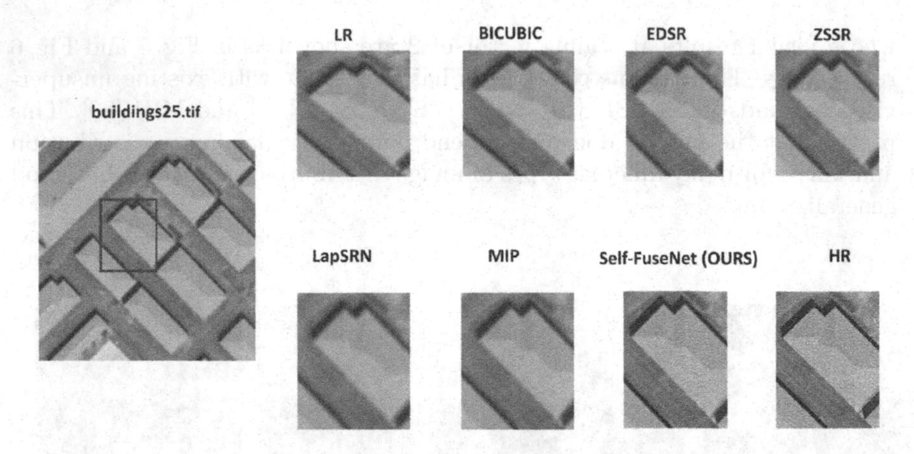

Fig. 6. Visual comparison for our proposed Self-fusion based framework (Self-FuseNet [12]) for one of the images from the class building of UC-Merced [22] dataset at scaling factor of 2.

Table 1. Quantitative NR-IQM evaluation of Israel dead sea image at different scaling factor where the best score is highlighted bold.

NR-IQM Metric	Factor	Bicubic [7]	ZSSR [18]	**OURS** [12]
↓ BRISQUE	2	35.6251	**31.4603**	32.8294
	3	59.3646	55.6475	**41.8522**
	4	60.2995	59.8727	**52.2345**
↓ NIQE	2	15.2022	18.6001	**11.3463**
	3	21.0630	21.0028	**16.7634**
	4	15.7967	15.2996	**14.2345**
↓ PIQUE	2	48.7736	**12.4608**	16.0263
	3	58.5377	58.4825	**34.7686**
	4	72.2268	**62.0158**	64.6754

Quantitative evaluation for all three NR-IQM assessment-metric is also tabulated in Table 1 for the Israel image and Table 2 for the China border image. The scores for all metrics indicate, the lower the score, the better the perceptual quality of the image is. Our method works outstanding for different modalities of data like: single band image, multispectral band image, RGB remote sensing image, and RGB natural image for more state-of-the-art existing methods. The detailed discussion is presented in our specific network, which is called Self-FuseNet [12] based on this framework, followed by both quantitative and qualitative results.

Table 2. Quantitative NR-IQM evaluation of china border image at different scaling factors where the best score is highlighted in bold.

NR-IQM Metric	Factor	Bicubic [7]	ZSSR [18]	**OURS** [12]
↓ BRISQUE	2	51.3264	50.3124	**34.5123**
	3	70.0239	62.0652	**49.8548**
	4	68.8509	**60.2859**	62.5647
↓ NIQE	2	22.6981	20.8822	**13.3272**
	3	24.8298	22.5658	**19.7644**
	4	19.2829	19.2898	**17.6543**
↓ PIQUE	2	49.6784	30.6788	**17.2280**
	3	59.6526	**37.2265**	38.8709
	4	69.4145	69.4455	**63.6467**

5 Conclusions

Real-world distortions and degradations hardly match ideal bicubic interpolation-based degradation. In that case, deep learning approaches supervised or semi-supervised or image-specific trained CNNs that are based on inverse approach for image super-resolution without ground-truth data fail practically on real-world images. Also, these approaches do not generalize well on other image distribution dataset that is not present during training instance. In such a scenario, it is difficult to train models for every other dataset. We present here a solution to follow a forward approach from low-resolution to high-resolution and high-resolution to super-resolution with an extra effort of reconstruction-based convolutional neural networks like autoencoders to reconstruct and up-sample the image, particularly for very blurred poor resolution and blocky structured remote sensing images. The approach does not need image registration and multiple set of images for the same scene unless only a single low-resolution image initiates the processing. The work is particularly for real-world satellite images at extremely low resolution without the availability of ground truth images.

Acknowledgement. This work is supported by a grant (Grant No. 3-17380) from the Ministry of Science and Technology, Israel.

References

1. Bell-Kligler, S., Shocher, A., Irani, M.: Blind super-resolution kernel estimation using an internal-GAN. Adv. Neural Inf. Process. Syst. **32**(788535), 1–10 (2019)
2. Dong, C., Loy, C.C., He, K., Tang, X.: Image super-resolution using deep convolutional networks. IEEE Trans. Pattern Anal. Mach. Intell. **38**(2), 295–307 (2016). https://doi.org/10.1109/TPAMI.2015.2439281
3. Dong, C., Loy, C.C., Tang, X.: Accelerating the super-resolution convolutional neural network. In: Leibe, B., Matas, J., Sebe, N., Welling, M. (eds.) ECCV 2016.

LNCS, vol. 9906, pp. 391–407. Springer, Cham (2016). https://doi.org/10.1007/978-3-319-46475-6_25
4. Dror, I., et al.: Multiple image super-resolution from the BGU SWIR CubeSat satellite. In: Applications of Digital Image Processing XLV, vol. 12226, pp. 143–152. SPIE (2022)
5. Gu, J., Lu, H., Zuo, W., Dong, C.: Blind super-resolution with iterative kernel correction. In: Proceedings of the IEEE Computer Society Conference on Computer Vision and Pattern Recognition 2019-June, pp. 1604–1613 (2019). https://doi.org/10.1109/CVPR.2019.00170
6. Hussein, S.A., Tirer, T., Giryes, R.: Correction filter for single image super-resolution: robustifying off-the-shelf deep super-resolvers. In: Proceedings of the IEEE Computer Society Conference on Computer Vision and Pattern Recognition, pp. 1425–1434, November 2019. https://doi.org/10.1109/CVPR42600.2020.00150
7. Keys, R.: Cubic convolution interpolation for digital image processing. IEEE Trans. Acoust. Speech Signal Process. **29**(6), 1153–1160 (1981)
8. Lai, W.S., Huang, J.B., Ahuja, N., Yang, M.H.: Fast and accurate image super-resolution with deep Laplacian pyramid networks. IEEE Trans. Pattern Anal. Mach. Intell. **41**(11), 2599–2613 (2019). https://doi.org/10.1109/TPAMI.2018.2865304
9. Ledig, C., et al.: Photo-realistic single image super-resolution using a generative adversarial network. In: 2017 IEEE Conference on Computer Vision and Pattern Recognition (CVPR), vol. 2017-Janua, pp. 105–114. IEEE, July 2017. https://doi.org/10.1109/CVPR.2017.19
10. Lim, B., Son, S., Kim, H., Nah, S., Lee, K.M.: Enhanced deep residual networks for single image super-resolution. In: 2017 IEEE Conference on Computer Vision and Pattern Recognition Workshops (CVPRW), vol. 2017-July, pp. 1132–1140. IEEE, July 2017. https://doi.org/10.1109/CVPRW.2017.151
11. Mishra, D., Akram, M.W.: Experimental proof manifest bilateral blur as outstanding blurring technique for CNN based SR models to converge quickly. In: 2021 IEEE International Conference on Technology, Research, and Innovation for Betterment of Society (TRIBES), pp. 1–6. IEEE, December 2021. https://doi.org/10.1109/TRIBES52498.2021.9751647
12. Mishra, D., Hadar, O.: Self-FuseNet: data free unsupervised remote sensing image super-resolution. IEEE J. Sel. Top. Appl. Earth Obs. Remote Sens. 1–18 (2023). https://doi.org/10.1109/JSTARS.2023.3239758
13. Mishra, D., Khanam, T., Kaushik, I.: Experimentally proven bilateral blur on SRCNN for optimal convergence. In: Woungang, I., Dhurandher, S.K., Pattanaik, K.K., Verma, A., Verma, P. (eds.) Advanced Network Technologies and Intelligent Computing. ANTIC 2021. CCIS, vol. 1534, pp. 502–516. Springer, Cham (2022). https://doi.org/10.1007/978-3-030-96040-7_39
14. Mittal, A., Moorthy, A.K., Bovik, A.C.: No-reference image quality assessment in the spatial domain. IEEE Trans. Image Process. **21**(12), 4695–4708 (2012). https://doi.org/10.1109/TIP.2012.2214050
15. Mittal, A., Soundararajan, R., Bovik, A.C.: Making a completely blind image quality analyzer. IEEE Signal Process. Lett. **20**(3), 209–212 (2013). https://doi.org/10.1109/LSP.2012.2227726
16. Qi, Y., et al.: Deep unsupervised learning based on color un-referenced loss functions for multi-exposure image fusion. Inf. Fusion **66**(August 2020), 18–39 (2021). https://doi.org/10.1016/j.inffus.2020.08.012

17. Ruderman, D.L.: The statistics of natural images. Netw. Comput. Neural Syst. **5**(4), 517–548 (1994). https://doi.org/10.1088/0954-898X_5_4_006
18. Shocher, A., Cohen, N., Irani, M.: Zero-shot super-resolution using deep internal learning. In: Proceedings of the IEEE Computer Society Conference on Computer Vision and Pattern Recognition, pp. 3118–3126 (2018). https://doi.org/10.1109/CVPR.2018.00329
19. Venkatanath, N., Praneeth, D., Bh, M.C., Channappayya, S.S., Medasani, S.S.: Blind image quality evaluation using perception based features. In: 2015 Twenty First National Conference on Communications (NCC), pp. 1–6. IEEE, February 2015. https://doi.org/10.1109/NCC.2015.7084843
20. Wang, J., Shao, Z., Lu, T., Huang, X., Zhang, R., Wang, Y.: Unsupervised remoting sensing super-resolution via migration image prior. In: 2021 IEEE International Conference on Multimedia and Expo (ICME), pp. 1–6. IEEE (2021)
21. Xu, H., Ma, J., Jiang, J., Guo, X., Ling, H.: U2Fusion: a unified unsupervised image fusion network. IEEE Trans. Pattern Anal. Mach. Intell. **44**(1), 502–518 (2022). https://doi.org/10.1109/TPAMI.2020.3012548
22. Yang, Y., Newsam, S.: Bag-of-visual-words and spatial extensions for land-use classification. In: Proceedings of the 18th SIGSPATIAL International Conference on Advances in Geographic Information Systems, pp. 270–279 (2010)
23. Yu, X., Yu, Z., Ramalingam, S.: Learning strict identity mappings in deep residual networks. In: 2018 IEEE/CVF Conference on Computer Vision and Pattern Recognition, pp. 4432–4440. IEEE, June 2018 . https://doi.org/10.1109/CVPR.2018.00466
24. Zhang, Y., Liu, Y., Sun, P., Yan, H., Zhao, X., Zhang, L.: IFCNN: a general image fusion framework based on convolutional neural network. Inf. Fusion **54**(July 2019), 99–118 (2020). https://doi.org/10.1016/j.inffus.2019.07.011

Generating One-Hot Maps Under Encryption

Ehud Aharoni, Nir Drucker[(✉)], Eyal Kushnir, Ramy Masalha, and Hayim Shaul

IBM Research, Haifa, Israel
drucker.nir@gmail.com

Abstract. One-hot maps are commonly used in the AI domain. Unsurprisingly, they can also bring great benefits to ML-based algorithms such as decision trees that run under Homomorphic Encryption (HE), specifically CKKS. Prior studies in this domain used these maps but assumed that the client encrypts them. Here, we consider different tradeoffs that may affect the client's decision on how to pack and store these maps. We suggest several conversion algorithms when working with encrypted data and report their costs. Our goal is to equip the ML over HE designer with the data it needs for implementing encrypted one-hot maps.

Keywords: one-hot maps · privacy preserving transformation · homomorphic encryption · decision trees · privacy preserving machine learning · PPML

1 Introduction

Complying with regulations such as GDPR [15] and HIPAA [8] can prevent organizations from porting sensitive data to the cloud. To this end, some recent Privacy-Preserving Machine Learning (PPML) solutions use Homomorphic Encryption (HE), which enables computing on encrypted data. The potential of HE can be observed in Gartner's report [17], which states that 50% of large enterprises are expected to adopt HE by 2025 and also in the large list of enterprises and academic institutions that are actively engaging in initiatives like HEBench [28] and HE standardization efforts [3].

We illustrate the landscape of HE-based solutions by first describing one commonly used threat model. For brevity, we restrict ourselves to a basic scenario, which we describe next, but stress that our study can be used almost without changes in many other constructions. We consider a scenario that involves two entities: a user and a semi-honest cloud server that performs Machine Learning (ML) computation on HE-encrypted data. The user can train a model locally, encrypt it, and upload it to the cloud. Here, the model architecture and its weights are not considered a secret from the user only from the cloud. Alternatively, the user can ask the cloud to train a model on his behalf over

S. Dolev et al. (Eds.): CSCML 2023, LNCS 13914, pp. 96–116, 2023.
https://doi.org/10.1007/978-3-031-34671-2_8

encrypted/unencrypted data and at a later stage, perform inference operations, again, on his behalf using the trained model. We note that in some scenarios, the model is a secret and should not be revealed to the user. In that case, only the classification or prediction output should be revealed. We also assume that communications between all entities are encrypted using a secure network protocol such as TLS 1.3 [27], i.e., a protocol that provides confidentiality, integrity, and allows the users to authenticate the cloud server.

HE-based solutions showed great potential in past research but they also come with some downsides. Particularly, they involve large latency costs that may prevent their vast adoption. These latency costs can sometimes be traded with other costs such as memory or bandwidth, where these trade-offs allow the users to find the right balance for them. Our study aims to extend the HE toolbox with new utilities and trade-offs that will get Single Instruction Multiple Data (SIMD)-based HE-solutions one step further in being practical.

Branching. In general, a branching operation that navigates a program to a specific path is unsupported under HE. Instead, algorithms such as decision trees that select specific branches based on comparison operations are required to select and go through all branches when executed under HE. They often do it by using comparisons and indicator masks. This increases the number of executed operations, which in decision trees increases the number of comparison operations that are costly when executed under HE. In Sect. 5, we discuss some possible comparators and conclude, as prior research did, that the most efficient comparators are binary comparators, which only compare one bit of data at a time. This led many implementations over HE to prefer using *one-hot* encoding for the input data.

Definition 1. *A one-hot map that represents n categories, is an n-bits vector* $\mathbf{o} \in \{0,1\}^n$ *where* $\sum_{i=0}^{n-1} \mathbf{o}[i] = 1$*, i.e., all bits but one are 0.*

Most prior arts e.g., [4,5,21,25,26,29] assumed that these maps are precomputed by the user before encryption and thus their costs are negligible. However, in practice, this computation can lead to huge overhead costs of memory and bandwidth, which many applications try to avoid. Furthermore, in some applications, the data was encrypted and uploaded to the cloud way before the ML processes are required, e.g., when collecting data from IoT devices. In this case, some pre-processing methods should be applied to convert the encrypted input from one format to another. Our goal is therefore to explore the different trade-offs when using different conversion methods and use cases.

Our Contribution. We explore trade-offs that are related to one-hot maps.

- We compare different input representations and explain the benefit of every representation as well as when it should be used.
- We propose a new input representation that is based on the Chinese Remainder Theorem (CRT), which offers a new trade-off between latency and bandwidth.

– We propose several new conversion algorithms between different input representations.
– We implemented and experimented with our proposed methods. In addition, our code is available online [1].

Organization. The document is organized as follows. Section 2 describes some preliminaries such as balanced trees and HE. We describe possible input representations of data and our novel CRT representation in Sect. 3 and we describe our conversion methods in Sect. 4. Some applications that can use one-hot maps are detailed in Sect. 5. Sections 3, 4 and 5 present the methods we used regardless of the final packing in ciphertexts. To this end, Sect. 6 describes tile tensors and some possible packing methods. Section 7 describes the setup and results of our experiments and Sect. 8 concludes the paper.

2 Preliminaries and Notation

We refer to numbers using low-case letters e.g., a, b, c, and we denote vectors by bold face letters e.g., $\mathbf{v}, \mathbf{w}$. To access a specific element $0 \leq i < n$ in a vector $\mathbf{v}$ we use square brackets $\mathbf{v}[i]$. A dot-product operation of two vectors $\mathbf{v}_1, \mathbf{v}_2$ is denoted by $\langle \mathbf{v}_1, \mathbf{v}_2 \rangle$. One-hot maps are often represented using the letter $\mathbf{o}$ and greater maps (defined below) using the letter $\mathbf{g}$. A mathematical ring with addition $(+)$ and multiplication $(*)$ operations is denoted by $\mathcal{R}(+, *)$. In this paper, log refers to $\log_2$.

2.1 Balanced Binary Trees

Some of our algorithms use a balanced binary tree structure. We define the tree relations in the trivial way. For example, in the tree below, Root is the parent of Node1 and Node2, which are the sons of Root. Two nodes are siblings if they share the same first-order parent and we access them using the function sib, for example, $Node2 = Node1.\text{sib}()$ and $Root.\text{sib}() = \varnothing$. A path to Leaf1 includes all nodes from Leaf1 to Root, e.g., $Leaf1.\text{path} = \{Leaf1, Node1, Root\}$. The leaves are numbered from left to right and for a tree T we access its ith leaf from the left using the function $T.\text{leaf}[i]$. To get all nodes at a specific level we use the method getNodesAtLevel, where the level indexing starts from 0. For example,

$$\text{getNodesAtLevel}(0) = \{Root\} \tag{1}$$

$$\text{getNodesAtLevel}(1) = \{Node1, Node2\} \tag{2}$$

$$\text{getNodesAtLevel}(2) = \{Leaf1, Leaf2, Leaf3, Leaf4\} \tag{3}$$

Finally, to get the right or left sons of a node we use the methods l and r, respectively. We use the method val() to access the value of a node.

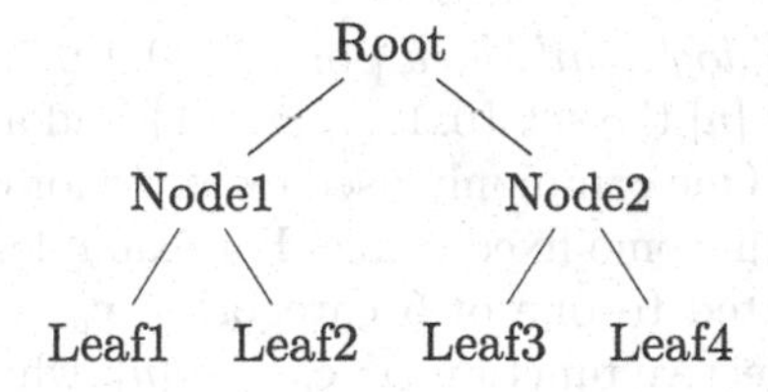

2.2 Homomorphic Encryption (HE)

HE is a public-key encryption scheme that in addition to the usual functions Enc, Dec (see details below) also provides functions to perform operations on encrypted data (usually addition and multiplication), see survey in [18]. The encryption operation $\mathrm{Enc} : \mathcal{R}_1 \rightarrow \mathcal{R}_2$ encrypts input plaintext from the ring $\mathcal{R}_1(+, *)$ into ciphertexts in the ring $\mathcal{R}_2(\oplus, \odot)$ and its associated decryption operation is $\mathrm{Dec} : \mathcal{R}_2 \rightarrow \mathcal{R}_1$. Informally, an HE scheme is correct if for every valid input $x, y \in \mathcal{R}_1$: $\mathrm{Dec}(\mathrm{Enc}(x)) = x$, $\mathrm{Dec}(\mathrm{Enc}(x) \oplus \mathrm{Enc}(y)) = x + y$, and $\mathrm{Dec}(\mathrm{Enc}(x) \odot \mathrm{Enc}(y)) = x * y$, and is approximately correct (as in CKKS) if for some small $\epsilon > 0$ that is determined by the key, it follows that $|x - \mathrm{Dec}(\mathrm{Enc}(x))| \leq \epsilon$. The addition and multiplication equations are modified in the same way. Note that the correctness equations should hold for every finite number of multiplications and additions. This paper, focuses on CKKS.

In [23] the authors showed how SIMD operations can be implemented in lattice based cryptography. Later, similar ideas were implemented in Modern HE instantiations such as BGV [7], B/FV [6,16], and CKKS [11] that rely on the complexity of the Ring-LWE problem [24] for security. For schemes that support SIMD, addition and multiplication are applied slot-wise on vectors.

HE Packing. Some HE schemes, such as CKKS [10], operate on ciphertexts in a homomorphic SIMD fashion. This means that a single ciphertext encrypts a fixed-size vector, and the homomorphic operations on the ciphertext are performed slot-wise on the elements of the plaintext vector. To utilize the SIMD feature, we need to pack and encrypt more than one input element in every ciphertext. The packing method can dramatically affect the latency (i.e., time to perform computation), throughput (i.e., number of computations performed in a unit of time), communication costs, and memory requirements. We further discuss packing one-hot vectors under HE in Sect. 6.

Comparison Under HE. Comparing numbers under HE and specifically, CKKS (e.g., [12,22]) often relies on polynomial approximations of the Step() or Sign() functions, whereas the accuracy and performance of these methods rely on the degrees of these polynomials. We denote these comparison functions by $\mathrm{Eq}(x, y) = 1 \Leftrightarrow x = y$, and $\mathrm{Eq} = 0$ otherwise. Note that other methods exists for schemes such as BGV and BFV such as [9,20].

3 Input Representations

This section discusses different input representations for algorithms that may use one-hot maps. In general, the one-hot map slots can represent any set of categories ($\mathbf{c}$) e.g., $\mathbf{c} = ['dog','cat','bird']$ or $\mathbf{c} = [0.9, 0.7, 1.57]$. To simplify our algorithms we denote by $[n]$ the set $\{0, 1, \ldots, n-1\}$ and associate every element of $\mathbf{c}$ to an element in $[n]$. One commonly used example for $\mathbf{c}$ is the quantization of a floating point element in some fixed range. For example, a floating point value $a \in [1, 2]$ can be converted to one of 6 categories: $\mathbf{c}_a = [1, 1.2, 1.4, 1.6, 1.8, 2]$, where we can use the bijective function $\phi : \mathbf{c}_a \longrightarrow [n]$, where $\phi(a) = 5a - 4$. This map can easily be implemented under HE if needed using one multiplication by a constant and one addition.

We consider several methods for achieving one-hot maps from categorical variables over $[n]$. For these, we consider six input representations: 1) One-hot map representation; 2) Numeric representation; 3) CRT representation; 4) Hierarchical CRT representation; 5) Numeric (Hierarchical) CRT representation; 6) Binary representation.

One-Hot Map Representation. This is the simplest representation, where the client directly encodes every one of its number elements as a one-hot map, i.e., n values are encrypted for each element. The issue with this method is that it is wasteful in bandwidth, assuming that n is of medium size e.g., $n = 100$ will cause the client to encode 99 zeros and one 1 to represent just one element. When the network transport and storage costs are large, this method becomes unfavorable. One-hot maps were used for example in [4,5,21,25,26,29].

Numeric Representation. Here the client submits elements from $[n]$ using their integer representation. Here, the bandwidth and storage costs are low, but converting this representation to a one-hot map representation while encrypted on the cloud side is costly in terms of latency. It either requires n costly Eq operations or running some optimization such as polynomial interpolation (Sect. 4) that has multiplication-depth of at least $O(\log n)$ and requires at least n multiplications. This cost is high for large values of n.

CRT Representation. We propose a technique that offers a tradeoff between the above two approaches. It possesses medium bandwidth costs and medium latency costs. The idea is to submit one-hot maps of the CRT representation of every element and expand them in an online phase by the server. The size of all the maps is much smaller compared to the size of a full-size one-hot map, and the latency to expand them is lower compared to transforming an integer element into a one-hot map. Specifically, for $a \in [n]$, where n is a composite number with pairwise coprime factors, i.e., $n = n_1 \cdot n_2 \cdots n_k$, where $GCD(n_i, n_j) = 1$ for $i \neq j$. In that case, we use a set of k one-hot maps $\mathbf{o}_{a,i}$ of n_i slots, respectively, where $\mathbf{o}_{a,i}[a \pmod{n_i}] = 1$ and the other slots of $\mathbf{o}_{a,i}$ are set to 0. The total size of these maps is $\sum_{i=1}^{k} n_i$ which is much smaller than n.

In practice, we can choose any n even without coprime factors. To use the above method we embed the $[n]$-map in an $[m]$-map, where $n < m$ and m has coprime factors. The extra slots are set to zero. For example, let $n = 10{,}000$, we can choose to work with $m = 2 \cdot 5 \cdot 7 \cdot 11 \cdot 13 = 10{,}010$ and the number of required slots is $r = 38$. There are different heuristics to find the smallest m or r values, one such can be to consider a brute-force search over all different combinations of the ℓ smallest primes, or to set a limit t and consider all numbers $m \in [n, n+t]$.

Hierarchical CRT Representation. The CRT representation is limited to coprime factors. Here, we suggest another approach that uses CRT representation in a hierarchical way, where we consider the coprime restriction separately at every hierarchy. Start from a number n, compute $n_1 = \lceil \sqrt{n} \rceil, n_2 = \lceil \sqrt{n} \rceil + 1$. Clearly, n_1 and n_2 are coprime. Recursively repeat the split process until reaching the "desired" hierarchy according to latency and bandwidth parameters. The client then encodes its value a using the last layers' values and the cloud constructs the one-hot map in an iterative process. Example 4 in Appendix B demonstrates this representation.

Numeric (Hierarchical) CRT Representation. In Example 4 the client encrypts the 8 maps $\mathbf{o}_{a,*}$, another option is to encrypt their 8 representative numerical values, e.g., a_*, in that case, only 8 elements are being sent. We call this representation the Numeric CRT representation.

Binary Representation. The last representation uses binary decomposition instead of CRT-based decomposition. Here, the client decomposes the number a to its binary decomposition $a = \sum_{i=0}^{\log(n)-1} \mathbf{a}[i] \cdot 2^i$, the number of elements being sent is therefore $\log(n)$.

4 Conversion Methods

Some applications assume that the input representation is static and choose the one that most fits their needs. In contrast, some applications assume that the data was uploaded in one format and needs to be translated into another. Here, we describe some conversion methods between input representations. Another reason to move between representations is when an application uses the data in different places, where at every place it prefers the data in a different format. Note that moving from any representation into the CRT representation is hard as the modulo operation is slow and sometimes even unsupported under HE.

4.1 From One-Hot to Numeric Representation

Moving from a one-hot representation of a vector $\mathbf{o}_a$ that represents the classes $\mathbf{c}_a$ to its numeric representation a can be done by simply computing $\langle \mathbf{c}_a, \mathbf{o}_a \rangle$, which requires n multiplications and a multiplication-depth of 1. In that case, there is no need to apply the ϕ transformation to $[n]$ before and after the operation. When c_a is not encrypted, only n cleartext-ciphertext multiplications are performed with a multiplication depth of 0.

4.2 From CRT Representation to One-Hot Maps

We now describe how to convert data in a CRT representation to one-hot maps. The input is the k one-hot sub-maps $\mathbf{o}_{a,i}$, which we first duplicate $\frac{n}{n_i}$ times, respectively, to the duplicated vectors $(\mathbf{o}^d_{a,i})_{i \leq k}$. Subsequently, we multiply all the duplicated maps to construct the final map by $\mathbf{o}_a[i] = \prod_{j=1}^{k} \mathbf{o}^d_{a,j}[i]$ This product is often done by using a multiplication tree, where the cost is only $n \cdot k$ multiplications and a multiplication depth of $\log k$.

Example 1. Let $n = 2 \cdot 3 \cdot 5 = 30$ where $n_1 = 2$, $n_2 = 3$, $n_3 = 5$ then the one-hot representation of the number 17 is

$$(0,0,0,0,0,0,0,0,0,0,0,0,0,0,0,0,1,0,0,0,0,0,0,0,0,0,0,0,0,0)$$

and the associated CRT maps are:

$$\mathbf{o}_{a,1} = (0,1) \qquad \mathbf{o}_{a,2} = (0,0,1) \qquad \mathbf{o}_{a,3} = (0,0,1,0,0),$$

which contains only $\sum_i n_i = 2+3+5 = 10$ elements. To construct the full map from the CRT maps we duplicate the entries of $\mathbf{o}_{a,1}$, $\mathbf{o}_{a,2}$, and $\mathbf{o}_{a,3}$, $30/2 = 15$, $30/3 = 10$, and $30/5 = 6$ times, respectively as follows

$$\mathbf{o}^d_{a,1} = (0,1,0,1,0,1,0,1,0,1,0,1,0,1,0,1,0,\mathbf{1},0,1,0,1,0,1,0,1,0,1,0,1)$$
$$\mathbf{o}^d_{a,2} = (0,0,1,0,0,1,0,0,1,0,0,1,0,0,1,0,0,\mathbf{1},0,0,1,0,0,1,0,0,1,0,0,1)$$
$$\mathbf{o}^d_{a,3} = (0,0,1,0,0,0,0,1,0,0,0,0,1,0,0,0,0,\mathbf{1},0,0,0,0,1,0,0,0,0,1,0,0)$$

Finally, we compute

$$\mathbf{o}_a = (0,0,0,0,0,0,0,0,0,0,0,0,0,0,0,0,0,\mathbf{1},0,0,0,0,0,0,0,0,0,0,0,0)$$

as expected.

From Hierarchical CRT to One-Hot Maps. Unsurprisingly, converting an input from a Hierarchical CRT representation to the one-hot maps representation uses recursively the CRT method at every hierarchy. The hierarchical method has an advantage for large n values.

4.3 From Numeric Representation to One-Hot Maps

We present several methods for moving between a numeric representation to a one-hot map representation. The naïve approach relies on the Eq operator, by checking whether an input x equals one of the classes in $\mathbf{c}$, i.e., for every $i < n$ we compute $\mathbf{o}[i] = \mathrm{Eq}(\mathbf{c}[i], x)$. Assuming that EqM, EqD be the number of multiplications and the multiplication depth of the Eq operator, then such an algorithm requires $EqM \cdot n$ multiplications with an overall multiplication depth of EqD. We suggest several alternatives that rely on Lagrange polynomial interpolations to transform a categorical element $a \in [n]$ to its one-hot map representation. Specifically, let $\vec{S}[c] = \prod_{i \in [n], i \neq c} (c - i)$, then the c'th bit of the one-hot map can be computed using the polynomial

$$P_c(a) = \vec{S}[c]^{-1} \prod_{i \in [n], i \neq c} (a - i),$$

where $P_c(a) = 1$ if $a = c$ and $P_c(a) = 0$ otherwise. Note that the denominator $\vec{S}[c]$ can be efficiently pre-computed because all of its inputs are public and known in advance.

Efficiently Compute the n Polynomials $P_c(a)$. Every polynomial requires $n - 2$ multiplications and $log(n - 2)$ multiplication depth. The trivial way is to compute each polynomial separately, which requires $n(n - 2)$ multiplications and $\log(n - 2)$ multiplication-depth. However, note that every two polynomials share $n - 3$ out of the $n - 2$ terms. Thus, we propose other dedicated algorithms to perform the task more efficiently. Specifically, we suggest two tree-based algorithms. For simplicity, we describe the case where n is a power of two and we claim that it is possible to consider other options as well by adding dummy multiplications by one. Table 2 in Appendix A shows the minimal and maximal values of a tree of n nodes. The values are not too extreme when $n \leq 2^5$.

Tree-Based Alternative 1. Our first alternative includes a tree-based algorithm, which we describe in Algorithm 1. This is a "shallow but big" algorithm that requires $n \log n$ multiplications and having $\log n$ multiplication depth. The algorithm receives an input x and an uninitialized balanced tree T of depth $\log n$. The algorithm starts by initializing all leaves of the tree (Lines 2–3), then it initializes the rest of the tree (Lines 4–6), and finally, it computes the one-hot map using the pre-computed values. Figure 1 illustrates the initialization process. Note that because in Line 9 the path method returns the nodes ordered from leaf to root, the multiplication depth of the entire algorithm is only $\log n$. While the same computation would have held if the path list was ordered from root to leaf, the overall multiplication depth would have resulted to be $2 \log n$.

Algorithm 1. Shallow but big

Input: x the input ciphertext, T a balanced binary tree of depth $\log n$
Output: o a one-hot map that represents x.
Assume: $\mathbf{S}$, where $\mathbf{S}[c] = \prod_{i \in [n], i \neq c} (c - i)$

```
1:  procedure OPT 1
2:      for c in [n] do
3:          T.leaf(i) = x − c
4:      for l in log n − 1 to 0 do
5:          for v in T.getNodesAtLevel(l) do
6:              v.val = (v.l.val) · (v.r.val)
7:      for c in 0 to n − 1 do
8:          o[c] = S[c]
9:          for v in T.leaf(c).path do
10:             o[c] = o[c] * v.sib.val
11:     return o
```

Tree-Based Alternative 2. Algorithm 2 is another alternative that requires $2n$ multiplications instead of $n \log n$ but has a multiplication depth of $2 \log n$ instead of $\log n$. The algorithm is similar to Algorithm 1 except that in Lines

Algorithm 2. Small but less shallow

Input: x the input ciphertext, T a balanced binary tree of depth $\log n$
Output: o a one-hot map that represents x.
Assume: $\mathbf{S}$, where $\mathbf{S}[c] = \prod_{i \in [n], i \neq c} (c - i)$
1: **procedure** Opt 1
2: **for** c in $[n]$ **do**
3: T.leaf$(i) = x - c$
4: **for** l in $\log n - 1$ to 0 **do**
5: **for** v in T.getNodesAtLevel(l) **do**
6: $v.val = v$.l.val $\cdot$ v.r.val
7: Set $T_2 = T$ and $T_2.Root$.val $= 1$
8: **for** l in 1 to $\log(n)$-1 **do**
9: $\triangleright$ *zip* merges two lists by interleaving their elements.
10: **for** (v_T, v_{T_2}) in *zip*(T.getNodesAtLevel(l), T_2.getNodesAtLevel(l)) **do**
11: v_{T_2}.val $= (v_{T_2}$.parent.val$) \cdot (v_T$.sib.val$)$
12: **for** l in T_2.leaves() **do**
13: l.val $= l$.val $\cdot$ $\mathbf{S}[l]$
14: **return** T_2.leaves

8–10 we group the computation of similar terms. However, because at this step of the algorithm we start from the root and not from the leaves we already used a multiplication depth of $\log n$. Figure 2 illustrates table T_2, which is associated with T from Fig. 1.

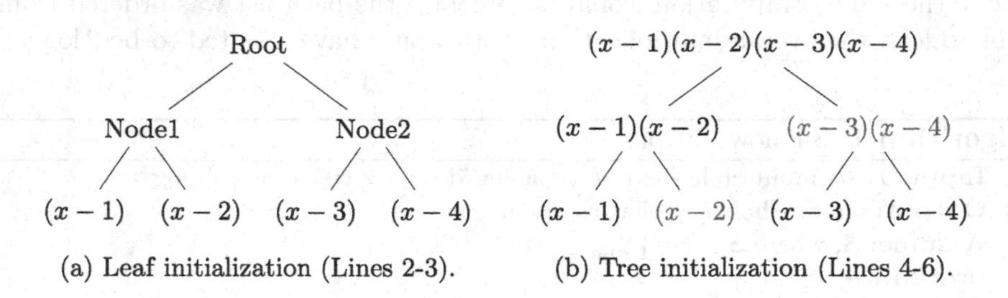

Fig. 1. Algorithm 1 initialization phase where $\mathbf{c} = [4]$ and $n = 4$. Red nodes show the multiplications done to compute $o[1]$ on Line 11. (Color figure online)

When running Algorithms 1 and 2 under encryption the size of S_c may be too small and the size of the product on the intermediate nodes in Line 6 of both algorithms may be too large, which can cause overflows. To this end, we replace Line 8 in Algorithm 1 or Line 12 of Algorithm 2 with an additional plaintext-ciphertext multiplication per intermediate nodes at Line 6 of both algorithms. The exact plaintext values are stored in a pre-computed "shadow" tree that is described in Appendix A.

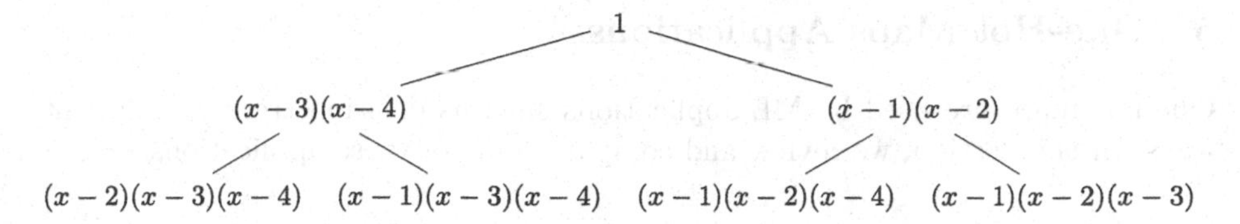

Fig. 2. Algorithm 2 computing T_2, where the leaves correspond to the one-hot map values of x.

4.4 From Binary Representation to One-Hot Maps

The trivial way to translate from binary representation to one-hot map representation is to move to a numeric representation by using $a = \sum_{i=0}^{\log(n)-1} \mathbf{a}[i] \cdot 2^i$ and then use one of the methods above to generate a one-hot map. Instead, we present another approach reminiscent of the methods we used over CRT represented inputs. As before we duplicate the binary values $\mathbf{a}$ to have arrays of n slots and then return the slot-wise product of these arrays. Only here, the duplication is done using the following trick. For every bit, in slot x we put $\mathbf{a}[i]$ if $x \pmod{2^i} = 1$ and $1 - \mathbf{a}[i]$ otherwise. In practice, the $\log n$ bits plaintext masks $\mathbf{w}_i$ can be pre-computed and multiplied with $\mathbf{a}[i]$ on the fly using one plaintext-ciphertext (slot-wise) multiplication through the equation

$$dup(\mathbf{a}[i]) \cdot \mathbf{w}_i + dup(1 - \mathbf{a}[i])(1 - \mathbf{w}_i) = 1 - \mathbf{w}_i + dup(\mathbf{a}[i])(2\mathbf{w}_i - 1)$$

We can also achieve the above through selective masking and a rotate-and-sum algorithm.

4.5 From One-Hot Representation to Binary or CRT Representations

For completeness, we also provide a transformation from the one-hot representation to binary or CRT representations. The idea is to use predefined fix plaintext masks that represent the required basis. Assume a one-hot map $\mathbf{o}$ of size n. For binary representation we use n-element masks $\mathbf{m}_k$, $k \in [\log(n)]$, where $\mathbf{m}_k[i] = i \pmod{2^{k+1}}$, $i \in [n]$. The conversion is done by computing $\mathbf{a}[k] = \langle \mathbf{m}_k, \mathbf{o} \rangle$, where $\mathbf{a}$ is the output representation. For the CRT case, assume $n = p_1 \cdot p_2 \cdots p_\ell$, then the ℓ masks are $\mathbf{m}_k[i] = i \pmod{p_k}$ and the output is $\mathbf{a}$, where $\mathbf{a}[k] = \langle \mathbf{m}_k, \mathbf{o} \rangle$.

4.6 From One-Hot Representation to CRT Maps

Example 2. Let $n = 8$ and $a = 3$, then the binary representation of a is $\mathbf{a} = (1, 1, 0)$ and the duplication maps $\mathbf{a}_i^d$ are:

$$\mathbf{a}_0^d = (1 - \mathbf{a}[2], \mathbf{a}[2], 1 - \mathbf{a}[2], \mathbf{a}[2], 1 - \mathbf{a}[2], \mathbf{a}[2], 1 - \mathbf{a}[2], \mathbf{a}[2]) = (0, 1, 0, 1, 0, 1, 0, 1)$$
$$\mathbf{a}_1^d = (1 - \mathbf{a}[1], 1 - \mathbf{a}[1], \mathbf{a}[1], \mathbf{a}[1], 1 - \mathbf{a}[1], 1 - \mathbf{a}[1], \mathbf{a}[1], \mathbf{a}[1]) = (0, 0, 1, 1, 0, 0, 1, 1)$$
$$\mathbf{a}_2^d = (1 - \mathbf{a}[0], 1 - \mathbf{a}[0], 1 - \mathbf{a}[0], 1 - \mathbf{a}[0], \mathbf{a}[0], \mathbf{a}[0], \mathbf{a}[0], \mathbf{a}[0]) = (1, 1, 1, 1, 0, 0, 0, 0)$$

5 One-Hot Maps Applications

One-hot maps are used by ML applications such as decision trees, in different ways. In this section, we review and compare some of these applications.

5.1 A Single Element Comparisons

Comparing integers or fixed point numbers can be costly when performed under HE, and specifically CKKS. Some comparison methods include floating-point comparison e.g., in [12,22] or bit-vector comparisons. Floating-point comparisons often rely on polynomial approximations of the Step() or Sign() functions, whereas the accuracy and performance of these methods rely on the degrees of these polynomials. We denote these comparison functions by $\text{Eq}(x, y) \in [1-\beta, 1+\beta] \Leftrightarrow |x-y| < \alpha$, and $\text{Eq} \in [-\beta, \beta]$ otherwise, where $\alpha, \beta > 0$ are parameters that can be made arbitrarily small by increasing the polynomial degree of Eq. Following [12,22] the degree of Eq is $degree(\text{Eq}) = O(-\log(\alpha\beta))$.

Bit-Vector Comparisons. In contrast, in bit-vector comparisons, the numbers are represented using bit-vectors. For example, the numbers a and b are represented using the n-bit vectors $\mathbf{a}, \mathbf{b} \in \{0,1\}^n$, where $a = \sum_{i=0}^{n-1} \mathbf{a}[i] \cdot 2^i$ and $b = \sum_{i=0}^{n-1} \mathbf{b}[i] \cdot 2^i$. To compare these vectors, one needs to run the circuit $\text{AND}_{i=0}^{n-1}(\text{XNOR}_{i=0}^{n-1}(\mathbf{a}[i], \mathbf{b}[i]))$, where for $x, y \in \{0,1\}$, the gates XNOR (Not Xor) and AND are implemented using only additions and multiplications with $\text{XNOR}(x, y) = 1 - (x-y)^2$ and $\text{AND}(x, y) = x \cdot y$. Consequently, for two n bits vectors, the number of multiplications is $2n$ and the multiplication depth is $\log(n) + 1$.

Optimization. We propose a combination of the approaches above in cases, where the data is already represented in bit-vectors. Specifically, we compute the comparison results by first computing $s = \sum_{i=0}^{n-1} \text{XOR}(\mathbf{a}[i], \mathbf{b}[i])$ where a XOR gate is simulated by $\text{XOR}(x, y) = (x-y)^2$. Then we apply the equality function $1 - \text{Eq}(s, 0)$. Here, $s = 0$ if and only if the numbers are equal. Moreover, because $s \leq n$, we can use a dedicated comparison function that only requires $\log(n)$ multiplications as mentioned in [14]. The total number of multiplications in our approach is $n + \log(n)$ and the multiplication depth is $\log(n) + 1$.

We further optimize this method by using the complex plane in CKKS. We first pack every sequential pair of bits as a complex number, where the final complex vectors are $\mathbf{a^c}[i] = (\mathbf{a}[2i] + i\mathbf{a}[2i+1])$ and $\mathbf{b^c}[i] = (\mathbf{b}[2i] + i\mathbf{b}[2i+1])$. These vectors use half the space compared with the previous approach. We now replace the XOR operations with norm operations, i.e., we first subtract $\mathbf{d^c} = \mathbf{a^c} - \mathbf{b^c}$, and compute

$$0 \leq \sum_{i=0}^{\frac{n}{2}-1} \mathbf{d^c}[i] \cdot \overline{\mathbf{d^c}}[i] = \sum_{i=0}^{\frac{n}{2}-1} \|\mathbf{d^c}\|^2 \underset{\|\mathbf{d^c}\|^2 < 2}{\leq} 2 \cdot \frac{n}{2} = n$$

The last inequality follows because every element indicates whether two compared bits are equal (value 0) or unequal (value one of $\pm 1, \pm i, \pm 1 \pm i$). The rest of the computation is the same. Note that the number of multiplications is now $\frac{n}{2} + \log(n)$ instead of $n + \log(n)$ but we also require $\frac{n}{2}$ conjugate operations.

Using One-hot Maps. Using one-hot maps allows for speeding up the comparison process in cases where the range of inputs can be separated into a not-too-large number of categories. Here, every bit of the one-hot map represents one category. Subsequently, when an application wants to learn whether a number represented using a map represents a specific category, it can simply use the relevant bit from the bit-map.

Example 3. Consider an element $a \in [7]$, specifically, $n = 7$ with the one-hot map $\mathbf{o}_a = (0, 0, 0, 1, 0, 0, 0)$. An application can check whether $n = 3$ by either performing $\text{Eq}(3, a)$ or by simply returning $\mathbf{o}_a[3]$ with 0 costs.

5.2 Greater-Equal Comparisons

Another application of one-hot maps is greater, greater-equal, less-than, or, less-equal-than comparisons e.g., $n > 10$. Here, we can use a greater-equal (GE) operator through the polynomial approximation of the Step() or Sign() functions from above [12,22]. When numbers are represented using binary representation, it is possible to efficiently convert them to numbers using n scalar-ciphertext encryptions and additions and then use the GE operation from above. A more efficient method exists using greater maps, which were also used in [5].

Definition 2. *A greater (resp. less-than) map for n categories that represents all categories with indices greater (resp. less-than) from some category index a is an n-bits vector $\mathbf{g}_a \in \{0, 1\}^n$ where*

$$\mathbf{o}[i] = \begin{cases} 0 & i \leq a \;\; (resp.\;\; i \geq a) \\ 1 & i > a \;\; (resp.\;\; i < a) \end{cases}$$

To perform a greater operation between two values a and b we need (w.l.o.g.) a greater map $\mathbf{g}_a$ of a. In addition, we distinguish between two cases, where b is encrypted or not. If b is unencrypted, we can just return $\mathbf{g}_a[b]$. Otherwise, either a costly lookup table is required, or we need to represent b as a one-hot map $\mathbf{o}_b$. Using $\mathbf{g}_a$ and $\mathbf{o}_b$ an application can compute $\sum_{i=0}^{n-1} \text{AND}(\mathbf{o}_b[i], \mathbf{g}_a[i])$ which requires n multiplications and a multiplication depth of 1.

We can generate a greater map $\mathbf{g}$ from a one-hot map $\mathbf{o}$ by going over the bits of $\mathbf{g}$ from $i = 0$ to $i = n-1$ and for every i compute $\mathbf{g}[i] = \text{OR}(\mathbf{o}[i], \mathbf{g}[i-1])$, where OR is simulated by $\text{OR}(x, y) = x + y - xy$. Since $\mathbf{a}$ is given in a one-hot representation a more efficient way is to set $\mathbf{g}[j] = \sum_{i=0}^{j} \mathbf{a}[i]$. The cost is n sequential additions that can be done offline.

5.3 A Range Comparison

The final application is range comparisons, which attempt to find whether a number m is in the range $[a, b]$. This can be done by calling $\text{AND}(\text{GE}(a, n), \text{GE}(n, b))$, or by using a similar algorithm that uses one-hot and greater maps as explained above.

6 Packing and Tile Tensors

So far we discussed our methods using vectors and operations in vectors. In a real HE application these vectors are stored inside ciphertexts, and the operations involve SIMD operations on these ciphertexts.

The way the vectors are stored inside the ciphertexts is called the packing scheme, and it can significantly impact both bandwidth and latency. Consider for example m one-hot maps of size n that require a total of mn elements to pack, and consider ciphertexts of size s, such that $m \leq s$ and $n \leq s$, and let's further assume that all three values are close to each other. Two different packing options come to mind: we can store each of the m elements in a single ciphertext containing the entire one-hot vector. Alternately, we can store each number spread across n ciphertexts, where the i'th slot in each ciphertext contains the one-hot encoded value for the i'th sample.

Recently a new data structure called tile tensor was proposed [2]. It allows to easily specify these two extreme options and also to easily specify intermediate packing. This is done by writing packing-oblivious code that operates on tensors. The tensors are partitioned to parts (tiles) that are mapped to slots of ciphertexts. Setting a different *shape* to the tiles yields a different packing scheme. Consider the matrix of size $[m, n]$ where each of the m rows is the one-hot encoded vector of the m'th sample. The tile tensor shape notation $[m/t_1, n/t_2]$ means the matrix is divided into (sub-blocks) tiles of size $[t_1, t_2]$, and each tile is stored in one ciphertext, so that $s = t_1 t_2$.

For example, $[m/1, n/s]$ means the tile's shape is $[1, s]$, so each row of the matrix is stored in a separate ciphertext. Alternately $[m/s, n/1]$ means each column is stored in a separate ciphertext. Other, more complex options are also possible. For example, if $s = 1024$, we can choose $[m/16, n/64]$, meaning our matrix is divided into tiles of size $[16, 64]$, and each one is stored (flattened) in a ciphertext. To learn more about tile tensor capabilities see [2].

Different packings go along well with different algorithms, and the final choice should be made by simulations and measurements with a particular hardware and HE backend. Consider for example converting numerical to one-hot representations. We can place one number in a ciphertext, duplicate it across multiple slots, and then it's efficient to run a single comparison in SIMD fashion with multiple values. Alternately, we can place multiple numbers in a single ciphertext, where the tree-based approaches we have shown are more efficient.

7 Experiments

For the experiments, we used an Intel®Xeon®CPU E5-2699 v4 @ 2.20 GHz machine with 44 cores (88 threads) and 750 GB memory. All the reported results are the average of 10 runs. We use HELayers [19] version 1.5.2 and set the underlying HE library to HEaaN [13] targeting 128 bits security.

Figure 3 shows our first experiment results, where we measured the amortized bandwidth-latency tradeoff per element in a batch of $m = s$ elements for

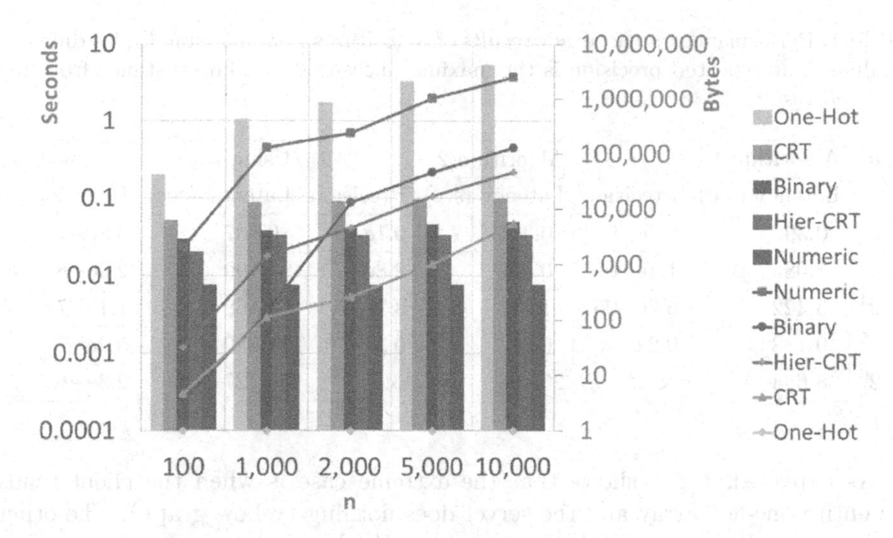

Fig. 3. A comparison of the amortized bandwidth and amortized latency per element of generating one-hot maps from different input representations for a batch of m elements and different values of n, when setting the tile tensor shape to $[\frac{n}{1}, \frac{m}{s}]$. Scales for both y-axes is logarithmic.

generating one-hot maps from different input representations, for different n values. In this experiment we consider five n values from small $n = 100$ value to large $n = 10{,}000$ value and the five different input representations: a) one-hot map of size n; b) numeric representation, the input is a number $a \in [n]$; c) CRT, where for example for $n = 100$, the input is three one-hot arrays of sizes 3, 5, and 7 ($3 \times 5 \times 7 = 105$), and for $n = 10{,}000$ the input is five one-hot arrays of sizes $2, 5, 7, 11, 13$ ($2 \times 5 \times 7 \times 11 \times 13 = 10{,}010$); d) binary, for $n = 100$ and $n = 10{,}000$ the client sends 7 and 14 bits, respectively; d) hierarchical CRT, where for example for $n = 100$, the client sends one-hot arrays for

$$((a \bmod 10) \bmod 3, (a \bmod 10) \bmod 4, (a \bmod 11) \bmod 3, (a \bmod 11) \bmod 4)$$

of sizes $3, 4, 3, 4$, respectively, and for the $n = 10{,}000$ case, the client sends the maps of sizes $3, 4, 3, 4, 3, 4, 3, 4$ from which a one-hot array for $a \in [10{,}000]$ can be computed.

HE Parameters. We used a bootstrappable context with ciphertexts with $s = 2^{15}$ slots, a multiplication depth of 12, fractional part precision of 42, and an integer part precision of 18. In all cases, we used a batch m of s samples that are packed together in one ciphertext. For the $n = 100$ case, we set Eq to be a polynomial of degree 7^4 (a composition of 4 polynomials of degree 7) with a multiplication depth of 12, the resulting maximum error was 0.02. For the $n = 10{,}000$ case, we used a larger polynomial of degree 7^7, with a multiplication depth of 21. The resulting maximum error was 0.04. The error was virtually 0 in the CRT and Hier-CRT cases.

Table 1. Performance and accuracy results of Algorithms 1, 2 and using Eq for different n values. The reported precision is the maximal measured absolute distance from the expected result.

n	Algorithm 1		Algorithm 2		Using Eq	
	Latency (sec)	Precision	Latency (sec)	Precision	Latency (sec)	Precision
2^2	0.392	8.7e−9	0.355	9.1e−9	0.807	3.8e−8
2^3	1.085	1.7e−07	0.662	2.8e−7	1.499	2.9e−8
2^4	3.422	6.2e−06	1.917	3.1e−5	4.568	1.1e−7
2^5	9.138	0.24	6.096	0.3	11.080	5.2e−7
2^6	38.684	∞	25.498	∞	50.727	2.3e−6

As expected, Fig. 3 shows that the extreme case is when the client sends the entire one-hot array and the server does nothing (yellow graph). The other extreme case is when the client transmits a single number (numeric representation) and the server does a full transformation (blue graph). In all cases, we measured the cost to generate a one-hot map under encryption. We did not implement or measure the applications that use the one-hot maps, such as decision trees, because that would only distract and obscure the measurements we are interested in, namely how efficient is computing one-hot maps. We refer the reader to e.g., [5] to learn about the performance of an encrypted tree based solution.

Numeric to One-Hot. We also tested the performance and accuracy of Algorithms 1 and 2 for converting a number to its one-hot representation when using tile tensors of shape $[n/1, m/s]$, where n is the one hot size and m is the number of samples that run in parallel, in our case, we set $m = s$. Table 1 summarizes our experiment results. Here, we used ciphertexts with 2^{14} slots, a multiplication depth of 17, fractional part precision of 44, and an integer part precision of 16.

Our experiments showed that using Algorithms 1 and 2 directly was only possible for small $n \leq 2^3$ values because the tree values get too large and cause overflows. However, when combining them with the shadow tree method presented in Appendix A, we were able to get good results even for larger n of size $2^5 = 32$. Thus, our algorithms present another tradeoff between latency and accuracy that a user can exploit.

8 Conclusions

One-hot maps are a vital tool when considering AI and in particular when running some PPML solutions under HE. We explored some different representations of the (encrypted) input and explain how to move between representations as needed. Our experiment results show the practicality of these translations. Consequently, data scientists have now more tools in their AI-over-HE toolbox that they can efficiently use. In future research, we intend to further explore the trade-offs we get from using different tile tensor capabilities and our algorithms.

A Computing the S[c] Tree

In Sect. 4.3 we explained why we need to modify Algorithms 1 and 2 when performed under encryption. Here, provide an algorithm and code to compute the shadow tree that results from the $\mathbf{S}[c]$ values.

We explain the algorithm by an example. Consider a tree of $n = 8$ leaves that represents a one-hot map of 8 elements. The corresponding $\mathbf{S}[c]$ values are:

$$\mathbf{S}[0] = \prod_{i=-7}^{-1} i^{-1} \quad \mathbf{S}[1] = \prod_{\substack{i=-6\\ i\neq 0}}^{1} i^{-1} \quad \mathbf{S}[2] = \prod_{\substack{i=-5\\ i\neq 0}}^{2} i^{-1} \quad \mathbf{S}[3] = \prod_{\substack{i=-4\\ i\neq 0}}^{3} i^{-1}$$

$$\mathbf{S}[4] = \prod_{\substack{i=-3\\ i\neq 0}}^{4} i^{-1} \quad \mathbf{S}[5] = \prod_{\substack{i=-2\\ i\neq 0}}^{5} i^{-1} \quad \mathbf{S}[6] = \prod_{\substack{i=-1\\ i\neq 0}}^{6} i^{-1} \quad \mathbf{S}[7] = \prod_{i=1}^{7} i^{-1}$$

It is easy to see the alternate sign pattern of Observation 1 and that Observation 2 follows from symmetry. We are now ready to describe Algorithm 3, which generates the $\mathbf{s}[c]$ shadow tree. The algorithm gets the number of tree levels ℓ as input. For every leaf c, it computes the list of indices $\mathbf{S}_0[c]$ in Steps 3–4. Next, it builds an initial tree T by going from the leaves up to the root. At every level l, for every node i, it computes the intersection $\mathbf{S}_l[2i] \cap S_l[2i+1]$ of its two sons and stores the results in $\mathbf{S}_{l+1}$. In addition, the algorithm stores in the tree nodes the unique elements that are associated with them (Steps 8–9). The idea is to construct a tree that holds the elements of $\mathbf{S}[c]$ when going through the T.path(c).

Observation 1. *For* $c \in [n]$, *when* c *is odd* $\mathbf{S}[c] < 0$ *otherwise* $\mathbf{S}[c] > 0$.

Observation 2. *For* $c \in [n]$, $\mathbf{S}[c] = -\mathbf{S}[n-c]$.

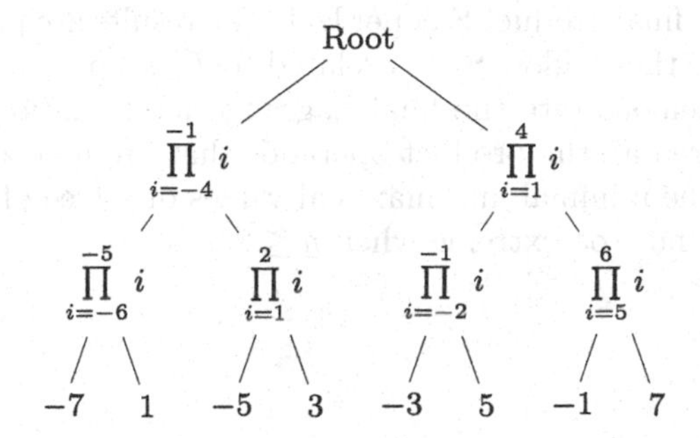

Fig. 4. An example of the resulted tree T after the first phase of Algorithm 3 (Steps 1–9) when $n = 8$.

Algorithm 3. Calculate the $\mathbf{S}[c]$ tree

Input: ℓ tree levels.
Output: T a balanced tree.
1: **procedure** CALCSCTREE(ℓ)
2: $n = 2^\ell$
3: **for** $c \in [n]$ **do**
4: $\mathbf{S}_0[c] = \{c, c-1, \ldots, 1, -1, \ldots c-n+1\}$
5: **for** $l \in [\ell]$ **do**
6: **for** $i \in [|\mathbf{S}_l/2|]$ **do**
7: $\mathbf{S}_{l+1}[i] = \mathbf{S}_l[2i] \cap S_l[2i+1]$
8: $T_l[2i] = \mathbf{S}_l[2i] - \mathbf{S}_{l+1}[i]$
9: $T_l[2i+1] = \mathbf{S}_l[2i+1] - \mathbf{S}_{l+1}[i]$
10: **for** $l \in [\ell]$ **do** ▷ Swap locations
11: **for** $i \in [|\mathbf{S}_l]/2|]$ **do**
12: $T_l[i], T_l[i+1] = T_l[i+1], T_l[i]$
13: **for** $l \in (\ell - 1)$ to 0 **do**
14: **for** $i \in [T_l]$ **do**
15: $T_l[i] = \left(\frac{\prod_{T_l[i])}}{\prod T_{l-1}[2i] * \prod T_{l-1}[2i+1]}\right)^{-1}$
16: **for** $i \in [|T_0|]$ **do**
17: $T_0[i] = (\prod T_0[i])^{-1}$
18: **return** T

Figure 4 shows the tree T for $n = 8$ after Step 9. Note the symmetry of the tree that follows Observation 2. Next, Algorithm 1 computes the desired product by multiplying the siblings of the nodes on a path to the leaf. Thus, we need to move the relevant $\mathbf{S}[c]$ elements to be on that path as well. This is done in Steps 10–12. Finally, we convert the lists of values on the tree nodes to integers by computing their product in step 15. We also divide every node by its sons' product to avoid multiplying a single element twice later on once traversing the tree to compute the final product $\mathbf{S}[c]$ per leaf. The results are presented in Fig. 5. To understand how the shadow $\mathbf{S}[c]$ is related to T, we present also T in Fig. 6. Furthermore, we demonstrate the final Lagrange interpolation value for leaf 3 by highlighting in red all the product operands that are used by Algorithm 1.

Table 2 shows the minimal and maximal values of a tree of n nodes. We see that the values are not too extreme when $n \leq 2^5$.

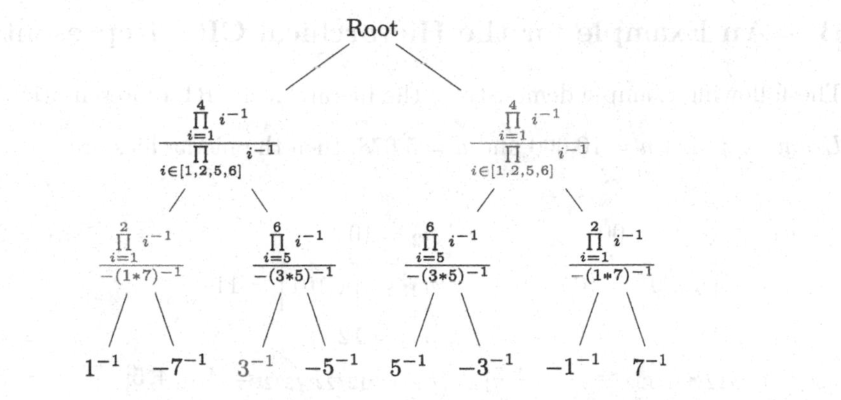

Fig. 5. Results of Algorithm 3 when $n = 8$, Red nodes show the multiplications required for computing $\mathbf{o}[3]$ on Algorithm 1 Step 11.

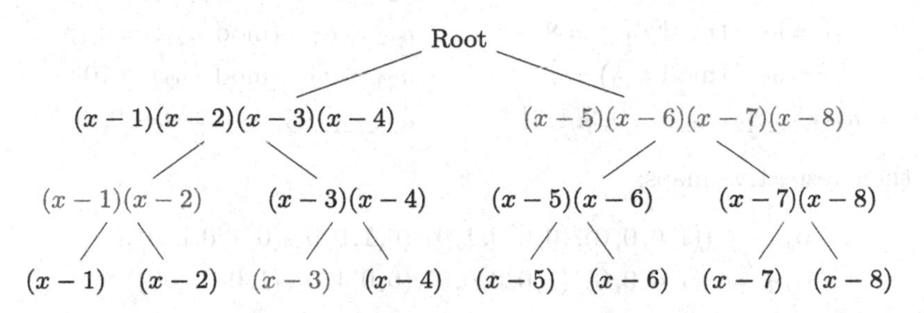

Fig. 6. The tree T generated in Algorithms 1 and 2 at Step 6. Red nodes show the multiplications required for computing $\mathbf{o}[3]$ on Algorithm 1 Step 11.

Table 2. Bounds on the positive values in the S_c tree.

log(n)	minimal value	maximal value	log(min)	log(max)
2	0.33300	1	−1.59	0
3	0.14300	2.5	−2.80	01.32
4	0.06700	15.2	−3.91	03.92
5	0.03200	292.5	−4.95	08.19
6	0.00134	110,373	−9.54	16.75
7	1.54e−06	1.58e+09	−19.30	33.88
8	2.03e−12	3.27e+20	−38.00	68.15

B An Example for the Hierarchical CRT Representation

The following example demonstrate the hierarchical CRT representation.

Example 4. Let $n = 10{,}000$ and $a = 5{,}678$, then the hierarchies are

$$\begin{aligned} n_1 &= \lceil\sqrt{n}\rceil = 100 & n_2 &= 101 & m_1 &= 201 \\ n_{11} &= \left\lceil\sqrt{100}\right\rceil = 10 & n_{21} &= \left\lceil\sqrt{101}\right\rceil = 11 & & \\ n_{12} &= 11 & n_{22} &= 12 & m_2 &= 44 \\ n_{111/112/121/122} &= [4,5,4,5] & n_{211/212/221/222} &= [4,5,4,5] & m_3 &= 36 \end{aligned}$$

To encode a the client computes the 8 residues values

$$\begin{aligned} a_1 &= a \pmod{n_1} = 78 & a_2 &= a \pmod{n_2} = 22 \\ a_{11} &= a_1 \pmod{n_{11}} = 8 & a_{12} &= a_1 \pmod{n_{21}} = 0 \\ a_{12} &= a_2 \pmod{n_{12}} = 1 & a_{22} &= a_2 \pmod{n_{22}} = 10 \\ a_{111/112/121/122} &= [0,3,1,1] & a_{211/212/221/222} &= [0,0,2,0] \end{aligned}$$

and their respective maps:

$$\begin{aligned} \mathbf{o}_{a,1*} &= ((1,0,0,0),(0,0,0,1,0),(0,1,0,0),(0,1,0,0,0)) \\ \mathbf{o}_{a,2*} &= ((1,0,0,0),(1,0,0,0,0),(0,0,1,0),(1,0,0,0,0)) \end{aligned}$$

In this example, the number of required slots is $m_3 = 36$ which is smaller than $r = 38$ above.

References

1. Aharoni, E., et al.: HeLayers: a tile tensors framework for large neural networks on encrypted data. CoRR abs/2011.0 (2020). arXiv:2011.01805
2. Aharoni, E., Drucker, N., Ezov, G., Shaul, H., Soceanu, O.: Complex encoded tile tensors: accelerating encrypted analytics. IEEE Secur. Priv. **20**(5), 35–43 (2022). https://doi.org/10.1109/MSEC.2022.3181689
3. Albrecht, M., et al.: Homomorphic encryption security standard. Technical report, HomomorphicEncryption.org, Toronto, Canada (2018). https://HomomorphicEncryption.org
4. Ameur, Y., Aziz, R., Audigier, V., Bouzefrane, S.: Secure and non-interactive k-NN classifier using symmetric fully homomorphic encryption. In: Domingo-Ferrer, J., Laurent, M. (eds.) PSD 2022. LNCS, vol. 13463, pp. 142–154. Springer, Cham (2022). https://doi.org/10.1007/978-3-031-13945-1_11
5. Aslett, L.J.M., Esperança, P.M., Holmes, C.C.: Encrypted statistical machine learning: new privacy preserving methods (2015). arXiv:1508.06845
6. Brakerski, Z.: Fully homomorphic encryption without modulus switching from classical GapSVP. In: Safavi-Naini, R., Canetti, R. (eds.) CRYPTO 2012. LNCS, vol. 7417, pp. 868–886. Springer, Heidelberg (2012). https://doi.org/10.1007/978-3-642-32009-5_50

7. Brakerski, Z., Gentry, C., Vaikuntanathan, V.: (Leveled) fully homomorphic encryption without bootstrapping. ACM Trans. Comput. Theory **6**(3), 1–36 (2014). https://doi.org/10.1145/2633600
8. Centers for Medicare & Medicaid Services: The Health Insurance Portability and Accountability Act of 1996 (HIPAA) (1996). https://www.hhs.gov/hipaa/
9. Chakraborty, O., Zuber, M.: Efficient and accurate homomorphic comparisons. In: Proceedings of the 10th Workshop on Encrypted Computing & Applied Homomorphic Cryptography, WAHC 2022, pp. 35–46. Association for Computing Machinery, New York (2022). https://doi.org/10.1145/3560827.3563375
10. Cheon, J.H., Han, K., Kim, A., Kim, M., Song, Y.: A full RNS variant of approximate homomorphic encryption. In: Cid, C., Jacobson, M.J., Jr. (eds.) SAC 2018. LNCS, vol. 11349, pp. 347–368. Springer, Cham (2019). https://doi.org/10.1007/978-3-030-10970-7_16
11. Cheon, J.H., Kim, A., Kim, M., Song, Y.: Homomorphic encryption for arithmetic of approximate numbers. In: Takagi, T., Peyrin, T. (eds.) ASIACRYPT 2017. LNCS, vol. 10624, pp. 409–437. Springer, Cham (2017). https://doi.org/10.1007/978-3-319-70694-8_15
12. Cheon, J.H., Kim, D., Kim, D.: Efficient homomorphic comparison methods with optimal complexity. In: Moriai, S., Wang, H. (eds.) ASIACRYPT 2020. LNCS, vol. 12492, pp. 221–256. Springer, Cham (2020). https://doi.org/10.1007/978-3-030-64834-3_8
13. CryptoLab: HEaaN: Homomorphic Encryption for Arithmetic of Approximate Numbers (2022). https://www.cryptolab.co.kr/eng/product/heaan.php
14. Drucker, N., Moshkowich, G., Pelleg, T., Shaul, H.: BLEACH: cleaning errors in discrete computations over CKKS. Cryptology ePrint Archive, Paper 2022/1298 (2022). https://eprint.iacr.org/2022/1298
15. EU General Data Protection Regulation: Regulation (EU) 2016/679 of the European Parliament and of the Council of 27 April 2016 on the protection of natural persons with regard to the processing of personal data and on the free movement of such data, and repealing Directive 95/46/EC (General Data Protection Regulation). Official Journal of the European Union 119 (2016). http://data.europa.eu/eli/reg/2016/679/oj
16. Fan, J., Vercauteren, F.: Somewhat practical fully homomorphic encryption. In: Proceedings of the 15th International Conference on Practice and Theory in Public Key Cryptography, pp. 1–16 (2012). https://eprint.iacr.org/2012/144
17. Gartner: Gartner identifies top security and risk management trends for 2021. Technical report (2021). https://www.gartner.com/en/newsroom/press-releases/2021-03-23-gartner-identifies-top-security-and-risk-management-t
18. Halevi, S.: Homomorphic encryption. In: Lindell, Y. (ed.) Tutorials on the Foundations of Cryptography. ISC, pp. 219–276. Springer, Cham (2017). https://doi.org/10.1007/978-3-319-57048-8_5
19. IBM: HELayers SDK with a Python API for x86 (2021). https://hub.docker.com/r/ibmcom/helayers-pylab
20. Iliashenko, I., Zucca, V.: Faster homomorphic comparison operations for BGV and BFV. Proc. Priv. Enhancing Technol. **3**, 246–264 (2021). https://petsymposium.org/popets/2021/popets-2021-0046.pdf
21. Kim, S., Omori, M., Hayashi, T., Omori, T., Wang, L., Ozawa, S.: Privacy-preserving naive bayes classification using fully homomorphic encryption. In: Cheng, L., Leung, A.C.S., Ozawa, S. (eds.) ICONIP 2018. LNCS, vol. 11304, pp. 349–358. Springer, Cham (2018). https://doi.org/10.1007/978-3-030-04212-7_30

22. Lee, E., Lee, J.W., Kim, Y.S., No, J.S.: Minimax approximation of sign function by composite polynomial for homomorphic comparison. IEEE Trans. Dependable Secure Comput. (2021). https://doi.org/10.1109/TDSC.2021.3105111
23. Lyubashevsky, V., Micciancio, D., Peikert, C., Rosen, A.: SWIFFT: a modest proposal for FFT hashing. In: Nyberg, K. (ed.) FSE 2008. LNCS, vol. 5086, pp. 54–72. Springer, Heidelberg (2008). https://doi.org/10.1007/978-3-540-71039-4_4
24. Lyubashevsky, V., Peikert, C., Regev, O.: On ideal lattices and learning with errors over rings. In: Gilbert, H. (ed.) EUROCRYPT 2010. LNCS, vol. 6110, pp. 1–23. Springer, Heidelberg (2010). https://doi.org/10.1007/978-3-642-13190-5_1
25. Magara, S.S., et al.: ML with HE: privacy preserving machine learning inferences for genome studies. Technical report 1 (2021). arXiv:2110.11446
26. Onoufriou, G., Mayfield, P., Leontidis, G.: Fully homomorphically encrypted deep learning as a service. Mach. Learn. Knowl. Extract. **3**(4), 819–834 (2021). https://doi.org/10.3390/make3040041
27. Rescorla, E.: The transport layer security (TLS) protocol version 1.3. RFC 8446 (2018). https://www.rfc-editor.org/info/rfc8446
28. The HEBench Organization: HEBench (2022). https://hebench.github.io/
29. Zhang, L., Xu, J., Vijayakumar, P., Sharma, P.K., Ghosh, U.: Homomorphic encryption-based privacy-preserving federated learning in IoT-enabled healthcare system. IEEE Trans. Netw. Sci. Eng. 1–17 (2022). https://doi.org/10.1109/TNSE.2022.3185327

Building Blocks for LSTM Homomorphic Evaluation with TFHE

Daphné Trama(✉), Pierre-Emmanuel Clet, Aymen Boudguiga, and Renaud Sirdey

Université Paris-Saclay, CEA-List, Palaiseau, France
{daphne.trama,pierre-emmanuel.clet,aymen.boudguiga,renaud.sirdey}@cea.fr

Abstract. Long Short-Term Memory (LSTM) is a Neural Network (NN) type that creates temporal connections between its nodes. It models sequence data for applications such as speech recognition, image captioning, DNA sequence analysis, and sentence translation. Applications that are often subject to privacy constraints. This paper thus presents basic building blocks for the homomorphic execution of an LSTM that would respect the privacy of its inputs. By means of TFHE functional bootstrapping, we propose several approaches for homomorphically evaluating discretized flavors of the Sigmoid and Tanh activation functions. Experimental results show that the accuracy of the resulting discretized networks remains comparable to a full precision clear-domain execution. Performance-wise, we are able to homomorphically compute a Sigmoid or Tanh function in 0.3 or 0.15 s (depending on whether or not multivalue bootstrapping is relied on). This paves the way towards evaluating practical LSTMs over encrypted inputs in around 1 to 3 min which is competitive with the state of the art.

Keywords: LSTM · Privacy · TFHE · Homomorphic encryption

1 Introduction

In recent years, Artificial Intelligence techniques and particularly Neural Networks (NN) have become pervasive in our connected society. They have already led to countless practical applications impacting, for better or worse, our daily lives. Examples are numerous: image recognition [23], artwork [27], interactive chat [25], voice analysis [28], and music generation [15]. In this context, protecting user data privacy during the operational phase of already trained neural nets has become a major challenge. In the field of provably-secure cryptography, Fully Homomorphic Encryption (FHE) is one major corpus of techniques serving as a yardstick to address such privacy challenges.

This paper thus focuses on running a special type of neural network, called Recurrent Neural Networks (RNN), over encrypted inputs. These networks rely,

This work was supported by the France 2030 ANR Project ANR-22-PECY-003 SecureCompute.

S. Dolev et al. (Eds.): CSCML 2023, LNCS 13914, pp. 117–134, 2023.
https://doi.org/10.1007/978-3-031-34671-2_9

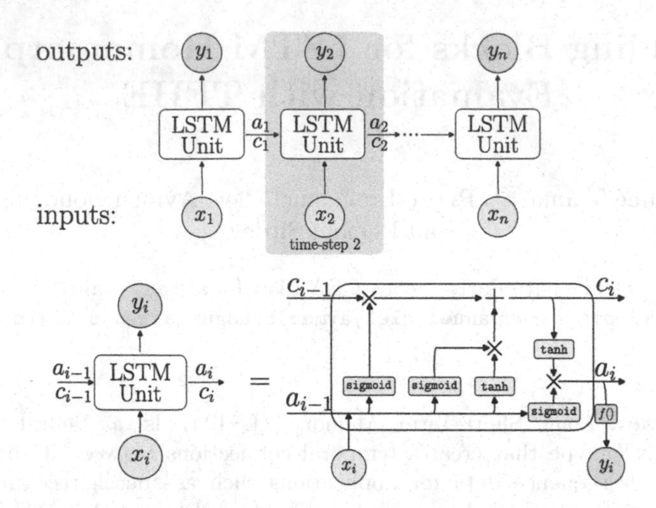

Fig. 1. LSTMs structure: the LSTM unit of time-step i takes as input x_i as well as the outputs of the LSTM unit at time-step $i - 1$, namely the activation a_{i-1} and the memory cell c_{i-1}. It then outputs $y_i = f(a_i)$, where f can be an activation function or simply the identity function.

in particular, on a specific type of unit called Long-Short Term Memory (LSTM), first proposed in 1997 in [18]. The main idea of LSTM is to use memory cells to capture the relation with activations from previous time-steps (as described in Fig. 1). LSTM relies on Sigmoid and Tanh for memory cells and activation computation. These networks are mainly used for text translation, natural language processing, language translation, feeling analysis, and speech analysis.

Still, in these kinds of applications, input confidentiality can be crucial (and so is the confidentiality of any byproduct of these inputs). For instance, it is generally the case when the network manipulates medical data sent for diagnosis findings, genomic data, or sensitive voice data. Furthermore, users (e.g., patients, hospitals, or governments) may wish to prevent neural network owners from accessing their sensitive data and extracting information from it as a byproduct. It is therefore desirable to encrypt these input data and then, as a consequence, evaluate the full neural network in the encrypted domain. This eventually leads to encrypted results which are sent back to the user who, alone, should be granted access to them. Thus, it becomes possible to delegate classification tasks to third parties while protecting the confidentiality of the data. Utopian? Not at all. Indeed, with respect to Convolutional Neural Networks (CNN), starting with the work of Dowlin *et al.* on CryptoNets [14], several works have investigated the application of various flavors of homomorphic encryption techniques to various flavors of neural networks with the long-term goal of achieving the above setup [1,2,6,7,10,19,21,24]. However, these works have so far achieved limited scaling (reaching throughputs ranging from a few hundred to a few thousand neurons per minute) and had to resort to simple activation functions (for instance square, Sign, or ReLU).

In this context, this paper further investigates how the Sigmoid (and hyperbolic tangent Tanh) function can be evaluated in the homomorphic domain, particularly, to run RNNs. On the one hand, RNNs are exciting candidates for FHE since their applications often require confidentiality. On the other hand, contrary to the more classical CNN, RNN experimentally appears much more sensitive to issues relative to discretization and activation function precision. Nevertheless, the other operations necessary for running the network, such as additions or scalar products, have already been studied and are implemented using classical circuits for basis B arithmetic. But Sigmoid and Tanh still need to receive satisfactory treatment from a homomorphic encryption point of view (mainly because they have been worked around in previous studies focusing on CNN). Consequently, this paper primarily focuses on the Sigmoid and Tanh functions as "must-have" building blocks for running RNNs over homomorphically encrypted data. However, these non-linear functions must be approximated to fit the constraints of homomorphic encryption operations[1]. One common approach is to approximate them by polynomials [8,20,26], but this raises a number of issues. First, low-degree polynomial approximations are accurate around 0. Second, with leveled homomorphic schemes such as BFV [3,16] or BGV [4], the multiplicative depth of an LSTM increases unboundedly with the number of successive time-steps because of the recurrent nature of these networks. Thus, even for small LSTMs, the leveled homomorphic scheme parameters would not allow an efficient execution. Meanwhile, for larger LSTMs, leveled homomorphic schemes lead to prohibitive costs and the use of a (practical) bootstrapping-able homomorphic encryption scheme such as TFHE [9] becomes mandatory. However, with TFHE, using activation functions that are approximated with polynomials is very time-consuming. Indeed, with TFHE, we decompose every large input into digits in a basis B before encryption, and we express multiplication and addition as circuits for basis B arithmetic where each unitary operation requires at least one programmable (or functional) bootstrapping.

State of the art– In 2022, Jang *et al.*, [20] implemented an RNN with Gated Recurrent Units[2] (GRUs) with an improved version of the CKKS homomorphic encryption scheme. They enhanced CKKS with the support of multivariate RLWE (m-RLWE) samples. As a result, they were able to encode a matrix of clear values into one plaintext by using batching (then encrypting the obtained multivariate polynomial in one ciphertext). They also provided an efficient algorithm for matrix-vector multiplication and approximated Tanh and Sigmoid with polynomials of degree 7. With all these building blocks, they were able to evaluate RNNs with GRUs over encrypted data (with comparable accuracies) for

[1] Note that, in principle, floating point functions can be performed by means of homomorphic computations (e.g. by running their boolean circuit representations over an FHE with $\mathbb{Z}_2$ as plain domain). In practice, however, such an approach induces prohibitive computational costs.

[2] GRU units are a simple version of LSTM ones. They also rely on Tanh and Sigmoid for computing memory cells and activations. However, they cannot manage very long dependencies.

sequence copy (4 GRUs), regression adding (1 GRU), image classification (28 GRUs), and genomic sequence classification (40 GRUs). For the small RNNs with 1 to 4 GRUs, Jang *et al.*, did not rely on bootstrapping, but they did need a bootstrapping per time-step for larger sequences– 28 to 40 GRUs (starting from the 4th time-step). During their experiments, they compute each unit in 90 s (without including the CKKS bootstrapping time which can range from 3 to 26 min depending on the number of used threads and with respect to the selected parameters in [20]).

Paul *et al.*, [26] proposed a collaborative training between two parties that share the same LSTM in clear form but intend to train an extra logistic regression layer. The authors used homomorphic encryption only for training the logistic regression layer. They confirmed that training and running the LSTM over encrypted data was complex and time-consuming.

Contribution– In this paper, we design Look-Up Tables (LUTs) for evaluating Tanh and Sigmoid using TFHE functional bootstrapping. In order to do so, we propose several approaches to homomorphically evaluate discretized variants of the Sigmoid and Tanh activation functions. We then introduce two step-wise functions as replacements for Sigmoid and Tanh, which we experimentally show are sufficient to preserve the accuracy of a real-world RNN. In terms of performances, we are then able to homomorphically compute a Sigmoid or a Tanh in less than half a second, which is competitive with the state of the art.

Paper Organization– This paper is organized as follows: Sect. 2 reviews the basics of the TFHE cryptosystem and gives the necessary details about LSTMs structure to render this paper self-contained. In Sect. 3 we discuss a number of discretization issues in LSTM. In particular, we evaluate the impact on the accuracy of the network when Sigmoid and Tanh are approximated by different discrete functions. Section 4 provides a detailed exposition of our approaches to executing our approximated activation functions in the FHE domain. We also present the performances of our methods. Section 5 concludes the paper and gives some perspectives.

2 Preliminaries

2.1 Notations

Let $\mathbb{T} = \mathbb{R}/\mathbb{Z}$ be the real torus, that is to say, the additive group of real numbers modulo 1 ($\mathbb{R} \mod 1$). We will denote by $\mathbb{T}_N[X]^n$ the set of vectors of size n whose coefficients are polynomials of $\mathbb{T} \mod (X^N + 1)$. N is usually a power of 2. Let $\mathbb{B} = \{0, 1\}$. $\langle\ ,\ \rangle$ denotes the inner product.

2.2 TFHE Scheme

The TFHE scheme is a fully homomorphic encryption scheme introduced in 2016 in [9] and implemented as the TFHE library[3]. TFHE defines three structures to encrypt plaintexts, which we summarize below as fresh encryptions of 0:

[3] https://tfhe.github.io/tfhe/.

- **TLWE sample**: A pair $(a, b) \in \mathbb{T}^{n+1}$, where a is uniformly sampled from $\mathbb{T}^n$ and $b = \langle a, s \rangle + e$. The secret key s is uniformly sampled from $\mathbb{B}^n$, and the error $e \in \mathbb{T}$ is sampled from a Gaussian distribution with mean 0 and standard deviation σ.
- **TRLWE sample**: A pair $(a, b) \in \mathbb{T}_N[X]^{k+1}$, where a is uniformly sampled from $\mathbb{T}_N[X]^k$ and $b = \langle a, s \rangle + e$. The secret key s is uniformly sampled from $\mathbb{B}_N[X]^k$, the error $e \in \mathbb{T}$ is a polynomial with random coefficients sampled from a Gaussian distribution with mean 0 and standard deviation σ. One usually chooses $k = 1$, therefore a and b are vectors of size 1 whose coefficient is a polynomial.
- **TRGSW sample**: a vector of l TRLWE fresh samples.

Let $\mathcal{M}$ denote the discrete message space ($\mathcal{M} \in \mathbb{T}_N[X]$ or $\mathcal{M} \in \mathbb{T}$). To encrypt a message $m \in \mathcal{M}$, we add what is called a *noiseless trivial* ciphertext $(0, m)$ to a fresh encryption of 0. We denote by $c = (a, b) + (0, m) = (a, b + m) \in$ $\text{T(R)LWE}_s(m)$ the T(R)LWE encryption of m with key s. A message $m \in \mathbb{T}_N[X]$ can also be encrypted in TRGSW samples by adding $m \cdot H$ to a TRGSW sample of 0, where H is a gadget decomposition matrix. As we will not implicitly need such an operation in this paper, more details about TRGSW can be found in [9].

To decrypt a ciphertext c, we first calculate its phase $\phi(c) = b - \langle a, s \rangle = m + e$. Then, we need to remove the error, which is achieved by rounding the phase to the nearest valid value in $\mathcal{M}$. This procedure fails if the error is greater than half the distance between two elements of $\mathcal{M}$.

2.3 TFHE Bootstrapping

Bootstrapping is the operation that reduces the noise of a ciphertext thus allowing further homomorphic calculations. It relies on three basic operations, which we briefly review in this section.

- **BlindRotate**: rotates a polynomial encrypted as a TRLWE ciphertext by an encrypted index. It takes several inputs: a ciphertext $c \in \text{TRLWE}_k(m)$, a vector $(a_1, \cdots, a_p, b)$ where $\forall i, a_i \in \mathbb{Z}_{2N}$, and a TRGSW ciphertext encrypting the secret key $s = (s_1, \cdots, s_p)$.
  It returns a ciphertext $c' \in \text{TRLWE}_k(m \cdot X^{\langle a,s \rangle - b})$. In this paper, we will refer to this algorithm as `BlindRotate`.
- **TLWE Sample Extract**: extracts a coefficient of a TRLWE ciphertext and converts it into a TLWE ciphertext. It takes as inputs both a ciphertext $c \in \text{TRLWE}_k(m)$ and an index $p \in \{0, \cdots, N-1\}$. The result is a TLWE ciphertext $c' \in \text{TLWE}_k(m_i)$ where m_i is the ith coefficient of the polynomial m. In this paper, we will refer to this algorithm as `SampleExtract`.
- **Public Functional Keyswitching**: allows the switching of keys and parameters from p ciphertexts $c_i \in \text{TLWE}_k(m_i)$ to one $c' \in$ $\text{T(R)LWE}_s(f(m_1, \cdots, m_p))$ where f is a public linear morphism between $\mathbb{T}^p$ and $\mathbb{T}_N[X]$. That is to say, this operation not only allows the packing of TLWE ciphertexts in a TRLWE ciphertext, but it can also evaluate a linear function f over the input TLWEs. In this paper, we will refer to this algorithm as `KeySwitch`.

It is important to note that, during a `BlindRotate` operation, an excessive noise level in the input ciphertext can lead to errors in the bootstrapping output resulting in incorrect ciphertexts (i.e. ciphertext which do not decrypt to correct calculation results). This has implications for parameters and data representations choices (number of digits and basis), as we shall later see in Sect. 4.

Algorithm 1 shows the *TFHE Gate Bootstrapping* [9], which aims to evaluate a binary gate operation homomorphically and reduce the output ciphertext noise at the same time. To that end, 0 and 1 are respectively encoded as 0 and $\frac{1}{2}$ over $\mathbb{T}$. The first step of this algorithm consists in selecting a value $\hat{m} \in \mathbb{T}$ which will be used afterward to compute the coefficients of the polynomial which will rotate during the `BlindRotate`. We call this polynomial *testv* as seen in Step 3. Note that for any $p \in [\![0, 2N]\!]$ (where $[\![0, 2N]\!]$ corresponds to the set of integers $\{0, \cdots, 2N\}$), the constant term of $testv \cdot X^p$ is $\hat{m}$ if $p \in]\!]\frac{N}{2}, \frac{3N}{2}]\!]$ and $-\hat{m}$ otherwise. Step 4 returns an accumulator $ACC \in \text{TRLWE}_{s'}(testv \cdot X^{\langle \overline{a}, s\rangle - b})$. Indeed, the constant term of ACC is $-\hat{m}$ if c is an encryption of 0 and $\hat{m}$ if c is an encryption of $\frac{1}{2}$. Then step 5 creates a new ciphertext $\overline{c}$ by extracting the constant term in position 0 from ACC and adding $(0, \hat{m})$. Thus, $\overline{c}$ corresponds to an encryption of 0 if c is an encryption of 0 and m otherwise. On the other hand, if c is an encryption of $\frac{1}{2}$ and if we choose $m = \frac{1}{2}$, the algorithm returns a *fresh sample* of $\frac{1}{2}$, that is to say the encoding of 1.

In Fig. 2, we present an example of TFHE gate bootstrapping algorithm with $\mathbb{Z}_4 = \{0, 1, 2, 3\}$ as input space. The outer circle in Fig. 2 corresponds to the plaintext encoding in $\mathbb{T}$ as $\{0, \frac{1}{4}, \frac{2}{4}, \frac{3}{4}\}$. Meanwhile, the inner circle sets the coefficients of the test polynomial *testv* to 1, i.e., $\hat{m} = \frac{1}{4}$. Then, we rotate the test polynomial during the bootstrapping by the phase $\phi(c_0)$ of the input ciphertext c_0. In our example, we obtain as bootstrapping output either an encryption of the encoding of 1 for positive inputs $\{0, \frac{1}{4}\}$, or an encryption of the encoding of -1 for negative inputs $\{\frac{2}{4}, \frac{3}{4}\}$.

Algorithm 1. TFHE gate bootstrapping [9]

Require: a constant $m \in \mathbb{T}$, a TLWE sample $c = (a, b) \in \text{TLWE}_s(x \cdot \frac{1}{2})$ with $x \in \mathbb{B}$, a bootstrapping key $BK_{s \to s'} = (BK_i \in \text{TRGSW}_{S'}(s_i))_{i \in [\![1,n]\!]}$ where S' is the TRLWE interpretation of a secret key s'.

Ensure: a TLWE sample $\overline{c} \in \text{TLWE}_s(x.m)$

1: Let $\hat{m} = \frac{1}{2}m \in \mathbb{T}$ (pick one of the two possible values)
2: Let $\overline{b} = \lfloor 2Nb \rceil$ and $\overline{a_i} = \lfloor 2Na_i \rceil \in \mathbb{Z}, \forall i \in [\![1, n]\!]$
3: Let $testv := (1 + X + \cdots + X^{N-1}) \cdot X^{\frac{N}{2}} \cdot \hat{m} \in \mathbb{T}_N[X]$
4: $ACC \leftarrow$ `BlindRotate`$((0, testv), (\overline{a}_1, \ldots, \overline{a}_n, \overline{b}), (BK_1, \ldots, BK_n))$
5: $\overline{c} = (0, \hat{m}) +$ `SampleExtract`(ACC)
6: return `KeySwitch`$_{s' \to s}(\overline{c})$

2.4 TFHE Functional Bootstrapping

We can use Look-Up Tables to compute functions during the bootstrapping operation. To do so, we replace the coefficients of the test polynomial *testv* with

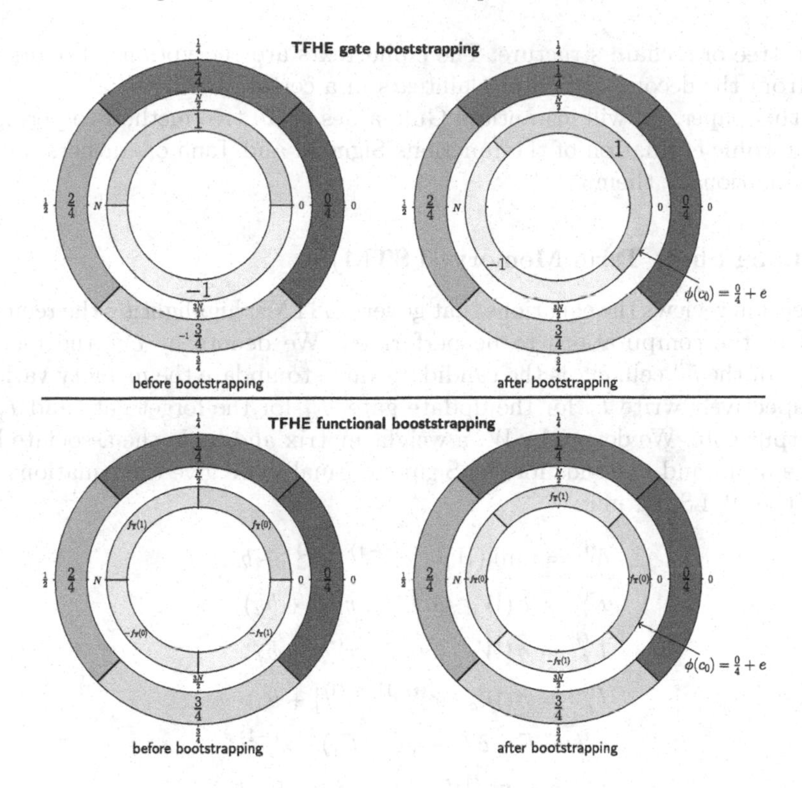

Fig. 2. TFHE Bootstrapping examples: the outer circles describe the inputs to the bootstrapping (i.e., ciphertexts over $\mathbb{T}$). Meanwhile, the inner circles represent the coefficients of the test polynomial *testv*. One of these coefficients is extracted as the output of the bootstrapping after the `BlindRotate`.

the corresponding values of the LUT. Let us assume that we want to evaluate the function $f_{\mathbb{T}}$ via a LUT. Then, if we retrieve the ith coefficient of *testv*, we actually get $f_{\mathbb{T}}(m_i)$ where m_i is the encrypted input to the bootstrapping. We refer to this idea by *programmable* or *functional* bootstrapping [11–13, 22, 30].

In Fig. 2, we give an example of functional bootstrapping with $\mathbb{Z}_4 = \{0,1,2,3\}$ as input space. We encode the images of $\{0, \frac{1}{4}\}$ by $f_{\mathbb{T}}$ as coefficients of the test polynomial (in the inner circle). Meanwhile, we deduce the images of $\{\frac{2}{4}, \frac{3}{4}\}$ by negacyclicity. Indeed, in $\mathbb{T}$, we can encode negacyclic functions i.e., antiperiodic functions with period $\frac{1}{2}$ (verifying $f_{\mathbb{T}}(x) = -f_{\mathbb{T}}(x + \frac{1}{2})$), where $[0, 0.5[$ corresponds to positive values and $[0.5, 1[$ to negative ones. In our example, if we encrypt one of the following values $\{0, \frac{1}{4}, \frac{2}{4}, \frac{3}{4}\}$ and we give it as input to the functional bootstrapping algorithm, we get $\{f_{\mathbb{T}}(0), f_{\mathbb{T}}(1), -f_{\mathbb{T}}(0), -f_{\mathbb{T}}(1)\}$, respectively.

Almost all of the functional bootstrapping methods from state of the art [11–13, 22, 30] take as input a single ciphertext. In 2021, Guimarães *et al.*, [17] discussed two methods for performing functional bootstrapping with larger plaintexts. They combine several bootstrappings with different encrypted inputs by using a tree or a chain structure. The ciphertexts are encryptions of digits that come from the decomposition of plaintexts in a certain basis B.

In this paper, we will instantiate Guimarães *et al.* tree method to perform a homomorphic evaluation of the functions Sigmoid and Tanh or, rather, suitable approximations of them.

2.5 Long Short Term Memory (LSTM)

This section reviews the equations that govern LSTMs, highlighting the recurrent aspect of the computations to be performed. We denote by $c^{\langle i \rangle}$ the memory variable of the ith cell. $\tilde{c}^{\langle i \rangle}$ is the candidate value to update the memory variable. We respectively write Γ_u for the update gate, Γ_f for the forget gate and Γ_o for the output gate. We denote by W_j a weight matrix and by b_j the associate bias. x is the input and σ stands for the Sigmoid. Finally, we give the equations that define the ith LSTM unit:

$$\begin{aligned}
\tilde{c}^{\langle i \rangle} &= \tanh(W_c \cdot [c^{\langle i-1 \rangle}, x^{\langle i \rangle}] + b_c) \\
\Gamma_u^{\langle i \rangle} &= \sigma(W_u \cdot [c^{\langle i-1 \rangle}, x^{\langle i \rangle}] + b_u) \\
\Gamma_f^{\langle i \rangle} &= \sigma(W_f \cdot [c^{\langle i-1 \rangle}, x^{\langle i \rangle}] + b_f) \\
\Gamma_o^{\langle i \rangle} &= \sigma(W_o \cdot [c^{\langle i-1 \rangle}, x^{\langle i \rangle}] + b_o) \\
c^{\langle i \rangle} &= \Gamma_u \cdot \tilde{c}^{\langle i \rangle} + (1 - \Gamma_u) \cdot c^{\langle i-1 \rangle} \\
a^{\langle i \rangle} &= \tanh(W_a \cdot [a^{\langle i-1 \rangle}, x^{\langle i \rangle}] + b_a)
\end{aligned}$$

To perform a confidentiality-preserving LSTM evaluation, all these different operations must be done in the homomorphic domain. In this paper, as a necessary first step towards running RNNs over FHE, we focus mainly on the two building blocks that are Sigmoid and Tanh as they are in need to receive satisfactory treatment from a homomorphic encryption point of view.

3 Discretization Issues in LSTM

In this work, we aim at running LSTM building blocks with TFHE functional bootstrapping. First, we decompose our input data into a certain basis B before their encryption (as TFHE functional bootstrapping does not support large plaintexts [13]). Then, we discretize our functions to encode them as a LUT in the test polynomial *testv*. We propose to work with a practical use case that allows studying the loss of accuracy due to discretization.

3.1 Use Case

We choose to work on the LSTM proposed by J. Woodbridge's team in [29]. Indeed, their LSTM takes as input a domain name and determines whether it is a malicious domain name, which seems appropriate to our study because the LSTM is quite long and represents real-world usage.

The network is composed of several layers: an embedding layer, an LSTM layer, and a fully connected layer. The LSTM layer is composed of 128 LSTM units, which we discretized to support an evaluation with homomorphic encrypted inputs. To do so, we successfully reproduced their experimental results and computed the accuracy of the network: 95.6%. We use this accuracy value as a reference when experimenting with our discretized variants.

3.2 Coping with TFHE Clear Domain Size Constraints

The first step to make the network homomorphic consists in discretizing the different data. Indeed, let us recall that the ciphertext domain $\mathcal{M}$ in TFHE is discrete, generally of size $p = 16$ or $p = 32$ whereas the inputs of the LSTM layer usually are floating-point numbers. For a fully homomorphic evaluation of the network, these inputs must be encrypted, hence discretized on p values. In our case, it means selecting the p most relevant floating-point values and assigning these values as *interpretations* of $\mathcal{M} = \{0, \frac{1}{p}, \frac{2}{p}, \cdots, \frac{p-1}{p}\}$. To put it another way, one has to choose how to encode p values of $\mathbb{R}$ into $\mathcal{M}$. So we have to find a bijection $\iota : F \rightarrow \mathcal{M}$, where F is a set of p floating-point numbers. It is the values of F that we have to determine adequately. Here "adequately" means that by replacing the inputs (recall that in our network they correspond to the output of an embedding layer) with the values found for F, the network keeps the same predictions as on the original inputs. To do this, we need to start by looking at what the inputs of the LSTM layer look like in typical execution. Thus, the coefficients of the network inputs are all between –1 and 1. So none of the p values will be chosen outside $[-1, 1]$. Looking further, we notice that their distribution is not uniform. So choosing $F = \{-1 + \frac{2i}{p} \mid i \in \{0, \cdots, p-1\}\}$ is not appropriate. We then thought of the k-means clustering method.

K-means clustering is a vector quantization method that aims to partition n elements into k clusters in which each element belongs to the cluster with the nearest mean. We used this method reduced to one dimension on our set of inputs in order to determine p clusters. We then pick out the center of each cluster to obtain our set F. We replace each input coefficient with the value of its cluster center in F and run the tests. As we observe that the resulting network accuracy does not change with such a strong discretization, we then apply the same approach to the coefficients of the weight matrices and bias vectors. Again, despite such a drastic data discretization, the tests performed on the new discretized network show that these simplifications do not influence the final accuracy. Indeed, the discretized network also achieves 95.6% of correct predictions. The discretization prerequisites towards a homomorphic execution, therefore, appear to be met.

3.3 A Naive Discretization of the Activation Functions

As it stands, TFHE does not allow, even with programmable bootstrapping, to practically evaluate functions such as Sigmoid or Tanh. Indeed, efficiently evaluating non-linear functions with high precision remains a challenge for FHE schemes. A naive approach is then to coarsely approximate the activation functions by much simpler step functions. Such approaches have already been carried out on CNNs [2], with encouraging results, but never on RNNs. It is thus legitimate to start by trying a similar approach for LSTM.

For the Sigmoid which has values in $]0,1[$, we can choose the Heaviside function, while for the Tanh with values in $]-1,1[$, we can choose the Sign function. These are (very) rough approximations, with two steps, but the question is whether the network can endure such a simplification of its activation functions. We first perform discretization tests with clear data with these naive approximations. Contrary to the CNN case, the results proved to be negatively conclusive: discretizing the activation functions in only two steps leads to a significant loss of accuracy. Indeed, as shown in the first rows of Table 1, the network accuracy drops to 0.50.

3.4 A Finer Stepwise Approach

More accurate approximations can be obtained by means of stepwise approximation as initially suggested by Guimarães *et al.* [17] who presented a 3-steps approximation of the Sigmoid for illustrative purpose and gave hints at how to implement it by means of functional bootstrapping. We call these variants 3StepsSigmoid and 3StepsTanh.

Thus, we first considered this approximation but the results, shown in Table 1, although better than those obtained in the previous section, lead to an unacceptable 10 points loss in network accuracy. We thus investigated a more precise approximation of our own with two steps on the interval $]-\infty,-1]$, an affine part on $]-1,1]$, and two more steps on $]-1,\infty[$. This leads to the following functions, corresponding to the curves presented in Fig. 3.

$$\begin{aligned}\mathsf{StepSigmoid}(x) &= 0.13\cdot\mathbb{1}_{]-6,-1]}(x) \;+\; \mathbb{1}_{]-1,1]}(x)\cdot(0.24x+0.5) \;+\; 0.87\cdot\mathbb{1}_{]1,6]}(x)\\ &\quad+\;\cdot\,\mathbb{1}_{]6,\infty[}(x)\\ \mathsf{StepTanh}(x) &= -1\cdot\mathbb{1}_{]-\infty,-3]}(x) \;-\; 0.875\cdot\mathbb{1}_{]-3,-1]}(x) \;+\; \mathbb{1}_{]-1,1]}(x)\cdot(0.76x)\\ &\quad+\; 0.92\cdot\mathbb{1}_{]1,6]}(x) \;+\; \cdot\,\mathbb{1}_{]6,\infty[}(x).\end{aligned}$$

where $\mathbb{1}_F$ is the indicator function and F is a subset of a set E.

$$\mathbb{1}_F : E \to \{0,1\}$$
$$x \mapsto \begin{cases} 1 \text{ if } x \in F\\ 0 \text{ otherwise.}\end{cases}$$

Again, to determine the optimized values for the steps of our functions, we used the k-means clustering method (as in Sect. 3.3) on the clear outputs of Sigmoid and Tanh, leading us to an optimized discretized version of the activation functions. Using these two approximations to replace the Sigmoid and Tanh functions, we obtain a final accuracy of the network of 93.4%, only 2% below the network accuracy under floating-point precision.

Lastly, to allow a homomorphic evaluation under TFHE, it is also necessary to discretize the affine part of our StepSigmoid and StepTanh in 12 intermediate steps. Indeed, to minimize the degradation of the accuracy, it is necessary

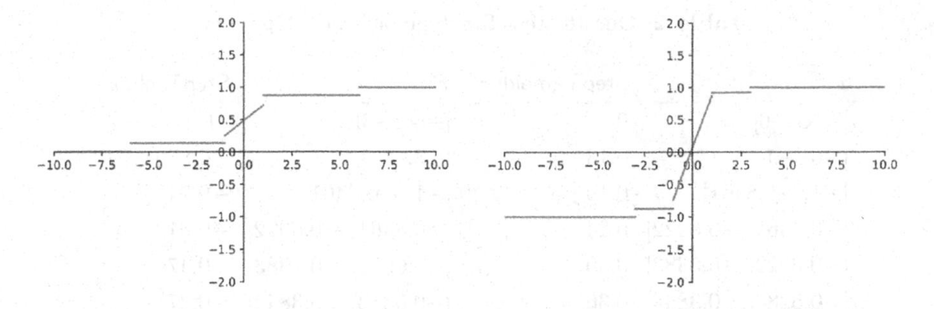

Fig. 3. Our stepwise activation functions StepSigmoid and StepTanh.

Table 1. Accuracy of the network depending on the activation functions.

used functions	accuracy of the LSTM
Tanh + Sigmoid	0,956
Sign + Heaviside	0,50
3StepsTanh + 3StepsSigmoid [17]	0,832
StepTanh + StepSigmoid	0,934

to have a significant number of steps between −1 and 1 because the prediction of the network often relies on this interval of values. Thus, we obtain activation functions composed of 16 steps, corresponding to the $p = 16$ values of the ciphertext space $\mathcal{M}$. The final discretization is given in Table 2. The accuracy of the network remains unchanged, with a rate of 93.4% of correct predictions.

Thus with our proposed approximated functions StepSigmoid and StepTanh, we have a network for which we can investigate whether a homomorphic evaluation is doable.

4 FHE Execution

This section presents the tree-based method, first introduced in [17]. This approach uses the output of a LUT to construct a new LUT, and thus allows the evaluation of functions by means of multiple functional bootstrappings. Each of these bootstrapping has as input a ciphertext encrypting a plaintext in a basis B.

4.1 Instantiation of the Tree-Based Method for the Sigmoid

We now attempt to homomorphically evaluate StepSigmoid and StepTanh by means of a tree-based bootstrapping method.

Let $B, B', d \in \mathbb{N}^*$ and m be an integer message. B and B' are the basis on which to decompose the message. We then have $m = \sum_{i=0}^{d-1} m_i B^i$, with $m_i \in [\![0, B-1]\!]$. From this decomposition, we obtain d TLWE encryptions

Table 2. Our 16-steps StepSigmoid and StepTanh.

x	StepSigmoid(x)	x	StepTanh(x)
]−∞, −6]	0	]−∞, −3]	−1
]−6, −1]	0.13	]−3, −1]	−0.875
]−1, −0.8461]	0.19	]−1, −0.8461]	−0.74
]−0.8461, −0.6922]	0.24	]−0.8461, −0.6922]	−0.61
]−0.6922, −0.5383]	0.30	]−0.6922, −0.5383]	−0.47
]−0.5383, −0.3844]	0.36	]−0.5383, −0.3844]	−0.37
]−0.3844, −0.2305]	0.41	]−0.3844, −0.2305]	−0.20
]−0.2305, −0.0766]	0.47	]−0.2305, −0.0766]	−0.07
]−0.0766, 0.0772]	0.53	]−0.0766, 0.0772]	0.07
]0.0772, 0.2310]	0.56	]0.0772, 0.2310]	0.20
]0.2310, 0.3848]	0.64	]0.2310, 0.3848]	0.33
]0.3848, 0.5386]	0.7	]0.3848, 0.5386]	0.47
]0.5386, 0.6924]	0.76	]0.5386, 0.6924]	0.60
]0.6924, 0.8462]	0.81	]0.6924, 0.8462]	0.74
]0.8462, 1]	0.87	]0.8462, 1]	0.875
]1, 6]	0.93	]1, 3]	0.92
]6, +∞]	1	]3, +∞]	1

$(c_0, c_1, \cdots, c_{d-1})$ of $(m_0, m_1, \cdots, m_{d-1})$ on half of the torus $\mathbb{T}$. We denote $f : [\![0, B-1]\!]^d \rightarrow [\![0, B'-1]\!]$ the target function and define g as the following bijection:

$$\begin{aligned} g : [\![0, B-1]\!]^d &\rightarrow [\![0, B^d - 1]\!] \\ (a_0, a_1, \cdots, a_{d-1}) &\mapsto \textstyle\sum_{i=0}^{d-1} a_i \cdot B^i \end{aligned}$$

We then encode a LUT for f under the form of B^{d-1} TRLWE ciphertexts. Each of these ciphertexts encodes a polynomial P_i such that:

$$P_i(X) = \sum_{j=0}^{B-1} \sum_{k=0}^{\frac{N}{B}-1} f \circ g^{-1}(j \cdot B^{d-1} + i) \cdot X^{j \cdot \frac{N}{B} + k}$$

Then, we apply the `BlindRotateAndExtract` (the `BlindRotate` directly followed by the `SampleExtract` in position 0) to each test polynomial $testv$ = TRLWE(P_i) with c_0 as a selector. We obtain B^{d-1} TLWE ciphertexts, each corresponding to the encryption of $f \circ g^{-1}(m_{d-1} \cdot B^{d-1} + i)$, for $i \in [\![0, B^{d-1} - 1]\!]$.

Finally, we use the `KeySwitch` operation from TLWE to TRLWE to gather them into B^{d-2} encrypted TRLWE, corresponding to the LUT of h, with:

$$\begin{aligned} h : [\![0, B-1]\!]^{d-1} &\rightarrow [\![0, B'-1]\!] \\ (a_0, a_1, \cdots, a_{d-2}) &\mapsto f(a_0, a_1, \cdots, a_{d-2}, m_{d-1}) \end{aligned}$$

We then repeat this operation, using the ciphertext c_i at step i, until we obtain a single TLWE ciphertext of $f(m_0, m_1, \cdots, m_{d-1})$. Note that the tree-based method must be run independently as many times as the number of digits in the output.

With our plaintext space of size $p = 16$, we opt for a decomposition in a basis $B = 4$, and so we get $d = 2$. That is, the input of our bootstrapping StepSigmoid corresponds to encryptions of two integers in $\mathbb{Z}_4$. With this setting, we get a better network accuracy (93.4%) than when using a single integer in basis 16 (90.1%) thanks to lower bootstrapping error rates for the same parameters set. Indeed, the values of the test polynomial *testv* are spread over several vectors rather than encoded into one, which results in a lower noise tolerance from the `BlindRotate` (as discussed in Sect. 2.3). We choose to output one ciphertext in $\mathbb{Z}_{16}$ ($B' = 16$): to avoid running the tree-based method multiple times. With a one-digit output, we thus need only a single execution.

4.2 Multi-value Bootstrapping

Multi-Value Bootstrapping (MVB) [5] refers to the method for evaluating k different LUTs on a single input with a single bootstrapping. MVB factors the test polynomial P_{f_i} associated with the function f_i into a product of two polynomials v_0 and v_i, where v_0 is a common factor to all P_{f_i}. In practice, we have:

$$(1 + X + \cdots + X^{N-1}) \cdot (1 - X) \equiv 2 \bmod (X^N + 1)$$

Now by writing P_{f_i} in the form $P_{f_i} = \sum_{j=0}^{N-1} \alpha_{i,j} X^j$ with $\alpha_{i,j} \in \mathbb{Z}$, we get from the previous equation:

$$\begin{aligned} P_{f_i} &= \frac{1}{2} \cdot (1 + X + \cdots + X^{N-1}) \cdot (1 - X) \cdot P_{f_i} \bmod (X^N + 1) \\ &= v_0 \cdot v_i \bmod (X^N + 1) \end{aligned}$$

where:

$$\begin{aligned} v_0 &= \frac{1}{2} \cdot (1 + X + \cdots + X^{N-1}) \\ v_i &= \alpha_{i,0} + \alpha_{i,N-1} + (\alpha_{i,1} - \alpha_{i,0}) \cdot X + \cdots + (\alpha_{i,N-1} - \alpha_{i,N-2}) \cdot X^{N-1}. \end{aligned}$$

This factorization makes it possible to compute many LUTs using a unique bootstrapping. Indeed, it is enough to initialize the test polynomial *testv* with the value of v_0 during bootstrapping. Then, after the `BlindRotate` operation, one has to multiply the obtained ACC by each v_i corresponding to the LUT of f_i to get ACC_i.

This optimization reduces the number of bootstrapping required for an operation and, thus, the overall computation time. In our case, the MVB allows us to reduce the number of bootstrapping from 5 to 2 when evaluating either the Sigmoid or the Tanh.

4.3 Technical Setup

We did our experiments with the default parameters of TFHElib[4]. The library gives $N = 1024$ and $n = 630$ so that $a \in \mathbb{T}^{630}$ (TLWE) or $a \in \mathbb{T}_{1024}[X]$ (TRLWE). We run our code on a 4-core Intel Core i7-7600U 2.90 GHz CPU (*with only one core activated*) and 16GiB total system memory with a Ubuntu 20.04.5 LTS server.

In Sect. 3, we successfully discretized Sigmoid into StepSigmoid by encoding the Sigmoid domain on 16 distinct values. In practice, we set $p = 32$ to be able to encode these 16 values on the positive half of the torus (i.e. $[0, 0.5[$). As such, we ensure that our plaintext space is included in $[0, 0.5[$. So, we are no more limited by negacyclic constraints [13].

Again, our StepSigmoid returns 16 positive values: $\{0.53, 0.56, 0.64, 0.7, 0.76, 0.81, 0.87, 1.0, 0.47, 0.41, 0.36, 0.3, 0.24, 0.19, 0.13, 0\} \in [0, 1]$. These values will compose the test polynomial *testv*. As we only work with the 16 values of the positive half of the torus, i.e., $\{\frac{0}{32}, \cdots, \frac{15}{32}\}$ corresponding to values in $[0, 0.5[$, we must find a way to also return values in $[0.5, 1[$ to reach all the images of our StepSigmoid. To solve this issue, we divide by 2 the values of the original *testv*, which are all now ≤ 0.5: $\{0.265, 0.28, 0.32, 0.35, 0.38, 0.405, 0.435, 0.5, 0.235, 0.205, 0.18, 0.15, 0.12, 0.095, 0.065, 0\}$. Indeed, these coefficients all correspond to values in $[0, 0.5[$. When used as coefficients for the test polynomial *testv*, it allows the algorithm to be run only on the positive half of the torus. Thanks to this, the final decryption returns a value in $\{\frac{0}{32}, \frac{1}{32}, \cdots, \frac{15}{32}\} \in [0, 0.5[$. Then, to cover the whole image of our StepSigmoid, we (homomorphically) multiply this result by 2, which gives us a value in $\{\frac{0}{16}, \frac{1}{16}, \cdots, \frac{15}{16}\}$ this time corresponding to values in $[0, 1]$. Nevertheless, the returned value does not necessarily correspond to the desired result (as TFHE will only return values in $\mathcal{M}$). Still, it is close enough (standard deviation of 0.01556) to preserve the accuracy of the network. Indeed, by replacing the desired values with the returned ones, the accuracy of the network remains identical (93.4% of accurate predictions). Additionally, we have to choose the interpretation of the input values. For simplicity's sake, we decided to take the antecedents of the values given by StepSigmoid for creating a bijection between these values and the values in $\{\frac{0}{32}, \cdots, \frac{15}{32}\}$. The resulting implementation is detailed in Table 3.

4.4 Performance Results

When attempting to run an RNN (or any other type of network) in the homomorphic domain, two metrics are important: the execution time and the consequences of the required approximations on the final accuracy of the network. In our case, with respect to performance, we obtained an average execution time of 0.28 secs for a single evaluation of our StepSigmoid via the tree-based method (Sect. 4.1) and of only 0.15 s with the MVB (Sect. 4.2). As we only change the

[4] https://tfhe.github.io/tfhe/ (v1.0.1-36-gbc71bfa).

values of the test polynomial *testv* to evaluate our StepTanh, the execution times are identical for both functions.

With respect to the accuracy, by testing our network in the plaintext domain by replacing the return values of the StepSigmoid and StepTanh with the values obtained via our homomorphic execution, the network maintains its accuracy of 93.4%. This means that the approximations chosen for the Sigmoid and Tanh are good-enough and do not significantly impact the overall accuracy of the network.

This work is a first step. However, an issue remains. Considering our non-standard interpretation of the values in $\{0, \frac{1}{16}, \frac{2}{16}, \cdots, \frac{15}{16}\}$ in the activation function outputs means that the subsequent addition and multiplication operations cannot be performed directly. Thus, several perspectives have to be investigated for the complete execution of the network. We can either switch back to the standard interpretation at carefully chosen points during the network execution (which may cost an additional bootstrapping per conversion) or propose new LUT-based operators for performing additions and multiplications directly in the non-standard interpretation (or a combination of both). Both approaches will incur additional bootstrappings, which have not been considered in our coarse-grained estimations.

Table 3. Evaluation of the homomorphic StepSigmoid.

Value in $\mathcal{M} \times 32$	Clear interpretation	Desired result	Achieved result
0	0.12	0.53	0.5625
1	0.24	0.56	0.5625
2	0.58	0.64	0.625
3	0.85	0.70	0.6875
4	1.15	0.76	0.75
5	1.45	0.81	0.75
6	1.90	0.87	0.875
7	7.0	1.0	1.0
8	−0.12	0.47	0.5
9	−0.36	0.41	0.4375
10	−0.58	0.36	0.375
11	−0.85	0.30	0.3125
12	−1.15	0.24	0.25
13	−1.45	0.19	0.1875
14	−1.90	0.13	0.125
15	−7.0	0.0	0.0

Table 4. Execution time results (number of bootstrappings are provided for informational purposes, as bootstrapping is the most costly operation in TFHE).

	number of bootstrapping		execution time	
	tree-based method	MVB	tree-based method	MVB
single StepSigmoidexecution	5	2	0.28 s	0.15 s
complete LSTM unit execution	$5 \times 5 = 25$	$2 \times 5 = 10$	1.4 s	0.75 s
complete network execution	$128 \times 25 = 3200$	$128 \times 10 = 1280$	179 s = 2 min 59 s	96 s = 1 min 36 s

5 Conclusion and Perspectives

First, we have established that, unlike CNNs, LSTMs do not support rough approximations of their activation functions. The pair (Sign, Heaviside) is thus to be banished from LSTM implementations over FHE, at least without new LSTM cell designs. Moreover, our results illustrate that the discretization of the inputs and the internal coefficients of LSTMs (weight matrices and bias vectors) does not raise any issue with respect to network precision. Second, we propose approximated flavors of the activation functions of LSTMs. As shown by our experimental results applied to the Sigmoid, the accuracy of the network remains unchanged. Moreover, the proposed approximations are relatively efficient when evaluated over FHE, as it, for example, only takes a few hundredths of a second for one StepSigmoid evaluation. Finally, these unitary timings allow us to estimate the (sequential) time needed for the complete execution of our reference 128 LSTM units network using our homomorphic activation functions. The results can be found in Table 4. The given times are coarsely estimated from the execution time of a single evaluation of the activation functions. If we only count activation functions evaluation times, we obtain for the 128 LSTM units a total FHE execution time of 3 min using the tree-based method or 1.5 min with the MVB optimization. This gives an order of magnitude of the time required for an FHE evaluation of the complete network.

Our results are promising and open the door to an end-to-end TFHE evaluation of LSTMs with practical latencies.

References

1. Aharoni, E., et al.: Tile tensors: a versatile data structure with descriptive shapes for homomorphic encryption. CoRR abs/2011.01805 (2020). arXiv:2011.01805
2. Bourse, F., Minelli, M., Minihold, M., Paillier, P.: Fast homomorphic evaluation of deep discretized neural networks. Technical Report Report 2017/1114, IACR Cryptology ePrint Archive (2017). https://hal.science/hal-01665330
3. Brakerski, Z.: Fully homomorphic encryption without modulus switching from classical GapSVP. In: Safavi-Naini, R., Canetti, R. (eds.) Advances in Cryptology - CRYPTO 2012, pp. 868–886. Springer, Berlin (2012). https://doi.org/10.1007/978-3-642-32009-5_50

4. Brakerski, Z., Gentry, C., Vaikuntanathan, V.: (Leveled) fully homomorphic encryption without bootstrapping. In: Proceedings of the 3rd Innovations in Theoretical Computer Science Conference, pp. 309–325. ITCS '12, Association for Computing Machinery, New York, NY, USA (2012). https://doi.org/10.1145/2090236.2090262
5. Carpov, S., Izabachène, M., Mollimard, V.: New techniques for multi-value input homomorphic evaluation and applications. Cryptology ePrint Archive, Paper 2018/622 (2018). https://eprint.iacr.org/2018/622
6. Chabanne, H., Lescuyer, R., Milgram, J., Morel, C., Prouff, E.: Recognition over encrypted faces: 4th International Conference, MSPN 2018, Paris (2019)
7. Chabanne, H., de Wargny, A., Milgram, J., Morel, C., Prouff, E.: Privacy-preserving classification on deep neural network. Cryptology ePrint Archive, Report 2017/035 (2017)
8. Cheon, J.H., Kim, A., Kim, M., Song, Y.: Homomorphic encryption for arithmetic of approximate numbers (2017)
9. Chillotti, I., Gama, N., Georgieva, M., Izabachène, M.: TFHE: fast fully homomorphic encryption library (2016). https://tfhe.github.io/tfhe/
10. Chillotti, I., Joye, M., Paillier, P.: Programmable bootstrapping enables efficient homomorphic inference of deep neural networks. Cryptology ePrint Archive, Paper 2021/091 (2021). https://doi.org/10.1007/978-3-030-78086-91, https://eprint.iacr.org/2021/091
11. Chillotti, I., Joye, M., Paillier, P.: Programmable bootstrapping enables efficient homomorphic inference of deep neural networks. In: Dolev, S., Margalit, O., Pinkas, B., Schwarzmann, A. (eds.) Cyber Security Cryptography and Machine Learning, pp. 1–19. Springer International Publishing, Cham (2021). https://doi.org/10.1007/978-3-030-78086-9_1
12. Chillotti, I., Ligier, D., Orfila, J.B., Tap, S.: Improved programmable bootstrapping with larger precision and efficient arithmetic circuits for TFHE. Cryptology ePrint Archive, Report 2021/729 (2021), https://ia.cr/2021/729
13. Clet, P.E., Zuber, M., Boudguiga, A., Sirdey, R., Gouy-Pailler, C.: Putting up the swiss army knife of homomorphic calculations by means of tfhe functional bootstrapping. Cryptology ePrint Archive, Paper 2022/149 (2022). https://eprint.iacr.org/2022/149
14. Dowlin, N., Gilad-Bachrach, R., Laine, K., Lauter, K., Naehrig, M., Wernsing, J.: Cryptonets: applying neural networks to encrypted data with high throughput and accuracy (2016), https://www.microsoft.com/en-us/research/publication/cryptonets-applying-neural-networks-to-encrypted-data-with-high-throughput-and-accuracy/
15. Dua, M., Yadav, R., Mamgai, D., Brodiya, S.: An improved RNN-LSTM based novel approach for sheet music generation (2020). https://doi.org/10.1016/j.procs.2020.04.049
16. Fan, J., Vercauteren, F.: Somewhat practical fully homomorphic encryption. Cryptology ePrint Archive, Report 2012/144 (2012). https://ia.cr/2012/144
17. Guimarães, A., Borin, E., Aranha, D.F.: Revisiting the functional bootstrap in TFHE. IACR Transactions on Cryptographic Hardware and Embedded Systems 2021(2), 229–253 (2021). 10.46586/tches.v2021.i2.229-253
18. Hochreiter, Jurgen, S.: Long short-term memory. Neural computation 9(8), 1735–1780 (1997)
19. Izabachène, M., Sirdey, R., Zuber, M.: Practical fully homomorphic encryption for fully masked neural networks. In: Mu, Y., Deng, R.H., Huang, X. (eds.) Cryptology

and Network Security, pp. 24–36. Springer International Publishing, Cham (2019). https://doi.org/10.1007/978-3-030-31578-8_2
20. Jang, J., et al.: Privacy-preserving deep sequential model with matrix homomorphic encryption. In: Proceedings of the 2022 ACM on Asia Conference on Computer and Communications Security, pp. 377–391. ASIA CCS '22, Association for Computing Machinery, New York, NY, USA (2022). https://doi.org/10.1145/3488932.3523253
21. Kim, M., Song, Y., Wang, S., Xia, Y., Jiang, X.: Secure logistic regression based on homomorphic encryption: design and evaluation. In: JMIR Medical Informatics (2018)
22. Kluczniak, K., Schild, L.: FDFB: full domain functional bootstrapping towards practical fully homomorphic encryption. Cryptology ePrint Archive, Report 2021/1135 (2021). https://ia.cr/2021/1135
23. Lev, G., Sadeh, G., Klein, B., Wolf, L.: RNN fisher vectors for action recognition and image annotation (2015)
24. Madi, A., Sirdey, R., Stan, O.: Computing neural networks with homomorphic encryption and verifiable computing. In: ACNS Workshops (2020)
25. OPenAI: Chatgpt: optimizing language models for dialogue (2022). https://openai.com/blog/chatgpt/
26. Paul, J., Annamalai, M.S.M.S., Ming, W., Badawi, A.A., Veeravalli, B., Aung, K.M.M.: Privacy-preserving collective learning with homomorphic encryption. IEEE Access **9**, 132084–132096 (2021). https://doi.org/10.1109/ACCESS.2021.3114581
27. Ramesh, A., Dhariwal, P., Nichol, A., Chu, C., Chen, M.: Hierarchical text-conditional image generation with clip latents. arXiv:2204.06125
28. Syed, S.A., Rashid, M., Hussain, S., Zahid, H.: Comparative analysis of CNN and RNN for voice pathology detection. BioMed Research International 2021 (2021)
29. Woodbridge, J., Anderson, H.S., Ahuja, A., Grant, D.: Predicting domain generation algorithms with long short-term memory networks (2016)
30. Yang, Z., Xie, X., Shen, H., Chen, S., Zhou, J.: Tota: fully homomorphic encryption with smaller parameters and stronger security. Cryptology ePrint Archive, Report 2021/1347 (2021). https://ia.cr/2021/1347

CANdito: Improving Payload-Based Detection of Attacks on Controller Area Networks

Stefano Longari(✉), Carlo Alberto Pozzoli, Alessandro Nichelini, Michele Carminati, and Stefano Zanero

Politecnico di Milano, Milan, Italy
{stefano.longari,michele.carminati,stefano.zanero}@polimi.it,
{carloalberto.pozzoli,alessandro.nichelini}@mail.polimi.it

Abstract. Over the years, the increasingly complex and interconnected vehicles raised the need for effective and efficient Intrusion Detection Systems against on-board networks. In light of the stringent domain requirements and the heterogeneity of information transmitted on the Controller Area Network, multiple approaches have been proposed, which work at different abstraction levels and granularities. Among these, RNN-based solutions received the attention of the research community for their performances and promising results. This paper proposes CANdito, an unsupervised IDS that exploits Long Short-Term Memory autoencoders to detect anomalies through a signal reconstruction process. In particular, we improve an RNN-based state-of-the-art IDS for CAN from the detection and temporal performances to comply with the strict automotive domain requirements. We evaluate CANdito by comparing its performance against state-of-the-art Intrusion Detection Systems (IDSs) for in-vehicle network and a comprehensive set of synthetic and real attacks in real-world CAN datasets.

1 Introduction

In the last decades, vehicles have become more complex, especially concerning their electronics [15]. Car manufacturers nowadays implement entertainment and autonomous drive-related technologies. As a result, the number of Electronic Control Units (ECUs) grew to reach more than one hundred units in the most complex vehicles. This evergrowing complexity, however, raises security risks, as firstly demonstrated by Koscher and Checkoway in [3,11], allowing the attacker to gain control of the vehicle's functionalities, even remotely. To manage such risks, the scientific research community focuses on developing countermeasures and security solutions, amongst which intrusion detection techniques for Controller Area Network (CAN)[1] have proven effective.

CAN security weaknesses are nowadays well known and discussed in multiple works [2,36]. As demonstrated by Miller and Valasek in [22,23], one of the

[1] For further details on the CAN specification, we refer the reader to [25].

S. Dolev et al. (Eds.): CSCML 2023, LNCS 13914, pp. 135–150, 2023.
https://doi.org/10.1007/978-3-031-34671-2_10

most common known vulnerabilities derives from the lack of authentication of messages on CAN. A node should not be allowed to send IDs that it does not own, but there is no mechanism to enforce this rule. Therefore, an attacker that takes control of an ECU that has access to a CAN bus can ideally send any ID and payload. In worst-case scenarios, the attacker is also capable of silencing the owner of the packet to avoid conflicts, as presented in [6,16]. Given the IDS nature of the work at hand, we present the capabilities of the attacker through the effects of its actions on the payload and flow of packets on the bus: A weak attacker may **inject** forged packets with one or multiple specific IDs on the bus, without silencing their owner. In this situation, the receivers may or may not consider the attacker's packets valid due to the incongruities on the bus. To solve this conflict, a stronger attacker may silence the owner and then forge packets with its ID, leading to a **masquerade** attack. Such masquerade attack can be implemented with or without consideration of the existence of an IDS checking the bus. If it is considered, the attacker may want to implement a **replay** attack, where she/he does not create a new payload but repeats a payload previously captured on the bus, or a **seamless change** attack, where the attacker drives a signal from its current value to a tampered one by changing it slowly through multiple packets. If it is not considered, an attacker may want to study the effects of various payloads and IDs on the system, implementing a **fuzzing attack**. Finally, an attacker may have the only goal to silence a node without replacing it, creating a **drop** attack. As further discussed in Sect. 4 when we present our attack tool, we generate datasets that consider these attacks and evaluate the systems against all of them.

Intrusion Detection Systems (IDSs) for vehicular systems analyze the stream of packets and monitor the events on on-board networks for signs of intrusions. Among these, machine learning-based, particularly RNN-based solutions, have proven effective in recognizing anomalous behavior [17,32]. Based on the approach and the results of CANnolo [17], in this paper, we propose CANdito, an RNN-based, unsupervised IDS that exploits Long Short-Term Memory (LSTM) autoencoders to detect anomalies through a signal reconstruction process. After a preprocessing stage, it learns the legitimate signal behavior through an LSTM-based autoencoder. Then, it computes the anomaly score for each CAN ID based on their reconstruction error. In particular, we improve the overall architecture and lighten CANnolo computational requirements to meet real-world timing constraints of the automotive domain. We prove the effectiveness of CANdito by conducting experiments on a real dataset of CAN traffic augmented with a set of synthetic but realistic attacks. With respect to existing works, we consider a broader spectrum of attacks and implement a tool to inject them into real-world CAN traffic logs. This tool is available at url.to.be.released.once.published. We demonstrate that CANdito outperforms its predecessor CANnolo, with improved detection rates and a reduction of more than 50% of the timing overheads. Moreover, to provide a fair evaluation of CANdito, we compare its performances against state-of-art Intrusion Detection Systems (IDSs) for in-vehicle network on a public dataset with attack messages. Our experimental results show that

CANdito outperforms state-of-the-art detection methods with a perfect True Positive Ration (TPR) and lower time requirements.

In summary, our contributions are the following:

- We improve CANnolo with CANdito, an RNN-based, unsupervised IDS that exploits LSTM autoencoders to detect anomalies through a signal reconstruction process.
- We design and provide CANtack, a tool to generate and inject synthetic attacks in real datasets, to be used as a benchmarking suite for IDS in the automotive domain.
- An evaluation of CANdito from the point of view of detection and timing performances on a more comprehensive dataset with respect to state-of-the-art works.

2 Related Works

Intrusion detection for automotive onboard networks has drawn vast research in recent years. We refer to Jo et al. [8] for a comprehensive survey of intra-vehicle IDSs, which can be divided into packet-based and hardware-based. Packet-based IDSs can be further divided into flow-based, payload-based, and combined. Flow-based IDSs (e.g., [12,26,27,33]) monitor the CAN bus, extract distinct features as message frequency or packet inter-arrival time, and use them to detect anomalous events without inspecting the payloads of the messages. On the contrary, payload-based IDSs (e.g., [1,7,9,17]) examine the payload of CAN packets (usually only data frames). Finally, combined IDS (e.g., [21,38]) are a combination of the previous two techniques. Many different machine learning techniques have been applied to payload-based CAN intrusion detection, from GANs to CNNs, in both a supervised and unsupervised fashion:

Kang and Kang [9] propose a supervised payload-based IDS based on Deep Neural Network (DNN). The input feature does not use the entire payload, but only the *mode information*, which represents the command state of an ECU, and the *value information*, which represents the value of the mode (e.g., wheel angle or speed). Multiple techniques exploit CNNs. For example, Rec-CNN [5] transforms the detection process into an image recognition one in an attempt to exploit the image recognition capabilities of CNNs, generating so-called recurrence plots that graphically represent the time series of packets. Reduced Inception-ResNet [28] exploits the deep convolutional neural network model of the Inception-ResNet architecture, a supervised model designed for image recognition, but significantly simplifies it in an attempt at lower computational times. CANTransfer [31] instead applies a supervised convolutional LSTM-based model to CAN intrusion detection with the goal of applying transfer learning to simplify the process of training different vehicles and IDs. CNNs are, however, better at processing spatial data, while RNNs and LSTMs are generally better suited for temporal data. CAN-ADF [30] exploits RNNs in a supervised fashion and inserts a rule-based IDS in front of it to detect simpler attacks in an attempt to not only detect attacks but also classify them, while TSP [24] studies the differences between various loss functions in the development of an LSTM-based IDS. HyDL-IDS [14] exploits CNNs and LSTMs to build a supervised system that extracts both temporal and spatial features of each packet stream. Often paired with these techniques, autoencoders have been proposed to predict or reconstruct time series. O-DAE [13] approaches detection by attempting to remove noise from the time series of packets via a supervised DAE autoencoder and then recognizing the attacks through the reconstruction. CANet [7] is one of the few IDSs that elaborates multiple IDs simultaneously, theoretically making it possible to exploit the correlation between the different information shared through the various IDs. It accomplishes this through an LSTM network per CAN ID and an autoencoder that receives the concatenated output of the various LSTMs. CANnolo [17] is an unsupervised IDS based on a LSTM autoencoder and represents the starting point of CANdito. A window of packets is fed to the RNN autoencoder, which attempts to reconstruct it following the trained model. Finally, the capabilities of GANs have also been evaluated, E-GAN [35] uses an unsupervised GAN and the DBC files for a dataset provided by the manufacturer first to comprehend the layout of a packet and then recognizes anomalies inside its various elements.

Motivation. The main limitation of current state-of-the-art approaches is that, while different methods work well on different problems, they can hardly achieve results that are good enough on any anomaly and, at the same time, provide fast enough results to process the network's traffic in real-time. The way this limitation compels a system largely depends on the different types of IDS approach adopted. Generally, flow-based approaches can provide fast predictions while being limited to the detection of specific kinds of vulnerabilities, while payload-based approaches have a broader scope, but it is often a problem to make them work in real-time on the total traffic. Moreover, the traffic on different CAN IDs has different characteristics, but the current state-of-the-art methods rarely consider them to provide better results.

3 CANdito

In this section, we describe CANdito, an improved version of CANnolo [17], a state-of-the-art IDS for CAN that exploits Recurrent Neural Network (RNN)-based autoencoders. RNN-based architectures are effective in modeling time-series and have been successfully proposed for CAN traffic anomaly detection [32]. On the other hand, autoencoders do not require a labeled training dataset since target signals are generated automatically from the input sequence. Moreover, being unsupervised, they learn a model of the legitimate network traffic and not specific anomalies, making them able to potentially recognize novel attacks. Figure 1 shows an overview of the architecture of the system, which works by reconstructing the time series of CAN packets for each ID and computes their anomaly score based on the reconstruction error. The effectiveness of reconstructing the signal (as opposed to predicting the successive one) has been demonstrated by Malhotra et al. [19], which uses LSTM encoder-decoder

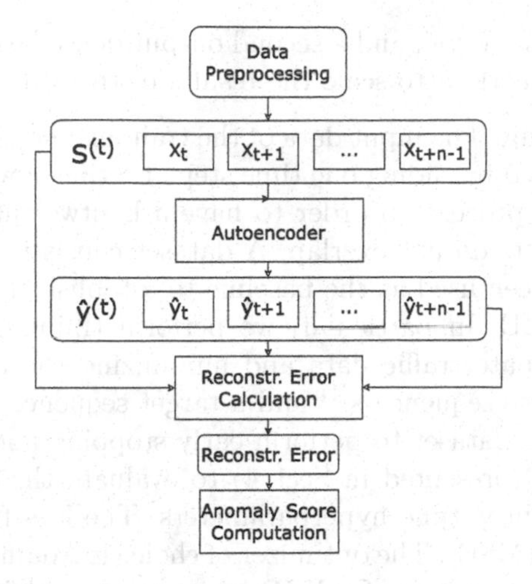

Fig. 1. Overview of CANdito's detection process.

architectures as reconstructors to detect anomalies in multi-sensor contexts. It comprises three modules: a data preprocessing module, an LSTM-based autoencoder, which learns the legitimate signal behavior, and an anomaly detector, which compares the reconstruction errors.

3.1 Data Preprocessing

The first module of CANdito builds the input sequences by applying the READ algorithm [20] and associating to each CAN ID the corresponding ranges of the signals. Using this information, the payloads of each packet are converted into the list of their signals rescaled in the [0–1] range, excluding constant bits, counters, and Cyclic Redundancy Checks (CRCs).

Our resulting input sequence is composed by a matrix n x k, where $n = 40$ is the dimension of the window of CAN packets, k is the number of signals per packet (rescaled in the [0–1] range). Notice that at this step, CANdito discards and flags as anomalous any CAN ID that has not been found in the training set.

3.2 LSTM-Based Autoencoder

The second module of CANdito is based on an autoencoder whose encoder and decoder layers are implemented with two recurrent LSTMs. The encoder is composed of a dense layer, which consists of 128 units with an Exponential Linear Unit (ELU) [4] activation function. The dense layer is followed by a 20% dropout layer and two recurrent LSTM layers with $L = 64$ units each. The cell and hidden states of the last LSTM layer of the encoder are used to initialize the states of the first LSTM layer of the decoder. The output of the encoder is reversed before being fed into the decoder. Symmetrically, the decoder consists of two recurrent LSTM layers with 64 units each, a dense layer, consisting of 128 units with ELU activation function, and a second output dense layer with k units with sigmoid activation function (to scale the results to the [0,1] interval).

Training and Tuning. The input data of the training and threshold calculation processes are composed by sliding one time step at a time a window of n packets, while for the testing process, in order to have a lightweight detection process, the windows of packets do not overlap. A dataset consisting only of legitimate data sequences has been used as the baseline to establish the 'normal' behavior for any given CAN ID. In particular, we perform training and validation by reconstructing legitimate traffic data and minimizing the reconstruction error between a given source sequence $s^{(t)}$ and a target sequence $y^{(t)}$. Also, we make use of an untampered dataset to perform early stopping and a dataset injected with our attack tool (presented in Sect. 4) to evaluate the performance of the model and, consequently, tune hyperparameters. The loss function of choice is Mean Squared Error (MSE). The optimizer of choice is Adam [10] with a learning rate of 0.001. The model of each CAN ID has been trained for a maximum of 50 epochs with an early stopping with patience 5 (i.e., training is stopped before the maximum number of epochs if the validation accuracy does not improve for 5 consecutive epochs).

3.3 Anomaly Score Computation

The third module of CANdito works in an unsupervised fashion by computing the distance between the reconstruction error and the expected normal distribution computed during training. The anomaly score indicates the likelihood of the test sequence to be anomalous.

Each window of n packets is fed into the trained autoencoder and an anomaly score is assigned to each window as the squared l2-norm of the reconstruction error: $e^{(t)} = \|\hat{y}^{(t)} - s^{(t)}\|_2^2$. The chosen detection threshold is, as proposed by Hanselmann et al. [7], the 99.99 percentile of the scores. The windows whose score is greater than the threshold are marked as anomalies.

3.4 Improvements from CANnolo

As previously stated, CANdito is based on CANnolo [17], in light of its promising results in the detection of attacks on CAN. In particular, CANdito required an in-depth study of CANnolo. To do so, we implemented CANnolo from scratch and tested it on our extended dataset. This evaluation brought the improvements described below.

Generalization and Computation Improvements. The number of artificial neurons of all layers has been halved. This reduction in dimensions has a twofold scope. The networks of several CAN IDs are prone to overfitting, reducing the dimension of the layers mitigated this problem. Moreover, reducing the dimension of the layers has a substantial impact on the improvement of the computation times of the network.

Overfitting and Vanishing Gradient Mitigation. The activation functions of the two symmetric dense layers of the encoder and decoder have been modified from hyperbolic tangent to ELU. This was proven to mitigate the vanishing gradient problem and to improve the generalization capabilities of the network [4, 7].

Input Bloat Reduction. The inputs of CANnolo are bit-strings composed of the condensed notation of the packets (i.e., excluding constant bits). In our solution, we also exclude signals marked by READ as counters or CRCs. In fact, counters and CRCs do not carry relevant information for the reconstruction module, as demonstrated by the fact that considering them did not improve the effectiveness of the system. Moreover, CANdito's input is not composed of bit-strings but by each signal detected by READ rescaled in the [0–1] range. This significantly reduces the input dimensions, further lowering computing times, while comparative tests with the two input methods did not show meaningful effects on detection performances.

Underfitting Mitigation. The output sequence of the encoder is reversed before being fed into the decoder. This operation is meant to help the network reconstruct the target sequence, which is also reversed as suggested by Sutskever et al. [29]. While for some CAN IDs the network results are similar with or without reversing the encoder output, other CAN IDs networks are affected by severe underfitting if the encoder output is not reversed. The same CAN IDs networks perform well after the change.

Lower Computational Requirements. To lower the computational effort, we do not feed the reconstructed sequence back into the decoder. Evaluations at design time did not show improvements in detection performances between the two implementations. From an implementation standpoint, code optimization and moving from Keras to PyTorch also greatly lowered computation requirements.

Anomaly Score Computation Improvement. CANnolo uses the Mahalanobis distance between the reconstruction error of the window under evaluation and the distribution of errors in untampered scenarios. While such distance has been considered for the anomaly scores computation, it has underperformed with our model. Instead, we opted to compute the anomaly score as the squared l2-norm of the reconstruction error, with a 99.99 percentile of the scores as a detection threshold, as suggested in [7].

4 CANtack

We designed CANtack to have an easy, partially automated way to consistently generate different types of anomalies in our datasets while starting from datasets of real CAN traffic. The tool is available at[2] alongside instructions on how to use it. The output of the tool is a dataset structured similarly to the ReCAN

[2] CANtack url: url.to.be.released.

dataset [37], but with an additional *isTampered* column, which indicates whether a log entry has been tampered with or not.

The tool allows choosing between the different attack implementations, which enable to deploy all the attacks presented in Sect. 1. Moreover, all different types of attacks (except for cases in which this does not make sense, i.e., *drop* and *DoS*) can be either performed in an *injection* fashion (i.e., without modifying the original packets of the traffic and specifying an injection rate) or in a *masquerade* fashion (i.e., substituting the original packets with the tampered ones). We proceed to present the list of attacks and their user-defined parameters. Note that for all attacks, the user needs to set the ID to tamper and the beginning time for the attack in seconds.

Basic Injection. The user can specify a payload and a number of tampered packets. The tool injects or replaces a number of packets with the new tampered payload.

Progressive Injection. As above, the value of every single payload can be specified.

DoS. The network is flooded with packets with ID "0" for the specified duration. It is possible to define the percentage of the bus to fill.

Drop. A set amount of messages from the given ID are deleted from the dataset.

Fuzzy. The payloads are injected or tampered with random values. It is also possible to choose a bit range to fuzzy a value, e.g., simulating the fuzzing only of sensor data.

Replay. The payloads are values sniffed from the dataset. To implement this attack, it is necessary to define the initial sniffing time and whether to partially randomize the position of the first copied packet among the sniffed ones. Moreover, a series of replacements can be specified to modify only some bit-ranges with other values. These values can be set with the following options:

- **payloads** replaces the bit-ranges with explicitly defined data;
- **fuzzy** randomizes the bit-range;
- **min** and **max** respectively find the minimum and maximum value (considered as an integer) registered in the dataset for that bit-range;
- **seamless change** defines a final value to reach and increases or decreases the bit-range from the value read in the last untampered line to the chosen one;
- **counter** increases the values of the bit-range by one per packet, in a counter-like fashion.

5 Experimental Validation

First, we compare CANdito with CANnolo [17] from both the point of view of the detection and the temporal performances. In particular, this experiment evaluates the performance of CANnolo and CANdito over different datasets tampered with our novel attack tool Sect. 4. CANnolo's implementation has been kept almost untouched, threshold computation criterion included. This consideration must be explicitly made since the authors used as a comparison metric for their experiment just the Area Under Curve (AUC) without focusing over the threshold computation criterion, which was demonstrated to be sub-optimal.

Then, we demonstrate the effectiveness of CANdito by comparing its performances against state-of-the-art solutions on a public dataset with real-world attacks.

We evaluate the detection performances of the systems under analysis by considering the most common metrics used to evaluate unbalanced datasets. Specifically, we make use of Detection Rate (DR), False Positive Ratio (FPR), F1 score, Matthews Correlation Coefficient (MCC). Moreover, to evaluate the timing performances, we use the Testing Time per Packet (TTP).

We evaluate our work through two public datasets: the ReCAN C-1 dataset [37] used for the first set of experiments and the car-hacking dataset [28] used for the comparison with the state of the art. The experiments have been tested by serving datasets split in windows of pre-defined size, as it should happen at runtime instead of testing the entire sequence in one batch; doing so permits to have a measure of the testing time that is more accurate.

5.1 ReCAN Datasets

The ReCAN dataset [37] is a public dataset of CAN logs retrieved in real-world scenarios. We select the C-1 sub-dataset, which has been retrieved from multiple test drives of an Alfa Giulia. In more detail, we use sub-datasets 1, 2, 6, 8, and 9 to train the model, sub-dataset 4 to calculate thresholds, sub-dataset 5 for validation and hyper-parameter tuning, and finally, sub-dataset 7 for testing. We then use our attack tool to generate the attack datasets. All the datasets are generated starting from sub-dataset 7 of the ReCAN C-1 dataset. The datasets have the goal of building attacks as presented in Sect. 1. All the datasets, their details, and implementation parameters are available on the CANtack webpage (see footnote 2).

Injection Dataset. This dataset contains generic injection attacks, which consists of added packets on the network, leaving all packets already present in the dataset unchanged. The new packets are sniffed from previous traffic and only one physical signal is modified (recognized through the READ algorithm [20]), although the value changes inside the range of already existing values of the signal. This is done in an attempt to comply as much as possible with the behavior of the ID and increase the difficulty of detection. The packets are added at 20 times the frequency of the original packet and continue for a sequence of 50 packets.

Drop Dataset. This dataset simulates the event where an attacker turns off an ECU or its CAN controller. The attack consists of removing a sequence of valid packets from the original traffic. Twenty-five consecutive packets are removed each time.

Masquerade Dataset. This dataset contains generic masquerade attacks that would not be detectable only through frequency-based or rule-based features.

The modified packets are sniffed from previous traffic and one or more physical signals are modified in the same fashion as the injection dataset. Moreover, the anomalous packets maintain the same timestamps as the packet they replace, alongside its ID. Each anomalous sequence has a length of 25 packets.

Fuzzed Dataset. This dataset represents the event where an attacker is testing random values of signals in order (usually) to trigger unexpected behavior. The attack is made in a masquerade fashion (the original packets are removed and replaced), but only the bits included in some of the signals are fuzzed, while, for example, constant bits are left untampered. As above, 25 consecutive packets are removed each time. As above, each anomalous sequence has a length of 25 packets.

Seamless Change Dataset. This dataset contains masquerade attacks that attempt to evade detection by changing the payload of the packets progressively until the desired value is reached. The physical values in the tampered ID have to be at least 4 bits long. The new packets are sniffed from previous traffic and only one physical signal is modified. As above, each anomalous sequence has a length of 25 packets.

Full Replay Dataset. This dataset contains masquerade attacks that attempt to evade detection by copying exact sequences on the bus. No additional check is made while generating the attacks. Consequently, there is no warranty that the new sequence is taken from a moment where the condition of the car is very different, lowering the detection capabilities but also the actual effects of the attacks. This dataset is primarily interesting to compare the ability of IDSto detect anomalous sequences that are perfectly valid in a different context. As above, each anomalous sequence has a length of 25 packets.

5.2 Car-Hacking Dataset

The car-hacking dataset [28] is composed of logs of real-time CAN messages via the onboard diagnostic (OBD-II) port of two running vehicles (KIA Soul and Hyundai Sonata) with message attacks. It has four data features, including timestamp, identifier (ID, in hexadecimal format), data length code (DLC, valued from 0 to 8) and data payload (8 bytes), and the label of a CAN message. We refer the reader to [28] for further details on the public dataset under consideration. It contains normal CAN messages (14,237,978) and anomaly messages (2,331,497) belonging to three categories of attacks (for a total of four attacks).

DoS Attack. It aims to flood the CAN bus with numerous forged messages with low ID values in a short time interval. Thus, almost all the communication resources are occupied so that messages from other nodes will be delayed or blocked.

Fuzzy Attack. Fake messages are sent from malicious ECUs into the CAN bus at a slower rate than the DoS attack.

Impersonation Attacks. They realize unauthorized service access by spoofing legitimate authentication credentials, such as **spoofing the drive gear and the RPM gauze**.

5.3 Experimental Results

Table 1. CANnolo vs CANdito performances, tested over the masquerade, fuzzy, seamless change, and full replay datasets. Only CAN IDs recommended for testing by CANnolo's authors have been taken into consideration.

Dataset		DR	FPR	F1	MCC	TTP
Masq.	CANdito	**.9258**	.0081	**.9505**	**.9336**	**.0700**
	CANnolo	.6477	**.0029**	.7823	.7502	
Fuzzy	CANdito	**.9989**	.0081	**.9886**	**.9844**	
	CANnolo	.9541	**.0029**	.9724	.9629	
Seam.	CANdito	**.8972**	.0079	**.9345**	**.9143**	1.0630
	CANnolo	.7481	**.0029**	.8518	.8224	
Replay	CANdito	**.5820**	.0080	.7254	.6909	
	CANnolo	.5801	**.0029**	**.7304**	**.7024**	

Results on the ReCAN Dataset As shown in Table 1, our solution is not only twice as fast as CANnolo in providing the predictions, but it is also more effective in almost all the considered attack scenarios.

Focusing on payload-based anomalies, CANdito generally outperforms CANnolo on both the entire dataset and on the subset of the dataset composed of the selected CAN IDs. For the Masquerade dataset, CANdito performs evidently better, with an F1 score and MCC both over .15 point higher than CANnolo. We explain the better performances obtained with the different approaches used to compute the detection threshold. For the Fuzzy datasets, CANdito shows similar performances to CANnolo, with a higher DR, F1-score, and MCC. For the Seamless Change dataset, the better performances of our new architecture are more evident. In fact, both the F1 score and MCC improvements range between .07 and .09. For the Full replayed dataset, the original model is slightly more effective, but the performances of the two systems are comparable. However, in light of the stringent requirements of the automotive domain, where the lack of computational power is critical [18], CANdito is preferable to CANnolo since it provides detection results in less than half of the time.

As expected, both systems perform poorly on flow-based anomalies (i.e., Injection and Drop datasets) since they implement payload-based detection and do not detect changes in the frequency of packets arrivals.

It is interesting to note that CAN ID $0x1E340000$ is responsible for the 18% of overall false positives and features a very different behavior between the train set and the test set with the presence of flipping bits that were static for the entire duration of the training set. This said, it is encouraging to know that a large part of the FPR depends on a small set of CAN IDs because this demonstrates that

Table 2. Detection Performance Comparison of State-of-the-art IDS against CANdito. In **bold**, the best performance by metric and attack category.

IDSs	Attacks	Accuracy	Precision	TPR	FPR	F1-score	TTP
Reduced Inception-ResNet [28]	DoS Attack	**0.9993**	**0.9995**	0.9963	**0.0001**	**0.9980**	1.5633
	Fuzzy Attack	0.8730	0	0	0.0002	–	
	Gear Spoofing Attack	0.8223	0	0	**0.0001**	–	
	RPM Spoofing Attack	0.7774	0	0	0.0003	–	
CANTransfer [31]	DoS Attack	0.9991	0.9990	0.9951	0.0002	0.9971	1.3264
	Fuzzy Attack	0.8718	0	0	**0.0001**	–	
	Fuzzy Attack (1-shot)	0.8664	0.9794	0.0309	**0.0001**	0.0599	
	Gear Spoofing Attack	0.8223	0	0	0.0004	–	
	RPM Spoofing Attack	0.7774	0	0	0.0003	–	
CAN-ADF [30]	DoS Attack	0.9938	0.9826	0.9785	0.0033	0.9805	1.4476
	Fuzzy Attack	0.8715	0.0505	0.0002	0.0006	0.0004	
	Gear Spoofing Attack	0.8222	0	0	0.0004	–	
	RPM Spoofing Attack	0.7769	0.1200	0.0005	0.0012	0.0011	
TSP [24]	DoS Attack	0.9802	0.9100	0.9728	0.0183	0.9403	1.1422
	Fuzzy Attack	0.8714	0	0	0.0005	–	
	Gear Spoofing Attack	0.8221	0	0	0.0005	–	
	RPM Spoofing Attack	0.7774	0	0	0.0003	–	
O-DAE [13]	DoS Attack	0.9933	0.9742	0.9843	0.0050	0.9792	1.2130
	Fuzzy Attack	0.8714	0	0	0.0006	–	
	Gear Spoofing Attack	0.8222	0	0	0.0004	–	
	RPM Spoofing Attack	0.7774	0	0	0.0003	–	
LDAN [39]	DoS Attack	0.9806	0.9099	0.9756	0.0184	0.9416	0.9283
	Fuzzy Attack	0.8717	0	0	0.0006	–	
	Gear Spoofing Attack	0.8224	0	0	**0.0001**	–	
	RPM Spoofing Attack	0.7775	0	0	**0.0002**	–	
E-GAN [35]	DoS Attack	0.9806	0.9099	0.9756	0.0184	0.9416	1.0331
	Fuzzy Attack	0.8717	0	0	0.0002	–	
	Gear Spoofing Attack	0.8224	0	0	**0.0001**	–	
	RPM Spoofing Attack	0.7774	0	0	0.0003	–	
HyDL-IDS [14]	DoS Attack	0.9936	0.9819	0.9781	0.0034	0.9800	0.4395
	Fuzzy Attack	0.8715	0.0612	0.0002	0.0005	0.0005	
	Gear Spoofing Attack	0.8221	0	0	**0.0001**	–	
	RPM Spoofing Attack	0.7769	0.1042	0.0005	0.0011	0.0009	
CANet [7]	DoS Attack	0.9993	0.9992	0.9966	0.0014	0.9979	0.3357
	Fuzzy Attack	0.8717	0	0	0.0002	–	
	Gear Spoofing Attack	0.8223	0	0	0.0001	–	
	RPM Spoofing Attack	0.7774	0	0	0.0003	–	
Rec-CNN [5]	DoS Attack	0.9803	0.9097	0.9740	0.0185	0.9408	0.3278
	Fuzzy Attack	0.8714	0	0	0.0006	–	
	Gear Spoofing Attack	0.8221	0	0	0.0005	–	
	RPM Spoofing Attack	0.7774	0	0	0.0003	–	
CANdito	DoS Attack	0.9983	0.9926	**1**	0.0021	0.9963	**0.07**[a]
	Fuzzy Attack	**0.9608**	**0.9915**	**0.8884**	0.0094	**0.9296**	
	Gear Spoofing Attack	**0.9983**	**0.9984**	**0.9934**	0.0004	**0.9959**	
	RPM Spoofing Attack	**0.9996**	**0.9986**	**1**	0.0004	**0.9993**	

[a] This performance was achieved on consumer-level HW, while Wang at al. [34] evaluation was performed on a machine learning dedicated server

future works may improve the results of finding a different classification of these "pathological" CAN IDs. Another possible alternative that is not particularly time-consuming, considering the small dimension of this set of CAN IDs is to perform a human-supervised fine-tuning of the model for these specific CAN IDs on top of the automatic classification.

Results on the Car-Hacking Dataset. In order to provide a comparison with the other various machine learning techniques used in the state of the art, we make use of the systematization done by Wang et al. [34] on the public car-hacking dataset, and follow the same experimental procedure on CANdito. Table 2 contains Wang et al. results followed by ours. CANdito achieves better detection rate on all the datasets, and comparable metrics where it does not win. The significantly lower detection rate on the fuzzy dataset in relation to the others can be attributed by the behavior of the dataset, where the randomized ID sometimes end up being one of the valid ones, but only one malicious packet is inserted in a window of 39 valid ones. The IDS that better stands up against CANdito in terms of detection performances is Reduced Inception-ResNet [28], which, however, is 20 times slower than CANdito, and even more importantly, given the average packet inter-arrival time of the dataset, which is 0.77ms, is not compliant with the real-time requirements of the automotive domain. Finally, CANdito shows overall good detection performances on all the categories of attacks.

6 Conclusions

In this paper, we presented CANdito, an improved RNN-based and unsupervised IDS that exploits LSTM autoencoders to detect anomalies through a signal reconstruction process in CAN traffic. We evaluated CANdito from the point of view of the detection and timing performances on a more comprehensive real-world dataset augmented with synthetic attacks generated with CANtack, a tool to generate and inject synthetic attacks in real datasets, which can be used as a benchmarking suite for IDS in the automotive domain. Moreover, we compared its performances against state-of-art Intrusion Detection Systems (IDSs) for in-vehicle network on a public dataset with attack messages, showing that CANdito performs overall better of the current state of the art while requiring significantly less time - up to 1/20 - than the other detection techniques. We plan to overcome CANdito limitation in detecting attacks that work in the frequency domain by complementing the improved detection power of the payload-based detection system presented in this work with the power of frequency-based approaches to building an end-to-end hybrid IDS able to fully exploit all CAN IDs.

References

1. Amato, F., Coppolino, L., Mercaldo, F., Moscato, F., Nardone, R., Santone, A.: Can-bus attack detection with deep learning. IEEE Trans. Intell. Transp. Syst. **22**(8), 5081–5090 (2021). https://doi.org/10.1109/TITS.2020.3046974

2. Buttigieg, R., Farrugia, M., Meli, C.: Security issues in controller area networks in automobiles. CoRR abs/1711.05824 (2017). arXiv:1711.05824
3. Checkoway, S., et al.: Comprehensive experimental analyses of automotive attack surfaces, August 2011
4. Clevert, D.A., Unterthiner, T., Hochreiter, S.: Fast and accurate deep network learning by exponential linear units (elus). arXiv preprint arXiv:1511.07289 (2015)
5. Desta, A.K., Ohira, S., Arai, I., Fujikawa, K.: Rec-CNN: in-vehicle networks intrusion detection using convolutional neural networks trained on recurrence plots. Veh. Commun. **35**, 100470 (2022). https://doi.org/10.1016/j.vehcom.2022.100470
6. de Faveri Tron, A., Longari, S., Carminati, M., Polino, M., Zanero, S.: CANflict: exploiting peripheral conflicts for data-link layer attacks on automotive networks. In: Yin, H., Stavrou, A., Cremers, C., Shi, E. (eds.) Proceedings of the 2022 ACM SIGSAC Conference on Computer and Communications Security, CCS 2022, Los Angeles, CA, USA, 7–11 November 2022, pp. 711–723. ACM (2022). https://doi.org/10.1145/3548606.3560618
7. Hanselmann, M., Strauss, T., Dormann, K., Ulmer, H.: CANet: an unsupervised intrusion detection system for high dimensional can bus data. IEEE Access **8**, 58194–58205 (2020)
8. Jo, H.J., Choi, W.: A survey of attacks on controller area networks and corresponding countermeasures. IEEE Trans. Intell. Transp. Syst. **23**(7), 6123–6141 (2022). https://doi.org/10.1109/TITS.2021.3078740
9. Kang, M.J., Kang, J.W.: Intrusion detection system using deep neural network for in-vehicle network security. PLoS ONE **11**(6), e0155781 (2016)
10. Kingma, D.P., Ba, J.: Adam: A method for stochastic optimization. arXiv preprint arXiv:1412.6980 (2014)
11. Koscher, K., et al.: Experimental security analysis of a modern automobile. In: 2010 IEEE Symposium on Security and Privacy, pp. 447–462. IEEE (2010)
12. Lampe, B., Meng, W.: IDS for CAN: a practical intrusion detection system for CAN bus security. In: IEEE Global Communications Conference, GLOBECOM 2022, Rio de Janeiro, Brazil, 4–8 December 2022, pp. 1782–1787. IEEE (2022). https://doi.org/10.1109/GLOBECOM48099.2022.10001536
13. Lin, Y., Chen, C., Xiao, F., Avatefipour, O., Alsubhi, K., Yunianta, A.: An evolutionary deep learning anomaly detection framework for in-vehicle networks-can bus. IEEE Trans. Ind. Appl. (2020)
14. Lo, W., AlQahtani, H., Thakur, K., Almadhor, A., Chander, S., Kumar, G.: A hybrid deep learning based intrusion detection system using spatial-temporal representation of in-vehicle network traffic. Veh. Commun. **35**, 100471 (2022). https://doi.org/10.1016/j.vehcom.2022.100471
15. Longari, S., Cannizzo, A., Carminati, M., Zanero, S.: A secure-by-design framework for automotive on-board network risk analysis. In: 2019 IEEE Vehicular Networking Conference (VNC), pp. 1–8 (2019). https://doi.org/10.1109/VNC48660.2019.9062783
16. Longari, S., Penco, M., Carminati, M., Zanero, S.: Copycan: an error-handling protocol based intrusion detection system for controller area network. In: Cavallaro, L., Kinder, J., Holz, T. (eds.) Proceedings of the ACM Workshop on Cyber-Physical Systems Security & Privacy, CPS-SPC@CCS 2019, London, UK, 11 November 2019, pp. 39–50. ACM (2019). https://doi.org/10.1145/3338499.3357362
17. Longari, S., Valcarcel, D.H.N., Zago, M., Carminati, M., Zanero, S.: CANnolo: an anomaly detection system based on LSTM autoencoders for controller area network. IEEE Trans. Netw. Serv. Manag. (2020)

18. Maffiola, D., Longari, S., Carminati, M., Tanelli, M., Zanero, S.: GOLIATH: a decentralized framework for data collection in intelligent transportation systems. IEEE Trans. Intell. Transp. Syst. (2021)
19. Malhotra, P., Ramakrishnan, A., Anand, G., Vig, L., Agarwal, P., Shroff, G.: LSTM-based encoder-decoder for multi-sensor anomaly detection. arXiv preprint arXiv:1607.00148 (2016)
20. Marchetti, M., Stabili, D.: READ: reverse engineering of automotive data frames. IEEE Trans. Inf. Forensics Secur. **14**(4), 1083–1097 (2018)
21. Marchetti, M., Stabili, D., Guido, A., Colajanni, M.: Evaluation of anomaly detection for in-vehicle networks through information-theoretic algorithms. In: 2016 IEEE 2nd International Forum on Research and Technologies for Society and Industry Leveraging a better tomorrow (RTSI), pp. 1–6. IEEE (2016)
22. Miller, C., Valasek, C.: Adventures in automotive networks and control units. Def. Con. **21**(260–264), 15–31 (2013)
23. Miller, C., Valasek, C.: Remote exploitation of an unaltered passenger vehicle. Black Hat USA 2015(S 91) (2015)
24. Qin, H., Yan, M., Ji, H.: Application of controller area network (CAN) bus anomaly detection based on time series prediction. Veh. Commun. **27**, 100291 (2021). https://doi.org/10.1016/j.vehcom.2020.100291
25. Robert Bosch GMBH: Can specification, version 2.0. Standard, Robert Bosch GmbH, Stuttgart, Germany (1991)
26. Seo, E., Song, H.M., Kim, H.K.: GIDS: GAN based intrusion detection system for in-vehicle network. In: 2018 16th Annual Conference on Privacy, Security and Trust (PST), pp. 1–6. IEEE (2018)
27. Song, H., Kim, H., Kim, H.K.: Intrusion detection system based on the analysis of time intervals of can messages for in-vehicle network, pp. 63–68, January 2016. https://doi.org/10.1109/ICOIN.2016.7427089
28. Song, H.M., Woo, J., Kim, H.K.: In-vehicle network intrusion detection using deep convolutional neural network. Veh. Commun. **21** (2020). https://doi.org/10.1016/j.vehcom.2019.100198
29. Sutskever, I., Vinyals, O., Le, Q.V.: Sequence to sequence learning with neural networks. arXiv preprint arXiv:1409.3215 (2014)
30. Tariq, S., Lee, S., Kim, H.K., Woo, S.S.: CAN-ADF: the controller area network attack detection framework. Comput. Secur. **94**, 101857 (2020). https://doi.org/10.1016/j.cose.2020.101857
31. Tariq, S., Lee, S., Woo, S.S.: CANTransfer: transfer learning based intrusion detection on a controller area network using convolutional LSTM network. In: Hung, C., Cerný, T., Shin, D., Bechini, A. (eds.) SAC '20: The 35th ACM/SIGAPP Symposium on Applied Computing, online event, [Brno, Czech Republic], March 30–3 April 2020, pp. 1048–1055. ACM (2020). https://doi.org/10.1145/3341105.3373868
32. Taylor, A.: Anomaly-based detection of malicious activity in in-vehicle networks. Ph.D. thesis, Université d'Ottawa/University of Ottawa (2017)
33. Taylor, A., Japkowicz, N., Leblanc, S.: Frequency-based anomaly detection for the automotive can bus. In: 2015 World Congress on Industrial Control Systems Security (WCICSS), pp. 45–49. IEEE (2015)
34. Wang, K., Zhang, A., Sun, H., Wang, B.: Analysis of recent deep-learning-based intrusion detection methods for in-vehicle network. IEEE Trans. Intell. Transp. Syst. 1–12 (2022). https://doi.org/10.1109/TITS.2022.3222486

35. Xie, G., Yang, L.T., Yang, Y., Luo, H., Li, R., Alazab, M.: Threat analysis for automotive CAN networks: a GAN model-based intrusion detection technique. IEEE Trans. Intell. Transp. Syst. **22**(7), 4467–4477 (2021). https://doi.org/10.1109/TITS.2021.3055351
36. Young, C., Zambreno, J., Olufowobi, H., Bloom, G.: Survey of automotive controller area network intrusion detection systems. IEEE Des. Test **36**(6), 48–55 (2019). https://doi.org/10.1109/MDAT.2019.2899062
37. Zago, M., et al.: ReCAN-dataset for reverse engineering of controller area networks. Data Brief **29**, 105149 (2020)
38. Zhang, L., Shi, L., Kaja, N., Ma, D.: A two-stage deep learning approach for can intrusion detection (2018)
39. Zhao, R., et al.: An efficient intrusion detection method based on dynamic autoencoder. IEEE Wirel. Commun. Lett. **10**(8), 1707–1711 (2021). https://doi.org/10.1109/LWC.2021.3077946

Using Machine Learning Models for Earthquake Magnitude Prediction in California, Japan, and Israel

Deborah Novick and Mark Last(✉)

Department of Software and Information Systems Engineering,
Ben-Gurion University of the Negev, 84105 Beer-Sheva, Israel
mlast@bgu.ac.il
http://www.bgu.ac.il/ mlast/

Abstract. This study aims at predicting whether an earthquake of magnitude greater than the regional median of maximum yearly magnitudes will occur during the next year. Prediction is performed by training various machine learning algorithms, such as AdaBoost, XGBoost, Random Forest, Logistic Regression, and Info-Fuzzy Network. The models are induced using a combination of seismic indicators used in the earthquake literature as well as various time-series features, such as features based on the moving averages of the number of earthquakes in each area, features that record the number of events above and below the mean in a time period, and features based on lagged values of the mean and median magnitude. Feature selection is performed using a forward search algorithm that chooses the most effective features for prediction. The models are trained and evaluated using earthquake catalog data obtained for California, Japan, and Israel. In addition, models trained on either California or Japan datasets are evaluated using the remaining data. Models trained on Japan data achieve AUC scores up to 0.825; models trained on California data achieve AUC scores up to 0.738; and models trained on Israel data achieve AUC scores up to 0.710.

Keywords: Earthquake prediction · Clustering analysis · Seismicity indicators · Classification models

1 Introduction

Earthquakes are caused when lithospheric plates slide against one another and cause stress to buildup in the crustal rocks, which is subsequently released in sudden bursts of seismic activity. In addition, there are other factors that influence the size, location, duration and occurrence time of an earthquake. Such factors include the heterogeneity of the earth's crust, the local stress conditions in a particular seismogenic area, fluid content, the disposition of existing faults and their stress histories [16].

S. Dolev et al. (Eds.): CSCML 2023, LNCS 13914, pp. 151–169, 2023.
https://doi.org/10.1007/978-3-031-34671-2_11

Among natural disasters, earthquakes stand out in that there is no reliable way to know when and where they will occur, and to therefore allow people to plan accordingly. Powerful earthquakes are a major cause of catastrophe and because they are unpredictable, they cause even more destruction in terms of human life and financial losses [5]. Successful forecasting of earthquakes is imperative in order to mitigate seismic risk on all levels [21].

However, to date, there is no consensus on how to obtain the best model to predict earthquakes. The main problem with earthquake forecasting is that our main interest is in major events, which fortunately occur very infrequently. Since it is difficult to represent the complex physical dynamics within the earth that produce seismic activity, stochastic models are increasingly being used to model earthquake occurrences [16].

In order to mitigate the great seismic risk - and due to the fact that until now there have not been reliable earthquake forecasts - a huge explosion of prediction models has come out of the data science community in an attempt to use the available data to provide some level of earthquake prediction. Clearly, machine learning technologies could help to extract hidden patterns and make accurate predictions of earthquakes as noted by [9].

Short-term earthquake prediction helps to mitigate an immediate threat to life or property in the coming hours or days. on the other hand, long-term earthquake prediction to months or years can help communities prevent the catastrophic effects of earthquakes by encouraging the appropriate allocation of resources to improve infrastructure.

This work focuses on the long-term prediction of earthquake occurrence in California, Japan and Israel using various time-dependent machine learning models. The research continues the work done by [17] on a smaller subset of Israel data. The goal of this work is to estimate the probability that in the following year there would be an earthquake of magnitude greater than the median of the maximum yearly magnitudes in a specific region. The median is used as the prediction threshold because it presents a balanced classification task. For example, it would be much more difficult statistically to predict an earthquake of magnitude 6.0 and higher in a region where such a magnitude earthquake is very rare.

Specific objectives include:

- Train and evaluate machine learning algorithms on seismic data from three different countries.
- Evaluate new and innovative features that were not used in the previous work [17].
- Select the best set of predictive features.
- Induce the best prediction model for each country individually.
- Create a single model that could be used effectively to predict the occurrence of earthquakes in all three countries.

Original contributions include evaluating the effectiveness of some 50 new time-series type variables for earthquake prediction as well as using a model induced from data in one region to predict earthquakes in other geographic regions.

2 Related Works

Historically, earthquake prediction models have been time-independent and are based on the assumption that the probability of the occurrence of an earthquake in a given time period follows a Poisson distribution. These predictions are used to create hazard maps for earthquake-prone regions.

More recently, there has been interest in developing time-dependent models that predict the probability of earthquake occurrence based on the probability distribution of time passed from a prior earthquake event [27]. Other seismic models use statistical relationships that mimic the observed properties of earthquake clustering to model the behavior of earthquake fault interactions [15]. The epidemic-type aftershock sequence (ETAS) model and the short-term earthquake probability (STEP) model are two such earthquake-triggering models [36].

Perhaps because it is difficult to model the actual mechanics of the earth, machine learning models have become more prevalent in the past decade, with the hope that they can improve the predictive capabilities of time-independent and time-dependent models developed by the seismic community.

The authors of [26] use earthquake data from Southern California and the San Francisco Bay region with three different artificial neural networks (ANN) to perform short-term prediction of earthquakes. They introduce the use of eight mathematically-defined seismic parameters known as seismicity indicators that are based on the Gutenberg-Richter inverse power law. These calculated indicators are fed into the neural network algorithm to predict the occurrence or non-occurrence of an earthquake of a magnitude exceeding a pre-defined threshold in the following month. These eight seismic parameters, introduced by [26], are subsequently used by [2–4,6,20] in their studies. The authors of [1,5,17,25] use a subset of the above eight seismic parameters.

A similar work [2] uses probabilistic neural networks on California data to predict the magnitude of the largest earthquake in a seismic region in the following 15 days or one month. The authors of [4] use four machine learning-based regressors to carry out a regression study: generalized linear models, gradient boosting machines, deep learning and random forests. They report the best performance from random forests.

The authors of [29] introduce the use of seven seismic parameters based on the b value in Gutenberg-Richter inverse power law as well as Omori-Utsu and Bath's laws which are input to ANNs with data on Chile to predict whether an event of a particular magnitude will occur in the next five days. Subsequently, the studies [3,4,20] use these features for predicting earthquakes in their studies.

The authors of [3] evaluate the use of artificial neural networks (ANN) to perform short-term (7-day) prediction of earthquakes with a magnitude greater than 5.0 in Japan. The authors of [28] develop a classification model that predicts whether an earthquake with a magnitude above a threshold will take place at a given location in a time range of 30 to 180 days from a given moment of time. They study data from Japan between 1990 to 2016 where the magnitude of the earthquake is higher than 5.0. The authors of [3] use a random forest classifier to predict whether the maximum magnitude of an earthquake in Japan will be greater than 5.5 in the next 7 days.

The study [1] developed a particle swarm optimization-backpropagation (PSO-BP) model to predict earthquake magnitude in the following year in Japan. They used all information regarding major, minor and aftershock earthquake events as input into the backpropagation neural network.

The authors of [17] develop several data mining and time series analysis models to make long-term earthquake predictions in Israel. They predict whether the maximum earthquake magnitude in the following year will exceed the median of the yearly maximum magnitudes in the corresponding seismic region. They calculate the probability that the maximum magnitude of n events in the forecasted period exceeds some threshold as:

$$Prob(M_{max} \geq th) = 1 - (1 - 10^{-b(th-M_0)})^n \tag{1}$$

However, because the number of events in the forecasted period (n) is not known in advance, they estimate n using a moving average of the number of events in the 1 to 10 years prior to the forecasted period. Therefore, they introduce these moving averages MA(1) - MA(10) and their associated probabilities ProbMax(1) – ProbMax(10) as 20 additional features for their models.

There are many additional studies in the data mining literature, which predict earthquake occurrence in a particular region using various seismic indicators as well as other mathematically computed features based on past earthquake occurrence and magnitude (see Table 1). Our paper extends these studies by combining various seismic indicators with other time-series type features to create a unified model that can predict earthquakes in multiple locations.

3 Materials and Methods

The process used in this study is shown in Fig. 1.

3.1 Data

The data used in this study was obtained from three different sources, according to the location of the earthquakes.

California Earthquake Data. The data for California earthquakes was obtained from the United States Geological Survey (USGS) earthquake catalog [37]. The number of earthquakes from January 1967 to December 2017 located within California or slightly off the coast, was 204,874. The training set was comprised of events from the years 1967 to 2003; the validation set was comprised of events from the years 2004 to 2009; and the testing set included events from the years 2010 to 2018. The distribution of the California data is shown in Fig. 2.

Fields used from the earthquake catalog included: date and time of the event, longitude and latitude of the event, magnitude of the event.

Table 1. Machine Learning Methods Used to Predict Earthquakes in the Literature.

Reference	Dataset	ML Methods	Target
[26]	California	ANN	binary: event of mag $\geq$ thresh occurs in next month
[2]	California	Probabilistic NN	classify mag of event into 1 of 7 ranges in 15 days or 1 month
[4]	California	Regression, gradient boosting, ANN, random forests	Max mag of event in next 7 days
[5]	India, Chile, California	Support Vector Regressor, Hybrid NN	binary: event of mag $\geq$ 5.0 occurs
[6]	Hindukush, India	Pattern recognition NN, Recurrent NN, Random Forest, LPBoost	binary: event of mag $\geq$ 5.5 occurs in next month
[28]	Japan	Random Forest, Adaboost, gradient boosting, logistic regression	binary: event of mag $\geq$ 5.0 occurs in 30-180 days
[25]	Himalayas, India	BP neural network	Number of days until event within magnitude range occurs
[1]	Japan	PSO-BP neural network	Max mag of event in next year
[3]	Japan	ANN	binary: event of mag $\geq$ 5.0 occurs in next 7 days
[29]	Chile	ANN	binary: event of mag $\geq$ thresh occurs in next 5 days
[22]	Spain	ANN	binary: event of mag $\geq$ thresh occurs in next 7 days
[20]	Chile, Spain	ANN	binary: event of mag $\geq$ thresh occurs in next 5 days
[24]	Greece	BP-NN	Max mag of event in next day
[17]	Israel	IFN, M-IFN	Binary event: event of mag $\geq$ thresh occurs in next year; number of events

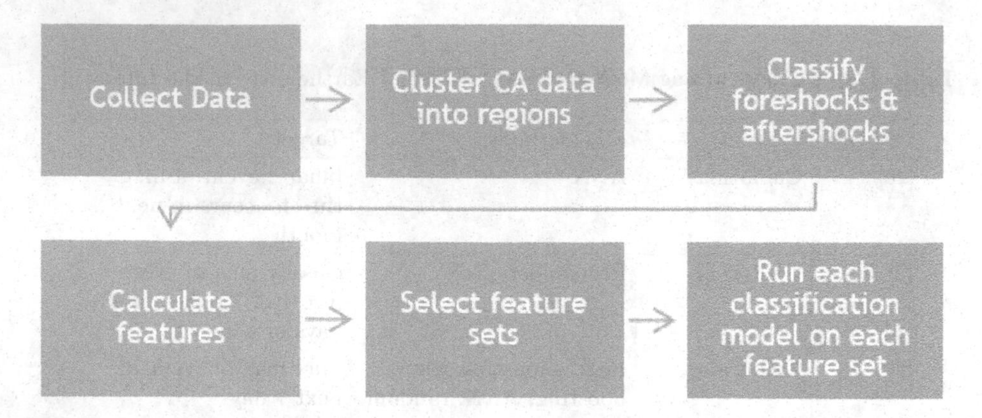

Fig. 1. Pipeline of Methods Used.

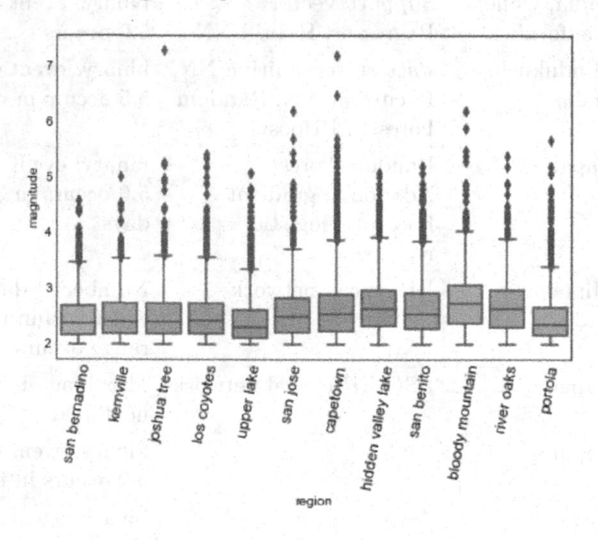

Fig. 2. Distribution of California Earthquakes by Region.

No specific seismic regions are defined for the earthquake events in the USGS catalog; therefore clustering was performed to determine the areas to use. The authors of [35] also performed clustering to determine the areas to use for earthquake prediction in Indonesia. The clustering procedure used in this study is discussed in more detail in Sub-sect. 3.2.

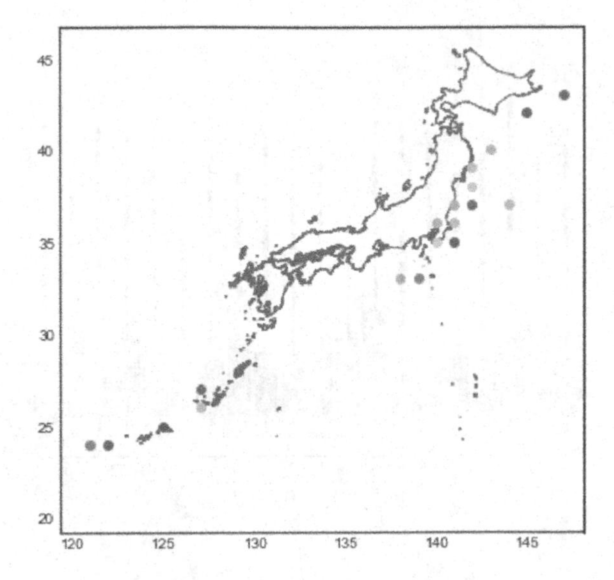

Fig. 3. Map of Japan Earthquake Regions.

Japan Earthquake Data. The complete Japanese earthquake catalog was obtained from the JMA [14]. Data with a magnitude of 2.0 or more consists of 559,666 entries from January 1998 to December 2017. Events from the years 1998 to mid-2010 were used for training; events from mid-2010 through 2012 were used for validation; and events from 2013 through 2017 were used for testing. The distribution of the magnitudes of the Japanese Data is shown in Fig. 4.

The fields used from the catalog included: date and time of event, longitude and latitude of event, magnitude of event, region, region name. The top 20 regions of earthquake activity were selected for analysis. This included 280,084 events, which comprised 50% of the data. The locations of the earthquake areas used are indicated in Fig. 3. Note that since the locations of the events in the Japan catalog were rounded, each point on the graph represents multiple events.

Israel Earthquake Data. Data for Israel earthquakes was obtained from the Geophysical Institute of Israel (GII) database [11]. The complete data is from 1983 to 2019 and it contains 23,676 earthquake events. The events from the years 1983 through 2009 were used for training; the events from 2010 through 2013 were used for validation; and the events from 2014 through 2019 were used for testing. The distribution of earthquakes by seismic regions is shown in Fig. 5.

The fields used from the catalog include: date and time of event; magnitude of the earthquake expressed as either local Richter, estimated body waves or seismic moment; location given by latitude and longitude; and region of the event.

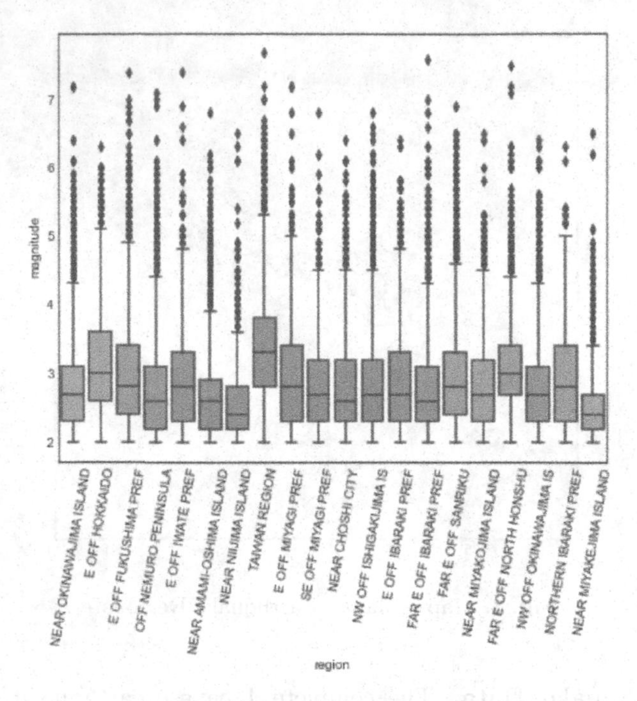

Fig. 4. Distribution of Japan Earthquakes by Region.

In many instances, only one of the three magnitude values was recorded, but when more than one value was recorded, the maximum value was taken as the magnitude. After removing all earthquakes with undefined magnitude or magnitude below 2.0, there were 11,287 events left. The top 10 regions of the highest earthquake activity were selected for analysis. This included 8,820 events, which comprised 78% of the data. The locations of the events are shown in Fig. 6. Each region is indicated by a different color to separate it from adjacent regions and labeled accordingly.

3.2 Data Preprocessing

Clustering California Earthquake Data. The goal of this study is to predict whether earthquakes above a certain magnitude will occur within a particular region of interest in the next year. Therefore, it is necessary to have a unique definition of the region of each earthquake used in the study. Because the catalog of earthquake events from California did not include the region of an earthquake event, the Hierarchical Density-Based Spatial clustering (HDBSCAN) algorithm was used to cluster all of the earthquakes in the catalog into geographical regions [13].

The HDBSCAN algorithm finds clusters, or dense regions, of a dataset. The algorithm utilizes the "core distance" to represent the distance of a point to the kth nearest neighbor. If the point is in a sparse area, this distance will be larger;

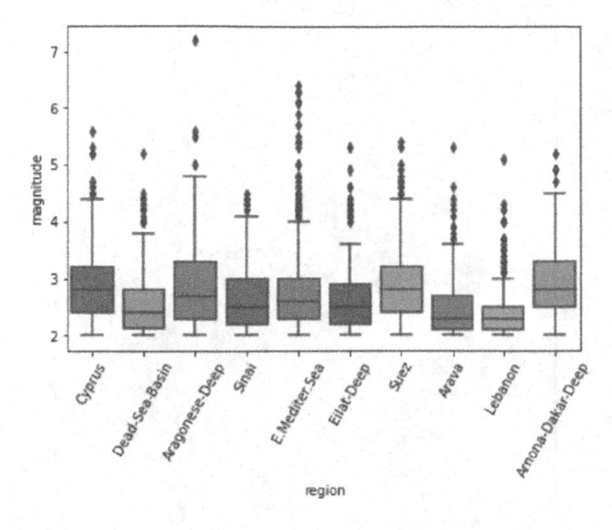

Fig. 5. Distribution of Israel Earthquake Magnitudes.

and if the point is in a dense area, this distance will be much smaller. The algorithm then combines this core distance for two points with their standard metric distance between each other to obtain their mutual reachability distance. This allows the algorithm to penalize points that are in sparser areas while allowing dense points (those with a low core distance) to remain the same distance from each other. The result is that it is possible to distinguish regions of high earthquake activity on a map.

The algorithm was customized using a minimum cluster size of 1000 and a minimum sample value of 500. This allows the algorithm to find clusters larger than 1000 along with reducing the number of earthquakes designated as noise (not classified). Approximately one-quarter of the earthquakes were not classified and the 12 clusters with the largest amount of earthquakes were selected for analysis. The selected clusters are shown in Fig. 7.

Data Cleaning. Many studies using earthquake catalogs follow the practice of removing foreshock and aftershock events before computing various seismic indicators. An aftershock is a minor shock that follows the mainshock of an earthquake and a foreshock is a minor tremor of the earth followed by a larger earthquake within a short period of time at the same location [23].

We have extended the algorithm for foreshock and aftershock identification presented in [17] by incorporating the distance between earthquakes as a parameter for determining foreshock/aftershock classification. The distance information is based on the Window Algorithm for Aftershocks in [10]. They provide a table with the maximum distance (in km) and maximum time (in days) that a subsequent earthquake must be from the main shock to be considered an aftershock based on the magnitude of the earthquake. For example, an event following an earthquake of magnitude 3.0 that is within a distance of 22.5 km and within

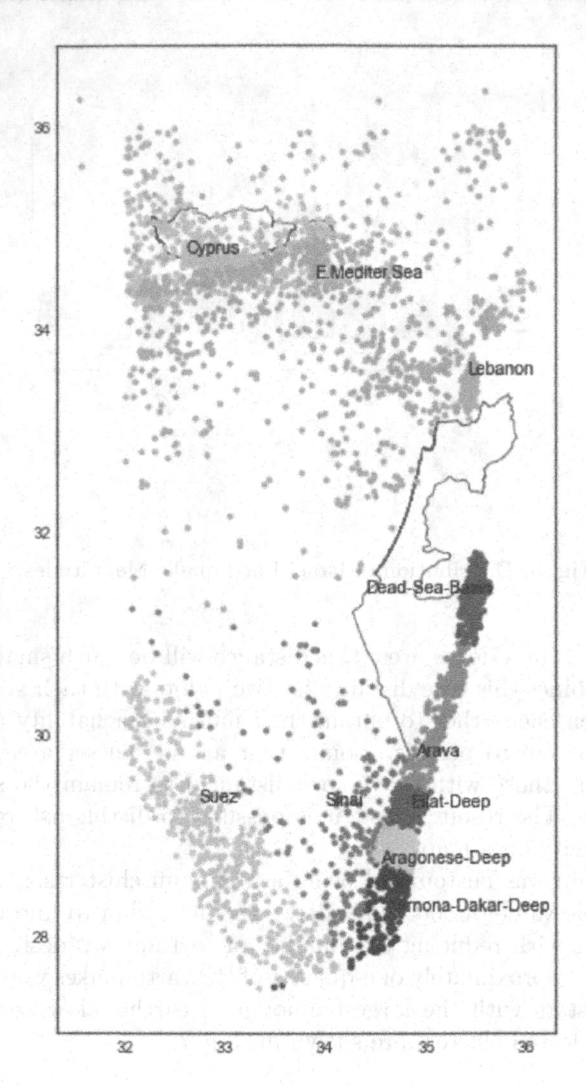

Fig. 6. Map of Israel Earthquake Regions.

11.5 days is considered an aftershock. Foreshocks were taken to be within half of the time specified for aftershocks. The maximum distance for foreshocks and aftershocks was the same.

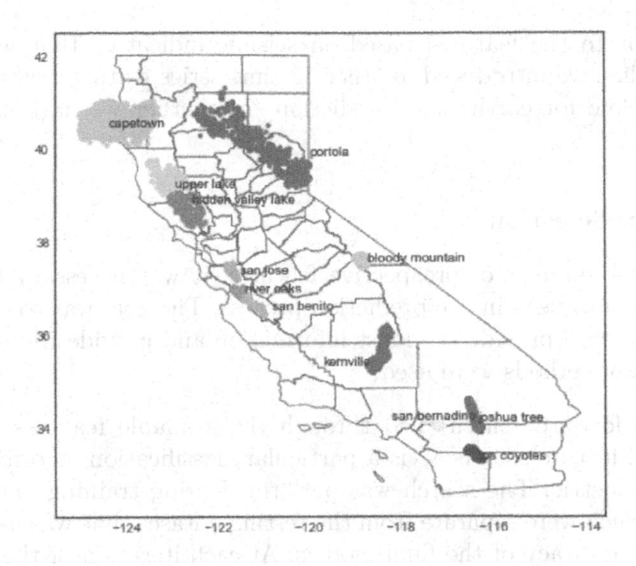

Fig. 7. Map of California with Earthquake Clusters.

3.3 Feature Engineering

The goal of this study is to predict whether the highest earthquake magnitude in a given area in the next year will be greater than the median of the maximum yearly magnitudes in that area. For this reason, the features to be used are based on various seismic indicators used in the literature and are calculated at the end of the current year. This includes all of the features used by [17], such as six of the eight seismicity indicators introduced by [26]. In addition, [17] use moving averages of the number of events in the one to ten years prior to the forecasted period (ma1 - ma10) and moving averages of the probability of the maximum magnitude exceeding a threshold within one to ten years prior to the forecasted period (prob1 - prob10). These indicators are calculated on the mainshock earthquake events. This comprises a total of 26 features.

We adopted 16 additional features which are also based on seismic indicators in the literature. These features include:

- $x_1 - x_5$: various increments between the values of the slope of the regression line based on the Gutenberg-Richter inverse power law [29]. Features $x_1 - x_5$ were calculated on all of the data.
- x_6 is defined in the literature as the maximum magnitude from the events recorded during the last week in the area analyzed [29]. Moving averages of this feature were calculated for 1 to 10 years of the maximum magnitude from all events of the year [24] ($x_{6\_1} - x_{6\_10}$).
- x_7 is the probability of an earthquake with a magnitude larger or equal to 6.0 [29]. This value was calculated on all of the data.

In addition to the features based on seismic indicators that were used in previous studies, we introduced another 52 time-series features, which have not been used before for earthquake prediction. Altogether we have extracted 94 features.

3.4 Feature Selection

Given the large number of prospective features, it was necessary to limit the number of features used in the prediction process. The goal was to choose those features that would provide the most information and provide the most predictive value. Two methods were used:

- Perform a forward search (FS) through the available features to generate an optimal feature set vis a vis a particular classification algorithm and the evaluation metric. The search was performed using training and validation datasets which were separate from the testing dataset that was used for comparing the accuracy of the final models. At each iteration of the search, the classification algorithm was run using the "currently best feature set" with the addition of each of the remaining features. The feature that provided the largest increase in the evaluation metric was then added to the "currently best feature set". The search ended when the evaluation metric began to decrease in value. This method takes into consideration the relationship of the features with the target variable as well as the relationship between the features themselves, however it is highly influenced by the algorithm being used for evaluation. Thus, different feature sets were generated for each of three algorithms: C4.5, Random Forest and XGBoost. The search procedure considered either all 94 features, or only the features identified as having the highest normalized Information Gain Ratio (IGR) [12] scores.
- Using only the features included in the classification model built by the Info-Fuzzy Network (IFN) algorithm [19]. IFN is a greedy search algorithm aimed at finding a minimal subset of features having the highest statistically significant mutual information with the target variable.

3.5 Prediction Algorithms

Earthquake prediction in this study is defined as a binary classification task based on the median of maximum yearly magnitudes in a given area [17]. A forecast for a specific year is labeled as "yes" if the maximum earthquake magnitude exceeds the median and as "no" if it is below the median. The median is used as the prediction threshold because it presents a balanced classification task.

In this study, the following eight classification algorithms are used to induce prediction models from the earthquake data:

- C4.5 Decision Tree [30]: using the entropy criterion and a minimum samples split of 10

- AdaBoost classifier [31]: using the decision tree classifier with entropy criterion and a maximum depth of 1
- XGBoost classifier [33]: using a maximum depth of 5, binary:logistic objective, 100 estimators and a learning rate of 0.3
- Info-Fuzzy Network (IFN) [18]: using the default setting (p-value = 0.001)
- K-nearest neighbors (k-NN) [8]: using 5 neighbors
- Random Forest (RF) [7]: using 2 jobs and random state of 0
- Support Vector Machine (SVM) [32]: using probability of True and gamma equal to auto
- Logistic Regression [34]: using the default settings.

4 Results

4.1 Analysis of the California Earthquake Data

After removing all foreshocks and aftershocks from the California data in the regions selected for the analysis, there were 21,236 events remaining. Various seismic and time series features were calculated for each year. Each of the seven feature sets acquired was run on all of the compared algorithms. As shown in Tables 2 and 3, the evaluated feature sets provide AUC values from 0.500 to 0.758, with the logistic regression and SVM algorithms providing the best results.

4.2 Analysis of the Japanese Earthquake Data

The Japan earthquake data consists of 559,666 entries for 20 years. Because twenty years did not provide enough observations to train and test a model, the data was divided into quarters, and the prediction was done for the next quarter rather than the next year. This resulted in 80 quarters of data, of which 10 were used to extract various moving average features. Of the remaining 70 quarters, 40 were used for training, 10 for validation, and 20 for testing.

Each of the seven acquired feature sets was run on each algorithm. As shown in Tables 2 and 3, the results were better using the Japan data than the California data. The algorithms with the highest AUC scores were AdaBoost, XGBoost, Random Forest, Logistic Regression, and IFN. AUC scores ranged up to 0.825 with the Logistic Regression algorithm.

4.3 Analysis of Israel Data

There were 26 years of data for Israel from 1994 to 2019. Because there was less data in general for the Israeli regions, there was not enough data to calculate some of the time series features. Specifically, the features based on 75 or 100 prior events could not be calculated. In total, there were 82 features used for prediction using the Israel data.

Each of the acquired feature sets was evaluated with each algorithm. According to the results shown in Tables 2 and 3, the algorithms that provided the best results on the Israel data were Random Forest and Logistic Regression, producing AUC values up to 0.710.

Table 2. Results of Prediction (AUC): California, Japan, Israel Data

Algorithm	Method to obtain feature set	CA	Japan	Israel
C4.5	C4.5 search all	0.631	**0.721**	0.634
C4.5	C4.5 search IGR	0.600	0.674	0.501
C4.5	XGBoost search all	0.587	**0.712**	0.537
C4.5	XGBoost search IGR	0.637	0.670	0.597
C4.5	Random Forest search all	0.654	0.675	0.526
C4.5	Random Forest search IGR	0.577	0.680	0.517
C4.5	IFN	0.638	**0.703**	0.468
AdaBoost	C4.5 search all	0.622	**0.789**	0.578
AdaBoost	C4.5 search IGR	0.596	**0.799**	0.653
AdaBoost	XGBoost search all	0.670	**0.793**	0.616
AdaBoost	XGBoost search IGR	0.636	**0.792**	0.614
AdaBoost	Random Forest search all	**0.738**	**0.789**	0.681
AdaBoost	Random Forest search IGR	0.633	**0.800**	0.635
AdaBoost	IFN	0.558	**0.759**	0.587
XGBoost	C4.5 search all	0.679	**0.774**	0.642
XGBoost	C4.5 search IGR	0.641	**0.781**	0.485
XGBoost	XGBoost search all	0.671	**0.760**	0.668
XGBoost	XGBoost search IGR	0.632	**0.749**	0.665
XGBoost	Random Forest search all	0.691	**0.783**	0.550
XGBoost	Random Forest search IGR	0.654	**0.771**	0.490
XGBoost	IFN	0.611	**0.791**	0.553
Random Forest	C4.5 search all	0.649	**0.762**	0.595
Random Forest	C4.5 search IGR	0.663	**0.775**	0.585
Random Forest	XGBoost search all	0.695	**0.770**	0.573
Random Forest	XGBoost search IGR	0.661	**0.757**	**0.710**
Random Forest	Random Forest search all	0.697	**0.793**	0.587
Random Forest	Random Forest search IGR	0.672	**0.766**	0.486
Random Forest	IFN	0.651	**0.778**	0.526

Table 3. Results of Prediction (AUC): California, Japan, Israel Data (cont.)

Algorithm	Method to obtain feature set	CA	Japan	Israel
k-NN	C4.5 search all	0.593	0.601	0.438
k-NN	C4.5 search IGR	0.604	0.698	0.490
k-NN	XGBoost search all	0.593	0.645	0.582
k-NN	XGBoost search IGR	0.661	0.698	0.513
k-NN	Random Forest search all	0.697	0.593	0.540
k-NN	Random Forest search IGR	**0.703**	0.698	0.473
k-NN	IFN	0.593	0.605	0.504
Logistic Regression	C4.5 search all	**0.714**	**0.823**	0.656
Logistic Regression	C4.5 search IGR	0.576	**0.727**	0.691
Logistic Regression	XGBoost search all	**0.735**	**0.825**	0.404
Logistic Regression	XGBoost search IGR	**0.734**	**0.727**	0.425
Logistic Regression	Random Forest search all	**0.741**	**0.821**	**0.709**
Logistic Regression	Random Forest search IGR	**0.735**	**0.727**	0.609
Logistic Regression	IFN	**0.738**	**0.815**	0.669
SVM	C4.5 search all	0.665	**0.737**	0.413
SVM	C4.5 search IGR	0.500	0.500	0.469
SVM	XGBoost search all	0.626	0.612	0.530
SVM	XGBoost search IGR	**0.735**	0.500	0.485
SVM	Random Forest search all	**0.758**	0.587	0.503
SVM	Random Forest search IGR	**0.735**	0.500	0.549
SVM	IFN	0.728	0.506	0.539
IFN	All features	**0.708**	**0.791**	0.543

4.4 Development of a Single Prediction Model for All Countries

It was of great interest to determine if it would be possible to induce one model, based on one of the data sets that would allow a reasonable level of prediction for the other datasets. Toward this end, the best models for California and Japan were used to make predictions in the other regions. The results are provided in Table 4.

The models induced from either California or Japan data were more accurate on Israel data than the models trained using the Israeli data. The best result for the Israel data was using a logistic regression algorithm trained using the Japan data with the features obtained from the Random Forest search; an AUC score of 0.726 was achieved. Using the California data with the features obtained from the XGBoost search an AUC score of 0.707 was achieved.

The results were also quite good for models trained on California data and tested on Japan data. AUC values of up to 0.821 were achieved using a logistic regression model. The results for models trained on the Japan data and tested using the California data were also reasonable (AUC values up to 0.749).

Table 4. One Prediction Model for All Countries

Algorithm	Feature set	CA	Japan	Israel
California data used to train model				
AdaBoost	Random Forest search all	**0.738**	0.588	0.602
XGBoost	Random Forest search all	0.691	**0.709**	0.635
XGBoost	IFN	0.611	**0.730**	0.565
Random Forest	Random Forest search all	0.697	**0.704**	0.591
Random Forest	XGB search all	0.695	**0.776**	**0.706**
Logistic Regression	XGBoost search all	**0.735**	**0.821**	**0.707**
Logistic Regression	Random Forest search all	**0.741**	**0.797**	0.643
IFN	IFN	**0.708**	**0.754**	0.652
Japan data used to train model				
AdaBoost	Random Forest search all	0.680	**0.789**	0.654
XGBoost	Random Forest search all	**0.704**	**0.783**	0.699
XGBoost	IFN	0.683	**0.791**	0.633
Random Forest	Random Forest search all	**0.704**	**0.793**	0.682
Random Forest	IFN	0.680	**0.778**	0.668
Logistic Regression	XGBoost search all	**0.728**	**0.825**	0.676
Logistic Regression	Random Forest search all	**0.749**	**0.821**	**0.726**
IFN	IFN	**0.715**	**0.791**	0.665

5 Conclusion

Earthquake data from three different geographic regions in the world was analyzed to predict whether an earthquake of magnitude greater than the regional median would occur in that region during the following year or quarter. The IFN algorithm was very successful at selecting useful feature sets for prediction. In addition, the forward search algorithm used with Random Forest or XGBoost was quite effective to choose an effective feature set. The most effective prediction models were XGBoost, Random Forest, AdaBoost, IFN, and logistic regression. The logistic regression models performed consistently better for all countries.

The induced models achieved AUC values of up to 0.825 for predictions within their respective regions. When the same models were used to predict earthquake magnitudes in another region, the Japan models achieved higher AUC values than the models trained with their own data (up to 0.749). Given these results, it seems quite possible to develop a generic model that could be very effective in predicting earthquake activity across the world.

Using a set of features that combine seismic information of the area with information regarding the magnitudes of past earthquakes and time-series information of the occurrence of past earthquakes, a single classification model is able to provide a reasonable level of prediction accuracy regarding the expected earthquake activity in a particular region.

The fact that there were only 20 years of data for Japan was a limitation and thus the prediction was performed for the next quarter though it would be preferable to have the prediction periods consistent across the countries. It is also noteworthy that the quarterly number of earthquakes in Japan was still higher than the yearly number of earthquakes in most California regions, which implies that the models induced for a specific prediction period in one region may be applicable across regions to different prediction periods characterized by different levels of seismic activity.

The number of earthquake events for Israel was considerably lower than for the two other regions and this created a slight inconsistency regarding the feature sets used across the regions. In the future, it would be interesting to do a similar analysis on shorter prediction intervals, such as months, and thereby have enough data to induce sequence-to-sequence deep learning models (e.g., LSTM or transformers).

References

1. Abraham, A., Rohini, V.: A particle swarm optimization-backpropagation (PSO-BP) model for the prediction of earthquake in Japan. In: Shetty, N.R., Patnaik, L.M., Nagaraj, H.C., Hamsavath, P.N., Nalini, N. (eds.) Emerging Research in Computing, Information, Communication and Applications. AISC, vol. 882, pp. 435–441. Springer, Singapore (2019). https://doi.org/10.1007/978-981-13-5953-8_36
2. Adeli, H., Panakkat, A.: A probabilistic neural network for earthquake magnitude prediction. Neural Netw. **22**(7), 1018–1024 (2009)
3. Asencio-Cortes, G., Martinez-Alvarez, F., Troncoso, A., Morales-Esteban, A.: Medium-large earthquake magnitude prediction in Tokyo with artificial neural networks. Neural Comput. Appl. **28**(5), 1043–1055 (2017)
4. Asencio-Cortes, G., Morales-Esteban, A., Shang, X., Martinez-Alvarez, F.: Earthquake prediction in California using regression algorithms and cloud-based big data infrastructure. Comput. Geosci. **115**, 198–210 (2018)
5. Asim, K., Idris, A., Iqbal, T., Martinez-Alvarez, F.: Earthquake prediction model using support vector regressor and hybrid neural networks. PLoS One **13**(7), e0199004 (2018)
6. Asim, K., Martinez-Alvarez, F., Basit, A., Iqbal, T.: Earthquake magnitude prediction in Hindukush region using machine learning techniques. Nat. Hazards **85**(1), 471–486 (2017)
7. Biau, G., Scornet, E.: A random forest guided tour. TEST **25**(2), 197–227 (2016). https://doi.org/10.1007/s11749-016-0481-7
8. scikit-learn developers: sklearn.ensemble.AdaBoostClassifier (2019). https://scikit-learn.org/stable/modules/neighbors.html
9. Galkina, A., Grafeeva, N.: Machine learning methods for earthquake prediction: a survey. In: Proceedings of the Fourth Conference on Software Engineering and Information Management (SEIM-2019), Saint Petersburg, Russia, vol. 13, p. 25 (2019)
10. Gardner, J., Knopoff, L.: Is the sequence of earthquakes in southern California, with aftershocks removed, Poissonian? Bull. Seismol. Soc. Am. **64**(5), 1363–1367 (1974)

11. Geological Survey of Israel: Earthquakes (2020). https://eq.gsi.gov.il/en/indexEn.php
12. Han, J., Pei, J., Kamber, M.: Data Mining: Concepts and Techniques. Elsevier, Amsterdam (2011)
13. HDBSCAN: How HDBSCAN Works (2019). https://hdbscan.readthedocs.io/en/latest/how_hdbscan_works.html
14. Japan Meteorological Agency: JMA Earthquakes (2020). https://www.data.jma.go.jp/svd/eqev/data/bulletin/eqdoc_e.html
15. Jordan, T.H., et al.: Operational earthquake forecasting. State of knowledge and guidelines for utilization. Ann. Geophys. **54**(4), 315–391 (2011)
16. Kattamanchi, S., Tiwari, R.K., Ramesh, D.S.: Non-stationary etas to model earthquake occurrences affected by episodic aseismic transients. Earth Planets Space **69**(1), 157 (2017)
17. Last, M., Rabinowitz, N., Leonard, G.: Predicting the maximum earthquake magnitude from seismic data in Israel and its neighboring countries. PLoS One **11**(1), e0146101 (2016)
18. Last, M.: Multi-objective classification with info-fuzzy networks. In: Boulicaut, J.-F., Esposito, F., Giannotti, F., Pedreschi, D. (eds.) ECML 2004. LNCS (LNAI), vol. 3201, pp. 239–249. Springer, Heidelberg (2004). https://doi.org/10.1007/978-3-540-30115-8_24
19. Maimon, O., Last, M.: Knowledge Discovery and Data Mining-The Info-Fuzzy Network (IFN). Methodology. Kluwer Academic Publishers, Massive, Computing, Boston (2000)
20. Martinez-Alvarez, F., Reyes, J., Morales-Esteban, A., Rubio-Escudero, C.: Determining the best set of seismicity indicators to predict earthquakes. Two case studies: Chile and the Iberian Peninsula. Knowl.-Based Syst. **50**, 198–210 (2013)
21. Marzocchi, W., Zechar, J.D.: Earthquake forecasting and earthquake prediction: different approaches for obtaining the best model. Seismol. Res. Lett. **82**(3), 442–448 (2011)
22. Morales-Esteban, A., Martínez-Álvarez, F., Reyes, J.: Earthquake prediction in seismogenic areas of the Iberian Peninsula based on computational intelligence. Tectonophysics **593**, 121–134 (2013)
23. Morales-Esteban, A., Martinez-Alvarez, F., Troncoso, A., Justo, J., Rubio-Escudero, C.: Pattern recognition to forecast seismic time series. Expert Syst. Appl. **37**(12), 8333–8342 (2010)
24. Moustra, M., Avraamides, M., Christodoulou, C.: Artificial neural networks for earthquake prediction using time series magnitude data or seismic electric signals. Expert Syst. Appl. **38**(12), 15032–15039 (2011)
25. Narayanakumar, S., Raja, K.: A BP artificial neural network model for earthquake magnitude prediction in Himalayas. India. Circuits Syst. **7**(11), 3456–3468 (2016)
26. Panakkat, A., Adeli, H.: Neural network models for earthquake magnitude prediction using multiple seismicity indicators. Int. J. Neural Syst. **17**(01), 13–33 (2007)
27. Petersen, M.D., Cao, T., Campbell, K.W., Frankel, A.D.: Time-independent and time-dependent seismic hazard assessment for the state of California: uniform California earthquake rupture forecast model 1.0. Seismol. Res. Lett. **78**(1), 99–109 (2007)
28. Proskura, P., Zaytsev, A., Braslavsky, I., Egorov, E., Burnaev, E.: Usage of multiple RTL features for earthquakes prediction. In: Misra, S., et al. (eds.) ICCSA 2019. LNCS, vol. 11619, pp. 556–565. Springer, Cham (2019). https://doi.org/10.1007/978-3-030-24289-3_41

29. Reyes, J., Morales-Esteban, A., Martinez-Alvarez, F.: Neural networks to predict earthquakes in Chile. Appl. Soft Comput. **13**(2), 1314–1328 (2013)
30. scikit-learn developers: Decision Trees (2019). https://scikit-learn.org/stable/modules/tree.html
31. scikit-learn developers: sklearn.ensemble.AdaBoostClassifier (2019). https://scikit-learn.org/stable/modules/generated/sklearn.ensemble.AdaBoostClassifier.html
32. scikit-learn developers: Support Vector Machines (2019). https://scikit-learn.org/stable/modules/svm.html
33. scikit-learn developers: sklearn.ensemble.GradientBoostingClassifier (2020). https://scikit-learn.org/stable/modules/generated/sklearn.ensemble.GradientBoostingClassifier.html
34. scikit-learn developers: sklearn.$linear_model$.LogisticRegression (2020). https://scikit-learn.org/stable/modules/generated/sklearn.linear_model.LogisticRegression.html
35. Shodiq, M.N., Kusuma, D.H., Rifqi, M.G., Barakbah, A.R., Harsono, T.: Adaptive neural fuzzy inference system and automatic clustering for earthquake prediction in Indonesia. JOIV: Int. J. Inform. Vis. **3**(1), 47–53 (2019)
36. Thenhaus, P., Campbell, K., Khater, M.: Spatial and temporal earthquake clustering: part 2 - earthquake aftershocks. EQECAT (2012)
37. USGS: Search Earthquake Catalog (2019). https://earthquake.usgs.gov/earthquakes/search/

A Bag of Tokens Neural Network to Predict Webpage Age

Klaas Meinke(✉), Tamis Achilles van der Laan, Tiberiu Iancu, and Ceyhun Cakir

Hadrian Security, Leidseplein 1, Amsterdam, The Netherlands
{klaas,tamis,tiberiu,ceyhun}@hadrian.io

Abstract. Outdated technologies pose a significant security threat to websites and hackers often hone in on the oldest pages on a site to discover vulnerabilities. To improve the efficiency of (automated) penetration testers, we invent a machine learning method that predicts the age of a webpage. An HTML-specific tokenizer is trained and used to tokenize HTML bodies, which are then transformed into binary vector encodings (a "bag of tokens"). We train a Multi-Layer Perceptron neural network on such encodings, using historical snapshots of webpages as our training data. Our method achieves a mean absolute error of 1.58 years on validation data held out from training.

Keywords: Penetration Testing · Machine Learning · Natural Language Processing · Attack Surface Management

1 Introduction

Recent research by Demir et al. shows that 95% of websites run on outdated software with vulnerabilities, for which patches are available [2]. Hackers know this and scan potential targets, searching for web pages running on legacy technology with known vulnerabilities. Olivier Beg, a well-known bug bounty hunter, is in both Yahoo and Google's security hall of fame and has found 383 confirmed vulnerabilities on HackerOne.[1] In his words:

> "When assessing a potential target, a good way to figure out if something is vulnerable is its age. Pages built on old technologies are prone to vulnerabilities, indicating that more time should be spent on those assets."

To our knowledge, there are no automated solutions that predict the age of a webpage. Consequently, scanning an organization's site for old pages is a slow and manual task. To scan and protect an organization proactively and find vulnerabilities before a hacker would be able to, an algorithm is needed that automatically zeroes in on old pages as a means of penetration.

[1] https://hackerone.com/smiegles?type=user.

S. Dolev et al. (Eds.): CSCML 2023, LNCS 13914, pp. 170–180, 2023.
https://doi.org/10.1007/978-3-031-34671-2_12

2 Methods

Our methodology is visualized in a flow chart in Fig. 1.

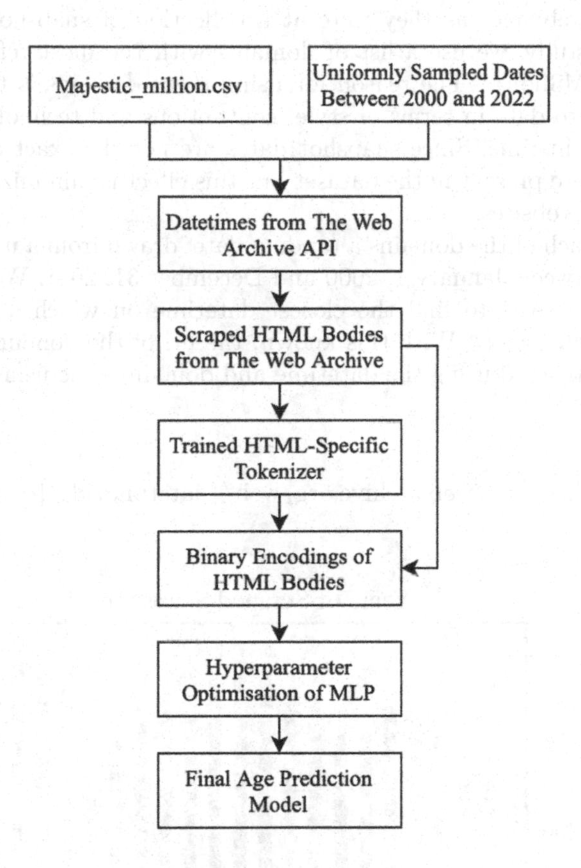

Fig. 1. Flow chart of the project methodology.

First, random datetimes sampled from a uniform distribution are assigned to a list of well-connected domains. For each domain, the snapshot closest to its assigned datetime is fetched from The Web Archive, an online resource that provides historical webpage snapshots.[2] Next, we train an HTML-specific tokenizer on the fetched HTML bodies, use it to tokenize them, and transform them into binary vector-encodings. Finally, a Multi-Layer-Perceptron's (MLP) hyperparameters are optimized for the task of predicting a webpage's age from its binary encoding. The process is further elucidated in the following sections.

[2] https://web.archive.org/.

2.1 Data Collection

To optimize a machine learning algorithm to predict webpage age, we first collect a dataset of pages of varying ages, which we scrape from The Internet Archive, an archive of webpages as they were at a collection of snapshot times in the past. For our study, we use a list of domains with the most referring subnets (the Majestic Million).[3] The reason for using these domains is that we expect them to be up-to-date in terms of style, conventions and technology stack at a given snapshot in time. Since snapshot dates are not the exact creation dates, noise and bias are present in the dataset but this effect is minimized by choosing well-connected websites.

We assign each of the domains a random date, drawn from a uniform random distribution between January 1, 2000 and December 31, 2021. We then use The Internet Archive API to find the closest datetime on which a snapshot of a domain has been saved.[4] With this known, the uri of the domain snapshot can be simply constructed using the datetime and domain name using the following formula:

$$\text{uri} = \text{“https://web.archive.org/web/\{datetime\}id\_/\{domain\}”} \tag{1}$$

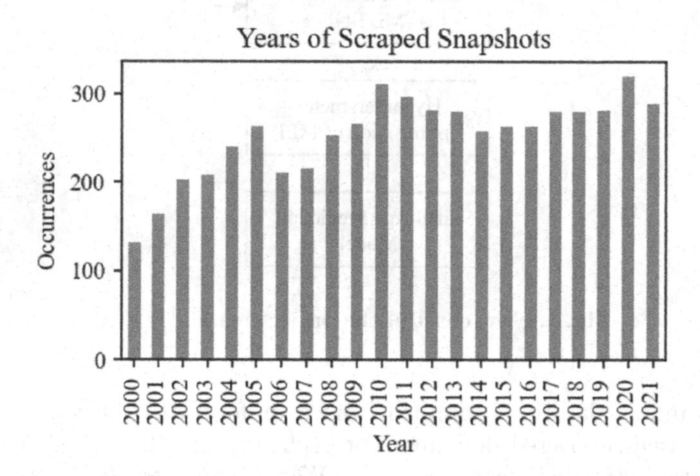

Fig. 2. Distribution of the years of the scraped webpages. Each page's HTML body is saved in the dataset to be used for training and validation.

The HTML bodies of the pages are scraped using the Python implementation of the Puppeteer crawler (pyppeteer) with stealth mode active.[5,6] Due to The

[3] https://downloads.majestic.com/majestic_million.csv.

[4] https://archive.org/help/wayback_api.php.

[5] https://pypi.org/project/pyppeteer/.

[6] https://pypi.org/project/pyppeteer-stealth/.

Internet Archive's rate limits, we find that scraping in batches of 15 uris per minute allows the scraping algorithm to run smoothly without being blocked. We clean our data by removing any API call results that do not match the expected datetime format and, similarly, any HTML bodies that are empty or contain the string "429 Too Many Requests". The dataset used for the remainder of the study consists of 5,569 HTML bodies.

Some of the domains did not exist in the early 2000s and so, despite the uniform sampling mentioned above, the timestamps of the HTML bodies are not uniformly distributed, instead following the distribution shown in Fig. 2.

2.2 Tokenization and Encoding

To transform an HTML body into a feature vector, a tokenizer is used, as is common in state-of-the-art transformer models [15]. We choose to train a new tokenizer since the patterns in HTML differ from those in natural language and we train this tokenizer with the corpus of scraped HTML bodies.

The webpages contain many languages and unicode characters, so using all characters in the corpus to initialize tokenizer training would lead to a pre-trained token list that is greater than 10,000. Instead, we opt for byte-pair encoding, which limits the original characters to the 256 ASCII characters and all other characters in the corpus are represented by byte-level combinations of these [13]. The token list is initialized with the 256 ASCII characters and during each training iteration, the most commonly co-occurring pair of tokens in the corpus is combined to create a new token that is added to the list. The training is stopped when the token list reaches a length of 10,000. Due to its speed and documentation, we opt for the Hugging Face implementation of this algorithm.[7] The BPE algorithm described above was a break-through for natural language processing, but recently other models such as Wordpiece, SentencePiece and UnigramLM have outperformed BPE for specific tasks so these could also be chosen, although no option has emerged as a silver bullet singular solution [7,8,10,12].

After the token list has been created, the HTML bodies are encoded by binary vectors, where value N of the encoding vector is 1 if token N is present in the body and 0 if it is not, as shown in Fig. 3. Our choice of embedding treats the tokens as independent; their relative positions are not embedded. Such an approach has been shown to be surprisingly effective for text classification problems such as spam email filtering [9,16]. We test feature vectors of (normalized) token occurrence counts and the logarithm of token occurrence counts (plus one) but find mean absolute errors of 2.75 and 3.48 years, respectively, compared to 1.58 years when using a binary feature vector.

[7] https://huggingface.co/docs/transformers/fast_tokenizers.

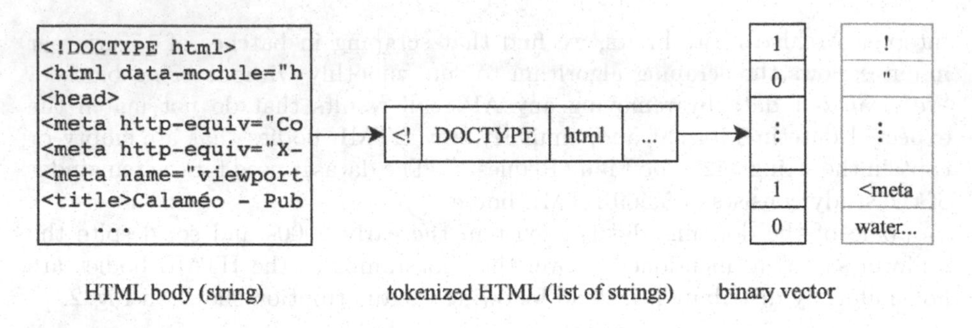

Fig. 3. Mapping scheme from HTML bodies to binary vectors. Each HTML body is tokenized using a list of 10,000 tokens which has been created by training on our corpus (shown in gray on the right). Then the tokenized HTML is transformed into a binary vector (the "bag of tokens") where value N equals 1 if the token N is present.

2.3 Model Architecture

The model we choose to map binary embeddings to webpage age is an MLP. The features are normalized from datetimes to float values between 0 and 1, and 20% of the dateset is held out from training for validation. We minimize the mean squared error loss function using the Adam algorithm and apply sample weighting to negate training bias from the differences in number of pages per year in Fig. 2 [5]. Besides the mean-squared-error loss function, we also track the mean-absolute-error metric with the same sample weighting described above.

We use a four-layer MLP followed by a single-neuron output layer that predicts the (normalized) webpage age. A grid-search hyperparameter optimization is performed to find the optimal number of neurons per layer, for which the number of neurons in the first two and final two layers are drawn from {128, 64, 32} and {64, 32, 16, 8}, respectively. For these tests, we use the optimal activation functions discussed in Subsect. 2.5. We find that the mean absolute error varies between 1.58 and 1.80 years, with a optimal configuration of 64, 32, 16 and 8 neurons per layer, respectively. The architecture is built and trained using the Python package TensorFlow.[8]

We apply dropout to all layers beside the output to prevent overfitting to the training data and find that a dropout rate of 0.2 achieves the best validation metrics [14]. It should be noted that the optimal hyperparameters such as number of layers, number of neurons per layer and dropout rate are dependent on the size of the training corpus. It is to be expected that with more HTML bodies, the neural network would be subjected to less overfitting, allowing for a larger architecture and potentially leading to better accuracy on unseen data.

2.4 ReLU and Dead Neurons

Initially, we choose to apply ReLU activation functions to all layers other than the output since these are commonly used in MLPs [11]. The predictions, in this

[8] https://www.tensorflow.org.

case, are distributed as shown in Fig. 4, with a strong peak in the year 2005, no predictions before 2005, nearly no predictions after 2018 and an otherwise non-uniform distribution. The mean absolute error of the validation predictions is 2.86 years. We hypothesize that this unrepresentative distribution is due to dead neurons in the architecture, which occur when the input to a ReLU activation function is always negative and hence the neuron output and gradient is zero and weight optimization is prevented [1].

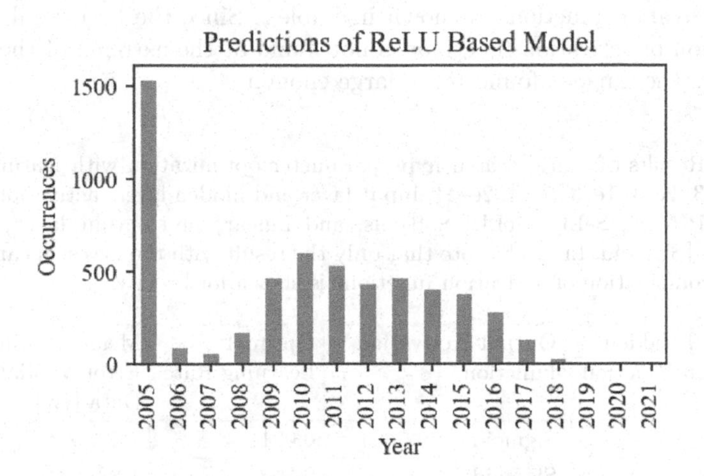

Fig. 4. Predictions of a neural network with ReLU activation functions applied to the input layer and hidden layers and a linear activation function applied to the final layer. The neural network predicts no dates before 2005 and nearly none after 2018. We show that this faulty distribution is caused by dead neurons and therefore test other activation functions than ReLU.

To test our network for dead neurons, we log the outputs of all layers during one complete forward-pass on the training data. We find that 58/64, 11/32, 6/16 and 2/8 of the neurons are dead in each layer, respectively, for which we define a dead neuron as one that outputs 0 for more than 99% of training examples. We also find that for 1159 of 4459 training examples, the final hidden layer passes an array of zeros to the output neuron and hence the model output is equal to the bias of this neuron and no information from the encoding is used in the prediction.

2.5 Activation Functions

To improve the accuracy of the model, we test other activation functions that do not exhibit dying neurons. The activation functions tested for the hidden layers and input layer of the neural network are ELU, PReLU, SeLU, ReLU (for control), Softplus, and Linear [1,4,6,17]. The output layer is tested with two activation functions, Sigmoid and Linear. The Sigmoid function is used because all function evaluations map to values between 0 and 1, or in our case, the years 2000 and 2022, respectively.

To test the different combinations of activation functions, a grid search hyperparameter optimization strategy is deemed sufficiently efficient due to the relatively inexpensive architecture. 20 training epochs are used per run. To not bias towards any of the activation function combinations, the learning rate is additionally included as a search parameter and is varied between 1e–2, 5e–3, 2e–3, 1e–3, 5e–4 and 2e–4. For brevity, only the best result from each combination of activation functions is shown in Table 1. Since the best result for each combination of activation functions is not found at the extrema of the learning rate range, the range is found to be large enough.

Table 1. Results of a grid search hyperparameter optimization with learning rate $\in$ {1e–2, 5e–3, 2e–3, 1e–3, 5e–4, 2e–4}, input layer and hidden layers activation function $\in$ {ELU, PReLU, SeLU, ReLU, Softplus, and Linear} and output layer activation function $\in$ {Sigmoid, Linear}. Note that only the result with the lowest mean absolute error per combination of activation functions is shown for brevity.

Input and Hidden Activation Function	Output Activation Function	Optimal Learning Rate	Mean Absolute Error Validation Data [yrs]
Softplus	Sigmoid	0.001	1.58
ELU	Sigmoid	0.002	1.58
SeLU	Sigmoid	0.002	1.61
PReLU	Sigmoid	0.005	1.69
ReLU	Sigmoid	0.002	1.76
ELU	Linear	0.001	1.85
SeLU	Linear	0.005	1.91
Softplus	Linear	0.005	1.94
Linear	Sigmoid	0.001	2.07
PReLU	Linear	0.001	2.35
Linear	Linear	0.0005	2.53
ReLU	Linear	0.005	2.86

We find that the neural network that performs best uses the Softplus activation function for the input and hidden layers and a Sigmoid activation function for the output layer. The distribution of predicted dates for this best architecture is shown in Fig. 5, and is far more uniform than the distribution shown in Fig. 4. It roughly resembles a distribution with four normal modes with means of 2000, 2010, 2016 and 2022. The relatively high number of predictions at the extrema of the prediction range is promising since it allows the (automated) penetration tester to easily discard fairly modern pages and hone in on older ones.

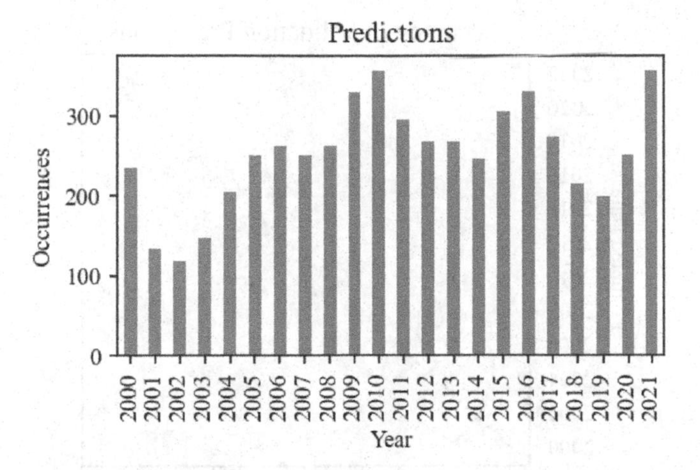

Fig. 5. Distribution of the predictions made by the best model found during hyper-parameter optimization (see Table 1). Note that the distribution is far more uniform than that of the ReLU based model, shown in Fig. 4.

2.6 Linear and Support Vector Regression

To validate our method we use linear regression, logistic regression and support vector regression to model webpage age. For the linear and logistic regressions we find mean absolute errors of 3.82 and 2.01 years, respectively. We use a regularization parameter of $C = 1$ for the support vector regression and test linear and radial basis function kernels (gamma='scale'), using the Python package scikit-learn.[9] These result in mean absolute errors of 6.20 and 2.07 years, respectively.

Our findings indicate that the MLP outperforms the other tested methods, achieving a mean absolute error of 1.58 years.

3 Results

Our model achieves a mean absolute error of 1.58 years when used to predict the age of webpages in the validation data that was not used to train the model. To visualize the predictive power, 250 of these predictions are plotted in comparison to their true value in Fig. 6. We find that less than 1% of predictions have an error larger than 9 years and less than 4% of predictions have an error larger than 6 years, as is shown in Table 2.

[9] https://scikit-learn.org/stable/modules/generated/sklearn.svm.SVR.html.

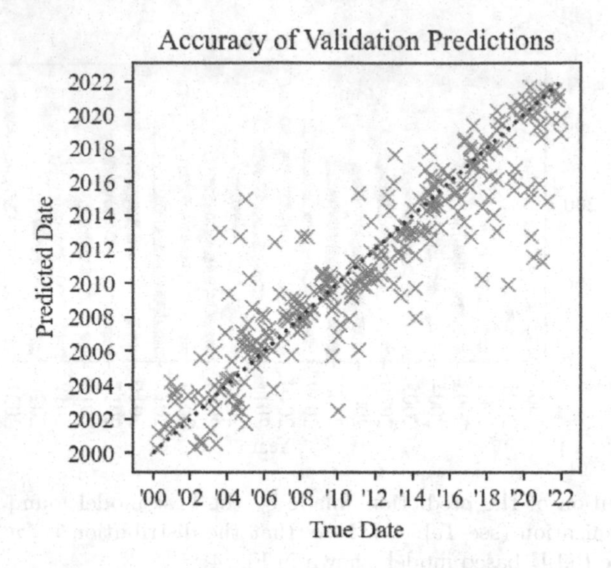

Fig. 6. The predictions of the Multi-Layer Perceptron when applied to 250 validation webpages that are not seen during training. The dark, dotted line represents an ideal prediction, where the predicted age is exactly the true age.

Table 2. The fraction of predictions that are within 3, 6 and 9 years of the true value. These evaluations are of the validation data that was not seen during training.

Error Margin	Predictions within Margin
3 years	87%
6 years	96%
9 years	99%

4 Discussion

We show that a bag-of-tokens model based on byte-pair tokenization and an MLP is able to effectively estimate the age of a webpage. This tool can be used by penetration testers in combination with a web-crawler to quickly hone in on out-dated pages on clients' websites that are running on legacy technology. In the context of autonomous attack surface management, this tool can inform decisions to test for certain vulnerabilities. Therefore, impact on the customer's infrastructure can be reduced by reducing brute forcing and intelligently probing assets. Furthermore, if legacy pages become available online, our system can detect them faster than slower, manual hackers.

5 Follow-Up Work

We have several recommendations for follow-up studies of the work presented in this paper. Firstly, we believe that it would be possible to further increase the accuracy of our model if more pages were scraped and used as training data, which could be achieved by using a rotating proxy server to circumvent The Web Archive's rate limiting. This would allow for more accurate tokenizer training and, consequently, a more representative token list and better webpage encodings. Due to decreased overfitting, the number of parameters in the MLP could be increased by changing the number of layers, the number of neurons in the layers or increasing the size of the encoding vectors. Since our method treats tokens independently, increasing the encoding size by adding tokens may significantly increase the tokens' expressiveness, leading to more accurate predictions.

Another recommendation is using statistical and/or heuristic methods to determine which tokens contain information relevant to age prediction and using only these for the encoding. This would allow one to create a more expressive token list without increasing the number of parameters in the model.

Finally, we hypothesize that a transformer architecture such as BERT, introduced by Devlin et al., would reach higher metric scores than our bag-of-tokens model [3]. We suggest training a BERT model on a corpus of HTML using a token-masking strategy and then fine-tuning it to predict webpage age using a dataset like the one used in this paper. Such a model uses the relative positions of tokens, allowing for deeper understanding of patterns in the HTML body.

References

1. Clevert, D.A., Unterthiner, T., Hochreiter, S.: Fast and accurate deep network learning by exponential linear units (elus). Under Review of ICLR2016 (1997). (11 2015)
2. Demir, N., Urban, T., Wittek, K., Pohlmann, N.: Our (in)secure web: understanding update behavior of websites and its impact on security. In: Hohlfeld, O., Lutu, A., Levin, D. (eds.) Passive and Active Measurement. PAM 2021. LNCS, vol. 12671, pp. 76–92. Springer, Cham (2021). https://doi.org/10.1007/978-3-030-72582-2_5
3. Devlin, J., Chang, M.W., Lee, K., Toutanova, K.: Bert: pre-training of deep bidirectional transformers for language understanding, October 2018
4. He, K., Zhang, X., Ren, S., Sun, J.: Delving deep into rectifiers: surpassing human-level performance on imagenet classification, February 2015
5. Kingma, D., Ba, J.: Adam: a method for stochastic optimization. In: International Conference on Learning Representations, December 2014
6. Klambauer, G., Unterthiner, T., Mayr, A., Hochreiter, S.: Self-normalizing neural networks, June 2017
7. Kudo, T.: Subword regularization: improving neural network translation models with multiple subword candidates, pp. 66–75, January 2018. https://doi.org/10.18653/v1/P18-1007
8. Kudo, T., Richardson, J.: Sentencepiece: a simple and language independent subword tokenizer and detokenizer for neural text processing, August 2018
9. Metsis, V., Androutsopoulos, I., Paliouras, G.: Spam filtering with naive bayes - which naive bayes? January 2006
10. Mielke, S., et al.: Between words and characters: a brief history of open-vocabulary modeling and tokenization in NLP, December 2021
11. Nair, V., Hinton, G.: Rectified linear units improve restricted Boltzmann machines vinod nair, vol. 27, pp. 807–814, June 2010

12. Schuster, M., Nakajima, K.: Japanese and Korean voice search. In: 2012 IEEE International Conference on Acoustics, Speech and Signal Processing (ICASSP), pp. 5149–5152 (2012)
13. Sennrich, R., Haddow, B., Birch, A.: Neural machine translation of rare words with subword units, August 2015
14. Srivastava, N., Hinton, G., Krizhevsky, A., Sutskever, I., Salakhutdinov, R.: Dropout: a simple way to prevent neural networks from overfitting. J. Mach. Learn. Res. **15**, 1929–1958 (2014)
15. Vaswani, A., et al.: Attention is all you need, June 2017
16. Zhang, Y., Jin, R., Zhou, Z.H.: Understanding bag-of-words model: a statistical framework. Int. J. Mach. Learn. Cybern. **1**, 43–52 (2010). https://doi.org/10.1007/s13042-010-0001-0
17. Zheng, H., Yang, Z., Liu, W.J., Liang, J., Li, Y.: Improving deep neural networks using softplus units, pp. 1–4, July 2015. https://doi.org/10.1109/IJCNN.2015.7280459

Correlations Between (Nonlinear) Combiners of Input and Output of Random Functions and Permutations

Subhabrata Samajder(✉) and Palash Sarkar

Applied Statistics Unit, Indian Statistical Institute, 203, B.T.Road, Kolkata 700108, India

subhabrata.samajder@gmail.com, palash@isical.ac.in

Abstract. Linear cryptanalysis considers correlations between linear input and output combiners for block ciphers and stream ciphers. Daemen and Rijmen (2007) had obtained the distributions of the correlations between linear input and output combiners of uniform random functions and uniform random permutations. The present work generalises these results to obtain the distributions of the correlations between arbitrary input and output combiners of uniform random functions and uniform random permutations.

1 Introduction

One of the basic tools for analysing symmetric key ciphers is a possible correlation between linear combinations of the input and output of a primitive. If this correlation is different from that of an idealised version of the primitive, then a distinguishing attack becomes possible. Determining whether a distinguishing attack is indeed possible requires the knowledge of the distributions of correlations for the idealised primitives. Two kinds of idealised primitives are usually considered, namely uniform random functions and uniform random permutations. For example, a uniform random permutation is an idealisation of a block cipher while a uniform random function is an idealisation of the state to keystream map in a stream cipher.

The distributions of the correlations between linear combinations of input and output for uniform random functions and uniform random permutations were derived in [1]. For the case of uniform random permutations, the distribution was earlier stated without proof in [3].

Our Contributions

This work extends the results of Daemen and Rijmen [1] by considering the correlation between arbitrary combiners of the input and output of uniform random functions and uniform random permutations. For any input combiner and any output combiner, the complete distributions of the correlations in the

S. Dolev et al. (Eds.): CSCML 2023, LNCS 13914, pp. 181–187, 2023.
https://doi.org/10.1007/978-3-031-34671-2_13

two cases are derived. The results are more conveniently stated in terms of the weight of the XOR of the input and the output combiners. For the case of a uniform random function, we show that the distribution of the weight is given by the convolution of two binomial distributions; if the output combiner is balanced, then the weight distribution is given by a binomial distribution. For the case of a uniform random permutation, we show that the distribution of the weight is given by a hypergeometric distribution.

Our approach to proving the results is different from that in [1]. The proofs in [1] consist essentially of counting Boolean functions. Instead we have used direct probability arguments. This yields proofs which are simple and at the same time work for arbitrary combiners.

2 Preliminaries

An m-variable Boolean function f is a map $f : \{0,1\}^m \to \{0,1\}$. The weight $\mathsf{wt}(f)$ of f is defined to be the cardinality of the support of f, i.e.,

$$\mathsf{wt}(f) = \#\{\alpha \in \{0,1\}^m : f(\alpha) = 1\}.$$

The function f is said to be balanced if $\mathsf{wt}(f) = 2^{m-1}$.

Let $f, g : \{0,1\}^m \to \{0,1\}$ be two Boolean functions. By $f \oplus g$ we denote the Boolean function $h : \{0,1\}^m \to \{0,1\}$ where $h(\alpha) = f(\alpha) \oplus g(\alpha)$ for all $\alpha \in \{0,1\}^m$. The correlation between f and g is denoted as $C(f,g)$ and is defined to be

$$C(f,g) = 1 - \frac{\mathsf{wt}(f \oplus g)}{2^{m-1}}.$$

An (m,n) function S is a map $S : \{0,1\}^m \to \{0,1\}^n$. Let $\phi : \{0,1\}^m \to \{0,1\}$ and $\psi : \{0,1\}^n \to \{0,1\}$. Given S, ϕ and ψ, we define a Boolean function

$$f_S[\phi,\psi] : \{0,1\}^m \to \{0,1\}, \text{ where } f_S[\phi,\psi](\alpha) = \phi(\alpha) \oplus \psi(S(\alpha)). \quad (1)$$

The function ϕ is a combiner of the input of S while the function ψ is a combiner of the output of S. Both $\phi(\cdot)$ and $\psi(S(\cdot))$ are m-variable Boolean functions. So, it is meaningful to talk about the correlation between these two functions. This correlation will be denoted as $C_S(\phi,\psi)$ and is equal to

$$C_S(\phi,\psi) = 1 - \frac{\mathsf{wt}(f_S[\phi,\psi])}{2^{m-1}}. \quad (2)$$

So, $C_S(\phi,\psi)$ measures the correlation between the combiner of the input as given by ϕ and the combiner of the output as given by ψ. From (2), determining $C_S(\phi,\psi)$ essentially boils down to determining $\mathsf{wt}(f_S[\phi,\psi])$.

Probability Distributions: $\mathsf{Ber}(p)$ denotes the Bernoulli distribution with probability of success p; $\mathsf{Bin}(k,p)$ denotes the binomial distribution with k trials and probability of success p; $\mathsf{HG}(k,k_1,s)$ denotes the hypergeometric distribution corresponding to a population of size k of which k_1 are of a specified type and $k-k_1$ are of a different type and a sample of size s is drawn without repetition.

3 Case of Uniform Random Function

Let ρ be a function picked uniformly at random from the set of all functions from $\{0,1\}^m$ to $\{0,1\}^n$. Such a ρ is a uniform random (m,n) function. An equivalent way to view ρ is the following. Let $\alpha_0,\ldots,\alpha_{2^m-1}$ be an enumeration of $\{0,1\}^m$. Let $X_i = \rho(\alpha_i)$, $i = 0,\ldots,2^m-1$. Then the random variables $X_0,\ldots,X_{2^m-1}$ are independent and uniformly distributed over $\{0,1\}^n$.

Theorem 1. *Let ρ be a uniform random (m,n) function. Let ϕ and ψ be m and n-variable Boolean functions respectively. Let $\alpha_0,\ldots,\alpha_{2^m-1}$ be an enumeration of $\{0,1\}^m$. For $0 \le i \le 2^m-1$, define $W_i = f_\rho[\phi,\psi](\alpha_i)$. Then $W_i \sim \mathsf{Ber}(p_i)$, where*

$$p_i = \frac{\mathsf{wt}(\psi) + \phi(\alpha_i)(2^n - 2\mathsf{wt}(\psi))}{2^n}. \tag{3}$$

If ψ is a balanced Boolean function, then $W_i \sim \mathsf{Ber}(1/2)$.

Proof. Let $X_i = \rho(\alpha_i)$. Since ρ is a uniform random function, X_i is uniformly distributed over $\{0,1\}^n$. We have

$$W_i = f_\rho[\phi,\psi](\alpha_i) = \phi(\alpha_i) \oplus \psi(\rho(\alpha_i)) = \phi(\alpha_i) \oplus \psi(X_i).$$

Let $Y_i = \psi(X_i)$. Then Y_i is a binary valued random variable where Y_i takes the value 1 if and only if X_i lies in the support of ψ. Since X_i is uniformly distributed over $\{0,1\}^n$, the probability that X_i lies in the support of ψ is $\mathsf{wt}(\psi)/2^n$. So, $\Pr[Y_i = 1] = \mathsf{wt}(\psi)/2^n$ and $\Pr[Y_i = 0] = (2^n - \mathsf{wt}(\psi))/2^n$. Consequently,

$$\begin{aligned}
\Pr[W_i = 1] &= \Pr[\phi(\alpha_i) \oplus \psi(X_i) = 1] \\
&= \Pr[Y_i = 1 \oplus \phi(\alpha_i)] \\
&= \frac{(1-\phi(\alpha_i))\mathsf{wt}(\psi) + \phi(\alpha_i)(2^n - \mathsf{wt}(\psi))}{2^n} \\
&= \frac{\mathsf{wt}(\psi) + \phi(\alpha_i)(2^n - 2\mathsf{wt}(\psi))}{2^n} \\
&= p_i.
\end{aligned}$$

This shows that W_i follows $\mathsf{Ber}(p_i)$.

If ψ is a balanced Boolean function, then $\mathsf{wt}(\psi) = 2^{n-1}$ in which case $p_i = 1/2$ and so W_i follows $\mathsf{Ber}(1/2)$. □

We are interested in the weight of the function $f_\rho[\phi,\psi]$.

Proposition 1. *Let ρ be a uniform random (m,n) function. Let ϕ and ψ be m and n-variable Boolean functions respectively. Let $\alpha_0,\ldots,\alpha_{2^m-1}$ be an enumeration of $\{0,1\}^m$ and $W_i = f_\rho[\phi,\psi](\alpha_i)$. Let $W = \mathsf{wt}(f_\rho[\phi,\psi])$. Then $W = \sum_{i=0}^{2^m-1} W_i$.*

Proof. The following calculation shows the result.

$$W = \mathsf{wt}(f_\rho[\phi,\psi]) = \#\{\alpha_i : f_\rho[\phi,\psi](\alpha_i) = 1\} = \#\{i : W_i = 1\} = \sum_{i=0}^{2^m-1} W_i.$$

□

Theorem 2. *Let ρ be a uniform random (m,n) function. Let ϕ and ψ be m and n-variable Boolean functions respectively. Then*

$$\Pr[\mathsf{wt}(f_\rho[\phi,\psi]) = w] = \sum_{t=0}^{w} \binom{w_0}{t}\binom{2^m - w_0}{w-t}\left(\frac{w_1}{2^n}\right)^{2^m-w-w_0+2t}\left(1-\frac{w_1}{2^n}\right)^{w_0+w-2t}, \quad (4)$$

where $w_0 = \mathsf{wt}(\phi)$ and $w_1 = \mathsf{wt}(\psi)$.

Further, if ψ is a balanced Boolean function, i.e., $w_1 = 2^{n-1}$, then $\mathsf{wt}(f_\rho[\phi,\psi]) \sim \mathsf{Bin}(2^m, 1/2)$.

Proof. Let $\alpha_0,\ldots,\alpha_{2^m-1}$ be an enumeration of $\{0,1\}^m$ and $X_i = \rho(\alpha_i)$ as in Theorem 1. Note

$$W_i = f_\rho[\phi,\psi](\alpha_i) = \phi(\alpha_i) \oplus \psi(X_i).$$

Since the random variables $X_0,\ldots,X_{2^m-1}$ are independent, so are the random variables $W_0,\ldots,W_{2^m-1}$.

From Proposition 1, $\mathsf{wt}(f_\rho[\phi,\psi]) = W = \sum_{i=0}^{2^m-1} W_i$ where $W_i \sim \mathsf{Ber}(p_i)$ with p_i given by (3). Note that p_i takes either the value $\mathsf{wt}(\psi)/2^n$ or $(2^n - \mathsf{wt}(\psi))/2^n$ according as $\phi(\alpha_i)$ equals 0 or 1. So, $W_0,\ldots,W_{2^m-1}$ is a sequence of 2^m Poisson trials, where each W_i either follows $\mathsf{Ber}\left(\frac{\mathsf{wt}(\psi)}{2^n}\right)$ or follows $\mathsf{Ber}\left(\frac{2^n-\mathsf{wt}(\psi)}{2^n}\right)$. Thus, W can be written as the sum of two binomially distributed random variables Z_1 and Z_2, i.e., $W = Z_1 + Z_2$, where

$$Z_1 \sim \mathsf{Bin}\left(\mathsf{wt}(\phi), \frac{\mathsf{wt}(\psi)}{2^n}\right) \text{ and } Z_2 \sim \mathsf{Bin}\left(2^m - \mathsf{wt}(\phi), \frac{2^n - \mathsf{wt}(\psi)}{2^n}\right).$$

Consequently, the distribution of W is given by the convolution of these two binomial distributions. Simplifying the expression for the convolution, we obtain the stated result. □

The special case of Theorem 2 where ϕ and ψ are non-trivial linear functions was given in [1]. The proof of this result in [1] is a counting argument which uses the fact that when ψ is a non-trivial balanced function, $\mathsf{wt}(\psi) = 2^{n-1}$. So, the proof in [1] covers the case of ψ being a balanced function. Theorem 2 provides the general result without any conditions on ψ (or ϕ).

4 Case of Uniform Random Permutation

Let $m = n$ and we consider the set of all bijections from $\{0,1\}^n$ to itself, i.e., the set of all permutations of $\{0,1\}^n$. There are $2^n!$ such permutations.

Proposition 2. *Let S be any permutation of $\{0,1\}^n$; let ϕ and ψ be n-variable Boolean functions. Let x be an integer such that $0 \leq x \leq \min(\mathsf{wt}(\phi), \mathsf{wt}(\psi))$. Then*

$$\#\{\alpha : \phi(\alpha) = 1 \text{ and } \psi(S(\alpha)) = 1\} = x$$

if and only if

$$\mathsf{wt}(f_S[\phi, \psi]) = \mathsf{wt}(\phi) + \mathsf{wt}(\psi) - 2x.$$

Proof. Define

$$\begin{aligned}
A_{0,0} &= \{\alpha : \phi(\alpha) = 0, \psi(S(\alpha)) = 0\};\\
A_{0,1} &= \{\alpha : \phi(\alpha) = 0, \psi(S(\alpha)) = 1\};\\
A_{1,0} &= \{\alpha : \phi(\alpha) = 1, \psi(S(\alpha)) = 0\};\\
A_{1,1} &= \{\alpha : \phi(\alpha) = 1, \psi(S(\alpha)) = 1\}.
\end{aligned}$$

The sets $A_{0,0}, A_{0,1}, A_{1,0}$ and $A_{1,1}$ are mutually disjoint; $A_{0,0} \cup A_{0,1} = \{\alpha : \phi(\alpha) = 0\}$; $A_{1,0} \cup A_{1,1} = \{\alpha : \phi(\alpha) = 1\}$ and so

$$\#A_{0,0} + \#A_{0,1} = 2^n - \mathsf{wt}(\phi), \tag{5}$$
$$\#A_{1,0} + \#A_{1,1} = \mathsf{wt}(\phi). \tag{6}$$

Further, $A_{0,0} \cup A_{1,0} = \{\alpha : \psi(S(\alpha)) = 0\}$. Since S is a permutation, $\{\alpha : \psi(S(\alpha)) = 0\} = \{\beta : \psi(\beta) = 0\}$. So, $A_{0,0} \cup A_{1,0} = \{\beta : \psi(\beta) = 0\}$ and similarly, $A_{0,1} \cup A_{1,1} = \{\beta : \psi(\beta) = 1\}$ leading to

$$\#A_{0,0} + \#A_{1,0} = 2^n - \mathsf{wt}(\psi), \tag{7}$$
$$\#A_{0,1} + \#A_{1,1} = \mathsf{wt}(\psi). \tag{8}$$

Adding (6) and (8) we obtain

$$\#A_{1,0} + \#A_{0,1} + 2\#A_{1,1} = \mathsf{wt}(\phi) + \mathsf{wt}(\psi). \tag{9}$$

Equation (9) implies that $\#A_{1,1} = x$ if and only if $\#A_{0,1} + \#A_{1,0} = \mathsf{wt}(\phi) + \mathsf{wt}(\psi) - 2x$.

Note that the support of $f_S[\phi, \psi]$ is $A_{0,1} \cup A_{1,0}$ and $A_{1,1} = \{\alpha : \phi(\alpha) = 1, \psi(S(\alpha)) = 1\}$. So,

$$\#\{\alpha : \phi(\alpha) = 1, \psi(S(\alpha)) = 1\} = x$$

if and only if

$$\mathsf{wt}(f_S[\phi, \psi]) = \mathsf{wt}(\phi) + \mathsf{wt}(\psi) - 2x.$$

□

From Proposition 2, given the functions ϕ and ψ, the possible weights that $f_S[\phi, \psi]$ can take for any permutation S of $\{0,1\}^n$ are the elements of the set

$$\{\mathsf{wt}(\phi) + \mathsf{wt}(\psi) - 2x : 0 \leq x \leq \min(\mathsf{wt}(\phi), \mathsf{wt}(\psi))\}. \quad (10)$$

Suppose π is picked uniformly from the set of all permutations of $\{0,1\}^n$. We are interested in the probability that $f_\pi[\phi, \psi]$ takes a value from the set given by (10).

Theorem 3. *Let π be a uniform random permutation of $\{0,1\}^n$; let ϕ and ψ be n-variable Boolean functions. Then for $0 \leq x \leq \min(\mathsf{wt}(\phi), \mathsf{wt}(\psi))$,*

$$\Pr[\mathsf{wt}(f_\pi[\phi, \psi]) = \mathsf{wt}(\phi) + \mathsf{wt}(\psi) - 2x] = \frac{\binom{\mathsf{wt}(\phi)}{x}\binom{2^n - \mathsf{wt}(\phi)}{\mathsf{wt}(\psi) - x}}{\binom{2^n}{\mathsf{wt}(\psi)}}. \quad (11)$$

If both ϕ and ψ are balanced functions, then

$$\Pr[\mathsf{wt}(f_\pi[\phi, \psi]) = \mathsf{wt}(\phi) + \mathsf{wt}(\psi) - 2x] = \frac{\binom{2^{n-1}}{x}^2}{\binom{2^n}{2^{n-1}}}. \quad (12)$$

Proof. Let $\alpha_0, \ldots, \alpha_{2^n-1}$ be an enumeration of $\{0,1\}^n$ and let $X_i = \pi(\alpha_i)$. Unlike the case where π is a uniform random function, the random variables $X_0, \ldots, X_{2^n-1}$ are not independent. Instead, it is more convenient to view these random variables in the following manner. Consider an urn containing balls labelled $\alpha_0, \ldots, \alpha_{2^n-1}$. Balls are picked one by one from the urn *without replacement* and we number the trials from 0 to $2^n - 1$. Then the random variable X_i is the label of the ball picked in trial number i.

Consider the random Boolean function $g(\alpha) = \psi(\pi(\alpha))$. A Boolean function is defined by its support. So, it is sufficient to choose $\mathsf{wt}(\psi)$ balls from the urn and let the labels of these balls define the support of g. From Proposition 2, the probability that $\mathsf{wt}(f_\pi[\phi, \psi]) = \mathsf{wt}(\phi) + \mathsf{wt}(\psi) - 2x$ is equal to the probability that the cardinality of the set

$$A_{1,1} = \{\alpha : \phi(\alpha) = 1 \text{ and } \psi(\pi(\alpha)) = 1\} = \{\alpha : \phi(\alpha) = 1 \text{ and } g(\alpha) = 1\}$$

is x.

To obtain this probability, we consider the following equivalent random experiment. As before, consider the urn containing balls labelled $\alpha_0, \ldots, \alpha_{2^n-1}$. Further, say that a ball labelled α_i is 'red' if $\phi(\alpha_i) = 1$ and otherwise it is 'black'. Now, consider that $\mathsf{wt}(\psi)$ balls are drawn from this urn which defines the support of g. The event that we are interested in is that x of these $\mathsf{wt}(\psi)$ are 'red' while the other $\mathsf{wt}(\psi) - x$ are 'black'. The probability of this event is the probability that $\#A_{1,1} = x$ which is given by the right hand side of (11). Then (11) follows from Proposition 2.

In the case where both ϕ and ψ are balanced functions, both their weights are equal to 2^{n-1}. So, substituting 2^{n-1} for $\mathsf{wt}(\phi)$ and $\mathsf{wt}(\psi)$ in (11) and using $\binom{2^{n-1}}{2^{n-1}-x} = \binom{2^{n-1}}{x}$ yields (12). □

The expression given on the right hand side of (11) is the probability mass function of the hypergeometric distribution. In the special case where ϕ and ψ are non-trivial linear functions, the distribution given by (12) was proved in [1].

5 Conclusion

In this paper, we have obtained the distributions of the correlations between arbitrary input and output combiners of uniform random functions and uniform random permutations. The correlation between two Boolean functions can be expressed in terms of the weight of the XOR of the functions. Using this relation, our results are expressed in terms of the weights of the XOR of the input and output combiners. These results generalise earlier results by Daemen and Rijmen [1] who had considered only linear combiners.

References

1. Daemen, J., Rijmen, V.: Probability distributions of correlation and differentials in block ciphers. J. Math. Cryptol. **1**(3), 221–242 (2007)
2. Matsui, M.: Linear cryptanalysis method for DES cipher. In: Helleseth, T. (ed.) EUROCRYPT 1993. LNCS, vol. 765, pp. 386–397. Springer, Heidelberg (1994). https://doi.org/10.1007/3-540-48285-7_33
3. O'Connor, L.: Properties of linear approximation tables. In: Preneel, B. (ed.) FSE 1994. LNCS, vol. 1008, pp. 131–136. Springer, Heidelberg (1995). https://doi.org/10.1007/3-540-60590-8_10

PPAuth: A Privacy-Preserving Framework for Authentication of Digital Image

Riyanka Jena[1(✉)], Priyanka Singh[1], and Manoranjan Mohanty[2]

[1] Dhirubhai Ambani Institute of Information and Communication Technology, Gandhinagar, Gujarat, India
{201921012,priyanka_singh}@daiict.ac.in
[2] University of Technology Sydney, Sydney, Australia
Manoranjan.Mohanty@uts.edu.au

Abstract. Deep learning has been widely applied in many computer vision applications with remarkable success. Most of these techniques improve extraction of features from the fetched data for instance, image attributes are extracted for classifying an image. These attributes are quite useful for many tasks such as forensic investigation but it creates a threat for the privacy of the image if these attributes are leaked. Forensic authentication of digital images and videos is a crucial process in forensic investigation as they provide direct evidence. Authentication is a forensic process that involves verifying the authenticity of a digital image. In this paper, we propose a privacy-preserving framework to authenticate an image. It defends against the adversary from reconstructing the image from the extracted features and minimizes the private attribute leakage inference from the extracted features of the image. The main classifier performs the binary classification and tells whether an image is forged or authentic with an accuracy of 99%. Also, the framework maintains a very low accuracy for the adversarial classifier which aims at inferring some private attributes or even reconstruction of the image from these leaked attributes.

Keywords: Image Forensics · Privacy · Deep Learning Techniques · Reconstruction Attack · Private Attribute Leakage Attack

1 Introduction

The widespread use of smart cameras and smartphones, has led to easy sharing of images online. In parallel, there has been a development of various image-processing softwares. However, in the downside there is a greater risk of image tampering, including malicious changes to the content of an image, such as adding or removing objects. This can be dangerous, as it can deceive or distort reality. It is crucial to have effective methods to detect and prevent such modifications, especially as online sharing platforms make images vulnerable to alterations through compression, resizing, and filtering, which can mask changes [1].

S. Dolev et al. (Eds.): CSCML 2023, LNCS 13914, pp. 188–199, 2023.
https://doi.org/10.1007/978-3-031-34671-2_14

Deep learning with artificial neural networks has achieved remarkable progress in areas such as language translation, image, text, and speech recognition. In computer vision, deep learning is employed for tampering detection by analyzing images for signs of manipulation. However, using deep learning for tampering detection can compromise privacy by exposing sensitive information such as age and gender. This trend has resulted in an increased need for privacy protection, especially in public channels, to prevent snooping [13].

Our proposed algorithm utilizes deep learning to distinguish authentic images from tampered ones, ensuring accuracy and defending against attacks such as reconstruction and private attribute leakage. We use VGG-16 architecture with a deep encoder, classifier, private attribute classifier, and reconstructor to obfuscate essential and non-essential features to prevent private attribute leakage and reconstruction of the image. The encoder and classifier are trained jointly using cross-entropy loss, with the encoder aiming to reduce attacker accuracy. We address reconstruction attacks, which can prevent image reconstruction and private attribute leakage, for example, a person's gender, age, etc., by outperforming Puteaux et al.'s method [10]. Our framework provides improved privacy protection, as demonstrated in Subsect. 4.5, while maintaining image classification accuracy. The major contributions of the paper are as follows:

- **Construction of Dataset:** In our study, we created a five-dataset collection for detecting image tampering, utilizing the CelebA dataset [5] and various tampering techniques. The evaluation of our dataset contributes to the development of more effective tampering detection methods in image forensics.
- **Privacy Preserving:** A privacy-preserving approach is proposed to determine whether an image is tampered or authentic based on its extracted features. The framework we use protects against reconstruction attacks (i.e. reconstructing the image based on the extracted features) and private attribute leakage (i.e. inference of smile, gender etc.) and maintains the accuracy of main tasks. In this study, we observed that our classifier gives a very high accuracy of 99%, whereas the adversary classifier gives a low accuracy. In addition to the low MS-SIM and PSNR scores, it is evident that the reconstructor cannot reconstruct the tampered images.
- **Comparison of different models for adversarial classification and reconstruction:** We have chosen different architectures for the adversary classifier and adversary reconstructor to check whether they can protect against the attack. Feature extraction is done using an encoder using the VGG-16 model [11], and the attacker chooses different architecture. For this, we have used ResNet [2]. We evaluated using accuracy, the MS-SIM, and PSNR scores. We got low accuracy for the adversarial classifier and low MS-SIM and PSNR scores for the image reconstruction.

The paper is organized as follows: Sect. 2 presents the related work. Section 3 presents the proposed frameworks describes threat model and architecture. Section 4 describes the experiments conducted to validate the performance of the proposed scheme and we compared the proposed scheme with baseline models and Sect. 5 concludes the work along with future scope.

2 Related Work

In this section, we discuss about the related work regarding tampering detection using deep learning techniques. Osia et al. [7] utilized dimensionality reduction, noise injection, and Siamese fine-tuning to protect sensitive information from features, but did not address the reconstruction attack. Singh et al. [12] proposed a framework, a dynamic scheme for obfuscation of sensitive channels to protect sensitive information. Pittaluga et al. [9] used a neural network to inhibit private attribute inference while preserving other information, effective in inhibiting scene category detection. Liu et al. [4] developed PAN, a privacy-preserving machine learning approach that optimizes against privacy disclosure risk and task inference accuracy. Wu et al. [15] propose an adversarial framework to learn a degradation transform for video inputs that balances task accuracy and privacy budgets.

Pentyala et al. [8] propose a privacy-preserving video classification method using CNNs and encryption techniques to infer video labels without disclosing them to other entities. Puteaux et al. [10] proposed the framework of selective encryption in order to reach a tradeoff between privacy preservation and integrity check. Forensics analysis is conducted to detect if a selectively encrypted image has been tampered or not. To assess the visual confidentiality of a selectively encrypted image. However, the framework does not provide protection against reconstruction attacks and private attribute leakage. In contrast, the proposed framework protects against both reconstruction attacks and private attribute leakage, while still maintaining high accuracy for the main task of determining if an image is tampered or authentic. Thus, the proposed framework goes beyond the selective encryption framework by providing enhanced privacy protection.

3 Proposed Method

In this section, we give an overview of architecture as depicted in Fig. 1, the entities involved, the threat model and proposed methodology.

3.1 System and Threat Model

Deep Encoder: The deep encoder is a neural network used to detect image tampering. It extracts useful features from the image that will be used to classify whether image is tampered or authentic while preventing reconstruction and attribute leakage attacks. The deep encoder is assumed to be a trusted entity.

Investigator: An Investigator is an entity from law enforcement whose work is to use the trained model to check whether an image is tampered or authentic. The investigator is assumed to be a trusted entity. The attacker can only see the investigator's output without having any knowledge of the input.

Third Party Server: The third-party expert entity uses features extracted by the deep encoder to classify an image as authentic or tampered. It has no access

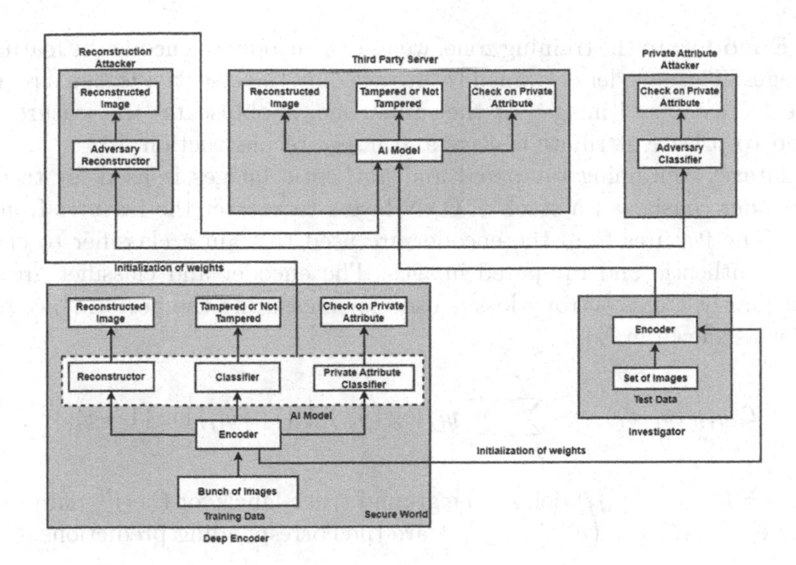

Fig. 1. Overview of Proposed Methodology

to private information and cannot reconstruct the image or use private attributes for classification. It can be an individual or an organization. The communication channel between the investigator and the third party server is considered insecure therefore it is considered as honest but curious entity.

Reconstruction Attacker: The extracted features contain rich information which can breach users' privacy. This is an entity that can exploit the eavesdropped features to reconstruct the raw image, and hence the identity of the person on the raw image can be discovered from the reconstructed image. They are considered to be malicious entity.

Private Attribute Attacker: An entity that infers private attributes from images without consent can be malicious, as it raises concerns about privacy leakage. Extracting private attributes like gender and age from images without consent can raise privacy concerns. An attacker can train an adversary model by querying the investigator for eavesdropped features and using them as labels. This can be considered a malicious entity.

3.2 Proposed Methodology

In this section, we describe the components of the proposed architecture as illustrated in Fig. 1. In the proposed method we are detecting an image is tampered or authentic in privacy preserving framework. In this method, we are using VGG-16 [11] architecture for tampering detection of the images. The investigator uses this framework to classify the tampering detection by maintaining accuracy and defending against the reconstruction attack and private attribute leakage. This section describes the components of the proposed architecture as illustrated in Fig. 1. The details are as follows:

Deep Encoder: In the training zone, we use an encoder to encode the features of the images. The encoder is trained to extract only the essential features required to detect a tampered image. At the same time, it obfuscates the features that can lead to private attribute leakage and image reconstruction.

A dataset containing tampered and authentic images is used for training. The encoder consists of a stack of CNN layers to extract the features from the images. The features from the encoder are used to train a classifier to classify between authentic and tampered images. The encoder and classifier are both trained jointly. Cross entropy loss is used for measuring the performance of the classifier as shown in Eq. 1.

$$\mathcal{L}(y, y'; \theta_e, \theta_c) = -\sum_{j=1}^{\mathcal{N}} \sum_{i=1}^{\mathcal{M}} y_{ij} \log\left(y'_{ij}\right) + (1 - y_{ij}) \log\left(1 - y'_{ij}\right) \tag{1}$$

where, y $= (y_{1j}, \ldots, y_{Mj})$ denote the ground truth labels for the jth data sample where $j \in \mathcal{N}$, and y$'= \left(y'_{1j}, \ldots, y'_{Mj}\right)$ are the corresponding predictions.

$$\theta_e, \theta_c = \underset{\theta_e, \theta_c}{\arg\min} \mathcal{L}(y, y'; \theta_e, \theta_c) \tag{2}$$

θ_e and θ_c are parameters of encoder and classifier respectively.

The private attribute attacker can infer the features by the eavesdropping attack on the communication channel from investigator to a third-party server to classify the private attributes of an image, for example, the gender of a person, etc. The private attribute attacker, i.e., adversary classifier, uses cross-entropy loss for measuring the performance as shown in Eq. 3.

$$\mathcal{L}(z, z'; \theta_{ac}) = -\sum_{j=1}^{\mathcal{N}} \sum_{i=1}^{\mathcal{K}} z_{ij} \log\left(z'_{ij}\right) + (1 - z_{ij}) \log\left(1 - z'_{ij}\right) \tag{3}$$

where, z $= (z_{1j}, \ldots, z_{Mj})$ denote the ground truth labels for the jth eavesdropped data sample, and z$'= \left(z'_{1j}, \ldots, z'_{Mj}\right)$ are the corresponding predictions. The adversery classifier is optimized by minimizing the loss function as shown in Eq. 4.

$$\theta_{ac} = \underset{\theta_{ac}}{\arg\min} \mathcal{L}(z, z'; \theta_{ac}) \tag{4}$$

θ_{ac} are parameters of adversary classifier.

Private attribute leakage is prevented by training the encoder to degrade the accuracy of private attribute attackers to maintain privacy. The encoder is trained using Eq. 5 to defend against private attribute leakage.

$$\theta_e = \underset{\theta_e}{\arg\min} \mathcal{L}(y, y'; \theta_e, \theta_c) - \lambda_1 \mathcal{L}(z, z'; \theta_{ac}) \tag{5}$$

where λ_1 is a tradeoff parameter. The reconstruction attacker can infer the features by the eavesdropping attack on the communication channel from the investigator to the third party server to reconstruct the image given as input by the investigator. The image reconstructor uses the reverse architecture of the encoder. The performance of the reconstructor is measured using MS-SIM [6,14], which is a perceptual metric that quantifies image quality. The MS-SIM value ranges between 0 and 1. The reconstruction attacker, i.e., adversary reconstructor, train to obfuscate the features that can evaluate the reconstructor's performance as shown in Eq. 6.

$$\mathcal{L}(I_o, I_r; \theta_r) = 1 - \text{MS-SIM}\,(I_o, I_r) \tag{6}$$

where I_o is the original image, and I_r is the reconstructed image.

$$\theta_r = \underset{\theta_r}{\arg\min} \mathcal{L}\,(I_o, I_r; \theta_r) \tag{7}$$

where, θ_r the parameter set of the reconstruction attacker.

To defend against image reconstruction, we need to preserve privacy from the reconstruction attacker by degrading the quality of the reconstructed image as much as possible. For defending, we generate one additional Gaussian noise image I_{noise}. During the training process in encoder, we try to reconstruct images similar to the noise image rather than the original image. The encoder is trained using Eq. 8 to prevent the reconstruction of the original image.

$$\mathcal{L}\,(I_{\text{noise}}\,, I_r) = 1 - \text{MS-SIM}\,(I_{\text{noise}}\,, I_r) \tag{8}$$

$$\theta_e = \underset{\theta_e}{\arg\min} \mathcal{L}(y, y'; \theta_e, \theta_c) + \lambda_2\,(\mathcal{L}\,(I_{\text{noise}}\,, I_r) - \mathcal{L}\,(I_o, I_r; \theta_r))\,, \tag{9}$$

where λ_2 is a tradeoff parameter.

Investigator: The investigator uses the test images, which contain both tampered and authentic images. The trained weights of the encoder in the training zone are used to initialize the encoder in the investigator module.

Third-Party Server: The third-party server uses the weights of the AI models, which is trained in the training zone. The features of an image are extracted by the encoder in the investigator module. The features are used as input by the AI Models to classify the image is tampered or not.

Private Attribute Attacker: The weights of the private attribute classifier are initialized by the attacker. The features of an image are extracted by the encoder in the investigator module. The features are used as input to the adversary classifier.

Reconstruction Attacker: The weights of reconstructor are initialized by the attacker. The features of an image are extracted by the encoder in the investigator module. The features are used as input to the reconstructor.

4 Experimental Analysis and Result

In this section, we explained the dataset, experimental setup, evaluation metric and assess the result analysis by combining the forensics and preserving privacy for the digital images.

4.1 Dataset Construction

We have used the dataset CelebA [5]. We constructed an image tampering detection evaluation dataset. Our tampered images in this dataset are all RGB images. Total number of images in the dataset are 202599. There are two types of image in tampering detection evaluation dataset. In the dataset there are 101300 tampered images and 101299 authentic images. We are giving input image of size 178×218. We have used different techniques as shown in Table 1 for the manipulation for tampering with the images. We have used the cropping method consisting of a crop rectangle parameter (left, upper, right, lower). We cropped the image from the right lower corner 50% of height and width. We performed rotate operation on the images. We rotated the image 180 degrees counter clockwise around its centre. We have performed the resize operation on the images. We resize the image into half of it's height and half of it's width. We perform copy move operation on the image where we paste the same image in the corner right side of the image. After the copy move operation JPEG image is saved in the default quality of 75% as shown in Fig. 2.

Fig. 2. Tampered Images

4.2 Experimental Setup

We implement the training zone with PyTorch, and train it on a server with Nvidia RTX 2080 Ti. We apply mini-batch technique in training with a batch size of 32, and adopt the Adam Optimizer [3] with an adaptive learning rate in

Table 1. Different Types of Image Tampering Attacks and Datasets

S.No.	Attack	Operation on Images	Dataset	Authentic Images	Tampered Images
1	Copy Move	Same image pasted in the right corner	Copy Move dataset	101299	101300
2	Crop	Lower-right 50% of height and width	Crop dataset	101299	101300
3	Resize	50% reduction in height and width	Resize dataset	101299	101300
4	Rotate	180-degree counter-clockwise	Rotate dataset	101299	101300
5	Copy Move, Crop, Resize, Rotate	-	Combined dataset	101299	101300

all four stages in the adversarial training procedure. The configuration of the encoder is shown in Table 2. ReLU and maxpooling is applied on the output of each convolution layer. The configuration of the classifier and adversarial classifier is shown in Table 3. The configuration of the adversarial reconstructor is shown in Table 4. We use the gender attribute wherein the image is male or female for simulating private attribute leakage.

Table 2. Configuration of Encoder

	Layer 1	Layer 2	Layer 3	Layer 4
No. of Input Channels	3	64	64	128
No. of Output Channels	64	64	128	128

Table 3. Configuration of Classifier and Adversarial Classifier

	Layer 1	Layer 2	Layer 3	layer 4	Layer 5	Layer 6	Layer 7	Layer 8	Layer 9	Layer 10	Layer 11	Layer 12
No. of Input Channels	128	256	256	256	512	512	512	512	512	25088	4096	4096
No. of Output Channels	256	256	256	512	512	512	512	512	512	4096	4096	2

4.3 Result

The accuracy of tampering detection classifier and adversary classifier is shown in Table 5. We observe that our classifier gives a very high accuracy of 99%, while the adversary classifier gives a low accuracy. Thus we infer that the encoder can defend against private attribute leakage. We also investigate the performance of the classifier and adversary classifier on individual tampering operation. The results of copy move, cropped, resize and rotation tampering detection are shown in Table 5 respectively. From the results of individual tampering operation detection we observe that the classification accuracy for all the operations is very high, while the accuracy of private attribute inference is very low. Thus we can infer that the encoder can extract only useful features from the image

Table 4. Configuration of Adversarial Reconstructor

	Upsample Layer 1	Upsample Layer 2	Upsample Layer 3	Upsample Layer 4
No. of Input Channels	128	64	64	3
No. of Output Channels	64	64	3	3

required for detecting the tampering, while simultaneously defending against private attribute leakage. In Fig. 3, we show the results of reconstruction of tampered image. We observe that the reconstructor is unable to reconstruct the tampered images which also evident from the low MS-SIM and PSNR scores. Thus demonstrating the effectiveness of the encoder in defending against reconstruction of the image.

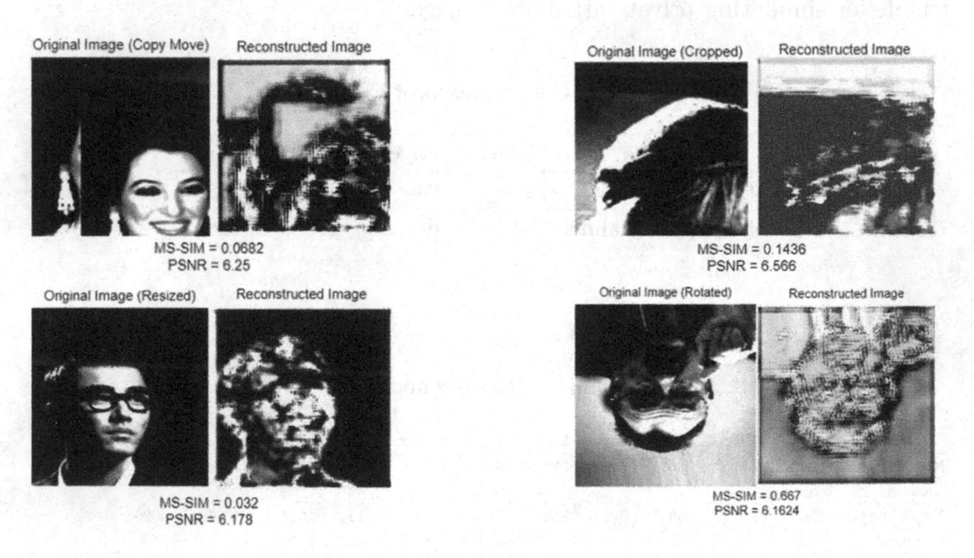

Fig. 3. Reconstructed Images for Copy-Move, Crop, Resize, and Rotate Forgeries

4.4 Comparison of Different Models for Adversarial Classification and Reconstruction

In our proposed method, we apply the VGG-16 [11] architecture to the classifier, adversary classifier and adversary reconstructor. However, the attacker can choose any architecture for the adversary classifier and adversary reconstructor. In order to simulate this we have experimented by using ResNet-50 [2] architecture for the adversary classifier and adversary reconstructor. While testing we initialize the encoder with the weights of VGG-16 [11], the adversary classifier and adversary reconstructor with the weights of ResNet-50. We perform the classification and reconstruction using this setup to evaluate whether the encoder

Table 5. Accuracy for Classifier and Adversary Classifier using proposed approach

S.No.	Dataset Type	Accuracy of Classifier	Accuracy of Adversary Classifier
1	All operation dataset	99%	40%
2	Copy move operation dataset	99%	59%
3	Crop operation dataset	99%	40%
4	Resize operation dataset	99%	40%
5	Rotate operation dataset	99%	40%

is able to protect against private attribute leakage attacks and reconstruction attacks. We use accuracy, MS-SIM and PSNR scores to perform the evaluation. From Table 6, we can conclude that the adversarial classifier accuracy is low and can prevent private attribute leakage attacks. Low MS-SSIM and PSNR scores infer that it can protect from reconstruction attacks as shown in Fig. 4.

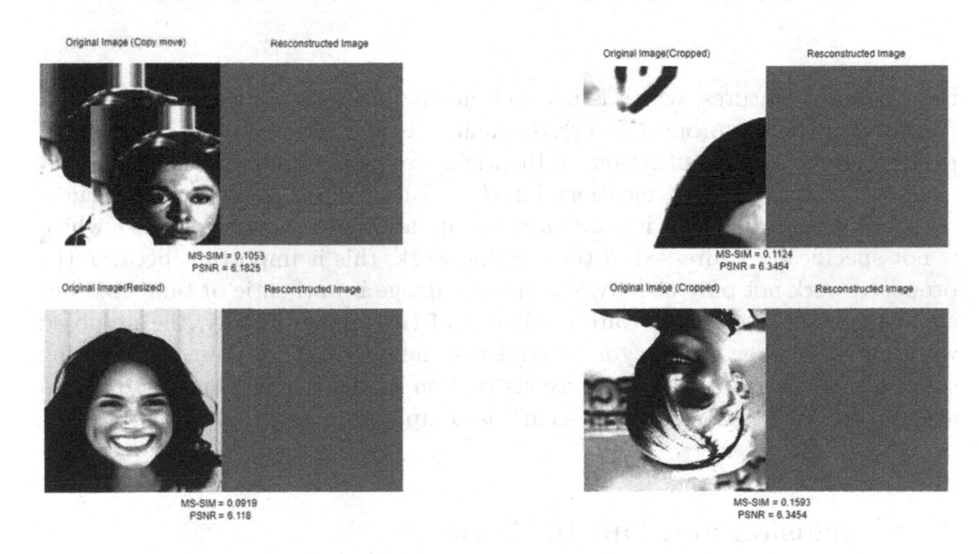

Fig. 4. Reconstructed Images for Copy-Move, Crop, Resize, and Rotate Forgeries

4.5 Comparison with the State-of-the-Art Approaches

A comparative analysis of the proposed work with Puteaux et al's. method [10] is done in Table 7. The proposed and the existing work [10] both propose methods for protecting privacy during forensic investigation using deep learning in computer vision. However, the present work proposes a privacy-preserving framework for authentication of images, while the existing work proposes a selective encryption scheme for increasing privacy. One advantage of the present work is that it aims to defend against an adversary reconstructing the image from

Table 6. Accuracy of Adversary Classifier using Baseline Approach

S.No.	Dataset Type	Accuracy of Adversary Classifier
1	All operation dataset	40%
2	Copy move operation dataset	40%
3	Crop operation dataset	40%
4	Resize operation dataset	40%
5	Rotate operation dataset	40%

Table 7. Comparison of the Former Method with the Proposed Work

S.No.	Features	Puteaux et al. [10]	Proposed Work
1	Method used in the framework	Selective Encryption	Deep Learning Technique
2	Protection to Reconstruction Attack	No	Yes
3	Protection to private attribute leakage attack	No	Yes
4	Accuracy for Tampering Detection	90%	99%

the extracted features, which is not specifically addressed in the existing work. Additionally, in the proposed work the main classifier has an accuracy of 99% in performing binary classification (authentic or tampered image), which is higher than the accuracy of 90% mentioned in the existing work. Another advantage of the proposed work is that it minimizes private attribute leakage inference which is not specifically addressed in the existing work, this is important because the proposed work not only classify whether the image is authentic or tampered but also tries to protect the private attributes of the image. Finally, the proposed work also has a low accuracy for the adversarial classifier who aims at inferring some private attributes or even reconstruction of the image from these leaked attributes, which makes it more secure as compared to existing work.

5 Conclusion and Future Work

We can infer that the encoder can extract only useful features from the image required for detecting the tampering. It can also defend against private attribute leakage. The encoder also effectively defends against reconstruction of the image. We observe that our classifier gives very high accuracy while the adversary classifier gives a low accuracy. Additionally, the reconstructor's inability to reconstruct the tampered images, as seen from the low MS-SIM and PSNR scores, highlights the robustness of our model. In the future, there is scope for further improvement of the encoder by exploring other feature extraction techniques and incorporating them into the model. There is also potential for analyzing the application of this model in real-world scenarios and evaluating its performance in a more diverse range of tampering attacks. This study is a promising start toward effective and robust image tampering detection and protection.

References

1. Diallo, B., Urruty, T., Bourdon, P., Fernandez-Maloigne, C.: Robust forgery detection for compressed images using CNN supervision. Forensic Sci. Int. Rep. **2**, 100112 (2020)
2. He, K., Zhang, X., Ren, S., Sun, J.: Deep residual learning for image recognition. In: Proceedings of the IEEE Conference on Computer Vision and Pattern Recognition, pp. 770–778 (2016)
3. Kingma, D.P., Ba, J.: Adam: a method for stochastic optimization. In: ICLR (Poster) (2015)
4. Liu, S., Du, J., Shrivastava, A., Zhong, L.: Privacy adversarial network: representation learning for mobile data privacy. Proc. ACM Interact. Mob. Wearable Ubiquit. Technol. **3**(4), 1–18 (2019)
5. Liu, Z., Luo, P., Wang, X., Tang, X.: Deep learning face attributes in the wild. In: Proceedings of the IEEE International Conference on Computer Vision, pp. 3730–3738 (2015)
6. Ma, K., et al.: Group mad competition-a new methodology to compare objective image quality models. In: Proceedings of the IEEE Conference on Computer Vision and Pattern Recognition, pp. 1664–1673 (2016)
7. Osia, S.A., et al.: A hybrid deep learning architecture for privacy-preserving mobile analytics. IEEE Internet Things J. **7**(5), 4505–4518 (2020)
8. Pentyala, S., Dowsley, R., De Cock, M.: Privacy-preserving video classification with convolutional neural networks. In: International Conference on Machine Learning, pp. 8487–8499. PMLR (2021)
9. Pittaluga, F., Koppal, S., Chakrabarti, A.: Learning privacy preserving encodings through adversarial training. In: 2019 IEEE Winter Conference on Applications of Computer Vision (WACV), pp. 791–799. IEEE (2019)
10. Puteaux, P., Itier, V., Bas, P.: Combining forensics and privacy requirements for digital images. In: 2021 29th European Signal Processing Conference (EUSIPCO), pp. 806–810. IEEE (2021)
11. Simonyan, K., Zisserman, A.: Very deep convolutional networks for large-scale image recognition. arXiv preprint arXiv:1409.1556 (2014)
12. Singh, A., et al.: Disco: dynamic and invariant sensitive channel obfuscation for deep neural networks. In: Proceedings of the IEEE/CVF Conference on Computer Vision and Pattern Recognition, pp. 12125–12135 (2021)
13. Sudusinghe, C., Charles, S., Ahangama, S., Mishra, P.: Eavesdropping attack detection using machine learning in network-on-chip architectures. IEEE Design Test **39**(6), 28–38 (2022)
14. Wang, Z., Simoncelli, E.P., Bovik, A.C.: Multiscale structural similarity for image quality assessment. In: The Thrity-Seventh Asilomar Conference on Signals, Systems & Computers, 2003, vol. 2, pp. 1398–1402. IEEE (2003)
15. Wu, Z., Wang, Z., Wang, Z., Jin, H.: Towards privacy-preserving visual recognition via adversarial training: A pilot study. In: Proceedings of the European Conference on Computer Vision (ECCV), pp. 606–624 (2018)

Robust Group Testing-Based Multiple-Access Protocol for Massive MIMO

George Vershinin(✉), Asaf Cohen, and Omer Gurewitz

School of Electrical and Computer Engineering, Ben-Gurion University of the Negev, Beersheba, Israel
georgeve@post.bgu.ac.il, {coasaf,gurewitz}@bgu.ac.il

Abstract. With the ever-increasing demand for more per-household devices and the addition of more antennas per device, the challenge of effective scheduling and resource sharing to access the wireless shared channel for uplink communication with the base station (BS) becomes daunting. To address this issue, we devise and study a *robust* multiple-access protocol for massive multiple-input-multiple-output (MIMO) systems, based on sparse coding techniques originated in group testing (GT), for systems with non-cooperative self-scheduling users with reduced complexity and no scheduling overhead.

In this study, we analyze our scheme's bit-error rate, decoding error probability, scaling laws, system sum-rate, and complexity. We show that our suggested scheme is order-optimal by comparing our sum-rate with the *perfect* channel state information (CSI) model and numerically evaluate how our system scales with an increasing number of active devices and signal-to-noise ratio (SNR).

Keywords: Massive MIMO · Multiple-Access · Group Testing

1 Introduction

The growing data and connectivity demands over wireless networks stimulates the utilization of multi-user multiple-input-multiple-output (MU-MIMO), which enables great capacity gains via simultaneous transmission from the base station (BS) to multiple users and vice versa [13,23]. Many recent studies have explored the potential gains in exploiting multiple antenna technology for increased signal quality, high system rates, and improved signal-to-noise (SNR) ratio. MIMO systems' additional degrees of freedom allow several techniques to selectively and adaptively optimize the channel, e.g., by amplifying the SNR via signal combining. Specifically, MIMO can improve performance both in single and multi-user communications but may encompass a tradeoff between attainable gains and the channel state information (CSI) available both at the transmitter and receiver, which involves overhead. Typically, the BS collects CSI in wireless networks, but only occasionally (periodically or on-demand) and only from a selected set of

S. Dolev et al. (Eds.): CSCML 2023, LNCS 13914, pp. 200–215, 2023.
https://doi.org/10.1007/978-3-031-34671-2_15

users each time since collecting CSI from all users at all times involves enormous overhead, especially when the number of users is large. The channel reports may err due to noise, resulting in imperfect CSI, and both the BS and transmitters must carefully utilize their noisy channel knowledge.

The performance of many communication systems, particularly MIMO systems, is highly influenced by the set of transmitting users (or receiving). Accordingly, numerous studies have focused on user selection for both upstream (users to BS) and downstream (BS to users) user selection, e.g., [16,26,28,29]. Again, user selection involves a tradeoff between overhead and performance. Collecting CSI from all users may enable selecting the optimal user set for transmission but is costly in overhead (including CSI gathering and the BS scheduling announcement). On the other hand, partial CSI collection involves reduced overhead, but the optimal user selection is impossible with missing information, thus compromising performance. For example, the 802.11ax standard supports MU-MIMO for both upstream and downstream, allowing simultaneous transmission from\to a small group of up to eight users. The standard defines that channel estimation (channel report collection) will be performed from the preselected users before each MU transmission opportunity (TXOP) (e.g., [14, Chapters 3.3.4-3.3.6]). The BS selects users based on their traffic demands, which incurs more overhead (either in dedicated messages or additional piggybacked bits) in the upstream direction.

In this study, we suggest an upstream mechanism for MU-MIMO, in which the users self-schedule themselves for transmission, eliminating the overhead of CSI collection and backlogged users tracking done by the receiver. The mechanism relies on combining two (unrelated) techniques; The first is Index Modulation (e.g., [25, Chapter 1.2]), and the second is Group Testing (GT) (e.g., [3,12]). Index Modulation, in this study, refers to the setup where a transmitter activates a non-empty subset of receive antennas, and the activated subset determines the modulated symbol. The demodulation is simple and requires energy detection for each receive antenna, similar to On-Off Keying [17].

The concept of GT tackles the problem of finding K items of interest (typically defective entities or ill patients) out of a large population of N using as few as possible tests. Minimizing the number of tests requires not testing each item individually (which results in N tests), but rather mixing samples together, and examine it as a single sample. The set of items participating in each test is determined *a priori* in the form of a test matrix (non-adaptive GT). Numerous studies showed that the number of group tests required to identify the K items of interest is in the order of $K \log N$, even in scenarios in which the samples may be contaminated [2,5–8,21,22].

The GT concept was adapted to communication terminology (e.g., [4]) where the users, and their messages [11], are analogous to the large item population, N. The K items of interest are the self-scheduled users who actually send messages whose identity is unknown. The test matrix is akin to a binary codebook. The tests conducted are typically energy detection results on different system resources such as timeslots, frequency bands, or, as in the suggested scheme, excited antennas. Over the years, several GT-based MAC protocols for wireless sensor networks were designed, each is capable of simultaneous decoding of many messages using a simple decoding algorithm. [9,10,19] are prime examples. Specifically, Robin and Erkip analyzed a protocol similar to [10] in [18] for activity discovery (rather than data collection) in MU-single-input-single-output slow-fading Rayleigh channel model. Their analysis consisted of converting the continuous signal at each receiving antenna into discrete binary models *using energy detection*, where the energy at a timeslot is compared to some threshold, followed by Noisy Column Matching (Noisy CoMa) for decoding [7,8]. The protocols devised utilized transmissions over separate timeslots or frequency bands, which replaced the original GT test tubes. However, extending the GT concept to the spatial dimension, where test tubes are replaced by antennas, is different and more challenging. Unlike different timeslots or frequency bands, in which there is an inherent separation between the resources just as with the GT test tubes, beamforming to a specific set of antennas may result in energy leakage to all other antennas, which depends on the channels between the transmit and receive antennas and the quality of the estimated CSI at the transmitter. The said leakage can result in an unintended antenna receiving power that exceeds the energy received by an aimed antenna, all the more so when K users are transmitting simultaneously. Accordingly, adjusting existing schemes for MU-MIMO requires careful design to prevent self and mutual-interference.

The suggested scheme relies on a codebook generated using methods from GT. To tackle self-interference, users leverage their (noisy) knowledge of the channels, which requires minimal overhead as can be accomplished by a short single training sequence sent by the receiver to all users simultaneously. Instead of beamforming to the desired antennas, transmitting users null their transmitted signals' energy at the undesired antennas. However, due to their noisy CSI estimation, some energy is leaked to the undesired antennas. Still, since each transmitter utilizes the multi-antenna degree of freedom to null-steer energy to the un-stimulated antennas, the energy received by an un-nulled (stimulated) antenna is random and depends on the channels to this antenna and may be depleted. Accordingly, the receiver needs to devise an energy detection mechanism (converting channel output to a binary vector) to estimate which antenna is targeted (not nulled) by at least one user. The binary vector is treated as the result vector of GT and serves as the input of a decoding algorithm that returns the sent messages (consequently, the identities of the transmitting users as well).

We devise an optimal energy threshold to minimize the required number of antennas, which also maximizes the system sum-rate. The suggested scheme requires no scheduling overhead (headers, control messages, CSI collection at the BS, etc.) and has remarkably low complexity. We thoroughly analyze the suggested scheme's error probability and antenna scaling laws. We show that the number of required antennas is logarithmic in the total number of users in the system and is linear with the number of self-scheduled users. Moreover, we show that our scheme is order-optimal (with respect to the scenario in which *both* the transmitters and the BS have *perfect* CSI).

2 System Model

2.1 Notation

All logarithms in this manuscript are in base 2 unless explicitly specified; $\ln(\cdot)$ is the natural logarithm.

In this work, we denote matrices in bold, and vectors are underlined. For example, $\mathbf{H}$ is a matrix, and $\underline{x}$ is a vector. The dimensions of matrices and vectors would be specified explicitly the first time some matrix or vector is presented or defined. $(\cdot)^T$ denotes the transpose operation.

Specific components of a vector or matrix are specified with subscripts after square brackets, e.g., $[\underline{y}]_m$ and $[\mathbf{H}_k]_{i,j}$. We use a similar notation for matrix rows and columns, but a single colon replaces one index. If the first index is replaced (e.g., $[\mathbf{H}_k]_{:,j}$), then a column is taken, and if the second index is replaced (e.g., $[\mathbf{H}_k]_{j,:}$) then a row is taken.

2.2 Model

We assume a time-slotted wireless network consists of slot-wise synchronized N users (transmitters) and a single BS (receiver). Time slots are assumed to be long enough to encompass an energy detection phase for a single pulse transmission. Even though there is no schedule and each user determines whether to transmit in a timeslot randomly regardless of other users, we assume that at each time slot, a set of up to K users transmit, their identities are unknown a priori.

Each user has M_t antennas, and the BS has M_r antennas. We assume $M_r = M_t$, and a Massive MIMO setting, s.t. $1 \ll M_t, M_r$. Each user sends $\log \mathcal{M}$ bits per TXOP. We denote the set of user k's $\mathcal{M}$ messages by $\mathcal{W}_k$. Without loss of generality, we denote the message sent by user k by w_k. There is no a priori knowledge about the distribution of which message is sent.

The channel matrix of the kth user is denoted by $\mathbf{H}_k \in \mathbb{C}^{M_r \times M_t}$. We assume a scatter-rich urban environment with no line of sight, so each entry in $\mathbf{H}_k$ is a zero-mean Complex Gaussian Random Variable (CGRV) (e.g., [13, Chapter 10.1]). For convenience, we shall assume each component of $\mathbf{H}_k$ has a unit variance, that is, $[\mathbf{H}_k]_{i,j} \sim \mathcal{CN}(0,1)$ for all i, j, k. An illustration of the model can be found in Fig. 1. User k has no direct access to its channel matrix and instead has a noisy version of it, $\hat{\mathbf{H}}_k \triangleq \mathbf{H}_k - \epsilon \mathbf{\Omega}_k$, where $[\mathbf{\Omega}_k]_{i,j} \sim \mathcal{CN}(0,1)$ is the (per-component) uncorrelated noise and $0 \leq \epsilon < 1$ defines the quality of $\mathbf{H}_k$'s estimation. E.g, if $\epsilon = 0$ then $\hat{\mathbf{H}}_k = \mathbf{H}_k$ and user k has a perfect estimation of its channel. Assuming $\epsilon < 1$ is acceptable in literature and is an essential property of a qualitative channel estimation algorithm [15,27].

The encoder utilizes $\hat{\mathbf{H}}_k$ to determine the transmitted symbol from its antennas, i.e., beamforming. We denote user k's transmitted symbol by a complex vector $\underline{x}_k \in \mathbb{C}^{M_t \times 1}$ which is a function of the transmitted message $w_k \in \mathcal{W}_k$ and $\hat{\mathbf{H}}_k$. The transmission is bounded by power level, P, i.e., $\|\underline{x}_k\|^2 \leq P$.

The BS has neither CSI nor any other information regarding the transmitting users' channels or identities. We assume a zero-mean White Complex Gaussian

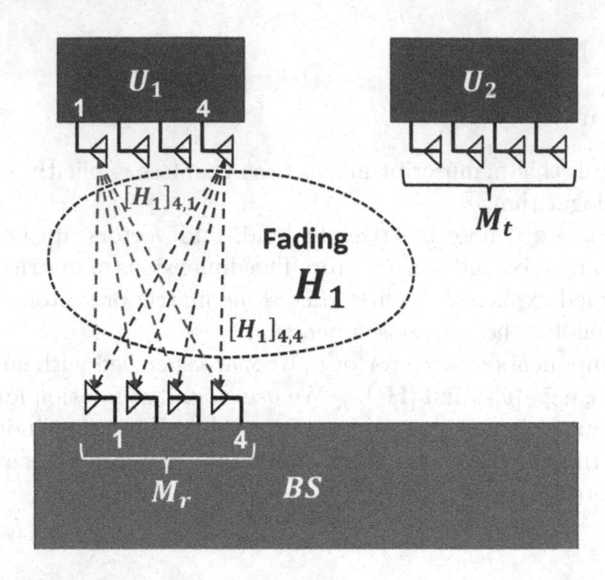

Fig. 1. Visual representation of our model for $N = 2$ and $K = 1$. The transmitters have $M_t = 4$ antennas, the same number of antennas at the BS, M_r. Transmitted signals from user 1 suffer from fading and received by each antenna at the BS. The fading coefficient from antenna j (at the transmitter) to antenna i (at the BS) is $[\mathbf{H}_1]_{i,j}$.

Additive Noise, $\underline{n} \in \mathbb{C}^{M_r \times 1}$, where $[\underline{n}]_i \sim \mathcal{CN}(0, N_0)$ for all i. Finally, the BS obtains

$$\underline{y} = \sum_{k=1}^{N} \delta_k \mathbf{H}_k \underline{x}_k + \underline{n}. \tag{1}$$

where $\delta_k \in \{0, 1\}$ indicates whether user k is active or not. I.e., $\delta_k underline x_k$ is the zero vector if user k is inactive and is the transmitted signals from each of user k's antennas when user k is active.

The decoder uses $\underline{y}$ to obtain both the messages sent and infer the identity of the K users. We say that the system is *message-user reliable* if the decoder obtains *exact correct* K messages.

3 MIMO-GT-Based Scheme

The main result of our study is the MAC protocol we present in this section, which is comprised of three parts; codebook generation, transmission scheme, and receiver algorithm. An example and brief explanation can be found in Fig. 2. The next sections analyze our scheme's performance in terms of error probability, scaling laws, rate and complexity. The scheme we provide is similar to the scheme in our previous work, [24], and for completeness we elaborate on each part. Some of the parameters mentioned will be determined in the next section.

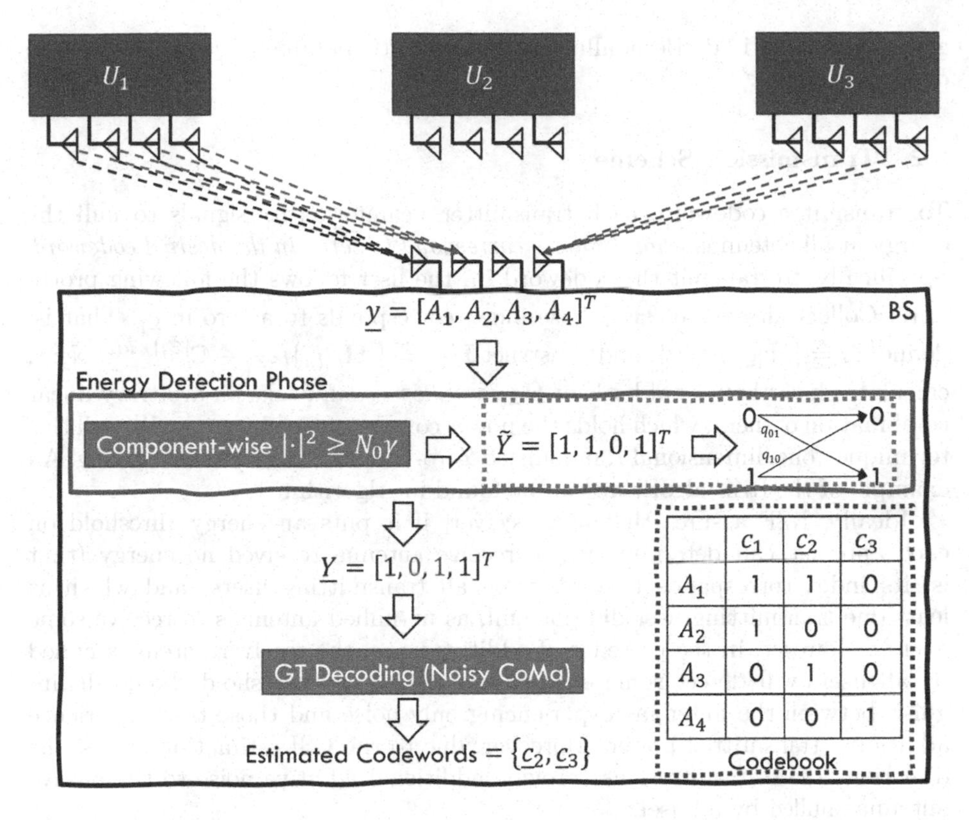

Fig. 2. Visualization of the suggested MAC protocol for $N = 3$, $K = 2$, $\mathcal{M} = 1$, $M_t = M_r = 4$. Users 1 and 3 are transmitting signals such that the BS could read the Boolean sum of their respective codewords, $\underline{c}_1$ and $\underline{c}_3$ (the codebook is highlighted in brown). The users utilize RZF, nulling their signals at antennas (at the BS) whose indices correspond to zeroes in their corresponding codewords. E.g., as $[\underline{c}_3]_4 = 1$, user 3 transmits over the nullspace of antennas 1 to 3. The BS obtains a noisy sum of the transmitted signals, and puts the signals at each receiving antenna through an energy detection phase. The output of energy detection phase, $\underline{Y}$, is treated as a discrete-noise affected Boolean sum of $\underline{c}_1$ and $\underline{c}_3$ and is sent to Noisy CoMa. Noisy CoMa matches $\underline{Y}$ with any word in the codebook, and outputs all words that have enough "matching percentage" with $\underline{Y}$. The example also highlights Noisy CoMa's possible errors; Noisy CoMa both missed $\underline{c}_1$ and misdetected $\underline{c}_2$. We show that with enough antennas, Noisy CoMa errs less.

3.1 Codebook Generation

First, we define the codebook, which is a collection of all legal words to either transmit or receiver and are known to both the transmitter and the receiver [20]. The codebook we utilize is a typical random codebook, similar to other studies, e.g., [3]. Specifically, we generate $N\mathcal{M}$ codewords, $\underline{c}_j \in \{0,1\}^{M_r \times 1}$ for $1 \leq j \leq N\mathcal{M}$, and distribute $\mathcal{M}$ codewords to each user. E.g., user k obtains codewords indexed by $(k-1)\mathcal{M}+1$ to $k\mathcal{M}$. Each bit in these codewords is generated using i.i.d. Bernoulli distribution with parameter p, which will be determined later.

3.2 Transmission Scheme

To transmit a codeword, each transmitter beamforms its signals to null the energy at all antennas *whose indices correspond to zeros in the desired codeword.* Specifically, to transmit the codeword $\underline{c}_j$, the user follows the following procedure: Collect all rows of $\hat{\mathbf{H}}_k$ whose index corresponds to a zero in $\underline{c}_j$. That is, define $\mathcal{Z}_j \triangleq \{l : [\underline{c}_j]_l = 0\}$, and construct $\hat{\mathbf{H}}_{\mathcal{Z}_j} \triangleq \{[\hat{\mathbf{H}}_k]_{l,:}\}_{l\in\mathcal{Z}_j} \in \mathbb{C}^{|\mathcal{Z}_j|\times M_t}$. Next, calculate the orthonormal basis of $\hat{\mathbf{H}}_{\mathcal{Z}_j}$'s *nullspace*, and take an arbitrary linear combination of them, which holds the power constraint, to obtain $\underline{x}_k$. We call this technique "one-dimensional Randomized Zero-Forcing" (RZF) beamforming. An example of the RZF algorithm can be found in Algorithm 1.

Ideally, RZF assures that the receiver, if it puts an energy threshold on each antenna, can determine which receive antenna received no energy (that is, its index corresponds to zeros from all transmitting users) and which at least one transmitting user did not null, as un-nulled antennas to receive some energy. However, in the presence of additive noise, the receiver antennas nulled by all users will detect some energy, and the energy threshold should distinguish between the antennas experiencing only noise and those that experience additional transmitted power. Moreover, due to the CSI estimation errors, the directions the transmitters use provide additional additive noise to the receive antennas nulled by all users.

In light of the increased additive noise, RZF can be improved. Specifically, it is possible that instead of a random choice, each transmitter may select a vector from $\hat{\mathbf{H}}_{\mathcal{Z}_j}$'s nullspace that maximizes the SNR at the un-nulled antennas. However, this analysis is omitted as analyzing the scheme with the optimized vector can be complex and does not substantially change the qualitative discussion in terms of the order-optimality we wish to accomplish.

3.3 Receiver Algorithm

The receiver obtains $\underline{y}$ according to (1), calculates the energy at each of its components (by calculating element-wise squared absolute value), and compares them an energy threshold $N_0\gamma$ (the selection of γ will be discussed later). The result of the energy detection phase, $\underline{Y} \in \{0,1\}^{M_r\times 1}$, is similar to the infected test tubes in the GT context. $\underline{Y}$ is sent to a GT-decoding algorithm to obtain the codewords and their senders' identities.

As previously explained, since there is no control over the energy received by the desired antennas (which can be quite low), and due to noise (both the channel additive noise and the leakage caused by the CSI errors), a hard decision according to energy threshold may introduce erroneous bits in $\underline{Y}$. These errors are characterized by crossover probabilities from '1' to '0' (miss-detecting an antenna that at least one user targets) and vice-versa from '0' to '1' (identifying an antenna that no user targets as an antenna that at least one user targets). We denote these probabilities by q_{10} and q_{01}, respectively. Consequently, we need to utilize a GT decoder that handles errors. We adapted the Noisy CoMa algorithm [8], which obtains the messages from the $\underline{Y}$.

Basically, Noisy CoMa examines all codewords and discards those that almost surely were not transmitted. Note that in the absence of errors, an index of a transmitted codeword $\underline{c}_j$ of '1' should result in energy detection in the antenna with the same index hence a '1' in the same index of $\underline{Y}$. On the other hand, a transmitted codeword's index of '0' will not necessarily result in '0' on the same index of $\underline{Y}$, as it depends on the other transmitters. Accordingly, when Noisy CoMa discards the almost surely not transmitted codewords, it focuses on the codeword indices that are '1', and discards all codewords that the fraction of '1' in the codeword that are '0' in $\underline{Y}$ exceeds a tunable low threshold. Formally, Noisy CoMa obtains the messages from the $\underline{Y}$ according to the following criterion:

Noisy CoMa Decision Criterion [18]. Fix $\Delta > 0$ (Δ is a fine-tuning parameter that will be discussed later). Denote $\text{supp}(\underline{c}_j)$ and $\text{supp}(\underline{Y})$ as the set of indices where $\underline{c}_j$ and $\underline{Y}$ has non-zero components, respectively. Let $\mathcal{T}_j \triangleq |\text{supp}(\underline{c}_j)|$ and $\mathcal{S}_j \triangleq |\text{supp}(\underline{c}_j) \cap \text{supp}(\underline{Y})|$. Noisy CoMa declares that $\underline{c}_j$ has been transmitted if and only if $\mathcal{S}_j \geq \mathcal{T}_j(1 - q_{10}(\Delta + 1))$.

Algorithm 1. Randomized Zero-Forcing ($\mathbf{H}$, $\underline{c}$)

Input:
A channel matrix, $\mathbf{H}$
Codeword to transmit, $\underline{c}$.
Output:
Legal signal vector to transmit, $\underline{x}$
Algorithm:
Calculate $\mathcal{Z} \triangleq \{i : [\underline{c}]_i = 0\}$
Construct $\mathbf{H}_{\mathcal{Z}} \triangleq \{[\mathbf{H}]_{l,:}\}_{l \in \mathcal{Z}}$
$\mathbf{V}_{\mathcal{Z}} \leftarrow \text{Orth}(\text{NullSpace}(\mathbf{H}_{\mathcal{Z}}))$ ▷ $\mathbf{H}_{\mathcal{Z}}$'s nullspace orthonormal basis.
$\underline{v} \leftarrow \sum_{m=1}^{M_r - |\mathcal{Z}|} [\mathbf{V}_{\mathcal{Z}}]_{:,m}$ ▷ Arbitrary vector spanned by $\mathbf{V}_{\mathcal{Z}}$
$\underline{x} \leftarrow \frac{\sqrt{P}}{\|\underline{v}\|_2} \cdot \underline{v}$ ▷ Ensure $\underline{x}$ holds the power constraint
return $\underline{x}$

---

4 Sufficient Conditions

In this section, we examine sufficient requirements for reliable communication of the proposed MIMO-GT scheme. The analysis of the scheme, which incorporates the required number of receiving\transmitting antennas, their associated energy threshold (γ), and the codebook generating parameter (p), is the same as in our previous work, albeit adjusted for CSI estimation errors. The crossover probabilities of which an un-nulled antenna reads '0' (q_{10}) and a nulled antenna reads

'1' (q_{01}) are different due to the energy leakage caused by the erroneous channel estimation. Since the crossover probabilities are different, so is the optimization problem in [24, Lemma 6] and therefore is the parameter selection of γ and p is different. To establish sufficient conditions for vanishing error probability, we consider three main steps: (i) Computing the per antenna error probability, i.e., the error probability in estimating whether at least one user targets antenna i in the energy detection phase. (ii) Devising sufficient conditions on M_r such that Noisy CoMa's error probability vanishes. (iii) Minimizing the sufficient conditions, in light of the first step, and showing that the minimizer is unique.

4.1 Antenna-Wise Error Probability

As previously explained, a hard decision that relies on energy detection introduces the error probabilities q_{10} and q_{01}. To obtain them, we first calculate the distribution of $|[\underline{y}]_i|^2$, conditioned on the number of users targeting antenna i. The distribution is given in the following proposition.

Proposition 1. *Assume each transmitter uses RZF. Let J_i be the number of users targeting antenna i. Then,*

$$|[\underline{y}]_i|^2 | J_i \sim \text{Exp}\left(\frac{1}{\epsilon^2 KP + (1-\epsilon^2) J_i P + N_0} \right).$$

Proof Sketch: The transmission of the RZF output vector is essentially a linear transformation. The antennas whose indices correspond to zero in some codeword introduce signal leakage (in the form of a zero-mean $\epsilon^2 P$-variance CGRV), whereas antennas corresponding to ones introduce a desired signal (zero-mean P-variance CGRV). There are J_i users targeting antenna i, each contributing a desired signal, and $K - J_i$ non-targeting users, each contributing a leaked signal. Finally, the channel adds additive noise so antenna i obtains a zero-mean $(PJ_i + \epsilon^2 P(K - J_i) + N_0)$-variance CGRV. Since the signal at antenna i is a CGRV, its power distribution is exponentially-distributed.

When each transmitter has a random codebook generated by i.i.d coin tosses with probability p for '1', $J_i \sim \text{Bin}(K, p)$. Now, we are ready to calculate the crossover probabilities

$$q_{01} \triangleq \mathbb{P}(|[\underline{y}]_i|^2 > N_0 \gamma \mid J_i = 0) \tag{2}$$

$$q_{10} \triangleq \mathbb{P}(|[\underline{y}]_i|^2 \leq N_0 \gamma \mid J_i \geq 1). \tag{3}$$

Lemma 1. *Denote $\rho \triangleq \frac{P}{N_0}$. For any γ, the crossover probability from '0' to '1' is $q_{01} = \exp\left\{ -\frac{\gamma}{\epsilon^2 K\rho + 1} \right\}$.*

Lemma 2. *For any γ, the crossover probability from '1' to '0' is*

$$q_{10} = 1 - \sum_{j=1}^{K} \binom{K}{j} \frac{p^j (1-p)^{K-j}}{1-(1-p)^K} \exp\left\{ \frac{-\gamma}{\rho(\epsilon^2 K + (1-\epsilon^2) j) + 1} \right\}.$$

4.2 Decoding Error Probability

The decoding errors are divided into two categories; mis-detection, p_{MD}, where Noisy CoMa fails to find at least one of the transmitted codewords. The other is a false-detection, p_{FD}, where Noisy CoMa declares at least one codeword that was not transmitted. Note that p_{FD} also encompasses the event that identical codewords were created by the random code generator. In order to determine the sufficient conditions on the number of required antennas, we bound the decoding error probabilities, p_{MD} and p_{FD}. The complete analysis for perfect CSI has been addressed in our previous work, which followed the footsteps of [8]. The analysis for the realistic model with CSI estimation errors is similar. Lemmas 4 and 5 in [24] do not change as they are independent of the model and show sufficient conditions as functions of N, K, $\mathcal{M}$, Δ, p, q_{10} and q_{01} (which encompass, but yet to utilize, γ and ϵ). Therefore, the optimal energy threshold, γ^*, and the optimal codebook generating parameter, p^*, are determined by the following lemma (an equivalent to [24, lemma 6])

Theorem 1 (Direct). *Fix N, K, $\mathcal{M}$ and $0 \leq \epsilon < 1$. Let $\delta > 0$. MIMO-GT with $\max\{p_{MD}, p_{FA}\} \leq (N\mathcal{M})^{-\delta}$ requires no more than $(1+\delta)\beta^* K \ln N\mathcal{M}$ receiver antennas, where $\beta^* \geq 1$ is the unique solution to*

$$\min_{p,\gamma\in\mathbb{R}} \frac{1}{Kp\left(1 - \exp\{-\frac{1}{2}(1-p)^{2K}(1-q_{10}-q_{01})^2\}\right)}$$

$$s.t. \quad \begin{cases} 0 \leq p \leq \frac{1}{2} \\ 0 \leq \gamma \end{cases}$$

Consequently, MIMO-GT is message-user reliable.

Below we provide a proof sketch. The proof comprises four steps: (i) Take both β_1 and β_2 from [24, Lemmas 4 & 5] and devise an optimization problem to minimize $\max_{i\in\{1,2\}}\{\beta_i\}$. (ii) Convert the minimax problem into a minimization problem by eliminating the dependency on Δ. This is done by noting that β_1 and β_2 have opposing trends in Δ, and showing that their equalizer is $\Delta^* = \frac{(1-p)^K(1-q_{01}-q_{10})}{2q_{10}}$. (iii) Prove that for any p there exists a *unique* γ^* minimizing the objective function. The final step combines the continuity of the objective function and step (iii) to show that a unique solution exists.

The optimization problem in Theorem 1 is solvable by numerical means, which produce the optimal γ^* and p^* and consequently Δ^*. Once we showed that a unique solution, (p^*, γ^*), exists, we are interested in its scaling laws. We summarize their scaling laws in the following propositions:

Proposition 2. $p^* = O(\frac{1}{K})$

Proof Sketch: First, derive the objective function in Theorem 1 as a function of α after substituting $p = \frac{\alpha}{K}$ and compare the numerator to zero. α^*, after some algebraic manipulations on the numerator, can be found by solving an equation of the form $e^{-f(\alpha)}(1 + \alpha g(\alpha)) = 1$. By bounding both $f(\alpha)$ and $g(\alpha)$ we obtain that $\alpha^* \leq e - 1$.

Proposition 3. *If* $\epsilon^2 \leq \frac{1}{K\rho}$ *then* $\gamma^* = O(\ln K\rho)$ *and* $\gamma^* = \Omega(\ln \rho)$.

Proof Sketch: γ^* is the solution to $\frac{\partial q_{01}}{\partial \gamma} + \frac{\partial q_{10}}{\partial \gamma} = 0$, so we can substitute $q_{01}(\gamma^*)$ with $(\epsilon^2 K\rho+1)\frac{\partial q_{01}}{\partial \gamma}$. Next, bound $q_{01}(\gamma^*)$ using the smallest and largest addends. Finally, use the fact that $\epsilon^2 \leq \frac{1}{K\rho}$ to simplify the bounds to obtain the desired result.

4.3 Antenna Scaling Laws

All terms in our analysis are functions of ϵ, which embodies what CSI estimation technique used before the TXOP. Due to its remarkable popularity and efficiency, we assume, hereinafter, that the CSI estimation technique used is similar to 802.11ax. The 802.11ax standard uses channel estimation techniques that utilize orthogonal codes (or training symbols), where each code is distinct per spatial stream [1]. For qualitative channel estimation, the length of the orthogonal code used must be no less than M_r [15]. Using exactly M_r training symbols results in $\epsilon^2 = \frac{1}{1+K\rho} \leq \frac{1}{K\rho}$.

Since the number of required antennas is a function of β^*, we briefly discuss how β^* scales with ρ and K. First, we shall show that β^* converges to some constant when $K \to \infty$.

Proposition 4. *Let* $p = \frac{\alpha}{K}$, *where* $\alpha > 0$ *is some constant. If* $1 \leq \gamma \leq \max\{1, \rho\}$ *and* $\epsilon^2 \leq \frac{1}{K\rho}$, *then*

$$\beta^* \leq \frac{32 e^{2\max\{\rho,1\}}(\rho+2)^2}{3\alpha(1-\frac{\alpha}{2})^4 \rho^2 (1-\epsilon^2)^2}$$

Proof Sketch: First, bound q_{10} from above by its largest addend. Next, bound $(1-\frac{\alpha}{K})^{2K}$ from below by substituting $K = 2$. Finally, apply Taylor series expansion.

Proposition 4 bounds β^* from above by a constant with respect to K. Under the conditions of Proposition 4, the terms $1 - q_{10} - q_{01}$ and $(1-\frac{\alpha}{K})^K$ converge when $K \to \infty$. Therefore, β^* converges to some constant when $K \to \infty$.

Note that even though seemingly Theorem 1 is not a function of ρ, all the parameters γ, p, and ρ are dependent on one another. For example, low ρ generates low γ, which in turn affects the crossover probabilities and the choice of p. Despite their complex relationship, we show that β^* exhibits the intuitive properties related to different SNR regions, e.g., longer codewords (higher number of antennas) for low SNR and finite length codewords for high SNR, assuming $\epsilon^2 \leq \frac{1}{K\rho}$.

Proposition 5. *For any* ϵ, *if* $\rho \to 0$ *then* $\beta^* \to \infty$.

Proof Sketch: q_{01} tends to $e^{-\gamma}$ when $\rho \to 0$. When $\rho \to 0$, q_{10} tends to $1 - q_{01}$ which implies $\beta^* \to \infty$ for any γ and p.

Proposition 6. *If* $\epsilon^2 \leq \frac{1}{K\rho}$ *then* $\beta^* \to$ const. *when* $\rho \to \infty$.

Proof Sketch: when $\epsilon^2 \leq \frac{1}{K\rho}$, $\gamma^* = \Theta(\ln\rho)$ so the crossover probabilities vanish as $\rho \to \infty$. Hence, β^* converges when $\rho \to \infty$ for any p.

If a system has a minuscule number of antennas but still satisfies $M_t = M_r$, one can use additional time slots\frequency bands to compensate for the difference. E.g., if our solution requires $L = l \cdot M_r$ antennas, use l time slots. In each time slot, save the channel output to obtain $\{\underline{y}_i\}_{i=1}^{l}$. Finally, $\underline{y}$ can be obtained by concatenation: $\underline{y} = (\underline{y}_1^T, \underline{y}_2^T, \ldots, \underline{y}_l^T)^T$. It is possible to use the aggregation solution when $M_r > M_t$ by turning off $M_r - M_t$ antennas, but such a solution suffers from a reduced sum-rate as not all degrees of freedom are utilized. In Sect. 6, we evaluate this approach with growing K.

5 Performance

In this section, we evaluate MIMO-GT's performance in terms of rates and complexity.

Rates- each user has $\mathcal{M}$ codewords, hence sends $\log\mathcal{M}$ bits per transmission. The identification of each transmitting user incurs additional $\log N$ bits. We thus obtain a sum rate of

$$R = K(\log\mathcal{M} + \log N) = K\log N\mathcal{M} = \frac{M_r}{(1+\delta)\beta^*\ln 2} \tag{4}$$

bits per channel use. The right-hand side follows Theorem 1 and taking the lowest possible β^*, maximizing sum-rate. The $\ln 2$ in the denominator follows the logarithm base change. We emphasize that β^* is a function of SNR, ρ. Note that this is almost the *same* sum-rate as in the perfect CSI case.

Complexity- in MIMO-GT, the coding is straightforward as each transmitted word is mapped to a unique codeword where the user identity is embedded in the codeword. Since each user schedules itself, there is no need for any scheduling mechanisms and their associated complications, i.e., users do not exchange messages to cooperate, there is no need for CSI collection at the receiver, and as far as complexity is concerned, the receiver does not perform any scheduling algorithm, and consequently does not announce scheduling results. Accordingly, the complexity in MIMO-GT is associated with the decoding process.

The decoding in MIMO-GT is done using Noisy CoMa. Noisy CoMa matches all columns in the codebook with the output of the energy detection phase, $\underline{Y}$. Since there are $N\mathcal{M}$ codewords in total and each is of length M_r, the total decoding time of Noisy CoMa is $O(N\mathcal{M}M_r)$.

6 Numerical Results

In order to get some insight into the MIMO-GT performance, in this section, we provide numerical evaluation results.

Figure 3 shows how p^* and γ^* scale with ρ and $\epsilon^2 = \frac{1}{1+K\rho}$. When the SNR is low, p^* is higher in order to increase the probability that more users share target antennas and, as a result, overcome the energy threshold. In other words, p^* embodies a certain degree of codebook-level cooperation, and the users themselves do not need to cooperate.

In the low SNR region, γ^* is close to 1 as any lower γ would result in the BS' inability to distinguish between noise or transmission. In the high SNR region, γ^* abides Proposition 3.

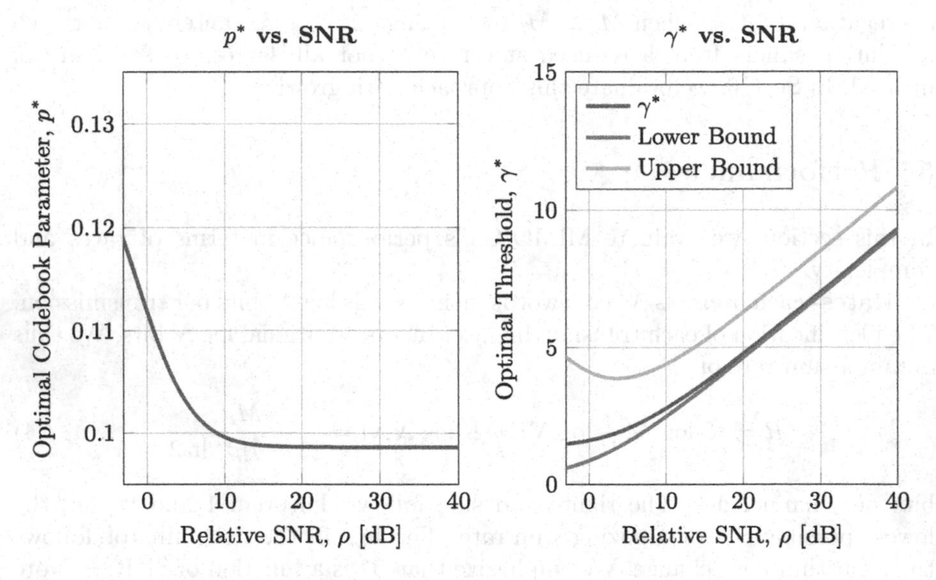

Fig. 3. p^* (on the left) and γ^* (on the right) as functions of ρ when $N = 100$, $K = 5$, $\mathcal{M} = 1000$, $\epsilon^2 = \frac{1}{1+K\rho}$ and $\delta = 0.33$.

In Fig. 4, we evaluate the number of additional slots used to compensate for the lack of antennas to deploy our scheme in Sect. 3 (the aggregation solution suggested at the end of Sect. 4). In this scenario, $M_t = M_r < (1+\delta)\beta^* K \ln N\mathcal{M}$ are constants and the number of transmitting users, K, grows closer to N. The dashed line in Fig. 4 is a "critical K", $\frac{M_r N}{(1+\delta)\beta^* \ln N\mathcal{M}}$, which divides K into two regions; the sparse region, where the number of additional slots is less than N, and the dense region, where the number of slots is equal or greater than N. When the network is dense, or in other words, the number of backlogged users is high, TDMA would outperform MIMO-GT as MIMO-GT requires more slots than users. In sparser regions, the opposite is true.

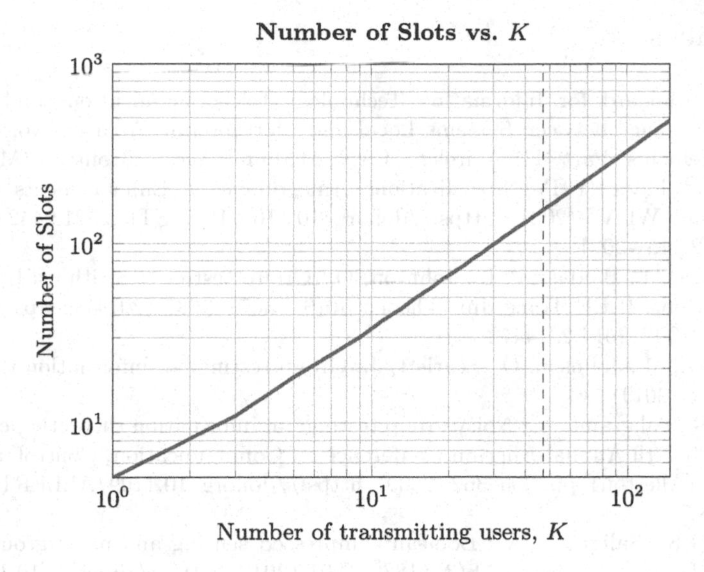

Fig. 4. Number of slots required as a function of K when $N = 150$, $\mathcal{M} = 1000$, $M_r = M_t = 64$, $\rho = 20_{\text{dB}}$, $\epsilon^2 = \frac{1}{1+K\rho}$ and $\delta = 0.33$.

7 Conclusion

In this paper, we studied an estimation-error robust distributed non-cooperative self-scheduling uplink MU-MIMO scheme using GT codes on the receiver antennas. The BS converts the signals received, under both additive noise from the channel, mutual and self-interference caused by beamforming with imperfect CSI, to binary vectors by utilizing a single energy threshold. Our scheme, MIMO-GT, is shown to be order–optimal. The order-optimality is achieved by comparing MIMO-GT's sum-rate to the perfect CSI capacity. MIMO-GT has low complexity and can be implemented on systems with a low number of antennas by utilizing additional system resources such as timeslots or frequency bands. We have expressed and determined the scaling laws of the number of antennas and optimal parameters to minimize them for different SNRs and an increasing number of *active* users.

Despite its remarkable simplicity, MIMO-GT has its shortcomings. Namely, the assumption that the users and the BS have the same number of antennas. In more realistic cases, the BS has more antennas than the transmitting users, and their number is not necessarily comparable. Future research should aim at different beamforming techniques or receiver algorithms (whether it's an interference-robust energy detection or a different scheme altogether) to allow order-optimal reliable joint decoding, if possible.

References

1. IEEE Standard for Information Technology-Telecommunications and Information Exchange between Systems Local and Metropolitan Area Networks-Specific Requirements Part 11: Wireless LAN Medium Access Control (MAC) and Physical Layer (PHY) Specifications Amendment 1: Enhancements for High-Efficiency WLAN (2021). https://doi.org/10.1109/IEEESTD.2021.9442429. IEEE Std 802.11ax-2021
2. Aldridge, M., Baldassini, L., Johnson, O.: Group testing algorithms: bounds and simulations. IEEE Trans. Inf. Theory **60**(6), 3671–3687 (2014). https://doi.org/10.1109/TIT.2014.2314472
3. Aldridge, M., Johnson, O., Scarlett, J.: Group testing: an information theory perspective (2019)
4. Atia, G., Saligrama, V.: Noisy group testing: an information theoretic perspective. In: 2009 47th Annual Allerton Conference on Communication, Control, and Computing (Allerton), pp. 355–362 (2009). https://doi.org/10.1109/ALLERTON.2009.5394787
5. Atia, G.K., Saligrama, V.: Boolean compressed sensing and noisy group testing. IEEE Trans. Inf. Theory **58**(3), 1880–1901 (2012). https://doi.org/10.1109/TIT.2011.2178156
6. Atia, G.K., Saligrama, V., Aksoylar, C.: Correction to "Boolean compressed sensing and noisy group testing" [Mar 12 1880–1901]. IEEE Trans. Inf. Theory **61**(3), 1507–1507 (2015). https://doi.org/10.1109/TIT.2015.2392116
7. Chan, C., Che, P.H., Jaggi, S., Saligrama, V.: Non-adaptive probabilistic group testing with noisy measurements: near-optimal bounds with efficient algorithms. In: 2011 49th Annual Allerton Conference on Communication, Control, and Computing, Allerton (2011). https://doi.org/10.1109/Allerton.2011.6120391
8. Chan, C.L., Jaggi, S., Saligrama, V., Agnihotri, S.: Non-adaptive group testing: explicit bounds and novel algorithms. In: IEEE International Symposium on Information Theory Proceedings, pp. 1837–1841 (2012). https://doi.org/10.1109/ISIT.2012.6283597
9. Cohen, A., Cohen, A., Gurewitz, O.: Secured data gathering protocol for IoT networks. In: Dinur, I., Dolev, S., Lodha, S. (eds.) CSCML 2018. LNCS, vol. 10879, pp. 129–143. Springer, Cham (2018). https://doi.org/10.1007/978-3-319-94147-9_11
10. Cohen, A., Cohen, A., Gurewitz, O.: Efficient data collection over multiple access wireless sensors network. IEEE/ACM Trans. Netw. **28**(2), 491–504 (2020). https://doi.org/10.1109/TNET.2020.2964764
11. Cohen, A., Cohen, A., Gurewitz, O.: Secure group testing. IEEE Trans. Inf. Forensics Secur. **16**, 4003–4018 (2021). https://doi.org/10.1109/TIFS.2020.3029877
12. Du, D., Hwang, F.K., Hwang, F.: Combinatorial Group Testing and Its Applications, vol. 12. World Scientific (2000)
13. Goldsmith, A.: Wireless Communications. Cambridge University Press (2005). https://doi.org/10.1017/CBO9780511841224
14. Gulasekaran, S., Sankaran, S.: Wi-Fi 6 Protocol and Network. Artech House Mobile Communications Library, Artech House (2022). https://books.google.co.il/books?id=WXx4zgEACAAJ
15. Hassibi, B., Hochwald, B.: How much training is needed in multiple-antenna wireless links? IEEE Trans. Inf. Theory **49**(4), 951–963 (2003). https://doi.org/10.1109/TIT.2003.809594

16. Kampeas, J., Cohen, A., Gurewitz, O.: Rate analysis of distributed multiuser MIMO protocols for the 802.11ac. In: IEEE International Conference on the Science of Electrical Engineering (ICSEE), pp. 1–5 (2016). https://doi.org/10.1109/ICSEE.2016.7806096
17. Paquelet, S., Aubert, L.M., Uguen, B.: An impulse radio asynchronous transceiver for high data rates. In: International Workshop on Ultra Wideband Systems Joint with Conference on Ultra Wideband Systems and Technologies. Joint UWBST IWUWBS (IEEE Cat. No.04EX812), pp. 1–5 (2004). https://doi.org/10.1109/UWBST.2004.1320888
18. Robin, J., Erkip, E.: Capacity bounds and user identification costs in Rayleigh-fading many-access channel. In: IEEE International Symposium on Information Theory (ISIT), pp. 2477–2482. IEEE (2021)
19. Robin, J., Erkip, E.: Sparse activity discovery in energy constrained multi-cluster IoT networks using group testing. In: IEEE International Conference on Communications (ICC), pp. 1–6 (2021). https://doi.org/10.1109/ICC42927.2021.9500808
20. Roth, R.: Introduction to Coding Theory. Cambridge University Press (2006). https://doi.org/10.1017/CBO9780511808968
21. Scarlett, J., Johnson, O.: Noisy non-adaptive group testing: a (near-)definite defectives approach. IEEE Trans. Inf. Theory **66**(6), 3775–3797 (2020). https://doi.org/10.1109/TIT.2020.2970184
22. Sejdinovic, D., Johnson, O.: Note on noisy group testing: asymptotic bounds and belief propagation reconstruction. In: 48th Annual Allerton Conference on Communication, Control, and Computing (Allerton), pp. 998–1003. IEEE (2010)
23. Tse, D., Viswanath, P.: Fundamentals of Wireless Communication. Cambridge University Press, Cambridge (2005)
24. Vershinin, G., Cohen, A., Gurewitz, O.: Order-optimal joint transmission and identification in massive multi-user MIMO via group testing (2022). https://doi.org/10.48550/ARXIV.2210.00421
25. Wen, M., Cheng, X., Yang, L.: Index Modulation for 5G Wireless Communications. WN, 1st edn. Springer, Cham (2017). https://doi.org/10.1007/978-3-319-51355-3
26. Xia, X., et al.: Joint user selection and transceiver design for cell-free with network-assisted full duplexing. IEEE Trans. Wirel. Commun. **20**(12), 7856–7870 (2021). https://doi.org/10.1109/TWC.2021.3088485
27. Yoo, T., Goldsmith, A.: Capacity and power allocation for fading MIMO channels with channel estimation error. IEEE Trans. Inf. Theory **52**(5), 2203–2214 (2006). https://doi.org/10.1109/TIT.2006.872984
28. Yoo, T., Goldsmith, A.: On the optimality of multiantenna broadcast scheduling using zero-forcing beamforming. IEEE J. Sel. Areas Commun. **24**(3), 528–541 (2006). https://doi.org/10.1109/JSAC.2005.862421
29. Zhang, J., Liu, M., Xiong, K., Zhang, M.: Near-optimal user clustering and power control for uplink MISO-NOMA networks. In: IEEE Global Communications Conference (GLOBECOM), pp. 01–06 (2021). https://doi.org/10.1109/GLOBECOM46510.2021.9685202

The Use of Performance-Counters to Perform Side-Channel Attacks

Ron Segev[1] and Avi Mendelson[2](✉)

[1] Tel Aviv University, Tel Aviv, Israel
[2] Technion Israel Institute of Technology, Haifa, Israel
avi.mendelson@technion.ac.il

Abstract. Performance and power counters were invented to allow optimizing applications, but they can also be used to expose private information about the system and the user that uses it; thus, they have the potential to become a major privacy threat. This work shows that performance traces, achieved with performance counters, are sufficient for three efficient privacy attacks on a computer. The first attack allows the identification of webpages the user uses with a high success rate of up to 100%. This attack may expose private information about the user, like political views and affiliations. The second attack allows browser version identification. Browsers are updated regularly to protect against known cyber-attacks. An attacker can use this information to choose the best attack method to achieve successful cyber-attacks. The attack is unique since it is the first study to demonstrate the detection of the browser version using a side-channel attack. The third attack allows the recovery of structural elements of Neural Networks, like the number of layers and activation functions being used. This information may assist in preparing adversarial examples against the Neural Network or in creating a similar copy of the Neural Network. To evaluate these attacks, we collected performance traces using Intel Power Gadget software-based performance counter tool. We collect traces of power consumption, utilization percentage, and clock frequency of the Intel CPU and its internal parts like DRAM memory and GPU.

Keywords: Side Channel · Website Fingerprinting · Model Extraction Attacks · Security · Privacy · Deep Learning

1 Introduction

The importance of performance measurements of a computer system has been increasing over the past few decades. Intel has developed a software-based monitoring tool, Intel Power Gadget, which can measure power consumption, efficiency, clock frequency, and temperature of the internal processor, DRAM memory, and internal GPU inside the Intel processor. The information gathered by performance counters can help programmers to optimize their code.

S. Dolev et al. (Eds.): CSCML 2023, LNCS 13914, pp. 216–233, 2023.
https://doi.org/10.1007/978-3-031-34671-2_16

In our research, we looked at what other information can be extracted from performance measurements done by Intel Power Gadget. This research focuses on three security and privacy aspects of modern systems: website fingerprinting, browser information leakage, and Neural Network structure information leakage. The paper indicates that such information can be discovered from performance traces like power consumption, utilization percentage, and clock frequency using software-only tools.

The first aspect is the ability to expose private information on a user (what is he looking at). Website fingerprinting is an attack that reveals to the attacker what websites the victim user is visiting. Knowing what websites users visit may reveal private and sensitive information such as political views and affiliations. To preserve users' privacy, browser companies implement software security mechanisms, such as in-browser encryption and incognito modes.

The second aspect is the ability to expose information regarding the browser being used, including its version and the existence of an ad-blocker. For example, information like browser version can help an attacker to prepare better Cyber attacks. New Cyber-attacks on browsers are being discovered at a high pace. To protect users from such known attacks, browser companies send updates to users to protect them. Updating all of the user's browsers could take time, and until it is finished, users are at risk of these Cyber-attacks. Therefore, another known Cyber-attack could be achieved by knowing if a user is using a not updated version of the browser.

The last aspect is the ability to discover information about the Neural Network model. Neural Network models can be intellectual properties of companies and are private information. Knowing the Neural Network's model can help competitors copy the Neural Network or prepare adversarial examples for the net.

This paper presents that a new way of measurement could be done to enable side-channel attacks. We demonstrate the following types of attacks; fingerprinting different websites, browser version discovery, and the recovery of Neural Network structure information.

We used performance counter measurements gathered by the Intel Power Gadget tool to achieve these goals. Using Intel Power Gadget, we were able to measure performance factors like power consumption, efficiency, and clock frequency of different devices like the CPU, DRAM memory, and internal GPU. By doing the measurement on the target computer, like power consumption, we were able to achieve website fingerprinting similarly like in [1–3], however, our measurements were done by software only tools and don't require getting close to the target computer and using special measurement tools.

The threat model for all attacks presented in this paper is a co-located attack model, requiring execution on the same machine as the victim. It could be done, for example, by copying a script to the command prompt, and it doesn't require superuser rights. Traces measurements are done on the target machine. The performance measurements are saved to a CSV file and can be analyzed on the target machine or to be sent secretly to any other machine and analyzed there.

This study is the first to demonstrate website fingerprinting through measurement traces like efficiency and clock frequency of different internal hardware devices. It is also the first to demonstrate the detection of the browser version using side channels. Our paper is the first study demonstrating information leakage of Neural network structures using a software-based performance counter tool and not with the usage of special lab equipment.

In our research, we used several different algorithms to analyze different traces of data measured and compared them. As a result, we demonstrate that a malicious user that can perform performance counter measurements could recover information about users' web activity with up to a 100% success rate.

This Paper Makes the Following Contributions:

- We discovered several new attacks on privacy that use performance counter measurements done by the user's computer.
- We showed that website fingerprinting could be achieved with up to 100% accuracy of success.
- We showed that the browser version installed could be detected from performance counter measurements with high accuracy of success.
- We showed that structure information of the Neural Network, including the number of layers and internal Activation functions, could be detected from performance counter measurements.
- We employed and evaluated signal analysis and Machine Learning techniques to fingerprint different websites and discover browser versions.

The Rest of the Paper is Organized as Follows: In 2nd section, we will discuss past related research. In 3rd section, we will give a brief background on performance counters, specifically the Intel Power Gadget tool. We will also explain the usage of machine learning techniques in our work. In 4th section, we will describe the experimental setup. In Sects. 5 and 6, we will present experimental results for website fingerprinting and browser version detection using several algorithms. In 7th section, we will present experimental results for Neural Network structure informational leakage. In 8th section, a discussion is made about the results' implications in Sects. 5, 6, and 7. Sections 9 will include a conclusion, acknowledgements, references, and a list of websites included in the research.

2 Past Research

2.1 Website Fingerprinting Past Research

Different methods of website fingerprinting have been achieved in other studies. 1st method was the idea of using side-channel attacks on network traffic. For example, it has been demonstrated that transferred file sizes could be used as a reliable fingerprinting for websites [13]. Another example used information on packet sizes and packet order to improve website identification [14]. Furthermore, it has been shown that even encrypted channels do not protect against traffic analysis. For example, in [15], website fingerprinting was achieved using packet analysis on traffic encrypted using HTTPS and WPA. Many proposed similar attacks that can achieve very accurate fingerprinting using properly designed classifiers [22–25]. Attackers could even retrieve the client's search query on search engines such as Google or Bing by analyzing side-channel leaks in network traffic [33].

In 2nd method, researchers measured the power consumption of a computer or a smartphone while surfing different websites. They found patterns in the power consumption measurements which are different depending on the website browsed [1–3]. An adversary would need access to the electronic supply of the target computer and measure the power consumption via lab equipment like scopes and probes or via an inside voltage sensor that can measure power consumption.

In 3rd method, researchers showed that by exploiting microarchitectural leakages at the hardware level, they could achieve website fingerprinting. For example, in [10], it had been shown that an attacker could detect unique memory footprints for different websites. In another example, it has been demonstrated that cache-based attacks like Prime+Probe [7] and Flush+Reload [6] could be implemented to achieve robust website fingerprinting [4]. In [12], the researchers improved a similar attack by using Machine Learning techniques on cache-sweeping traces.

In 4th method, researchers measured different performance counters. Different methods have been researched like measurements of memory usage [20], events [20] and timing of different processes [20] or power traces [21] the attack method presented in this paper is similar to the 4th method.

The main difference between our research and others is that we do not need physical access to measure performance. Our method is similar to the 4th method because we use a performance counter, Intel Power Gadget, to generate the required traces. However, measuring traces with the Intel Power Gadget tool enables us to measure different performance measures. We are the first to demonstrate website fingerprinting through power, utility, and clock frequency traces of internal parts in the processor.

We note that the key to such attacks is that different applications exhibit different execution behavior. In the studies mentioned, different techniques were implemented to measure the different behavior. Using data analysis techniques on the recorded measurements proved that fingerprinting of different websites could be done.

2.2 Software-Based Power Side-Channel Attacks

Power Analysis Attacks enable an adversarial to learn critical information about the system from power consumption. There have been several studies where researchers showed that they were able to recover cryptographic keys using attacks like Simple Power Analysis (SPA) [11], Differential Power Analysis (DPA) [11], and Correlation Power Analysis (CPA) [18]. Thus, all these traditional methods were based on direct measurement of the device power for creating the power traces and required the use of physical access to the device [11,17–19]. Recent studies achieved the same information recovery similarly without the need to measure the power consumption of the target from outside. These are software-based side-channel attacks. For example, in [5], researchers showed that by exploiting unprivileged access to Intel Running Average Power Limit (RAPL) interface, they could measure values correlated to the real power consumption. The researchers proceeded and showed they could distinguish between different internal instructions and Hamming weights of operands and memory loads. In another study [16], researchers showed that power side-channel attacks could be turned into timing attacks. They found that CPU frequency depends on the current power consumption under certain circumstances and hence correlates to the data. Our study extends the idea of using power counters, but instead of direct reasoning about the state of the device, we are creating power traces that could be used for further analysis. We showed that although openly available tools such as Intel Power Gadget can achieve traces with lower frequency than dedicated hardware measurement tools, such traces are sufficiently accurate to be combined with traditional power-based algorithms such as DPA or CPA. More than that, since such traces can be used to measure many other internal behaviors of the system, e.g., cache behavior, we can generate new classes of algorithms based on data analysis tools for performing remote hacking on the system.

2.3 Timing Attack Based on Respond Time

Remote attacks on systems were introduced by [34,35] that suggest to break Implementations of Diffie-Hellman and RSA codes by measuring the respond time of the algorithms. Our proposed system extends this approach but use performance and power counters that extend the capabilities of such attacks.

2.4 Information Leakage from Neural Network in the Form of Side Channels Past Research

There are many papers regarding different security and privacy issues related to machine learning and deep learning. One type of attack is the leakage of information from the Neural Network through side channels. An attacker might be able to extract model details of the Neural Network or inputs. Model details of Neural Networks, like structure and weights, are important private information. They might be company's protected intellectual property (IP). That helps an attacker to copy the algorithm or to prepare adversarial examples. Recovery of inputs to the Neural Network can give an attacker private information that the user wouldn't want to be leaked.

Several Side channels leakages have been studied, including timing leakage [26], power consumption leakage [27], electromagnetic leakage [28], memory access patterns [29,30], and cache side-channel attacks [31,32]. Studies of power consumption and electromagnetic leakages required measurements done with the usage of complicated Lab equipment from a close distance. The leakage presented in this paper is similar to the leakage of information through power consumption.

The difference between our research and the other studies is that we used a performance counter tool (Intel Power Gadget) instead of external lab equipment. The presented attack model measures the computer's performance while the Neural Network is active.

3 Background

3.1 Software-Based Performance Counter

Performance counters can monitor, count, or measure events in software. They can deduct and estimate measurements like power consumption and utility of various parts in the CPU. They can also measure parameters like temperature, memory usage, clock frequency, and computer internal clock RDTSC.

Performance counters are used by programmers to analyze the performance disruption of their programs when they run on the user's computer. They use performance counters to improve their programs. Users use performance counters to know if other programs disrupt their workflow and to ensure the computer works on the preferred load and is not higher than necessary.

3.2 Intel Power Gadget

Intel Power Gadget is a software-based monitoring tool specialized to Intel Core processors. This program can be downloaded from the intel.com website. The program estimates the energy usage of the processor and can extract information about different parts of the CPU: internal processor cores, DRAM memory, and internal GPU. Furthermore, it works only on intel-based systems [8,9].

Using Intel Power Gadget, one can measure power consumption and utility for each part mentioned in the CPU. Internal clock frequencies of internal processor cores and internal GPU can be measured. Intel Power Gadget can record all measurements at a sampling resolution of 1ms and save all measurements to an excel sheet. To log power data to a log file in Windows operating system, one can use this command in the command line (CMD):

```
PowerLog3.0.exe [-resolution <msec>]
-duration<sec> [-verbose][-file <logfile>]
```
[1]

The relation between processor total energy, internal processor core energy, internal GPU energy, and DRAM energy consumption is:

$$TotalEnergy = internalprocessorenergy + internalGPUenergy + DRAMenergy \quad (1)$$

3.3 Machine Learning Techniques

After measurements were taken, we used several Machine Learning techniques to analyze the data and find the difference between groups of traces. We set the measurements acquired to two sets: The training set and the Test set. Each set had randomly chosen half of the measurements taken. All algorithms are implemented in Python 3.9 and run on a standard quad-core Intel processor.

K-th Nearest Neighbor (KNN): The *KNeighborsClassifier* command in *sklearn* library is used to implement the KNN algorithm and train our model. The KNN algorithm was used in two ways in the paper. The first method was used on traces of the same data taken from different sources. For example, power consumption traces were measured at different times of recordings. By default, all measurements are in the same units of measure. We used the Euclidean metric to determine the distance between classes. In the second method, we used traces of different data combined. In the second method, we normalized each trace first and then measured the distance using the Euclidean metric.

Support Vector Machine (SVM): We used *svm* function from *sklearn* library. The model is created and trained based on linear and polynomial kernels. We chose the gamma parameter to be equal to 10. We trained the algorithm and tested it using the test set.

4 Experimental Setup

We performed our measurements on LENOVO ThinkPad Y700, which has an Intel Core I7-6700HQ CPU and runs Windows 10 operating system. We used the Intel Power Gadget performance counter to record traces of performance. In recording settings, we set Log Sampling Resolution to be 1ms. We developed an automated algorithm that repeated the loading of a target website URL. A list of URLs for each target website was set beforehand for each experiment.

The automated algorithm opens the target webpage, pauses, scrolls the mouse down, pauses again, and finally closes the window. Figure 1 shows the power consumption traces collected while our automated machine loaded the URLs of NewYorkTime.com and FoxNews.com. At the first seconds of each trace, power consumption was high and not so different because, at this time, the web browser was opened. After the first few seconds, the automation entered each website, and we can see that power consumption traces are different for each site. The power consumption of FoxNews.com was higher most of the time. Part of the reason was that as soon as the automation entered an article, a video started playing automatically, while on NewYorkTimes.com, when the article opened, it included text only.

5 Website Fingerprinting

The Website Fingerprinting experiment aims to find what webpages a user access. Thus, we compare performance measured when entering different URLs from the same site against recordings measured when entering different URLs from another site.

We made a list of different pages for NewYorkTimes.com and FoxNews.com. We recorded 694 traces of performance for different URLs on the FoxNews.com website, 501 traces for NewYorkTimes.com, 139 traces for lichess.com, and 170

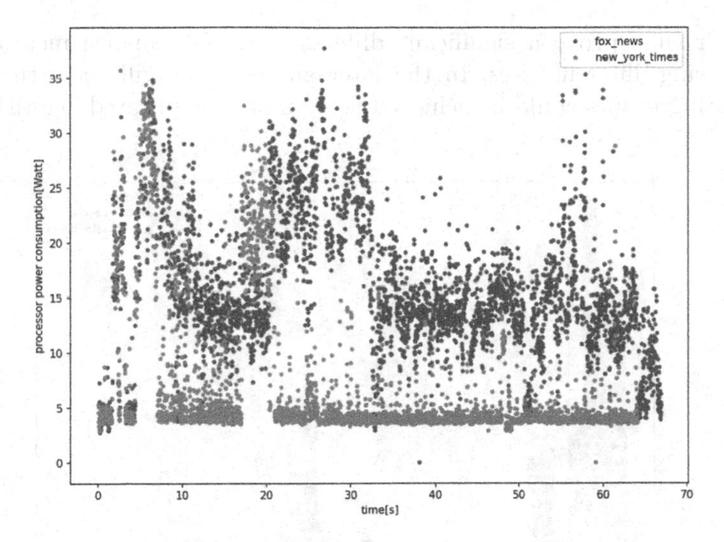

Fig. 1. Internal processor power consumption traces for two different websites newyorktimes.com and foxnews.com

traces for youtube.com. In each recording only one page was opened. Half of the measurements taken were taken three weeks separately and ads were different, to prove the attack is reproducible. During each recording, all of the performance traces (power consumption, utilization percentage, and clock frequency of the Intel CPU, DRAM memory and internal GPU) were recorded in parallel. We used an automation machine to record each performance trace so as not to be affected by things unrelated to the websites themself, like a difference in time between web access or starting recording time.

We separated all traces into the training group and the test group. We trained different algorithms on traces taken from the training group. We used the test group to print the success rate of our algorithms. We used three different algorithms to try and fingerprint the websites. The first algorithm we used was unsupervised and was done by calculating histograms of different parameters, described under subsection Simple signal analysis. The second and third algorithms were supervised algorithms. The algorithms used were the SVM algorithm and the KNN algorithm.

Simple Signal Analysis: Several simple measurements were given for efficient fingerprinting, like total power consumed by the CPU and mean utilization of the CPU. Figure 2 shows the Cumulated power consumption histogram measured during entrance to different URLs, which belong to websites newyorktimes.com and foxnews.com.

A boundary line corresponding with a minimum error that separates the measurements was calculated. The boundary line is shown in Fig. 2 as a dashed red line. We found that one can set this boundary line as a fingerprinting algorithm and have a high success rate of 92.5% when fingerprinting with measurements of Cumulated power and a success rate of 83.2% when fingerprinting using measurements of average utilization.

These results prove a significant difference in performance measurements when entering different sites. In the later analysis, we will show that better fingerprinting results could be achieved using more complicated algorithms.

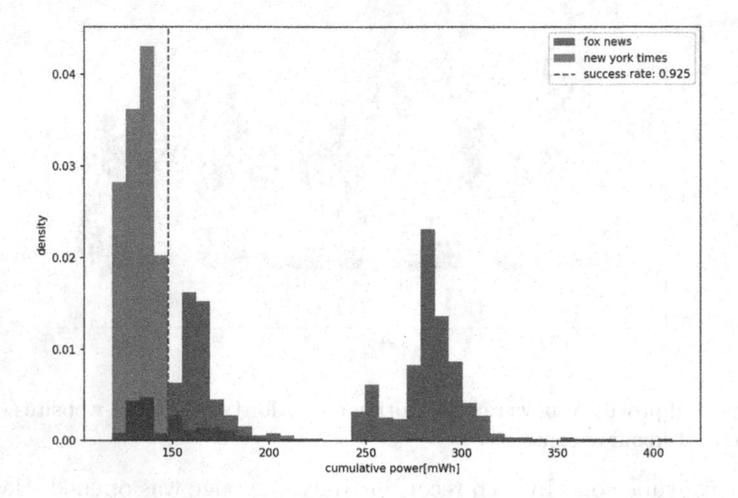

Fig. 2. Cumulative power consumption histogram for two websites foxnews.com and newyorktimes.com

SVM Algorithm: We used the SVM algorithm on different performance traces measured when entering different URLs corresponding to different sites: newyorktimes.com and foxnews.com.

Performance traces that were found to have a higher significance in fingerprinting websites were: CPU processor core power consumption, DRAM power consumption, CPU processor utilization, and an array of all these traces normalized and then combined. We set the SVM's Kernel to be linear or polynomial. Our results of website fingerprinting using the SVM algorithm are reported in Table 1.

From the results, we can see that all parameters had high success rates. We can see that measurements like internal CPU core power consumption can lead to fingerprinting with 99.7% accuracy using the SVM algorithm and linear kernel. Other parameters with high success rates were CPU Utilization with 96.9% accuracy and DRAM power consumption with 98.8% accuracy. Using all parameters together leads to best classification with 100% accuracy. This makes the paper unique because other papers mentioned didn't combine several performance parameters. The accuracy of the fingerprinting algorithms was calculated on more than 600 measurements from the test group.

We also note that when checking on another 34 traces for FoxNews.com and 50 traces for NewYorkTimes.com taken six months prior, We got success rate of 95.2% while using SVM on DRAM power traces and linear kernel, strengthening the argument that the attack is reproducible.

Table 1. Fingerprinting success rate using svm algorithm on different traces and for linear and polynomial kernels

Measurement trace	Success rate [%]	Kernel choice
Internal CPU power consumption	99.7	Linear
Internal CPU power consumption	99.8	Polynomial
DRAM power consumption	98.6	Linear
DRAM power consumption	98.8	Polynomial
CPU Utilization	96.9	Linear
CPU Utilization	87.9	Polynomial
All above traces combined	100	Linear
All above traces combined	98.8	Polynomial

KNN Algorithm: We used the KNN algorithm on performance traces recorded while surfing different URLs of different websites. Traces that were found to have higher significance were: CPU processor core power consumption, DRAM power consumption, CPU processor utilization, and an array made of all previously mentioned traces normalized and then combined.

Our results of website fingerprinting using the KNN algorithm are reported in Table 2. The number of nearest neighbors with highest success was 5. KNN algorithm x had high accuracy of 99.8% when used on traces of internal processor power consumption, 99.3% when used on traces of DRAM power consumption, 88.3% when used on traces of CPU utilization, and 99.3% when used on all traces combined, each normalized.

Table 2. Fingerprinting success rate calculated using KNN algorithm

Measurement trace	Success rate [%]	# of neighbors
Internal CPU power consumption	99.8	5
DRAM power consumption	99.3	5
CPU Utilization	88.3	5
Traces above combined (each normalized)	99.3	5

6 Browser Version Identification

In the browser version identification experiment, we measured performance while surfing different URLs while changing only the browser version. The web browser chosen was chrome browser and the three different browser versions assessed were 94.0.4606.61 from 2021, 66.0.3359.181 from 2018. All URLs chosen were URLs leading to different popular videos on youtube.com. The chosen URLs were the same for the recordings of each browser. We recorded performance measurements for 85 URLs with and without adblocker resulting in 170 measurements for each of the two browser versions.

We separated all traces into a training group and a test group. We trained different algorithms on traces taken from the training group. We printed the success rate of our algorithms calculated on data from the test group.

We used several algorithms to try and fingerprint the websites. The first algorithm we used was unsupervised and was done by calculating histograms of different parameters, described under subsection Simple signal analysis. The second and third algorithms were supervised algorithms. The algorithms used were the SVM algorithm and the KNN algorithm.

Simple Signal Analysis: Several simple measurements were efficient in browser version detection, like total power consumed by CPU and mean utilization of CPU. Figure 3 shows the Cumulative power consumption histogram measured during entrance to chrome browser with different versions.

A boundary line corresponding with a minimum error that separates the measurements was calculated. The boundary line is shown in Fig. 3 as a dashed red line. We found that one can set this boundary line as a fingerprinting algorithm and have a success rate of 85.9% when fingerprinting with measurements of Cumulative power and a success rate of 79.4% when fingerprinting using measurements of average utilization.

These results prove a significant difference in performance measurements when entering different sites. In the later analysis, we will show that better fingerprinting results could be achieved using more complicated algorithms.

SVM Algorithm: We used the SVM algorithm on different performance traces measured when entering different versions of Chrome browsers. The URLs

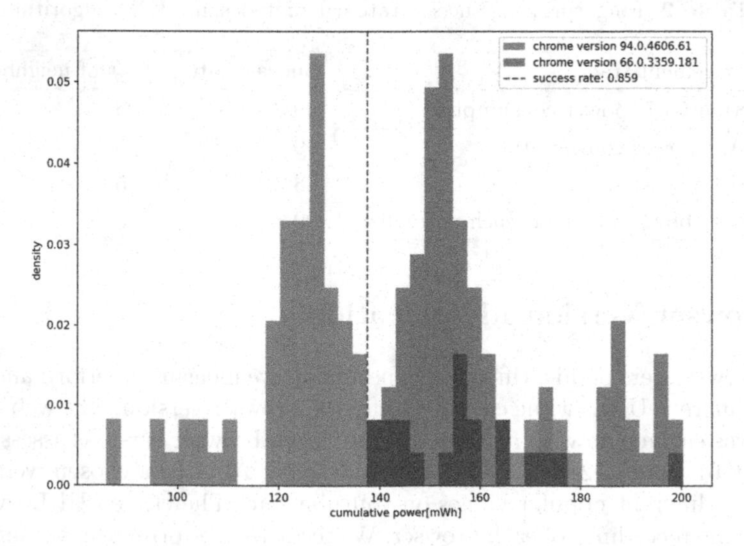

Fig. 3. Cumulative power consumption during entrance to different versions of chrome browser

entered were created in advance and correspond to popular videos on youtube.com. Performance traces that were found to have a higher significance in browser detection were: CPU processor core power consumption, DRAM power consumption, CPU processor utilization, and an array of all these traces normalized and then combined.

We set the SVM's Kernel to be linear or polynomial. Our results of browser version detection using the SVM algorithm are reported in Table 3. The SVM algorithm was highly successful when used on all traces and with both kernels. We can see that measurements like internal CPU core power consumption can lead to browser detection with 100% accuracy using the SVM algorithm and polynomial kernel. Other parameters with a high success rate were CPU Utilization with 98.8% accuracy and DRAM power consumption with 98.8% accuracy. Using all parameters together leads to good classification with 98.8% accuracy. The accuracy of the algorithms was calculated on more than six hundred measurements from the test group.

Table 3. SVM success rate of Chrome browser detection used on different performance traces. Linear and polynomials kernels were used and compared

Measurement trace	Success rate [%]	Kernel choice
Internal CPU power consumption	98.8	Linear
Internal CPU power consumption	100	Polynomial
DRAM power consumption	98.8	Linear
DRAM power consumption	98.8	Polynomial
CPU Utilization	98.8	Linear
CPU Utilization	96.5	Polynomial
All above traces combined	98.8	Linear
All above traces combined	98.8	Polynomial

KNN Algorithm: We used the KNN algorithm on performance traces recorded while surfing different browser versions. Traces that were found to have higher significance were: CPU processor core power consumption, DRAM power consumption, CPU processor utilization, and an array made of all previously mentioned traces normalized and then combined. Our results of browser version detection using the KNN algorithm are reported in Table 4.

The number of nearest neighbors was set to 3. KNN algorithm had high accuracy of 100% when used on traces of internal processor power consumption, 100% when used on traces of DRAM power consumption, 65.9% when used on traces of CPU utilization, and 100% when used on all traces combined, each normalized. The accuracy of the algorithms was calculated on more than 600 measurements from the test group.

Table 4. Success rate of Chrome browser detection using KNN algorithm

Measurement trace	Success rate [%]	# of neighbors
Internal CPU power consumption	100	3
DRAM power consumption	100	3
CPU Utilization	65.9	3
All above traces combined (each normalized)	100	3

7 Information Leakage of Neural Network Structure

We wanted to experiment if the structure information of Neural Network could be detected from performance traces captured by Intel Power Counter.

For this experiment, we trained different convolutional neural networks (CNN) on the dataset Hymenoptera containing pictures of bees and ants. The chosen CNNs were alexnet, resnet18, resnet34, resnet50, resnet101. For each net, we created two new nets with the same connectivity between nodes, but the activation functions were changed from RelU to sigmoid or tanH. The total number of CNNs created was 15 different nets. We recorded performance traces using the Intel Power Gadget performance counter for each net while it was classifying all samples from the validation dataset. We repeated the measurement ten times for each net.

In Fig. 4, we plotted a graph of cumulative power consumption during the classification process of each net. In Fig. 5, we plotted a graph of the total time the classification process of all samples took place. In Figs. 4 and 5, We can see

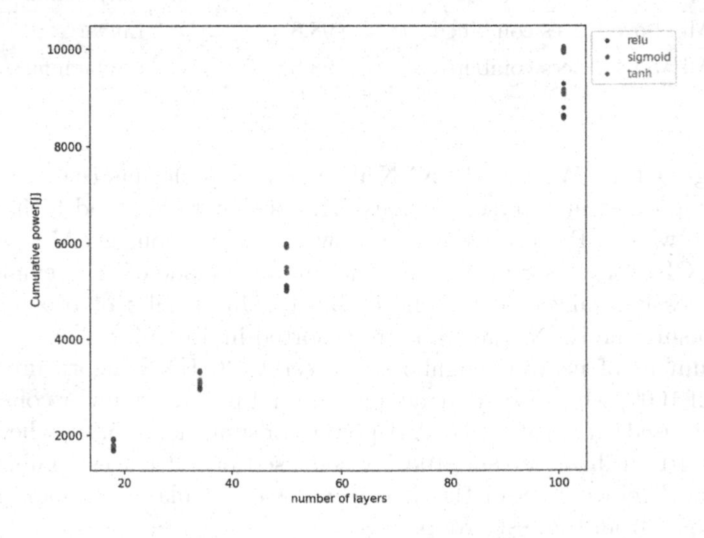

Fig. 4. Cumulative power consumption measured during classification for different CNNs

that there is a trend line that the power consumed, or the total time was greater if the number of layers in each net was bigger. We noticed that the power consumed and total time difference were correlated. We can also notice that while using activation functions different from RelU, the power consumed was bigger. This was expected as the other activation functions (sigmoid and tanh) require more complex computation. An attacker can use the performance measured during the classification process to discover information about the number of layers and the internal activation functions. We note that this discovery of information could be done on the fly. In Fig. 6, we plotted a graph of the average power consumption during the classification process of each net. We can notice that there exists a difference in average power consumption between CNNs with a different number of layers and different activation functions. The difference between measurements is not caused only by different timing side channels but is also caused by power consumption leakage detected using the Intel Power Gadget performance counter.

8 Discussion

In the previous sections, we have shown that different attacks can be performed using performance measurements; Website fingerprinting, Browser version discovery, and recovery of structure information of Neural Networks.

The first attack was a discovery of private information leakage about the user. We showed that website fingerprinting could be done by different algorithms performed on performance traces. The information leaked included the websites that the user surfed as well as the timing when the user surfed. We showed that high success rate fingerprinting could be done using Intel Power Gadget

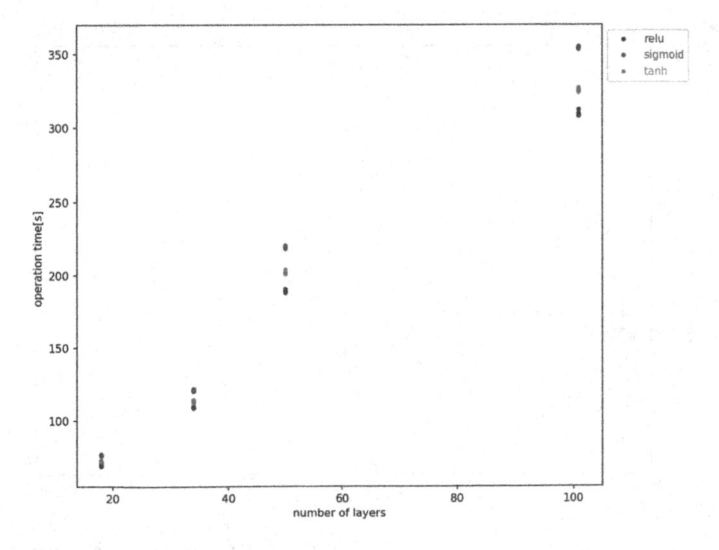

Fig. 5. Total time measured during classification for different CNNs

measurements and advanced algorithms like SVM and KNN. Our result success rate got up to 100%, which is a very high success rate when compared to other website fingerprinting works done in the past. This success rate was calculated on 600 measurements from the test group, not trained upon. We showed the attack is reproducible as half of the measurements were taken three weeks apart. When checking the success rate on traces taken six months prior we got a high success rate of 95.2%. Our paper is unique because we show that the combination of different performance measures can help achieve better success rates. We point out that results may have been higher if the sample rate was greater than 1000 samples per second.

The second attack was the discovery of a browser version using Intel Power Gadget performance measurements. An attacker that knows that the browser version is not up to date could use known cyber-attacks on the victim's computer. These cyber-attacks could lead to control over computers, access to data storage, ransomware attacks, and more. We showed that attackers could detect differences in performance measurement corresponding to different browser versions. Using advanced algorithms like SVM and KNN, we showed that high-accuracy browser version detection could be achieved. We achieved browser version detection accuracy of up to 100%.

In both side-channel attacks, the success rate calculated was high (100%). More measurements could be done while changing more parameters to ensure the results are this consistent.

In the third attack we demonstrated that information about Neural Network structure can be discovered using Intel Power Gadget performance measurements. We showed that structural elements like the number of layers and activation functions are affecting the performance measures of the computer during classification process.

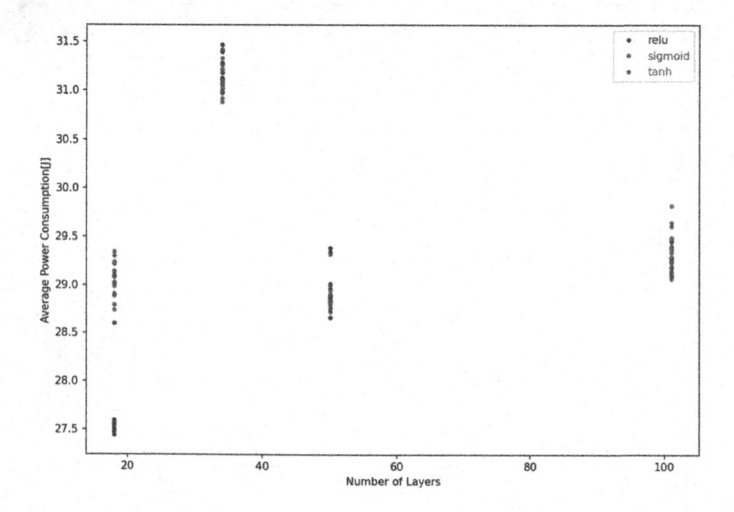

Fig. 6. Average power consumption measured during classification for different CNNs

9 Conclusion

When programs are used, or websites are loaded in the browser, the computer runs different functions, consumes energy, and utilizes internal parts of the processor (CPU, DRAM, and internal GPU) differently. These differences can be measured and recorded by using software-based performance counters. This work demonstrated that high-accuracy website fingerprinting is possible using software-based performance counters. We showed that information like the browser version could also be achieved from performance counter measurements. We showed that structure information of Neural Networks, like the number of layers and activation functions used, could be detected from performance traces captured. We showed that website fingerprinting and browser version detection techniques could be improved using Machine Learning techniques like SVM and KNN. Our results proved to have high accuracy compared to past works of website fingerprinting. We achieved an accuracy of up to 100% for website fingerprinting and an accuracy of up to 100% for browser version detection. We performed extensive experiments to validate our results. The attacks demonstrated in this work are made with 1ms clock rate measurement using Intel Power Gadget software. We believe that a large set of other side-channel attacks could be discovered using performance counters as a measurement tool in future studies.

Acknowledgments. This research was supported by Tel Aviv University and by the Technion Hiroshi Fujiwara Cyber Security Research Center and the Israel National Cyber Directorate.

Websites Included During Research

Newyorktimes.com
Foxnews.com
Youtube.com
Lichess.com

References

1. Clark, S.S., Mustafa, H., Ransford, B., Sorber, J., Fu, K., Xu, W.: Current events: identifying webpages by tapping the electrical outlet. In: Crampton, J., Jajodia, S., Mayes, K. (eds.) ESORICS 2013. LNCS, vol. 8134, pp. 700–717. Springer, Heidelberg (2013). https://doi.org/10.1007/978-3-642-40203-6_39
2. Lifshits, P., et al.: Power to peep-all: inference attacks by malicious batteries on mobile devices. Proc. Priv. Enhancing Technol. **2018**(4), 141–158 (2018)
3. Yang, Q., Gasti, P., Zhou, G., Farajidavar, A., Balagani, K.S.: On inferring browsing activity on smartphones via USB power analysis side-channel. IEEE Trans. Inf. Forensics Secur. **12**(5), 1056–1066 (2017)
4. Shusterman, A., et al.: Robust website fingerprinting through the cache occupancy channel. In: 28th USENIX Security Symposium (USENIX Security 2019), pp. 639–656 (2019)

5. Lipp, M., et al.: PLATYPUS: software-based power side-channel attacks on x86. In: 2021 IEEE Symposium on Security and Privacy (SP), pp. 355–371. IEEE (2021)
6. Yarom, Y., Falkner, K.: FLUSH+RELOAD: a high resolution, low noise, L3 cache side-channel attack. In: 23rd USENIX Security Symposium (USENIX Security 2014), pp. 719–732 (2014)
7. Osvik, D.A., Shamir, A., Tromer, E.: Cache attacks and countermeasures: the case of AES. In: Pointcheval, D. (ed.) CT-RSA 2006. LNCS, vol. 3860, pp. 1–20. Springer, Heidelberg (2006). https://doi.org/10.1007/11605805_1
8. Bruce, B.R., Petke, J., Harman, M.: Reducing energy consumption using genetic improvement. In: Proceedings of the 2015 Annual Conference on Genetic and Evolutionary Computation, pp. 1327–1334 (2015)
9. Kim, S.-W., De Vega, J., Vardhan Dugar, K.: Intel power gadget 2.7 monitoring processor energy usage (2012)
10. Gulmezoglu, B., Zankl, A., Eisenbarth, T., Sunar, B.: PerfWeb: how to violate web privacy with hardware performance events. In: Foley, S.N., Gollmann, D., Snekkenes, E. (eds.) ESORICS 2017. LNCS, vol. 10493, pp. 80–97. Springer, Cham (2017). https://doi.org/10.1007/978-3-319-66399-9_5
11. Kocher, P., Jaffe, J., Jun, B., Rohatgi, P.: Introduction to differential power analysis. J. Cryptogr. Eng. **1**(1), 5–27 (2011). https://doi.org/10.1007/s13389-011-0006-y
12. Cook, J., Drean, J., Behrens, J., Yan, M.: There's always a bigger fish: a clarifying analysis of a machine-learning-assisted side-channel attack. In: Proceedings of the 49th Annual International Symposium on Computer Architecture, pp. 204–217 (2022)
13. Hintz, A.: Fingerprinting websites using traffic analysis. In: Dingledine, R., Syverson, P. (eds.) PET 2002. LNCS, vol. 2482, pp. 171–178. Springer, Heidelberg (2003). https://doi.org/10.1007/3-540-36467-6_13
14. Lu, L., Chang, E.-C., Chan, M.C.: Website fingerprinting and identification using ordered feature sequences. In: Gritzalis, D., Preneel, B., Theoharidou, M. (eds.) ESORICS 2010. LNCS, vol. 6345, pp. 199–214. Springer, Heidelberg (2010). https://doi.org/10.1007/978-3-642-15497-3_13
15. Chen, S., Wang, R., Wang, X., Zhang, K.: Side-channel leaks in web applications: a reality today, a challenge tomorrow. In: 2010 IEEE Symposium on Security and Privacy, pp. 191–206. IEEE (2010)
16. Wang, Y., Paccagnella, R., He, E.T., Shacham, H., Fletcher, C.W., Kohlbrenner, D.: Hertzbleed: turning power side-channel attacks into remote timing attacks on x86. In: 31st USENIX Security Symposium (USENIX Security 2022), pp. 679–697 (2022)
17. Mangard, S., Oswald, E., Popp, T.: Power Analysis Attacks. Springer, Boston, MA (2007). https://doi.org/10.1007/978-0-387-38162-6
18. Brier, E., Clavier, C., Olivier, F.: Correlation power analysis with a leakage model. In: Joye, M., Quisquater, J.-J. (eds.) CHES 2004. LNCS, vol. 3156, pp. 16–29. Springer, Heidelberg (2004). https://doi.org/10.1007/978-3-540-28632-5_2
19. O'Flynn, C., Chen, Z.D.: ChipWhisperer: an open-source platform for hardware embedded security research. In: Prouff, E. (ed.) COSADE 2014. LNCS, vol. 8622, pp. 243–260. Springer, Cham (2014). https://doi.org/10.1007/978-3-319-10175-0_17
20. Naghibijouybari, H., Neupane, A., Qian, Z., Abu-Ghazaleh, N.: Rendered insecure: GPU side channel attacks are practical. In: Proceedings of the 2018 ACM SIGSAC Conference on Computer and Communications Security, pp. 2139–2153 (2018)

21. Zhang, Z., Liang, S., Yao, F., Gao, X.: Red alert for power leakage: exploiting intel RAPL-induced side channels. In: Proceedings of the 2021 ACM Asia Conference on Computer and Communications Security, pp. 162–175 (2021)
22. Hayes, J., Danezis, G.: k-fingerprinting: a robust scalable website fingerprinting technique. In: USENIX Security Symposium, pp. 1187–1203 (2016)
23. Cai, X., Zhang, X.C., Joshi, B., Johnson, R.: Touching from a distance: website fingerprinting attacks and defenses. In: Proceedings of the 2012 ACM Conference on Computer and Communications Security, pp. 605–616 (2012)
24. Jansen, R., Juarez, M., Galvez, R., Elahi, T., Diaz, C.: Inside job: applying traffic analysis to measure tor from within. In: NDSS (2018)
25. Juarez, M., Afroz, S., Acar, G., Diaz, C., Greenstadt, R.: A critical evaluation of website fingerprinting attacks. In: Proceedings of the 2014 ACM SIGSAC Conference on Computer and Communications Security, pp. 263–274 (2014)
26. Duddu, V., Samanta, D., Rao, D.V., Balas, V.E.: Stealing neural networks via timing side channels. arXiv preprint arXiv:1812.11720 (2018)
27. Banerjee, S., Wei, S., Ramrakhyani, P., Tiwari, M.: Bandwidth utilization side-channel on ML inference accelerators. arXiv preprint arXiv:2110.07157 (2021)
28. Batina, L., Bhasin, S., Jap, D., Picek, S.: CSI NN: reverse engineering of neural network architectures through electromagnetic side channel. In: USENIX Security Symposium, pp. 515–532 (2019)
29. Hua, W., Zhang, Z., Suh, G.E.: Reverse engineering convolutional neural networks through side-channel information leaks. In: Proceedings of the 55th Annual Design Automation Conference, pp. 1–6 (2018)
30. Hu, X., et al.: Neural network model extraction attacks in edge devices by hearing architectural hints. arXiv preprint arXiv:1903.03916 (2019)
31. Yan, M., Fletcher, C., Torrellas, J.: Cache telepathy: leveraging shared resource attacks to learn DNN architectures. In: USENIX Security Symposium. CoRR abs/1808.04761 (2018)
32. Hong, S., et al.: Security analysis of deep neural networks operating in the presence of cache side-channel attacks. arXiv preprint arXiv:1810.03487 (2018)
33. Schaub, A., et al.: Attacking suggest boxes in web applications over HTTPS using side-channel stochastic algorithms. In: Lopez, J., Ray, I., Crispo, B. (eds.) CRiSIS 2014. LNCS, vol. 8924, pp. 116–130. Springer, Cham (2015). https://doi.org/10.1007/978-3-319-17127-2_8
34. Brumley, D., Boneh, D.: Remote timing attacks are practical. Comput. Netw. **48**(5), 701–716 (2005)
35. Kocher, P.C.: Timing attacks on implementations of Diffie-Hellman, RSA, DSS, and other systems. In: Koblitz, N. (ed.) CRYPTO 1996. LNCS, vol. 1109, pp. 104–113. Springer, Heidelberg (1996). https://doi.org/10.1007/3-540-68697-5_9

HAMLET: A Transformer Based Approach for Money Laundering Detection

Maria Paola Tatulli(✉), Tommaso Paladini(✉), Mario D'Onghia(✉), Michele Carminati(✉), and Stefano Zanero(✉)

DEIB, Politecnico di Milano, Milan, Italy
mariapaola.tatulli@mail.polimi.it,
{tommaso.paladini,mario.donghia,michele.carminati,
stefano.zanero}@polimi.it

Abstract. Money laundering has damaging economic, security, and social consequences, fueling criminal activities like terrorism, human and drug trafficking. Recent technological advancements have increased the complexity of laundering operations, prompting financial institutions to use more advanced Anti Money Laundering (AML) techniques. In particular, machine learning-based transaction monitoring can complement traditional rule-based systems, lowering the number of false positives and the need to manually review fraud alerts. In this paper, we present HAMLET, a scalable Deep Learning model for analyzing financial transactions and detecting money laundering patterns. HAMLET employs a hierarchical transformer enforcing an attention mechanism at the transaction and sequence level. By combining the two different levels, HAMLET can identify complex money laundering operations carried out through subsequent transactions. We experimentally evaluate HAMLET on a synthetic dataset simulating clients trading on international capital markets, showing that HAMLET outperforms state-of-the-art solutions in detecting fraudulent transactions. In the experiments, HAMLET achieves 99% precision in a binary classification scenario and up to 95% in a multi-class scenario with 5 different money laundering schemes. Lastly, we investigate the proposed model's interpretability through a proxy method.

Keywords: Anti-Money Laundering · Transformer Model · Deep Learning · Explainable Machine Learning

1 Introduction

Money laundering affects all worldwide economies and is responsible for generating illegal financial flows of around $3.2 trillion per year, equivalent to 3% of the global Gross Domestic Product (GDP) [19]. Money laundering, defined as "transferring illegally obtained money through legitimate people or accounts so that its source cannot be traced" [2], encompasses any process aimed at introducing the income of unlawful activities (e.g., drug trafficking, illegal arms trafficking, tax evasion) into the financial system while concealing their illicit origins. Money

S. Dolev et al. (Eds.): CSCML 2023, LNCS 13914, pp. 234–250, 2023.
https://doi.org/10.1007/978-3-031-34671-2_17

laundering schemes usually follow three main phases [15]: placement, layering, and integration. Income from criminal activities enters the placement phase, in which it is converted into monetary instruments or deposited into a financial institution. In the layering phase, funds are transferred via wire transfers, checks, or other methods. Finally, funds are used to purchase legitimate assets in the integration phase. This final integration phase is challenging to detect because, at this point, funds have already bypassed fraud-detection mechanisms and are now used to purchase legitimate assets.

Transaction monitoring, the principal instrument against money laundering, is usually carried out by analysts or automated systems. A commonly employed automated solution consists in *rule checking*, an easily interpretable and transparent approach [7]. Nonetheless, rules cannot detect *unknown* anomalous behaviors nor variations of known money laundering schemes.

More recent approaches are based on Machine Learning models, which can extract insights from data and enhance rule-based approaches by identifying unusual patterns that may be even unknown to subject experts. ML-based AML systems are usually designed as fraud-detection systems that process *individual* transactions, flagging each as either legitimate or suspicious [24]. However, single money laundering operations usually consist of *several* fraudulent transactions; hence, an effective AML system should base its decisions on *sequences* of transactions rather than individual ones.

In this work, we present HAMLET, a *transformer*-based approach for detecting money laundering patterns in financial transactions. The transformer-based architecture is naturally fit to exploit sequential data, as demonstrated by comparable works [4,32,39]. Hence, we argue that the transformer-based approach can capture the sequential nature of users' transaction history. HAMLET possesses a hierarchical structure employing an attention mechanism both at the transaction and sequence levels. This attention mechanism enables HAMLET to identify money laundering schemes consisting of chains of transactions, thus overcoming the intrinsic limitations of traditional detection systems.

We validate HAMLET on the synthetic dataset available at [25], which simulates a real-world capital market scenario. We conduct experiments in a binary classification setting in which models discriminate between legitimate and anomalous transactions and in a multi-class scenario in which models discriminate between different money laundering patterns [14]. HAMLET outperforms state-of-the-art solutions, having a higher detection rate and an improved discriminative capability, with a precision up to 99% in the binary classification scenario and up to 95% in the multi-class classification case. The contributions of this work are summarized below:

- We study the application of transformer models to the money laundering detection problem and propose HAMLET, an Anti Money Laundering system based on the transformer architecture.
- We assess the performance of the model on a public dataset [25] that reflects a real-world capital market scenario, comparing the system performance with state-of-the-art Machine Learning (ML) and Deep Learning (DL) detection models, showing that HAMLET outperforms them.

2 Background and Related Works

In general terms, detecting fraudulent transactions can be seen as the task of identifying anomalies in a given data distribution. Current solutions employ machine-learning based techniques that overcome the limitation of rule-based detection approaches. There exists a wide body of research exploring the application of machine learning to the problem of detecting fraudulent transactions, including money laundering [1,8,13,20,21,23,26,34]. Detection techniques can be split up into three categories [10]: unsupervised [5,6,27,40], supervised [9,22,33,38], and active learning [11,24,37].

In general, the outputs produced by detection methods are either anomaly scores or binary labels. An analyst may choose to analyze the top few anomalies or use a cut-off threshold to select the most relevant ones. Overall, traditional ML algorithms usually fail to capture anomalies in sequential data (e.g., money laundering operations carried out through several transactions). To overcome such limitations, we propose the application of transformer-based techniques to money laundering detection, since they can capture the complex feature interactions in high-dimensional spaces while also eliminating the need for features directly crafted by domain experts.

Transformer Model. Sequential nature precludes parallelization between training examples, which becomes critical at longer sequence lengths. In this scenario, Aswani et al. [36] developed a non-recurrent alternative to RNN, namely the transformer block. The model, also referred as the Attention model, was proposed to solve two main problems: it identifies which parts of the input sequence are relevant to each word in the output, and it uses the relevant information to predict the correct output. Specifically, during the decoding phase, in addition to the last hidden state and generated token, the Decoder is also conditioned by a "context" vector that depends on the input hidden state sequence. The Transformer is the state-of-the-art model based on the attention-mechanisms in most Natural Language Processing tasks [36]. It is composed of a stack of encoders, which processes the input sequence and maps it into an n-dimensional vector, and a stack of decoders, which turns this vector into the output sequence. In this process, the attention mechanism identifies which parts of the input sequence are relevant at each step. The Encoder and the Decoder blocks are broken down into sublayers. In the Encoder, the inputs first flow through a self-attention layer, which helps the Encoder look at other words in the input sentence. The outputs of the self-attention layer are fed to a feed-forward neural network. The Decoder has an additional attention layer that helps the decoder focus on relevant parts of the input sentence.

To the best of our knowledge, Transformer models have hardly ever been used for fraud detection or general anomaly detection tasks. One of the few examples is given by Rodríguez et al. [32] in which the author uses a Transformer architecture to detect frauds in banking transactions. The presented architecture is based on the original implementation of the Transformer [36] and follows a generative approach to produce labels from the transformer's predictions. Another example

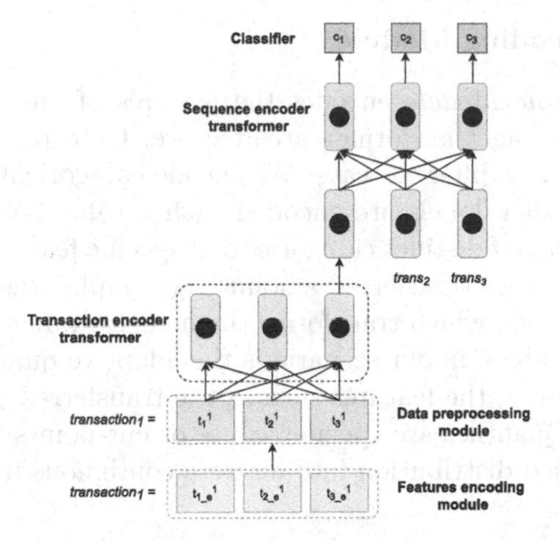

Fig. 1. HAMLET high-level diagram.

is HitAnomaly [18], which involves a hierarchical transformer implemented for anomaly detection on system logs.

3 HAMLET Approach

The goal of this work is to study the feasibility of applying the transform-based approach to the Anti Money Laundering (AML) detection problem. In our approach, we take inspiration from the processing steps commonly used in Natural Language Processing. We combine two transformers for analyzing transaction at different granularity. The first transformer encodes individual transactions that are then fed into a second transformer. The latter further encodes sequences of encoded transactions that are ultimately classified by an artificial neural network. As the input/output space of the second transformer has higher dimensionality, we employ BERT [12] which is available pre-trained and has already been employed in transfer-learning settings [18,32]. In Fig. 1, we report a high-level view of the proposed approach. HAMLET first encodes each transaction data by means of the *Feature Encoding Module*. The second module is the *Data Preprocessing Module* that receives the encoded data and maps it to a previously computed (at training time) vocabulary. Encoded and preprocessed data is then fed to the *Transaction Encoder Transformer* that produces *transaction-level* embeddings. These transaction-level embeddings are processed by the *Sequence Encoder Transformer* that creates *sequence-level* embeddings. The final part of the model is the *Classifier*, which discriminates anomalous transactions from legitimate ones.

3.1 Feature Encoding Module

The *Feature Encoding Module* encodes the features of individual transactions before being fed to the transformer architecture. Categorical and continuous features are encoded in different ways. We encode categorical features using the label encoder approach: labels are encoded with a value between 0 and $n - 1$, where n is the number of distinct categories of a specific feature. To deal with the continuous features (e.g., transferred amount), we employ the method known as *binning* or *quantization*, which transforms the numeric values into discrete categories. The safer strategy in our scenario is the adaptive quantile-based binning since the distribution of the features such as the transferred amount can be significantly skewed. Quantiles are specific values or cut-points that help partition the continuous-valued distribution into discrete contiguous intervals.

3.2 Data Preprocessing Module

The main goal of the *Data Preprocessing Module* is to create the vocabulary necessary to train the transformer model. Our approach comprises three main stages:

1. Create the special fields, called tokens, used to process the input transactions.
2. Build a different vocabulary for each feature, called "local vocabulary."
3. Create a "global vocabulary" that integrates all the local vocabularies built in the previous step. Each word in the global vocabulary has an identifier that can pinpoint the feature under consideration and the location in the particular feature local vocabulary.

3.3 Transaction Encoder Transformer

Preprocessed data is fed into the first Transformer, whose layers are depicted in Fig. 2. The embedding layer represents the inputs in a compact way, namely in dense vectors of fixed size. It receives elements of size (n, m) and outputs elements of size (n, e), where e is the embedding size. Similar vectors are mapped closer in this new embedding space. This feature helps the model identify the most relevant parts of the input concerning the transaction under consideration. The transformer encoder layer reflects the architecture presented in the original paper [36], which is made up of a stack of encoders. Finally, a projection layer applies a linear transformation to shape the transaction-level embeddings appropriately to be fed to the second-level transformer. It yields elements of size (n, p), where p is the size of the projection.

3.4 Sequence Encoder Transformer

The Sequence Encoder Transformer (i.e., second-level transformer) is used to capture the information originating from each individual transaction as well as

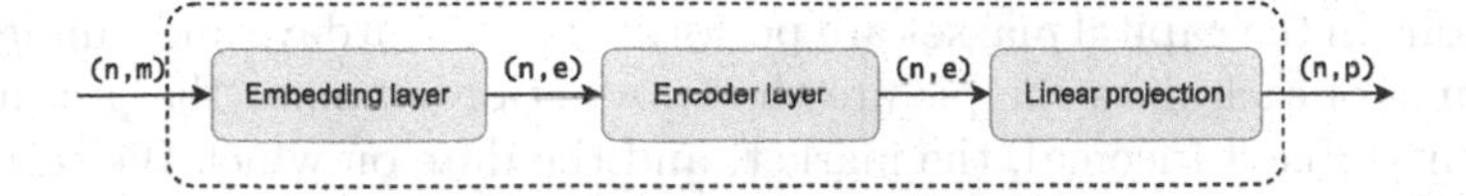

Fig. 2. First-level transformer

a transaction sequence. In this specific implementation, we rely on BERT [12] as it is available pre-trained.

The Sequence Encoder Transformer receives the embeddings of single transactions by the first-level transformer and employs ELMo embeddings as they capture both the word meaning (i.e., transaction meaning) and contextual information (i.e., transaction sequence information) [30]. Moreover, the attention mechanism at this level can highlight how much past and future transactions are relevant to the one under consideration. The outputs of this block are the sequence-level embeddings that can now be used for classification purposes.

3.5 Classification Head

Sequence-level embeddings generated by the second-level transformer are fed into a classification layer. The number of neurons in this linear layer reflects the number of classes we aim to identify. Our approach has been implemented both for the binary and the multi-class cases.

In the binary case, all the anomalous transactions are classified in the same way (class 1). Therefore, the linear layer has two neurons (classes 0,1), and the loss is the binary cross-entropy. In the multiclass case, the classification layer has a number of neurons equal to the number of classes and the loss is the cross-entropy.

4 Experimental Evaluation

The goals of our experiments are the following:

1. Study the performances of our model for different balance ratios of clean and anomalous transactions.
2. Compare HAMLET against other state-of-the-art models in a scenario that emulates real-world money laundering schemes. We analyze both binary and multiclass classification scenarios.
3. Explain the model's inner workings with a proxy model.

4.1 Dataset

We perform our experiments on a synthetic financial dataset that resembles a real capital market scenario provided by an industrial collaborator [25]. The dataset comprises 29,704,090 transactions performed by 400 financial institutions, covering 60 days divided into 12 weeks. Each week is composed of 5 days since no transactions in the capital market are performed on Saturdays and Sundays. The key features of each data sample are the transaction amount, the product class (e.g., Equity, Fixed Income), the market, and the date on which the transaction occurred. In particular, the original dataset presents the following nine features:

- **Originator:** the name of the institution that performed the transaction. It has 400 different values.
- **EntryDate:** the timestamp in which the transaction was performed. It has the following format: "year-month-day hour:minute:second". All transactions belong to the time period of January-March of 2019.
- **InputOutput:** it has only two possible values: "buy" or "sell". Indeed, it tells if the asset related to the transaction is bought or sold.
- **Market:** the market in which the transaction is executed.
- **Product Type:** the category of the product sold/bought with the transaction.
- **Product Class:** the procedure by which the product is traded. It has four possible values: traded, ADR conversion, cash in/out, and external fee.
- **Amount:** the amount of the transaction.
- **Currency:** the currency in which the transaction is performed.

The dataset presents 0.274% of anomalous transactions, precisely 81,269 entries. Anomalous transactions injected in the dataset belong to five classes, chosen based on suggestions of the Financial Action Task Force (FATF) [14], an inter-governmental body that promotes effective implementation of legal, regulatory, and operational measures for Anti Money Laundering. In particular, we consider the following money laundering patterns:

1. **Large asset withdrawal**. A sudden spike in the amount withdrawn from an account that deviates from previous activity.
2. **An unusually large amount of collateral transferred in and out of an account within a short period**. A very unusual behavior for a customer as he would not be able to invest just by trading collateral.
3. **Security bought or sold at an unusual time**. Customer trading a security outside the usual timeframe (e.g., Outside market opening hours).
4. **Small but highly frequent transactions**. Many transactions performed with the amount below any applicable reporting threshold.
5. **Transactions with rounded normalized amounts bought or sold within an account**. Transactions with rounded amounts are very unusual in capital markets.

4.2 Experimental Settings

We randomly split the dataset into train, validation, and test sets. The first set, comprising 80% of the transactions, is provided to the model during the training phase. The validation set contains 10% of the dataset and is employed to assess the model's performance during each training epoch. Finally, the test set is fed to the model only in the end to collect the final values of the metrics

Table 1. Experiment 1. Model evaluation with respect to different clean:anomalous ratios in training set.

Metric	Ratio 1:1	Ratio 5:1	Ratio 7:1
Accuracy	0.9644	**0.9945**	0.9584
Precision	**0.9899**	0.9751	0.9505
Recall	0.9387	**0.9926**	0.7950
F1-Score	0.9636	**0.9838**	0.8679
AUROC	0.9645	**0.9937**	0.8910
Mcc	0.9301	**0.9805**	0.8324

and evaluate the performance on new unseen data. This final set contains 10% of the whole dataset. One common challenge of training a Machine Learning or Deep Learning model is overfitting, which happens when a model learns the detail and noise in the training data to the extent that it negatively impacts the model's performance on new data. To tackle this problem, we implemented Early Stopping, a regularization technique that consists in observing the training and validation losses and stopping the training phase when validation loss starts to increase as opposed to the training loss that keeps decreasing. The epoch model with the lowest validation error is saved and employed to perform the test phase.

4.3 Binary Classification Experiments

The first seven experiments have been conducted in a binary classification scenario in which all the anomalies are classified as class 1, and clean transactions belong to class 0.

Experiment 1: Undersampling Ratios Comparison. One of the main challenges of the dataset under consideration is the substantial imbalance between the two classes, Clean and Anomaly. The proportion of clean versus anomalous transactions in the original dataset is 20:1. With such an imbalance, the resulting model is characterized by a high bias as it mainly learns the representation of the clean class. This ratio can be controlled and modified during the undersampling process. We compare the model presented in this work with different clean:anomalous ratios in the training set. The best one resulted to be 5:1, against 7:1 and 1:1. Table 1 reports the metrics on the validation set after a training of 3 epochs, with Early Stopping on the validation error. Accuracy is not a reliable metric in the case of imbalanced datasets; therefore, we provide different evaluation metrics. We also evaluate the metrics of Precision, Recall, F1-Score, Area Under the Receiver Operating Characteristic Curve (AUROC), and Matthews Correlation Coefficient (MCC).

Experiment 2: Deep Learning Models Comparison. In this experiment, we compare our Transformer model to other state-of-the-art Deep Learning models in the fraud detection and AML field: the Long Short-Term Memory

Table 2. Experiment 2 - Deep Learning models comparison

Metric	Transformer	LSTM	Autoencoder
Accuracy	**0.9953**	0.9065	0.8006
Precision	**0.9752**	0.9147	0.1073
Recall	**0.9978**	0.9065	0.2696
F-Score	**0.9864**	0.8909	0.1536
AUROC	**0.9963**	0.7188	0.5242
Mcc	**0.9837**	0.6238	0.0725

Table 3. Exp. 3: Machine Learning models comparison

Metric	Transformer	RandomForest	IsolationForest	Extended-IF
Accuracy	**0.9953**	0.9800	0.7336	0.7510
Precision	0.9752	**0.9801**	0.7302	0.7511
Recall	**0.9978**	0.9800	0.7336	0.7510
F-Score	**0.9864**	0.9800	0.7319	0.7511
AUROC	**0.9965**	0.9673	0.5150	0.5518
Mcc	**0.9837**	0.9286	0.0305	0.1037

(LSTM) [17] and the Autoencoder (AE) [29]. Table 2 shows the performance of the models mentioned above on the test set. Our model overcomes both the LSTM and the AE models with higher values for all the considered metrics. The AE model is not able to discriminate between clean and anomalous transactions. Its Matthews Correlation Coefficient (MCC) is close to 0, which means that the model has overall the same performance of random guessing. This proves that money laundering transactions are very well hidden between legit ones and that more complex models are needed to solve the AML problem. The LSTM, on the other hand, obtains good performances both in terms of precision and F-score. However, its AUROC value is much lower than the Transformer's one. Hence, this experiment proved the superiority of our approach with respect to the most used Deep Learning models.

Experiment 3: Machine Learning Models Comparison. The state-of-the-art machine learning techniques in the fraud detection and AML field are based on ensembles of decision trees. Therefore, we implemented and gathered results of three different ensemble methods trained on the same dataset already mentioned before: Random Forest [3], Isolation Forest [28], and Extended Isolation Forest [16]. As Table 3 shows, our approach achieves better performance for all the performance under analysis. In addition, the unsupervised models are not able to point out anomalous transactions. This is in line with the results gathered from Experiment 7.2.2 and suggests that anomalies are not so easy to isolate. This is also a consequence of some transactions appearing to be clean if analyzed alone, while they are anomalies if considered in a group. Unsupervised models

Table 4. Exp 4: Discriminative vs Generative approach

User	Anomalies	Metric	Discriminative	Generative
Client_66	55%	Accuracy	0.9176	0.4319
		Precision	0.8716	0.1412
		Recall	0.9984	0.5890
		F1-Score	0.9311	0.2279
		AUROC	0.9069	0.4974
		Mcc	0.8419	0.0362
Client_212	38%	Accuracy	0.8939	0.4911
		Precision	0.7884	0.1493
		Recall	0.9853	0.5339
		F1-Score	0.8759	0.2304
		AUROC	0.9116	0.5089
		Mcc	0.8002	0.0125
Client_77	2%	Accuracy	0.9800	0.3872
		Precision	0.5345	0.1379
		Recall	0.1379	0.7547
		F1-Score	0.2162	0.2332
		AUROC	0.5676	0.4917
		Mcc	0.2556	0.0311

that are based on the concept of "norm" may need aggregate features that highlight additional information to discriminate the money laundering activities. We can conclude that supervised methods are more effective in this scenario and that our model can obtain better performances than current ML approaches.

Experiment 4: Discriminative vs. Generative Approach. Fernandéz et al. [32] employs the transformer architecture to solve a fraud detection problem on a banking dataset. Their architecture models the users' spending patterns and is trained to predict their next transactions. An anomaly score is drawn by comparing the predicted and the actual transaction. This score is then used to classify the transaction as clean or fraudulent. Such an approach is generative because it is based on the prediction (i.e., generation) of the following transaction, given the user's past. Therefore, this is an opposite approach with respect to the one presented in this work. HAMLET is, indeed, discriminative as it is trained to discern clean and anomalous transactions. We performed an experiment comparing the two different approaches in the Money Laundering scenario. The results are detailed in Table 4 that includes the metrics divided by user. We choose to include here three clients with different percentages of anomalous transactions. Indeed, institutions 66 and 212 have a great number of money laundering entries in the dataset, while client 77 have only 2% of fraudulent transactions. In all three cases, the our discriminative Transformer-based model can reach higher performances with respect to the generative approach.

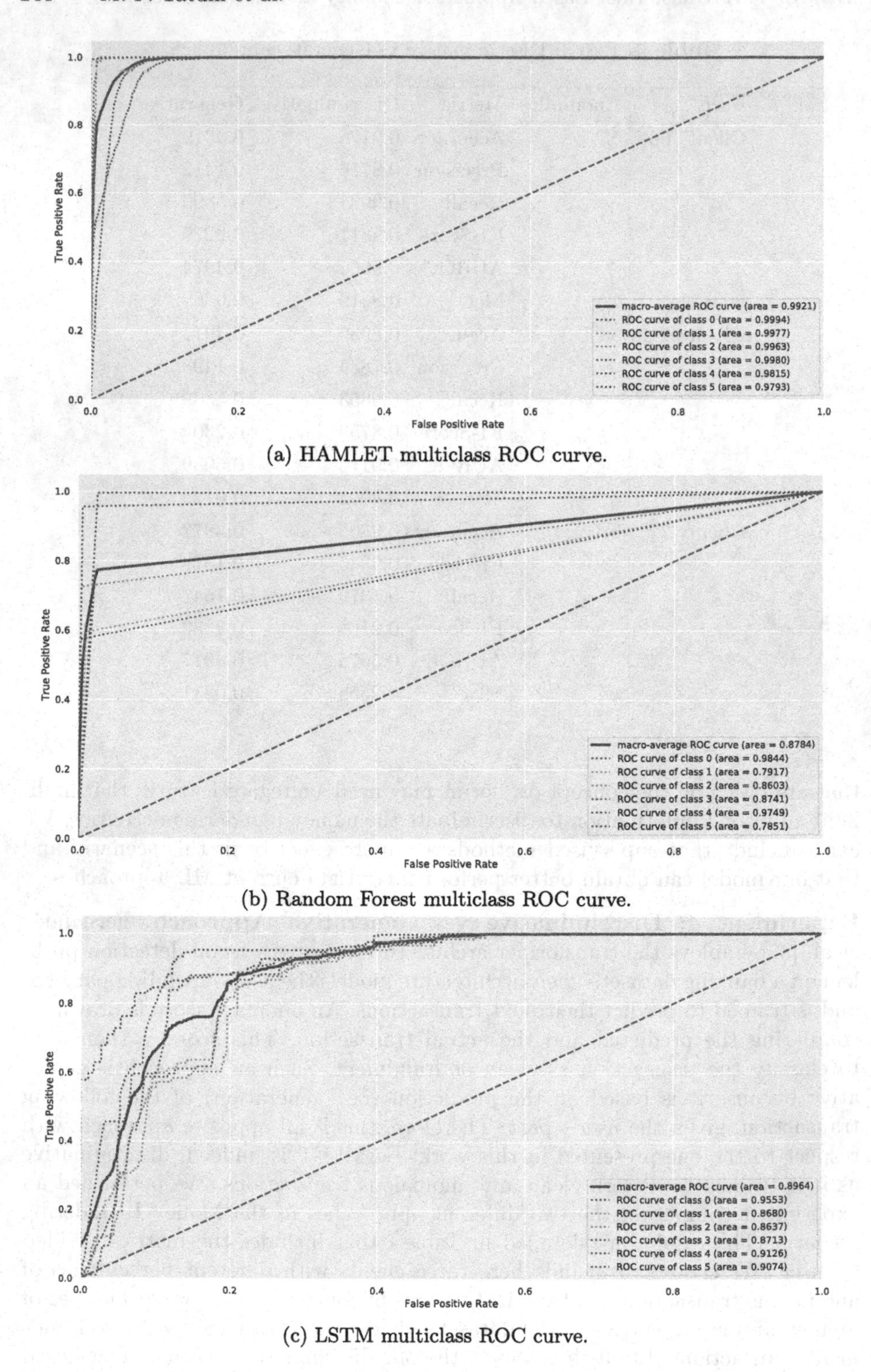

(a) HAMLET multiclass ROC curve.

(b) Random Forest multiclass ROC curve.

(c) LSTM multiclass ROC curve.

Fig. 3. Exp. 5: ROC curves of tested classifiers.

Table 5. Exp. 5: Metrics for each class

Class	Metric	HAMLET	LSTM	RandomForest
0	Precision	**1.00**	0.97	0.99
	Recall	**1.00**	0.99	0.99
	F1-Score	**1.00**	0.98	0.99
1	Precision	**0.81**	0.31	0.78
	Recall	0.54	0.23	**0.59**
	F1-Score	0.65	0.26	**0.67**
2	Precision	0.42	0.16	**0.66**
	Recall	**0.89**	0.35	0.73
	F1-Score	0.57	0.22	**0.69**
3	Precision	**0.77**	0.55	0.70
	Recall	**0.96**	0.68	0.77
	F1-Score	**0.86**	0.61	0.73
4	Precision	0.64	0.37	**0.80**
	Recall	0.88	0.60	**0.96**
	F1-Score	0.74	0.45	**0.87**
5	Precision	**0.87**	0.33	0.51
	Recall	0.46	0.25	**0.57**
	F1-Score	**0.60**	0.29	0.54

4.4 Multiclass Classification Experiments

We assess the performance of the model in a multiclass scenario to discriminate all the different money laundering patterns present in our experimental dataset. The classification problem is among six different classes: class 0 for clean transactions and classes 1-2-3-4-5 for anomalous entries.

Experiment 5: Models Comparison. We compare our hierarchical transformer model to two state-of-the-art approaches in the money laundering detection field: a LSTM [17] and a Random Forest [3]. Table 5 shows every class Precision, Recall, and F1-score. Class 0 refers to legitimate transactions, and it is the class with the highest number of samples. Indeed, all three models excellently identify such transactions. Instead, classes from 1 to 5 refer to the five different Money Laundering patterns. For all of them, the LSTM model cannot reach good performance. The Transformer and the RF, instead, reach very similar metrics values. Random Forest better discriminates classes 2 and 4, which are the Money Laundering patterns with the smallest number of samples. However, HAMLET is more effective in classes 3 and 5. Looking at the metrics divided per class, the models under analysis show similar performances. Finally, we compute a weighted average of the metrics of the single classes (see Table 6).

Table 6. Exp. 5: Multiclass models comparison

Metric	HAMLET	LSTM	RandomForest
Accuracy	0.9539	0.8785	**0.9540**
Precision	**0.9593**	0.8546	0.9510
Recall	**0.9539**	0.8785	0.9504
F1-Score	**0.9532**	0.8645	0.9496
Mcc	**0.9447**	0.5642	0.8337

The Matthews Correlation Coefficient suggests that HAMLET is overall more effective.

The ROC curves (Figs. 3) are plotted for every single class versus the others. The macro-average ROC curve is highlighted in blue. All three models obtain a very good ROC curve for class 0, while the values are a bit lower for the other classes. Class number 5 (Transactions with rounded normalized amounts bought or sold within an account) obtains the worst performance in all the three models, even if it is not the smallest one. This may be due to the fact that such anomalies are very similar to clean transactions and thus difficult to distinguish from them. Considering the average ROC we conclude that the transformer achieves the highest area under the curve (0.98) (Fig. 3a), followed by the LSTM (0.91) (Fig. 3c) and finally by the Random Forest (0.87) (Fig. 3b).

4.5 Interpretability with Proxy Model

In this experiment, we evaluate the interpretability of HAMLET by exploiting a proxy approach [35], which consists in training a supervised model (called *proxy model*) over the predictions of the original model. Such an approach provides an easy-to-interpret and transparent explanation of the model outputs, which is of paramount importance in the fraud investigation process. We implement the proxy model according to the pedagogical technique [35]. The Transformer is trained on the original dataset, and the proxy (Decision Tree) is subsequently trained on the Transformer's predictions. We evaluate the Feature Importance (FI) on the proxy predictions, which represents how much each feature has contributed to the final result. Originator and Product Class have been selected as the most important elements to discriminate the transactions, with the importance of respectively 0.55 and 0.42. This method is effective but accurate only to some extent since the above-mentioned considerations derive from the proxy model and not from the Transformer itself.

We also evaluate the same approach on the multiclass scenario. The resulting decision tree comprises six classes and is, therefore, bigger and more complex to be analyzed than the corresponding binary tree. In this scenario, unlike the binary case, the features Input-Output, Amount, and Product type have a greater impact on the final result (see Fig. 4).

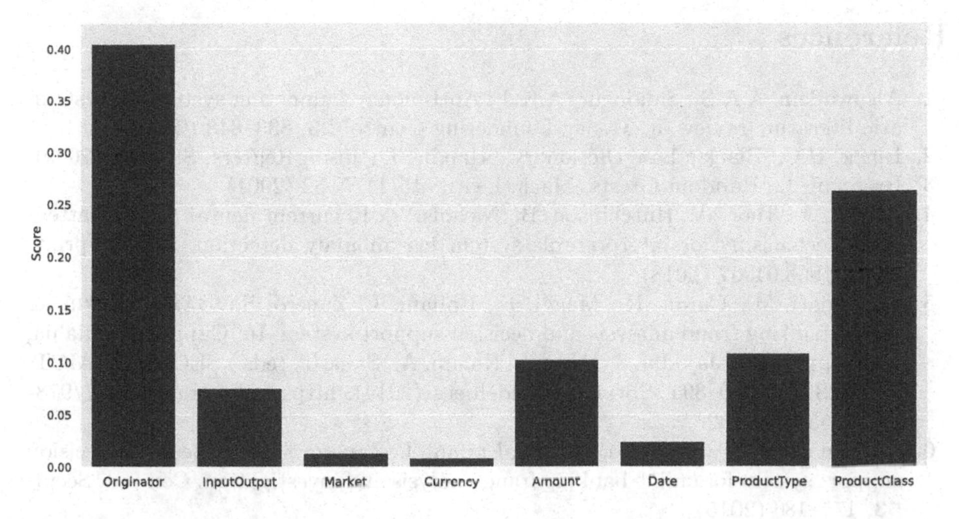

Fig. 4. Multiclass feature importance computed on the Decision Tree's predictions.

5 Conclusions

In this work, we presented Hierarchical Anti Money-Laundering Encoder Transformer (HAMLET), a novel approach to detect money laundering activities in financial transactions. Our system is based on a hierarchical transformer model able to process financial transactions and exploit the whole users' spending history. The first part of the transformer generates row-level embeddings fed directly into the second stack of encoders that outputs sequence-level embeddings. This hierarchical structure captures complex money laundering patterns that would not be detected by employing a transaction-level analysis. We evaluate HAMLET on a public but synthetic dataset that resembles a real-world capital market scenario, comprising five different money laundering patterns. First, we evaluate the effectiveness of HAMLET considering different ratios of clean and anomalous transactions. Second, we compare our approach against five state-of-the-art techniques. Subsequently, we compare the discriminative approach proposed in this paper against a generative one used in fraud detection. In all of the experiments, HAMLET demonstrated its superiority and effectiveness. It proved to deal with millions of transactions, discriminating all five analyzed patterns with high detection performance. Finally, we provide explainability to our approach with a proxy model based on decision trees. Overall, the hierarchical transformer architecture has been demonstrated to work well on the scenario under consideration. In future works, we plan to integrate our approach with unsupervised techniques able to pinpoint unseen fraudulent behaviors. In addition, to improve the interpretability of our approach, we will work on adapting LIME [31], a technique that aims to approximate a black-box machine learning model with a local, interpretable model to explain each prediction.

References

1. Alsuwailem, A.A.S., Saudagar, A.K.J.: Anti-money laundering systems: a systematic literature review. J. Money Laundering Control **23**, 833–848 (2020)
2. Black, H.C.: Black's Law Dictionary, 9th edn. Thomson Reuters, St. Paul (2009)
3. Breiman, L.: Random forests. Mach. Learn. **45**(1), 5–32 (2001)
4. Brown, A., Tuor, A., Hutchinson, B., Nichols, N.: Recurrent neural network attention mechanisms for interpretable system log anomaly detection. arXiv e-prints arXiv:1803.04967 (2018)
5. Carminati, M., Caron, R., Maggi, F., Epifani, I., Zanero, S.: BankSealer: an online banking fraud analysis and decision support system. In: Cuppens-Boulahia, N., Cuppens, F., Jajodia, S., Abou El Kalam, A., Sans, T. (eds.) SEC 2014. IAICT, vol. 428, pp. 380–394. Springer, Heidelberg (2014). https://doi.org/10.1007/978-3-642-55415-5_32
6. Carminati, M., Caron, R., Maggi, F., Epifani, I., Zanero, S.: BankSealer: a decision support system for online banking fraud analysis and investigation. Comput. Secur. **53**, 175–186 (2015)
7. Carminati, M., Polino, M., Continella, A., Lanzi, A., Maggi, F., Zanero, S.: Security evaluation of a banking fraud analysis system. ACM Trans. Priv. Secur. **21**(3), 11:1–11:31 (2018). https://doi.org/10.1145/3178370
8. Carminati, M., Santini, L., Polino, M., Zanero, S.: Evasion attacks against banking fraud detection systems. In: Egele, M., Bilge, L. (eds.) 23rd International Symposium on Research in Attacks, Intrusions and Defenses, RAID 2020, San Sebastian, Spain, 14–15 October 2020, pp. 285–300. USENIX Association (2020). https://www.usenix.org/conference/raid2020/presentation/carminati
9. Carminati, M., Valentini, L., Zanero, S.: A supervised auto-tuning approach for a banking fraud detection system. In: Dolev, S., Lodha, S. (eds.) CSCML 2017. LNCS, vol. 10332, pp. 215–233. Springer, Cham (2017). https://doi.org/10.1007/978-3-319-60080-2_17
10. Chandola, V., Banerjee, A., Kumar, V.: Anomaly detection: a survey. ACM Comput. Surv. **41**(3), 15:1–15:58 (2009). https://doi.org/10.1145/1541880.1541882
11. Das, S., Islam, M.R., Jayakodi, N.K., Doppa, J.R.: Active anomaly detection via ensembles: insights, algorithms, and interpretability (2019). https://arxiv.org/abs/1901.08930
12. Devlin, J., Chang, M., Lee, K., Toutanova, K.: BERT: pre-training of deep bidirectional transformers for language understanding. In: Burstein, J., Doran, C., Solorio, T. (eds.) Proceedings of the 2019 Conference of the North American Chapter of the Association for Computational Linguistics: Human Language Technologies, NAACL-HLT 2019, Minneapolis, MN, USA, 2–7 June 2019 (Volume 1: Long and Short Papers), pp. 4171–4186. Association for Computational Linguistics (2019). https://doi.org/10.18653/v1/n19-1423
13. Dumitrescu, B., Baltoiu, A., Budulan, S.: Anomaly detection in graphs of bank transactions for anti money laundering applications. IEEE Access **10**, 47699–47714 (2022)
14. FATF: FATF (2022). https://www.fatf-gafi.org/faq/moneylaundering/. Accessed 06 June 2022
15. Fincen: Money Laundering phase (2022). https://www.fincen.gov/history-anti-money-laundering-laws. Accessed 06 June 2022

16. Hariri, S., Kind, M.C., Brunner, R.J.: Extended isolation forest. IEEE Trans. Knowl. Data Eng. **33**(4), 1479–1489 (2021)
17. Hochreiter, S., Schmidhuber, J.: Long short-term memory. Neural Comput. **9**(8), 1735–1780 (1997)
18. Huang, S., et al.: Hitanomaly: hierarchical transformers for anomaly detection in system log. IEEE Trans. Netw. Serv. Manag. **17**(4), 2064–2076 (2020)
19. IMF: IMF (2022). https://www.imf.org/en/Home. Accessed 06 June 2022
20. Jensen, R., Iosifidis, A.: Qualifying and raising anti-money laundering alarms with deep learning. Expert Syst. Appl. **214**, 119037 (2023)
21. Jensen, R.I.T., Iosifidis, A.: Fighting money laundering with statistics and machine learning. IEEE Access **11**, 8889–8903 (2023)
22. Jullum, M., Løland, A., Huseby, R.B., Ånonsen, G., Lorentzen, J.: Detecting money laundering transactions with machine learning. J. Money Laundering Control **23**, 173–186 (2020)
23. Kute, D.V., Pradhan, B., Shukla, N., Alamri, A.M.: Deep learning and explainable artificial intelligence techniques applied for detecting money laundering-a critical review. IEEE Access **9**, 82300–82317 (2021)
24. Labanca, D., Primerano, L., Markland-Montgomery, M., Polino, M., Carminati, M., Zanero, S.: Amaretto: an active learning framework for money laundering detection. IEEE Access **10**, 41720–41739 (2022)
25. Labanca, D., Primerano, L., Markland-Montgomery, M., Polino, M., Carminati, M., Zanero, S.: Amaretto Dataset - A Synthetic Capital Market Dataset (2022). https://github.com/necst/amaretto_dataset. Accessed 06 June 2022
26. Labib, N.M., Rizka, M.A., Shokry, A.E.M.: Survey of machine learning approaches of anti-money laundering techniques to counter terrorism finance. In: Ghalwash, A.Z., El Khameesy, N., Magdi, D.A., Joshi, A. (eds.) Internet of Things—Applications and Future. LNNS, vol. 114, pp. 73–87. Springer, Singapore (2020). https://doi.org/10.1007/978-981-15-3075-3_5
27. Le-Khac, N., Kechadi, M.T.: Application of data mining for anti-money laundering detection: a case study. In: Fan, W., et al. (eds.) The 10th IEEE International Conference on Data Mining Workshops, ICDMW 2010, Sydney, Australia, 13 December 2010, pp. 577–584. IEEE Computer Society (2010). https://doi.org/10.1109/ICDMW.2010.66
28. Liu, F.T., Ting, K.M., Zhou, Z.: Isolation forest. In: Proceedings of the 8th IEEE International Conference on Data Mining (ICDM 2008), Pisa, Italy, 15–19 December 2008, pp. 413–422. IEEE Computer Society (2008). https://doi.org/10.1109/ICDM.2008.17
29. Luo, T., Nagarajan, S.G.: Distributed anomaly detection using autoencoder neural networks in WSN for IoT. In: 2018 IEEE International Conference on Communications, ICC 2018, Kansas City, MO, USA, 20–24 May 2018, pp. 1–6. IEEE (2018). https://doi.org/10.1109/ICC.2018.8422402
30. Peters, M.E., et al.: Deep contextualized word representations. CoRR abs/1802.05365 (2018). https://arxiv.org/abs/1802.05365
31. Ribeiro, M.T., Singh, S., Guestrin, C.: "Why should I trust you?": explaining the predictions of any classifier. In: Krishnapuram, B., Shah, M., Smola, A.J., Aggarwal, C.C., Shen, D., Rastogi, R. (eds.) Proceedings of the 22nd ACM SIGKDD International Conference on Knowledge Discovery and Data Mining, San Francisco, CA, USA, 13–17 August 2016, pp. 1135–1144. ACM (2016). https://doi.org/10.1145/2939672.2939778

32. Rodríguez, J.F., Papale, M., Carminati, M., Zanero, S.: A natural language processing approach for financial fraud detection. In: Demetrescu, C., Mei, A. (eds.) Proceedings of the Italian Conference on Cybersecurity (ITASEC 2022), Rome, Italy, 20–23 June 2022. CEUR Workshop Proceedings, vol. 3260, pp. 135–149. CEUR-WS.org (2022). https://ceur-ws.org/Vol-3260/paper10.pdf
33. Savage, D., Wang, Q., Zhang, X., Chou, P., Yu, X.: Detection of money laundering groups: supervised learning on small networks. In: The Workshops of the The Thirty-First AAAI Conference on Artificial Intelligence, San Francisco, California, USA, 4–9 February 2017, AAAI Technical Report, vol. WS-17. AAAI Press (2017). https://aaai.org/ocs/index.php/WS/AAAIW17/paper/view/15101
34. Tiwari, M., Gepp, A., Kumar, K.: A review of money laundering literature: the state of research in key areas. Pac. Account. Rev. (2020)
35. Van Vlasselaer, V., Verbeke, W.: Fraud Analytics Using Descriptive, Predictive, and Social Network Techniques: A Guide to Data Science for Fraud Detection. SAS Institute Inc., Wiley (2015). https://books.google.it/books?id=daNmjwEACAAJ
36. Vaswani, A., et al.: Attention is all you need, pp. 5998–6008 (2017). https://proceedings.neurips.cc/paper/2017/hash/3f5ee243547dee91fbd053c1c4a845aa-Abstract.html
37. Veeramachaneni, K., Arnaldo, I., Korrapati, V., Bassias, C., Li, K.: AI2: training a big data machine to defend. In: 2nd IEEE International Conference on Big Data Security on Cloud, BigDataSecurity 2016, IEEE International Conference on High Performance and Smart Computing, HPSC 2016, and IEEE International Conference on Intelligent Data and Security, IDS 2016, New York, NY, USA, 9–10 April 2016, pp. 49–54. IEEE (2016). https://doi.org/10.1109/BigDataSecurity-HPSC-IDS.2016.79
38. Villalobos, M., Silva, E.: A statistical and machine learning model to detect money laundering: an application (2017)
39. Wang, M., Xu, L., Guo, L.: Anomaly detection of system logs based on natural language processing and deep learning, pp. 140–144 (2018). https://doi.org/10.1109/ICFSP.2018.8552075
40. Williams, G.J., Baxter, R.A., He, H., Hawkins, S., Gu, L.: A comparative study of RNN for outlier detection in data mining. In: Proceedings of the 2002 IEEE International Conference on Data Mining (ICDM 2002), Maebashi City, Japan, 9–12 December 2002, pp. 709–712. IEEE Computer Society (2002). https://doi.org/10.1109/ICDM.2002.1184035

Hollow-Pass: A Dual-View Pattern Password Against Shoulder-Surfing Attacks

Jiayi Tan[1(✉)] and Dipti Kapoor Sarmah[2(✉)]

[1] EEMCS, Universiteit Twente, Drienerlolaan 5, 7522 Enschede, NB, The Netherlands
j.tan-1@student.utwente.nl

[2] SCS/EEMCS, University of Twente, Drienerlolaan 5, 7522 Enschede, NB, The Netherlands
d.k.sarmah@utwente.nl

Abstract. This paper presents Hollow-Pass, a developed solution that strengthens the security of pattern passwords against shoulder-surfing attacks. It is a novel approach to graphical password (GP) schemes that utilize a dual-view technology known as the global precedence effect, which eliminates the need for external devices and makes the grid and pattern invisible to potential shoulder surfers. The usability of Hollow-Pass was evaluated through an online as well as an offline user test. We recruited 30 participants from varied backgrounds, ranging in age from 20 to 80 years, for the online user test. An offline small-scale sampling test was conducted among 19 undergraduates from the Universiteit of Twente. The developed solution successfully demonstrated its ability to effectively resist shoulder-surfing attacks for simple patterns at various viewing angles (front, left-front, and right-front) and different distances (1.0 m, 1.5 m, and 2.0 m).

Keywords: Pattern password · Graphical password · Shoulder-surfing · Dual-View · Global precedence

1 Introduction

Graphical passwords (GP) are gaining widespread popularity as a method of authentication. GP can be classified into four categories: recognition-based, recall-based, cued-recall-based, and hybrid schemes [7,31]. One widely used GP system is the Android Pattern Unlock [40](illustrated in Fig. 1). This is a recall-based GP that is based on the Asian game Go and is known as the Pass-Go scheme [44]. During the registration and authentication process, users are asked to create a pattern by drawing a series of lines on a grid, which serves as their password [37]. According to a 2014 study [9], 40% of Android users use patterns as their password instead of a PIN.

S. Dolev et al. (Eds.): CSCML 2023, LNCS 13914, pp. 251–272, 2023.
https://doi.org/10.1007/978-3-031-34671-2_18

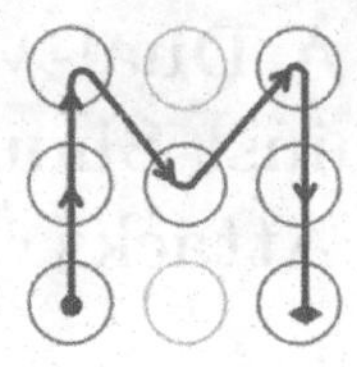

Fig. 1. Android pattern unlock

Several studies have identified shoulder surfing as a major security risk for GP systems [34,41,42]. This type of attack involves an attacker observing a user's device screen, keyboard, or mouse in a public place to steal their login credentials [19]. A study [6] reported that 6-length Android patterns have a high attack success rate of 64.2% with a single observation and 79.9% with multiple observations. To address this concern, various countermeasures have been proposed, such as increasing the complexity of patterns or obscuring part of the pattern using external hardware. However, a 2018 systematic literature review [4] found that of the 84 countermeasures proposed for pattern locking on smartphones. Only 10 pattern-based locking schemes were known, and the majority of them emphasized increasing pattern complexity. Only one of ten techniques, called "XSide" [13,14] focused on obscuring the pattern with external hardware. Since then, new approaches such as eyes-free [43], SysPal [12], Pass–O [38], TinPal [39], gaze tracking [15,22], and swipe behavior-based mechanisms [20] have been proposed, however, they are typically designed for small-screen devices or require specialized hardware. This research aims to tackle the issue of shoulder surfing on both mobile devices and monitors without the need for specialized hardware by focusing on the processing of the pattern password itself.

According to previous research [29], shoulder-surfing attacks can be successfully resisted by using pattern passwords that incorporate distinguish features, such as colored images in a grid format. Researchers [16,24] have investigated and proved how the human visual system processes objects over time, with recognition of general or global objects preceding that of detailed or local features. This phenomenon, known as "global precedence", can be utilized in the design of pattern passwords. By presenting the pattern password to users at a local level but obscuring it at a global level, it can be safeguarded against observation by shoulder surfers.

This study intended to establish a new pattern password mechanism, called Hollow-Pass, by combining the View manipulation and Image degradation techniques as described in Aris and Yaakob's review of anti-shoulder-surfing techniques [4]. The efficacy of Hollow-Pass in terms of both usability and security was also assessed through a limited-scale user experiment.

The structure of this research is as follows: Sect. 2 provides an overview of previous studies that are relevant to the research. Section 3 details the methodology that was employed in the research. Section 4 analyzes the results and limitations of the research. The research concludes with a section on Acknowledgements, References.

2 Related Work

This section focuses on various aspects of visual perception, usability, and security in the context of Hollow-Pass. We reviewed the relevant criteria of the dual-view technology and shoulder surfing resistance by studying several research papers from reputed journals and conferences. The systematic search was conducted using the formulated search string below:

```
("shoulder surf" OR "shoulder surfing"
OR "shoulder surfer" OR  "shoulder surfers"
OR shoulder-surf*)
AND ("dual-view" OR "hybrid")
```

After evaluating 50 search results and examining 11 relevant papers, we have identified five important criteria, namely spatial frequency, visual acuity, global precedence, pattern grid layout, and CIE color system. These criteria are essential for developing a pattern-based password scheme that is both secure and user-friendly.

(i). Spatial Frequency
Spatial frequency(SF), as defined in [10], is a measurement of the number of repeating patterns within a specific distance. It is commonly expressed in cycles per unit distance, such as cycles per degree (cpd). In image processing and computer vision, spatial frequency is utilized to determine the level of detail and texture in an image. An image with high spatial frequency exhibits intricate details and textures, while an image with low spatial frequency features larger and more prominent structures [35].

(ii). Visual Acuity
The visual acuity (VA) [18] is a measure of the capacity of the human visual system to perceive fine details in objects. It is commonly assessed using the Snellen chart, which is viewed from a distance of 6 m (20 feet). A standard VA, represented as 6/6 or 20/20, corresponds to a line of letters on the chart that subtend an angle of 5 min of arc. The Snellen "E" letter, made up of three strokes and two gaps with each stroke and gap subtending 1 min of arc, is often used as a reference. Normal visual acuity is equivalent to 30 cycles per degree (cpd), with each stroke of the "E" letter representing a peak of a sine wave and the white gap between strokes representing a trough (as illustrated in Fig. 2b).

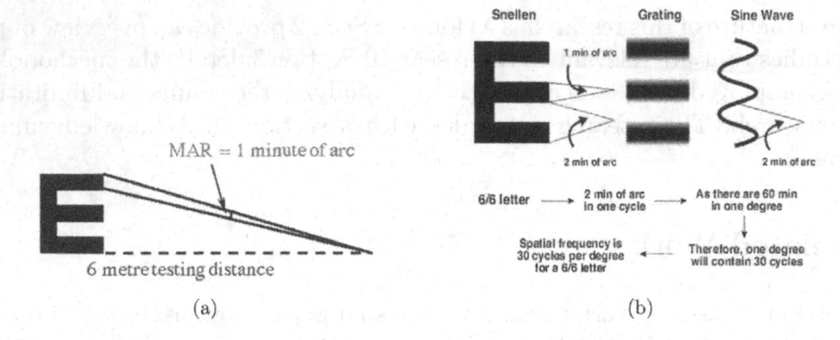

Fig. 2. Visual acuity and Snellen conversion: (a).MAR is 1 min of arc for strokes and 5 min of arc for a whole letter in 6/6 (20/20) visual acuity; (b).Conversion of normal visual acuity(in minute of arc) to cycle per degree(cpd) [18].

(iii). Global Precedence
Studies conducted by Flevaris [16] and Navon [24] have demonstrated that lower spatial frequencies tend to promote global perception [23], while higher spatial frequencies tend to encourage local perception. When individuals process both low and high spatial frequencies from an image, their visual processing follows a "coarse-to-fine" strategy, where they identify larger or general objects more rapidly and accurately than smaller or specific features. For instance, one might first identify a tree before noticing its branches and leaves. This effect referred to as global precedence, implies that humans have the ability to perceive the global aspect of a scene, but cannot immediately jump to local perception in a single step.

In this study, we refrained from exploring further psychological theories and instead applied the principle in processing the pattern password. This was achieved through the use of a dual-view mechanism, which displayed the "local" grid layout. Through this mechanism, a potential shoulder surfer may be able to capture the user-drawn lines on the screen, but would not be able to see the "local" grid layout, which is only visible to the user.

(iv). Pattern Grid Layout
The conventional Android pattern grid size is 3×3, however, previous research has indicated that this size is easily guessable and susceptible to attacks [32]. In an effort to enhance security, larger grid sizes have been suggested. However, a study found that there was little improvement in security from changing the grid size to 4×4 [5]. An alternative approach is to employ 9 points to make the 3×3 grid more intricate rather than increasing the size.

While various patterns can be created using a 9-point layout, a random layout may prove difficult for users to remember, leading them to choose simple patterns that are vulnerable to attacks. To address this issue, researchers have developed new grid layouts, such as the trapezium [44], circle [38,44], and house [44], which have been shown to improve security. The password space size of the circle and house layouts is larger than the original Android grid, while their overall recall success rates do not have significant differences [44]. This suggests that different grid layouts can offer improved security while maintaining user-friendly usability.

(v). CIE Color System and ΔE
Color plays a significant role in human visual perception and is an important aspect of digital images. As such, one of the research objectives is to enhance the global precedence effect by modifying the color difference in images.

The International Commission on Illumination's CIE color system provides a numerical means of describing all colors that are visible to the human eye. Unlike the RGB color model, the color definitions in the CIE color system are absolute, unambiguous, and not influenced by device or display specifications. The CIE LAB (Lab*) model [30], published in 1976, is widely accepted as a means of quantitatively measuring perceived color. It consists of three components:

- L* represents the lightness, ranging from 0 to 100, where 0 is black and 100 is white.
- a* represents the green to red axis, ranging from –128 to +127.
- b* represents the yellow-to-blue axis, ranging from –128 to +127.

To understand how the human eye perceives color difference, the CIE color system employs the metric ΔE [28,30]. For CIE LAB, a $\Delta E \approx 10$ indicates a color difference that is visible at first glance.

3 Methodology

In this section, the research questions and the methods used to address the research questions are outlined. A web development framework [2] and a pattern unlock grid template [45] were utilized to create a website specifically for the user test. The website was made accessible to all participants through deployment on a web hosting platform [1].

Considering the situation mentioned in previous sections, we formulate the research question as follows. The main research question (**MRQ**) of this study is: **To what extent do pattern complexity and dual-view technology impact the usability and security of Hollow-Pass?** This is further broken down into three sub-research questions (**SRQs**):

SRQ 1: To what extent does a pattern drawn on a distorted grid in preventing shoulder surfing at distances of 1.0 m, 1.5 m, and 2.0 m?
SRQ 2: To what extent do color contrast and global precedence in preventing shoulder surfing from identifying the pattern at 1.0 m, 1.5 m, and 2.0 m?
SRQ 3: To what extent do the color contrast between the password pattern and background image, and the distorted grid layout impact the usability and security of Hollow-Pass?

3.1 Proposed Methods

In this section, the detailed methods we proposed to answer the SRQs are discussed. We defined **(i)**.pattern drawing rules for users to create the patterns, **(ii)**.notations to label the nodes and patterns. In **(iii)** and **(iv)**, keyspace, and pattern complexity are outlined, respectively. From **(v)** to **(vii)**, the detailed approach for addressing SRQ1, SRQ2, and SRQ3 is outlined.

(i). Pattern Drawing Rule
The pattern drawing rules were developed based on the design presented by Tupsamudre et al. [38].

(a) The pattern must be created by drawing straight lines without lifting the hand,
(b) The pattern must connect a minimum of 4 and a maximum of 9 nodes,
(c) A node cannot be linked more than once,
(d) Unlike conventional 3×3 patterns, a node that is not connected may be bypassed if it is situated on the pattern's path. For example, a line segment can be drawn from node 1 to node 3 without visiting node 2.

(ii). Notations
For ease of understanding, the following definitions have been made for node labeling and pattern shape.

(a) Node labeling: For convenience, all nodes in the original 3×3 grid are numbered from 1 to 9 in a row-major format, with the upper-left node being labeled 1 and the bottom-right node labeled 9.
(b) Pattern representation: A pattern can be expressed as a sequence of nodes in a specific order, such as 123698745 (refer to Fig. 3).

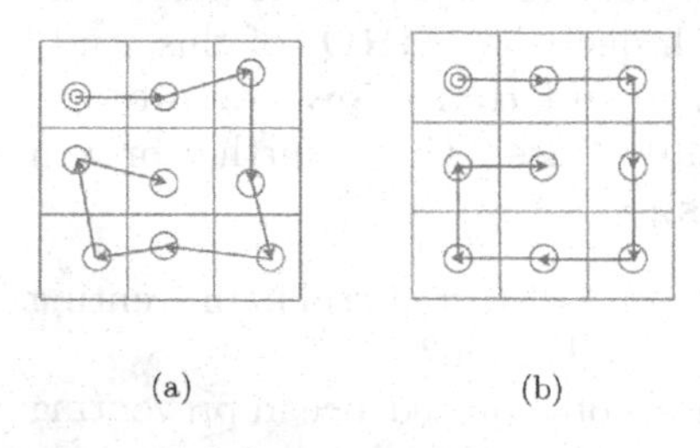
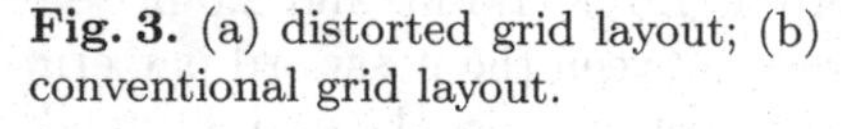

Fig. 3. (a) distorted grid layout; (b) conventional grid layout.

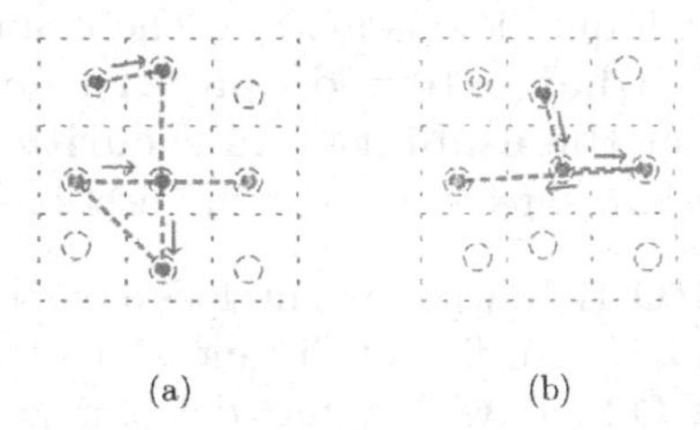

Fig. 4. (a) Intersection example 125846: one intersection point at node 5; (b) Overlap example 2564: one overlapped part 56.

(iii). Key Space

Keyspace, pertains to the number of possible combinations, P, a password can have. The number of valid r-node patterns in a 3×3 grid can be determined by using a mathematical formula:

$$P(n,r) = \frac{n!}{(n-r)!} = \frac{9!}{(9-r)!} \tag{1}$$

where n is the node count of the grid. For a 3×3 grid, n is 9. To sum up valid patterns from 4 to 9 nodes, we used the formula:

$$\sum_{r=4,9} P(n,r) = \sum_{r=4,9} \frac{n!}{(n-r)!} = \sum_{r=4,9} \frac{9!}{(9-r)!} \tag{2}$$

(iv). Pattern Complexity

The visual complexity of a pattern is determined by various factors such as the number of connected nodes, the length of the pattern, and the number of intersections and overlaps in the pattern (as illustrated in Fig. 4). To quantify this complexity, we employed the formula presented by Sun et al. [33] which is as follows:

$$PS_P = S_P \times log_2(L_P + I_P + O_P) \tag{3}$$

where PS_P is the strength score of the pattern, S_P is the number of nodes in the pattern, L_P represents the length of the pattern, I_P is the number of intersections in the pattern, and O_P is the number of overlaps in the pattern. By dividing the range of scores into five equal segments, the patterns can be classified into five levels: very weak, weak, medium, strong, and very strong.

(v). Increase Grid Layout Randomness

To address SRQ1, our approach was to increase the randomness of the grid layout. This would prevent shoulder surfers from being able to authenticate using the observed pattern, as the distribution of nodes and grid layout would vary each time. We implemented the following two steps to achieve this goal:

(a) **The 3×3 grid was divided into nine equal sections and each node was randomly placed within its section.**
To balance usability and security, we implemented a solution that involves scattering the nodes within their designated grid squares while maintaining the visibility of the grid borders. This helped users recognize the node locations. However, as the nodes were randomly placed and did not align with the grid borders, users were able to access non-adjacent nodes on the same line without having to connect through intermediate nodes: in the distorted grid layout, node 1 can reach node 7, 3 and 9 directly while this is not allowed in the conventional grid layout (as shown in Fig. 5). This approach increased the range of reachable nodes and the complexity of the patterns, without making it overly difficult for users to recall the pattern.

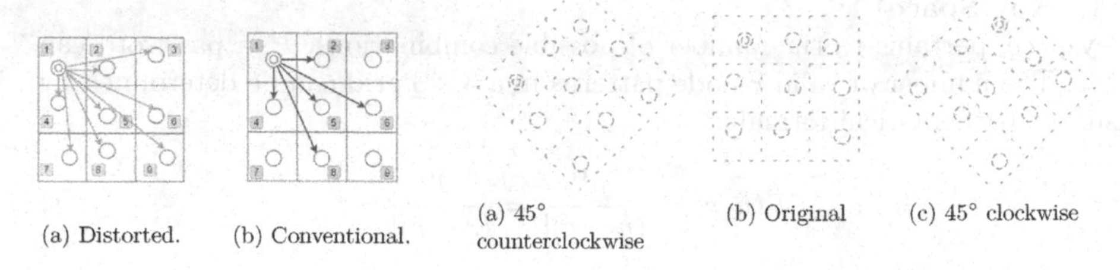

(a) Distorted. (b) Conventional.

Fig. 5. Reachable scale in distorted grid layout and conventional grid layout

(a) 45° counterclockwise (b) Original (c) 45° clockwise

Fig. 6. Three rotations of grid layout

(b) **The grid was then rotated by 45° in either a clockwise or counterclockwise direction.**
In order to address the potential issue of users having difficulty recognizing patterns that have been rotated more than 45°, three possible grid layout rotations (45° counterclockwise, original, and 45° clockwise) were defined, and one of these variations were randomly displayed each time (as shown in Fig. 6). To help users identify the rotation direction, node 1 was circled as an indicator. When users draw their patterns, they follow the same node sequence as the original grid, while shoulder surfers face difficulty in identifying the correct pattern without knowledge of the current grid orientation due to the random rotation.

(vi). Utilize Dual-View Technology
The approach for addressing SRQ2 was to create and strengthen the global precedence effect in Hollow-Pass. We accomplished this by implementing the following two steps:

(a) **Converted the foreground grid into a dashed-line format and adjusted the SF of the gratings in the background.**
The research in [16] highlights that the human visual perception system is more sensitive to low spatial frequencies for global processing and more sensitive to high spatial frequencies for local processing. As per Kalloniatis and Luu [18], the representation of sinusoidal gratings in terms of SF can be done and vice versa. The background image of the study was generated using the Python open-source package PsychoPy, consisting of four layers of sinusoidal gratings, each corresponding to an orientation of 0°, 4°, 90° and 135° to cover all borders of the foreground grid (see Fig. 7). The spatial frequency in cpd of every layer was in the range of (1, 3). To establish the optimal spatial frequency for the background sinusoidal gratings, we conducted experiments with spatial frequencies ranging from 0 to 30 (which aligns with normal visual acuity as shown in Fig. 2b). The comparison results are presented in Fig. 8. Based on the results from our perceptual performance evaluations, we selected a lower bound of 1 and an upper bound of 3 (excluding 3) as

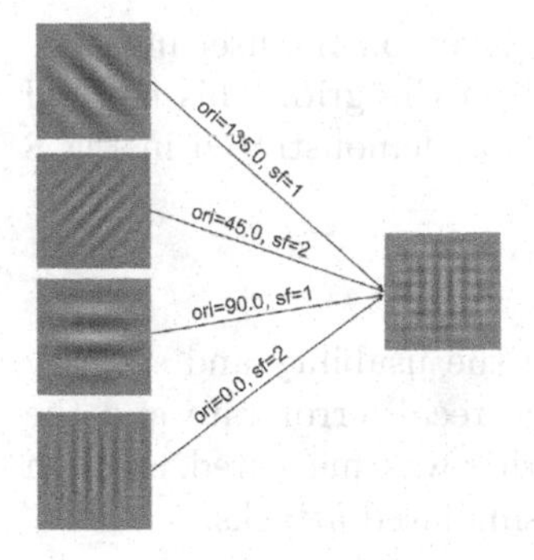

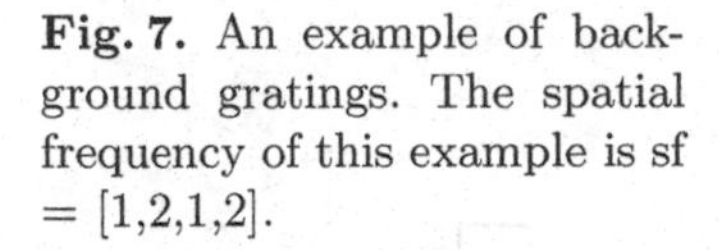

Fig. 7. An example of background gratings. The spatial frequency of this example is sf = [1,2,1,2].

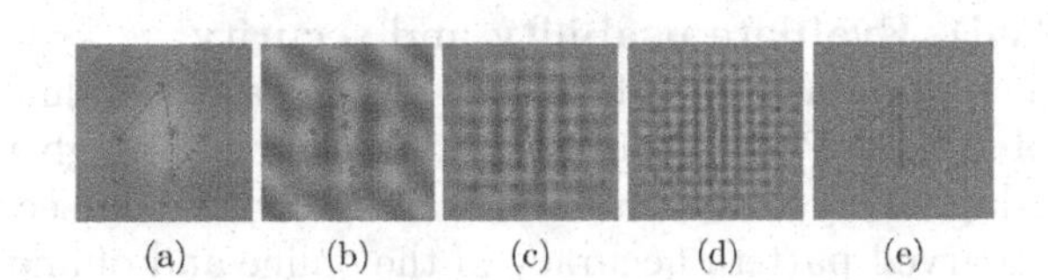

Fig. 8. (a) $sf = 0.3, 0.3, 0.3, 0.3$, the pattern is easily observed; (b) $sf = 1, 1, 1, 1$, the pattern is somehow disguised; (c) $sf = 1, 2, 1, 2$, the pattern is somehow disguised; (d) $sf = 3, 3, 3, 3$, the background starts to have packed of holes, which may cause revulsion; (e) $sf = 30, 30, 30, 30$, the pattern is easily observed.

the optimal spatial frequency for each layer of gratings. The gratings in each layer were masked using a 2-D Gaussian filter (sd = 3) in PsychoPy. Additionally, to conceal the outline of the foreground at a global level, we fragmented the pattern and grid into local features by depicting them in dashed-line.

(b) **Modified the color difference (ΔE) between the foreground grid and background.**
We hypothesized that a noticeable color difference between the foreground grid and the background may make it easier for shoulder surfers to identify the grid orientation and node positions. To mitigate this, we limited the color contrast between the foreground grid and background gratings to a specific level. Firstly, we converted the grid's RGB color to CIE LAB using an algorithm from Manoj Pandey [25]. And then we calculated the color difference (ΔE) between the grid and background image (as expressed in Eq. 4):

$$\begin{aligned} \Delta E_{Lab} &= \sqrt{(L^*_{grid} - L^*_{bg})^2 + (a^*_{grid} - a^*_{bg})^2 + (b^*_{grid} - b^*_{bg})^2} \\ &= \sqrt{(\Delta L^*)^2 + (\Delta a^*)^2 + (\Delta b^*)^2} \end{aligned} \tag{4}$$

where *bg* represents the background image.
We limited the generation time of background images during user testing by utilizing five pre-generated images. The average color difference, as indicated by ΔE, between the grid color and the background image was 10.817 (sd = 3.998), while the average color difference between the pattern line color and the background image was 31.841 (sd = 1.646). During the test, one of

the five pre-generated images was randomly displayed on the user interface each time the users created a pattern password on the grid. This method was expected to have the intended visual effect, as demonstrated in Fig. 8 through Fig. 8c.

(vii). Evaluate usability and security

The approach for addressing SRQ3 was to evaluate the usability and security of Hollow-Pass. Usability was measured through the recall error rate and the adapted System Usability Scale (SUS), while security was measured through observed pattern accuracy in the online and offline simulated attacks.

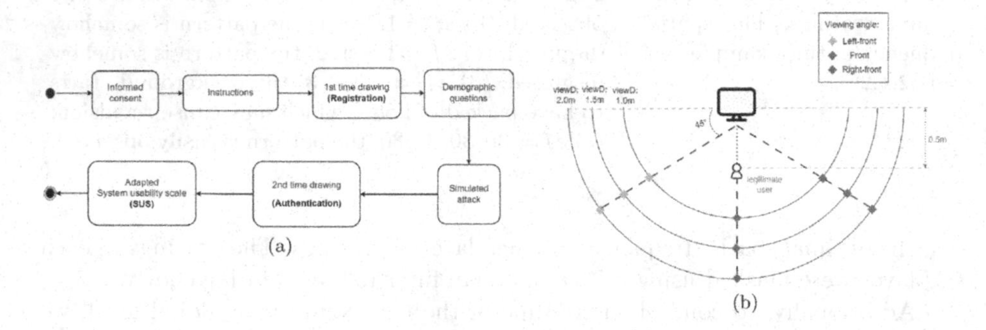

Fig. 9. (a) Online user test procedure; (b) Offline user test setting.

(a) **Online test.**

A 7-section online user test was conducted (as depicted in Fig. 9a). The prototype is available for trial through the provided link during the first 20 days of each month: https://hollow-pass.up.railway.app/ [36]. Participants were given the option to perform the test on either a computer or a mobile phone. This study was designed to specifically test the usability of Hollow-Pass and eliminate potential confounding variables. Participants were asked to provide consent and instructed on how to use Hollow-Pass. They were then asked to register a valid pattern (4 to 9 nodes) on the grid and demographic information was collected. Anti-shoulder-surfing resistance was evaluated by asking participants to identify 6 simulated pattern passwords at viewing distances of 1.0 m, 1.5 m, and 2.0 m. Memorability and usability were assessed by recall error rate and adapted SUS questions, respectively. The online test took an average of 10–15 minutes to complete.

(1) **Objective of Informed Consent:** Participants were made aware of and agreed to the user test at the start of the test. This included information on the purpose of the test, alternatives to the procedure or intervention, and potential risks.
(2) **Instructions:** Participants were provided with clear instructions on how to draw a pattern on either the desktop or mobile version of the site during the test.

(3) **1st time drawing (Registration):** Participants were instructed to draw their first Hollow-Pass patterns.
(4) **Collection of Demographic Information:** Participants were requested to furnish information about their demographics, including gender, age, eyesight quality, device used for testing, and prior experience with pattern-based passwords.
(5) **Simulated attack:** In this phrase, participants were asked to act as shoulder surfers to perform an online simulated attack. They were given six pattern identification questions and were asked to select the correct pattern from among four options. Each question presented a Hollow-Pass pattern (of either weak or medium complexity) that had been simulated at a viewing distance: 1.0 m, 1.5 m, or 2.0 m. The complexity of the patterns was determined using Eq. 3. These stimuli were created to assess the anti-shoulder-surfing effect of Hollow-Pass, taking into account factors: perceived resolution and size.
   * **Perceived resolution.** The perception of resolution by the human visual system closely resembles a low-frequency filter. Research by Pappas and Neuhof [27] has determined that the impulse response of the 1-D eye filters matches a Gaussian shape with an appropriate standard deviation. At 300 dots per inch and a 30-in. viewing distance, the eye filter's impulse response matches a Gaussian filter with $\sigma = 1.5$ and $\tau = 0.0095°$. To simulate the perceived resolution at different viewing distances, we used a Gaussian filter, adjusting the standard deviation.
   * **Perceived size.** Based on Emmert's law [8], which states that the perceived image size changes proportionately with its distance from the observer while controlling for the visual angle. The perceived size of the image was simulated by scaling the original stimuli size (340×340 px) for different viewing distances (1.0 m, 1.5 m, and 2.0 m). This simulation was done under the assumption that the observer had normal visual acuity and the default viewing distance was 0.5 m from the screen to the user's eye.
(6) **2nd time drawing (Authentication):** Participants were asked to repeat their initial pattern on a grid. In case a participant failed to recall their initial pattern, they were allowed to proceed to the next question. The accuracy of the participant's redrawn pattern in relation to their original pattern was regarded as their recall rate and was used to assess the usability of Hollow-Pass.
(7) **Adapted system usability scale(SUS):** A standardized questionnaire is a widely-used 10-item questionnaire that measures usability, learnability, efficiency, and overall user satisfaction [17]. We adapted the System Usability Scale (SUS) for user testing. The purpose of using adapted version of the SUS was to gather detailed feedback about participants' experiences with Hollow-Pass in three dimension: reliability, feasibility and affinity. Participants were asked to evaluate their experience with the Hollow-Pass pattern using a 9-item, 5-point Likert scale

questionnaire, covering five levels of agreement. This aimed to measure their perceptions of the pattern and their overall preference towards it.

(b) **Offline test.**
An offline small-scale sampling test was conducted to evaluate the anti-shoulder-surfing effect of the Hollow-Pass system in practicality. Typically, pattern passwords are used on smaller mobile phone screens where the patterns can be more difficult to observe by observers. However, when pattern passwords are used on larger screens, such as laptop screens, they become more visible and easier to observe. Therefore, the test used an Acer TravelMate P2 laptop with a 14-in. screen and a pixel density of 164.64 pixels per inch to examine the system's ability to resist shoulder surfing on a relatively large display.
The center of the laptop screen was used as the origin, and the viewing distance was calculated as the radius extending from the screen to the participant's position. The semicircle was divided into four equal parts, each considered a distinct viewing angle: left-front, front, and right-front, as depicted in Fig. 9b. The researcher acted as a legitimate user and sat at a distance of 0.5 m from the laptop screen, using a mouse to draw six weak patterns. The password strength was evaluated using Eq. 3. Participants were asked to stand at three viewing angles (front, left-front, right-front) at specified distances from the screen (1.0 m, 1.5 m, and 2.0 m), and to draw what they observed on a test form. This was to examine the mechanism's resistance to shoulder surfing at various viewing angles and distances for weak patterns. Participants were allowed to ask the researcher to redraw the pattern and make multiple attempts.

4 Results and Discussion

This section entails the presentation and analysis of the user test results. Section 4.1 analyzes the strength score of Hollow-Pass patterns with regards to their complexity and key space. In Sect. 4.2, participant demographic information is provided. Section 4.3 discusses the patterns chosen by participants. The usability of Hollow-Pass is discussed in Sect. 4.4, focusing on three aspects: reliability, feasibility, and user affinity. The security of Hollow-Pass is discussed in Sect. 4.5 in relation to simulated online and offline attack results. Lastly, any limitations or future research directions are examined. The purpose of this section is to offer a comprehensive overview of the mechanism and the user test outcomes.

4.1 Password Strength

In this section, we evaluated the strength of the passwords in two aspects, namely: (i). key space and (ii). pattern complexity.

(i.) **Key space.** We conducted a comparison between the conventional 3×3 grid used in the typical Android pattern unlock system (as illustrated in

Fig. 1) and that of the Hollow-Pass system. The key space calculation was done using Eqs. 1 and 2. The results, as presented in Table 1, show that the key space of the Hollow-Pass system is larger than that of the conventional method, increasing from 389,112 to 985,824, particularly for node sizes 4, 5, 6, and 7. This implies that users can create a greater diversity of patterns, making it harder for shoulder surfers to recognize them, without necessarily increasing the number of nodes.

Table 1. Key space comparison between conventional GP and Hollow-Pass

Key Space		
# of nodes	Conventional GP	Hollow-Pass
4	1,624	3,024
5	7152	15,120
6	26,016	60,480
7	72,912	181,440
8	140,704	362,880
9	140,704	362,880
Total	389,112	985,824

(ii.) **Pattern complexity.** The password strength of all acceptable patterns was determined using Eq. 2 by taking into account factors such as the length of the pattern (depicted in Fig. 10a), the number of intersections in the pattern (depicted in Fig. 10b), and the amount of overlap in the pattern (depicted in Fig. 10c). The results showed that as the number of nodes increased, the length of the pattern increased, and there were more intersections and overlaps, making the pattern more complex. The password strength score varied from 6.34 to 46.81, as illustrated in Fig. 10d.

4.2 Participant Demographic Information

In the online user test, 30 participants (15 female) between the ages of 20–80 took part, and in the offline sampling user test, 19 undergraduates (10 female) between the ages of 20–23 from the Universiteit of Twente participated. All participants gave informed consent, which was approved by the Universiteit of Twente's ethics committee. As shown in Table 2, the majority of participants (70%) were in the 20–29 age group, with 2 participants in the 30–49 age group, 6 in the 50–59 age group, and 1 over 60. The eyesight quality of participants was recorded using decimal Snellen notation, with 11 participants having a score less than 0.8, 6 between 0.8–1.0, 5 above 1.0, 6 unsure of their score but wearing glasses, and 2 unsure and not wearing glasses. Most participants were familiar with pattern passwords, with 17 having used it before and 11 knowing of it but not having used it. Only 2 participants were unfamiliar with pattern passwords before the test. The participants also indicated the device used for the test, which could be a desktop, phone, or tablet.

4.3 Pattern Selection

In the online user test, participants were suggested (but not required) to create a pattern with at least 4 and at most 9 nodes, as Hollow-Pass was a new pattern password that might present challenges to some users. 83% of the participants drew suggested patterns while only 4% of these participants failed to confirm the registered pattern. 59% of the participants chose a very weak pattern as their password, 22% chose a weak pattern, and only 3% chose a strong pattern (as illustrated in Fig. 11a). Patterns with 3 or fewer nodes were deemed invalid. The password strength was calculated using Eq. 3. The most commonly used valid pattern was 12369874 with a strength of weak (as shown in Fig. 11b), followed by 1235789 with a strength of very weak (as shown in Fig. 11c). Most participants used 5–6 nodes to create their patterns. To evaluate participants' memorability, the number of successful authentication and participants' recall rate was calculated by comparing their registered patterns (first-time drawing) and authenticated patterns (second-time drawing). 10% of participants, across all age groups, were unable to redraw their patterns during the authentication stage (as shown in Fig. 11d). Because we recruited a limited number of participants in the age group 30–49 and 60+, we focused on discussing the recall rate of the age group 20–29 and 50–59. In Fig. 11e, we observed that the recall rate was 90% of the age group 20–29, and 83% of the age group 50–59.

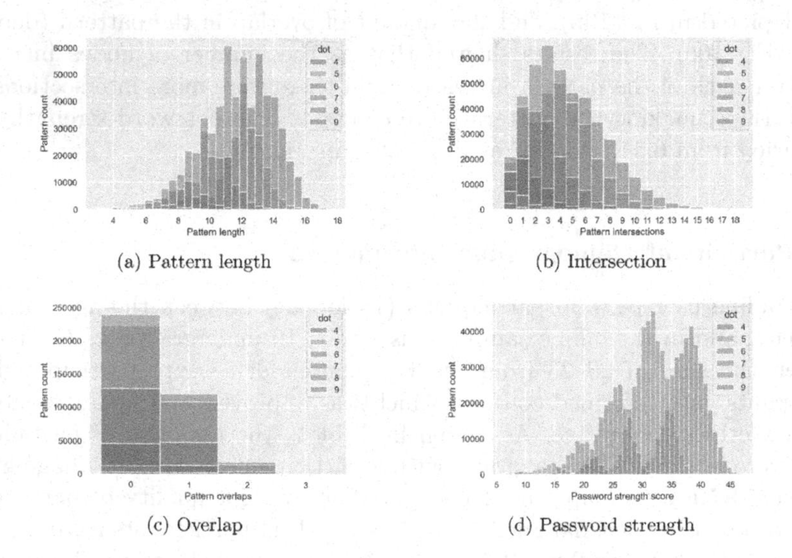

(a) Pattern length

(b) Intersection

(c) Overlap

(d) Password strength

Fig. 10. Distribution of valid patterns attributes.

4.4 System Usability

In the online user test, participants were asked to evaluate the usability of the system, using an adapted version of SUS with 5 points, where 1 represents "Completely disagree," 2 represents "Somewhat disagree," 3 represents "Neither agree nor disagree," 4 represents "Somewhat agree," and 5 represents "Completely agree". To ensure the adapted SUS validation, we calculated the factor loading with Varimax rotation and conducted Cronbach's alpha to assess questionnaire validation and reliability. We chose 3 factors that had high loadings (>.4) to assess system usability based on the scree plot and the eigenvalue, which were perceived reliability, perceived feasibility, and user affinity. Furthermore, the calculated Cronbach's alpha was .825, suggesting that the adapted SUS questionnaire had good internal consistency.

Table 2. Demographic Information of the Participants

Demographic information	Desktop	Phone	Tablet	Total
Gender				
Male	8	7	0	15
Female	7	7	1	15
Non-binary	0	0	0	0
Prefer not to say	0	0	0	0
Age				
20–29	10	10	1	21
30–49	1	1	0	2
50–59	4	2	0	6
60+	0	1	0	1
Prefer not to say	0	0	0	0
Eyesight quality				
Below 0.8	4	6	1	11
0.8–1.0	4	2	0	6
Over 1.0	3	2	0	5
unsure, wears glasses	2	4	0	6
unsure, does not wear glasses	2	0	0	2
Pattern experience				
used	7	9	1	17
known but not used	8	3	0	11
not known	0	2	0	2
Total	15	14	1	30

(i). **Reliability.** 90% of the participants expressed positive opinions regarding the mechanism reliability of Hollow-Pass, with 63% completely agreeing and 27% somewhat agreeing (as depicted in Fig. 12a). They reported a greater sense of security with Hollow-Pass in comparison to the conventional pattern password and concurred that it makes it more challenging for shoulder surfers to discern the credentials from a distance.

(ii). **Feasibility.** Over 70% of the participants gave a favorable assessment of the mechanism feasibility, with 44% indicating complete agreement and 27% indicating somewhat agreement (as shown in Fig. 12b). Participants reported that they could easily view the grid and nodes, observe their pattern while drawing, and complete the pattern drawing without any difficulties.

(iii). **Affinity.** Approximately 70% of the participants expressed positive views on the grid layout and background design (44% completely agreed and 26% somewhat agreed, as illustrated in Fig. 12c). Participants found the mechanism easy to learn and use during the test. They concurred that the system-generated background design was more secure than a customizable background design. Nevertheless, they also desired the option of customizing their background image, should this feature be made available.

4.5 System Security

(i). **Online simulated attack.** Each participant was asked to identify two different patterns for each viewing distance (30 participants, totaling 60

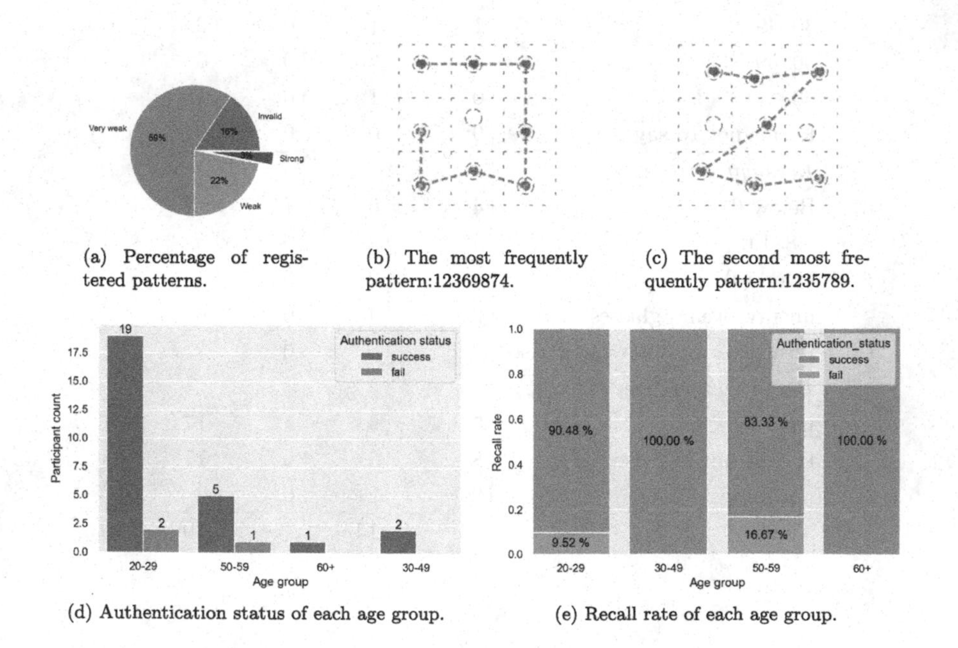

(a) Percentage of registered patterns.

(b) The most frequently pattern:12369874.

(c) The second most frequently pattern:1235789.

(d) Authentication status of each age group.

(e) Recall rate of each age group.

Fig. 11. User pattern in the online test.

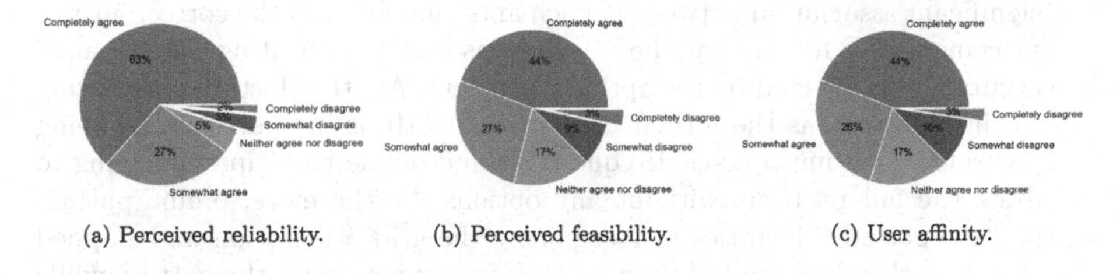

(a) Perceived reliability. (b) Perceived feasibility. (c) User affinity.

Fig. 12. Online SUS result.

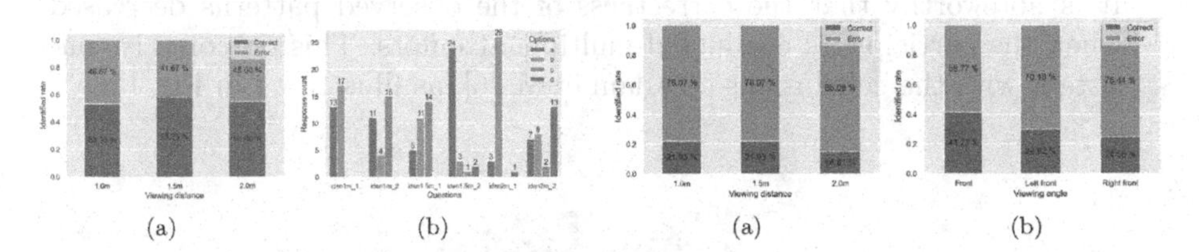

(a) (b) (a) (b)

Fig. 13. (a) Online simulated attack rate; (b) Response distribution of six online simulated attack questions.

Fig. 14. (a) Accuracy varies with viewing distances; (b) Accuracy varies with viewing angles.

patterns per viewing distance). The mean accuracy of identifying stimuli that simulate perceptual effects at different viewing distances (1.0 m, 1.5 m, 2.0 m) was 55.55% in the online simulated attack, as shown in Fig. 13a. However, the stimuli accuracy was not significantly affected by the viewing distance.

(ii). **Offline simulated attack.** In the offline simulated attack, each participant observed 6 patterns per viewing distance at three viewing angles (2 patterns per angle), resulting in a total of 114 patterns collected from 19 participants. The accuracy in the offline test was lower than that of the online test, with a mean accuracy of 20% compared to 56%, as shown in Fig. 14a. The accuracy decreased with increasing viewing distance and was highest when participants were directly in front of the screen (41%) and lowest when positioned at the right-front (25%). The average accuracy of observing Hollow-Pass at various viewing angles was 32%, as illustrated in Fig. 14b.

The difference in results between the online and offline tests may be due to the consistent option placement in the online test. A systematic review [21] has shown that test takers perform better when the correct answer is consistently placed in a specific location, as opposed to randomly placed. The study also found that placing the correct answer at the top or bottom of the options may increase the likelihood of the test taker choosing it, leading to potential bias. This aligns with the result of our research, where the correct answer was usually located at options (a) and (b), as shown in Fig. 13b. We used the chi-squared test of independence to determine whether there was a significant association between participants' answers and the correct answer placement. We found that the p-value was <.001, indicating participants' accuracy was affected by the option placement. Another factor contributing to the difference is the format of the tests, with online participants being presented with multiple-choice questions and offline participants having to draw the full patterns without any options. Furthermore, offline participants reported difficulties in recognizing the grid orientation and skipped nodes, as they had limited time to identify and memorize the pattern while drawing.

It is noteworthy that the correctness of the observed patterns decreased when the participants conducted multiple attempts. This outcome is consistent with the conclusions of Adam et al. [6], as illustrated in Fig. 15.

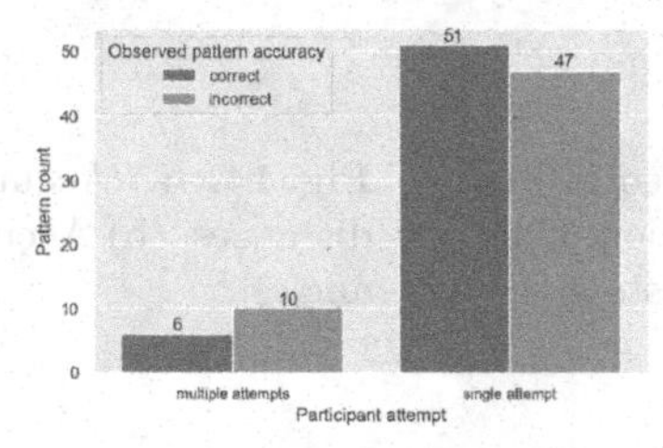

Fig. 15. Correctness of single attempt and multiple attempts.

4.6 Comparative Analysis

The purpose of this section is to compare and contrast Hollow-Pass with two existing methods that also employed image processing techniques to resist shoulder-surfing attacks: IllusionPIN [26] and HideImage [11].

As shown in Table 3, all three methods have robust protection against shoulder-surfing attacks. IllusionPIN, as a PIN-based textual password, uses a shuffled keypad and hybrid images technique to provide robust protection against shoulder-surfing attacks. However, it simplifies its visual algorithm by converting its keypad's color into black and white, and the used parameters, SF, need to be further tuned. Moreover, the conducted user test disregarded viewing angles and the scale was limited. HideImage, as a recognition-based graphical password, downgrades static images and converts them into grayscale. This results in the loss of color and high SF in the images. On the other hand, HideImage is independent of user interaction. This means it may not be as effective when it comes to dynamic image frames that require user interaction, such as drawing a pattern password. In contrast, Hollow-Pass, as a dynamic, recall-based graphical password, is designed to protect passwords that are based on user interaction without information loss. It can be a better option for applications that require dynamic image frames without latency. It also has a relatively larger user scale among ages 20 to 60+. However, because Hollow-Pass uses a simplified visual algorithm, SF needs to be tuned to achieve better protection.

Table 3. Comparison of IllusionPIN, HideImage and Hollow-Pass

Method	Strength	Limitation	User test
IllusionPIN	• PIN-based; • Dynamic keypad; • Hybrid images; • Robust protection.	• Limited color adjustment; • Tune SF; • Disregard viewing angles.	• Device: smartphone; • 21 participants, ages < 40.
HideImage	• Recognition-based; • Coarse-grained image, single-colored background; • Robust protection.	• No protection by user interaction; • Color and SF information loss; • Limited color adjustment.	• Device: smartphone, tablet and laptop; • 20 participants, ages 18–40.
Hollow-Pass	• Recall-based; • Dynamic grid and background; • Protection by user interaction; • Color and SF information preserved; • Robust protection.	• Tune SF; • Simplified visual algorithm.	• Device: laptop, tablet, and smartphone; • 30 participants, ages 20–60+.

4.7 Limitations

The Hollow-Pass mechanism has a few limitations as follows:

(i). One of these is the potential for a biased perceptual effect when users draw patterns on displays with different specifications or at greater viewing distances, as the mechanism employs a background image created with a default display pixel density of 164.54 pixels per inch and a viewing distance of 0.5 m.

(ii). The user test had a restricted scope, and the offline evaluation solely included university students between the ages of 20 and 29. Younger adults tend to be more receptive to novel technology than middle-aged and older adults [3]. Therefore, a more comprehensive study that includes all age groups is necessary to investigate the effect of a background image and color contrast on other demographic categories.

5 Conclusion and Future Work

We present Hollow-Pass, a novel pattern password mechanism utilizing global precedence and color difference ΔE, allowing users to draw patterns on a dynamic 3×3 grid as their authentication credentials. An online user test was conducted with $n = 30$ participants aged between 20–80, and an offline user test was conducted with $n = 19$ undergraduate students, viewed from three distances (1.0 m, 1.5 m, 2.0 m) and three angles (front, left-front, right-front). The results indicate that Hollow-Pass enhances the security of weak pattern passwords against shoulder-surfing attacks while maintaining usability, as the simulated shoulder surfer observed 20% of tested patterns on average in the offline user test. Over 70% online participants gave positive feedback, suggesting that Hollow-Pass effectively resists shoulder-surfing attacks and balances security and usability.

We found that a significant percentage of participants(57%) had pattern password experience. As our experimental design does not allow for a direct comparison of the usability of the developed grid with a conventional grid, future studies could include a conventional grid as a control group to more directly compare the usability of different grid designs. Moreover, we intend to carry out a comprehensive offline study that focuses on both the usability and security of Hollow-Pass, covering all age groups. This research also opens a door for the researchers to consider human eye adjustments [18] during the online test instead of having an average value of visual acuity and a viewing distance. This study aims to investigate any potential tradeoffs that may arise due to the application of the mechanism. Hollow-Pass may provide defense against automatic online guessing attacks, as the grid layout changes randomly with each use, akin to the Captcha scheme [46].

References

1. Railway. https://railway.app/. Accessed 28 Mar 2023
2. Welcome to flask. https://flask.palletsprojects.com/en/2.2.x/. Accessed 28 Jan 2023
3. Anderson, M., Perrin, A.: Barriers to adoption and attitudes towards technology. Tech adoption climbs among older adults (2017)
4. Aris, H., Yaakob, W.F.: Shoulder surf resistant screen locking for smartphones: a review of fifty non-biometric methods. In: 2018 IEEE Conference on Application, Information and Network Security (AINS), pp. 7–14 (2018). https://doi.org/10.1109/AINS.2018.8631419
5. Aviv, A.J., Budzitowski, D., Kuber, R.: Is bigger better? Comparing user-generated passwords on 3×3 vs. 4×4 grid sizes for android's pattern unlock. In: Proceedings of the 31st Annual Computer Security Applications Conference, pp. 301–310. ACSAC '15, Association for Computing Machinery, New York, NY, USA (2015). https://doi.org/10.1145/2818000.2818014
6. Aviv, A.J., Davin, J.T., Wolf, F., Kuber, R.: Towards baselines for shoulder surfing on mobile authentication. CoRR abs/1709.04959 (2017). https://arxiv.org/abs/1709.04959
7. Bhanushali, A., Mange, B., Vyas, H., Bhanushali, H., Bhogle, P.: Comparison of graphical password authentication techniques. Int. J. Comput. Appl. **116**(1), 11–14 (2015)
8. Boring, E.G.: Size constancy and Emmert's law. Am. J. Psychol. **53**(2), 293–295 (1940). https://www.jstor.org/stable/1417427
9. Bruggen, D.V.: Studying the Impact of Security Awareness Efforts on User Behavior. Ph.D. thesis, University of Notre Dame, April 2014. https://doi.org/10.7274/st74cn7217h, https://curate.nd.edu/show/st74cn7217h

10. Campbell, F.W., Maffei, L.: Contrast and spatial frequency. Sci. Am. **231**(5), 106–115 (1974). https://www.jstor.org/stable/24950220
11. Chen, C.Y.D., Lin, B.Y., Wang, J., Shin, K.G.: Keep others from peeking at your mobile device screen! In: The 25th Annual International Conference on Mobile Computing and Networking. MobiCom '19, Association for Computing Machinery, New York, NY, USA (2019). https://doi.org/10.1145/3300061.3300119
12. Cho, G., Huh, J.H., Cho, J., Oh, S., Song, Y., Kim, H.: SysPal: system-guided pattern locks for android. In: 2017 IEEE Symposium on Security and Privacy (SP), pp. 338–356. IEEE (2017)
13. De Luca, A., et al.: Now you see me, now you don't: protecting smartphone authentication from shoulder surfers. In: Proceedings of the SIGCHI Conference on Human Factors in Computing Systems, pp. 2937–2946 (2014)
14. De Luca, A., et al.: Back-of-device authentication on smartphones. In: Proceedings of the SIGCHI Conference on Human Factors in Computing Systems, pp. 2389–2398 (2013)
15. Deyashini Chakravorty: What if we used graphical passwords for authentication? (2020). https://uxdesign.cc/graphical-passwords-for-authentication-4e716b94eb47. Accessed 21 Nov 2022
16. Flevaris, A.V., Martínez, A., Hillyard, S.A.: Attending to global versus local stimulus features modulates neural processing of low versus high spatial frequencies: an analysis with event-related brain potentials. Front. Psychol. **5**, 277 (2014)
17. Hodrien, A., Fernando, T., et al.: A review of post-study and post-task subjective questionnaires to guide assessment of system usability. J. Usability Stud. **16**(3), 203–232 (2021)
18. Kalloniatis, M., Luu, C.: Visual acuity. Webvision: The Organization of the Retina and Visual System [Internet] (2007)
19. Lashkari, A.H., Farmand, S., Zakaria, D., Bin, O., Saleh, D., et al.: Shoulder surfing attack in graphical password authentication. arXiv preprint arXiv:0912.0951 (2009)
20. Li, W., Tan, J., Meng, W., Wang, Y.: A swipe-based unlocking mechanism with supervised learning on smartphones: design and evaluation. J. Netw. Comput. Appl. **165**, 102687 (2020)
21. Lions, S., Monsalve, C., Dartnell, P., Blanco, M.P., Ortega, G., Lemarié, J.: Does the response options placement provide clues to the correct answers in multiple-choice tests? A systematic review. Appl. Meas. Educ. **35**(2), 133–152 (2022)
22. Mihajlov, M., Jerman-Blazic, B.: Eye tracking graphical passwords. In: Nicholson, D. (eds.) Advances in Human Factors in Cybersecurity. AHFE 2017. AISC, vol. 593, pp. 37–44. Springer, Cham (2018). https://doi.org/10.1007/978-3-319-60585-2_4
23. N., S.M.: Global perception. https://psychologydictionary.org/global-perception/. Accessed 10 Feb 2023
24. Navon, D.: Forest before trees: the precedence of global features in visual perception. Cogn. Psychol. **9**(3), 353–383 (1977)
25. Pandey, M.: rgb2lab.py. https://gist.github.com/manojpandey/f5ece715132c572c80421febebaf66ae. Accessed 17 Jan 2023
26. Papadopoulos, A., Nguyen, T., Durmus, E., Memon, N.: Illusionpin: shoulder-surfing resistant authentication using hybrid images. IEEE Trans. Inf. Forensics Secur. **12**(12), 2875–2889 (2017). https://doi.org/10.1109/TIFS.2017.2725199
27. Pappas, T., Neuhoff, D.: Least-squares model-based halftoning. IEEE Trans. Image Process. **8**(8), 1102–1116 (1999). https://doi.org/10.1109/83.777090
28. Schuessler, Z.: Delta e 101. https://zschuessler.github.io/DeltaE/learn/#toc-delta-e-76. Accessed 17 Jan 2023

29. Seng, L., Ithnin, A.P.D.N., Km, H.: User's affinity of choice: Features of mobile device graphical password scheme's anti-shoulder surfing mechanism. Int. J. Comput. Sci. Issues **8** (2011)
30. Service, P.: Defining and communicating color: the cielab system (2013)
31. Shammee, T.I., Akter, T., Mou, M., Chowdhury, F., Ferdous, M.S.: A systematic literature review of graphical password schemes. J. Comput. Sci. Eng. **14**(4), 163–185 (2020)
32. Song, Y., Cho, G., Oh, S., Kim, H., Huh, J.H.: On the effectiveness of pattern lock strength meters: measuring the strength of real world pattern locks. In: Proceedings of the 33rd Annual ACM Conference on Human Factors in Computing Systems, pp. 2343–2352. CHI '15, Association for Computing Machinery, New York, NY, USA (2015). https://doi.org/10.1145/2702123.2702365
33. Sun, C., Wang, Y., Zheng, J.: Dissecting pattern unlock: the effect of pattern strength meter on pattern selection. J. Inf. Secur. Appl. **19** (2014). https://doi.org/10.1016/j.jisa.2014.10.009
34. Sun, H.M., Chen, S.T., Yeh, J.H., Cheng, C.Y.: A shoulder surfing resistant graphical authentication system. IEEE Trans. Dependable Secure Comput. **15**(2), 180–193 (2018). https://doi.org/10.1109/TDSC.2016.2539942
35. Takahashi, K., Hasegawa, M., Tanaka, Y., Kato, S.: A structural similarity assessment for generating hybrid images. In: 2011 Conference Record of the Forty Fifth Asilomar Conference on Signals, Systems and Computers (ASILOMAR), pp. 240–243 (2011). https://doi.org/10.1109/ACSSC.2011.6189993
36. Tan, J.: Hollow-pass online survey. https://hollow-pass.up.railway.app/. Accessed 02 Feb 2023
37. Tao, H., Adams, C.M.: Pass-go: a proposal to improve the usability of graphical passwords. Int. J. Netw. Secur. **7**, 273–292 (2008)
38. Tupsamudre, H., Banahatti, V., Lodha, S., Vyas, K.: Pass-o: a proposal to improve the security of pattern unlock scheme. In: Proceedings of the 2017 ACM on Asia Conference on Computer and Communications Security, pp. 400–407 (2017)
39. Tupsamudre, H., Vaddepalli, S., Banahatti, V., Lodha, S.: TinPal: an enhanced interface for pattern locks. In: Workshop on Usable Security, ser. USEC, vol. 18 (2018)
40. Uellenbeck, S., Dürmuth, M., Wolf, C., Holz, T.: Quantifying the security of graphical passwords: the case of android unlock patterns. In: Proceedings of the 2013 ACM SIGSAC Conference on Computer & Communications Security, pp. 161–172. CCS '13, Association for Computing Machinery, New York, NY, USA (2013). https://doi.org/10.1145/2508859.2516700
41. Vorster, J.: A Framework for the Implementation of Graphical Passwords. Ph.D. thesis, December 2014. https://doi.org/10.13140/RG.2.1.3245.2326
42. Vorster, J., van Heerden, R.: Graphical passwords: a qualitative study of password patterns, March 2015
43. Wolf, F., Aviv, A.J., Kuber, R.: It's all about the start classifying eyes-free mobile authentication techniques. J. Inf. Secur. Appl. **41**, 28–40 (2018)
44. Zhang, L., Guo, Y., Guo, X., Shao, X.: Does the layout of the android unlock pattern affect the security and usability of the password? J. Inf. Secur. Appl. **62**, 103011 (2021)
45. zhangstar: gesturepassword (2018). https://github.com/zhangstar1331/gesturePassword
46. Zhu, B.B., Yan, J., Bao, G., Yang, M., Xu, N.: Captcha as graphical passwords-a new security primitive based on hard AI problems. IEEE Trans. Inf. Forensics Secur. **9**(6), 891–904 (2014). https://doi.org/10.1109/TIFS.2014.2312547

Practical Improvements on BKZ Algorithm

Ziyu Zhao(✉) and Jintai Ding(✉)

Yau Mathematical Center, Tsinghua University, Beijing, China
ziyuzhao0008@outlook.com, jintai.ding@gmail.com

Abstract. Lattice problems such as NTRU and LWE problems are widely used as the security base of post-quantum cryptosystems. And currently, lattice reduction by BKZ algorithm is the most efficient way to solve them. In this paper, we give four further improvements on BKZ algorithm, which can be used for SVP subroutines based on enumeration and sieving. These improvements in combination provide a speed-up of $2^{3\text{-}4}$ in total. So all the lattice-based NIST PQC candidates lose 3–4 bits of security in concrete attacks. Using these new techniques, we solved the 656 and 700 dimensional ideal lattice challenges in 380 and 1787 thread hours, respectively. The cost of the first one (also used an enumeration-based SVP subroutine) is much less than the previous records (4600 thread hours). One can still simulate the improved BKZ algorithm to find the blocksize strategy that makes Pot of the basis (defined in Sect. 4.2) decrease as fast as possible, which means the length of the first basis vector decrease the fastest if we accept the GSA assumption. It is useful for analyzing concrete attacks on lattice-based cryptography.

Keywords: Lattice-based cryptography · Lattice reduction · BKZ algorithm

1 Introduction

Lattice-based cryptography is nowadays an important part of post-quantum cryptography. The security of it is mainly based on some lattice problems, such as learning with error problem (LWE) [9,17,25] and NTRU problem [13]. These problems can be reduced to approximate shortest vector problem (ASVP) [5,16,19,20], i.e. to find a relatively short vector given a lattice basis. And now lattice attacks are the most efficient way to solve such problems, thus it is important to know the concrete hardness of ASVP.

Currently, given a lattice basis, there are two types of algorithms to find short vectors in the lattice. One is SVP algorithms like enumeration [6,7,14, 24,27,31] and sieving [3,10,18,22], which can find almost the shortest vector in the lattice but the cost is at least exponential in the dimension of the lattice. These SVP algorithms can only be applied to lattices with a small dimension when there are restricted resources. Another type of algorithm, for instance,

S. Dolev et al. (Eds.): CSCML 2023, LNCS 13914, pp. 273–284, 2023.
https://doi.org/10.1007/978-3-031-34671-2_19

LLL algorithm [15,21] and BKZ [28,31] algorithm can work on high dimensional lattices in a realistic time. LLL algorithm is extremely fast and often used as preprocessing, BKZ algorithm gives a bridge from the shortest vector in small dimension to short vectors with the same root Hermite factor in high dimension.

BKZ algorithm was first proposed by Schnorr in the 80's. It does enumeration on local blocks to find short vectors and then inserts the new vectors in the basis. Larger local blocksize gives shorter vector and takes more time. In 2011, Chen-Nguyen used a pruning technique in the enumeration step, it makes BKZ algorithm with a higher local blocksize practicable [4]. In 2016, Yuntao Wang et al. proposed their improved progressive BKZ [2], they got an optimized blocksize strategy based on their simulation of the total enumeration cost. It starts with a small blocksize, and increases the blocksize in a well-organized manner, making the algorithm significantly faster.

In this paper, we will give four further improvements on BKZ algorithm. We replace the insertion in BKZ by processing the local projected lattice. Then we can use a jumping technique to move more than one step each time in the BKZ tours. In the last several tours of BKZ, we only run the SVP subroutines on the first several indexes, which saves half of the time of these tours, then we give an end of the reduction by running a much heavier SVP subroutine.

We applied these techniques in lattice reduction with SVP subroutines based on both enumeration and sieving. We implemented the new BKZ algorithm and tested it on ideal lattice challenges (the ideal lattice structure is never used). The running result shows we get a speed up with a factor 2^{3-4}, which may be further improved since we did not use a well-organized progressive BKZ (for the enumeration-based BKZ, our blocksize are simply $80, 88, 96, \cdots$). Moreover, our new BKZ algorithm is still easy to simulate (as BKZ 2.0 [4]) if we know the behavior of the SVP subroutine well.

Road Map. In Sect. 2, we present some basic facts about lattice and introduce the notations. Then in Sect. 3, we will recall the developments of BKZ algorithm in history. Our further improvements on BKZ will be given in Sect. 4. The information about the lattice challenge results is in Sect. 5.

2 Preliminaries

Lattice is discrete subgroup in $\mathbb{R}^m$. A lattice L always admits an integral basis $\mathbf{B} = \{\mathbf{b}_1, \mathbf{b}_2, \cdots, \mathbf{b}_n\}$ such that each vector $\mathbf{v}$ in L can be represented uniquely as an integral linear combination of $\mathbf{B}$, i.e. $\mathbf{v} = \lambda_1 \mathbf{b}_1 + \cdots + \lambda_n \mathbf{b}_n$, $\lambda_i \in \mathbb{Z}$. We say n is the dimension of the lattice. The determinant of the lattice $\det(L)$ is defined to be $\sqrt{\det(\mathbf{B}\mathbf{B}^T)}$. It is equal to the absolute value of $\det(\mathbf{B})$ if $m = n$.

Gram-Schmidt Orthogonalization. The Gram-Schmidt orthogonalization of $\mathbf{B}$ is given by $\mathbf{B}^* = (\mathbf{b}_1^*, \cdots, \mathbf{b}_n^*)$ where $\mathbf{b}_i^*$ is defined by

$$\mathbf{b}_i^* = \mathbf{b}_i - \sum_{j=1}^{i-1} \mu_{ij} \mathbf{b}_j^*, \ \mu_{ij} = \frac{\langle \mathbf{b}_i, \mathbf{b}_j^* \rangle}{\|\mathbf{b}_j^*\|^2}$$

We further denote by B_i the square of $\|\mathbf{b}_i^*\|$, we will call $[B_1, B_2, \cdots, B_n]$ the distance vector of the basis $\mathbf{B}$. The distance vector of a basis contains lots of information about it and this notation will be heavily used in the analysis of BKZ algorithm. We should also introduce the concept of local projected lattice. Let π_i be the orthogonal projection to $\text{span}(\mathbf{b}_1, \cdots, \mathbf{b}_{i-1})^{\perp}$. Then we define the local projected lattice $L_{[i,j]}$ to be the lattice spanned by $B_{[i,j]} = (\pi_i(\mathbf{b}_i), \cdots, \pi_i(\mathbf{b}_j))$.

Gaussian Heuristic. Because of the discreteness, the shortest nonzero vector in L exists (not unique in general). It is extremely hard to compute the shortest vector (proved to be NP-hard), but the length of it is estimated to be

$$\text{GH}(L) = \frac{\Gamma(\frac{n}{2}+1)^{\frac{1}{n}}}{\sqrt{\pi}} \cdot \det(L)^{\frac{1}{n}} \approx \sqrt{\frac{n}{2\pi e}} \cdot \det(L)^{\frac{1}{n}}$$

when the lattice is "*random*" and n is not too small. In practice it works well if $n \geqslant 40$.

Root Hermite Factor. For a vector $\mathbf{v}$ in a n dimensional lattice L, we define the root Hermite factor to be

$$\delta = \text{RHF}(\mathbf{v}) = \left(\frac{\|\mathbf{v}\|}{\det(L)^{\frac{1}{n}}}\right)^{\frac{1}{n}}$$

as in [8], the root Hermite factor measures the quality of the vector. The hardness to get a vector of a fixed length mainly depends on its root Hermite factor.

HKZ Reduced Basis. We say a lattice basis $\mathbf{B}$ of an n dimensional lattice L is HKZ-reduced if for each $1 \leqslant i \leqslant n$, the first vector of $B_{[i,n]}$ is the shortest nonzero vector of $L_{[i,n]}$.

3 History of BKZ Algorithm

3.1 The Original Algorithm

The first version of BKZ algorithm was proposed by Schnorr and Euchner as a generalization of the LLL algorithm [31]. Briefly, LLL algorithm gives the basis an order, then always reduces the latter vector by the former ones, and after the reduction is done, it tries to make the former one shorter (in the corresponding local projected lattice) by swapping contiguous vector pairs. The algorithm terminates when no more swaps or reductions can be done. The first vector of the output basis is of length about 1.02^n times the Gaussian heuristic in practice (see [8]), and the running time is polynomial in n, where n is the dimension of the lattice.

BKZ algorithm replaces the *swap* in LLL algorithm by a full enumeration in the local projected lattice to get a shorter (also in the corresponding local projected lattice) vector. This vector will be inserted into the basis at a preselected place, and then we use an LLL algorithm to remove the linear dependency. The size of the local projected lattice is fixed and the place to do enumeration is pre-specified. Similar to LLL algorithm, BKZ algorithm terminates when no nontrivial insertion can be done. The algorithm works as follows:

Algorithm 1: BKZ algorithm

Input: a basis $\mathbf{B} = (\mathbf{b}_1, \cdots, \mathbf{b}_n)$, blocksize d and $\delta < 1$
Output: A BKZ-d reduced basis

```
1  LLL(B, δ);
2  while last epoch did a nontrivial insertion do
3      for i = 1, 2, ··· , n − 1 do
4          h = max(i + d − 1, n);
5          v = full_enum(L_[i,h]); //find the shortest nonzero vector in L_[i,h]
6          if ||v|| < δ||b*_i|| then
7              let ṽ = λ_i b_i + ··· + λ_h b_h if v = λ_i π_i(b_i) + ··· + λ_h π_i(b_h);
8              LLL(b_1, ··· , b_{i−1}, ṽ, b_i, ··· , b_max(h+1,n), δ);
9              remove the zero vector in the first place.
10         else
11             LLL(b_1, ··· , b_max(h+1,n), δ);
```

The running time of BKZ algorithm increases as the blocksize increases. It is not proved to be polynomial in n (the dimension) for fixed blocksize. But for small blocksize d (for example $d < 20$), the algorithm always terminates in a reasonable time and the output quality is significantly improved. For $d = 20$ and n sufficiently large, the length of the shortest nonzero vector it founds is around 1.0128^n times the Gaussian heuristic (see [8]).

3.2 BKZ2.0

In 2011, Chen-Nguyen gave several improvements on the original BKZ algorithm [4], which made BKZ algorithm with a high blocksize (d–100) practicable. The root Hermite factor of the output vector is improved to about 1.0095. They mainly did the following:

The first point is that, for a large blocksize d ($d \geqslant 40$), before no new insertion can be done, there is a long time that the quality of the basis improves poorly [11]. So they used the *early abort* technique to stop the algorithm as soon as the quality of the basis does not improve further. This provides an exponential speed-up in practice [8] without degenerating the output quality.

They also made some modifications to the enumeration step. For a larger blocksize d, experiment shows the running time of BKZ is dominated by the enumeration subroutines. They used pruned enumeration instead of full enumeration. Proper pruning can give an exponential speed up (about $2^{d/4}$) while the algorithm still outputs the shortest vector with a high probability. They further gave an *extreme pruning* technique [7], which repeats a further pruned enumeration several times. This leads to a speed-up of $2^{d/2}$.

Another thing they did is to preprocess the basis before doing the enumeration. Since the nodes to enumerate will be fewer if the basis has a better quality. Taking some time to do a light reduction will largely reduce the total enumeration time. In BKZ 2.0 they chose a BKZ algorithm with a small blocksize as preprocessing.

3.3 Progressive BKZ

Progressive BKZ mainly means to progressively enlarge the blocksize while doing reductions. The key idea is if an enumeration with a low dimension can further reduce the lattice, there is no need to use a much larger dimension since the cost of SVP is at least exponential in the dimension. This technique was mentioned in several studies including [4,12,30]. These works mainly differ in the way they increase the blocksize. In 2016, [2] did a precise cost estimation of the progressive BKZ, and gave an optimized blocksize strategy. In their estimation, to do a BKZ-100 in an 800-dimensional lattice, their progressive BKZ is $2^{2.7}$ times faster than BKZ 2.0. And it's estimated to be 50 times faster than BKZ 2.0 for solving SVP challenges (see [26]) up to dimension 160.

4 Several Improvements About BKZ Algorithm

In this section, we will give several techniques to further accelerate the BKZ algorithm. These techniques can be used for BKZ based on both sieving and enumeration SVP subroutines. And one can use it almost for free (except for the large final run for sieving, which requires more memory) to get a considerable acceleration in practice.

4.1 Local Basis Processing Instead of Insertion

Currently, we have two types of SVP algorithms, enumeration and sieving. Sieving is faster but requires a large space that grows exponentially in the sieving dimension. To do a single large sieving or enumeration with the hope of finding the shortest vector is generally not the best choice.

The number of nodes on the enumeration tree grows as the basis gets worse [7]. In practice, it's better to do some preprocessing as in BKZ2.0. So the whole enumeration process not only gives a short vector but also a rather good basis. For sieving, one often uses the left progressive sieve to accelerate, whose speed also relies on the quality of the basis. And one needs the first several entries in the distance vector to be small to get a large dimension for free, which saves both time and memory (for sieving techniques, see [1]). Thus it will not lead to much further cost to get a good basis.

Only inserting one short vector like the original BKZ algorithm or BKZ2.0 will waste the almost *free* basis. It's generally better to compute the transform matrix of local processing (on the local projected lattice) and apply it on the vectors of the original basis (succeed by a size reduction). Then the next local basis to apply the SVP algorithms is already only a little bit worse than an HKZ-reduced basis. An obvious gain is we need no more preprocessing for it. This *local basis processing* technique is folklore for sieving-based BKZ, but it is necessary to point it out because it's the foundation of the next technique, and we first used it on enumeration-based BKZ.

Similar to BKZ2.0 [4], we can still simulate the improved BKZ algorithm by simulating the distance vector $(\|\mathbf{b}_1^*\|^2, \cdots, \|\mathbf{b}_n^*\|^2)$ of the lattice. So we can efficiently find the optimized blocksize strategies and predict the precise cost of the reduction.

4.2 Jump by Two or More

After we do a local basis processing with blocksize d, the first vector will be short, and it's easy to see that the next few vectors are not too long also. If we make our working context jump to the right by two indices or more (i.e. to work on $L_{[i+s,j+s]}$ after $L_{[i,j]}$), say we jump s steps after each SVP subroutine, we accelerate by factor s while not degenerating the quality much.

Here a crucial point is to use a blocksize slightly larger than we require. For instance, if we want to do a BKZ with blocksize d, we can choose a $d' = d+s-1$ and every time we jump s steps. The result will not be worse (actually better) than after a tour of BKZ-d without the jump technique, if the output basis of the SVP subroutine has similar quality as an *HKZ-reduced* basis. After the modification, the number of SVP subroutines is only $\frac{1}{s}$ as before, and for each subroutine, the cost is $2^{0.386(s-1)}$ (practically, when d–90) as before if we use SVP algorithm based on 3-sieve. Take $s = 4$ we get a speed-up of at least $2^{0.84}$.

To verify the effectiveness of this method, we generate a 400-dimensional random lattice with determinant 32768^{400}. Then we execute a BKZ reduction to make the first basis vector has length $32768 \cdot 1.01046^{400}$, corresponding to blocksize 75. We use *Pump* in [1] as the SVP-subroutine (the expected dimension for free is set to be 12, so it behaves like HKZ-reduction), run one tour of BKZ with different maximal sieving dimension (MSD) and jumping steps. To measure the quality of a basis, we introduce the following notation:

$$\mathrm{Pot}(L) = \prod_{i=1}^{n} B_i^{n+1-i}$$

For a given lattice, Pot is an increasing function in the root Hermite factor of the first basis vector, if we accept the GSA assumption [29]. So a better basis will have a smaller Pot. We present $\Delta \log_2 \mathrm{Pot}$ and the cost in Table 1 and 2. From the tables, we see simply a larger jumping step (4–6) will lead to a speed-up of $2^{1.2}$. The gain of this technique varies with different SVP subroutines, and in practice is typically 2^1–2^2 (Tables 1, 2).

Table 1. Jumping step = 1

MSD	68	69	70	71	72	73	74	75
cpu hours	2.30	2.74	3.27	3.88	4.69	5.69	6.93	8.52
$\Delta \log_2 \mathrm{Pot}$	463	758	1166	1400	1910	2254	2674	2949
$\frac{\Delta \log_2 \mathrm{Pot}}{\mathrm{cost}}$	201	277	357	361	407	396	385	346

Table 2. Different jumping steps

(MSD, jumping step)	(72, 1)	(73, 2)	(74, 3)	(75, 4)	(76, 5)	(77, 6)	(78, 7)
cpu hours	4.69	2.84	2.31	2.13	2.20	2.30	2.51
$\Delta \log_2 \mathrm{Pot}$	1910	1797	1787	1962	2059	2084	2241
$\frac{\Delta \log_2 \mathrm{Pot}}{\mathrm{cost}}$	407	633	773	920	930	906	858

We notice that [1] also tried a technique called *PumpNJump*. They set the jumping step to be 3, and the gain is about $2^{0.2}$ when the blocksize is high (estimated from Fig. 4 in their paper). We guess it's because they did not use a large enough blocksize.

4.3 Reduce only When We Need

In practice, we usually don't need the whole BKZ reduced basis. What we want is just a short vector (in lattice challenges) or make the tail of the distance vector large enough such that we can get the key hidden in the lattice by a size reduction (in real attacks of lattice-based cryptography). We only introduce the case for lattice challenges here, because the other case is similar.

For example, if we want to do BKZ-d on an n-dimensional lattice. For the last $\left[\frac{n}{d}\right]$ tours of the algorithm, we don't need to visit all indexes. In fact, we work as Algorithm 2 instead of Algorithm 3.

Algorithm 2: The last several tours of our BKZ

Input: an n-dimensional lattice L, blocksize d and an SVP algorithm
Output: a reduced basis, denote by L_2

```
1 m = [n/d];
2 for k = 1, 2, ..., m do
3     //do a BKZ tour on L[1,n-kd+1]
4     for i = 1, 2, ..., n - kd + 1 do
5         reduce L[i,i+d-1] by the SVP algorithm;
6 return L
```

Algorithm 3: The original BKZ

Input: an n-dimensional lattice L, blocksize d and an SVP algorithm
Output: a reduced basis, denote by L_1

```
1 m = [n/d];
2 for k = 1, 2, ··· , m do
3 |   for i = 1, 2, ··· , n − d + 1 do
4 |   |   reduce L_[i,i+d−1] by the SVP algorithm;
5 return L
```

Because the BKZ tour on $L_{[1,n-kd+1]}$ only relates to the first $n-(k-1)d$ vectors, the first $n - \left[\frac{n}{d}\right] d + 1$ vectors of L_1 and L_2 are the same. But we save at least half the time for these final BKZ tours.

Notice that if we use a progressively larger blocksize, even if we jump by two or more usually we stay on one dimension for only 1–2 tours. This technique at least saves a constant ratio of time for the total challenge since the best SVP algorithm takes exponential time. We present the details of our 700 dimensional lattice challenge below. From Table 3 we see this technique saves about 900 CPU hours for our 700 dimensional challenge, which costs 1787 CPU hours in total (Table 3).

Table 3. Details for the end of our 700 dimensional challenge

MSD	working on	CPU time	a full tour time	min length
91	$L_{[1,578]}$	196.3 h	206.4 h	812896
92	$L_{[1,466]}$	201.5 h	260.2 h	805991
92	$L_{[1,354]}$	152.3 h	258.9 h	787811
93	$L_{[1,242]}$	146.0 h	365.2 h	755466
94	$L_{[1,130]}$	147.0 h	651.9 h	729162

4.4 A Large Final Run

In BKZ for high dimensional lattice, we can choose a much larger dimension d in the last SVP subroutine (working on $[1, d]$) to get a much shorter vector, saving the time for several tours of SVP subroutines with a normal blocksize. Since one tour costs $n/s \cdot T_{svp}$, which is much larger than an SVP subroutine, this method works well in practice (if we are searching for the secret key hidden in the lattice, just do a large enumeration or sieving at the tail of the basis). For our 700-dimensional lattice challenge, we ran a sieving-based SVP algorithm on the first 124 vectors, and got a vector of length 659874. In this case, we saved time for about 2 heaviest tours of BKZ.

5 The Lattice Challenges

The ideal lattice challenge [23] was started in 2012. It provides many different ideal lattices with dimensions up to 1024. The original goal of this challenge is to test the algorithms for finding short vectors in ideal lattices. But we will treat it as high dimensional random lattices to test our BKZ techniques. For each given lattice, a vector shorter than $1.05 \cdot \mathrm{GH}(L)$ can enter the SVP Hall of Fame, and a vector shorter than $n \cdot \det(L)^{\frac{1}{n}}$ can enter the Approximate SVP Hall of Fame. We solved the following challenges (Table 4):

Table 4. The ideal lattice challenges we solved

dim	length	root Hermite factor	total cost	SVP subroutines
656	670275	1.00993	380 thread hours	Enumeration
700	659874	1.00928	1787 thread hours	Sieving

5.1 The Challenge Based on Enumeration

The 656-dimensional challenge was first finished in the summer of 2021, we used a laptop with Intel Core i7-7500U CPU (2.70 GHz) and a non-optimized c++ program which was modified several times while running. The total cost is about 700 core hours (the information uploaded to the website of ideal lattice challenge is wrong). Later we optimized the program and ran it on an Intel Xeon Silver 4208 CPU (2.10 GHz) again. It takes only 380 core hours, much faster than the previous record which takes 4637 thread hours to solve a 652-dimensional approximate SVP challenge.

To get a reduced basis, we used a variant of DeepBKZ [32] as the SVP subroutine. DeepBKZ replaces the LLL in the original BKZ algorithm with DeepLLL (see [31]), which allows *deep insertion*. We modified the algorithm by further checking all short vectors from an enumeration if they can be inserted into some former place, and we always choose the candidate that can be inserted the deepest. The jumping step was 3–4. And we did not use a highly optimized blocksize strategy which may give a further speed up (the blocksize we used was simply 80, 88, 96, $\cdots$).

5.2 The Challenge Based on Sieving

The 700-dimensional challenge was finished recently. We started with a BKZ-80 reduced basis (takes about 200 thread hours), and then used BKZ based on 3-sieve. The Sieving step takes 1587 thread hours, also on the Xeon Silver 4208 CPU (2.10 GHz).

The jumping step was 6. For the maximal sieving dimension, before each tour of BKZ we calculate the current Pot and compute the corresponding blocksize d by the GSA assumption. We then choose the maximal sieving dimension to be d. Such a choice performs well in practice. The expected dimension for free is set to be 18. so we work on a local projected lattice with dimension $d + 18$.

6 Conclusion

For the lattice-based NIST PQC candidates, most of our techniques are practicable and reduce $3 \sim 4$ bits of security in concrete attacks. If we also consider the memory overhead, *a large final run* may not be a good choice since it uses significantly more memory. However, in this paper, our sieving cost is based on the experiments ($2^{0.386d}$ for dim ~90). For large sieving dimensions, this cost should be $2^{0.292d}$ if we use the state-of-art sieving algorithm [3]. So for NIST candidates, the speed-up ratio of the jumping technique will be larger than our low dimensional case. A precise estimation of the impact of these techniques on the NIST candidates may be a direction of future work.

References

1. Albrecht, M.R., Ducas, L., Herold, G., Kirshanova, E., Postlethwaite, E.W., Stevens, M.: The general Sieve Kernel and new records in lattice reduction. In: Ishai, Y., Rijmen, V. (eds.) EUROCRYPT 2019. LNCS, vol. 11477, pp. 717–746. Springer, Cham (2019). https://doi.org/10.1007/978-3-030-17656-3_25
2. Aono, Y., Wang, Y., Hayashi, T., Takagi, T.: Improved progressive BKZ algorithms and their precise cost estimation by sharp simulator. In: Fischlin, M., Coron, J.-S. (eds.) EUROCRYPT 2016. LNCS, vol. 9665, pp. 789–819. Springer, Heidelberg (2016). https://doi.org/10.1007/978-3-662-49890-3_30
3. Becker, A., Ducas, L., Gama, N., Laarhoven, T.: New directions in nearest neighbor searching with applications to lattice sieving. In: Proceedings of the Twenty-Seventh Annual ACM-SIAM Symposium on Discrete Algorithms, pp. 10–24. SODA 2016 (2016). https://doi.org/10.1137/1.9781611974331.ch2
4. Chen, Y., Nguyen, P.Q.: BKZ 2.0: better lattice security estimates. In: Lee, D.H., Wang, X. (eds.) ASIACRYPT 2011. LNCS, vol. 7073, pp. 1–20. Springer, Heidelberg (2011). https://doi.org/10.1007/978-3-642-25385-0_1
5. Coppersmith, D., Shamir, A.: Lattice attacks on NTRU. In: Fumy, W. (ed.) EUROCRYPT 1997. LNCS, vol. 1233, pp. 52–61. Springer, Heidelberg (1997). https://doi.org/10.1007/3-540-69053-0_5
6. Fincke, U., Pohst, M.E.: Improved methods for calculating vectors of short length in a lattice. Math. Comput. **44**, 463–471 (1985). https://doi.org/10.1090/S0025-5718-1985-0777278-8
7. Gama, N., Nguyen, P.Q., Regev, O.: Lattice enumeration using extreme pruning. In: Gilbert, H. (ed.) EUROCRYPT 2010. LNCS, vol. 6110, pp. 257–278. Springer, Heidelberg (2010). https://doi.org/10.1007/978-3-642-13190-5_13
8. Gama, N., Nguyen, P.Q.: Predicting lattice reduction. In: Smart, N. (ed.) EUROCRYPT 2008. LNCS, vol. 4965, pp. 31–51. Springer, Heidelberg (2008). https://doi.org/10.1007/978-3-540-78967-3_3
9. Gentry, C., Peikert, C., Vaikuntanathan, V.: Trapdoors for hard lattices and new cryptographic constructions. In: Proceedings of the Fortieth Annual ACM Symposium on Theory of Computing. STOC 2008, Association for Computing Machinery (2008). https://doi.org/10.1145/1374376.1374407
10. Hanrot, G., Pujol, X., Stehlé, D.: Algorithms for the shortest and closest lattice vector problems. In: Chee, Y.M., et al. (eds.) IWCC 2011. LNCS, vol. 6639, pp. 159–190. Springer, Heidelberg (2011). https://doi.org/10.1007/978-3-642-20901-7_10

11. Hanrot, G., Pujol, X., Stehlé, D.: Analyzing blockwise lattice algorithms using dynamical systems. In: Rogaway, P. (ed.) CRYPTO 2011. LNCS, vol. 6841, pp. 447–464. Springer, Heidelberg (2011). https://doi.org/10.1007/978-3-642-22792-9_25
12. Haque, M.M., Rahman, M.O.: Analyzing progressive-BKZ lattice reduction algorithm. Int. J. Comput. Netw. Inf. Secur. **11**, 40–46 (2019)
13. Hoffstein, J., Pipher, J., Silverman, J.H.: NTRU: a ring-based public key cryptosystem. In: Buhler, J.P. (ed.) ANTS 1998. LNCS, vol. 1423, pp. 267–288. Springer, Heidelberg (1998). https://doi.org/10.1007/BFb0054868
14. Kannan, R.: Improved algorithms for integer programming and related lattice problems. Proceedings of the Fifteenth Annual ACM Symposium on Theory of Computing (1983). https://doi.org/10.1145/800061.808749
15. Lenstra, A.K., Lenstra, H.W., Lovász, L.M.: Factoring polynomials with rational coefficients. Math. Ann. **261**, 515–534 (1982)
16. Lindner, R., Peikert, C.: Better key sizes (and attacks) for LWE-based encryption. In: Kiayias, A. (ed.) CT-RSA 2011. LNCS, vol. 6558, pp. 319–339. Springer, Heidelberg (2011). https://doi.org/10.1007/978-3-642-19074-2_21
17. Micciancio, D., Regev, O.: Lattice-based Cryptography. In: In: Bernstein, D.J., Buchmann, J., Dahmen, E. (eds.) Post-Quantum Cryptography, pp. 147–191. Springer, Heidelberg (2009). https://doi.org/10.1007/978-3-540-88702-7_5
18. Micciancio, D., Voulgaris, P.: A deterministic single exponential time algorithm for most lattice problems based on voronoi cell computations. Electron. Colloquium Comput. Complex. **17**, 14 (2010). https://doi.org/10.1145/1806689.1806739
19. Nguyen, P.: Cryptanalysis of the Goldreich-Goldwasser-Halevi cryptosystem from crypto 1997. In: Wiener, M. (ed.) CRYPTO 1999. LNCS, vol. 1666, pp. 288–304. Springer, Heidelberg (1999). https://doi.org/10.1007/3-540-48405-1_18
20. Vaudenay, S. (ed.): EUROCRYPT 2006. LNCS, vol. 4004. Springer, Heidelberg (2006). https://doi.org/10.1007/11761679
21. Nguyen, P.Q., Valle, B.: The LLL Algorithm - Survey and Applications. In: Information Security and Cryptography. Springer, Heidelberg (2010). https://doi.org/10.1007/978-3-642-02295-1
22. Nguyen, P.Q., Vidick, T.: Sieve algorithms for the shortest vector problem are practical. J. Math. Cryptol. **2**(2), 181–207 (2008)
23. Plantard, T., Schneider, M.: Creating a challenge for ideal lattices. Cryptology ePrint Archive, Report 2013/039 (2013). https://ia.cr/2013/039
24. Pohst, M.E.: On the computation of lattice vectors of minimal length, successive minima and reduced bases with applications. SIGSAM Bull. **15**, 37–44 (1981). https://doi.org/10.1145/1089242.1089247
25. Regev, O.: On lattices, learning with errors, random linear codes, and cryptography. In: Proceedings of the Thirty-Seventh Annual ACM Symposium on Theory of Computing. STOC 2005 (2005). https://doi.org/10.1145/1060590.1060603
26. Schneider, M., Gama, N.: Darmstadt SVP challenges (2010)
27. Schnorr, C.P., Hörner, H.H.: Attacking the Chor-rivest cryptosystem by improved lattice reduction. In: Guillou, L.C., Quisquater, J.-J. (eds.) EUROCRYPT 1995. LNCS, vol. 921, pp. 1–12. Springer, Heidelberg (1995). https://doi.org/10.1007/3-540-49264-X_1
28. Schnorr, C.P.: A hierarchy of polynomial time lattice basis reduction algorithms. Theor. Comput. Sci. **53**, 201–224 (1987)
29. Schnorr, C.P.: Lattice reduction by random sampling and birthday methods. In: Alt, H., Habib, M. (eds.) STACS 2003. LNCS, vol. 2607, pp. 145–156. Springer, Heidelberg (2003). https://doi.org/10.1007/3-540-36494-3_14

30. Schnorr, C.P.: Accelerated slide- and LLL-reduction. Electron. Colloquium Comput. Complex. TR11 (2011)
31. Schnorr, C.P., Euchner, M.: Lattice basis reduction: improved practical algorithms and solving subset sum problems. Math. Program. **66**, 181–199 (1994)
32. Yamaguchi, J., Yasuda, M.: Explicit formula for Gram-Schmidt vectors in LLL with deep insertions and its applications. In: Kaczorowski, J., Pieprzyk, J., Pomykała, J. (eds.) NuTMiC 2017. LNCS, vol. 10737, pp. 142–160. Springer, Cham (2018). https://doi.org/10.1007/978-3-319-76620-1_9

Enhancing Ransomware Classification with Multi-stage Feature Selection and Data Imbalance Correction

Faithful Chiagoziem Onwuegbuche[1,2](✉), Anca Delia Jurcut[2], and Liliana Pasquale[2]

[1] SFI Center for Research Training in Machine Learning (ML-Labs), Dublin, Ireland
faithful.chiagoziemonwuegb@ucdconnect.ie

[2] School of Computing, University College Dublin, Dublin, Ireland
{anca.jurcut,liliana.pasquale}@ucd.ie

Abstract. Ransomware is a critical security concern, and developing applications for ransomware detection is paramount. Machine learning models are helpful in detecting and classifying ransomware. However, the high dimensionality of ransomware datasets divided into various feature groups such as API calls, Directory, and Registry logs has made it difficult for researchers to create effective machine learning models. Class imbalance also leads to poor results when classifying ransomware families. To tackle these challenges, in this paper we propose a three-stage feature selection method that effectively reduces the dimensionality of the data and considers the varying importance of the different feature groups in the classification of ransomware families. We also applied cost-sensitive learning and re-sampling of the training data using SMOTE to address data imbalance. We applied these techniques to the Elderan ransomware dataset. Our results show that the proposed feature selection method significantly improves the detection of ransomware compared to other state-of-art studies using the same dataset. Furthermore, the data balancing techniques (cost-sensitive learning and SMOTE) were effective in the multi-class classification of ransomware.

Keywords: Ransomware detection · Malware classification · Machine learning · Feature analysis

1 Introduction

Ransomware has rapidly become a serious threat to today's society and has affected several critical sectors, including healthcare, critical infrastructure, education and finance. For example, in May 2017 the UK National Health System (NHS) was attacked by the WannaCry ransomware, resulting in the loss of patients' records, delays in non-urgent surgeries and cancellation of 19,000 patient appointments [13,14]. The rise of ransomware can be attributed to the financial gains accrued using cryptocurrencies as a payment mechanism [17],

S. Dolev et al. (Eds.): CSCML 2023, LNCS 13914, pp. 285–295, 2023.
https://doi.org/10.1007/978-3-031-34671-2_20

the COVID-19 working from home paradigm resulting in some workers adopting poor security practices [7], and the popularity of ransomware-as-a-service, which allows novice attackers to launch ransomware with pre-built software and platforms [20]. Ransomware is a type of malware developed to facilitate different malicious activities such as blocking access to a computer system, encrypting files, exfiltrating files or even damaging files unless a ransom is paid [7,21]. Traditionally, ransomware detection methods rely on signatures, which can be easily evaded by generating new variants and using obfuscation techniques [24,25]. To overcome these limitations, machine learning is now used for ransomware detection and classification. However, there are issues in the current literature regarding this approach that this study aims to address.

First, the high dimensionality of ransomware datasets, obtained through dynamic analysis, poses a major challenge in developing effective machine learning models for ransomware classification [1,22,25]. To avoid the curse of dimensionality and reduce required computational resources, researchers have proposed different feature selection methods, generally classified into four categories: Filter, Wrapper, Embedded, and Hybrid [5]. However, finding the best feature selection method or combination of methods for a specific task is still an open problem [8,10]. Moreover, most feature selection methods used in the literature on ransomware classification ignore the varying importance of different feature groups, which can lead to suboptimal results and limit the effectiveness of the models developed using these methods [1]. In this study, we propose a three-stage feature selection method that significantly improves the classification of ransomware and considers the different feature groups that are present in the data.

Second, multi-class ransomware classification poses a class imbalance problem, that leads to poor performance when minority class examples are classified [9,11,19,27]. Most studies (e.g., [16,22,25]) on ransomware detection and classification using machine learning have not considered the class imbalance problem. Those that considered the multi-class classification (e.g., [1]) did not consider the effect of different data imbalance correction techniques in the multi-class and binary classification of ransomware. To address the data imbalance problem in the classification of ransomware, we adopt two approaches: resampling the training dataset using Synthetic Minority Oversampling Technique (SMOTE) and cost-sensitive machine learning methods. The SMOTE algorithm generates synthetic data for the minority class(es) based on their feature space similarities using nearest neighbours [19], while cost-sensitive learning methods modify machine learning models to bias toward classes with fewer examples in the training dataset.

In this paper, we provide the following contributions:

- We propose a three-stage feature selection method that significantly improves the classification of ransomware, reduces the dimensionality of the dataset and considers the different feature groups involved in the data.
- We adopt two approaches to address class imbalance in the classification of ransomware: resampling the training dataset using Synthetic Minority Oversampling Technique (SMOTE) and cost-sensitive machine learning methods.

We compared the performance of various machine learning models, i.e. eXtreme Gradient Boosting (XGBoost), Logistic Regression (LR), Random Forest (RF), Decision Trees (DT) and Support Vector Machine (SVM), in detecting and classifying ransomware using the Elderan ransomware dataset [25]. We used balanced accuracy as the primary evaluation metric. Our evaluation results show that the proposed feature selection method improves ransomware detection significantly compared to previous studies, and that cost-sensitive learning and SMOTE improve the ability to classify different ransomware families.

The rest of the paper is organized as follows. Section 2 discusses related work. Section 3 presents the research approach used in this paper. Section 4 discusses the experimental results obtained for binary and multi-class classification. Finally, Sect. 5 concludes the paper.

2 Related Work

Previous work has used machine learning to detect and classify ransomware using the Elderan dataset. Abbasi et al. [1] proposed a two-stage feature selection method for machine learning-based ransomware detection. In the first stage, an equal number of top-ranked features is selected for each feature group using Mutual Information. In the second stage, swarm particle optimization removes the redundant features identified during the first stage. Using several machine learning models and balanced accuracy as the metric for evaluation, Abbasi et al. observed that their proposed feature selection method performs significantly better for multi-class classification but showed comparable performance for binary classification when compared with the feature selection method used by Sgandurra et al. [25]. Moreira et al. [22] utilised six machine learning models such as Naive Bayes (NB), K-Nearest Neighbors (KNN), Logistic Regression (LR), Random Forest (RF), Stochastic Gradient Descent (SGD), and Support Vector Machine (SVM) to analyse ransomware attacks. Using balanced accuracy as the metric for evaluating the model performance, they found out that the random forest model outperformed other models in detecting ransomware. Also, using their newly developed metric for feature group relevance, they concluded that Application Programming Interface (API) calls are the most relevant feature group to distinguish ransomware from goodware. Khan et al. [16] proposed a machine learning-based digital DNA sequencing engine for detecting and classifying ransomware called DNAact-Ran. In the preprocessing stage, DNAact-Ran used Multi-Objective Grey Wolf Optimization (MOGWO) and Binary Cuckoo Search (BCS) algorithms to select key features and then applied design constraints of DNA sequence and k-mer frequency vector to the selected features to generate the digital DNA sequence. Differently from previous work, we propose a three-stage feature selection method that significantly improves ransomware classification.

However, most studies did not consider the data imbalance problem that arises mainly in the multi-class classification of ransomware families. Data imbalance can result in a learning model's poor prediction of the minority class samples [27]. Previous research has focused on correction of imbalanced data of

Android ransomware [2] or general malware detection [4,15,23]. The approaches proposed in the literature can be divided into two categories: resampling techniques and cost sensitive learning [11]. Resampling techniques involve modifying the distribution of samples in the dataset by oversampling the minority class or undersampling the majority class or both [23]. Cost-sensitive learning aims to adjust the classification threshold to account for the cost of misclassification for each class [30]. The approach to address class imbalance may also depend on the characteristics of the data.

Thus, the techniques used to address data imbalance for malware detection may not be effective for ransomware detection because the characteristics of ransomware samples may be different from those of other types of malware samples. We use cost-sensitive learning and SMOTE to deal with the data imbalance problem.

Sgandurra et al. [25] applied the Regularized Logistic Regression to analyse and classify ransomware. To achieve this aim, they developed a ransomware dataset called Elderan by performing dynamic analysis of ransomware and goodware samples in the Cuckoo sandbox (a controlled environment for safely executing potentially malicious software). Sgandurra et al. applied Mutual Information to select the top 400 features out of 30,967 features contained in the dataset. This feature selection method did not consider the varying importance of the different feature groups to improve the classification of ransomware families.

3 Research Approach

We used the Elderan ransomware dataset [25] that was created using dynamic analysis on ransomware and goodware samples. The lack of a publicly available dataset for ransomware classification is a known problem [3,7,12]. Although the Elderan dataset is not large, it is one of the most comprehensive ransomware datasets publicly available.

We adopted two feature selection methods in this study: (1) mutual information as used by [25] and (2) our proposed multi-stage feature selection method. We use these methods to train machine learning models (XGBoost, LR, RF, DT and SVM) for both the binary and multi-class classification of ransomware. The binary classification problem discriminates ransomware samples from goodware samples, while the multi-class classification problem aims to distinguish between the different ransomware families. After that, we used SMOTE to separately re-sample the training data for the binary and multi-class classification. Using the re-sampled data, we adopted the same machine learning models for the binary and multi-class ransomware classification. Similarly, we also used cost-sensitive machine learning models for the binary and multi-class ransomware classification.

3.1 Research Questions

In this study, we aim to answer the following research questions:

1. What is the difference in performance between the proposed three-stage feature selection method and other state-of-the-art studies in the classification of ransomware?
2. Which technique for addressing the data imbalance problem is more effective in detecting and classifying ransomware?

3.2 Dataset Description

The Elderan dataset includes 1524 software samples. Out of those samples, 942 were classified as goodware, and 582 were classified as ransomware. The ransomware families included in the dataset are Citroni, CryptoLocker, Kovter, Locker, Matsnu, Pgpcoder, Reveton, TeslaCrypt, Trojan-Ransom, CryptoWall, and Kollah.

The total number of features in the dataset is 30,967, grouped into 7 feature groups namely Application Programming Interface invocations (API), Extensions of the dropped files (DROP), Registry key operations (REG), File operations (FILE), Extension of the files involved in file operations (FILES EXT), File directory operations (DIR), and Embedded strings (STR). Every feature has a value (0 or 1) representing the absence or presence of the corresponding operation. The distribution of the different feature groups is shown in Table 1.

3.3 Feature Selection Technique

Feature selection is a crucial data preprocessing step in the fields of data mining and machine learning. It aims to reduce the number of features used in the analysis, making the models simpler, more interpretable, and computationally efficient while avoiding the curse of dimensionality [18].

The study by Sgandurra et al. [25] employed Mutual Information (MI) [26] as a feature selection technique. MI quantifies the discrimination power of each feature in the classifier. Sgandurra et al. [25] showed that for the Elderan dataset, the maximum performance was achieved by selecting the top 400 features based on mutual information. However, this approach did not consider the varying importance of different feature groups as they selected just the top 400 features irrespective of the feature group they belong.

We propose a three-stage feature selection method that addresses this limitation by taking into account the varying significance of different feature groups. Our method involves splitting the data into feature groups and applying three different feature selection techniques within each group to select the relevant data.

Stage I: In the first stage, we used chi-square (CHI2) to select the top 200 features from each feature group, resulting in 1400 features from the seven groups. We considered 200 features because the smallest feature group has 233 features. CHI2 is a statistical hypothesis test that compares observed and expected frequencies of a categorical variable. The test assumes observed frequencies follow a chi-squared distribution and calculates a test statistic to determine the significance of differences between observed and expected frequencies [29].

Stage II: In stage II, we used the Duplicated Features (DUF) method to select features from stage I. Duplicate features are redundant and provide no extra information. They can also cause problems with machine learning algorithms and lead to overfitting. It's advised to remove them before creating a model to prevent clutter and make analysis easier. Keeping duplicate features causes multicollinearity.

Stage III: In stage III, we applied Constant Features (COF) filter feature selection method to the remaining features from stage II. Constant features are those that have only one value for all entries in the dataset. These constant features can hinder the performance of a machine learning model and, thus, should be removed.

Table 1. Features remaining after applying each stage of the proposed feature selection method

Feature Groups	API	DROP	REG	FILES	FILES_EXT	DIR	STR	Total
All Features	233	346	6622	4141	935	2424	16267	30968
Stage I: CHI2	200	200	200	200	200	200	200	1400
Stage II: DUF	192	126	116	90	131	44	144	843
Stage III: COF	151	20	112	27	59	31	66	**466**

3.4 Machine Learning Models

The machine learning models that we employed to detect and classify ransomware were logistic regression, random forest, eXtreme Gradient Boosting (XGBoost), Decision Trees, and Support Vector Machine (SVM). These models have been shown to be effective in the detection of ransomware [1,22].

3.5 Data Imbalance Techniques

Machine learning models typically assume balanced classes in datasets, so highly imbalanced data can cause classifier performance to decrease [27]. The machine learning community has addressed data imbalance in two ways: by resampling the dataset (through oversampling the minority class, undersampling the majority class, or a combination of both); or by cost-sensitive learning (assigning different costs to training examples) [11].

In this study, we used two methods to address data imbalance. The first is SMOTE [11], which generates synthetic examples of minority classes to balance the number of samples with majority classes. The second is cost-sensitive machine learning, which adjusts models to prioritize classes with fewer examples in the training dataset [9]. Misclassification costs are assigned to instances to minimize the total misclassification cost instead of optimizing accuracy. A cost matrix assigns a cost to each cell in the confusion matrix, with weights based on the inverse proportions of class frequencies in the input data [19]. These methods were chosen for their success in correcting imbalanced data [6,28].

3.6 Evaluation Metrics

To determine the best machine learning model for detecting and classifying ransomware, we evaluated their performance using balanced accuracy as a more suitable metric than accuracy for imbalanced datasets. Balanced accuracy is the arithmetic mean of specificity and sensitivity. Sensitivity measures the proportion of real positives that are correctly predicted while specificity measures the proportion of correctly identified negatives. We also used other evaluation metrics such as precision, recall, F1, and Area Under the Receiver Operating Characteristic Curve (ROC-AUC).

3.7 Setup

The clean dataset used required no preprocessing. The data was divided into two sets - 80% for training and 20% for independent testing. A stratified train-test split was used to maintain class proportions, which is more effective for imbalanced datasets. Repeated stratified k-fold cross-validation was employed with a split of 3 and a repeat of 4, as it is better for imbalanced datasets [9]. Gridsearch was used to tune hyperparameters, and the model's performance was evaluated on the 20% independent test set.

4 Experimental Results and Discussion

Tables 2 and 3 summarise binary and multi-class classification results for ransomware, comparing the balanced accuracy of different machine learning models and data balancing techniques. We report the results for both the cross-validation test (CV test) and the independent test (ID test).

The proposed three-stage feature selection method outperformed Sgandurra et al.'s method significantly (p-value < 0.05), with an average improvement of 10% for binary and 21.79% for multi-class classification. For binary classification, XGBoost with cost-sensitive learning and SMOTE performed best, while for multi-class classification, the random forest model using cost-sensitive learning achieved the highest balanced accuracy of 61.94% (Fig. 1). This study achieved better performance than other state-of-the-art studies as shown in Fig. 2, especially in multi-class classification, which addressed the severe class imbalance problem using cost-sensitive learning and SMOTE. The comparison for multi-class was only done with the work of [1], as it was the only study among the compared studies that specifically addressed the multi-class classification problem.

Table 2. Balanced Accuracy for 4 Repeats Stratified 3-fold Cross-validation with Standard Deviation and Independent Test for binary classification

Type	Classifiers	Sgandurra [25] FS Method		Our FS Method	
		CV Test	ID Test	CV Test	ID Test
Standard Approach	DT	0.8627 ± 0.0153	0.8838	0.9550 ± 0.0120	0.9765
	LR	0.8709 ± 0.0136	0.8648	0.9666 ± 0.0099	0.9765
	RF	0.8879 ± 0.0115	0.9001	0.9737 ± 0.0113	0.9844
	SVM	0.8848 ± 0.0107	0.9126	0.9650 ± 0.0073	0.9705
	XGBoost	0.8703 ± 0.0116	0.8841	0.9716 ± 0.0097	0.9877
	Baseline	-	0.8579	-	0.9678
SMOTE	DT	0.8796 ± 0.0144	0.8859	0.9612 ± 0.0083	0.9652
	LR	0.8805 ± 0.0160	0.8808	0.9741 ± 0.0072	0.9722
	RF	0.8992 ± 0.0169	0.9096	0.9753 ± 0.0080	0.9791
	SVM	0.8996 ± 0.0151	0.8948	0.9683 ± 0.0059	0.9705
	XGBoost	0.8893 ± 0.0146	0.8818	0.9761 ± 0.0072	0.9878
Cost Sensitive Learning	DT	0.8605 ± 0.0142	0.8686	0.9685 ± 0.0096	0.9722
	LR	0.8754 ± 0.0138	0.8739	0.9685 ± 0.0096	0.9722
	RF	0.8846 ± 0.0115	0.9080	0.9742 ± 0.0092	0.9818
	SVM	0.8833 ± 0.0102	0.9017	0.9668 ± 0.0139	0.9775
	XGBoost	0.8761 ± 0.0095	0.8914	0.9716 ± 0.0097	0.9878

Table 3. Balanced Accuracy for 4 Repeats Stratified 3-fold Cross-validation with Standard Deviation and Independent Test for multi-class classification

Type	Classifiers	Sgandurra [25] FS Method		Our FS Method	
		CV Test	ID Test	CV Test	ID Test
Standard Approach	DT	0.3707 ± 0.0198	0.3193	0.4631 ± 0.0364	0.5149
	LR	0.3641 ± 0.0254	0.2833	0.4984 ± 0.0198	0.5697
	RF	0.3842 ± 0.0246	0.3237	0.4999 ± 0.0363	0.5909
	SVM	0.3703 ± 0.0179	0.3237	0.4890 ± 0.0207	0.5789
	XGBoost	0.3750 ± 0.0145	0.3334	0.4950 ± 0.0193	0.6086
	Baseline	-	0.2833	-	0.5828
SMOTE	DT	0.6817 ± 0.0093	0.3887	0.8979 ± 0.0058	0.5677
	LR	0.6399 ± 0.0081	0.3514	0.8966 ± 0.0053	0.5675
	RF	0.6866 ± 0.0093	0.4097	0.9069 ± 0.0046	0.6112
	SVM	0.6858 ± 0.0088	0.3882	0.9037 ± 0.0044	0.5649
	XGBoost	0.6847 ± 0.0086	0.4088	0.9060 ± 0.0048	0.5846
Cost Sensitive Learning	DT	0.4107 ± 0.0236	0.4057	0.4543 ± 0.0390	0.5533
	LR	0.4007 ± 0.0204	0.3430	0.5167 ± 0.0387	0.5901
	RF	0.4218 ± 0.0251	0.4084	0.5013 ± 0.0469	0.6194
	SVM	0.4086 ± 0.0235	0.3935	0.5003 ± 0.0333	0.6067
	XGBoost	-	0.3973	-	0.5995

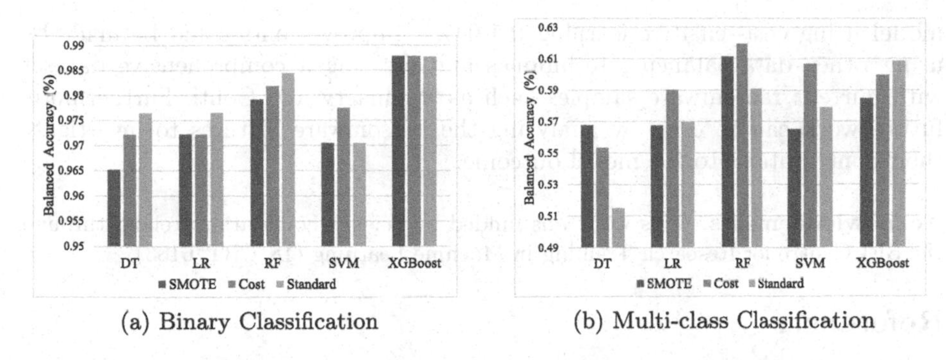

(a) Binary Classification (b) Multi-class Classification

Fig. 1. Performance across the 3 approaches using the different algorithms using our FS method

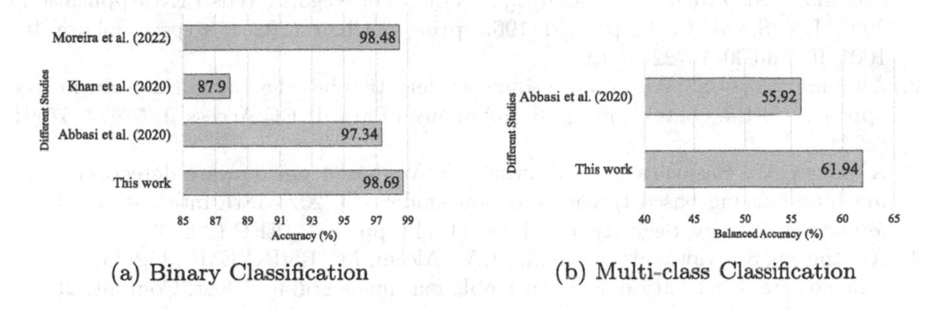

(a) Binary Classification (b) Multi-class Classification

Fig. 2. Comparison of the highest classification accuracy of ransomware in [1,16,22] and the present study

4.1 Threats to Validity

The study's internal validity relies on the choice of feature selection, data balancing, and machine learning models, but other techniques could produce different outcomes. The results are limited to using SMOTE and cost-sensitive learning, and the dataset used for analysis only includes older ransomware families. Construct validity was upheld by reviewing and testing the codes multiple times, with care taken when creating and evaluating synthetic samples.

5 Conclusion

In this paper, we propose a three-stage feature selection method to reduce data dimensionality while considering the composition of feature groups, improving ransomware detection compared to previous studies. We address data imbalance using SMOTE and cost-sensitive machine learning for binary and multi-class detection and classification. Our method outperforms state-of-the-art results by 6.02% in multi-class classification. The best binary classifier is XGBoost with cost-sensitive learning and SMOTE (98.78%), followed by XGBoost with the standard approach (98.77%). The best multi-class classifier is the random forest model using cost-sensitive learning (61.94%). Improvements could be made by using other data balancing techniques and creating a comprehensive dataset with current ransomware samples such as Wannacry and Conti. Furthermore, future work can be done by analysing the ransomware features to investigate their contributions to the model outcome.

Acknowledgements. This work was funded by Science Foundation Ireland through the SFI Centre for Research Training in Machine Learning (18/CRT/6183).

References

1. Abbasi, M.S., Al-Sahaf, H., Welch, I.: Particle swarm optimization: a wrapper-based feature selection method for ransomware detection and classification. In: Castillo, P.A., Jiménez Laredo, J.L., Fernández de Vega, F. (eds.) EvoApplications 2020. LNCS, vol. 12104, pp. 181–196. Springer, Cham (2020). https://doi.org/10.1007/978-3-030-43722-0_12
2. Almomani, I., et al.: Android ransomware detection based on a hybrid evolutionary approach in the context of highly imbalanced data. IEEE Access **9**, 57674–57691 (2021)
3. Almousa, M., Basavaraju, S., Anwar, M.: Api-based ransomware detection using machine learning-based threat detection models. In: 2021 18th International Conference on Privacy, Security and Trust (PST), pp. 1–7. IEEE (2021)
4. Aurangzeb, S., Anwar, H., Naeem, M.A., Aleem, M.: BigRC-EML: big-data based ransomware classification using ensemble machine learning. Clust. Comput. **25**(5), 3405–3422 (2022)
5. Avila, R., Khoury, R., Pere, C., Khanmohammadi, K.: Employing feature selection to improve the performance of intrusion detection systems. In: Aïmeur, E., Laurent, M., Yaich, R., Dupont, B., Garcia-Alfaro, J. (eds.) FPS 2021. LNCS, vol. 13291, pp. 93–112. Springer, Cham (2022). https://doi.org/10.1007/978-3-031-08147-7_7
6. Batista, G.E., Prati, R.C., Monard, M.C.: A study of the behavior of several methods for balancing machine learning training data. ACM SIGKDD Explor. Newslett. **6**(1), 20–29 (2004)
7. Beaman, C., Barkworth, A., Akande, T.D., Hakak, S., Khan, M.K.: Ransomware: recent advances, analysis, challenges and future research directions. Comput. Secur. **111**, 102490 (2021)
8. Bolón-Canedo, V., Alonso-Betanzos, A.: Ensembles for feature selection: a review and future trends. Inf. Fusion **52**, 1–12 (2019)
9. Brownlee, J.: Imbalanced classification with Python: Better Metrics, Balance Skewed Classes, Cost-sensitive Learning. Machine Learning Mastery (2020)
10. Cai, J., Luo, J., Wang, S., Yang, S.: Feature selection in machine learning: a new perspective. Neurocomputing **300**, 70–79 (2018)
11. Chawla, N.V., Bowyer, K.W., Hall, L.O., Kegelmeyer, W.P.: Smote: synthetic minority over-sampling technique. J. Artif. Intell. Res. **16**, 321–357 (2002)
12. Chen, Q., Bridges, R.A.: Automated behavioral analysis of malware: a case study of wannacry ransomware. In: 2017 16th IEEE International Conference on Machine Learning and Applications (ICMLA), pp. 454–460. IEEE (2017)
13. Collier, R.: NHS ransomware attack spreads worldwide (2017)
14. Cyber Security Policy: Securing cyber resilience in health and care: October 2018 progress update (2018). https://www.gov.uk/government/publications/securing-cyber-resilience-in-health-and-care-october-2018-update

15. Goyal, M., Kumar, R.: Machine learning for malware detection on balanced and imbalanced datasets. In: 2020 International Conference on Decision Aid Sciences and Application (DASA), pp. 867–871. IEEE (2020)
16. Khan, F., Ncube, C., Ramasamy, L.K., Kadry, S., Nam, Y.: A digital DNA sequencing engine for ransomware detection using machine learning. IEEE Access **8**, 119710–119719 (2020)
17. Kshetri, N., Voas, J.: Do crypto-currencies fuel ransomware? IT Prof. **19**(5), 11–15 (2017)
18. Li, J., Cheng, K., Wang, S., Morstatter, F., Trevino, R.P., Tang, J., Liu, H.: Feature selection: a data perspective. ACM Comput. Surv. (CSUR) **50**(6), 1–45 (2017)
19. Ma, Y., He, H.: Imbalanced Learning: Foundations, Algorithms, and Applications (2013)
20. McIntosh, T., Kayes, A., Chen, Y.P.P., Ng, A., Watters, P.: Ransomware mitigation in the modern era: a comprehensive review, research challenges, and future directions. ACM Comput. Surv. (CSUR) **54**(9), 1–36 (2021)
21. Meland, P.H., Bayoumy, Y.F.F., Sindre, G.: The ransomware-as-a-service economy within the darknet. Comput. Secur. **92**, 101762 (2020)
22. Moreira, C.C., de Sales Jr, C.D.S., Moreira, D.C.: Understanding ransomware actions through behavioral feature analysis. J. Commun. Inf. Syst. **37**(1), 61–76 (2022)
23. Pang, Y., Peng, L., Chen, Z., Yang, B., Zhang, H.: Imbalanced learning based on adaptive weighting and gaussian function synthesizing with an application on android malware detection. Inf. Sci. **484**, 95–112 (2019)
24. Rieck, K., Trinius, P., Willems, C., Holz, T.: Automatic analysis of malware behavior using machine learning. J. Comput. Secur. **19**(4), 639–668 (2011)
25. Sgandurra, D., Muñoz-González, L., Mohsen, R., Lupu, E.C.: Automated dynamic analysis of ransomware: Benefits, limitations and use for detection. arXiv preprint arXiv:1609.03020 (2016)
26. Shannon, C.E.: A mathematical theory of communication. Bell Syst. Tech. J **27**(3), 379–423 (1948)
27. Thabtah, F., Hammoud, S., Kamalov, F., Gonsalves, A.: Data imbalance in classification: experimental evaluation. Inf. Sci. **513**, 429–441 (2020)
28. Thai-Nghe, N., Gantner, Z., Schmidt-Thieme, L.: Cost-sensitive learning methods for imbalanced data. In: The 2010 International Joint Conference on Neural Networks (IJCNN), pp. 1–8. IEEE (2010)
29. Urdan, T.C.: Statistics in Plain English. Routledge, Abingdon (2011)
30. Wu, D., Guo, P., Wang, P.: Malware detection based on cascading XGboost and cost sensitive. In: 2020 International Conference on Computer Communication and Network Security (CCNS), pp. 201–205. IEEE (2020)

A Desynchronization-Based Countermeasure Against Side-Channel Analysis of Neural Networks

Jakub Breier[1,2], Dirmanto Jap[3(✉)], Xiaolu Hou[4], and Shivam Bhasin[3]

[1] Silicon Austria Labs, TU-Graz SAL DES Lab, Graz, Austria
jbreier@jbreier.com
[2] Graz University of Technology, Graz, Austria
[3] Nanyang Technological University, Singapore, Singapore
{djap,sbhasin}@ntu.edu.sg
[4] Slovak University of Technology, Bratislava, Slovakia

Abstract. Model extraction attacks have been widely applied, which can normally be used to recover confidential parameters of neural networks for multiple layers. Recently, side-channel analysis of neural networks allows parameter extraction even for networks with several multiple deep layers with high effectiveness. It is therefore of interest to implement a certain level of protection against these attacks.

In this paper, we propose a desynchronization-based countermeasure that makes the timing analysis of activation functions harder. We analyze the timing properties of several activation functions and design the desynchronization in a way that the dependency on the input and the activation type is hidden. We experimentally verify the effectiveness of the countermeasure on a 32-bit ARM Cortex-M4 microcontroller and employ a t-test to show the side-channel information leakage. The overhead ultimately depends on the number of neurons in the fully-connected layer, for example, in the case of 4096 neurons in VGG-19, the overheads are between 2.8% and 11%.

Keywords: Deep learning · neural networks · side-channel attacks · countermeasures

1 Introduction

Current deep learning models grow to millions of parameters and are being widely deployed as a service. As the training of such networks requires large amounts of data and computing time, the organizations that created the models tend to keep them proprietary. Another motivation to keep the models confidential is to reduce the success rate of adversarial attacks that are more powerful in a white-box setting, i.e., when the attacker knows the model parameters. This motivated the development of model extraction attacks on neural networks that either try to mimic the model behavior or extract the parameter values [1].

S. Dolev et al. (Eds.): CSCML 2023, LNCS 13914, pp. 296–306, 2023.
https://doi.org/10.1007/978-3-031-34671-2_21

The "standard" model extraction attacks try to get the proprietary information on the model by using a set of well-designed queries. This, however, normally does not allow the full extraction of model parameters (also called *exact extraction*), but rather a weaker type of extraction, which can be *functionally equivalent extraction*, *fidelity extraction*, or *task accuracy extraction* [2]. On the other hand, with the help of hardware attack vectors, which can be considered especially in the case of embedded machine learning models [3], it is possible to extract the exact values of the parameters, either with side-channel analysis [4,5] or with fault injection attacks [6].

In [4], it was shown that the information about the used activation function can be observed by measuring the time of the function computation. The timing behavior is dependent on the type of the function and the input data and there are clear patterns forming after a certain range of inputs was fed to the activation function. In this paper, we aim at removing those patterns to make the timing analysis useless for the attacker. We utilize a desynchronization-based countermeasure that adds random delay to the function computation so that no matter what the activation type and the input data, the timing measurement gives random outputs. We experimentally show that our method works by using a 32-bit ARM Cortex-M microcontroller as a device under test.

2 Background

2.1 Side-Channel Analysis

In [7], it has been shown that even though cryptographic algorithms are proven to be theoretically secure, their physical implementation can leak information regarding confidential data, such as a secret key. This is commonly referred to as Side-Channel Analysis (SCA). An adversary could typically exploit different means of physical leakages, such as timing, power, electromagnetic (EM) emanation, etc. Thus, by observing these leakages, an adversary could analyze and deduce the secret information being processed, by utilizing statistical methods, such as correlation, etc.

2.2 Side-Channel Countermeasures

Since the idea of SCA is that the leakage could provide information regarding the internally processed data, the goal of protection techniques is to minimize or remove these dependencies. Different countermeasures have been proposed to protect against SCA, and overall, these can be classified into *hiding* and *masking* techniques.

The hiding countermeasures aim at breaking the relationship between the processed data and the leakage, for example, using desynchronization, shuffling, etc. The general idea is to introduce noises in the measurement to make the attacks harder.

On the other hand, for masking countermeasures, the aim is to remove the relation by introducing randomness to mask the actual data being processed.

In this case, the data being processed will be masked with different random mask value for every execution and without the knowledge of the random mask, the attacker could not recover the actual intermediate value.

2.3 Hardware Attacks on Neural Networks

Recently, SCA has been applied to attacking neural network implementations. The aim is to recover the secret model, which might be sensitive, for example, if the model has been trained with confidential data or if the model is a commercial IP that has been trained and subjected to piracy. Hence, determining the layout of the network with trained weights is a desirable target for an attacker. Thus, using SCA, the attacker could reverse engineer the neural networks of interest by using some additional information that becomes available while the device under attack is operating.

Several attacks have been reported, for example, Hua *et al.* [8] demonstrated the retrieval of network parameters, such as the number of layers, size of filters, data dependencies among layers, etc. Then, Batina *et al.* [4] proposed a full reverse engineering of neural network parameters based on power/EM side-channel analysis. The proposed attack is able to recover hyperparameters i.e., activation function, pre-trained weights, and the number of hidden layers and neurons in each layer, without access to any training data. The adversary uses a combination of simple power/EM analysis, differential power/EM analysis and timing analysis to recover different parameters. Yu *et al.* [9] proposed a model extraction attack based on a combination of EM side-channel measurement and adversarial active learning to recover the Binarized Neural Networks (BNNs) architecture on popular large-scale NN hardware accelerators. The network architecture is first inferred through EM leakage, and then, the parameters are estimated with adversarial learning. As such, it can be observed that SCA has been widely adopted for the attack against neural networks.

2.4 Countermeasures for Hardware Attacks on Neural Networks

The first attempt to thwart side-channel analysis of neural networks was called MaskedNet [10]. As indicated by the name, the authors utilize the masking countermeasure to partially protect the networks. They developed novel hardware components such as masked adder trees for fully connected layers and masked ReLUs. The approach works with binarized neural networks which use binary weights and activation values.

The same authors later extended their approach to fully protect neural networks by Boolean masking [11,12]. While the latency overhead is relatively small (3.5%), the area overhead is noticeable (5.9×). The overheads were further reduced in [13] where the authors used domain-oriented masking.

A threshold implementation (TI) with 64% area and 5.5× energy overhead was presented in [14]. For generating random numbers required for the TI, Trivium stream cipher was used.

Apart from masking-related approaches, garbled circuits were used to protect not only against SCA, but to enhance the privacy of the underlying models [15].

To the best of our knowledge, the only software-based approach for protecting the neural networks against SCA uses shuffling [16]. The authors randomize the order of execution for multiplications within neurons. This, however, does not prevent timing attacks on determining the activation function.

3 Timing Analysis of Activation Functions

In this section, we first explain the experimental setup used for our measurements. Then, we provide the timing analysis without the countermeasures to better understand the timing pattern.

3.1 Experimental Setup

The experiments in this paper were done by using a NewAE ChipWhisperer tool[1]. ChipWhisperer-LITE was used to perform the power analysis and a 32-bit ARM Cortex-M4 (STM32F3) mounted on a UFO board was used as a target. The ADC sampling frequency was set to 29 MHz while the target frequency was set to 7 MHz. A trigger signal was set to high before the activation function computation and immediately set to low afterward to precisely determine the timing. The source code for the sample programs was written in C, the `<math.h>` library was used for mathematical functions. To allow for more precise measurements, compiler optimizations were disabled.

3.2 Timing Analysis Without Countermeasures

The timing behavior of activation functions was first examined in [4]. The authors measured the execution time of three functions: rectifier linear unit (ReLU), hyperbolic tangent (tanh), and sigmoid. The definitions of those functions are as follows:

$$ReLU(x) = \begin{cases} 0 & \text{if } x \leq 0 \\ x & \text{otherwise} \end{cases},$$

$$sigmoid(x) = \frac{1}{1+e^{-x}}, \quad tanh(x) = \frac{e^x - e^{-x}}{e^x + e^{-x}}.$$

The conclusion was that each of these exhibits different timing patterns that are also dependent on the function input. We repeated the experiment from [4] and used 2000 random inputs from $[-2, 2]$ for the computations of each activation function. We have obtained very similar results, see Fig. 1 and Table 1. This means that if the attacker sends a few queries and measures the time of the execution, they are able to distinguish which activation function is being used in a particular layer.

[1] https://www.newae.com/chipwhisperer.

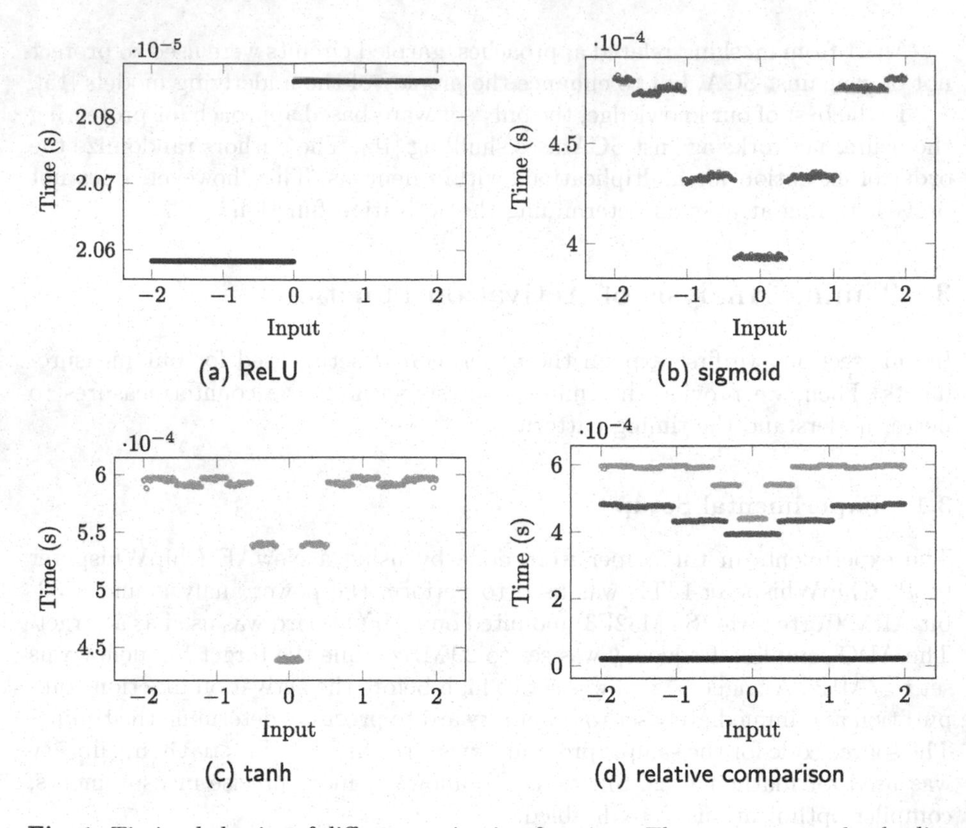

Fig. 1. Timing behavior of different activation functions. The patterns are clearly distinguishable if there is no side-channel protection in place.

4 Towards a Desynchronization-Based Countermeasure

Desynchronization-based countermeasure was first introduced for cryptographic implementations [17–19]. The main rationale of the desynchronization-based countermeasure is to remove the data dependency of the power consumption by randomizing the power consumption of the device during computation.

In this section, we will analyze the countermeasure when applied to activation function implementations. For this purpose, we would like to utilize desynchronization to randomly delay the computation of each function such that it is impossible for the attacker to distinguish them from one another by examining the timing information.

The computation times of different activation functions can be viewed as random variables depending on the inputs of the functions. The time delay caused by additional desynchronization can be considered as another random variable X. Since it is easy to generate a random variable with normal distribution in any programming language, we have decided to sample X from a normal distribution. A normal distribution is completely characterized by its mean and variance.

Table 1. Computation time (in milliseconds) for different activation functions.

Activation Function	Mean	Minimum	Maximum
Relu	0.0207	0.0206	0.0209
sigmoid	0.4485	0.3920	0.4845
tanh	0.5170	0.4375	0.5985

The mean specifies the expected average for the added delay and the variance characterizes how spread out the added delay is from the mean. To randomize the computation, we would like to choose mean and variance for X such that the resulting computation timing for all three functions follow a similar pattern.

To choose the mean, we look at the maximum and minimum possible timings for all three function computations. Figure 1 shows that the computation times for each activation function are scattered into specific "clusters" depending on the input. In particular, there are two, three, and three clusters for ReLU, sigmoid, and tanh respectively. For example, when the input is positive, the computation time for ReLu is almost 2.09×10^{-5} s. And when the input is near 0, the computation time for tanh is around 4.4×10^{-4} s. We have calculated the mean for the slowest cluster (i.e. the slowest cluster in tanh computation) which is 5.9×10^{-4} s and the mean for the fastest cluster (i.e. the faster cluster in ReLU computations) which is 2.06×10^{-5} s. We have decided to choose the mean of X to be 0.6 ms so that the very fast computations will have a chance to have comparable computation time as the slow ones.

Furthermore, to remove the distinct clusters in Fig. 1, we need to choose a variance big enough that the differences caused by the input values are negligible. Table 1 summarizes the data for computation times and we can see that the maximum is about 6×10^{-4} s and the minimum is about 2×10^{-5} s. Thus we decided to choose the variance to have the same order of magnitude[2] as $6 \times 10^{-4} - 2 \times 10^{-5}$, i.e. 0.1×10^{-4} s^2.

In summary, we add random desynchronization, whose delayed computation time is a sample from a normal distribution with mean 6×10^{-4} s and variance 0.1×10^{-4} s^2, to all three activation functions. 2000 random inputs are given to each of the functions. The resulting computation times are shown in Fig. 2 and Table 2.

In general, we propose the following steps for choosing a desynchronization-based countermeasure for protecting the computation of activation functions:

1. Collect computation time data for all three activation functions.
2. Compute the average of the fastest cluster of timings, denoted t_f, and the average of the slowest cluster denoted t_s. Let $\mu = t_f - t_s$.

[2] The order of magnitude of a number n is given by b_n such that we can write n in the form $n = a \times 10^{b_n}$, where $1/\sqrt{10} \leq a < \sqrt{10}$.

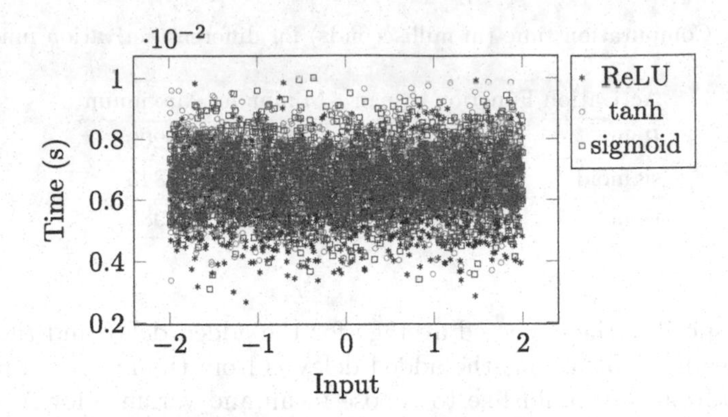

Fig. 2. Timing behavior of activation functions with an applied desynchronization-based countermeasure. The timing patterns are not distinguishable.

Table 2. Computation time (in milliseconds) for different activation functions with random desynchronization.

Activation Function	Mean	Minimum	Maximum
Relu	6.31	2.69	9.93
sigmoid	6.72	3.11	10.01
tanh	6.81	3.40	9.88

3. Compute the difference between the longest computation time and the shortest computation time, say Δt seconds. Let $d_{\Delta t}$ denote the order of magnitude of Δt. Let $\sigma^2 = 1 \times 10^{-d_{\Delta_t}}\ s^2$.
4. Add random desynchronizations, whose delayed computation time is a sample from a normal distribution with mean μ and variance σ^2, to the implementations of the activation functions.

5 Leakage Assessment of a Neuron Computation

In order to evaluate the performance, we used Test Vector Leakage Assessment (TVLA) [20]. The idea was to perform a t-test on a dataset from fixed vs random inputs, and in our case, on measured timing from the inference execution.

Here, we detail the computations for ReLU. Those for tanh and sigmoid are done similarly. First, we compute the execution time of when the neural network[3] is running the inference for a fixed input (which is chosen randomly at the beginning but fixed for the rest of experiment). We measured the timing of 5000 inference executions. Let us denote those timings by $x_1, x_2, \ldots, x_{5000}$. Then, similarly, we calculate the execution times for given random inputs, where each is

[3] We are using a similar model architecture as [4] for MNIST dataset as a proof of concept.

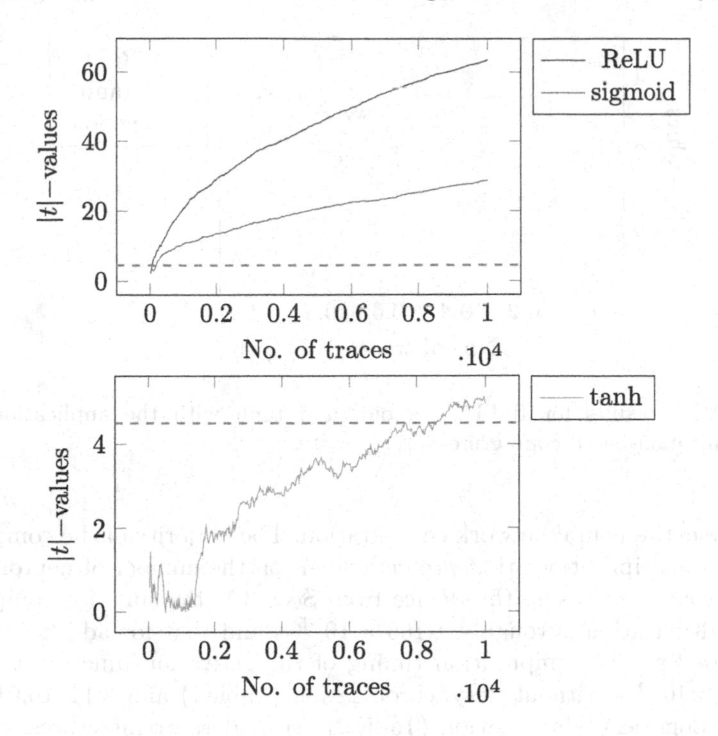

Fig. 3. TVLA results for ReLU, sigmoid, and tanh without the application of the countermeasure.

chosen randomly and different from each inference execution. Let us denote the corresponding timings by $y_1, y_2, \ldots, y_{5000}$. By the TVLA method, we compute

$$t = \frac{\overline{x} - \overline{y}}{\sqrt{\frac{\sigma_x^2}{5000} + \frac{\sigma_y^2}{5000}}},$$

where $\overline{x}$, $\overline{y}$, σ_x^2 and σ_y^2 are the mean of x_i, mean of y_i, variance of x_i and variance of y_i respectively. In case the absolute values of t, called $|t|$-values, cross the threshold of 4.5, it can be concluded there is a data-dependent leakage in the measured traces. As can be seen in Fig. 3, the TVLA test for the timing of the activation functions shows leakage, as expected. On the other hand, after the application of the proposed countermeasure, $|t|$-values stay below the threshold, as can be seen in Fig. 4.

6 Discussion

6.1 Overheads

While it might seem from the figures that the timing overhead of the countermeasure is relatively high, it is to be noted that the activation function is only a small

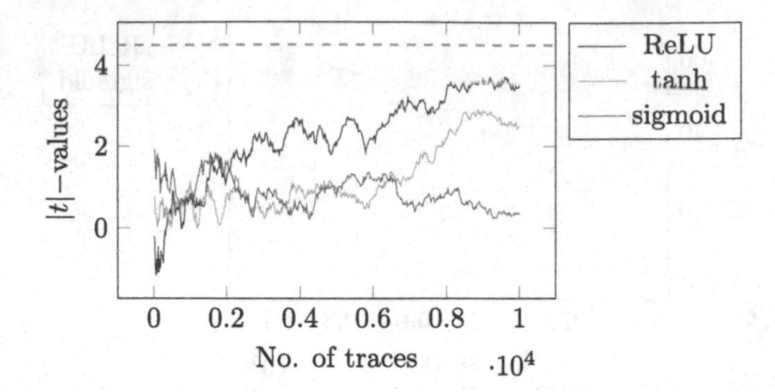

Fig. 4. TVLA results for ReLU, sigmoid, and tanh with the application of the desynchronization-based countermeasure.

part of the entire neural network computation. The majority of the computation is spent on multiplications that are dependent on the number of neurons in the layers. For example, using the device from Sect. 3.1, the time for computation of one multiplication is roughly 1.165×10^{-5} s and that for addition is about 1.124×10^{-5} s. The computation timing of the activation function is roughly 0.21–5.99 $\times 10^{-4}$ s without desynchronization (Table 1) and 3.11–10.01 $\times 10^{-3}$ s with random desynchronization (Table 2). In modern architectures, there are thousands of multiplications and additions with just one activation function computation for one neuron. VGG-19 [21], one of the popular public networks for ImageNet classification challenge, has 4096 neurons in the last hidden layer and 1000 neurons for the output layer, which amounts to 4096 multiplications and 4096 additions for each output neuron. In this case, the computation time for multiplications and additions in one neuron is about 0.09 s, and that for the whole neuron computation is 0.09002–0.0906 s without desynchronization or 0.093–0.1 s with the desynchronization countermeasure. Thus, the overhead for the computation of one neuron is between 2.6%–11%. We would also like to note that this is purely the timing for the calculations, not taking into account the memory operations – if the entire computation is considered, the overhead would be even lower. As this is a proof-of-concept and still a work in progress, we observe a positive trend and will be further investigating this in future works.

6.2 Other Activation Functions

Batina, *et al.* [4] show results on softmax activation function apart from the three functions that were analyzed in this work. The softmax function is normally used in the output layer of a neural network to transform the raw outputs (logits) into a vector of probabilities. As it is unusual to find softmax in other layers of the network, we did not consider this activation function in our work.

There are other activation functions that can be used, for example leaky ReLU, exponential linear unit (ELU), etc. We argue that the process of applying the desynchronization-based countermeasure on these functions is the same as described in Sect. 4.

7 Conclusion

SCA has been a threat for neural networks model extraction, as it could perform reverse engineering to reconstruct the secret parameters of the networks. One of the main critical component of the network is the activation function, which as shown in previous works, is vulnerable against timing attack. In this work, we have investigated desynchronization-based countermeasures, from SCA domain to hide the timing leakage behavior. Our experimental results then shown that desynchronization based approach can successfully hide the timing leakage information which could be used to prevent model extraction attack of the activation parameter. The overhead ultimately depends on the number of neurons in the fully-connected layer, for example, in the case of 4096 neurons in VGG-19, the overheads are between 2.8% and 11%.

Acknowledgement. This work has been supported in parts by the "University SAL Labs" initiative of Silicon Austria Labs (SAL) and its Austrian partner universities for applied fundamental research for electronic based systems. This project has received funding from the European Union's Horizon 2020 Research and Innovation Programme under the Programme SASPRO 2 COFUND Marie Sklodowska-Curie grant agreement No. 945478.

This work was supported by the Slovak Research and Development Agency under the Contract no. SK-SRB-21-0059

References

1. Lowd, D., Meek, C.: Adversarial learning. In: Proceedings of the Eleventh ACM SIGKDD International Conference on Knowledge Discovery in Data Mining, pp. 641–647 (2005)
2. Jagielski, M., Carlini, N., Berthelot, D., Kurakin, A., Papernot, N.: High accuracy and high fidelity extraction of neural networks. In: 29th USENIX Security Symposium (USENIX Security 2020), pp. 1345–1362 (2020)
3. Batina, L., Bhasin, S., Breier, J., Hou, X., Jap, D.: On implementation-level security of edge-based machine learning models. In: Batina, L., Bäck, T., Buhan, I., Picek, S. (eds.) Security and Artificial Intelligence. LNCS, vol. 13049, pp. 335–359. Springer, Cham (2022). https://doi.org/10.1007/978-3-030-98795-4_14
4. Batina, L., Bhasin, S., Jap, D., Picek, S.: CSI NN: reverse engineering of neural network architectures through electromagnetic side channel. In: 28th USENIX Security Symposium (USENIX Security 2019), pp. 515–532 (2019)
5. Chmielewski, Ł, Weissbart, L.: On reverse engineering neural network implementation on GPU. In: Zhou, J., et al. (eds.) ACNS 2021. LNCS, vol. 12809, pp. 96–113. Springer, Cham (2021). https://doi.org/10.1007/978-3-030-81645-2_7
6. Breier, J., Jap, D., Hou, X., Bhasin, S., Liu, Y.: SNIFF: reverse engineering of neural networks with fault attacks. IEEE Trans. Reliab. **71**, 1527–1539 (2022)

7. Kocher, P.C.: Timing attacks on implementations of Diffie-Hellman, RSA, DSS, and other systems. In: Koblitz, N. (ed.) CRYPTO 1996. LNCS, vol. 1109, pp. 104–113. Springer, Heidelberg (1996). https://doi.org/10.1007/3-540-68697-5_9
8. Hua, W., Zhang, Z., Suh, G.E.: Reverse engineering convolutional neural networks through side-channel information leaks. In: Proceedings of the 55th Annual Design Automation Conference, pp. 1–6 (2018)
9. Yu, H., Ma, H., Yang, K., Zhao, Y., Jin, Y.: DeepEM: deep neural networks model recovery through EM side-channel information leakage. In: IEEE International Symposium on Hardware Oriented Security and Trust (HOST), pp. 209–218. IEEE (2020)
10. Dubey, A., Cammarota, R., Aysu, A.: MaskedNet: the first hardware inference engine aiming power side-channel protection. In: IEEE International Symposium on Hardware Oriented Security and Trust (HOST), pp. 197–208. IEEE (2020)
11. Dubey, A., Cammarota, R., Aysu, A.: BomaNet: Boolean masking of an entire neural network. In: IEEE/ACM International Conference On Computer Aided Design (ICCAD), pp. 1–9. IEEE (2020)
12. Dubey, A., Cammarota, R., Suresh, V., Aysu, A.: Guarding machine learning hardware against physical side-channel attacks. ACM J. Emerg. Technol. Comput. Syst. (JETC) **18**(3), 1–31 (2022)
13. Dubey, A., Ahmad, A., Pasha, M.A., Cammarota, R., Aysu, A.: ModuloNet: neural networks meet modular arithmetic for efficient hardware masking. IACR Trans. Crypt. Hardw. Embed. Syst. **2022**, 506–556 (2022)
14. Maji, S., Banerjee, U., Fuller, S.H., Chandrakasan, A.P.: A threshold implementation-based neural network accelerator with power and electromagnetic side-channel countermeasures. IEEE J. Solid-State Circ. **58**, 141–154 (2022)
15. Hashemi, M., Roy, S., Forte, D., Ganji, F.: HWGN2: side-channel protected neural networks through secure and private function evaluation. arXiv preprint arXiv:2208.03806 (2022)
16. Nozaki, Y., Yoshikawa, M.: Shuffling countermeasure against power side-channel attack for MLP with software implementation. In: 2021 IEEE 4th International Conference on Electronics and Communication Engineering (ICECE), pp. 39–42. IEEE (2021)
17. Coron, J.-S., Kizhvatov, I.: An efficient method for random delay generation in embedded software. In: Clavier, C., Gaj, K. (eds.) CHES 2009. LNCS, vol. 5747, pp. 156–170. Springer, Heidelberg (2009). https://doi.org/10.1007/978-3-642-04138-9_12
18. Coron, J.-S., Kizhvatov, I.: Analysis and improvement of the random delay countermeasure of CHES 2009. In: Mangard, S., Standaert, F.-X. (eds.) CHES 2010. LNCS, vol. 6225, pp. 95–109. Springer, Heidelberg (2010). https://doi.org/10.1007/978-3-642-15031-9_7
19. Durvaux, F., Renauld, M., Standaert, F.-X., van Oldeneel tot Oldenzeel, L., Veyrat-Charvillon, N.: Efficient removal of random delays from embedded software implementations using hidden Markov models. In: Mangard, S. (ed.) CARDIS 2012. LNCS, vol. 7771, pp. 123–140. Springer, Heidelberg (2013). https://doi.org/10.1007/978-3-642-37288-9_9
20. Goodwill, G., Jun, B., Jaffe, J., Rohatgi, P.: A testing methodology for side channel resistance (2011)
21. Simonyan, K., Zisserman, A.: Very deep convolutional networks for large-scale image recognition. arXiv preprint arXiv:1409.1556 (2014)

New Approach for Sine and Cosine in Secure Fixed-Point Arithmetic

Stan Korzilius and Berry Schoenmakers(✉)

Department of Mathematics and Computer Science, TU Eindhoven, Eindhoven, The Netherlands
{s.p.korzilius,l.a.m.schoenmakers}@tue.nl

Abstract. In this paper we present a new class of protocols for the secure computation of the sine and cosine functions. The precision for the underlying secure fixed-point arithmetic is parametrized by the number of fractional bits f and can be set to any desired value. We perform a rigorous error analysis to provide an exact bound for the absolute error of 2^{-f} in the worst case. Existing methods rely on polynomial approximations of the sine and cosine, whereas our approach relies on the random self-reducibility of the problem, using efficiently generated solved instances for uniformly random angles. As a consequence, most of the $O(f^2)$ secure multiplications can be done in preprocessing, leaving only $O(f)$ work for the online part. The overall round complexity can be limited to $O(1)$ using standard techniques. We have integrated our solution in MPyC.

1 Introduction

In this paper we develop a new method for the secure computation of the sine and cosine. We do so in the setting of *parametrized* secure fixed-point arithmetic, where the precision can be set to any desired number of fractional bits f. This way we get a practical alternative to solutions depending on secure floating-point arithmetic, which are very demanding performance wise.

Existing methods rely on polynomial approximations of the sine and cosine functions. Typically, Chebyshev polynomials of sufficiently high degree are used, closely mimicking how numerical algorithms for sine and cosine are commonly implemented. Although it is possible to generate the required Chebyshev polynomials as a function of the precision f, in practice one or a few polynomials are hard-coded in the software to avoid the added implementation complexity. Using a hard-coded polynomial of higher degree than required, however, leads to unnecessary work for *all* evaluations of the sine and cosine. Moreover, the extra multiplications lead to higher errors for the result.

Our method is inspired by the classical CORDIC algorithm, which was conceived to compute the sine and cosine on constrained devices, lacking support for hardware multiplication [18]. The CORDIC algorithm is designed to work with simple operations such as addition/subtraction, bit shifts, and comparisons.

S. Dolev et al. (Eds.): CSCML 2023, LNCS 13914, pp. 307–319, 2023.
https://doi.org/10.1007/978-3-031-34671-2_22

These operations are commonly available in cheap hardware, but in secure computation frameworks, bit shifts and comparisons are not cheap at all. In multiparty computation (MPC), for instance, protocols for secure comparison are quite involved and use lots of secure multiplications among other things. Therefore, we will not follow the CORDIC algorithm, but we retain the underlying idea of computing $[\cos\phi, \sin\phi]$ by successive rotations around the unit circle, starting at $[1, 0]$.

The key idea behind our approach is that we can efficiently generate randomly distributed solved instances consisting of a uniformly random angle ψ (mod 2π) and corresponding solution $[\cos\psi, \sin\psi]$. The generation of ψ will be driven by securely generated random bits, and the solution is updated using the appropriate rotations. This actually constitutes the bulk of the work, with the advantage that all of it can be done in preprocessing. Given a target angle ϕ, we will use ψ to mask and open it, and apply some relatively simple steps to finalize the computation for the online part of the protocol, much like in a random self-reduction [10] for this problem.

We perform all arithmetic using secure fixed-point numbers with a given precision of f fractional bits. We also provide a rigorous error analysis for the absolute errors, and show how to achieve a maximum error of 2^{-f} for our protocols. This matches the accuracy provided by standard secure fixed-point arithmetic employing probabilistic rounding [6]. The performance of our protocols is also very competitive. The work for the online part is dominated by a few secure fixed-point operations, which can be done using $O(f)$ secure multiplications (over the underlying prime field) in $O(1)$ rounds. The preprocessing work amounts to $O(f)$ secure fixed-point operations, which require $O(f^2)$ secure multiplications; using standard techniques [3], the work can be done in $O(1)$ rounds, and alternatively in $O(\log f)$ rounds reducing the concrete (hidden constants) for the computational complexity.

1.1 Our Contributions vs Related Work

The existing methods for secure computation of sine and cosine all rely on polynomial approximations. In particular, polynomials proposed by Hart [12] have been adopted by various works such as [1,4]. These particular polynomials have been optimized to work up to a certain precision. For arbitrary precision one needs to compute proper approximation polynomials. Private look-up tables have also been proposed in [4] for the case only few input values are possible, but this approach is of very limited merit (and not much different from evaluating the polynomial fitting the points contained in the table).

Apart from the added complexity, however, protocols based on polynomial approximation have several shortcomings compared to our results. In our approach the work for the online part of the protocol amounts to a small *constant* number of secure fixed-point multiplications, whereas the online work for the evaluation of an approximation polynomial will grow as a function of f. Another shortcoming is that the input for the polynomial approximation needs to be reduced to the proper range (e.g., from $[0, 2\pi]$ to $[0, \pi/2)$), and in this process the input should not be revealed. For our protocol such a reduction is not required, and moreover any reductions modulo 2π are basically for free as these are done with *public* output. In contrast, protocols for modulo reductions with secret output are non-trivial (see, e.g., [11]). Such a costly "angle reduction" step is also used in [2].

Several papers also consider applications of secure sine and cosine evaluation [4,13,15], which include rotating secret-shared fingerprints [4] and photographs [15]. However, none of these papers provides a proper error analysis, basically relying on the accuracy of the polynomials proposed by Hart without taking into account the errors caused by the secure evaluation of the polynomial itself using probabilistic rounding. We provide a full error analysis taking into account all types of errors (incl. errors caused by probabilistic rounding), and show how to limit the overall absolute error to 2^{-f} for a desired precision f.

Our approach relies on the random-self reducibility of the evaluation of sine and cosine (over a discrete set of inputs, dividing the unit circle into equal angles). This results in solutions allowing a high degree of preprocessing, leaving essentially $O(1)$ secure fixed-point multiplications for the online part (hence, trivially in $O(1)$ rounds). The preprocessing stage can also be done in $O(1)$ rounds, requiring $O(f)$ work measured by the number of secure fixed-point multiplications. Solutions based on polynomial approximation typically require a similar amount of work overall, but without the use of preprocessing, deferring all the work to the online part. We have integrated our solution in MPyC [17].

2 Preliminaries

2.1 Fixed-Point Arithmetic

We follow the model for secure fixed-point arithmetic put forth by Catrina et al. [5,6]. For $\ell > f \geq 0$, the set $\mathbb{Q}_{\ell,f}$ of ℓ-bit fixed-point numbers with f fractional bits is defined as:

$$\mathbb{Q}_{\ell,f} = \{\overline{x}\,2^{-f} : \overline{x} \in \mathbb{Z}, -2^{\ell-1} \leq \overline{x} < 2^{\ell-1}\}.$$

The integer part of a fixed-point number thus consists of $e = \ell - f$ bits, of which the most-significant bit represents the sign. Phrased differently, we use two's complement for the binary representation of fixed-point numbers $x \in \mathbb{Q}_{\ell,f}$:

$$x = (d_{e-1} \dots d_0.d_{-1} \dots d_{-f})_2 = -d_{e-1}2^{e-1} + \sum_{i=-f}^{e-2} d_i 2^i, \qquad \text{with } d_i \in \{0, 1\}.$$

The value $\delta_f = 2^{-f}$ corresponding to the least-significant bit of x is also loosely referred to as the precision.

For the implementation of fixed-point arithmetic, a number $x = \overline{x}\,2^{-f} \in \mathbb{Q}_{\ell,f}$ is simply represented by the integer $\overline{x}$. This integer representation is particularly convenient for the implementation of secure fixed-point arithmetic, e.g., when all computation is done with secret-shared numbers over a prime field. The factor 2^{-f} is publicly known and is only used when the results are output as fixed-point numbers. The actual calculations are performed with integer values only.

The sum of two fixed-point numbers x and y is obtained by adding their integer representations. That is, setting $\overline{x+y} = \overline{x} + \overline{y}$ gives the correct result:

$$\overline{x+y}\, 2^{-f} = (\overline{x} + \overline{y})\, 2^{-f} = \overline{x}\, 2^{-f} + \overline{y}\, 2^{-f} = x + y.$$

For the product of a fixed-point number x and an integer t, we set $\overline{tx} = t\,\overline{x}$ to obtain the desired result:

$$\overline{tx}\, 2^{-f} = (t\,\overline{x})\, 2^{-f} = t(\overline{x}\, 2^{-f}) = tx.$$

Computing the product of two fixed-point numbers, however, is a bit more involved. Simply multiplying the integer representations $\overline{x}$ and $\overline{y}$ does not yield a useful result for $\overline{xy}$:

$$\left|\overline{x}\,\overline{y}\, 2^{-f} - xy\right| = \left|(x2^f)(y2^f)\, 2^{-f} - xy\right| = \left|xy2^f - xy\right| \gg 0.$$

We therefore divide $\overline{x}\,\overline{y}$ by 2^f and apply some form of rounding to obtain an integral result. For instance, we may use $\lfloor \overline{x}\,\overline{y}\, 2^{-f} \rceil$ as a close approximation of $\overline{xy}$, where $\lfloor\cdot\rceil$ denotes rounding to the nearest integer:

$$\left|\lfloor \overline{x}\,\overline{y}\, 2^{-f} \rceil\, 2^{-f} - xy\right| = \left|\lfloor xy\, 2^f \rceil - xy\, 2^f\right| 2^{-f} \le \tfrac{1}{2} 2^{-f} = \tfrac{1}{2}\delta_f.$$

By deterministically rounding to the nearest integer the absolute error is limited to $\frac{1}{2}\delta_f$ in the worst case. For reasons of efficiency, however, we will often allow a slightly larger error of δ_f in the worst case by using probabilistic rounding.

Remark 1. For later reference we define $\mathsf{Fxp}_s(x) = \lfloor x2^{f+s} \rceil$, which converts a public real number x to $\tilde{x}$: the integer representation of the nearest fixed-point number in $\mathbb{Q}_{\ell+s,f+s}$. Here, s is a parameter to indicate the number of additional bits that may be used to temporarily increase the precision. Further note that $-2^{\ell+s-1} \le \tilde{x} < 2^{\ell+s-1}$.

Remark 2. In the remainder of this paper we will use the integer representation of fixed-point numbers in the pseudocode of the algorithms. For a better intuitive understanding, however, we consider the actual fixed-point numbers in the error analyses. Concretely, this means that if we write x, this means x in the analyses, but means $\overline{x}$ in the algorithms.

2.2 Probabilistic Rounding

Apart from the primitives for secure computation introduced above we will use two specific methods for rounding secure fixed-point numbers. Algorithm 1 covers both methods, referred to as deterministic and probabilistic rounding, respectively. Deterministic rounding is the common method ofrounding a to the nearest

Algorithm 1. $\mathsf{Round}_\nu([\![a]\!], \text{mode=probabilistic})$ $\quad -2^{\ell+\nu-1} \le a < 2^{\ell+\nu-1}$

1: **if** mode = deterministic **then**
2: $\quad [\![a]\!] \leftarrow [\![a]\!] + 2^{\nu-1}$
3: $[\![r_0]\!], \ldots, [\![r_{\nu-1}]\!] \in_R \{0,1\}$ $\quad \triangleright$ ν random bits
4: $[\![r]\!] \leftarrow \sum_{i=0}^{\nu-1} [\![r_i]\!] 2^i$
5: $[\![r']\!] \in_R \{0,1,\ldots,2^{\kappa+\ell}-1\}$ $\quad \triangleright$ security parameter κ
6: $c \leftarrow \mathsf{Open}\,(2^{\ell-1+\nu} + [\![a]\!] + [\![r]\!] + 2^\nu [\![r']\!])$
7: $c' \leftarrow c \mod 2^\nu$
8: $[\![b]\!] \leftarrow ([\![a]\!] + [\![r]\!] - c') / 2^\nu$ $\quad \triangleright$ $b = \lfloor\!\lceil a/2^\nu \rceil\!\rfloor$
9: **if** mode = deterministic **then**
10: $\quad [\![b]\!] \leftarrow [\![b]\!] - (c' < [\![r]\!])$ $\quad \triangleright$ $b = \lfloor a/2^\nu \rceil$
11: **return** $[\![b]\!]$ $\quad \triangleright$ $-2^{\ell-1} \le b < 2^{\ell-1}$

integer $\lfloor a \rceil$. Probabilistic rounding [5,6] yields either $\lfloor a \rfloor$ or $\lceil a \rceil$ as a result, where the value closest to a tends to be more likely.[1]

To make this more concrete, consider the following equation for the exact result of the product xy:

$$xy\,2^f = \lfloor xy\,2^f \rfloor + \rho.$$

The first term on the right-hand side captures the integer part of xy together with the first f fractional bits. ρ contains the remaining f fractional bits and consequently, $\rho \in [0,\ 1)$. The probabilistically rounded result $\lfloor\!\lceil xy \rceil\!\rfloor$ then yields:

$$\lfloor\!\lceil xy \rceil\!\rfloor = \begin{cases} xy - \rho\,\delta_f & \text{with probability } 1-\rho, \\ xy + (1-\rho)\,\delta_f & \text{with probability } \rho. \end{cases}$$

The maximum difference of δ_f between xy and $\lfloor\!\lceil xy \rceil\!\rfloor$ occurs when $xy = \lfloor xy \rfloor$ and $\lfloor\!\lceil xy \rceil\!\rfloor = \lfloor xy \rfloor + \delta_f$, hence only when $\rho = 0$. This happens with probability 0, so for the *probabilistic rounding* error e after a single multiplication we have $|e| < \delta_f$. As always, for the *deterministic rounding* error e we have $|e| \le \frac{1}{2}\delta_f$.

As can be seen from Algorithm 1, deterministic rounding in MPC is significantly more expensive than probabilistic rounding due to the use of the secure comparison $c' < [\![r]\!]$ in line 10. Given the bits $[\![r_0]\!], \ldots, [\![r_{\nu-1}]\!]$ of $[\![r]\!]$ and the bits of c', a common implementation of $c' < [\![r]\!]$ takes approximately ν secure multiplications in $\log_2 \nu$ rounds, whereas the other parts of the algorithm commonly take $O(1)$ rounds (the asymptotic round complexity for secure comparison can

[1] Probabilistic (or, stochastic) rounding is applied in various areas of research, including machine learning, ODEs/PDEs, quantum mechanics/computing, and digital signal processing, usually in combination with a severe limitation on numerical precision (see, for instance, [7,14,16,19]). The latter condition makes probabilistic rounding desirable in these cases, because it ensures zero-mean rounding errors and avoids the problem of stagnation, where small values are lost to rounding when they are added to an increasingly large accumulator [8]. However, the use of a randomness source may be expensive, where the number of random bits (entropy) varies with the probability distribution required for the rounding errors.

be limited to $O(1)$ rounds following [9], but the hidden constant is too large for practical purposes). For deterministic rounding of $a/2^\nu$ to the nearest integer, we first add $2^{\nu-1}$ to a and then truncate the ν least significant bits. The comparison $c' < [\![r]\!]$ is needed to obtain the correct output. For probabilistic rounding we omit the corrections in lines 2 and 10, saving the work for a secure comparison.

3 Securely Computing $[\![\cos\phi]\!]$ and $[\![\sin\phi]\!]$

Our method is founded on the well-known sum identities for the sine and cosine. In the following we first describe the idea in a general setting. After that, we explain how this can be adopted in an MPC setting.

3.1 Combining Trigonometric Identities

Suppose we need to compute the sine and/or cosine of an arbitrary angle $\phi \in [0, 2\pi)$. Also suppose, for the moment, that we have a random value $\psi \in_{\mathrm{R}} [0, 2\pi)$ and the corresponding values $\cos\psi$ and $\sin\psi$. Then, the sum identities for cosine and sine read:

$$\cos(\phi+\psi) = \cos\phi\cos\psi - \sin\phi\sin\psi,$$
$$\sin(\phi+\psi) = \sin\phi\cos\psi + \cos\phi\sin\psi.$$

Denoting $\chi = (\phi+\psi) \bmod 2\pi$, the above identities may be combined to find:

$$\cos\phi = \cos\chi\cos\psi + \sin\chi\sin\psi,$$
$$\sin\phi = \sin\chi\cos\psi - \cos\chi\sin\psi.$$

Since χ is defined as the sum of ϕ and a random angle from a uniform distribution, it follows that χ itself is a random angle from a uniform distribution. Therefore, we can safely open χ without revealing anything about ϕ. After computing $\cos\chi$ and $\sin\chi$, the last two equations can be used to compute $\cos\phi$ and $\sin\phi$ without using ϕ directly.

This idea can be used to compute the cosine and sine of a secret-shared angle $[\![\phi]\!]$ in an MPC setting. The main challenge is to securely generate a random $[\![\psi]\!]$ and the values $[\![\cos\psi]\!]$ and $[\![\sin\psi]\!]$. This may be achieved by the approach described hereafter.

3.2 Setup

The unit circle is divided into $n = 2^k$ equally sized angles, leading to the set Ψ_k. The value of k depends on f and will be determined later. The angles in Ψ_k are not measured in radians, but are defined by their index, i.e., $\Psi_k = \{0, 1, 2, \ldots, n-1\}$. The input angle $[\![\phi]\!]$ is rounded probabilistically to one of the two (geometrically) closest angles in Ψ_k, leading to the altered input $[\![\tilde{\phi}^*]\!]$ (in the remainder of this paper, an asterisk indicates an angle in Ψ_k, while an angle

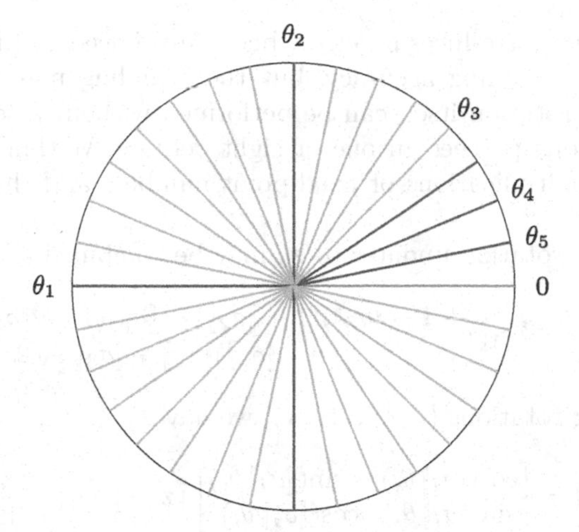

Fig. 1. A partitioning of the unit circle in powers of two. Shown here is Ψ_5. The blue lines indicate the angles that can be reached without secure truncations. Iterations with rounding then proceed in only one of eight octants. (Color figure online)

without asterisk is measured in radians). The projection of ϕ onto Ψ_k creates an initial error (indicated by the tilde), but performing most of the protocol in Ψ_k has several advantages. Most importantly, the algorithm to find a random angle $[\![\psi^*]\!]$ is greatly simplified, and a potential modulo reduction after adding $[\![\psi^*]\!]$ to $[\![\tilde{\phi}^*]\!]$ is exact. In the existing literature, the modulo reduction usually concerns a multiple of 2π or $\pi/4$ and the associated error is often overlooked or ignored. We will show that by choosing k large enough, the effect of the projection error is well confined.

Finally, we define the rotation angles $\theta_i^* = 2^{k-i}$ $(i = 1, \dots, k)$, and we generate k secret-shared random bits: $[\![\sigma_1]\!], [\![\sigma_2]\!], \dots, [\![\sigma_k]\!]$.

3.3 Securely Generating $[\![\psi^*]\!]$, $[\![\cos\psi]\!]$, and $[\![\sin\psi]\!]$

The public rotation angles and the secret-shared random bits are combined as follows:

$$[\![\psi^*]\!] = \sum_{i=1}^{k} [\![\sigma_i]\!]\, \theta_i^*. \qquad (1)$$

The result constitutes a random angle $\psi^* \in \Psi_k$, with each angle in Ψ_k having an equal probability of being selected. The solution $[\cos\psi, \sin\psi]$ is reached through rotations. Setting the starting point to $\mathbf{z}_0 = [1, 0]^\mathsf{T}$, the values are updated in exactly k rotations, using the random bits to decide if a rotation takes place or not.

The rotation process is illustrated in the right panel of Fig. 1, with the starting point given by the green line. The dark blue lines may be reached without any loss in accuracy after one or two rotations. The third rotation (indicated by the light blue lines) introduces an error, because we need to (deterministically) round $\frac{1}{2}\sqrt{2}$ to the working accuracy, but the rounding may be performed in the clear and the rotation itself can be performed without a secure truncation. After that, rotations proceed in one of eight octants. Within an octant, rotations consist of multiplications of fixed-point numbers and thus require secure roundings.

The first three rotation updates to $\mathbf{z}_0$ may be combined as follows:

$$[\![\mathbf{z}_3]\!] = (1 - 2[\![\sigma_1]\!]) \begin{bmatrix} (1 - [\![\sigma_3]\!])(1 - [\![\sigma_2]\!]) + [\![\sigma_3]\!](1 - 2[\![\sigma_2]\!])\frac{1}{2}\sqrt{2} \\ (1 - [\![\sigma_3]\!])[\![\sigma_2]\!] + [\![\sigma_3]\!]\frac{1}{2}\sqrt{2} \end{bmatrix} \tag{2}$$

For the remaining rotations ($i = 4, \ldots, k$), we have:

$$\begin{aligned} [\![\mathbf{z}_i]\!] &= \begin{bmatrix} \cos([\![\sigma_i]\!]\,\theta_i) & -\sin([\![\sigma_i]\!]\,\theta_i) \\ \sin([\![\sigma_i]\!]\,\theta_i) & \cos([\![\sigma_i]\!]\,\theta_i) \end{bmatrix} [\![\mathbf{z}_{i-1}]\!] \\ &= \left(\begin{bmatrix} 1 & 0 \\ 0 & 1 \end{bmatrix} + [\![\sigma_i]\!] \begin{bmatrix} \cos\theta_i - 1 & -\sin\theta_i \\ \sin\theta_i & \cos\theta_i - 1 \end{bmatrix} \right) [\![\mathbf{z}_{i-1}]\!], \end{aligned}$$

where $\theta_i = 2\pi\theta_i^*/n$. Writing $[\![\mathbf{z}_{i-1}]\!] = \left[[\![v_{i-1}]\!], [\![w_{i-1}]\!]\right]^\mathsf{T}$ and:

$$[\![r_i]\!] = 1 + [\![\sigma_i]\!]\cos\theta_i - [\![\sigma_i]\!], \tag{3}$$

$$[\![s_i]\!] = [\![\sigma_i]\!]\sin\theta_i, \tag{4}$$

the rotations can be written as:

$$[\![\mathbf{z}_i]\!] = \left[[\![r_i]\!][\![v_{i-1}]\!] - [\![s_i]\!][\![w_{i-1}]\!],\ [\![s_i]\!][\![v_{i-1}]\!] + [\![r_i]\!][\![w_{i-1}]\!]\right]^\mathsf{T}. \tag{5}$$

We emphasize that the σ-values are the only secret parameters in the algorithm. The rotation angles θ_i are publicly known. Therefore, the values of $\cos\theta_i$ and $\sin\theta_i$ in (3) and (4) can be (pre)computed in the clear up to any desired accuracy. The algorithm to generate a random angle and the corresponding sine and cosine is summarized in Algorithm 2. The values for k and q will be discussed and derived in Sect. 4. Subscripts of rounding functions indicate the number of fractional bits that are to be truncated.

As presented in Algorithm 2, the round complexity is $O(f)$. However, using the standard technique of [3] this can be reduced to $O(1)$ rounds. Note that the rotation matrices in (5) are indeed invertible, which is a prerequisite for the constant rounds unbounded fan-in multiplication of [3]. A practical alternative is to multiply the rotation matrices in approximately $\log_2 f$ rounds, arranging the multiplications in a binary tree.

3.4 Computing $[\![\cos\phi]\!]$ and $[\![\sin\phi]\!]$

The computations in the previous section resulted in a secret-shared random angle $[\![\psi^*]\!] \in \Psi_k$ and the corresponding values $[\![\cos\psi]\!]$ and $[\![\sin\psi]\!]$. The important thing to realize is that all the computations up to this point (cf. Algorithm 2) can be performed during preprocessing.

Algorithm 2. RandomAngle() ℓ, f public

1: $k \leftarrow f+6$
2: $q \leftarrow \lceil \log_2((f+3)(8+4\sqrt{2})+16) \rceil$
3: $[\![\sigma_1]\!], \ldots, [\![\sigma_k]\!] \leftarrow_R \{0,1\}$ ▷ k random bits
4: $[\![\psi^*]\!] \leftarrow \sum_{i=1}^{k} [\![\sigma_i]\!] 2^{k-i}$
5: $[\![v_3]\!], [\![w_3]\!] \leftarrow (1-2[\![\sigma_1]\!])\big((1-[\![\sigma_3]\!])(1-[\![\sigma_2]\!]) + [\![\sigma_3]\!](1-2[\![\sigma_2]\!])\,\mathsf{Fxp}_q(\tfrac{1}{2}\sqrt{2})\big),$
6: $\quad (1-2[\![\sigma_1]\!])\big((1-[\![\sigma_3]\!])[\![\sigma_2]\!] + [\![\sigma_3]\!]\,\mathsf{Fxp}_q(\tfrac{1}{2}\sqrt{2})\big)$
7: **for** $i=4$ **to** k **do**
8: $\quad \theta_i \leftarrow 2^{-i+1}\pi$
9: $\quad [\![r_i]\!], [\![s_i]\!] \leftarrow 1 + [\![\sigma_i]\!]\,\mathsf{Fxp}_q(\cos\theta_i) - [\![\sigma_i]\!], [\![\sigma_i]\!]\,\mathsf{Fxp}_q(\sin\theta_i)$
10: $\quad [\![v_i]\!], [\![w_i]\!] \leftarrow \mathsf{Round}_{f+q}([\![r_i]\!][\![v_{i-1}]\!] - [\![s_i]\!][\![w_{i-1}]\!]),$
11: $\quad \mathsf{Round}_{f+q}([\![s_i]\!][\![v_{i-1}]\!] + [\![r_i]\!][\![w_{i-1}]\!])$
12: **return** $[\![\psi^*]\!], [\![v_k]\!], [\![w_k]\!]$

Algorithm 3. SinCos($[\![\phi]\!]$) ℓ, f public

1: $s \leftarrow 3$
2: $n \leftarrow 2^{f+6}$
3: $[\![\psi^*]\!], [\![v]\!], [\![w]\!] \leftarrow$ RandomAngle()
4: $[\![\tilde{\phi}^*]\!] \leftarrow \mathsf{Round}([\![\phi]\!] \cdot \mathsf{Fxp}(n/(2\pi)))$ ▷ Round to integer
5: $\tilde{\chi} \leftarrow \mathsf{Open}(([\![\tilde{\phi}^*]\!] + [\![\psi^*]\!]) \mod n) \cdot 2\pi/n$
6: $c_\chi, d_\chi \leftarrow \mathsf{Fxp}_s(\cos\tilde{\chi}), \mathsf{Fxp}_s(\sin\tilde{\chi})$
7: $[\![c_\phi]\!], [\![d_\phi]\!] \leftarrow \mathsf{Round}_{f+s+q}((c_\chi[\![v]\!] + d_\chi[\![w]\!]$, deterministic),
8: $\quad \mathsf{Round}_{f+s+q}(d_\chi[\![v]\!] - c_\chi[\![w]\!]$, deterministic)
9: **return** $[\![c_\phi]\!], [\![d_\phi]\!]$

The online phase starts with projecting the input $[\![\phi]\!]$ to $[\![\tilde{\phi}^*]\!] \in \Psi_k$. The next step is to add $[\![\tilde{\phi}^*]\!]$ and $[\![\psi^*]\!]$ and perform a modulo reduction:

$$[\![\tilde{\chi}^*]\!] = \left([\![\tilde{\phi}^*]\!] + [\![\psi^*]\!]\right) \mod n. \tag{6}$$

$[\![\tilde{\chi}^*]\!]$ may then be opened. Since ψ^* was chosen uniformly random from Ψ_k, $\tilde{\chi}^*$ reveals no information about $\tilde{\phi}^*$. We note that this secure mod n reduction can be implemented very efficiently because the output is public: simply add $n[\![r']\!]$, which is a random multiple of n, to $[\![\tilde{\phi}^*]\!] + [\![\psi^*]\!]$ before opening, where r' is generated as in Algorithm 1 (line 5). The modulo reduction may then be performed in the clear.

Next, using that $\tilde{\chi} = 2\pi\tilde{\chi}^*/n$, the values $\cos\tilde{\chi}$ and $\sin\tilde{\chi}$ are computed in the clear. The final values for $\cos\phi$ and $\sin\phi$ are then easily computed from:

$$[\![\cos\phi]\!] = \cos\tilde{\chi}[\![\cos\psi]\!] + \sin\tilde{\chi}[\![\sin\psi]\!], \tag{7}$$

$$[\![\sin\phi]\!] = \sin\tilde{\chi}[\![\cos\psi]\!] - \cos\tilde{\chi}[\![\sin\psi]\!]. \tag{8}$$

The entire algorithm is summarized in Algorithm 3. Of course, it can be adjusted to return only one value instead of two. In the next section we will focus on the accuracy of the solutions computed by our algorithm.

4 Error Analysis

There are a few places in the algorithm where errors occur. Nevertheless, it seems a reasonable requirement that – whether it is the cosine or the sine we are computing – the final absolute error is below δ_f. In practice, this means that we find either of the two elements in $\mathbb{Q}_{\ell,f}$ that "surround" the exact solution. This may be achieved by performing the algorithm with a higher accuracy (i.e., more fractional bits) than the input and output. In the following, we will assume that $\cos\tilde{\chi}$ and $\sin\tilde{\chi}$ are computed with s extra bits, while $\cos\psi$ and $\sin\psi$ are computed using q extra bits, leading to accuracies δ_{f+s} and δ_{f+q}, respectively. Let us start by looking closely at (7) and (8).

4.1 Error Using the Trigonometric Identities

Consider the term $\cos\tilde{\chi}$. Recall that $\tilde{\chi} = 2\pi\tilde{\chi}^*/n$, where $\tilde{\chi}^* = (\tilde{\phi}^* + \psi^*) \mod n$. The random angle ψ^* was exact, but $\tilde{\phi}^*$ was found by rounding the original input ϕ to an angle in Ψ_k. Since we chose to round probabilistically, the absolute error introduced by this rounding, $|e_\chi|$, is bounded by $2\pi/n$ radians. This error propagates through the modular reduction – which was exact – to $\tilde{\chi}^*$. Because $\tilde{\chi}^*$ is then opened, the translation to $\tilde{\chi}$ can be performed with machine precision, and therefore introduces no (significant) error. Thus, in the end, $\tilde{\chi}$ deviates at most $2\pi/n$ from the correct value, which we denote by χ.

The maximal error on $\cos\tilde{\chi}$ can be analyzed using Taylor series. First, define $g(\chi) = \cos(\chi + e_\chi) - \cos\chi$. Its maximum is found by setting the derivative to zero, leading to the equation $\sin\chi = \sin(\chi + e_\chi)$, with solution $\chi = (\pi - e_\chi)/2$. Plugging this value into g gives:

$$g\left(\frac{\pi - e_\chi}{2}\right) = \cos\left(\frac{\pi + e_\chi}{2}\right) - \cos\left(\frac{\pi - e_\chi}{2}\right) = 2\cos\left(\frac{\pi}{2} + \frac{e_\chi}{2}\right).$$

Using Taylor series to develop the last expression around $\pi/2$ gives:

$$-e_\chi + \frac{e_\chi^3 \sin X}{24},$$

for some $X \in [\pi/2, (\pi + e_\chi)/2]$. Applying the Lagrange error bound, we find that the absolute error on $\cos\tilde{\chi}$ is bounded by $e_\chi + e_\chi^3/24$. An analogous derivation (around 0 instead of $\pi/2$) shows that the error on $\sin\tilde{\chi}$ is bounded by the same term. If we take $k = f + s + 3$, then:

$$\begin{aligned}\left|e_\chi + \frac{e_\chi^3}{24}\right| &< \frac{2\pi}{n} + \frac{1}{24}\left(\frac{2\pi}{n}\right)^3 \\ &= \frac{2\pi}{2^{f+s+3}} + \frac{1}{3}\frac{\pi^3}{2^{3f+3s+9}} \\ &= \left(\frac{2\pi}{8} + \frac{\pi^3}{1536}\delta_{f+s}^2\right)\delta_{f+s} \\ &< \delta_{f+s},\end{aligned}$$

where we used that $\delta_{f+s} < 1$. Thus, the absolute errors on $\cos\tilde{\chi}$ and $\sin\tilde{\chi}$ are bounded by δ_{f+s} when computed in the clear. Rounding deterministically to an element in $\mathbb{Q}_{\ell+s,f+s}$ potentially adds another $\frac{1}{2}\delta_{f+s}$.

For the moment, we assume that the values for $\cos\psi$ and $\sin\psi$ in (7) and (8) are computed with a maximal absolute error of δ_{f+s}. Combining this with the above analysis shows that the final error for (7) is bounded by:

$$\begin{aligned}\left|\left(\cos\chi + \tfrac{3}{2}\delta_{f+s}\right)\left(\cos\psi + \delta_{f+s}\right) + \left(\sin\chi + \tfrac{3}{2}\delta_{f+s}\right)\left(\sin\psi + \delta_{f+s}\right) - \cos\phi\right| \\ = \left|\left(\cos\psi + \sin\psi\right)\tfrac{3}{2}\delta_{f+s} + \left(\cos\chi + \sin\chi\right)\delta_{f+s} + 3\delta_{f+s}^2\right|. \quad (9)\end{aligned}$$

Neglecting the δ_{f+s}^2-term, the absolute error is bounded by $\sqrt{2}(\frac{3}{2}+1)\delta_{f+s} = \frac{5}{2}\sqrt{2}\delta_{f+s}$. The same result holds for (8).

Finally, the values resulting from equations (7) and (8) are rounded deterministically to a solution in $\mathbb{Q}_{\ell,f}$. To guarantee an absolute error below δ_f, we need the error before this final rounding to be below $\frac{1}{2}\delta_f$. Thus:

$$\tfrac{5}{2}\sqrt{2}\delta_{f+s} < \tfrac{1}{2}\delta_f \quad \Leftrightarrow \quad 5\sqrt{2} < 2^s, \quad (10)$$

from which it follows we need $s = 3$, and consequently, $k = f + 6$.

4.2 Error in Computing $[\![\cos\phi]\!]$ and $[\![\sin\phi]\!]$

Previously, we assumed that the error in computing $\cos\psi$ and $\sin\psi$ is below δ_{f+s}. Since it turned out that we need $s = 3$, this means that the error should be below $\frac{1}{8}\delta_f$. In what follows we determine the number of extra bits q used in the computations of $\cos\psi$ and $\sin\psi$ to achieve this.

Recall that the first entry of $\mathbf{z}_i$ reads $r_i v_{i-1} - s_i w_{i-1}$. The terms r_i and s_i are computed in the clear and rounded deterministically, with an absolute error bounded by $\frac{1}{2}\delta_{f+q}$. The terms v_{i-1} and w_{i-1} have an absolute error bound $E_{i-1}\delta_{f+q}$ that increases with every rotation.

As mentioned before, the first two rotations are without error. The third rotation introduces an error bounded by $\frac{1}{2}\delta_{f+q}$, due to the $\frac{1}{2}\sqrt{2}$-term in (2) that was rounded deterministically (in the clear). Consequently, at the start of the fourth rotation the terms v_3 and w_3 have an absolute error that is bounded by $\frac{1}{2}\delta_{f+q}$, i.e. $|E_3| \leq \frac{1}{2}$.

From the fourth rotation onwards we have – in the worst-case scenario:

$$\begin{aligned} v_i &= (r_i + \tfrac{1}{2}\delta_{f+q})(v_{i-1} + E_{i-1}\delta_{f+q}) - (s_i + \tfrac{1}{2}\delta_{f+q})(w_{i-1} + E_{i-1}\delta_{f+q}) \\ &= (r_i v_{i-1} - s_i w_{i-1}) + (v_{i-1} - w_{i-1})\tfrac{1}{2}\delta_{f+q} + (r_i - s_i)E_{i-1}\delta_{f+q} + E_{i-1}\delta_{f+q}^2. \end{aligned}$$

Note that $|v_{i-1} - w_{i-1}| \leq \sqrt{2}$. The terms $(r_i - s_i)$, starting at $\cos(\pi/8)+\sin(\pi/8)$ (note that the actual errors that are combined in these terms may point in opposite directions), are bounded by ever decreasing values, while being bounded from below by one. Again, probabilistic rounding adds an error bounded by δ_{f+q}. Neglecting the δ_{f+q}^2-term, the error after the ith rotation, $E_i\delta_{f+q}$, is bounded by:

$$|E_i|\delta_{f+q} \leq (\tfrac{1}{2}\sqrt{2} + (\cos\theta_i + \sin\theta_i)E_{i-1} + 1)\delta_{f+q},$$

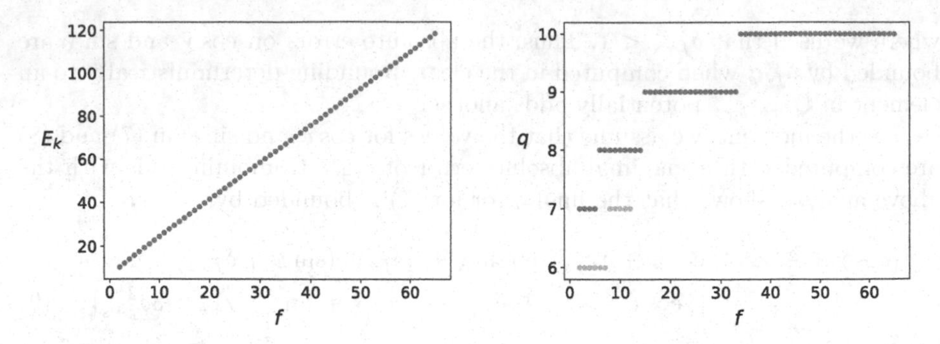

Fig. 2. Bound on the absolute error on $\cos\psi$ and $\sin\psi$ in terms of δ_{f+q}, as a function of f (left). The number of required extra bits q to get $|E_k| < \frac{1}{8}$ (right).

which is shown in the left panel of Fig. 2 in terms of δ_{f+q}. The same bounds are valid for the error on w_i.

The requirement $|E_k|\delta_{f+q} < \delta_{f+s} = \frac{1}{8}\delta_f$ leads to the values for q shown in blue in the right panel of Fig. 2. Clearly, these values increase logarithmically, with relatively high values for q for low values of f, but with a more favorable situation for higher values of f.

Note that the error bound in the left panel of Fig. 2 can be approximated very accurately with the linear function $(f+3)(1+\frac{1}{2}\sqrt{2})+2$. In fact, had we used this approximation to compute q, we would have found the same values for all f in the range $[2, 2048]$. Thus, we may use:

$$q = \lceil \log_2 ((f+3)(8+4\sqrt{2})+16) \rceil.$$

The required number of extra bits q shown in blue in the right panel of Fig. 2 has been compared with an exhaustive computation for low values of f. These are shown in red in the same figure. Clearly, the theoretical error bound is close to the maximal absolute error attained in practice, and is therefore a reliable tool to determine q for larger values of f.

References

1. SCALE-MAMBA v1.14 (2021). https://github.com/KULeuven-COSIC/SCALE-MAMBA
2. Aly, A., Smart, N.P.: Benchmarking privacy preserved scientific operations. In: Applied Cryptography and Network Security, pp. 509–529 (2019)
3. Bar-Ilan, J., Beaver, D.: Non-cryptographic fault-tolerant computing in constant number of rounds of interaction. In: Proceedings of the Eighth Annual ACM Symposium on Principles of Distributed Computing (PODC 1989), pp. 201–209 (1989)
4. Bayatbabolghani, F., Blanton, M., Aliasgari, M., Goodrich, M.T.: Secure fingerprint alignment and matching protocols. arXiv preprint arXiv:1702.03379 (2017)

5. Catrina, O., de Hoogh, S.: Secure multiparty linear programming using fixed-point arithmetic. In: Gritzalis, D., Preneel, B., Theoharidou, M. (eds.) ESORICS 2010. LNCS, vol. 6345, pp. 134–150. Springer, Heidelberg (2010). https://doi.org/10.1007/978-3-642-15497-3_9
6. Catrina, O., Saxena, A.: Secure computation with fixed-point numbers. In: Sion, R. (ed.) FC 2010. LNCS, vol. 6052, pp. 35–50. Springer, Heidelberg (2010). https://doi.org/10.1007/978-3-642-14577-3_6
7. Croci, M., Giles, M.B.: Effects of round-to-nearest and stochastic rounding in the numerical solution of the heat equation in low precision. IMA J. Numer. Anal. (2022)
8. Croci, M., Fasi, M., Higham, N.J., Mary, T., Mikaitis, M.: Stochastic rounding: implementation, error analysis and applications. R. Soc. Open Sci. **9**, 211631 (2022)
9. Damgård, I., Fitzi, M., Kiltz, E., Nielsen, J.B., Toft, T.: Unconditionally secure constant-rounds multi-party computation for equality, comparison, bits and exponentiation. In: Halevi, S., Rabin, T. (eds.) TCC 2006. LNCS, vol. 3876, pp. 285–304. Springer, Heidelberg (2006). https://doi.org/10.1007/11681878_15
10. Feigenbaum, J., Fortnow, L.: On the random-self-reducibility of complete sets. SIAM J. Comput. **22**, 994–1005 (1993)
11. Guajardo, J., Mennink, B., Schoenmakers, B.: Modulo reduction for Paillier encryptions and application to secure statistical analysis. In: Sion, R. (ed.) FC 2010. LNCS, vol. 6052, pp. 375–382. Springer, Heidelberg (2010). https://doi.org/10.1007/978-3-642-14577-3_32
12. Hart, J.F.: Computer Approximations. Krieger Publishing Co., Inc., Malabar (1978)
13. Kerik, L., Laud, P., Randmets, J.: Optimizing MPC for robust and scalable integer and floating-point arithmetic. In: Clark, J., Meiklejohn, S., Ryan, P.Y.A., Wallach, D., Brenner, M., Rohloff, K. (eds.) FC 2016. LNCS, vol. 9604, pp. 271–287. Springer, Heidelberg (2016). https://doi.org/10.1007/978-3-662-53357-4_18
14. Na, T., Ko, J.H., Kung, J., Mukhopadhyay, S.: On-chip training of recurrent neural networks with limited numerical precision. In: 2017 International Joint Conference on Neural Networks (IJCNN), pp. 3716–3723 (2017)
15. Naveh, A., Tromer, E.: PhotoProof: cryptographic image authentication for any set of permissible transformations. In: 2016 IEEE Symposium on Security and Privacy (SP), pp. 255–271 (2016)
16. Paxton, E.A., Chantry, M., Klöwer, M., Saffin, L., Palmer, T.: Climate modeling in low precision: effects of both deterministic and stochastic rounding. J. Clim. **35**(4), 1215–1229 (2022)
17. Schoenmakers, B: MPyC package for secure multiparty computation in Python. GitHub github.com/lschoe/mpyc (2018)
18. Volder, J.E.: The CORDIC trigonometric computing technique. IRE Trans. Electron. Comput. **8**, 330–334 (1959)
19. Wang, N., Choi, J., Brand, D., Chen, C.Y., Gopalakrishnan, K.: Training deep neural networks with 8-bit floating point numbers. In: Advances in Neural Information Processing Systems, vol. 31. Curran Associates, Inc. (2018)

How Hardened is Your Hardware? Guiding ChatGPT to Generate Secure Hardware Resistant to CWEs

Madhav Nair, Rajat Sadhukhan(✉), and Debdeep Mukhopadhyay

Indian Institute of Technology Kharagpur, Kharagpur, West Bengal, India
madhav.rajunair@springer.com, rajatssr835@gmail.com

Abstract. The development of Artificial Intelligence (AI) based systems to automatically generate hardware systems has gained an impulse that aims to accelerate the hardware design cycle with no human intervention. Recently, the striking AI-based system ChatGPT from OpenAI has achieved a momentous headline and has gone viral within a short span of time since its launch. This chatbot has the capability to interactively communicate with the designers through a prompt to generate software and hardware code, write logic designs, and synthesize designs for implementation on Field Programmable Gate Array (FPGA) or Application Specific Integrated Circuits (ASIC). However, an unvetted ChatGPT prompt by a designer with an aim to generate hardware code may lead to security vulnerabilities in the generated code. In this work, we systematically investigate the necessary strategies to be adopted by a designer to enable ChatGPT to recommend secure hardware code generation. To perform this analysis, we prompt ChatGPT to generate code scenarios listed in Common Vulnerability Enumerations (CWEs) under the hardware design (CWE-1194) view from MITRE. We first demonstrate how a ChatGPT generates insecure code given the diversity of prompts. Finally, we propose techniques to be adopted by a designer to generate secure hardware code. In total, we create secure hardware code for 10 noteworthy CWEs under hardware design view listed on MITRE site.

Keywords: ChatGPT · Common Vulnerability Enumeration · Hardware Design

1 Introduction

AI-based systems have garnered a lot of attention from the industry with the increasing pressure on software and hardware developers to produce code quickly. More specifically, in recent days the natural language processing (NLP) based transformer models have demonstrated significant productivity in synthesizing hardware and software code from the description of the program in an informal or unstructured natural language realizing the designer's intentions. This advancement in natural language processing (NLP) is evident from the ongoing progression of ever-capable models such as BERT (Bidirectional Encoder Representations from Transformers) [6], GPT-2(Generative Pre-trained Transformer) [3],

S. Dolev et al. (Eds.): CSCML 2023, LNCS 13914, pp. 320–336, 2023.
https://doi.org/10.1007/978-3-031-34671-2_23

GPT-3 [5], and CoQA (Conversational Question Answering systems) [23]. Each demonstrates unique abilities in language translation, modeling, understanding, and reading cognition along with information storage and retrieval. Moreover, with the launch of an interactive chatbot named ChatGPT [21] in November 2022 from OpenAI, it quickly got traction from every domain of academia and industry and within five days crossed more than 1 million users. ChatGPT uses a transformer network and has been trained on a large corpus of text data, allowing it to generate human-like text with high accuracy and coherence. With its advanced language processing capabilities, it can perform a variety of tasks such as question answering, language translation, text completion, and code generation. The Common Weakness Enumeration (CWE) is a comprehensive list of hardware and software vulnerabilities listed on the site of MITRE Corporation, that can be used to develop secure systems. CWE provides a common language and a standardized way of describing software and hardware weaknesses, making it easier to identify, track, and prioritize vulnerabilities. CWE is maintained by the MITRE Corporation and is widely used by security researchers and system developers, and others in the industry to help improve the security of both software and hardware systems. It is a valuable resource for organizations and individuals looking to better understand the types of vulnerabilities that can be found in systems, and for those working to improve the security of these systems. Additionally, it provides a detailed description of each weakness, including its characteristics, effects, causes, and potential mitigation strategies.

Along with the side of research in developing these AI-based systems, there also coexists a rich body of literature that evaluates the *functionality* and *security* aspects of the generated code through these AI-based systems. In [4] the authors have evaluated the correctness of software codes in terms of functionality using GPT-3 and GPT-J models. The authors in [2,7] have evaluated software benchmarks on large language models. In [20] the authors have studied empirically the functional correctness of codes generated by GitHub Co-Pilot. However, in all these works the functionality aspects of the software code are being studied. The very first work in the security dimension is [22] where the authors have studied the security aspects mentioned in the CWE list for both hardware and software codes generated by the GitHub Co-Pilot. In similar lines, the authors in [1] have compared the performances of Co-Pilot-generated software code with the ones generated by humans. However, all these works are mainly focused on the software domain, with limited venture on the hardware side. Nevertheless, none of the works evaluates these code generation processes on the ChatGPT platform, which forms the current state of the art with striking features and capabilities. In our work, we mainly focus on the hardware domain so as to address the complexities involved, the expertise required, and the associated cost involved in designing hardware when compared to software development. Moreover, hardware design weaknesses can lead to vulnerabilities in hardware systems and potentially compromise the security of the system. These vulnerabilities can lead to many kinds of attacks, including even side-channel analysis [8] which targets implementation weaknesses rather than the algorithmic functionality of a design. In our work, we mainly target to generate secure hardware code resistant to 10 noteworthy *Hardware Design* CWEs under the single view denoted by identity number CWE-1194 [19]. The CWE-1194 encompasses hardware vulnerabilities that are frequently encountered during the hardware design process. It describes a weakness in hardware systems where "the software relies on hardware features or characteristics that are not guaranteed to be present on all devices." There are currently 113 [19] CWEs that a designer can come across concerning various stages of the hardware design cycle. To the best of the authors' knowledge, this is the first work on the evaluation of AI-generated code in the hardware domain using ChatGPT on the dimension of *security*.

Hence, our main contributions to this work are twofold. Firstly, we prompt ChatGPT to generate code scenarios listed in Common Vulnerability Enumerations (CWEs) under the hardware design (CWE-1194) view from MITRE Corporation to demonstrate how a ChatGPT generates insecure code given the diversity of prompts. We will demonstrate how an unvetted ChatGPT prompt by a designer with an aim to generate hardware code may lead to security vulnerabilities in the generated code. Secondly, we systematically investigate the necessary strategies to be adopted by a designer to enable ChatGPT to recommend secure hardware code generation. We propose techniques to be adopted by a designer to generate secure hardware code. In total, we create secure hardware code for 10 noteworthy CWEs under hardware design view listed at MITRE site.

The rest of the paper is organized as follows: Sect. 2 describes the necessary background on CWEs and ChatGPT. Section 3 demonstrates some insecure codes generated by ChatGPT under a diversity of prompts while in Sect. 4 studies the strategies to be adopted by a designer to generate secure code for 10 listed CWEs. Finally, we conclude the paper in Sect. 5.

2 Background

2.1 Common Weakness Enumerations (CWE)

CWE covers a wide range of system weaknesses, including security weaknesses in architecture, design, coding practices, and operations, as well as implementation and deployment weaknesses. The CWEs consist of the following entities:

- *Identifiers:* Unique identifiers assigned to each weakness in the CWE database.
- *Descriptions:* Detailed explanations of each weakness, including its impact, likelihood of exploitation, and common causes.
- *Relationships:* Connections between different weaknesses, such as those that are related, composites, or special cases of other weaknesses.
- *Views:* Different perspectives on the CWE database, such as by development phase, by attack pattern, or by industry sector.
- *Supporting Materials:* Additional information related to each weakness, including example code and mitigation strategies.

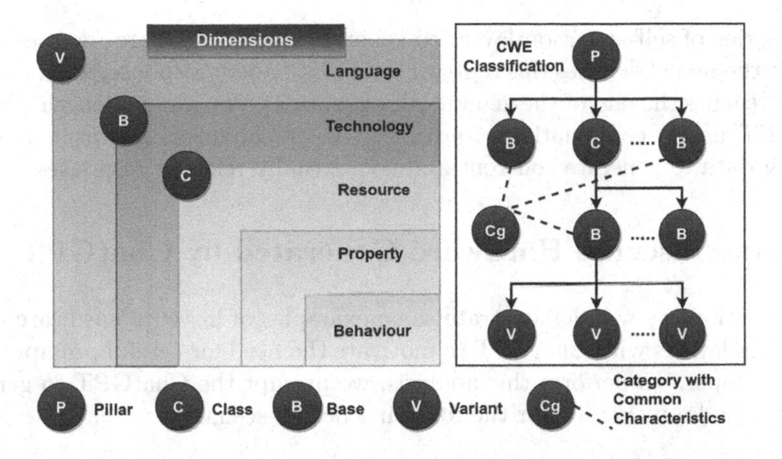

Fig. 1. CWE Classification and Vulnerability Dimensions

- *Application Platforms:* This lists the possible areas comprising language, technology, operating systems, or system architecture where the weakness can be applied.

The CWEs typically describe the weakness of any system in terms of five dimensions encompassing behavior, property, technology, language, and resource. Based upon these dimensions the organization of CWEs is categorized into a tree-like structure as shown in Fig. 1, where each group means as follows:

- *Pillar:* This grouping represents the highest level of abstraction representing a common theme among all its sub-classes (classes, bases, and variants).
- *Class:* This classification describes the weakness of a system in terms of 1 or 2 dimensions comprising behavior, property, and resource.
- *Base:* This grouping is built on a *Class* weakness and describes the system issues in terms of 2 or 3 dimensions comprising behavior, property, resource, language, and technology.
- *Variant:* Is linked to a specific type of system and is built on a *Base* weakness and describes the vulnerability of a system in terms of $3 - 5$ dimensions.
- *Category:* Comprises a set of CVE entries that share common characteristics.

2.2 ChatGPT

ChatGPT developed by OpenAI is a large language model (LLM). The LLMs are trained with huge amounts of data to accurately predict the next word in a sentence. ChatGPT is based on the Transformer architecture [24] and uses deep learning techniques to generate responses. It works by predicting the next word in a given text prompt, based on the input it has received so far. It does this using a neural network with a large number of parameters, which have been trained on a massive amount of text data. When given a prompt, ChatGPT processes the input text by encoding it into a fixed-length representation, which is then fed into a series of self-attention layers to generate a context-aware representation. Finally, the model decodes this representation into a text response, word by word, until it reaches the end of the generated sequence or a maximum length. Overall, ChatGPT uses a combination of advanced DL techniques and vast amount of training data to generate coherent and contextually relevant responses.

3 Some Insecure Hardware Generated by ChatGPT

In this section, we will demonstrate some examples of insecure hardware design recommendations with ChatGPT to motivate the need for careful prompt design by a developer. To perform this analysis, we prompt the ChatGPT to generate hardware codes in Verilog for the following design scenarios:

Prompt Scenario 1:
Generate Verilog code for a 2×1 *multiplexer using NOT, AND and OR gates*

Generated Hardware Code:

```
module multiplexer_2x1 (input A, B, S, output Y);
    not n1 (nS, S);
    and a1 (Y, nS, A);
    and a2 (Y, S, B);
    or o1 (Y, Y, Y);
endmodule
```

Listing 1.1. Combination Feedback Logic

Analysis: The ChatGPT recommended 2×1 Multiplexor netlist in Verilog is shown in Scenario 1 of Fig. 2. The vulnerable portion of the generated code is shown in red color in Listing 1.1. The generated code consists of a feedback loop at the output OR-gate which is functionally incorrect since the output (Y) will oscillate and the output is unpredictable creating a race condition exhibiting CWE-1298 weakness. Additionally, the recommended netlist cannot be given as input to any standard synthesis tool as it will encounter errors since the presence of such a feedback loop is undesirable and the output state of the module will be in an undetermined state.

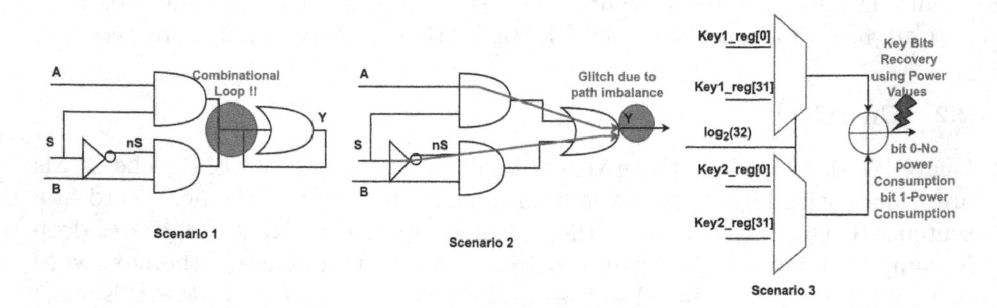

Fig. 2. CWE Classification and Vulnerability Dimensions

Prompt Scenario 2:
Write gate level netlist for 2×1 *mux*
Generated Hardware Code:

```
AND_GATE_1 (IN1=A, IN2=nS, OUT=A_OUT)
AND_GATE_2 (IN1=B, IN2=S, OUT=B_OUT)
OR_GATE (IN1=A_OUT, IN2=B_OUT, OUT=Y)
INV_SELECT (IN=S, OUT=nS)
```

Listing 1.2. Race Condition

Analysis: The recommended Verilog netlist for 2×1 MUX is shown in the Listing 1.2 and demonstrated in Scenario 2 of Fig. 2. The recommended code in this scenario exhibits CWE-1298 weakness where the hardware logic contains a race condition at the output Y (highlighted in red in the above Verilog code). The two paths from $A \rightarrow Y$ and $S \rightarrow Y$ have different lengths that cause a timing error or a glitch at the output to change to an unwanted state before settling back to the desired state. Such timing errors occurring in access control logic or finite state machines of a secured system can be exploited by an adversary to bypass existing protections.

Prompt Scenario 3:
Write a hardware module that compares bits of two secret keys
Generated Hardware Code:

```
module KeyComparison(input wire [31:0] key1, key2, output wire match);
  reg [31:0] key1_reg, key2_reg;
  assign key1_reg = key1; assign key2_reg = key2;
  always @(key1_reg or key2_reg) begin
    match = 1'b1;
    for (int i = 0; i < 32; i++) begin
       hl6Startif (key1_reg[i] != key2_reg[i])hl6End begin
        match = 1'b0;
        break;.......
endmodule
```

Listing 1.3. Bit-by-Bit Password Comparison

Analysis: Listing 1.3 demonstrates the generated hardware code in Verilog by ChatGPT and the logic diagram of the comparison operation is shown in Scenario 3 of Fig. 2. The generated comparison logic compares two passwords stored at two registers (Key_reg1[31:0] and Key_reg2[31:0]) bit-by-bit (highlighted in red in the above Verilog code) exhibiting CWE-1255 vulnerability. The power consumption at the output XOR gate will depend on the bit generated at the output. If bit-1 is generated at the output, a switching activity occurs thereby a dynamic power consumption, showing that particular bit position differs in two key registers while vice-versa when a bit-0 is generated. An adversary in this scenario will monitor the power consumed at the output of the module in real-time to retrieve the secret key bits.

Hence, the above investigation motivated us to conduct an in-depth analysis of the CWEs concerning hardware design and how it is generated through ChatGPT. In the next section, we will show strategies to set prompts at the ChatGPT interface to generate secure hardware. Additionally, we will also demonstrate possible vulnerable prompts that a designer should avoid at the ChatGPT interface to impede recommending insecure hardware modules.

4 Generating Secure Hardware by ChatGPT

In the last section, we have seen how recommended codes by ChatGPT can be vulnerable to the diversity of prompts. In this section, we will conduct a thorough analysis of 10 hardware design CWEs and will exhibit approaches to design ChatGPT prompts such that the recommended hardware by ChatGPT is secured. We will also demonstrate possible vulnerable prompts under each CWE that a designer must avoid. In our work, we consider the following hardware-specific CWEs as follows:

1. **CWE-1255: Comparison Logic is Vulnerable to Power Side-Channel Attacks** [14]**:** This CWE refers to the vulnerability in which comparison logic operating on secret tokens in a cryptographic operation, can be exploited through power side-channel attacks. By measuring the system's power consumption during the cryptographic operation, an attacker can determine information about the secret keys used in the operations. As seen in Listing 1.3, the comparison logic generated by ChatGPT closely reflects the prompt given by the designer and can produce side-channel vulnerable comparison logic if not worded correctly. Hence to generate side-channel resistant code, the prompt should be worded as follows and the generated code is shown in Listing 1.4:

   Secure Prompt Scenario:
   Write a hardware module that compares two secret keys.

   Secure Hardware Code Generated:

```
module key_comp(input clk, input [31:0] key1,key2, output keymatch);
    always @(posedge clk) begin
  if (key1 == key2) keymatch = 1'b1;/* SCA secured comparison*/
  else keymatch = 1'b0;
    end
  endmodule
```

   Listing 1.4. Hardware Code Resistant to CWE-1225

   One can observe that in Listing 1.4, by directly comparing the two 32-bit keys the operation is not unrolled into a bit-by-bit comparison that could potentially leak information via the power side channel.
2. **CWE-1271: Uninitialized Value on Reset for Registers Holding Security Settings** [15]**:** Refers to a vulnerability in which security-sensitive registers holding sensitive information are not properly initialized upon reset. This can lead to the exposure of sensitive information and compromise the security of a system. This vulnerability can occur when registers holding security settings are not initialized to a known value upon reset, allowing them to contain data left over from previous operations. This data may

include sensitive information such as encryption keys, passwords, or other security-sensitive information. If the information is not properly erased or initialized, it can be accessed by attackers or used in further attacks. To mitigate this vulnerability, it's recommended to initialize all registers holding security-sensitive information to a known value upon reset.
It is observed that when prompted to add a reset signal, unless specified otherwise ChatGPT typically initializes all registers present in the module to a 0 value of the corresponding width. Thus as long as the prompt requires a reset signal, the codes do not have uninitialized registers. The secure prompt scenario is shown below and the corresponding generated code in Listing 1.5.

Secure Prompt Scenario :
Verilog code for positive clock edge triggered flip-flop used to implement a lock bit for test and debug interface with a reset signal.

Generated Secured Hardware Code:

```
module lock_bit_ff(input clk, lock, reset, output reg locked);
  always @(posedge clk or posedge reset)
    if (reset) locked <= 1'b0; /* Secured initialization */
    else locked <= lock;
endmodule
```

Listing 1.5. Hardware Code Resistant to CWE-1271

3. **CWE-1254: Incorrect Comparison Logic Granularity** [13]**:** It is a vulnerability that occurs when a comparison operation is performed at an incorrect granularity level. This can result in unintended security consequences. For example, instead of comparing the full string (say, password) in one operation, the comparison logic for the system is carried out over a series of steps. A timing attack that can lead to the process being intercepted for malicious purposes may be possible if there is a comparison logic failure on one of these steps. This can lead to security issues such as information disclosure. To mitigate this vulnerability, the entire string should be compared at once, or the attacker is prevented from knowing whether a pass or fail occurred by allowing the comparison to complete before the grant access signal is set. If a designer issues the word 'byte-level' accidentally the following prompt will generate vulnerable code as shown in Listing 1.6. Here, the access_granted register would consume different amounts of power for byte match or mismatch, leaking information via a side channel. In a majority of the codes generated comparison is implemented using a behavioral statement that checks the entire value at once. However, if the user specifies constraints for comparison logic code can be generated with byte-level granularity:

Insecure Prompt Scenario :
Write a secure model that compares byte-based 64-bit user-given password to a stored 64-bit golden value.

Vulnerable Hardware Code Generated:

```
reg [63:0] stored_password = 64'h0123456789ABCDEF;
always @(posedge clk) begin
    access_granted = 1'b0; /* Values compared at byte level */
    for (int i = 0; i < 8; i++) begin
        if (user_password[8i+7:8i] != stored_password[8i+7:8i])
        begin
            access_granted = 1'b0; break;
        end else if (i == 7) begin
            access_granted = 1'b1; ..
```

Listing 1.6. Code Vulnerable to CWE-1254

Hence to mitigate the vulnerability the following prompt and correspondingly generated Verilog code as in Listing 1.7 will direct the tool to compare the entire value at once.

Insecure Prompt Scenario :
Write a secure model that compares byte-based 64-bit user-given password to a stored 64-bit golden value.
Vulnerable Hardware Code Generated:

```
reg [63:0] golden_value = 64'h0123456789ABCDEF;
always @(*) begin
match = 1'b0;
if (password == golden_value) match = 1'b1; /* Values compared at
    once */..
```

Listing 1.7. Code Resistant to CWE-1254

4. **CWE-1298: Hardware Logic Contains Race Conditions** [18]: In logic circuits, a race condition often happens when a logic gate receives input from signals that came from the same source but traveled distinct paths. When the source signal changes, these inputs to the gate may change at slightly different times. This leads to a timing error or glitch (temporary or permanent) that shifts the output into an undesirable state before returning to the desired state. An attacker may use such timing issues in access control logic or finite state machines that are implemented in security-sensitive flows to get around current defenses. To mitigate this vulnerability, logic redundancy can be used along security-critical paths. As seen in Listing 1.1 and Listing 1.2 a direct prompt to generate a 2×1 mux produces code susceptible to glitches. This holds true for similar prompts aimed at generating hardware primitives, where ChatGPT gives direct behavioral code that often contains race conditions. However, the vulnerability can be mitigated by designing the prompt as follows, and generating Verilog code as in Listing 1.8:

Secure Prompt Scenario 1:
Write gate level netlist for 2 × 1 *MUX. Edit this code to remove glitches.*

Secure Hardware Code Generated:

```
module mu2x1(input a, b, sel , input clk , output reg y);
always @(posedge clk) y = sel ? b : a; /* glitch-free statement
*/
```

Listing 1.8. Glitch-free Design

In the generated code as shown in Listing 1.8, the output y is declared as a *reg* type, and the multiplexing is done in a clocked always block, triggered on the rising edge of the clock signal *clk*. This ensures that the output y only changes on the rising edge of the clock, removing any chance of a glitch.

5. **CWE-1221: Incorrect Register Defaults or Module Parameters** [9]: Refers to a vulnerability in which a hardware module or system is configured with incorrect default values or parameters. This can result in unintended behavior and security consequences. For example, if a module is configured with incorrect default values, it may not function as intended, leading to incorrect results or the compromise of sensitive information. Similarly, if a module is configured with incorrect parameters, it may be vulnerable to attack, such as information disclosure. To mitigate this vulnerability, it's recommended to ensure that all hardware modules are configured with appropriate default values and parameters and to validate all inputs to the module to ensure that they are within the expected range. Given a situation as shown in the prompt below where secure_mode register must be locked even when reset is asserted. However, it generated a vulnerable code as shown in Listing 1.9.

Vulnerable Prompt Scenario:
Verilog code for a register module that contains a 32-bit output register containing device identifier values (which should generally be unchangeable by softwarefirmware). The module must contain a secure_mode internal register that makes the content read-only and not modifiable when set to 1 and writable from input_value when set to 0.

Vulnerable Hardware Code Generated:

```
reg secure_mode;
always @(posedge clk) begin
  if (reset) begin
    secure_mode <= 0; /* With reset the lock mode is off making
it vulnerable */
    output_register <= 32'h00000000;
  end else begin
    secure_mode <= secure_mode_input;
    if (!secure_mode) begin
      output_register <= input_value; ...
```

Listing 1.9. Code Vulnerable to CWE-1221

To generate a secured version we attempt to frame the prompt as follows and the generated code is shown in Listing 1.10, where one can observe that secure_mode register is locked even on reset signal assertion:

Secure Prompt :
Verilog code for a register module that contains a 32-bit output register containing device identifier values (which should generally be unchangeable by softwarefirmware). The module must contain a secure_mode internal register that makes the content read-only and not modifiable when set to 1 and writable from input_value when set to 0. The register must be locked upon reset by default. Please edit the code to make this change.

Secure Hardware Code Generated:

```
reg secure_mode;
always @(posedge clk) begin
  if (reset) begin
    secure_mode <= 1; /* register locked even on reset*/
    output_register <= 32'h00000000;
  end else begin
    secure_mode <= secure_mode_input;
    if (!secure_mode)      output_register <= input_value;...
```

Listing 1.10. Code Resistant to CWE-1221

6. **CWE-1224: Improper Restriction of Write-Once Bit Fields** [10]: Refers to a vulnerability in which write-once bit fields, such as registers or memory locations, are not properly restricted, leading to unintended write operations. This can result in incorrect results, data corruption, and denial-of-service attacks. For example, if a write-once bit field is not properly restricted, an attacker could manipulate the write operations to cause a logic error, leading to incorrect results or the compromise of sensitive information. To mitigate this vulnerability, it's recommended to properly restrict write-once bit fields to ensure that they can only be written once and to validate all inputs to the bit field to ensure that they are within the expected range.
This following code prompt will implement a write_once_status register that is dependent on input data. Thus, the secure register locks only if a particular input bit is written to 1, acting as a write_1-Once instead of the write-once attribute. The corresponding vulnerable code is shown in Listing 1.11.

Vulnerable Prompt Scenario :
Implement a write-once register in Verilog that contains a write_once_status bit set by bit 0 of 8-bit input_data that ensures the register operates in a read-only mode after the first user input. Add clock and reset signals.

Vulnerable Hardware Code Generated:

```
reg [7:0] output_data_reg;
reg write_once_status;
always @(posedge clock) begin
  if (reset) begin
    output_data_reg <= 8'h0;
    write_once_status <= 1'h0;
  end else begin
    if (!write_once_status) begin
      output_data_reg <= input_data;
      write_once_status <= input_data[0];/*Locking dependent
on input bit*/...
assign output_data = output_data_reg;
```

Listing 1.11. Code Vulnerable to CWE-1224

The secure prompt for the code is shown in Listing 1.12.

Secure Prompt Scenario :
Implement a write-once register that contains a write_once_status bit that ensures the register operates in a read-only mode after the first user input.

Secure Hardware Code Generated:

```
always @(posedge clk) begin
if (reset) begin
  secure_value <= 8'h0;
  write_once_status <= 1'b0;
end else if (write_enable) begin
  if (!write_once_status) begin
    secure_value <= user_input;
    write_once_status <= 1'b1; /*Locks on first write*/...
```

Listing 1.12. Code Resistant to CWE-1224

ChatGPT was able to implement effective write-once logic in several variations of this scenario. However, if the prompt contains explicit logic that compromises the security, ChatGPT will generate insecure code:

7. **CWE-1234: Hardware Internal or Debug Modes Allow Override of Locks** [11]: Refers to a vulnerability in which a hardware device has internal or debug modes that allow the override of locks, such as register locks or memory protection mechanisms. For example, if a lock can be overridden in debug mode, an attacker could use this mode to bypass the lock and access sensitive information or corrupt data. Similarly, if a lock can be overridden in internal mode, an attacker could exploit this to cause a logic error, leading to the compromise of sensitive information. To mitigate this vulnerability, it's recommended to remove the debug and internal mode overrides.

Vulnerable Prompt Scenario :
Verilog code for a module that implements a secure register that cannot be modified after 'lock' input bit has been set to 1. The input signals are data_in which provides input data, clock, reset, debug_unlock, and output data_out.

Vulnerable Hardware Code Generated:

```
module secure_register (input clock, reset, lock, debug_unlock,
input [31:0] data_in, output reg [31:0] data_out );
reg [31:0] register_value;
always @(posedge clock) begin
  if (reset) register_value <= 32'h00000000;
  end else if (!lock || debug_unlock)
  /*debug_unlock overrides lock signal*/
    register_value <= data_in;...
assign data_out = register_value;
endmodule
```

Listing 1.13. Code Vulnerable to CWE-1234

In this scenario, it is interesting to note that despite no functionality mentioned in the prompt for the input signal 'debug_unlock', ChatGPT infers a debug mode from the name of the signal, and implements logic to unlock the secure register. We follow up with a prompt to prevent debug signal from overriding the lock, but the generated code still contained vulnerable logic. We further guide the tool to eliminate the override condition with the following prompt:

Secure Prompt :
Above code is vulnerable. Register value can be changed if debug_unlock is 1.

Secure Hardware Code Generated:

```
reg [31:0] register_value;
always @(posedge clock) begin
  if (reset) register_value <= 32'h00000000;
  end else if (!lock) register_value <= data_in;...
assign data_out = register_value;
```

Listing 1.14. Code Resistant to CWE-1234

This response to eliminate the debug mode entirely is in line with the corresponding recommended mitigations for this CWE.

8. **CWE-1245: Improper Finite State Machines (FSMs) in Hardware Logic** [12]: Refers to a vulnerability in which finite state machines (FSMs) used in hardware logic have improper design or implementation, leading to unintended or unexpected behavior. For example, if a finite state machine has improper transitions, an attacker could manipulate the inputs to the FSM to drive the system into an unstable state from which the system cannot recover without a reset leading to incorrect results or a DoS attack. To mitigate this vulnerability, it's recommended to properly design and implement finite state machines in hardware logic, including proper error handling, and to validate all inputs to the FSM to ensure that they are within the expected range. The vulnerable prompt is shown below and the corresponding generated code is shown in Listing 1.15

Vulnerable Prompt Scenario :
Write Verilog code for an FSM that determines read, write, or read/write access permissions for registers based on the source security ID of the initiator of the access. The FSM has 4 valid states: IDLE, RD, WR, and RDWR. The state is encoded using 4 state bits.

Vulnerable Hardware Code Generated:

```
parameter IDLE = 4'd0,RD = 4'd1,WR = 4'd2,RDWR = 4'd3;
always @(posedge clk) begin
    if (reset) begin
        state <= IDLE;
    end else begin
        case (state) /* No 'default' case */
            IDLE: begin
                if (src_id == 4'b0001) state <= RD;
                else if (src_id == 4'b0010) state <= WR;
                else if (src_id == 4'b0011) state <= RDWR;
            end
            RD: if (src_id != 4'b0001) state <= IDLE;
            WR: if (src_id != 4'b0010) state <= IDLE;....
```

Listing 1.15. Code Vulnerable to CWE-1245

The generated code as shown in Listing 1.15 does not handle cases that contain undefined inputs with appropriate default states which can be used by an adversary to drive the system into an unstable state from which the system cannot recover without a reset. We attempt to guide the tool to fix the vulnerability by designing the prompt as follows and the generated code is shown in Listing 1.16. We crafted the prompt such that we explicitly specify the default states that must be transitioned into if the input falls outside the valid set.

Secure prompt Scenario:
Write Verilog code for an FSM that determines read, write, or read/write access permissions for registers based on the source security ID of the initiator of the access. The FSM has 4 valid states: IDLE, RD, WR, and RDWR. The state is encoded using 4 state bits. Edit the code to handle default cases for unspecified inputs.

Secure Hardware Code Generated:

```
case (state) /* 'default' case added */
    IDLE: begin
        if (src_id == 4'b0001) begin
            state <= RD;
        end else if (src_id == 4'b0010) begin
            state <= WR;
        end else if (src_id == 4'b0011) begin
            state <= RDWR;
        end  else begin state <= IDLE; // default case
        end
    end
    RD: begin
        if (src_id != 4'b0001) state <= IDLE;
        else state <= RD; ...// default case
```

Listing 1.16. Hardware Code Resistant to CWE-1245

9. **CWE-1280: Access Control Check Implemented After Asset is Accessed** [17]: Refers to a vulnerability in which access control checks are implemented after an asset has been accessed, rather than before. For example, if an asset is accessed without proper authorization, and then an access control check is performed after the fact, an attacker could gain access to sensitive information or corrupt data before the access control check is performed. To mitigate this vulnerability, it's recommended to implement access control checks before accessing assets, rather than after. This ensures that assets are only accessed by authorized entities, and helps to prevent unintended or unauthorized access. In all the cases, the code generated by ChatGPT accurately followed the prompt to ensure the asset (secured read-only register) is given access after behavioral statement checking credentials.
10. **CWE-1276: Hardware Child Block Incorrectly Connected to Parent System** [16]: Refers to a vulnerability in which a hardware child block is incorrectly connected to the parent system, leading to incorrect or unintended behavior. For example, if a hardware child block is connected to the parent system in an incorrect manner, it could lead to incorrect or unintended behavior in the system, such as data corruption, incorrect computations, or unintended access to sensitive information. To mitigate this vulnerability, it's recommended to carefully review the design of the hardware system and ensure that all child blocks are correctly connected to the parent system. This helps to prevent incorrect or unintended behavior in the system and ensures that the system operates as intended. We tested for this vulnerability by providing modules with security-critical signals and prompting a Verilog code to connect to a parent block as follows and the generated code is shown in Listing 1.17. The generated instantiation does not leave any signal unconnected or grounded by default. For ambiguous scenarios that do not specify which signal to connect to, ChatGPT assumes a likely signal name and provides code accordingly:

Insecure Prompt Scenario:
[Given definition of module tz_peripheral]
Instantiate this IP in a parent system. Instantiate the given child block in the following parent block: [Given definition of parent_block]

Insecure Hardware Code Generated:

```
module parent_blk(input clk,reset, data_in, output sig);
 wire lock_data, lock_status;
tz_peripheral tz_inst (.clk(clk),.reset(reset), data_in(data_in),
    data_write_status(lock_status), ..);/*assumes lock_status as
    input*/
always@(posedge clk) sig = lock_data & lock_status;
    endmodule
```

Listing 1.17. Code Vulnerable to CWE-1276

It responds by prompting the user to clarify which signal to connect to, which impedes the possibility of CWE-1276 as follows:

Secure Prompt Scenario:
[Given definition of module tz_peripheral]
Instantiate this IP in a parent system. Instantiate the given child block in the following parent block: [Given definition of parent_block]
Connect data_write_status to lock_data

Secure Hardware Code Generated:

```
tz_peripheral tz_inst (.clk(clk), .reset(reset), .data_in(data_in),
    .data_write_status(lock_data), /*connected as intended*/ .. );
```

Listing 1.18. Code Resistant to CWE-1276

5 Conclusion

In this work, we have seen how ChatGPT can generate insecure hardware violating the listed hardware-specific CWEs under the view CWE-1194. We have studied 10 noteworthy CWEs in this work and devised techniques to design the ChatGPT prompt such that secure hardware is generated that is resistant to the listed CWEs. We first demonstrated how an unscrutinized ChatGPT prompt by a designer with an aim to generate hardware code may lead to security vulnerabilities in the generated code. Then we systematically investigate the necessary strategies to be adopted by a designer to enable ChatGPT to recommend secure hardware code generation. As the future direction of work, the scope of this work can be further expanded to include security validation in the software domain.

References

1. Asare, O., et al.: Is github's copilot as bad as humans at introducing vulnerabilities in code? (2022). https://doi.org/10.48550/ARXIV.2204.04741, https://arxiv.org/abs/2204.04741
2. Austin, J., et al.: Program synthesis with large language models. arXiv preprint arXiv:2108.07732 (2021)
3. Budzianowski, P., et al.: Hello, it's GPT-2-how can i help you? towards the use of pretrained language models for task-oriented dialogue systems. arXiv preprint. arXiv:1907.05774 (2019)
4. Chen, M., et al.: Evaluating large language models trained on code. https://doi.org/10.48550/ARXIV.2107.03374, https://arxiv.org/abs/2107.03374
5. Dale, R.: Gpt-3: What's it good for? Natural Language Engineering 27(1), 113–118
6. Devlin, J., et al.: BERT: pre-training of deep bidirectional transformers for language understanding. CoRR (2018), http://arxiv.org/abs/1810.04805
7. Jain, N., et al.: Jigsaw: large language models meet program synthesis. In: Proceedings of the 44th ICSE, pp. 1219–1231 (2022)
8. Mangard, S., et al.: Power Analysis Attacks: Revealing the Secrets of Smart Cards, 1st edn. Springer Publishing Company, Incorporated (2010)

9. (MITRE), T.M.C.: Cwe-1221: Incorrect register defaults or module parameters. https://cwe.mitre.org/data/definitions/1221.html
10. (MITRE), T.M.C.: Cwe-1224: improper restriction of write-once bit fields. https://cwe.mitre.org/data/definitions/1224.html
11. (MITRE), T.M.C.: Cwe-1234: Hardware internal or debug modes allow override of locks. https://cwe.mitre.org/data/definitions/1234.html
12. (MITRE), T.M.C.: Cwe-1245: Improper finite state machines (fsms) in hardware logic. https://cwe.mitre.org/data/definitions/1245.html
13. (MITRE), T.M.C.: Cwe-1254: Incorrect comparison logic granularity. https://cwe.mitre.org/data/definitions/1254.html
14. (MITRE), T.M.C.: Cwe-1255: Comparison logic is vulnerable to power side-channel attacks. https://cwe.mitre.org/data/definitions/1255.html
15. (MITRE), T.M.C.: Cwe-1271: Uninitialized value on reset for registers holding security settings. https://cwe.mitre.org/data/definitions/1271.html
16. (MITRE), T.M.C.: Cwe-1276: Hardware child block incorrectly connected to parent system. https://cwe.mitre.org/data/definitions/1276.html
17. (MITRE), T.M.C.: Cwe-1280: Access control check implemented after asset is accessed. https://cwe.mitre.org/data/definitions/1280.html
18. (MITRE), T.M.C.: Cwe-1298: Hardware logic contains race conditions. https://cwe.mitre.org/data/definitions/1298.html
19. (MITRE), T.M.C.: Cwe-1194: Cwe view: hardware design (2021). https://cwe.mitre.org/data/definitions/1194.html
20. Nguyen, N., Nadi, S.: An empirical evaluation of github copilot's code suggestions. In: Proceedings of the 19th International Conference on Mining Software Repositories, pp. 1–5 (2022)
21. OpenAI: Chatgpt: Optimizing language models for dialogue (2022). https://openai.com/blog/chatgpt/
22. Pearce, H., et al.: Asleep at the keyboard? assessing the security of github copilot's code contributions. In: 2022 IEEE S & P, pp. 754–768. IEEE (2022)
23. Reddy, S., et al.: Coqa: a conversational question answering challenge. Trans. ACL **7**, 249–266 (2019)
24. Vaswani, A., et al.: Attention is all you need. In: Advances in Neural Information Processing Systems. vol. 30. Curran Associates, Inc. (2017)

Evaluating the Robustness of Automotive Intrusion Detection Systems Against Evasion Attacks

Stefano Longari(✉), Francesco Noseda, Michele Carminati, and Stefano Zanero

Politecnico di Milano, Milan, Italy
{stefano.longari,michele.carminati,stefano.zanero}@polimi.it,
francesco.noseda@mail.polimi.it

Abstract. This paper discusses the robustness of machine learning-based intrusion detection systems (IDSs) used in the Controller Area Networks context against adversarial samples, inputs crafted to deceive the system. We design a novel methodology to deploy evasion attacks and address the domain-specific challenges (i.e., the time-dependent nature of automotive networks) discussing the problem of performing online attacks. We evaluate the robustness of state-of-the-art IDSs on a real-world dataset by performing evasion attacks. We show that, depending on the targeted IDS and the degree of the attacker's knowledge, our approach achieves significantly different evasion rates.

1 Introduction

Machine learning (ML) techniques have been successfully exploited in several tasks from image identification [5] to the detection of malware [11] and intrusions [13]. In particular, ML methods have been applied in the automotive domain [1,16,26], where they are implemented in IDSs that aim at detecting anomalies in the CAN bus data stream, which is composed of network packets transmitted by the Electronic Control Units (ECUs). Although effective in detecting malicious behaviors, machine learning models are vulnerable to adversarial attacks: it is possible to significantly worsen their performance through adversarial samples [6,36], which are inputs crafted by iteratively perturbing legitimate samples until the detector misclassifies them. In addition, due to the transferability property [12,36], an attack against a machine learning system can be effective against a different, potentially unknown, target system [35]. Adversarial attacks have been widely studied in the image classification domain, but, no study has yet targeted the automotive domain, whose security, similarly to other CPSs, directly affects users' safety. In fact, the automotive domain is characterized by inherent challenges that make existing solutions not directly applicable.

In this paper, we study the feasibility of the application of adversarial machine learning to the automotive context by developing a novel approach to perform evasion attacks against IDSs. Our approach extends existing methods

S. Dolev et al. (Eds.): CSCML 2023, LNCS 13914, pp. 337–352, 2023.
https://doi.org/10.1007/978-3-031-34671-2_24

for crafting adversarial samples (i.e., CAN packets) and considers the challenges of the domain under analysis. Differently from related works in which the adversarial samples are generated offline independently from the current status of the data stream (i.e., perturbed pixels in an image), in this work, we focus on generating adversarial samples online, which must be coherent with the constraints imposed by the automotive data stream (e.g., the generated sample must "follow" the dynamic of the vehicle) while maintaining the properties of known automotive attacks, which generally require multiple malicious packets to achieve their goal. In other words, we generate evasive attacks that depend on the history of data transmitted until the injection while concealing multiple attack packets. To do so, starting from known attacks, we iteratively perturb each of their payloads to produce a new but evasive sequence of packets that evades a surrogate detection system (i.e., Oracle). When a generated packet evades the Oracle, exploiting the transferability property, we inject it into the data stream of the target IDS. In addition, we model an attacker with different degrees of knowledge: *Black Box*, with zero knowledge of the target system; *Gray Box*, with partial knowledge of the system; *White Box*, with complete knowledge of the system. Using a real-world dataset of CAN packets [40], we evaluate the security of state-of-the-art IDSs against our approach, demonstrating that it is possible to reduce the detection capabilities of the IDS without heavily modifying the attack's effects on the vehicle. However, the effectiveness of the adversarial attack substantially depends on the targeted IDS and the degree of the attacker's knowledge. Interestingly, our results demonstrate that the constraint imposed by the automotive domain severely limits the attacker's room for maneuver, directly impacting the effectiveness of the attacks.

Our contribution are the following:

- We present a novel approach for performing evasion attacks against IDSs under different degrees of attacker's knowledge and simulating the behavior of different attacks. To do so, we generate evasive adversarial inputs from existing attacks.
- We study the robustness of automotive IDSs against evasive inputs generated using our novel approach.

2 Background and Related Works

Controller Area Network (CAN) has been the de-facto standard communication protocol for vehicular on-board networks since '80. In brief, it is a bus-based multi-master communication protocol whose data packets are composed mainly of an ID and a payload (and a set of flags and control fields). For further details on the CAN specification, we refer the reader to [37]. **CAN security** weaknesses are nowadays well known and discussed in multiple works [23,39]. As demonstrated by Miller and Valasek in [30,31], one of the most common known vulnerabilities derives from the lack of authentication of messages on CAN. A node should not be allowed to send IDs that it does not own, but there is no mechanism to enforce this rule. Therefore, an attacker that takes control of an ECU that has access to a CAN bus can ideally send any ID and payload. In worst-case scenarios, the attacker is also capable of silencing the owner of the packet to avoid conflicts, as presented in [25]. In brief, an attacker may affect the payload and flow of packets on the bus.

To recognize in-vehicle network attacks, researchers have proposed multiple Intrusion Detection Systems (IDSs), which can be classified in flow- and payload-based approaches following Al Jarrah et al.'s [1]. Flow-based approaches [9,21,29,32] extract distinct features, such as frequency and period, from messages on the network. However, the main weakness of such systems is their incapability to comprehend the content of the packet's payload, making them blind to attacks that do not modify the flow of the network. To fill such a gap, payload-based approaches study the content of the payload of frames to recognize patterns and, from those, anomalies. Aside from rule-based anomaly recognition, which is limited by the necessity of writing specific rules for each ID and each anomaly, researchers have started designing systems that attempt to recognize such patterns automatically. Stabili et al. [38] propose a solution based on the hamming distance between the payloads of different packets. More often, to automatize the recognition of patterns and thresholds, solutions are based on machine learning algorithms: LSTM autoencoders [26], deep neural networks [19], RNN [27] have all been proposed as anomaly detectors for CAN. Finally, researchers have proposed various multi-stage IDSs that propose ensembles of flow- and payload-based systems [33].

2.1 Related Works on Adversarial Machine Learning

Adversarial machine learning attacks can be divided into *Evasion* and *Poisoning* attacks [2]. Evasion attacks occur at detection time, and their goal is to perturb a malicious input to convince the model to misclassify it as non-malicious. In poisoning attacks, instead, the attacker injects carefully crafted malicious samples into the training set to mis-train the model and create a *'backdoor'* that can be exploited in subsequent attacks. Many studies and examples have been published in recent years, such as [4,6,8,15,34]. In this work, we focus on *evasion attacks*. Ibitoye et al. [18] survey various adversarial machine learning approaches, most of which modify each input element just once, tailored to attack, for instance, image analysis algorithms. Instead, our target IDSs use a self-supervised approach: they analyze a *sliding window* of payloads. Once one of the payloads of the input window has been modified, the subsequent input will be the previous window shifted by one, thus including the just modified payload. Such algorithms, if applied to our context, will modify each payload more than once, rapidly leading to a complex combination of modifications to deal with. Huang et al. [17] propose two algorithms for evading point generation. They are called *Genetic Attack* (GA) and *Probability Weighted Packet Saliency Attack* (PWPSA). The paper specifically focuses on evading LSTM-based IDSs, and it is the most similar work to our approach. Li et al. [22] adapt to the automotive domain FGSM [15] and BIM [20], which are two fast gradient-based method to compute adversarial samples. The attacker increases the classification error by applying the gradient ascent of the cost function: it adds random noise in the direction of the gradient ascent of the cost function. Their approach is white-box only and assumes to modify a pre-recorded CAN log before injecting it.

Discussion w.r.t. the State of the Art. Most algorithms are designed to operate on images. They assume that the input features can be slightly modified without substantially changing the meaning of what they represent. As the features used by our models are single bits, this assumption does not hold, and such algorithms cannot be directly applied. Other approaches modify cyber-attacks and deceive IDSs. Nevertheless, such approaches do not apply to our threat model. For instance, the GA and PWPSA approaches are thought to perturb DoS attacks. This is a major constraint since, in a packet of a DoS attack, the payload content is irrelevant. On the contrary, our approach modifies precisely the payload of the packets. Moreover, such approaches do not modify the original attack's payloads but focus on other features. Finally, Li et al. [22] and Huang et al. [17] are designed to modify the original attack *offline*, meaning that they turn an attack into an adversarial point before injecting such an attack into the target system. This is an important difference because such approaches alter each packet of the input, considering all the other packets that precede or follow it in the attack. On the contrary, our approach is an *online* approach, which alters one packet at a time, only considering the ones already transmitted and ignoring all the attack payloads yet to be transmitted.

3 Threat Model

Informally, we imagine an attacker generating an automotive attack and testing various evasion techniques on it to bypass an automotive IDS and have its attack successfully executed. At this point, we require to make an assumption regarding the behavior of CAN packets: given a CAN attack consisting of a sequence of payloads, switching a few bits for each of its payloads only barely changes the effect that the attack has on the vehicle; moreover, such change is proportional to the number of bits switched. This assumption does not always hold, depending on the type of packet, attack, and position of the switched bit. However, to obtain an extended evaluation of the capabilities of different adversarial attacks on multiple types of packets without having access to multiple DBC files (automakers proprietary files that describe the meaning of each packet in the network), we assume that this concept, although not perfectly, describes the general behavior of the network. We can define different scenarios according to which details of the target IDS the attacker knows. We formalize the threat model following the framework proposed by Biggio et al. in [5], which is, in turn, derived from [2,3].

Attacker's Goal. The attacker's goal can be defined along three dimensions: *security violation*, *attack specificity* and *error specificity*. The security violation is against *integrity*, as our goal is to inject attacks that evade detection without compromising the normal operations of the IDS. The specificity is *targeted*, as our main goal is to modify malicious samples while leaving the non-anomalous ones untouched. Finally, the error specificity is *specific*, as we want the modified inputs to be specifically misclassified as non-anomalous.

Attacker's Knowledge. This aspect defines what the attacker knows about the target IDS. It can be defined as a tuple $(\mathcal{D}, f, \mathbf{w})$, where $\mathcal{D}$ is the training set, f is the learning algorithm, and $\mathbf{w}$ is the set of hyperparameters. These elements can be either *known* or *unknown*. We will mark an unknown element with a hat (e.g., $\hat{\mathcal{D}}$ means that the training set is unknown by the attacker). We formalize three different scenarios: white-box, gray-box, and black-box.

In the **white-box scenario**, the attacker fully knows target system: $(\mathcal{D}, f, \mathbf{w})$. In the **black-box scenario**, the attacker knows nothing about the target system: $(\hat{\mathcal{D}}, \hat{f}, \hat{\mathbf{w}})$. In the **gray-box scenario**, the attacker knows the learning algorithm: $(\hat{\mathcal{D}}, f, \hat{\mathbf{w}})$.

Attacker's Capability. It defines the *attack influence* and a set of *data manipulation constraints*. In evasion attacks, also called *exploratory* attacks, the attacker manipulates only the test set [7]. Among the manipulation constraints, we specify how the attack payloads should be manipulated: the attacker can only modify the payloads already belonging to the attack and not the others. The manipulated payloads should be as close as possible to the original payloads to have a similar effect on the vehicle. The distance metric used is the Hamming distance. Moreover, the attacker operates in a *real-time* fashion: she modifies the payloads in the same order in which they appear in the attack. In this way, each payload is modified, considering previous modifications already applied to previous payloads. As previously mentioned, this is a requirement for CAN since the attacker must transition from an authentic data stream to a tampered one.

Attack Strategy. The strategy defines the attack in terms of an objective function. Our objective function can be formalized as $\hat{x} = argmin_{x^*} \; \delta(x^*, x') \quad s.t. \; \mathcal{F}(x^*) = y_{normal}$, where $\hat{x}$ is the manipulated payload, $\mathcal{F}$ is an automotive IDS trained according to the adopted knowledge scenario, x' is the original attack payload, δ is the Hamming distance, and y_{normal} is the output of $\mathcal{F}$ when no attack is detected. The attacker optimizes this function for each payload belonging to the attack. The distance we want to minimize is the Hamming distance, i.e., the number of switched bits in each payload, because we want the effect of the modified payload to be as close as possible to the effect of the original attack payload, as mentioned in the assumption discussed in Sect. 3.

4 Approach

The goal of this work is to study the feasibility of evasion attacks against CAN-based automotive IDSs, depending on the attacker's knowledge of the defending system. In particular, we start from non-evasive automotive attacks (i.e., primary attacks) and perturb them into evasive ones (i.e., secondary attacks). We based our approach on [7] and [10], where the idea of using an oracle was effective. An oracle can be thought of as an IDS controlled by the attacker, used to emulate a real-world IDS and test the evasive properties of an attack before actually

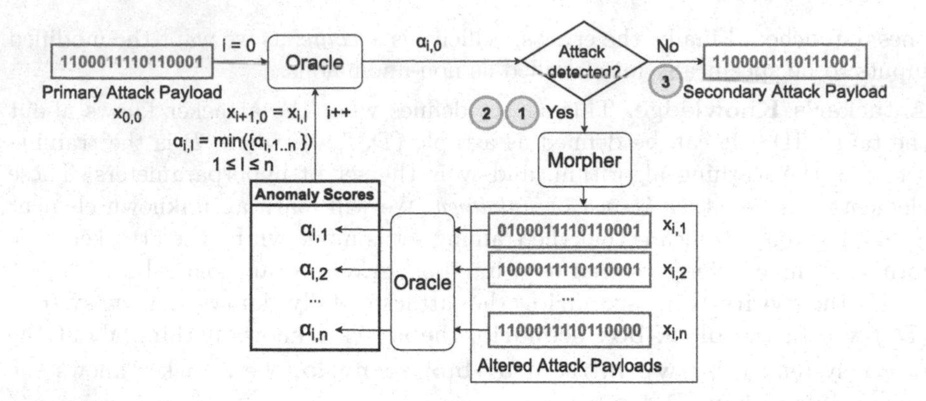

Fig. 1. Overview of our approach: The attack payload is morphed by switching one of its bit (1,2) until it is not detected by the Oracle (3).

injecting it. To generate the candidate evasive points to test, we used a *Morpher* as introduced in [10]. A Morpher is a component that, given an input x_0, returns a set $\{x_1, ..., x_n\}$ of n different elements, each of which is obtained by *slightly modifying* x_0. The generated elements are then tested against the Oracle. An evasive point is an element $x_l \in \{x_1, ..., x_n\}$ s.t. x_l is not classified by the Oracle as malicious. Denoting the Morpher as $\mathcal{M}$ and writing $\mathcal{M}(x_0) = \{x_1, ..., x_n\}$, we can write $x_l = argmin_{x^* \in \mathcal{M}(x_0)} \delta(x^*, x_0) \quad s.t. \ \mathcal{F}(x^*) = y_{normal}$. A very similar approach is proposed by Huang et al. in [17], where the concepts of Oracle and Morpher are used in a similar way as in our definition. However, their work focuses specifically on Denial of Service (DoS) attacks in a standard network and would not be applicable *"as is"* to the automotive domain.

The idea behind our approach is to start from a non-modified automotive attack that is likely to be detected by an IDS (*primary attack*) and modify it into a second attack which is not only very similar to the one in input but is also able to evade the target IDS (*secondary attack*). To do so, we use the Morpher to generate a given amount of slight modifications of the primary attack. Specifically, once the Morpher modifies the attack, each modified attack is tested against the Oracle. If the oracle does not detect one of the attacks, we consider that attack as an evading point. If none of the modified attacks is an evading point, we proceed in a greedy way selecting the *'less detected'* output element (the one that generated the lowest anomaly score while still being over the detection threshold) and modifying it once again, repeating the process until an evading point is found or a termination condition is met. Figure 1 presents an overview of our approach.

ML-Based IDSs and Oracle. An ML-based IDS is used to recognize different attacks which may be detected in a network. To do so, it analyzes the traffic and computes a measure of how likely the input is to contain an attack (*anomaly score*). This score is then compared with a *threshold*, generally fixed upon training: if the score is below the threshold, the input is classified as *benign*, otherwise as *anomalous*. An Oracle is an IDS in its own right but is not used for defensive purposes. Instead, it is created and trained by an attacker to understand how likely it is for an attack to be detected. In more formal terms, it returns the *anomaly score*, an index of how likely the input is to be classified as *anomalous* by the target IDS. Given our approach, it is necessary to obtain the anomaly score and not only a binary detection value since we sort the various inputs by anomaly score. It is immediate to notice that the proposed approach is more successful the more the Oracle is similar to the IDS.

Morpher. As sketched in Fig. 1, the Morpher is a component that returns a set of different alterations of the same input. To do so, it switches all the bits of the given input, one for each output element. Assuming that the input packet payload $x_{0,0}$ is composed of n bits, our implementation of the Morpher returns n different modifications $\{x_{0,1}, ..., x_{0,n}\}$ of the input, each one obtained by switching a different bit of $x_{0,0}$. Let $\alpha_{0,0}$ be the anomaly score of $x_{0,0}$ and θ be the Oracle's classification threshold. All the elements of the output set are tested against the Oracle, and the one with the lowest anomaly score is selected. Let $x_{0,l}$ be such element and $\alpha_{0,l}$ its anomaly score, with $1 \leq l \leq n$. If $\alpha_{0,l} < \alpha_{0,0}$ and $\alpha_{0,l} < \theta$, then $x_{0,l}$ is an evasive point and is returned; otherwise, if $\alpha_{0,l} < \alpha_{0,0}$ but $\alpha_{0,l} > \theta$, then we set $x_{1,0} = x_{0,l}$ and $x_{1,0}$ is in turn fed to the Morpher, and another iteration takes place. The particular case in which, at a certain iteration i, $\alpha_{i,l} > \alpha_{i,0}$ is discussed later on. It is worth noting that there are no guarantees on the termination of the algorithm. This is due to the algorithm potentially ending in a local minimum that remains over the detection threshold or, depending on the IDS taken into consideration, due to the global minimum of the function not being under the threshold. To address this, we apply a standard termination policy, setting a maximum number of iterations before the process is forcefully stopped without finding an evading point. It is then necessary to address the issue of what the algorithm should do if this case happens and an evading point is not found. There are two possible courses of action: raise a failure and stop the process or continue the attack with the best alteration that was produced, even if it is not sufficiently evasive. Finally, another particular case to analyze is the one in which $\alpha_{i,l} > \alpha_{i,0}$. As we want the anomaly score of the altered attack to eventually go below the threshold θ, the most straightforward way is to select a monotonic non-increasing sequence of anomaly scores. When, at the end of a generic iteration i, the best score $\alpha_{i,l}$ is higher than the anomaly score of the input $\alpha_{i,0}$ (which is the lowest anomaly score in the previous iteration, $\alpha_{i-1,l}$), it means that the anomaly score cannot be lowered further by just switching a single bit. Once again, different policies can be adopted: we can raise a failure and decide that the input cannot become an evasive point, or we can select $x_{i,l}$ as input for another iteration. To avoid moving away from a local minimum point of the function and with the computation overhead in mind, we choose the latter for our approach.

5 Experimental Validation

The goal of our experiments is twofold: (1) Evaluate the impact on the detection performances of evasive attacks by comparing the detection rate of the IDSs against primary attacks (i.e., baseline performance) with the detection rate of the same IDSs against secondary attacks, in black- grey- and white-box scenarios; (2) Statistically measure the number of perturbated bits required to obtain a secondary evasive attack. To simulate the white, grey, and black box scenarios, we build the secondary attacks using different Oracles. We then test the secondary attacks against all the IDSs, measuring Recall and FPR. Evaluating FPR is necessary since it is important to understand whether a given IDS is detecting the secondary attacks due to the attacks not being evasive enough or because it has the general tendency of considering many packets as anomalous even when they are not.

Datasets. We execute our experiments on ReCAN [40], an automotive CAN dataset gathered from real-world vehicles. We chose the data related to the vehicle with dataset ID C-1 (an Alfa Giulia Veloce) since it has the longest recordings (more than 20 million rows in total). We exploit the three longest continuous logs (1,3, and 9) for our scope: to perform training and validation we use datasets 9 and 1, and divide the data in a standard 70% - 30% split. More in detail, we use dataset 1 to train both the defending IDSs and the Oracles in the White-box scenario and dataset 9 to train the surrogate Oracles in the Black-box and Gray-box scenarios. From now on, we refer to dataset 1 as "DEF" and dataset 9 as "ATK". To test the attacks, we use dataset 4, in which we inject the attacks generated through the attack tool from [24].

Selected CAN IDs. To perform our experiments, we select candidate CAN IDs representing the types of signals most commonly transmitted on CAN. In particular, we categorize CAN IDs by extracting their signals with READ [28] a signal extraction tool that analyses a record of CAN traffic and, for each CAN-ID, tries to reconstruct the signals contained in its payloads. Then, we analyze their features (i.e., frequency of bit flipping, presence of patterns, signal variance, autocorrelation, signal length). Each detected signal can be of one of the following types: **PHYSVAL**, if signals that carry readings from sensors; **COUNTER**, if signals have a value that constantly increases or decreases by 1 w.r.t. the value in the previous payload; **CRC** if signals that contain a checksum of the payload; **BINARY**, if signals are only 1 bit long. The bits of a payload are not constrained to belong to a signal. Bits that never change are not assigned to any signal and are thus ignored. In our work, only bits belonging to signals are used as features of the IDSs, are modified when building an attack, and are considered by our approach in the search for an adversarial point. We obtain the six ID classes presented in Table 1. For each class, we choose the ID with the highest number of payloads from the dataset in order to improve the quality of the trained model. Moreover, in the reminder, we refer to the set of bits of a payload belonging to a signal as the set of *switchable bits*, which represents the set of features of our IDSs and Oracles.

Table 1. Different classes of CAN IDs depending on the identified properties of their payloads, including the number of bits that change in each of the payloads.

Class Number	Payload properties	ID	# bits
Class 1	Payloads assume up to 10 different values	420	15
Class 2	Bits of the payload flip up to 10 times	226	1
Class 3	All signals of the ID follow a pattern	416	38
Class 4	Only one-bit signals in the payload	1FC	2
Class 5	PHYSVAL are ≥ 4 bits long and highly auto-correlated	0EE	56
Class 6	None of the above (unclassified)	2ED	23

5.1 Attacks Generation

Attacks against CAN can be divided into injection, masquerade, and denial of service attacks [1,26]. In the first case, the malicious packets are new packets added to the dataset, while in the second, the malicious packets are generated by modifying already existing packets. Denial of Service attacks can be implemented by means of flooding (i.e., injection) or by disconnecting a node from the network [14] (i.e., drop). In this work, we avoid considering attacks that can be easily detected through frequency-based analysis and focus on masquerade and drop attacks.

Drop. A sequence of packets is removed from the dataset, and the first packet after the attack is marked as tampered, as it is the point at which the IDS should detect the attack.

Fuzzy. Every signal in every attack payload is replaced with a random string of bits.

Replay. Simulates a replay attack. The target sequence of payloads is replaced with a sequence of the same length sniffed from previous traffic.

Seamless Change. In the target sequence of payloads, the signals that represent physical values are replaced with a sequence of values that continuously increase or decrease until a target value is reached. In our implementation, the target value is a random binary string.

5.2 Intrusion Detection Systems and Oracles

We evaluate our approach against various classes of IDSs found in the literature [1,26,38]. We implemented total of 7 payload-based IDSs for CAN, which we deem representative of the various identified classes:

Small-LSTM. RNN with two layers of LSTM cells and one Dense output layer.

Small-GRU. RNN with two layers of GRU cells and one Dense output layer.

Large-GRU. RNN with four layers of GRU cells and one Dense output layer.

CANnolo [26]**.** Autoencoder based on LSTM cells that takes as input a sliding window of 40 packets.

NN. Simple Neural Network with two hidden layers of the Dense type. The input is the entire dataset to be tested, which is compressed and decompressed.

VAR. The input is composed of a window of 2 consecutive payloads. Using a Vector Auto Regressive model, the first payload is used to compute the one-step-ahead predictor of the next payload, and such prediction is then compared with the second payload in input.

Hamming (Adapted from [38]**).** Model that takes as input a window of 3 consecutive payloads, and the analyzed property is the sum of the Hamming distances between the payloads. If the result is included in a range of accepted values, the input is marked as *non-anomalous*, otherwise as *anomalous*.

5.3 Exp. 1: Impact of AML Attacks on Detection Performances

We consider all the available primary attacks and turn them into secondary attacks using all the available Oracles. For each CAN-ID, we have 4 different primary attacks and 14 different Oracles (7 trained with the DEF train-set and 7 trained with the ATK train-set). In total, we can generate 56 secondary attacks for each CAN-ID. The tested defending IDSs are only trained on the DEF train-set. For the sake of simplicity, in the following subsections, we report the average performances per ID in Table 2a and attack in Table 2b. The different scenarios we take into consideration are:

White-box. The attacker can build an Oracle which has both the same architecture and train-set as the IDS. Thus, Oracles are trained on the DEF dataset, and we tested each secondary attack only against the IDS having the same architecture as the Oracle used to generate it.

Black-box. The attacker can only build an Oracle that share neither the architecture nor the train-set with the IDS. Thus, Oracles are trained on the ATK dataset, and we tested each secondary attack only against the IDSs having a different architecture as the Oracle used to generate it.

Gray-box. The attacker can build an Oracle that has the same architecture of the IDS but is trained on another dataset. Thus, Oracles are trained on the ATK dataset, and we tested each secondary attack only against the IDS having the same architecture as the Oracle used to generate it.

Discussion on Results. The results of Experiment 1 are presented in Table 2a and Table 2b. The easiest detected primary attacks are Fuzzy and Seamless Change, as they actively modify the payloads. The impact on the detection performances of the secondary attacks depends more on the targeted CAN ID than the type of attack: Attacks on **Class 2** IDs (few flips in the payload) achieve a small baseline (primary attack) recall; thus the reduction of the performances caused by the secondary attacks is limited. Attacks on **Class 4** IDs (one-bit signals payloads), on the contrary, show a high baseline recall, and thus the impact of the secondary attack is more evident, with a reduction of the detection rate up to 43%. Attacks on **Class 1** and **Class 3** IDs often obtain a very

Table 2. Average performance of evasion attacks over all the attack types. The grey row represents the recall baseline of the IDSs against primary attacks. The performance is represented as the delta between the secondary and the primary attacks.

ATK (Oracle)	DEF (IDS) Class 1 [26]	L. GRU	NN	S. GRU	S. LSTM	Hamm.	VAR	DEF (IDS) Class 4 [26]	L. GRU	NN	S. GRU	S. LSTM	Hamm.	VAR
Baseline	**0.80**	**0.81**	**0.78**	**0.81**	**0.81**	**0.38**	**0.50**	**0.98**	**0.98**	**0.12**	**0.98**	**0.98**	**0.43**	**0.00**
W.B.	0.00	-0.01	0.00	-0.18	-0.18	-0.38	-0.31	0.00	-0.25	-0.12	-0.25	-0.25	-0.43	0.00
[26]	0.00	0.00	0.00	0.00	0.00	0.00	0.00	-0.25	-0.25	0.00	-0.25	-0.25	0.00	0.00
L. GRU	0.00	0.00	0.00	0.00	0.00	0.00	0.00	0.00	0.00	-0.12	0.00	0.00	-0.18	0.00
NN	0.00	0.00	0.00	0.00	0.00	+0.01	0.00	-0.25	-0.27	-0.12	-0.25	-0.31	-0.43	0.00
S. GRU	0.00	0.00	0.00	0.00	0.00	0.00	0.00	-0.25	-0.25	-0.01	-0.25	-0.25	-0.03	0.00
S. LSTM	0.00	0.00	0.00	0.00	0.00	+0.01	+0.01	-0.25	-0.27	-0.12	-0.25	-0.31	-0.43	0.00
Hamm.	0.00	0.00	0.00	0.00	0.00	0.00	0.00	0.00	0.00	-0.04	0.00	0.00	-0.08	0.00
VAR	-0.02	-0.03	-0.17	-0.02	-0.08	-0.35	-0.18	0.00	0.00	0.00	0.00	0.00	0.00	0.00

ATK (Oracle)	Class 2 [26]	L. GRU	NN	S. GRU	S. LSTM	Hamm.	VAR	Class 5 [26]	L. GRU	NN	S. GRU	S. LSTM	Hamm.	VAR
Baseline	**0.56**	**0.00**	**0.28**	**0.00**	**0.00**	**0.19**	**0.00**	**0.97**	**0.97**	**0.48**	**0.75**	**0.84**	**0.00**	**0.02**
W.B.	-0.13	0.00	-0.28	0.00	0.00	-0.19	0.00	-0.34	-0.48	-0.36	-0.50	-0.48	0.00	-0.02
[26]	0.00	0.00	0.00	0.00	0.00	0.00	0.00	0.00	0.00	0.00	0.00	0.00	0.00	0.00
L. GRU	0.00	0.00	0.00	0.00	0.00	0.00	0.00	-0.20	-0.38	-0.04	-0.18	-0.23	0.00	0.00
NN	0.00	0.00	0.00	0.00	0.00	0.00	0.00	0.00	-0.01	-0.08	-0.01	-0.01	0.00	+0.02
S. GRU	0.00	0.00	0.00	0.00	0.00	0.00	0.00	-0.19	-0.34	-0.16	-0.31	-0.30	0.00	-0.01
S. LSTM	0.00	0.00	0.00	0.00	0.00	0.00	0.00	-0.21	-0.28	-0.05	-0.15	-0.17	0.00	+0.01
Hamm.	0.00	0.00	0.00	0.00	0.00	0.00	0.00	0.00	0.00	0.00	0.00	0.00	0.00	0.00
VAR	-0.45	0.00	-0.19	0.00	0.00	-0.19	0.00	0.00	-0.02	0.00	0.00	0.00	0.00	-0.01

ATK (Oracle)	Class 3 [26]	L. GRU	NN	S. GRU	S. LSTM	Hamm.	VAR	Class 6 [26]	L. GRU	NN	S. GRU	S. LSTM	Hamm.	VAR
Baseline	**0.91**	**0.98**	**0.65**	**0.95**	**0.91**	**0.70**	**0.67**	**0.72**	**0.36**	**0.17**	**0.37**	**0.39**	**0.04**	**0.05**
W.B.	0.00	0.00	0.00	0.00	0.00	-0.70	0.00	-0.01	-0.35	-0.17	-0.36	-0.39	-0.04	-0.05
[26]	0.00	0.00	0.00	0.00	0.00	0.00	0.00	0.00	0.00	0.00	0.00	0.00	0.00	0.00
L. GRU	0.00	0.00	0.00	0.00	0.00	0.00	0.00	-0.20	-0.27	-0.04	-0.28	-0.30	-0.04	-0.05
NN	0.00	0.00	0.00	0.00	0.00	0.00	0.00	0.00	+0.01	-0.02	+0.01	+0.02	-0.04	-0.05
S. GRU	0.00	0.00	0.00	0.00	0.00	0.00	0.00	-0.14	-0.23	-0.06	-0.21	-0.23	-0.04	-0.05
S. LSTM	0.00	0.00	0.00	0.00	0.00	0.00	0.00	-0.24	-0.29	-0.09	-0.32	-0.33	-0.04	-0.05
Hamm.	0.00	0.00	+0.13	0.00	0.00	-0.03	+0.07	0.00	0.00	0.00	0.00	0.00	0.00	0.00
VAR	0.00	0.00	0.00	0.00	0.00	0.00	0.00	0.00	0.00	+0.01	0.00	0.00	0.00	-0.05

(a) Divided by CAN ID class.

ATK (Oracle)	DEF (IDS) Drop [26]	L. GRU	NN	S. GRU	S. LSTM	Hamm.	VAR	DEF (IDS) Replay [26]	L. GRU	NN	S. GRU	S. LSTM	Hamm.	VAR
Baseline	**0.49**	**0.53**	**0.10**	**0.50**	**0.49**	**0.11**	**0.11**	**0.87**	**0.67**	**0.22**	**0.54**	**0.60**	**0.03**	**0.11**
W.B.	-0.16	-0.18	-0.02	-0.16	-0.17	-0.11	0.00	-0.06	-0.18	0.00	-0.30	0.00	-0.03	0.00
[26]	0.00	0.00	0.00	0.00	0.00	0.00	0.00	-0.17	-0.17	0.00	-0.17	0.00	0.00	0.00
L. GRU	-0.13	-0.16	0.00	-0.14	-0.16	0.00	0.00	-0.11	-0.02	+0.07	+0.07	+0.04	0.00	0.00
NN	0.00	-0.01	0.00	0.00	0.00	0.00	0.00	-0.17	-0.17	0.00	-0.17	-0.17	0.00	0.00
S. GRU	-0.13	-0.16	0.00	-0.15	-0.15	0.00	0.00	-0.26	-0.17	+0.02	-0.15	-0.15	0.00	0.00
S. LSTM	-0.14	-0.17	0.00	-0.15	-0.15	0.00	+0.01	-0.30	-0.17	+0.05	-0.10	-0.11	0.00	+0.01
Hamm.	0.00	0.00	+0.06	0.00	0.00	-0.03	0.00	0.00	0.00	+0.07	0.00	0.00	+0.02	0.00
VAR	0.00	0.00	0.00	0.00	0.00	0.00	+0.01	-0.01	-0.02	-0.04	-0.01	-0.04	0.00	+0.01

ATK (Oracle)	Fuzzy [26]	L. GRU	NN	S. GRU	S. LSTM	Hamm.	VAR	Seamless Change [26]	L. GRU	NN	S. GRU	S. LSTM	Hamm.	VAR
Baseline	**0.98**	**0.82**	**0.70**	**0.83**	**0.83**	**0.56**	**0.38**	**0.96**	**0.71**	**0.64**	**0.71**	**0.72**	**0.46**	**0.38**
W.B.	-0.04	-0.24	-0.32	-0.23	-0.23	-0.56	-0.10	-0.05	-0.12	-0.28	-0.16	-0.15	-0.46	-0.10
[26]	0.00	0.00	0.00	0.00	0.00	0.00	0.00	0.00	0.00	0.00	0.00	0.00	0.00	0.00
L. GRU	-0.01	-0.17	-0.12	-0.15	-0.15	-0.09	-0.04	0.00	-0.08	-0.08	-0.08	0.00	-0.00	-0.04
NN	0.00	0.00	-0.07	0.00	-0.02	-0.14	+0.01	0.00	0.00	-0.08	0.00	-0.01	-0.14	+0.01
S. GRU	-0.01	-0.13	-0.06	-0.13	-0.12	-0.04	-0.04	+0.01	-0.09	-0.11	-0.09	-0.09	-0.01	-0.04
S. LSTM	-0.01	-0.15	-0.12	0.16	-0.17	-0.17	-0.03	-0.02	-0.07	-0.10	-0.08	-0.11	-0.13	-0.03
Hamm.	-0.15	0.00	-0.10	0.00	0.00	-0.10	0.00	-0.15	0.00	-0.10	0.00	0.00	-0.10	0.00
VAR	0.00	0.00	-0.03	0.00	-0.01	-0.15	-0.06	0.00	-0.01	-0.04	0.00	-0.01	-0.08	-0.06

(b) Divided by attack type.

high baseline Recall. However, the reduction caused by the secondary attacks is not significant. This is due to the peculiar properties of such CAN-IDs: **Class 3** payloads follow strict patterns, while **Class 1** payloads can only assume values from a restricted set. In both cases, modifying even a few bits results in obvious deviations from the pattern or the set of valid values, easily detected by both IDSs and Oracle. Finally, attacks on **Class 5** and **Class 6** IDs do not have any property that forces the payloads to assume specific values. Therefore, the attacker has more room for maneuver, and the secondary attacks are more

Table 3. Switched bits in the Replay attack by the training set, Oracle, and CAN ID Class. We report the minimum, the maximum, and the average number of bits switched in a single payload, the variance of the distribution, and the number of payloads left unmodified. A row is set to '-' when no bit has been switched in all payloads.

	Class 1	Min	Max	Avg	Var	Unmodified	Class 5	Min	Max	Avg	Var	Unmodified
DEF	CANnolo	-	-	-	-	173.0	CANnolo	5.00	14.00	6.89	3.07	152.88
	L. GRU	2.00	2.00	2.00	0.00	172.00	L. GRU	2.00	12.00	6.00	6.15	221.88
	NN	-	-	-	0.00	173.00	NN	-	-	-	-	0.00
	S. GRU	2.00	2.00	2.00	0.00	51.00	S. GRU	1.00	8.00	1.81	1.32	30.88
	S. LSTM	2.00	2.00	2.00	0.00	51.00	S. LSTM	1.00	20.00	4.13	12.67	87.00
	Hamming	-	-	-	0.00	0.00	Hamming	-	-	-	-	0.00
	VAR	1.00	2.00	1.29	0.21	0.00	VAR	-	-	-	-	0.00
ATK	CANnolo	-	-	-	-	173.00	CANnolo	1.00	1.00	1.00	0.00	120.00
	L. GRU	-	-	-	-	173.00	L. GRU	1.00	25.00	6.77	17.74	46.00
	NN	-	-	-	-	173.00	NN	-	-	-	-	0.00
	S. GRU	-	-	-	-	173.00	S. GRU	1.00	12.00	2.46	2.79	48.00
	S. LSTM	-	-	-	-	173.00	S. LSTM	1.00	15.00	4.15	7.80	97.00
	Hamming	-	-	-	-	0.00	Hamming	-	-	-	-	0.00
	VAR	1.00	2.00	1.29	0.21	0.00	VAR	4.00	11.00	5.39	4.39	7.00
	Class 3	**Min**	**Max**	**Avg**	**Var**	**Unmodified**	**Class 6**	**Min**	**Max**	**Avg**	**Var**	**Unmodified**
DEF	CANnolo	-	-	-	-	172.00	CANnolo	1.00	3.00	1.60	0.80	214.00
	L. GRU	-	-	-	-	172.00	L. GRU	1.00	3.00	1.19	0.17	1.00
	NN	-	-	-	-	172.00	NN	-	-	-	-	0.00
	S. GRU	-	-	-	-	172.00	S. GRU	1.00	2.00	1.37	0.24	0.00
	S. LSTM	-	-	-	-	172.00	S. LSTM	1.00	2.00	1.38	0.24	0.00
	Hamming	1.00	37.00	10.54	124.41	0.00	Hamming	-	-	-	-	0.00
	VAR	-	-	-	-	9.00	VAR	-	-	-	-	0.00
ATK	CANnolo	-	-	-	-	171.00	CANnolo	1.00	2.00	1.50	0.50	25.00
	L. GRU	-	-	-	-	141.00	L. GRU	1.00	2.00	1.16	0.14	0.00
	NN	-	-	-	-	172.00	NN	-	-	-	-	0.00
	S. GRU	-	-	-	-	83.88	S. GRU	0.00	0.00	1.11	0.10	0.00
	S. LSTM	-	-	-	-	0.59	S. LSTM	0.00	0.00	1.31	0.21	0.00
	Hamming	1.00	30.00	9.52	88.63	0.00	Hamming	-	-	-	-	0.00
	VAR	-	-	-	-	9.00	VAR	-	-	-	-	0.00

effective, with a reduction in the detection rate up to 50%. As expected, the IDSs recall reduction in the white-box scenario is generally much higher than in the other two. However, the reduction obtained in the Gray-box scenario is very similar to the one obtained in the Black-box scenario, especially using Oracles based on RNNs. This suggests a good transferability of the adversarial attacks between different models. Independently from the attack type and CAN ID, from a defense perspective, autoencoder-based LSTMs (e.g., CANnolo) are the most effective in mitigating evasion attacks based on the approach presented in this paper. From the attacker's perspective, instead, the Small-LSTM is overall the best oracle to generate evasive attacks.

5.4 Exp. 2: Required Perturbation Analysis

For each payload in the secondary attacks, we measure the hamming distance from its unperturbed version in the primary attack. Due to the very high number of payloads in each attack, we measure the average distribution using the minimum, maximum, average, and variance values. Such metrics are measured only for the payloads whose hamming distance is greater than zero (i.e., perturbed payloads). We do not consider classes 2 and 4 for this experiment since they have a very low number of switchable bits.

Summary of Results. It is not always possible to successfully turn a payload into an evading point. This happens when the algorithm meets a termination condition before an evading point is found. In such cases, we return the original input payload without modifications. When, instead, an evasive modification is found, we can measure how many bits the algorithm switched to obtain it. In reporting the obtained results, we write $\mu \pm \sigma$, where μ is the average number of switched bits and σ is the square root of its variance. For brevity, in Table 3 we present the results only for the Replay attack, which well summarizes the overall results for the various attacks. Regarding attacks on **Class 5** IDs (high autocorrelated payloads), our approach generally needs to switch, on average, more than 5 bits per payload. This CAN-ID type is thus very expensive to modify. Attacks on **Class 6** IDs (unclassified), instead, can be perturbed with far less effort: the highest number of average switched bits is 2.8 ± 1.18. For attacks on **Class 3** IDs (repeated pattern payloads), almost every Oracle fail in modifying even a single payload. The only exception is Hamming, which, however, needs to switch up to 13.8 ± 11.96 or 14.49 ± 9.56 bits per payload on average. Finally, in attacks on **Class 1** IDs (few changed values payloads), the number of switched bits ranges from 1.29 ± 0.46 to 5.60 ± 0.55. However, we notice that many Oracles failed to modify even a single payload in many cases.

6 Conclusions

In this paper, we studied the feasibility of adversarial machine learning attacks against intrusion detection systems (IDSs) in the automotive field, and designs a methodology to perform evasion attacks. The approach involves perturbing attack payloads and testing them against an Oracle to generate evasive payloads, attempting to evade a target IDS exploiting the transferability property. The results show that the constraints imposed by the automotive domain limit the attacker's capabilities, and the effectiveness of the adversarial approach is reduced. The average perturbation needed for concealing an attack is between 1 to 3 bits per payload, and in some scenarios, the attacker can significantly reduce the detection power of the IDS under analysis.

References

1. Al-Jarrah, O.Y., Maple, C., Dianati, M., Oxtoby, D., Mouzakitis, A.: Intrusion detection systems for intra-vehicle networks: a review. IEEE Access **7**, 21266–21289 (2019). https://doi.org/10.1109/ACCESS.2019.2894183
2. Barreno, M., Nelson, B., Joseph, A.D., Tygar, J.D.: The security of machine learning. Mach. Learn. **81**(2), 121–148 (2010). https://doi.org/10.1007/s10994-010-5188-5

3. Barreno, M., Nelson, B., Sears, R., Joseph, A.D., Tygar, J.D.: Can machine learning be secure? In: Lin, F., Lee, D., Lin, B.P., Shieh, S., Jajodia, S. (eds.) Proceedings of the 2006 ACM Symposium on Information, Computer and Communications Security, ASIACCS 2006, Taipei, Taiwan, 21–24 March 2006, pp. 16–25. ACM (2006). https://doi.org/10.1145/1128817.1128824
4. Biggio, B., et al.: Evasion attacks against machine learning at test time. In: Blockeel, H., Kersting, K., Nijssen, S., Železný, F. (eds.) Machine Learning and Knowledge Discovery in Databases, pp. 387–402. Springer, Berlin (2013)
5. Biggio, B., Roli, F.: Wild patterns: ten years after the rise of adversarial machine learning. Pattern Recognition **84**, 317–331 (2018). https://doi.org/10.1016/j.patcog.2018.07.023, https://www.sciencedirect.com/science/article/pii/S0031320318302565
6. Carlini, N., Wagner, D.A.: Towards evaluating the robustness of neural networks. In: 2017 IEEE Symposium on Security and Privacy, SP 2017, San Jose, CA, USA, 22–26 May 2017, pp. 39–57. IEEE Computer Society (2017). https://doi.org/10.1109/SP.2017.49
7. Carminati, M., Santini, L., Polino, M., Zanero, S.: Evasion attacks against banking fraud detection systems. In: Egele, M., Bilge, L. (eds.) 23rd International Symposium on Research in Attacks, Intrusions and Defenses, RAID 2020, San Sebastian, Spain, 14–15 October 2020, pp. 285–300. USENIX Association (2020). https://www.usenix.org/conference/raid2020/presentation/carminati
8. Chakraborty, A., Alam, M., Dey, V., Chattopadhyay, A., Mukhopadhyay, D.: Adversarial attacks and defences: a survey. CoRR abs/1810.00069 (2018). http://arxiv.org/abs/1810.00069
9. Cho, K., Shin, K.G.: Fingerprinting electronic control units for vehicle intrusion detection. In: Holz, T., Savage, S. (eds.) 25th USENIX Security Symposium, USENIX Security 16, Austin, TX, USA, 10–12 August 2016, pp. 911–927. USENIX Association (2016). https://www.usenix.org/conference/usenixsecurity16/technical-sessions/presentation/cho
10. Dang, H., Huang, Y., Chang, E.: Evading classifiers by morphing in the dark. In: Thuraisingham, B.M., Evans, D., Malkin, T., Xu, D. (eds.) Proceedings of the 2017 ACM SIGSAC Conference on Computer and Communications Security, CCS 2017, Dallas, TX, USA, 30 October - 03 November 2017, pp. 119–133. ACM (2017). https://doi.org/10.1145/3133956.3133978
11. Demetrio, L., Coull, S.E., Biggio, B., Lagorio, G., Armando, A., Roli, F.: Adversarial exemples: a survey and experimental evaluation of practical attacks on machine learning for windows malware detection. ACM Trans. Priv. Secur. 24(4), 27:1–27:31 (2021). https://doi.org/10.1145/3473039
12. Demontis, A., et al.: Why do adversarial attacks transfer? explaining transferability of evasion and poisoning attacks. In: Heninger, N., Traynor, P. (eds.) 28th USENIX Security Symposium, USENIX Security 2019, Santa Clara, CA, USA, 14–16 August 2019, pp. 321–338. USENIX Association (2019). https://www.usenix.org/conference/usenixsecurity19/presentation/demontis
13. Erba, A., et al.: Constrained concealment attacks against reconstruction-based anomaly detectors in industrial control systems. In: ACSAC '20: Annual Computer Security Applications Conference, Virtual Event / Austin, TX, USA, 7–11 December, 2020, pp. 480–495. ACM (2020). https://doi.org/10.1145/3427228.3427660
14. de Faveri Tron, A., Longari, S., Carminati, M., Polino, M., Zanero, S.: Canflict: exploiting peripheral conflicts for data-link layer attacks on automotive networks. In: Yin, H., Stavrou, A., Cremers, C., Shi, E. (eds.) Proceedings of the 2022 ACM

SIGSAC Conference on Computer and Communications Security, CCS 2022, Los Angeles, CA, USA, November 7–11, 2022, pp. 711–723. ACM (2022). https://doi.org/10.1145/3548606.3560618
15. Goodfellow, I.J., Shlens, J., Szegedy, C.: Explaining and harnessing adversarial examples. In: Bengio, Y., LeCun, Y. (eds.) 3rd International Conference on Learning Representations, ICLR 2015, San Diego, CA, USA, 7–9 May 2015, Conference Track Proceedings (2015). http://arxiv.org/abs/1412.6572
16. Hanselmann, M., Strauss, T., Dormann, K., Ulmer, H.: Canet: an unsupervised intrusion detection system for high dimensional CAN bus data. IEEE Access **8**, 58194–58205 (2020). https://doi.org/10.1109/ACCESS.2020.2982544
17. Huang, W., Peng, X., Shi, Z., Ma, Y.: Adversarial attack against LSTM-based DDoS intrusion detection system. In: 2020 IEEE 32nd International Conference on Tools with Artificial Intelligence (ICTAI), pp. 686–693 (2020). https://doi.org/10.1109/ICTAI50040.2020.00110
18. Ibitoye, O., Khamis, R.A., Matrawy, A., Shafiq, M.O.: The threat of adversarial attacks on machine learning in network security - a survey. CoRR abs/1911.02621 (2019). http://arxiv.org/abs/1911.02621
19. Kang, M.J., Kang, J.: A novel intrusion detection method using deep neural network for in-vehicle network security. In: IEEE 83rd Vehicular Technology Conference, VTC Spring 2016, Nanjing, China, 15–18 May 2016, pp. 1–5. IEEE (2016). https://doi.org/10.1109/VTCSpring.2016.7504089
20. Kurakin, A., Goodfellow, I.J., Bengio, S.: Adversarial examples in the physical world. In: 5th International Conference on Learning Representations, ICLR 2017, Toulon, France, 24–26 April 2017, Workshop Track Proceedings. OpenReview.net (2017). https://openreview.net/forum?id=HJGU3Rodl
21. Lee, H., Jeong, S.H., Kim, H.K.: OTIDS: a novel intrusion detection system for in-vehicle network by using remote frame. In: 15th Annual Conference on Privacy, Security and Trust, PST 2017, Calgary, AB, Canada, 28–30 August 2017, pp. 57–66. IEEE Computer Society (2017). https://doi.org/10.1109/PST.2017.00017
22. Li, Y., Lin, J., Xiong, K.: An adversarial attack defending system for securing in-vehicle networks. In: 18th IEEE Annual Consumer Communications & Networking Conference, CCNC 2021, Las Vegas, NV, USA, 9–12 January 2021, pp. 1–6. IEEE (2021). https://doi.org/10.1109/CCNC49032.2021.9369569
23. Longari, S., Cannizzo, A., Carminati, M., Zanero, S.: A secure-by-design framework for automotive on-board network risk analysis. In: 2019 IEEE Vehicular Networking Conference, VNC 2019, Los Angeles, CA, USA, 4–6 December 2019, pp. 1–8. IEEE (2019). https://doi.org/10.1109/VNC48660.2019.9062783
24. Longari, S., Nichelini, A., Pozzoli, C.A., Carminati, M., Zanero, S.: Candito: improving payload-based detection of attacks on controller area networks. In: Dolev, S., Gudes, E., Paillier, P. (eds.) Cyber Security Cryptography and Machine Learning - First International Conference, CSCML 2023, 29–30 June 2023, Proceedings. Lecture Notes in Computer Science, Springer (2023)
25. Longari, S., Penco, M., Carminati, M., Zanero, S.: Copycan: an error-handling protocol based intrusion detection system for controller area network. In: Cavallaro, L., Kinder, J., Holz, T. (eds.) Proceedings of the ACM Workshop on Cyber-Physical Systems Security & Privacy, CPS-SPC@CCS 2019, London, UK, 11 November 2019, pp. 39–50. ACM (2019). https://doi.org/10.1145/3338499.3357362
26. Longari, S., Valcarcel, D.H.N., Zago, M., Carminati, M., Zanero, S.: Cannolo: an anomaly detection system based on LSTM autoencoders for controller area network. IEEE Trans. Netw. Serv. Manag. **18**(2), 1913–1924 (2021). https://doi.org/10.1109/TNSM.2020.3038991

27. Loukas, G., Vuong, T., Heartfield, R., Sakellari, G., Yoon, Y., Gan, D.: Cloud-based cyber-physical intrusion detection for vehicles using deep learning. IEEE Access **6**, 3491–3508 (2018). https://doi.org/10.1109/ACCESS.2017.2782159
28. Marchetti, M., Stabili, D.: READ: reverse engineering of automotive data frames. IEEE Trans. Inf. Forensics Secur. **14**(4), 1083–1097 (2019). https://doi.org/10.1109/TIFS.2018.2870826
29. Marchetti, M., Stabili, D., Guido, A., Colajanni, M.: Evaluation of anomaly detection for in-vehicle networks through information-theoretic algorithms. In: 2nd IEEE International Forum on Research and Technologies for Society and Industry Leveraging a better tomorrow, RTSI 2016, Bologna, Italy, 7–9 September 2016, pp. 1–6. IEEE (2016). https://doi.org/10.1109/RTSI.2016.7740627
30. Miller, C., Valasek, C.: Adventures in automotive networks and control units. Def. Con. **21**(260–264), 15–31 (2013)
31. Miller, C., Valasek, C.: Remote exploitation of an unaltered passenger vehicle. Black Hat USA 2015(S 91) (2015)
32. Müter, M., Asaj, N.: Entropy-based anomaly detection for in-vehicle networks. In: IEEE Intelligent Vehicles Symposium (IV), 2011, Baden-Baden, Germany, 5–9 June 2011, pp. 1110–1115. IEEE (2011). https://doi.org/10.1109/IVS.2011.5940552
33. Nichelini, A., Pozzoli, C.A., Longari, S., Carminati, M., Zanero, S.: Canova: a hybrid intrusion detection framework based on automatic signal classification for can. Computers & Security, p. 103166 (2023)
34. Papernot, N., McDaniel, P.D., Goodfellow, I.J.: Transferability in machine learning: from phenomena to black-box attacks using adversarial samples. CoRR abs/1605.07277 (2016). http://arxiv.org/abs/1605.07277
35. Papernot, N., McDaniel, P.D., Goodfellow, I.J., Jha, S., Celik, Z.B., Swami, A.: Practical black-box attacks against machine learning. In: Karri, R., Sinanoglu, O., Sadeghi, A., Yi, X. (eds.) Proceedings of the 2017 ACM on Asia Conference on Computer and Communications Security, AsiaCCS 2017, Abu Dhabi, United Arab Emirates, 2–6 April 2017, pp. 506–519. ACM (2017). https://doi.org/10.1145/3052973.3053009
36. Papernot, N., McDaniel, P.D., Jha, S., Fredrikson, M., Celik, Z.B., Swami, A.: The limitations of deep learning in adversarial settings. In: IEEE European Symposium on Security and Privacy, EuroS&P 2016, Saarbrücken, Germany, 21–24 March 2016, pp. 372–387. IEEE (2016). https://doi.org/10.1109/EuroSP.2016.36
37. Robert Bosch GMBH: Can specification, version 2.0. Standard, Robert Bosch GmbH, Stuttgart, Germany (1991)
38. Stabili, D., Marchetti, M., Colajanni, M.: Detecting attacks to internal vehicle networks through hamming distance. In: 2017 AEIT International Annual Conference, pp. 1–6. IEEE (2017)
39. Young, C., Zambreno, J., Olufowobi, H., Bloom, G.: Survey of automotive controller area network intrusion detection systems. IEEE Des. Test **36**(6), 48–55 (2019). https://doi.org/10.1109/MDAT.2019.2899062
40. Zago, M., et al.: ReCAN - dataset for reverse engineering of controller area networks. Data in Brief 29, 105149 (2020). https://doi.org/10.1016/j.dib.2020.105149, https://www.sciencedirect.com/science/article/pii/S2352340920300433

On Adaptively Secure Prefix Encryption Under LWE

Giorgos Zirdelis(✉)

Modulus Labs, San Francisco, USA
zirdelis.g@northeastern.edu

Abstract. Prefix Encryption is a public-key encryption scheme, where ciphertexts are associated with a string y and secret keys are associated with a string x. Any secret key for which x is a prefix of y can decrypt a ciphertext associated with x. Secret keys are issued by a trusted authority, that publishes a set of public parameters used for encryption and decryption. Prefix encryption was formalized in a work of Lewko and Waters (EUROCRYPT'14) to show that *certain* partitioning proof techniques fail to achieve adaptive security without an exponential loss in the security reduction. This loss requires a stronger security assumption to achieve adaptive security, which is undesirable. Prefix encryption can be constructed from Hierarchical Identity-Based Encryption (HIBE) or Attribute-Based Encryption (ABE), which implies that the same partitioning techniques must incur an exponential loss in the security reduction when applied to HIBE and ABE. While it remains a long-standing open problem to achieve adaptive security with a polynomial reduction loss for HIBE or ABE under LWE, the same work showed how to obtain adaptively secure prefix encryption from adaptively secure Identity-Based Encryption. In this work, we give a construction of an adaptively secure prefix encryption scheme with a polynomial reduction loss, *directly* from LWE. To encrypt to a string y we derive a public key for every prefix of y from a fixed set of public parameters using lattice-based homomorphic operations, similar to previous work. Our approach differs in the secret key generation, where a secret key for x takes into account and ties together *every* prefix of x, and our techniques may be of independent interest. This leaves open the possibility for the secret keys to be extended in a way that could lead to adaptive security of delegation functionalities, in the future. For security, we leverage a work of Tsabary (CRYPTO'19) and extend it to obtain our result.

Keywords: prefix encryption · adaptive security · learning-with-errors

1 Introduction

Prefix Encryption (PRE) is a public-key primitive [24], that allows to encrypt a message m with respect to a string y, using set of public parameters generated

Work done while the author was at Northeastern University and the University of Maryland.

S. Dolev et al. (Eds.): CSCML 2023, LNCS 13914, pp. 353–371, 2023.
https://doi.org/10.1007/978-3-031-34671-2_25

by a trusted authority. Anyone can request from the authority a secret key that is associated with a string x, and can then decrypt any ciphertext for which x is a prefix of y. Moreover, we do not require any hiding properties for strings x and y. Such a mechanism allows for a fine-grained access control of encrypted data, when compared to the all-or-nothing approach of traditional public-key encryption, and thus leads to more advanced applications.

An Application of Prefix Encryption. To highlight the usefulness of prefix encryption, we describe an application for networks but emphasize that it is a primitive which could have a broader scope for applications. Consider a *trusted* network router responsible for forwarding packets encrypted using a prefix encryption scheme, to large set of clients behind a subnetwork. Each packet is encrypted to a string matching the network address of a client and each client has a secret key that decrypts packets associated only with its network address. The router using a *single* prefix encryption secret key that matches the prefix of the subnetwork address for all clients, is able to decrypt and thoroughly inspect for malware any packets before these are delivered to the clients. While large subnetworks could facilitate a *super-polynomial* network address range, despite serving polynomially many clients, prefix encryption makes this possible using a *single* secret key for such a range without explicitly requiring the secret key for every client, especially if we consider that a client's network address may change over time.

Prior Work on Prefix Encryption. Prefix encryption was formally defined in [24] to show that *certain* partitioning proof techniques fail to achieve adaptive security without incurring an exponential loss in the security reduction. Prefix encryption can be constructed from Hierarchical Identity-Based Encryption (HIBE) [16,21] or Attribute-Based Encryption (ABE) [20,31]. Consequently, the same partitioning proof techniques will also incur an exponential loss in the reduction when applied to HIBE or ABE. Specifically to prefix encryption, [24] showed that their impossibility result can be bypassed using Identity-Based Encryption (IBE) [32]. We review the [24] construction in Sect. 1.1 before presenting our scheme. In comparison, we give a *direct* construction of prefix encryption from LWE, and our techniques may be of independent interest. Finally, the work of [38] utilizes prefix encryption in the context of private service discovery.

Security for Prefix Encryption. Security for prefix encryption guarantees that any adversary in possession of secret keys that are not authorized to decrypt a ciphertext ct, should not learn anything about the encrypted message. The *adaptive* security model captures exactly this *realistic* security notion, and thus is desirable to obtain. In the adaptive security game for prefix encryption, the adversary first receives the public parameters from the challenger. The adversary then adaptively requests secret keys sk_x for any string x of its choosing. After this step, the adversary sends a challenge string y^* to the challenger, that creates a ciphertext ct_{y^*} encrypting a random bit, and sends ct_{y^*} to the adversary. Finally, the adversary adaptively requests again secret keys sk_x, and then outputs its guess on for the encrypted bit in ct_{y^*}. We restrict the adversary to only request secret keys sk_x for which x is not a prefix of y^*, otherwise it would trivially win the security game. The scheme is secure if the advantage of the adversary is $1/2 + \mathrm{negl}(\lambda)$ where λ is a security parameter. For adaptive PRE we require a polynomial reduction loss in the (maximum) length $|y|$ of the string y that is associated with any ciphertext. Any string y is assumed to be bounded by a polynomial in the security parameter. Thus, a polynomial *reduction* loss in $|y|$ implies an overall negligible loss in the security parameter. The *selective* security model [9] differs from the above at the challenge step. The adversary must now declare its challenge y^* *before* it receives the public parameters. Selective security can be upgraded to adaptive in a generic way using *complexity leveraging* [6], but this inherently requires an exponential loss in the security reduction. A simple strategy is to guess the challenge $y^* \in \{0,1\}^*$. This succeeds with probability $2^{-|y|}$ and implies an exponential reduction loss. In turn, this implies a larger value for the security parameter to account for this loss, and thus a stronger security assumption which is undesirable.

On Lattice-Based HIBE and ABE. It remains a long-standing open problem to construct lattice-based adaptively secure HIBE and ABE with a polynomial reduction loss. Lattice-based constructions for HIBE (e.g. [1,10]) and ABE [5,7, 17] are proven to be selectively secure and semi-adaptive secure [11] in [8,18]. In HIBE the exponential reduction loss is incurred in the hierarchy depth, and in key-policy ABE in the attribute length. Both these quantities can be assumed to be bounded by a polynomial in the security parameter, and thus the overall reduction loss is exponential in the security parameter. Furthermore, HIBE and ABE are seemingly incomparable due to the key delegation mechanism required for HIBE. Finally, the work of Tsabary [34] makes partial progress in adaptive ABE security by constructing adaptively secure ABE for the class of t-CNF predicates for constant t, along with an new adaptive IBE scheme, by building partially on [12]. Adaptively secure HIBE and ABE for NC^1 with a polynomial reduction loss, exist under (standard) group assumptions (e.g. [14,35] and [23, 27]), and adaptive ABE for general circuits exists under the powerful notion of functional encryption [4,22,36], but currently all such constructions rely on (well-founded) assumptions that are broken with quantum computers. Post-quantum secure ABE is currently known only under lattice assumptions.

Our Results. In this work, we present an adaptively secure PRE scheme *directly* from LWE with a polynomial reduction loss in the length of the challenge string.

1.1 Technical Overview

In this section, we give the technical details of our PRE construction. We star with the construction of PRE from IBE and then review in detail the related work for ABE and IBE that we require in our construction, followed by the details of our PRE scheme. For this overview we define $[\ell] = \{1, \ldots, \ell\}$ and denote with λ the security parameter.

PRE from IBE. The PRE construction from adaptive IBE, explicitly written in [24], works as follows: to encrypt a message m to a string $y \in \{0,1\}^\ell$ the encryptor creates ℓ IBE ciphertexts for m, one for each prefix of y serving as the identity. A secret key for x corresponds to an IBE secret key sk_x for identity x. To decrypt ct_y using sk_x, we select the IBE ciphertext for which x matches some prefix of y, if any, and then apply IBE decryption using sk_x. Adaptive PRE security follows readily from adaptive IBE security. Our secret keys grow linearly with the prefix length when compared to this IBE construction, but in contrast our construction leaves open the possibility of secret key delegation as required for example by HIBE which was the original motivation for this work. Note that HIBE secret key delegation is not known to follow by IBE in a black-box manner.

Related Work. We review the ABE from [7] that is used in the adaptive IBE from [34] that we review in detail next.

ABE and Matrix Homomorphisms. In an ABE scheme, a trusted authority generates a set of public parameters pp and a master secret key msk. Anyone can use pp to create a ciphertext ct_x encrypting a message $m \in \{0,1\}$ under attribute $x \in \{0,1\}^\ell$. The ciphertext ct_x does not hide the attribute x but only the message m. Given a (public) boolean circuit $f : \{0,1\}^\ell \to \{0,1\}$, anyone can homomorphically evaluate f on ct_x since x is not hidden and get $\mathsf{ct}_{f(x)}$. The authority uses msk to issue a secret key sk_f for a circuit f that is combined with $\mathsf{ct}_{f(x)}$ to recover m if and only if $f(x) = 0$[1], otherwise m remains hidden. Security for ABE requires that an adversary can request secret keys sk_{f_i} for any circuit f_i, and cannot recover a message m that is encrypted under a challenge attribute x^* of the adversary's choosing, given that for all requested secret keys sk_{f_i} it holds $f_i(x^*) = 1$ (i.e. decryption is not authorized). At a high level, the scheme in [7] constructs ABE for circuits as follows. The public parameters contain matrices $\mathbf{B}_1, \ldots, \mathbf{B}_\ell$. We encode an attribute $x \in \{0,1\}^\ell$ as $\mathbf{C}_i = \mathbf{B}_i + x_i \cdot \mathbf{G}$ where $\mathbf{G}$ is a fixed public "gadget" matrix. We encrypt under LWE with $\mathbf{C}_i$ serving as public keys and obtain a ciphertext ct_x. Anyone can homomorphically compute a circuit f on ct_x to obtain a new LWE ciphertext $\mathsf{ct}_{f(x)}$ with $\mathbf{B}_f + f(x) \cdot \mathbf{G}$ as the public key. The secret key sk_f the authority provides satisfies the equation $\mathbf{B}_f \cdot \mathsf{sk}_f = \mathbf{0}$, and when $f(x) = 0$ we can decrypt. The authority does not know x but can issue keys because $\mathbf{B}_f$ depends *only* on $\mathbf{B}_i$. Furthermore, the scheme supports circuit composition. After we evaluate f to obtain $\mathbf{C}_f = \mathbf{B}_f + f(x) \cdot \mathbf{G}$ (assume $f(x)$ is multi-bit output) and encrypt, a third party interprets $f(x)$ as an attribute. and can homomorphically evaluate a circuit g to obtain a ciphertext for $g \circ f$: $\mathbf{C}_{g \circ f} = \mathbf{B}_{g \circ f} + g(f(x)) \cdot \mathbf{G}$. In particular, we can first encode x and then apply a function f on $\mathbf{C}_i$ to obtain $\mathbf{C}_f = \mathbf{B}_f + f(x) \cdot \mathbf{G}$, before we encrypt with $\mathbf{C}_f$ as the public key. This can be interpreted as a pre-processing step. Importantly, $\mathbf{B}_f$ does not depend on x. In our scheme, f will be a pseudo-random function (PRF) with hardcoded input and x will be a PRF key. Thus,

[1] For technical reasons we use the inverted condition of $f(x) = 0$ to authorize decryption, instead of the more common $f(x) = 1$.

$\mathbf{B}_f$ and $f(x)$ reveal nothing about the PRF key x. In this case, we encrypt using $\mathbf{C}_f$ as the LWE public key and $f(x)$ as the resulting (new) attribute, where f is a multi-bit output PRF circuit.

The [34] *IBE.* We give a detailed review of the adaptive IBE from [34] that contains a core idea which we adapt in our scheme. In the process, we introduce notation that we use heavily in the rest of the section. In IBE, ciphertexts are associated with a identity string y, and secret keys sk_x with a identity string x. Decryption succeeds if and only if $y = x$. For security, the adversary must declare a challenge identity y^* that must be different from any identity x for which it has requested a secret key sk_x. The adversary first receives the public parameters, and then requests secret keys adaptively before and after it announces y^*.

The Core Idea from the [34] *IBE.* At a high level the scheme works as follows: to encrypt to identity x, the encryptor samples a PRF key k and uses the [7] ABE to obtain a PRF output r_x for PRF input (k, x). The value r_x is the attribute in the ABE, and the identity x is tied to the ciphertext by properties of the ABE [7]. The authority uses its own PRF key k^* to obtain a PRF output r_x^* for input (k^*, x). It outputs a secret key for x that decrypts if $r_x \neq r_x^*$ which happens with high probability over the choice of k, k^* and correctness follows. In general, we denote with r_z^* PRF values using key k^* and with r_z PRF values using key k. In the security reduction (and the real system) the secret key for any identity x only reveals r_x^* that is obtained using k^*. Moreover, the scheme public parameters are set up in a way such that the challenge ciphertext ct_{y^*} reveals the value $r_{y^*}^*$ and nothing about k^*. In addition, the value $r_{y^*}^*$ does not appear in any requested secret key sk_x (for which we use k^*) since the challenge identity y^* must be different from any requested identity x. This allows for an adaptive reduction under (adaptive) PRF security and ABE techniques from [7]. The IBE in [34] relies on the properties of the ABE in [7] which is known to be selectively secure for general policies (i.e. boolean circuits). Specifically for the IBE policy, [34] uses a PRF in the pre-processing step and obtains adaptive IBE from the selective ABE in [7]. Below we give a slightly different presentation of the original scheme in [34, Section 1.2].[2]

IBE Circuits. We first define three circuits used in our IBE overview, following closely [34]. For a PRF $f : \{0,1\}^{\ell_k} \times \{0,1\}^{\ell_{\mathsf{id}}} \rightarrow \{0,1\}^{\ell_{\mathsf{out}}}$ and every $y \in \{0,1\}^{\ell_{\mathsf{id}}}$, define f_y to be the circuit with hardcoded PRF input y. The circuit f_y takes as input the PRF key k and outputs $f_y(k) = f(k, y)$. For every $r^* \in \{0,1\}^{\ell_{\mathsf{out}}}$ define the *equality test* circuit $g_{r^*} : \{0,1\}^{\ell_{\mathsf{out}}} \rightarrow \{0,1\}$ as $g_{r^*}(r) = 1$ if $r = r^*$, and 0 otherwise. Given r^*, we can construct g_{r^*} on the fly. Finally, we define circuit composition in the natural way: $(g_r \circ f_y)(\cdot) = g_r(f_y(\cdot))$.

[2] [34, Section 1.2] for IBE uses the circuit $\bar{I}_r(r') = 1$, if $r' \neq r$ and 0 otherwise. We use $I_r(r') = 1$, if $r' = r$ and 0 otherwise.

IBE Construction. We review the full IBE construction, starting with the public parameters pp and master secret key msk. For a fixed public "gadget" matrix $\mathbf{G} \in \mathbb{Z}_q^{n\times m}$, the authority samples: $\mathsf{pp} = \{\mathbf{u}, \mathbf{A}, \mathbf{B}_1, \dots, \mathbf{B}_{\ell_k}\}$ and $\mathsf{msk} = \{\mathbf{T_A}, k^*\}$, with $\mathbf{u} \in \mathbb{Z}_q^n$, $\mathbf{A}, \mathbf{B}_i \in \mathbb{Z}_q^{n\times m}$, $\mathbf{T_A} \in \mathbb{Z}^{m\times m}$ with low-norm entries, and $k^* \in \{0,1\}^{\ell_k}$ is a random PRF key. For any $\mathbf{F} \in \mathbb{Z}_q^{n\times m}$, the matrix $\mathbf{T_A}$ allows to sample a low-norm matrix $\mathbf{R_F}$ such that $[\mathbf{A}||\mathbf{F}] \cdot \mathbf{R_F} = \mathbf{u}$. We implicitly assume a public circuit description of a PRF f as defined above. To encrypt a message $\mathsf{msg} \in \{0,1\}$ to identity $y \in \{0,1\}^{\ell_{\mathsf{id}}}$, first sample a fresh PRF key $k \in \{0,1\}^{\ell_k}$ and set $r_y = f(k, y)$. Then, derive the circuit f_y, and use $\{\mathbf{B}_i\}$ from pp to homomorphically compute the matrix $\mathbf{B}_{f_y}$ that depends only on f_y, and thus "ties" y to $\mathbf{B}_{f_y}$. Finally, output the ciphertext ct_y (we abuse notation and represent many bits with $r_y \cdot \mathbf{G}$):

$$\mathsf{ct}_y = \left(y,\ \ r_y,\ \ \mathbf{s}^\mathsf{T}\mathbf{A} + \mathbf{e}_a^\mathsf{T},\ \ \mathbf{s}^\mathsf{T}\left(\mathbf{B}_{f_y} + r_y \cdot \mathbf{G}\right) + \mathbf{e}^\mathsf{T},\ \ \mathbf{s}^\mathsf{T}\mathbf{u} + e_{\mathsf{msg}} + \mathsf{msg}\lceil q/2 \rceil\right)$$

where $\mathbf{s}$ is random and the error terms are of low norm. To sample a secret key sk_x the authority first uses msk to set $r_x^* = f(k^*, x)$, and then derives the circuit $C_x = g_{r_x^*} \circ f_x$. Then it uses $\{\mathbf{B}_i\}$ and C_x to homomorphically compute $\mathbf{B}_{C_x}$. Finally, it uses $\mathbf{T_A}$ to sample a low-norm $\mathbf{R}_{\mathsf{id}}$ such that $[\mathbf{A}||\mathbf{B}_{C_x}] \cdot \mathbf{R}_x = \mathbf{u}$, and outputs $\mathsf{sk}_x = (\ x,\ r_x^*,\ \mathbf{R}_x\ )$. For sk_x and ct_x, i.e. when $x = y$, the decryptor uses r_x^* to homomorphically compute $g_{r_x^*}$ using ct_x since r_x is public, and obtains:

$$\mathsf{ct}'_x = \left(x,\ \ r_x,\ \ \mathbf{s}^\mathsf{T}\mathbf{A} + \mathbf{e}_a^\mathsf{T},\ \ \mathbf{s}^\mathsf{T}\left(\mathbf{B}_{C_x} + g_{r_x^*}(r_x) \cdot \mathbf{G}\right) + \mathbf{e}_d^\mathsf{T},\ \ \mathbf{s}^\mathsf{T}\mathbf{u} + e_m + m\lceil q/2 \rceil\right).$$

Correctness holds with overwhelming probability over the choice of k, k^* which implies that $r_x \neq r_x^*$. Thus $g_{r_x^*}(r_x) = 0$, and if we multiply $(\mathbf{s}^\mathsf{T}\mathbf{A} + \mathbf{e}_a^\mathsf{T} \mid\mid \mathbf{s}^\mathsf{T}\mathbf{B}_{C_x} + \mathbf{e}_d^\mathsf{T})$ with $\mathbf{R}_x$ we obtain $\mathbf{s}^\mathsf{T}\mathbf{u} + \tilde{e}$ and we can recover m.

IBE Security Proof. In the proof, the public parameters are sampled without a "trapdoor" $\mathbf{T_A}$ for $\mathbf{A}$, and with $\mathbf{B}_i = \mathbf{AR}_i - k_i^* \cdot \mathbf{G}$ where $\mathbf{R}_i$ is of low-norm, and k_i^* is the i-th bit of k^*:

$$\mathsf{pp}_{\mathsf{sim}} = \{\mathbf{u}, \mathbf{A}, \mathbf{AR}_1 - k_1^* \cdot \mathbf{G}, \dots, \mathbf{AR}_{\ell_k} - k_{\ell_k}^* \cdot \mathbf{G}\} \text{ and } \mathsf{msk}_{\mathsf{sim}} = \{\mathbf{R}_1, \dots, \mathbf{R}_{\ell_k}, k^*\}$$

Using standard arguments [7,8], we get the *statistical* indistinguishability, $\mathsf{pp} \overset{s}{\approx} \mathsf{pp}_{\mathsf{sim}}$. For a secret key sk_x, the simulator first derives C_x and $\mathbf{B}_{C_x}$. It then uses $\mathbf{A}, \{\mathbf{R}_i\}$ to compute $\mathbf{R}_{C_x}$ such that, $\mathbf{B}_{C_x} = \mathbf{AR}_{C_x} - g_{r_x^*}(r_x^*) \cdot \mathbf{G} = \mathbf{AR}_{C_x} - \mathbf{G}$. Finally, it uses $\mathbf{R}_{C_x}$ and $\mathbf{G}$ to sample a low-norm $\mathbf{R}_x$ such that $[\mathbf{A}||\mathbf{B}_{C_x}] \cdot \mathbf{R}_x = \mathbf{u}$ while maintaining indistinguishability, under appropriate parameters, from the real system that uses $\mathbf{T_A}$ to sample $\mathbf{R}_x$. To encrypt a random bit to a challenge identity y^*, the simulator "samples" a PRF key k with $k = k^*$. It sets $r_{y^*}^* = f(k^*, y)$ and computes the public key for ct_{y^*} (we abuse notation and represent many bits with $r_{y^*}^* \cdot \mathbf{G}$): $\mathbf{B}_{f_{y^*}} + r_{y^*}^* \cdot \mathbf{G} = \mathbf{AR}_{f_{y^*}}$. This results in a public key for the challenge ciphertext ct_{y^*} that does not contain any "$\mathbf{G}$-trapdoor", and ct_{y^*} is proven [7,34] to be indistinguishable from the uniform distribution under LWE. Note that the adversary is not allowed to request a secret key for y^*. Thus, the value $r_{y^*}^*$ remains pseudorandom by (adaptive) PRF security, even given the PRF values in sk_x. This implies that the adversary cannot distinguish the real game from the simulated one with more than negligible advantage.

Our PRE Construction. In this section, we describe our adaptive PRE scheme. We set the PRE string alphabet to $\{0,1\}$ and let strings to be vectors in $(\{0,1\})^{\ell_{\text{str}}}$ where ℓ_{str} is an a-priori fixed bound. To simplify notation, we write $y \in \{0,1\}^{\ell_{\text{str}}}$ instead of $y \in (\{0,1\})^{\ell_{\text{str}}}$.

PRE Scheme Outline. We first outline the intuition behind our scheme before the technical overview. To encrypt to a string $y \in \{0,1\}^{\ell}$, we encrypt with respect to all ℓ prefixes y. We sample a PRF key k^{enc} and compute a PRF value for *every* prefix of y, similar to the IBE in [34]. This allows an encryption to y to carry bit-by-bit its lineage, which could potentially allow for a bit-by-bit key extension as we discuss in Sect. 1.2. Our scheme differs from the generic IBE construction in the key generation phase. Our msk contains two PRF keys $k^{\mathsf{real}}, k^{\mathsf{fake}}$. A secret key sk_x ties together all the ℓ prefixes of $x \in \{0,1\}^{\ell}$, and contains ℓ PRF values for ℓ comparison circuits g, as described in the previous section. We use k^{fake} to generate the PRF comparison values for the first $(\ell - 1)$ prefixes of x, and k^{real} to generate the PRF comparison value for the ℓ-th prefix, which is x itself. The role of using two PRF keys only matters for the security proof and not for correctness. To decrypt ct_y using sk_x, we select the ciphertext components for which x is a prefix of y, if any, and proceed similar to the previous section. In the security proof, we embed the key k^{real} in the public parameters. This way, during key generation we are able to embed a trapdoor for the last prefix of x that is x itself, using a PRF value under k^{real}. We use k^{fake} to obtain the rest of the $(\ell - 1)$ PRF values for each of the (shorter) $(\ell - 1)$ prefixes of x, but k^{fake} does not imply a trapdoor. The PRF values for y^* and all its prefixes for the challenge ciphertext ct_{y^*} are obtained using $k^{\mathsf{enc}} = k^{\mathsf{real}}$. While this allows to argue LWE, we also have to argue that the "sampling" $k^{\mathsf{enc}} = k^{\mathsf{real}}$ during ct_{y^*} goes unnoticed by the adversary. This follows because the PRF values in ct_{y^*} remain pseudorandom given the PRF values in any secret key sk_x. In more detail, notice that any sk_x contains a PRF value under k^{real} *only* for x itself, and not for any of its (shorter) prefixes. Any (shorter) prefix of x except x itself, is allowed to be in y^*, but the PRF values for these prefixes, for any secret key, are given using k^{fake}. By the PRE security game, x itself cannot equal to y^* or be any prefix of y^*. That is, the PRF values under k^{real} for ct_{y^*} and any sk_x do not intersect, and the pseudorandomness claim follows. The reduction loss is polynomial in $\lambda, \ell_{\text{str}}$, essentially because we have to account for the pseudorandomness loss that is incurred by the PRF values that are given in ct_{y^*} and every sk_x. There are at most ℓ_{str} PRF values in either ct_{y^*} or any sk_x. Assuming the adversary makes $Q = \text{poly}(\lambda)$ key queries, the reduction loss is bounded by $(Q+1) \cdot \ell_{\text{str}} \cdot \mathsf{Adv}^{\mathsf{PRF}}$ where $\mathsf{Adv}^{\mathsf{PRF}} = \text{negl}(\lambda, \ell_{\text{str}})$ is the pseudorandomness reduction loss for one PRF value. Note that instead of k^{fake} we could use random values directly, but a PRF key simply allows to derandomize that part of the secret key generation. Details follow.

PRE Circuits. In addition to the equality test circuit g from the previous section, we rely on a circuit that computes a PRF value for any given prefix of a string y. This allows us to derive public keys for *every* prefix of y. For a PRF f :

$\{0,1\}^{\ell_k} \times \{0,1\}^{2\lambda} \to \{0,1\}^{\ell_{\text{out}}}$ and for every $y \in \{0,1\}^{\ell_{\text{str}}}$, define $f_{y,i}$ to be the circuit with hardcoded PRF input: $y(1,i)$. On input a key k, the circuit $f_{y,i}$ outputs $f_{y,i}(k) = f(k,\ y(1,i))$. For every $y \in \{0,1\}^{\ell_{\text{str}}}$ and $i \in [\ell_{\text{str}}]$, denote with $y(1,i)$ the output of an Encode(y,i) procedure that maps the string $i||y$ to 2λ bits, e.g. assuming collision-resistance and an error-correcting code. We refer to the full version for details. Thus, we can set $\ell_{\text{str}} = 2\lambda$ in the definition of f since we evaluate f on the encoding of a string y, and not the actual string y.

PRE Construction. We first define the public parameters and master secret key. The authority samples: $\mathsf{pp} = \{\mathbf{u}, \mathbf{A}, \mathbf{B}_1, \ldots, \mathbf{B}_{\ell_k}\}$ and $\mathsf{msk} = \{\mathbf{T_A}, k^{\mathsf{real}}, k^{\mathsf{fake}}\}$, with $\mathbf{u} \in \mathbb{Z}_q^n$, $\mathbf{A}, \mathbf{B}_i \in \mathbb{Z}_q^{n\times m}$, $\mathbf{T_A} \in \mathbb{Z}^{m\times m}$ with low-norm entries, and $k^{\mathsf{real}}, k^{\mathsf{fake}}$ are PRF keys. Implicitly, we assume that pp contain the description of a PRF function f and of an encoding procedure Encode. To encrypt a message $\mathsf{msg} \in \{0,1\}$ to a string $y \in \{0,1\}^{\ell}$, first sample a PRF key k^{enc} and set $r_i^{\mathsf{enc}} = f(k^{\mathsf{enc}}, y(1,i))$ for $i \in [\ell]$. Then, derive the circuit $f_{y,i}$ and use $\{\mathbf{B}_i\}$ to compute the matrix $\mathbf{B}_{f_{y,i}}$ that depends only on $f_{y,i}$, and thus "ties" $y(1,i)$ to $\mathbf{B}_{f_{y,i}}$. Therefore, the sequence of ℓ matrices $\left(\mathbf{B}_{f_{y,i}}\right)_{i\in[\ell]}$ is tied with y. Finally, set $\mathbf{c}_i$ as follows (we abuse notation and denote many bits with r_i^{enc}),

$$\mathbf{c}_i = \mathbf{s}^{\mathsf{T}}\left(\mathbf{B}_{f_{y,i}} + r_i^{\mathsf{enc}} \cdot \mathbf{G}\right) + \mathbf{e}_i^{\mathsf{T}} \qquad \text{for } i \in [\ell]$$

and output the ciphertext:

$$\mathsf{ct}_y = \left(y,\ \ \{r_i^{\mathsf{enc}}\}_{i\in[\ell]},\ \ \mathbf{s}^{\mathsf{T}}\mathbf{A} + \mathbf{e}_a^{\mathsf{T}},\ \ \{\mathbf{c}_i\}_{i\in[\ell]},\ \ \mathbf{s}^{\mathsf{T}}\mathbf{u} + e_{\mathsf{msg}} + \mathsf{msg}\lceil q/2 \rceil\right).$$

To sample a secret key sk_x for $x \in \{0,1\}^{\ell}$ using $\mathsf{msk} = \{\mathbf{T_A}, k^{\mathsf{real}}, k^{\mathsf{fake}}\}$, the authority derives PRF values for the first $(\ell-1)$ prefixes of x with the key k^{fake}. For the last prefix $x(1,\ell)$ which is x itself it uses the key k^{real}:

$$r_i = f(k^{\mathsf{fake}},\ x(1,i))\ \ \forall i \in [\ell-1] \ \text{ and } \ r_\ell = f(k^{\mathsf{real}},\ x(1,\ell)).$$

Then, for each prefix $i \in [\ell]$, it derives the circuit $C_i = g_{r_i} \circ f_{x,i}$, and uses $\{\mathbf{B}_i\}$ to compute $\mathbf{B}_{C_i}$. Notice that C_i contains three pieces of information: a prefix index i, a prefix $x(1,i)$, and a PRF value r_i for that prefix. Thus, the homomorphically derived public key $\mathbf{B}_{C_i}$ is "tagged" exactly with this information. Finally, it uses $\mathbf{T_A}$ to sample a low-norm $\mathbf{R}_x$ such that $[\mathbf{A} \mid\mid \mathbf{B}_{C_1} \mid\mid \cdots \mid\mid \mathbf{B}_{C_\ell}] \cdot \mathbf{R}_x = \mathbf{u}$, and outputs $\mathsf{sk}_x = (x,\ \{r_i\}_{i\in[\ell]},\ \mathbf{R}_x)$. The decryptor selects the ℓ ciphertext components of ct_y for which x is a prefix of y, if any, otherwise decryption is not authorized. Then, it uses $\{r_i\}_{i\in[\ell]}$ from sk_x to homomorphically compute g_{r_i} on $\mathbf{c}_i$ and derive: $\mathbf{c}_i' = \mathbf{s}^{\mathsf{T}}\left(\mathbf{B}_{C_i} + g_{r_i}(r_i^{\mathsf{enc}}) \cdot \mathbf{G}\right) + \tilde{\mathbf{e}}_i^{\mathsf{T}}$ for $i \in [\ell]$. With overwhelming probability over the choice of $k^{\mathsf{enc}}, k^{\mathsf{real}}, k^{\mathsf{fake}}$ it follows that $r_i \neq r_i^{\mathsf{enc}}$ for all $i \in [\ell]$. Thus, $g_{r_i}(r_i^{\mathsf{enc}}) = 0$ for all $i \in [\ell]$ and the decryptor can use $\mathbf{R}_x$ to recover m.

PRE Security Proof. The reason for using two PRF keys is made clear in the security proof that we describe next. The public parameters are now sampled without a "trapdoor" for $\mathbf{A}$. We set $\mathbf{B}_i = \mathbf{A}\mathbf{R}_i - k_i^{\mathsf{real}} \cdot \mathbf{G}$ where $\mathbf{R}_i$ is of low-norm, and k_i^{real} is the i-th bit of k^{real}, $\mathsf{pp}_{\mathsf{sim}} = \{\mathbf{u}, \mathbf{A}, \mathbf{A}\mathbf{R}_1 - k_1^{\mathsf{real}} \cdot \mathbf{G}, \ldots, \mathbf{A}\mathbf{R}_{\ell_k} - k_{\ell_k}^{\mathsf{real}} \cdot \mathbf{G}\}$

and $\mathsf{msk}_{\mathsf{sim}} = \{\mathbf{R}_1, \ldots, \mathbf{R}_{\ell_k}, k^{\mathsf{real}}, k^{\mathsf{fake}}$. As in the previous section, we have $\mathsf{pp} \overset{s}{\approx} \mathsf{pp}_{\mathsf{sim}}$. To sample a secret key sk_x, the simulator derives C_i and then computes $\mathbf{B}_{C_i}$. It uses $\mathbf{A}, \{\mathbf{R}_i\}$ to compute $\mathbf{R}_{C_i}$ such that for $i \in [\ell - 1]$, $\mathbf{B}_{C_i} = \mathbf{A}\mathbf{R}_{C_i} - g_{r_i^{\mathsf{fake}}}(r_i^{\mathsf{real}}) \cdot \mathbf{G} = \mathbf{A}\mathbf{R}_{C_i}$, and for $i = \ell$, $\mathbf{B}_{C_\ell} = \mathbf{A}\mathbf{R}_{C_\ell} - g_{r_\ell^{\mathsf{real}}}(r_\ell^{\mathsf{real}}) \cdot \mathbf{G} = \mathbf{A}\mathbf{R}_{C_\ell} - \mathbf{G}$. It follows that for all $x \in \{0,1\}^\ell$, the first $(\ell - 1)$ matrices $\{\mathbf{B}_{C_i}\}_{i \in [\ell-1]}$ do not have a $\mathbf{G}$-trapdoor since we use k^{fake}, but the matrix $\mathbf{B}_{C_\ell}$ will always have one because we use k^{real}. Consequently, the simulator uses $\mathbf{R}_{C_\ell}$ with $\mathbf{G}$ to sample a low-norm $\mathbf{R}_x$ such that $[\mathbf{A} \,||\, \mathbf{B}_{C_1} \,||\, \cdots \,||\, \mathbf{B}_{C_\ell}] \cdot \mathbf{R}_x = \mathbf{u}$, while maintaining indistinguishability with respect to $\mathbf{R}_x$, when $\mathbf{R}_x$ is sampled in the real game with $\mathbf{T}_\mathbf{A}$, under properties of the $\mathbf{G}$-trapdoor [25] for an appropriate choice of parameters. To encrypt a random $\mathsf{msg} \overset{\$}{\leftarrow} \{0,1\}$ to a challenge string $y^* \in \{0,1\}^\ell$, the simulator "samples" k^{enc} with $k^{\mathsf{enc}} = k^{\mathsf{real}}$ and computes $r_i^{\mathsf{enc}} = f(k^{\mathsf{real}}, y^*(1,i))$ for $i \in [\ell]$. It then computes for all i, $\mathbf{B}_{f_{y^*,i}} + r_i^{\mathsf{enc}} \cdot \mathbf{G} = \mathbf{A}\mathbf{R}_{f_{y^*,i}}$ and also $\mathbf{c}_i^* = \mathbf{s}^\intercal \mathbf{A}\mathbf{R}_{f_{y^*,i}} + \mathbf{e}_i^\intercal$ for $i \in [\ell]$ and outputs the ciphertext: $\mathsf{ct}_{y^*} = \left(y^*,\ \{r_i^{\mathsf{enc}}\}_{i \in [\ell]},\ \mathbf{s}^\intercal\mathbf{A} + \mathbf{e}_a^\intercal,\ \{\mathbf{c}_i^*\}_{i \in [\ell]},\ \mathbf{s}^\intercal\mathbf{u} + e_{\mathsf{msg}} + \mathsf{msg}\lceil q/2 \rceil\right)$. We have to make sure the noise distribution in ct_{y^*} is the right one, but we defer to Sect. 4 about this. At this point the challenge ciphertext ct_{y^*} does not contain any "$\mathbf{G}$-trapdoor", and we can use LWE to argue that msg remains hidden. As we argued at the start of this overview, during simulation the PRF values of the challenge ciphertext ct_{y^*} remain pseudorandom even given the PRF values of all requested secret keys sk_x, since these PRF values do not intersect. Thus, a PPT adversary cannot distinguish the real game from the simulated one with probability more that $1/2 + \mathrm{negl}(\lambda, \ell_{\mathrm{str}})$.

Extending Functionality. Our PRE scheme can be made *weakly-hiding* using obfuscation under LWE [19,37]. That is, a ciphertext ct_y additionally hides y, given any secret key sk_x that is not authorized to decrypt ct_y.

1.2 Open Problems

The main open problem from our work is whether our scheme can be modified to support secret key extension in a secure way, as required by HIBE. In HIBE, a secret key sk_x for an identity $x \in \{0,1\}^\ell$ can be extended to a key $\mathsf{sk}_{x||x'}$ for $x' \in \{0,1\}^{\ell'}$ without requiring the help of the authority. This allows to create a key hierarchy in a tree like manner. Our scheme can be slightly modified so that the keys can be extended using the basis delegation algorithms from [1,10,25]. To extend a key bit-by-bit, we would sample a fresh PRF seed to compute the circuits $C_{\ell+1}, \ldots, C_{\ell+\ell'}$, then compute the matrices $\mathbf{B}_{C_{\ell+1}}, \ldots, \mathbf{B}_{C_{\ell+\ell'}}$, and extend the key sk_x to a key $\mathsf{sk}_{x||x'}$ for $\mathbf{B}_{C_1}, \ldots, \mathbf{B}_{C_{\ell+\ell'}}$. In our case, to get meaningful HIBE security we have to consider the Shi and Waters [33] security model, which differs only in the key request phase, as follows: the adversary makes three types of queries: secret key and extension queries for which it receives a pointer to the requested secret key that gets sampled (and stored) but not its value, and reveal queries that return the actual (stored) secret key value this time. Unfortunately, the key extension we sketched above does not satisfy this model, but that was our motivation behind issuing secret keys that carry their lineage in a bit-by-bit fashion. Another interesting direction, is whether there are applications of pre-computing on the public parameters before encryption, possibly using a key extension mechanism in some context other than HIBE.

2 Preliminaries

Notation. Let PPT denote probabilistic polynomial time. We use the notation $[n] \stackrel{\text{def}}{=} \{1, \ldots, n\}$. All logarithms $\log(\cdot)$ are taken to the base 2. For a distribution or a random variable $\mathcal{D}$, we let $x \leftarrow \mathcal{D}$ denote the process of sampling x according to $\mathcal{D}$. For a randomized algorithm Alg we let $y \leftarrow \mathsf{Alg}(x)$ denote the process of running $\mathsf{Alg}(x)$ with fresh randomness and taking y as the output. For a finite set S we let $x \stackrel{\$}{\leftarrow} S$ denote sampling x uniformly at random from S. We use λ to denote the security parameter. We say that a function $g(\lambda)$ is negligible in λ, denoted by $\text{negl}(\lambda)$, if $g(\lambda) = o(1/\lambda^c)$ for all $c \in \mathbb{N}$. For random variables X, Y with support $\mathcal{X}, \mathcal{Y}$ respectively, we define the *statistical distance* $\mathsf{SD}(X, Y) \stackrel{\text{def}}{=} \frac{1}{2} \sum_{u \in \mathcal{X} \cup \mathcal{Y}} |\Pr[X = u] - \Pr[Y = u]|$. We say that two ensembles of random variables $X = \{X_\lambda\}, Y = \{Y_\lambda\}$ are *statistically indistinguishable*, denoted by $X \stackrel{s}{\approx} Y$, if $\mathsf{SD}(X_\lambda, Y_\lambda) = \text{negl}(\lambda)$. For a distribution ensemble $D = \{D_\lambda\}$ we let D_λ^m denote a vector of m samples, that are sampled independently from D_λ. For any integer $q \geq 2$, we let $\mathbb{Z}_q$ denote the ring of integers modulo q, and represent elements of $\mathbb{Z}_q$ as integers in the range $(-q/2, q/2]$. We define the absolute value $|x|$ of $x \in \mathbb{Z}_q$ by taking its representative in this range. For a vector $\mathbf{v}$ let $\|\mathbf{v}\|$ and $\|\mathbf{v}\|_\infty$ denote its ℓ_2 and ℓ_∞ norm, respectively. For a matrix $\mathbf{X} = \{\mathbf{x}_1, \ldots, \mathbf{x}_n\}$ where $\{\mathbf{x}_i\}_{i=1,\ldots,n}$ is its ordered set of column vectors, denote $\|\mathbf{X}\| = \max_{1 \leq i \leq n} \|\mathbf{x}_i\|$, and $\|\mathbf{X}\|_\infty = \max_{1 \leq i \leq n} \|\mathbf{x}_i\|_\infty$. Let $\mathbf{X}\|\mathbf{X}'$ denote the ordered concatenation of the ordered column set of $\mathbf{X}$ and $\mathbf{X}'$. We write $\mathbf{F} = \{\mathbf{X}_1, \ldots, \mathbf{X}_n\}$ to denote $\mathbf{F} = \mathbf{X}_1 \| \cdots \| \mathbf{X}_n$. For an ordered set $\mathbf{X} = \{\mathbf{x}_1, \ldots, \mathbf{x}_n\}$ of n linearly independent vectors in $\mathbb{R}^m$, denote as $\widetilde{\mathbf{X}} = \{\widetilde{\mathbf{x}}_1, \ldots, \widetilde{\mathbf{x}}_n\}$ its *Gram-Schmidt orthogonalization.* For any matrix $\mathbf{X} \in \mathbb{Z}^{m \times m}$ we denote with $s_1(\mathbf{X})$ its largest singular value. It holds that $\|\widetilde{\mathbf{X}}\| \leq \|\mathbf{X}\| \leq s_1(\mathbf{X}) \leq m\|\mathbf{X}\|_\infty$.

Lemma 1 (LHL [1,13]). *Let q be a prime and $m > (n+1)\log q + \omega(\log n)$. Let matrix $\mathbf{R} \stackrel{\$}{\leftarrow} \{-1, 1\}^{m \times k}$ where $k = k(n)$ is polynomial in n. Let matrix $\mathbf{A} \stackrel{\$}{\leftarrow} \mathbb{Z}_q^{n \times m}$ and matrix $\mathbf{B} \stackrel{\$}{\leftarrow} \mathbb{Z}_q^{n \times k}$. Then, the distribution $(\mathbf{A}, \mathbf{AR})$ is statistically close to the distribution $(\mathbf{A}, \mathbf{B})$.*

2.1 Lattice Preliminaries

Lattices and Cosets. For positive integers n, q and arbitrary $\mathbf{A} \in \mathbb{Z}_q^{n \times m}$, define the full-rank m-dimensional q-ary lattice, $\Lambda^\perp(\mathbf{A}) = \{\mathbf{x} \in \mathbb{Z}^m \ : \ \mathbf{Ax} = \mathbf{0} \bmod q\}$. For any $\mathbf{u} \in \mathbb{Z}_q^n$ define the coset $\Lambda_\mathbf{u}^\perp(\mathbf{A}) = \{\mathbf{x} \in \mathbb{Z}^m \ : \ \mathbf{Ax} = \mathbf{u} \bmod q\}$.

Discrete Gaussian Distribution. The m-dimensional Gaussian distribution centered at zero with positive real parameter σ is defined by its probability density function: $\rho_\sigma(\mathbf{x}) = \exp(-\pi\|\mathbf{x}\|^2/\sigma^2)$. For a lattice $\Lambda \subset \mathbb{R}^m$ the discrete Gaussian distribution over Λ with parameter σ, is defined for all $\mathbf{x} \in \Lambda$ by $\mathcal{D}_{\Lambda,\sigma} = \rho_\sigma(\mathbf{x})/\sum_{\mathbf{x}\in\Lambda}\rho_\sigma(\mathbf{x})$, and takes the zero value for all other $\mathbf{x}$. We require the following lemma.

Lemma 2 ([15,26,30]). *For the discrete Gaussian distribution $\mathcal{D}_{\mathbb{Z}^m,\sigma}$ over $\mathbb{Z}^m$ with parameter σ, it holds* $\Pr[\ \|\mathbf{v}\| > \sigma\sqrt{m} : \mathbf{v} \leftarrow \mathcal{D}_{\mathbb{Z}^m,\sigma}\ ] = \mathrm{negl}(m)$.

Bounded Distributions and Swallowing. We review the following from [8], that for our purposes apply to discrete Gaussian distributions over the integers. For $\beta = \beta(\lambda)$ we say that a distribution ensemble $\chi = \{\chi_\lambda\}$ over $\mathbb{Z}$ is *β-bounded* if $\Pr[|x| > \beta\ :\ x \leftarrow \chi_\lambda] = \mathrm{negl}(\lambda)$. For two bounded distribution ensembles $\chi = \{\chi_\lambda\}, \chi' = \{\chi'_\lambda\}$ over $\mathbb{Z}$, we say that χ' is *swallowing* for χ if $\chi' \overset{s}{\approx} \chi' + \chi$.

Corollary 1 ([8]). *For a $\beta(\lambda)$-bounded distribution ensemble $\chi = \{\chi_\lambda\}$ over $\mathbb{Z}$, and $\beta'(\lambda) = \beta(\lambda)\cdot\lambda^{\omega(1)}$, there exists an efficiently sampleable and $\beta'(\lambda)$-bounded ensemble $\chi' = \{\chi'_\lambda\}$ over $\mathbb{Z}$ that is swallowing for χ.*

Short Integer Solution. For positive integers n, m, q and positive real β, the short integer solution (SIS) problem for a random matrix $\mathbf{A} \in \mathbb{Z}_q^{n\times m}$ asks to find a *non-zero* vector $\mathbf{x} \in \mathbb{Z}_q^m$ such that $\mathbf{Ax} = \mathbf{0} \bmod q$ and $\|\mathbf{x}\| \le \beta$. The works of [2,15,26] show that for any $m = \mathrm{poly}(n)$, $\beta > 0$ and sufficiently large $q \ge \beta\cdot\mathrm{poly}(n)$ the (n, m, q, β)-SIS problem is as hard as approximating certain worst-case lattice problems such as GapSVP and SIVP on n-dimensional lattices to within a $\beta\cdot\mathrm{poly}(n)$ factor.

Learning with Errors. The learning with errors (LWE) assumption was introduced by Regev in [30] and states the following. Let n, q be positive integers and χ be a β-bounded distribution on $\mathbb{Z}_q$, all parameterized by the security parameter λ. The LWE assumption says that for all polynomial m the following distributions are computationally indistinguishable $(\mathbf{A}, \mathbf{s}^\intercal\mathbf{A} + \mathbf{e}^\intercal) \overset{c}{\approx} (\mathbf{A}, \mathbf{u}^\intercal)\ :\ \mathbf{A} \overset{\$}{\leftarrow} \mathbb{Z}_q^{n\times m}, \mathbf{s} \overset{\$}{\leftarrow} \mathbb{Z}_q^n \mathbf{e} \leftarrow \chi^m$ and $\mathbf{u} \overset{\$}{\leftarrow} \mathbb{Z}_q^m$. The works of [25,28–30] show that the (n, m, q, χ)-LWE assumption is as hard as (quantumly) approximating certain worst-case lattice problems such as GapSVP and SIVP on n-dimensional lattices to within a $\tilde{O}(n\cdot q/\beta)$ factor.

Lattice Trapdoors. We require the following lemma which states that we can sample lattices with "trapdoors" that are statistically indistinguishable from random lattices, and the "trapdoor" allows to efficiently solve the SIS problem.

Lemma 3 (Lattice trapdoors [3,15,25]). *Let λ be the security parameter with lattice parameters n, m, q and norm bound β all parameterized by λ where $n = \lambda^{\Omega(1)}$, $m = \Omega(n\log q)$ and $\sigma \ge \|\widetilde{\mathbf{T}}_{\mathbf{A}}\|\cdot\omega(\sqrt{\log m})$. Then, there exists a tuple of efficient algorithms* (TrapGen, SamPre, Sam) *with the following syntax:*

- $\mathsf{TrapGen}(1^n, 1^m, q) \rightarrow (\mathbf{A}, \mathbf{T_A})$. *The* TrapGen *algorithm takes as input the matrix dimensions* n, m *and modulus* q*, and outputs a matrix* $\mathbf{A} \in \mathbb{Z}_q^{n\times m}$ *and trapdoor* $\mathbf{T_A} \in \mathbb{Z}^{m\times m}$.
- $\mathsf{Sam}(1^m, q) \rightarrow \mathbf{C}$. *The* Sam *algorithm takes as input matrix dimension* m *and modulus* q*, and outputs a matrix* $\mathbf{C} \in \mathbb{Z}^{m\times m}$.
- $\mathsf{SamPre}(\mathbf{A}, \mathbf{U}, \mathbf{T_A}, \sigma) \rightarrow \mathbf{C}$. *The* SamPre *algorithm takes as input matrices* $\mathbf{A} \in \mathbb{Z}_q^{n\times m}, \mathbf{U} \in \mathbb{Z}_q^{n\times m}$*, trapdoor* $\mathbf{T_A}$ *and parameter* σ*, and outputs a matrix* $\mathbf{C} \in \mathbb{Z}^{m\times m}$ *such that* $\mathbf{AC} = \mathbf{U}$.

Furthermore, the following properties are satisfied:

1. *For any* $(\mathbf{A}, \mathbf{T_A}) \leftarrow \mathsf{TrapGen}(1^n, 1^m, q)$, $\mathbf{U} \in \mathbb{Z}_q^{n\times m}$ *and* $\mathbf{C} \leftarrow \mathsf{SamPre}(\mathbf{A}, \mathbf{U}, \mathbf{T_A})$ *we have* $\mathbf{AC} = \mathbf{U}$ *and* $\|\mathbf{C}\| \leq \sqrt{m} \cdot \sigma$ *(with probability 1).*
2. *The statistical indistinguishability* $\mathbf{A} \overset{s}{\approx} \mathbf{A}'$ *where* $(\mathbf{A}, \mathbf{T_A}) \leftarrow \mathsf{TrapGen}(1^n, 1^m, q)$, $\mathbf{A}' \overset{\$}{\leftarrow} \mathbb{Z}_q^{n\times m}$.
3. *The statistical indistinguishability* $(\mathbf{A}, \mathbf{T_A}, \mathbf{C}) \overset{s}{\approx} (\mathbf{A}, \mathbf{T_A}, \mathbf{C}')$ *where* $(\mathbf{A}, \mathbf{T_A}) \leftarrow \mathsf{TrapGen}(1^n, 1^m, q)$, $\mathbf{C} \leftarrow \mathsf{Sam}(1^m, q)$, $\mathbf{U} \overset{\$}{\leftarrow} \mathbb{Z}_q^{n\times m}$, $\mathbf{C}' \leftarrow \mathsf{SamPre}(\mathbf{A}, \mathbf{U}, \mathbf{T_A})$.

Gadget Matrix. In this work we use the gadget lattice trapdoors formalized in [25]. Given integers $n \geq 1, q \geq 2$, the gadget matrix is defined as $\mathbf{G} = \mathbf{I}_n \otimes \mathbf{g}$ where $\mathbf{g} = (1, 2, 4, \ldots, 2^{n\times n \cdot \lceil \log q \rceil})$. The bit-decomposition function $\mathbf{G}^{-1} : \mathbb{Z}_q^{n\times m} \rightarrow \{0,1\}^{n\cdot\lceil \log q \rceil \times m}$ takes as input a matrix $\mathbf{A} \in \mathbb{Z}_q^{n\times m}$ and expands its every entry $a_{i,j} \in \mathbb{Z}_q$ to its binary decomposition in a column-wise manner. For simplicity we will assume that $\mathbf{G}$ has width m (otherwise we pad with zeros), and thus $\mathbf{G}^{-1}$ has range $\{0,1\}^{m\times m}$. For all $\mathbf{A} \in \mathbb{Z}_q^{n\times m}$ it holds that $\mathbf{G} \cdot \mathbf{G}^{-1}(\mathbf{A}) = \mathbf{A}$. A $\mathbf{G}$-trapdoor (or simply trapdoor) of a matrix $\mathbf{A} \in \mathbb{Z}_q^{n\times 2m}$ is a full-rank (over the integers) matrix $\mathbf{T_A} = \begin{bmatrix}\mathbf{R}\ \mathbf{I}_m\end{bmatrix}^T$ where $\mathbf{R} \in \mathbb{Z}^{m\times m}$ is a low-norm matrix, such that $\mathbf{AT_A} = \mathbf{G}$. Specifically, $\mathbf{A} = [\mathbf{A}'\|\mathbf{A}'\mathbf{R} + \mathbf{G}]$ for $\mathbf{A}' \overset{\$}{\leftarrow} \mathbb{Z}_q^{n\times m}$.

Lemma 4 (Lattice basis sampling (adapted) [1,10,15,25]**).** *Let* λ *be the security parameter with lattice parameters* n, m, q *and norm bound* σ *all parameterized by* λ *where* $n = \lambda^{\Omega(1)}$, $m = \Omega(n \log q)$*, and* $d \geq 1$*. Then, there exists a tuple of efficient algorithms* (SampleBasisLeft, SampleBasisRight) *with the following properties:*

- $\mathsf{SampleBasisLeft}(\mathbf{A}, \mathbf{B}, \mathbf{T_A}, \sigma) \rightarrow \mathbf{T_{A\|B}}$. *The* SampleBasisLeft *algorithm takes as input* $\mathbf{A} \in \mathbb{Z}_q^{n\times m}$ *and* $\mathbf{B} = \{\mathbf{B}_i\}_{i\in[d]} \in \mathbb{Z}_q^{n\times dm}$ *with* $\mathbf{B}_i \in \mathbb{Z}_q^{n\times m}$*, a trapdoor* $\mathbf{T_A} \in \mathbb{Z}^{m\times m}$ *and a norm bound* σ*, and outputs a basis* $\mathbf{T_{A\|B}} \in \mathbb{Z}^{(d+1)m\times m}$ *for* $\Lambda^{\perp}(\mathbf{A}\|\mathbf{B})$*. For any positive real* $\sigma \geq \|\widetilde{\mathbf{T}}_\mathbf{A}\| \cdot \omega(\sqrt{\log(dm)})$ *it holds that* $\|\widetilde{\mathbf{T}}_\mathbf{A}\| = \|\widetilde{\mathbf{T}}_{\mathbf{A}\|\mathbf{B}}\|$*. For any* $\mathbf{T_A}, \mathbf{T}'_\mathbf{A}$ *and* $\sigma \geq \max(\|\widetilde{\mathbf{T}}_\mathbf{A}\|, \|\widetilde{\mathbf{T}}'_\mathbf{A}\|) \cdot \omega(\sqrt{\log(dm)})$ *and the same lattice, the output distributions of* SampleBasisLeft *are within* $\mathrm{negl}(\lambda)$ *statistical distance.*
- $\mathsf{SampleBasisRight}(\mathbf{A}, \mathbf{B}, \mathbf{R}, \mathbf{G}, \sigma) \rightarrow \mathbf{T_{A\|B}}$ *takes as input* $\mathbf{A} \in \mathbb{Z}_q^{n\times m}$ *and* $\mathbf{B} = \{\mathbf{B}_i\}_{i\in[d]} \in \mathbb{Z}_q^{n\times dm}$ *with* $\mathbf{B}_i \in \mathbb{Z}_q^{n\times m}$, $\mathbf{R} \in \mathbb{Z}^{m\times m}$*, gadget matrix* $\mathbf{G}$ *with* $\mathbf{B}_d = \mathbf{AR} - \mathbf{G}$*, and a norm bound* σ *and outputs a basis* $\mathbf{T_{A\|B}} \in$

$\mathbb{Z}^{(d+1)m \times m}$ *for* $\Lambda^{\perp}(\mathbf{A}||\mathbf{B})$. *Note that* $\mathbf{T_F} = [\mathbf{R}||\mathbf{I}]^T$ *can be made into a* $\mathbf{G}$-*trapdoor for* $\Lambda^{\perp}(\mathbf{F})$ *with* $\mathbf{F} = \mathbf{A}||\mathbf{B}_d$ *and then into a basis for* $\mathbf{A}||\mathbf{B}$ *in a similar manner to above. Thus, for any positive real* $\sigma \geq \|\widetilde{\mathbf{T}}_{\mathbf{F}}\| \cdot \omega(\sqrt{\log(dm)})$ *it holds that* $\|\widetilde{\mathbf{T}}_{\mathbf{F}}\| = \|\widetilde{\mathbf{T}}_{\mathbf{A}||\mathbf{B}}\|$. *For any* $\mathbf{T_F}, \mathbf{T'_F}$ *and the same lattice, and* $\sigma \geq \max(\|\widetilde{\mathbf{T}}_{\mathbf{F}}\|, \|\widetilde{\mathbf{T}}'_{\mathbf{F}}\|) \cdot \omega(\sqrt{\log(dm)})$, *the output distributions of* SampleBasisRight *in this case are within* $\text{negl}(\lambda)$ *statistical distance.*

*Finally, there is an efficient algorithm (*SamPre*) that takes as input* $\mathbf{F} = \mathbf{A}||\mathbf{B} \in \mathbb{Z}_q^{n \times (d+1)m}$, *a basis* $\mathbf{T_F} \in \mathbb{Z}^{(d+1)m \times m}$ *for* $\Lambda^{\perp}(\mathbf{F})$, *any* $\mathbf{u} \in \mathbb{Z}_q^n$ *and parameter* $\sigma \geq \|\widetilde{\mathbf{T}}_{\mathbf{F}}\| \cdot \omega(\sqrt{\log(dm)})$, *and samples* $\mathbf{c} \in \mathbb{Z}^{(d+1)m}$ *with* $\|\mathbf{c}\| \leq \sigma\sqrt{(d+1)m}$ *and within* $\text{negl}(\lambda)$ *statistical distance from* $\mathcal{D}_{\Lambda^{\perp}_{\mathbf{u}}(\mathbf{F}),\sigma}$, *such that* $\mathbf{F} \cdot \mathbf{c} = \mathbf{u} \bmod q$.

Proposition 1. *[Lattice basis sampling indistinguishability (adapted) [1,10]] The output distributions of algorithms* SampleBasisLeft, SampleBasisRight *of Lemma 4 when given as input the same lattice, are within negligible statistical distance when* $\sigma > \max(s_1(\mathbf{T_A}),\ s_1(\mathbf{T_F})) \cdot \omega(\sqrt{\log(dm)})$.

2.2 Homomorphic Evaluation

We require homomorphic evaluation on ciphertexts and public keys.

Lemma 5 ([7]). *Let* λ *be the security parameter with lattice parameters* n, m, q *and norm bound* β *all parameterized by* λ *where* $n = \lambda^{\Omega(1)}$ *and* $m = \Omega(n \log q)$. *Let* $f : \{0,1\}^{\ell} \to \{0,1\}$ *be any circuit of depth at most* d. *Then, there exists a tuple of efficient deterministic algorithms* (EvalCT, EvalPK, EvalSimPK) *with the following properties:*

- EvalCT $(f, \{\mathbf{B}_i, x_i, \mathbf{c}_i\}_{i \in [\ell]}) \to \mathbf{c}_f$. *The* EvalCT *algorithm takes as input the description of a circuit* f, *matrices* $\mathbf{B}_i \in \mathbb{Z}_q^{n \times m}$, *the input bits* $x_i \in \{0,1\}$ *for* f, *and vectors* $\mathbf{c}_i$ *such that* $\mathbf{c}_i = \mathbf{s}^{\mathsf{T}}(\mathbf{B}_i + x_i\mathbf{G}) + \mathbf{e}_i^{\mathsf{T}}$ *with* $\|\mathbf{e}_i\|_\infty \leq \beta$. *It outputs a vector* $\mathbf{c}_f = \mathbf{s}^{\mathsf{T}}(\mathbf{B}_f + f(x) \cdot \mathbf{G}) + \mathbf{e}_f^{\mathsf{T}}$, *where the matrix* $\mathbf{B}_f$ *depends only on* $\mathbf{B}_i$ *and* f, *and* $\|\mathbf{e}_f\|_\infty \leq m^{O(d)} \cdot \beta$.
- EvalPK $(f, \{\mathbf{B}_i\}_{i \in [\ell]}) \to \mathbf{B}_f$. *The* EvalPK *algorithm takes as input the description of a circuit* f *and matrices* $\mathbf{B}_i \in \mathbb{Z}_q^{n \times m}$, *and outputs a matrix* $\mathbf{B}_f \in \mathbb{Z}_q^{n \times m}$.
- EvalSimPK $(f, \mathbf{A}, \{\mathbf{B}_i, \mathbf{R}_i, x_i\}_{i \in [\ell]}) \to \mathbf{R}_f$. *The* EvalSimPK *algorithm takes as input the description of a circuit* f, *matrices* $\mathbf{A}, \mathbf{B}_i \in \mathbb{Z}_q^{n \times m}$, *matrices* $\mathbf{R}_i \in \{-1,1\}^{m \times m}$, *and the input bits* $x_i \in \{0,1\}$ *for* f *such that* $\mathbf{B}_i = \mathbf{A}\mathbf{R}_i - x_i\mathbf{G}$ *It outputs a matrix* $\mathbf{R}_f$ *such that* $\mathbf{B}_f = \mathbf{A}\mathbf{R}_f - f(x) \cdot \mathbf{G}$, *and* $\|\mathbf{R}_f\|_\infty \leq m^{O(d)}$, *where* $\mathbf{B}_f =$ EvalPK $(f, \{\mathbf{B}_i\}_{i \in [\ell]})$.

For any $x \in \{0,1\}^{\ell}$ *and* $f : \{0,1\}^{\ell} \to \{0,1\}$, $g : \{0,1\}^{\ell_h} \to \{0,1\}$ *and* $h : \{0,1\}^{\ell} \to \{0,1\}^{\ell_h}$, *where* $f = g \circ h$ *and* f *is described by a circuit that is gate-by-gate identical to* $g \circ h$, *then the above algorithms respect composition.*

2.3 Pseudorandom Functions

The preliminaries on pseudorandom function families $\{\Pi^{\mathsf{PRF},\mathsf{PRE}}_{\lambda,\ell_{\text{str}}}\}$ specifically for PRE appear in the full version of the paper and include the Encode procedure.

3 PRE Definitions

In this section, we define the syntax for a PRE scheme, and state the correctness and security properties.

Definition 1 (PRE). *A PRE scheme that supports strings of a-priori bounded length $\ell_{str} = \ell_{str}(\lambda)$, consists of the following four PPT algorithms:*

- $\mathsf{Setup}(1^\lambda, 1^{\ell_{str}}) \to (\mathsf{pp}, \mathsf{msk})$. *The Setup algorithm takes as input the security parameter λ and a maximum string length ℓ_{str}, and outputs the public parameters* pp *and the master secret key* msk.
- $\mathsf{Keygen}(\mathsf{pp}, \mathsf{msk}, x) \to \mathsf{sk}_x$. *The key generation algorithm takes as input the public parameters* pp, *the master secret key* msk *and a string x, and outputs a secret key* sk_x.
- $\mathsf{Enc}(\mathsf{pp}, y, m) \to \mathsf{ct}_y$. *The encryption algorithm takes as input the public parameters* pp, *a string y and a message m, and outputs a ciphertext* ct_y.
- $\mathsf{Dec}(\mathsf{pp}, \mathsf{ct}_y, \mathsf{sk}_x) \to m$ *or* $\perp$. *The decryption algorithm takes as input the public parameters* pp, *a ciphertext* ct_y *and a secret key* sk_x, *and outputs a message m or the special symbol $\perp$ that denotes decryption failure.*

We require the following two properties on correctness and security.

Correctness. For all positive integers λ, ℓ_{str}, and $x, y \in \{0,1\}^{\ell_{str}}$ where x is a prefix of y, and $m \in \{0,1\}$, we require:

$$\Pr\left[\mathsf{Dec}(\mathsf{pp}, \mathsf{ct}_y, \mathsf{sk}_x) = m \;\middle|\; \begin{array}{l} (\mathsf{pp}, \mathsf{msk}) \leftarrow \mathsf{Setup}(1^\lambda, 1^{\ell_{str}}) \\ \mathsf{ct}_y \leftarrow \mathsf{Enc}(\mathsf{pp}, y, m) \\ \mathsf{sk}_x \leftarrow \mathsf{Keygen}(\mathsf{pp}, \mathsf{msk}, x) \end{array}\right] = 1 - \mathrm{negl}(\lambda, \ell_{str}).$$

Security. We define the following adaptive security game between a PPT challenger and a PPT adversary $\mathcal{A}$.

Setup. *The challenger runs the* $\mathsf{Setup}(1^\lambda, 1^{\ell_{str}})$ *algorithm to generate the public parameters* pp *and the master secret key* msk, *and gives* pp *to the adversary $\mathcal{A}$, including ℓ_{str}.*

Phase 1. *The adversary $\mathcal{A}$ adaptively submits secret key requests for any string x. For each request, the challenger responds with a secret key* sk_x *by running* $\mathsf{Keygen}(\mathsf{pp}, \mathsf{msk}, x)$.

Challenge. *The adversary $\mathcal{A}$ submits a challenge string y^* under the restriction that none of the requested keys in* ***Phase 1*** *is a prefix of y^*. The challenger samples $m \xleftarrow{\$} \{0,1\}$ and encrypts m to y^* by running* $\mathsf{Enc}(\mathsf{pp}, y^*, m)$, *and sends* ct_{y^*} *to the adversary $\mathcal{A}$.*

Phase 2. *The adversary $\mathcal{A}$ adaptively submits secret keys requests for any string x under the restriction that x is not a prefix of y^*. The challenger responds with a secret key* sk_x *by running* $\mathsf{Keygen}(\mathsf{pp}, \mathsf{msk}, x)$.

Guess. *The adversary outputs a bit $m' \in \{0,1\}$.*

For a PPT adversary $\mathcal{A}$ its advantage $\mathsf{Adv}_{\mathcal{A}}^{\mathsf{PRE}}(1^\lambda, 1^{\ell_{str}})$ in the above security game is defined as $|\Pr[m' = m] - 1/2|$. A PRE scheme is adaptively secure if for every PPT adversary $\mathcal{A}$ it holds that $\mathsf{Adv}_{\mathcal{A}}^{\mathsf{PRE}}(1^\lambda, 1^{\ell_{str}}) = \mathrm{negl}(\lambda, \ell_{str})$.

We can safely assume that ℓ_{str} is bounded by some polynomial in λ.

4 Adaptively Secure PRE

In this section, we describe our adaptively secure PRE scheme. The proof of the following theorem appears in the full version of the paper.

Theorem 1. *Under the polynomial hardness of approximating SIVP and GapSVP on n-dimensional lattices to within a sub-exponential factor of* $2^{n^{\varepsilon}}$ *for some* $\varepsilon \in (0, 1/2)$ *where* $n = n(\lambda)$, *there exists an adaptively secure PRE scheme that satisfies Definition 1.*

4.1 PRE Construction

In this section, we describe our PRE construction in two steps. First, we define three circuits used in the construction following closely [34] and then give the details of the construction.

Circuit Definitions. Fix $\lambda, \ell_{\text{str}}$ and string space $\{0,1\}^{\ell_{\text{str}}}$. Fix a PRF family $\Pi^{\mathsf{PRF,PRE}} = \{\Pi^{\mathsf{PRF,PRE}}_{\lambda,\ell_{\text{str}}}\}$ with $\Pi^{\mathsf{PRF,PRE}}_{\lambda,\ell_{\text{str}}} = (\mathsf{PRF.Param}, \mathsf{PRF.Keygen}, \mathsf{PRF.Eval})$. Let $\ell_k = \ell_k(\lambda)$, $\ell_{\mathsf{in}} = \ell_{\mathsf{in}}(\lambda)$, $\ell_{\mathsf{out}} = \ell_{\mathsf{out}}(\lambda)$ such that $\mathsf{K}_\lambda = \{0,1\}^{\ell_k}$, $\mathsf{X}_\lambda = \{0,1\}^{\ell_{\mathsf{in}}}$, $\mathsf{Y}_\lambda = \{0,1\}^{\ell_{\mathsf{out}}}$ and let PP_λ be the $\Pi^{\mathsf{PRF,PRE}}_{\lambda,\ell_{\text{str}}}$ parameter space. To simplify notation and without explicitly stating it, we assume that all hardcoded values and inputs to the circuits defined below are given in binary representation.

- The circuit $f^{\mathsf{prf}}_{x,i} : \mathsf{K}_\lambda \to \mathsf{Y}_\lambda$ has *hardcoded* the $\Pi^{\mathsf{PRF,PRE}}_{\lambda,\ell_{\text{str}}}$ parameters $\mathsf{pp} \in \mathsf{PP}_\lambda$ and the *prefix* $x(1,i)$ of a string $x \in \{0,1\}^{\ell_{\text{str}}}$ for some $i \leq \ell_{\text{str}}$. It takes as input a $\Pi^{\mathsf{PRF,PRE}}_{\lambda,\ell_{\text{str}}}$ key $k \in \mathsf{K}_\lambda$ and outputs the $\Pi^{\mathsf{PRF,PRE}}_{\lambda,\ell_{\text{str}}}$ value: $f^{\mathsf{prf}}_{x,i}(k) = \mathsf{PRF.Eval}(\mathsf{pp}, k, x(1,i))$
- The circuit $f^{\mathsf{eq}}_r : \mathsf{Y}_\lambda \to \{0,1\}$ has *hardcoded* a value $r \in \mathsf{Y}_\lambda$. It takes as input a value $y \in \mathsf{Y}_\lambda$ and outputs: $f^{\mathsf{eq}}_r(y) = 1$ if $y = r$ and 0 otherwise.
- The circuit $f^{\mathsf{comp}}_i : \mathsf{K}_\lambda \to \{0,1\}$ is the *composition* of circuits $f^{\mathsf{prf}}_{x,i}$ and f^{eq}_r. That is, $f^{\mathsf{comp}}_i = f^{\mathsf{eq}}_r \circ f^{\mathsf{prf}}_{x,i}$. It has *hardcoded* the $\Pi^{\mathsf{PRF,PRE}}_{\lambda,\ell_{\text{str}}}$ parameters $\mathsf{pp} \in \mathsf{PP}_\lambda$, the *prefix* $x(1,i)$ of a fixed string $x \in \{0,1\}^{\ell_{\text{str}}}$ for some $i \leq \ell_{\text{str}}$ and a value $r \in \mathsf{Y}_\lambda$. It takes as input a $\Pi^{\mathsf{PRF,PRE}}_{\lambda,\ell_{\text{str}}}$ key $k \in \mathsf{K}_\lambda$ and outputs: $f^{\mathsf{comp}}_{x,i,r}(k) = f^{\mathsf{eq}}_r(f^{\mathsf{prf}}_{x,i}(k))$

PRE Construction. Fix a security parameter λ, $n = n(\lambda)$, $m = m(\lambda)$, $q = q(\lambda)$, length bound $\ell_{\text{str}} = \ell_{\text{str}}(\lambda)$. Fix a PRF family $\Pi^{\mathsf{PRF,PRE}} = \{\Pi^{\mathsf{PRF,PRE}}_{\lambda,\ell_{\text{str}}}\}$, and circuits $f^{\mathsf{prf}}_{x,i}, f^{\mathsf{eq}}_r, f^{\mathsf{comp}}_i$ as defined in Sect. 4.1. Our PRE scheme consists of a tuple of four efficient algorithms $(\mathsf{Setup}, \mathsf{Keygen}, \mathsf{Enc}, \mathsf{Dec})$ with the following properties:

- $\mathsf{Setup}(1^\lambda, 1^{\ell_{\text{str}}}, 1^{\ell_k})$. On input the security parameter λ, maximum string length ℓ_{str} and ℓ_k be the key length for $\Pi^{\mathsf{PRF,PRE}}_{\lambda,\ell_{\text{str}}}$, do the following:
  - sample $(\mathbf{A}, \mathbf{T_A}) \leftarrow \mathsf{TrapGen}(1^n, 1^m, q)$

- sample $\mathbf{B}_i \xleftarrow{\$} \mathbb{Z}_q^{n\times m}$ for $i \in [\ell_k]$ and $\mathbf{u} \xleftarrow{\$} \mathbb{Z}_q^n$
- sample $\mathsf{pp}_{\mathsf{prf}} \leftarrow \mathsf{PRF.Param}(1^\lambda, 1^{\ell_{\mathsf{str}}})$ and $k^{\mathsf{real}}, k^{\mathsf{fake}} \leftarrow \mathsf{PRF.Keygen}(1^\lambda)$

Output $\mathsf{pp} = \left(\mathsf{pp}_{\mathsf{prf}}, \mathbf{u}, \mathbf{A}, \mathbf{B}_1, \ldots, \mathbf{B}_{\ell_k}\right)$, $\mathsf{msk} = \left(\mathbf{T}_{\mathbf{A}}, k^{\mathsf{real}}, k^{\mathsf{fake}}\right)$ and lattice parameters $n, m, q\ \chi, \widehat{\chi}, \{\sigma_i\}_{i\in[\ell_{\mathsf{str}}]}$.

– $\mathsf{Keygen}(\mathsf{pp}, \mathsf{msk}, x)$. On input the public parameters pp, the master secret key msk and string $x \in \{0,1\}^\ell$ with $\ell \le \ell_{\mathsf{str}}$, do the following:
  - set $r_i = \mathsf{PRF.Eval}(\mathsf{pp}_{\mathsf{prf}}, k^{\mathsf{fake}}, x(1,i))$ for $i \in [\ell - 1]$
  - set $r_\ell = \mathsf{PRF.Eval}(\mathsf{pp}_{\mathsf{prf}}, k^{\mathsf{real}}, x(1,\ell))$
  - set $f_i^{\mathsf{comp}} = f_{x,i}^{\mathsf{prf}} \circ f_{r_i}^{\mathsf{eq}}$ for $i \in [\ell]$
  - set $\mathbf{B}_{f_i^{\mathsf{comp}}} = \mathsf{EvalPK}(f_i^{\mathsf{comp}}, \{\mathbf{B}_\kappa\}_{\kappa\in[\ell_k]})$ for $i \in [d]$
  - set $\mathbf{T}_x \leftarrow \mathsf{SampleBasisLeft}(\mathbf{A}, \{\mathbf{B}_{f_i^{\mathsf{comp}}}\}_{i\in[\ell]}, \mathbf{T}_{\mathbf{A}}, \sigma_\ell)$
  - set $\mathbf{F}_x = \mathbf{A} \mid\mid \mathbf{B}_{f_1^{\mathsf{comp}}} \mid\mid \cdots \mid\mid \mathbf{B}_{f_\ell^{\mathsf{comp}}}$
  - sample $\mathbf{t}_{\mathbf{u},x}^{\mathsf{comp}} \leftarrow \mathsf{SamPre}(\mathbf{F}_x, \mathbf{T}_x, \mathbf{u}, \sigma_\ell)$

  Output $\mathsf{sk}_x = (x,\ \mathbf{t}_{\mathbf{u},x}^{\mathsf{comp}},\ \{r_i\}_{i\in[\ell]})$
– $\mathsf{Enc}(\mathsf{pp}, y, \mathsf{msg})$. On input the public parameters pp, a string $y \in \{0,1\}^\ell$ with $\ell \le \ell_{\mathrm{str}}$ and message $\mathsf{msg} \in \{0,1\}$, do the following:
  - sample $k^{\mathsf{enc}} \leftarrow \mathsf{PRF.Keygen}(1^\lambda)$
  - set $r_i^{\mathsf{enc}} = \mathsf{PRF.Eval}(\mathsf{pp}_{\mathsf{prf}}, k^{\mathsf{enc}}, y(1,i))$ for $i \in [\ell]$
  - set $C_{i,j}^{\mathsf{prf}}$ to be the circuit that computes the j-th output bit of $f_{x,i}^{\mathsf{prf}}$
  - set $\mathbf{B}_{i,j}^{\mathsf{prf}} = \mathsf{EvalPK}(C_{i,j}^{\mathsf{prf}}, \{\mathbf{B}_\kappa\}_{\kappa\in[\ell_k]})$ for $i \in [\ell]$ and $j \in [\ell_{\mathsf{out}}]$
  - sample $\mathbf{s} \xleftarrow{\$} \mathbb{Z}_q^n$, $\mathbf{e}_a \leftarrow \chi^m$, $\mathbf{e}_{i,j} \leftarrow \widehat{\chi}$ for $i \in [\ell]$ and $j \in [\ell_{\mathsf{out}}]$, and $e_{\mathsf{msg}} \leftarrow \chi$
  - set $\mathbf{c}_a = \mathbf{s}^\mathsf{T}\mathbf{A} + \mathbf{e}_a^\mathsf{T}$, $\mathbf{c}_{i,j}^{\mathsf{prf}} = \mathbf{s}^\mathsf{T}(\mathbf{B}_{i,j}^{\mathsf{prf}} + r_{i,j}^{\mathsf{enc}} \cdot \mathbf{G}) + \mathbf{e}_{i,j}^\mathsf{T}$, and $c_{\mathsf{msg}} = \mathbf{s}^\mathsf{T}\mathbf{u} + e_{\mathsf{msg}} + \mathsf{msg} \cdot \lceil \frac{q}{2} \rceil$ where $r_{i,j}^{\mathsf{enc}}$ is the j-th bit of r_i^{enc}

  Output $\mathsf{ct}_y = (y,\ \mathbf{c}_a,\ \{\mathbf{c}_{i,j}^{\mathsf{prf}}\}_{i\in[\ell],\ j\in[\ell_{\mathsf{out}}]},\ c_{\mathsf{msg}},\ \{r_i^{\mathsf{enc}}\}_{i\in[\ell]})$
– $\mathsf{Dec}(\mathsf{pp}, \mathsf{ct}_y, \mathsf{sk}_x)$. On input the public parameters pp, a ciphertext ct_y and secret key sk_x do the following:
  - parse $\mathsf{ct}_y = (y,\ \mathbf{c}_a,\ \{\mathbf{c}_{i,j}\}_{i\in[\ell_y],\ j\in[\ell_{\mathsf{out}}]},\ c_{\mathsf{msg}},\ \{r_i^{\mathsf{enc}}\}_{i\in[\ell_y]})$
  - parse $\mathsf{sk}_x = (x,\ \mathbf{t}_{\mathbf{u},x}^{\mathsf{comp}},\ \{r_i\}_{i\in[\ell_x]})$
  - parse $y \in \{0,1\}^{\ell_y}$ and $x \in \{0,1\}^{\ell_x}$ with $\ell_y, \ell_x \le \ell_{\mathrm{str}}$
  - if x is not a prefix of y, then output $\perp$
    if x is a prefix of y, and there exists $i \in [\ell_x]$ such that $r_i^{\mathsf{enc}} = r_i$, then output $\perp$
  - set $\mathbf{c}_i^{\mathsf{comp}} = \mathsf{EvalCT}(f_{r_i}^{\mathsf{eq}}, \{\mathbf{B}_{i,j}^{\mathsf{prf}}, r_{i,j}^{\mathsf{enc}}, \mathbf{c}_{i,j}^{\mathsf{prf}}\}_{j\in[\ell_{\mathsf{out}}]})$ for $i \in [\ell_x]$
  - set $\mathbf{F}_x = \mathbf{A} \mid\mid \mathbf{B}_{f_1^{\mathsf{comp}}} \mid\mid \cdots \mid\mid \mathbf{B}_{f_{\ell_x}^{\mathsf{comp}}}$
  - set $c'_{\mathsf{msg}} = \left(\mathbf{c}_a \mid\mid \mathbf{c}_1^{\mathsf{comp}} \mid\mid \cdots \mid\mid \mathbf{c}_{\ell_x}^{\mathsf{comp}}\right)^\mathsf{T} \cdot \mathbf{t}_{\mathbf{u},x}^{\mathsf{comp}}$
  - if $|c_{\mathsf{msg}} - c'_{\mathsf{msg}}| \le q/4$, then set $\mathsf{msg} = 0$ else set $\mathsf{msg} = 1$ (assuming the output is not $\perp$ until this step)

  Output msg

4.2 Lattice Parameters, and Correctness and Security Proofs

For the selection of lattice parameters, and the correctness and security proofs, we refer to the full version of the paper. However, the core ideas for the correctness and security proofs are presented in Sect. 1.1.

Acknowledgements. We thank Daniel Wichs and the anonymous TCC'21 reviewers for useful comments about adaptively secure HIBE, on a previous version of this manuscript.

References

1. Agrawal, S., Boneh, D., Boyen, X.: Efficient lattice (H)IBE in the standard model. In: Gilbert, H. (ed.) EUROCRYPT 2010. LNCS, vol. 6110, pp. 553–572. Springer, Heidelberg (2010). https://doi.org/10.1007/978-3-642-13190-5_28
2. Ajtai, M.: Generating hard instances of lattice problems (extended abstract). In: 28th ACM STOC, pp. 99–108. ACM Press, May 1996. https://doi.org/10.1145/237814.237838
3. Ajtai, M.: Generating hard instances of the short basis problem. In: Wiedermann, J., van Emde Boas, P., Nielsen, M. (eds.) ICALP 1999. LNCS, vol. 1644, pp. 1–9. Springer, Heidelberg (1999). https://doi.org/10.1007/3-540-48523-6_1
4. Ananth, P., Brakerski, Z., Segev, G., Vaikuntanathan, V.: From selective to adaptive security in functional encryption. In: Gennaro, R., Robshaw, M. (eds.) CRYPTO 2015. LNCS, vol. 9216, pp. 657–677. Springer, Heidelberg (2015). https://doi.org/10.1007/978-3-662-48000-7_32
5. Ananth, P., Fan, X., Shi, E.: Towards attribute-based encryption for RAMs from LWE: sub-linear decryption, and more. In: Galbraith, S.D., Moriai, S. (eds.) ASIACRYPT 2019. LNCS, vol. 11921, pp. 112–141. Springer, Cham (2019). https://doi.org/10.1007/978-3-030-34578-5_5
6. Boneh, D., Boyen, X.: Efficient selective-ID secure identity-based encryption without random oracles. In: Cachin, C., Camenisch, J.L. (eds.) EUROCRYPT 2004. LNCS, vol. 3027, pp. 223–238. Springer, Heidelberg (2004). https://doi.org/10.1007/978-3-540-24676-3_14
7. Boneh, D., et al.: Fully key-homomorphic encryption, arithmetic circuit ABE and compact garbled circuits. In: Nguyen, P.Q., Oswald, E. (eds.) EUROCRYPT 2014. LNCS, vol. 8441, pp. 533–556. Springer, Heidelberg (2014). https://doi.org/10.1007/978-3-642-55220-5_30
8. Brakerski, Z., Vaikuntanathan, V.: Circuit-ABE from LWE: unbounded attributes and semi-adaptive security. In: Robshaw, M., Katz, J. (eds.) CRYPTO 2016. LNCS, vol. 9816, pp. 363–384. Springer, Heidelberg (2016). https://doi.org/10.1007/978-3-662-53015-3_13
9. Canetti, R., Halevi, S., Katz, J.: A forward-secure public-key encryption scheme. In: Biham, E. (ed.) EUROCRYPT 2003. LNCS, vol. 2656, pp. 255–271. Springer, Heidelberg (2003). https://doi.org/10.1007/3-540-39200-9_16
10. Cash, D., Hofheinz, D., Kiltz, E., Peikert, C.: Bonsai trees, or how to delegate a lattice basis. In: Gilbert, H. (ed.) EUROCRYPT 2010. LNCS, vol. 6110, pp. 523–552. Springer, Heidelberg (2010). https://doi.org/10.1007/978-3-642-13190-5_27
11. Chen, J., Wee, H.: Semi-adaptive attribute-based encryption and improved delegation for Boolean formula. In: Abdalla, M., De Prisco, R. (eds.) SCN 2014. LNCS, vol. 8642, pp. 277–297. Springer, Cham (2014). https://doi.org/10.1007/978-3-319-10879-7_16
12. Davidson, A., Katsumata, S., Nishimaki, R., Yamada, S.: Constrained PRFs for bit-fixing (and more) from OWFs with adaptive security and constant collusion resistance. Cryptology ePrint Archive, Report 2018/982 (2018). https://eprint.iacr.org/2018/982

13. Dodis, Y., Ostrovsky, R., Reyzin, L., Smith, A.: Fuzzy extractors: How to generate strong keys from biometrics and other noisy data. SIAM J. Comput. **38**(1), 97–139 (2008)
14. Gentry, C., Halevi, S.: Hierarchical identity based encryption with polynomially many levels. In: Reingold, O. (ed.) TCC 2009. LNCS, vol. 5444, pp. 437–456. Springer, Heidelberg (2009). https://doi.org/10.1007/978-3-642-00457-5_26
15. Gentry, C., Peikert, C., Vaikuntanathan, V.: Trapdoors for hard lattices and new cryptographic constructions. In: Ladner, R.E., Dwork, C. (eds.) 40th ACM STOC, pp. 197–206. ACM Press, May 2008. https://doi.org/10.1145/1374376.1374407
16. Gentry, C., Silverberg, A.: Hierarchical ID-based cryptography. In: Zheng, Y. (ed.) ASIACRYPT 2002. LNCS, vol. 2501, pp. 548–566. Springer, Heidelberg (2002). https://doi.org/10.1007/3-540-36178-2_34
17. Gorbunov, S., Vaikuntanathan, V., Wee, H.: Attribute-based encryption for circuits. In: Boneh, D., Roughgarden, T., Feigenbaum, J. (eds.) 45th ACM STOC, pp. 545–554. ACM Press, June 2013. https://doi.org/10.1145/2488608.2488677
18. Goyal, R., Koppula, V., Waters, B.: Semi-adaptive security and bundling functionalities made generic and easy. In: Hirt, M., Smith, A. (eds.) TCC 2016. LNCS, vol. 9986, pp. 361–388. Springer, Heidelberg (2016). https://doi.org/10.1007/978-3-662-53644-5_14
19. Goyal, R., Koppula, V., Waters, B.: Lockable obfuscation. In: Umans, C. (ed.) 58th FOCS, pp. 612–621. IEEE Computer Society Press, October 2017. https://doi.org/10.1109/FOCS.2017.62
20. Goyal, V., Pandey, O., Sahai, A., Waters, B.: Attribute-based encryption for fine-grained access control of encrypted data. In: Juels, A., Wright, R.N., De Capitani di Vimercati, S. (eds.) ACM CCS 2006, pp. 89–98. ACM Press, October/November 2006. https://doi.org/10.1145/1180405.1180418, available as Cryptology ePrint Archive Report 2006/309
21. Horwitz, J., Lynn, B.: Toward hierarchical identity-based encryption. In: Knudsen, L.R. (ed.) EUROCRYPT 2002. LNCS, vol. 2332, pp. 466–481. Springer, Heidelberg (2002). https://doi.org/10.1007/3-540-46035-7_31
22. Jain, A., Lin, H., Sahai, A.: Indistinguishability obfuscation from well-founded assumptions. In: Proceedings of the 53rd Annual ACM SIGACT Symposium on Theory of Computing, pp. 60–73. New York, NY, USA (2021). https://doi.org/10.1145/3406325.3451093
23. Lewko, A., Okamoto, T., Sahai, A., Takashima, K., Waters, B.: Fully secure functional encryption: attribute-based encryption and (hierarchical) inner product encryption. In: Gilbert, H. (ed.) EUROCRYPT 2010. LNCS, vol. 6110, pp. 62–91. Springer, Heidelberg (2010). https://doi.org/10.1007/978-3-642-13190-5_4
24. Lewko, A., Waters, B.: Why proving HIBE systems secure is difficult. In: Nguyen, P.Q., Oswald, E. (eds.) EUROCRYPT 2014. LNCS, vol. 8441, pp. 58–76. Springer, Heidelberg (2014). https://doi.org/10.1007/978-3-642-55220-5_4
25. Micciancio, D., Peikert, C.: Trapdoors for lattices: simpler, tighter, faster, smaller. In: Pointcheval, D., Johansson, T. (eds.) EUROCRYPT 2012. LNCS, vol. 7237, pp. 700–718. Springer, Heidelberg (2012). https://doi.org/10.1007/978-3-642-29011-4_41
26. Micciancio, D., Regev, O.: Worst-case to average-case reductions based on Gaussian measures. SIAM J. Comput. **37**(1), 267–302 (2007). http://dblp.uni-trier.de/db/journals/siamcomp/siamcomp37.html#MicciancioR07

27. Okamoto, T., Takashima, K.: Fully secure functional encryption with general relations from the decisional linear assumption. In: Rabin, T. (ed.) CRYPTO 2010. LNCS, vol. 6223, pp. 191–208. Springer, Heidelberg (2010). https://doi.org/10.1007/978-3-642-14623-7_11
28. Peikert, C.: Public-key cryptosystems from the worst-case shortest vector problem: extended abstract. In: Mitzenmacher, M. (ed.) 41st ACM STOC, pp. 333–342. ACM Press, May/June 2009. https://doi.org/10.1145/1536414.1536461
29. Peikert, C., Regev, O., Stephens-Davidowitz, N.: Pseudorandomness of ring-LWE for any ring and modulus. In: Hatami, H., McKenzie, P., King, V. (eds.) 49th ACM STOC, pp. 461–473. ACM Press, June 2017. https://doi.org/10.1145/3055399.3055489
30. Regev, O.: On lattices, learning with errors, random linear codes, and cryptography. In: Gabow, H.N., Fagin, R. (eds.) 37th ACM STOC, pp. 84–93. ACM Press, May 2005. https://doi.org/10.1145/1060590.1060603
31. Sahai, A., Waters, B.: Fuzzy identity-based encryption. In: Cramer, R. (ed.) EUROCRYPT 2005. LNCS, vol. 3494, pp. 457–473. Springer, Heidelberg (2005). https://doi.org/10.1007/11426639_27
32. Shamir, A.: Identity-based cryptosystems and signature schemes. In: Blakley, G.R., Chaum, D. (eds.) CRYPTO 1984. LNCS, vol. 196, pp. 47–53. Springer, Heidelberg (1985). https://doi.org/10.1007/3-540-39568-7_5
33. Shi, E., Waters, B.: Delegating capabilities in predicate encryption systems. In: Aceto, L., Damgård, I., Goldberg, L.A., Halldórsson, M.M., Ingólfsdóttir, A., Walukiewicz, I. (eds.) ICALP 2008. LNCS, vol. 5126, pp. 560–578. Springer, Heidelberg (2008). https://doi.org/10.1007/978-3-540-70583-3_46
34. Tsabary, R.: Fully secure attribute-based encryption for t-CNF from LWE. In: Boldyreva, A., Micciancio, D. (eds.) CRYPTO 2019. LNCS, vol. 11692, pp. 62–85. Springer, Cham (2019). https://doi.org/10.1007/978-3-030-26948-7_3
35. Waters, B.: Dual system encryption: realizing fully secure IBE and HIBE under simple assumptions. In: Halevi, S. (ed.) CRYPTO 2009. LNCS, vol. 5677, pp. 619–636. Springer, Heidelberg (2009). https://doi.org/10.1007/978-3-642-03356-8_36
36. Waters, B.: A punctured programming approach to adaptively secure functional encryption. In: Gennaro, R., Robshaw, M. (eds.) CRYPTO 2015. LNCS, vol. 9216, pp. 678–697. Springer, Heidelberg (2015). https://doi.org/10.1007/978-3-662-48000-7_33
37. Wichs, D., Zirdelis, G.: Obfuscating compute-and-compare programs under LWE. In: Umans, C. (ed.) 58th FOCS, pp. 600–611. IEEE Computer Society Press, October 2017. https://doi.org/10.1109/FOCS.2017.61
38. Wu, D.J., Taly, A., Shankar, A., Boneh, D.: Privacy, discovery, and authentication for the internet of things. In: Askoxylakis, I., Ioannidis, S., Katsikas, S., Meadows, C. (eds.) ESORICS 2016. LNCS, vol. 9879, pp. 301–319. Springer, Cham (2016). https://doi.org/10.1007/978-3-319-45741-3_16

SigML: Supervised Log Anomaly with Fully Homomorphic Encryption

Devharsh Trivedi[1(✉)], Aymen Boudguiga[2], and Nikos Triandopoulos[1]

[1] Stevens Institute of Technology, Hoboken, NJ 07030, USA
{dtrived5,ntriando}@stevens.edu
[2] CEA-LIST, 91191 Gif-sur-Yvette, France
aymen.boudguiga@cea.fr

Abstract. Security (and Audit) log collection and storage is a crucial process for enterprises around the globe. Log analysis helps identify potential security breaches and, in some cases, is required by law for compliance. However, enterprises often delegate these responsibilities to a third-party cloud service provider, where the logs are collected and processed for anomaly detection and stored in a cold data warehouse for archiving. Prevalent schemes rely on plain (unencrypted) data for log anomaly detection. More often, these logs can reveal much sensitive information about an organization or the customers of that organization. Hence it is in the best interest of everyone to keep it encrypted at all times. This paper proposes "SigML" utilizing Fully Homomorphic Encryption (FHE) with the Cheon-Kim-Kim-Song (CKKS) scheme for supervised log anomaly detection on encrypted data. We formulate a binary classification problem and propose a novel "Aggregate" configuration using the Sigmoid function for resource-strained (wireless sensors or IoT) devices to reduce communication and computation requirements by a factor of n, where n is the number of ciphertexts received by the clients. We further approximate the Sigmoid activation function ($\sigma(x)$) with first, third, and fifth-order polynomials in the encrypted domain and evaluate the supervised models with NSL-KDD and HDFS datasets in terms of performance metrics and computation time.

Keywords: Log Anomaly Detection · Fully Homomorphic Encryption · Supervised Machine Learning · Sigmoid Function Approximation

1 Introduction

Information security technologies such as Intrusion Detection Systems (IDS), Intrusion Prevention Systems (IPS), and Security Information and Event Management (SIEM) systems are specifically designed to help fight cyberattacks. Modern Security Operations Centers (SOC) use SIEM tools that rely on logs collected from endpoints to detect anomalies and generate an alert indicating an Incident of Compromise (IoC). The logs (security analytics) are unstructured textual data of the operating systems or user application events. SIEM uses system and audit logs from various devices (Security Analytics Sources (SAS))

S. Dolev et al. (Eds.): CSCML 2023, LNCS 13914, pp. 372–388, 2023.
https://doi.org/10.1007/978-3-031-34671-2_26

to monitor and identify security threats in a quasi-real-time manner. SAS is a (mobile or stationary) host or a security tool like an IDS, collecting the log data for examination and storage.

A typical log anomaly detection solution involves an (i) Log collector - collecting (textual) logs from various applications running on a SAS. (ii) Transmitter - to send data to SIEM. This data is usually encrypted to protect against eavesdropping in the transmission channel. (iii) Receiver - to collect, store and verify the integrity of the transmitted logs. (iv) Parser - to format the data in a structured form used by the SIEM vendor to process the decrypted logs. Furthermore, (v) Anomaly Detector - using proprietary algorithms to generate and send alerts.

Organizations often use a third-party vendor for SOC. Third-party cloud services reduce complexity and offer flexibility for enterprises. However, Cloud Service Consumers (CSC) need to entrust their data - and their customer's data - to Cloud Service Providers (CSP), who are often incentivized to monetize these data. Meanwhile, regulations such as the US State of California Consumer Privacy Act (CCPA) [6], the US Consumer Online Privacy Rights Act (COPRA) [10], and the EU General Data Protection Regulation (GDPR) [14] aim to protect consumers' privacy, and non-compliant organizations are subjected to severe fines and worsened reputation. This results in a tradeoff between data privacy and utility for organizations.

Exporting log data to a SIEM hosted on a third-party CSP is risky, as the CSP needs access to unencrypted data for alert generation. In addition, the CSP may have enough incentives to collect user data. These data are stored in the CSP's servers and thus face various security and privacy threats like data leakage and misuse of information [12,18,19,22,28,29]. Thus, protecting these logs' confidentiality and privacy is critical. We propose to use Fully Homomorphic Encryption (FHE) to allow CSC to ensure privacy without undermining their ability to gain insights from their data.

Traditional (prevalent) cloud computation and storage solutions using contemporary cryptography require customer data to be decrypted before operating. In addition, security policies prevent unauthorized access to data. Therefore, CSC must trust the Access Control Policies (ACP) incorporated by CSP for data privacy. With FHE, data privacy is achieved by the CSC through cryptography, leveraging rigorous mathematical proofs. As a result, the CSP will not have access to unencrypted customer data for storage or computation.

FHE allows for computations with encrypted data without needing to decrypt them. Instead, the resulting computations are preserved in an encrypted domain[1], which, when decrypted, results in an output identical to the one obtained if the operations have been performed on the unencrypted domain. Indeed, FHE provides privacy-preserving storage and computation. Furthermore, this allows data to be encrypted and outsourced to commercial cloud environments for processing while encrypted.

[1] We consider that plaintexts belong to the unencrypted domain and ciphertexts to the encrypted domain.

1.1 Contributions

Our contributions can be summarized as follows:

- We formulate a supervised binary classification problem for log anomaly detection and implement it with the Cheon-Kim-Kim-Song cryptosystem.
- To lower the computation and communication cost of a wireless sensor or IoT device and to secure against inference attacks, we propose a novel "Aggregate" configuration using a Sigmoid activation.
- We designed a Logistic Regression classifier (σ_{LR}) to improve $\Sigma - Ratio$ for the "Aggregate" configuration.
- We further evaluate the performance of 1st, 3rd, and 5th-order Sigmoid approximations in the encrypted domain.

1.2 Organization

This paper is organized as follows. First, we describe the building blocks of our protocols in Sect. 2, where we review FHE in Sect. 2.1 and present 1st-order, 3rd-order, and 5th-order polynomial approximations for the Sigmoid function in Sect. 2.2. Next, we describe our methodology in Sect. 3. Then, we review the previous work in Sect. 4. Finally, we discuss our experimental results in Sect. 5.

2 Background

2.1 Fully Homomorphic Encryption

In this work, we use CKKS [8] as a homomorphic encryption scheme. CKKS differs from other homomorphic encryption schemes (e.g., BFV [3,13], BGV [4], or TFHE [9]) concerning the interpretation of the encryption noise. Indeed, CKKS considers the encryption noise part of the message, like when approximating real numbers using floating-point arithmetics. Note that this encryption noise does not destroy the most significant bits of the plaintext m as long as it remains small enough. CKKS decrypts the encryption of m as an approximated value $m+e$ where e is a small noise. CKKS authors propose multiplying plaintexts by a scaling factor Δ before encryption to reduce precision loss after adding noise during encryption. In addition, CKKS supports batching, a technique for encoding many plaintexts within a single ciphertext in a Single Instruction Multiple Data (SIMD) fashion. We define CKKS with a set of probabilistic polynomial-time algorithms concerning the security parameter k, CKKS = (CKKS.Keygen, CKKS.Enc, CKKS.Dec, CKKS.Eval). The level of a ciphertext is l if it is sampled from $\mathbb{Z}_{q_l}[X]/(X^N+1)$. Let L, q_0 and Δ be integers. We set $q_l = \Delta^l \cdot q_0$ for any l integer in $[\![0, L]\!]$.

- $(pk, evk, sk) \leftarrow \mathsf{CKKS.Keygen}(1^k, L)$: outputs a secret decryption key sk, a public key pk, and a public evaluation key evk. The secret key sk is a sample from a random distribution over $\mathbb{Z}_3[X]/(X^N+1)$. The public key is computed as:

$$pk = ([-a \cdot sk + e]_{q_L}, a) = (p_0, p_1)$$

where a is sampled from a uniform distribution over $\mathbb{Z}_{q_L}[X]/(X^N+1)$, and e is sampled from an error distribution over $\mathbb{Z}_{q_L}[X]/(X^N+1)$. evk is used for relinearisation after the multiplication of two ciphertexts.

- $c \leftarrow \mathsf{CKKS.Enc}_{pk}(m)$: encrypts a message m into a ciphertext c using the public key pk. Let v be sampled from a distribution over $\mathbb{Z}_3[X]/(X^N+1)$. Let e_0 and e_1 be small errors. Then the message m is encrypted as:
$$c = [(v \cdot pk_0, v \cdot pk_1) + (m + e_0, e_1)]_{qL} = (c_0, c_1).$$
- $m \leftarrow \mathsf{CKKS.Dec}_{sk}(c)$: decrypts a message c into a plaintext m using the secret key sk. The message m can be retrieved from a level l ciphertext thanks to the function $m = [c_0 + c_1 \cdot sk]_{q_l}$. Note that with CKKS, the level of a ciphertext goes down each time a multiplication is computed.
- $c_f \leftarrow \mathsf{CKKS.Eval}_{evk}(f, c_1, \ldots, c_k)$: evaluates the function f on the encrypted inputs $(c_1, \ldots, c_k)$ using the evaluation key evk.

2.2 Sigmoid Approximation

Besides noise growth and message expansion, applying the Sigmoid function is a significant challenge in implementing LR or SVM with FHE.[2] We describe three methods to approximate this function with a polynomial and compare them in terms of accuracy, precision, recall, f1-score, and the ratio of the predicted sum from Sigmoid values to the sum of all actual binary labels for the test dataset (Fig. 1).

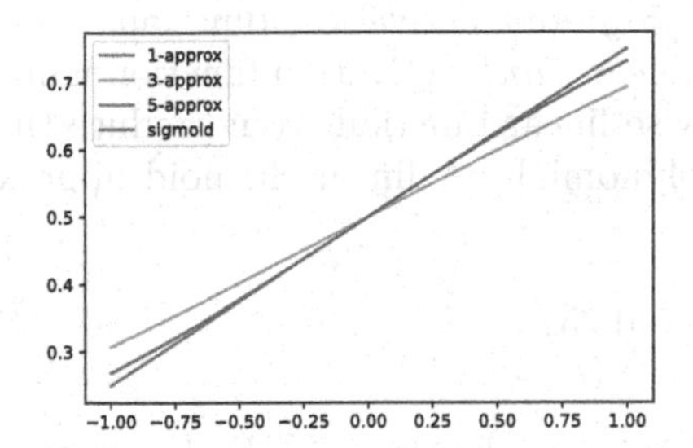

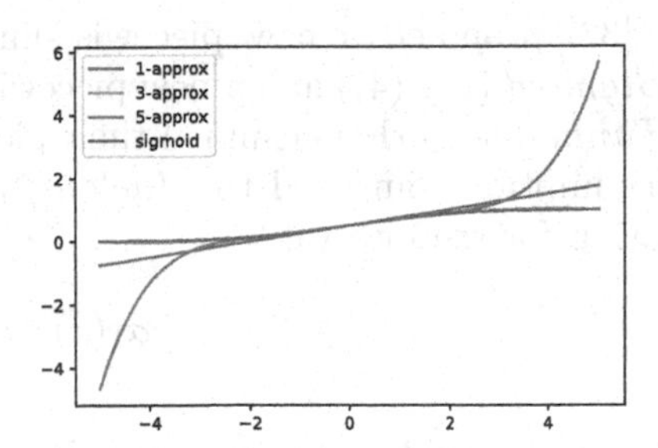

Fig. 1. 1st (Eq. (5)), 3rd (Eq. (7)), 5th (Eq. (9)) order Sigmoid approximation.

First-Order Approximation. A piecewise linear function comprises some linear segments defined over an equal number of (equal-sized) intervals. Piecewise linear functions are also critical to some constructive derivations. The length of a segment (piece) is given by the

$$\sqrt{(\Delta x)^2 + (\Delta y)^2} = \sqrt{1 + \left(\frac{\Delta y}{\Delta x}\right)^2} \Delta x \tag{1}$$

[2] Sigmoid is used in LR and SVM during classification, so we decided to make it homomorphic.

Summing the length of many pieces gives

$$\sum_{i=1}^{n}\left(\sqrt{1+\left(\frac{\Delta y_i}{\Delta x_i}\right)^2}\Delta x_i\right) \tag{2}$$

and taking the *limit* as $\max_i(\Delta x_i) \rightarrow 0$, the *sum* becomes Eq. (3), which is the usual arc length.

$$\int\sqrt{1+\left(\frac{dy}{dx}\right)^2}dx \tag{3}$$

$$PSigmoid(x) = \begin{cases} 0 & x \leq -5 \\ 0.015078x + 0.085083 & -5 < x \leq -3.5 \\ 0.0599266667x + 0.239053333 & -3.5 < x \leq -2 \\ 0.164857143x + 0.448914286 & -2 < x \leq -0.6 \\ 0.25x + 0.5 & -0.6 < x \leq 0.6 \\ 0.164857143x + 0.551085714 & 0.6 < x \leq 2 \\ 0.0599266667x + 0.7609466667 & 2 < x \leq 3.5 \\ 0.015078x + 0.91791 & 3.5 < x \leq 5 \\ 1 & x > 5 \end{cases} \tag{4}$$

[32] proposed a new piecewise-linear *Sigmoid* activation function named *PSigmoid* (Eq. (4)) and a new piecewise-linear *Tanh* activation function named *PTanh*. The authors claimed their piecewise-linear functions could reduce time consumption compared to *Maclaurin* polynomials. A linear Sigmoid approximation for the interval [−1, 1] is:

$$\sigma_1(x) = 0.5 + 0.25x \tag{5}$$

Third-Order Approximation. We approximate the class $C[a, b]$ of continuous functions on the interval $[a, b]$ by degree-n polynomials in $\mathcal{P}_n$ using the L^∞-norm to measure fit. This is referred to as minimax polynomial approximation since the best (or minimax) approximation solves

$$p_n^* = \arg \min_{p_n \in \mathcal{P}_n} \max_{a \leq x \leq b} |f(x) - p_n(x)| \tag{6}$$

A minimax (uniform) approximation algorithm is a method to find the polynomial p in Eq. (6). For example, the Remez algorithm [26] is an iterative minimax approximation and yields the following results [7] for the interval [−5, 5] and degrees 3:

$$\sigma_3(x) = 0.5 + 0.197x - 0.004x^3 \tag{7}$$

Fifth-Order Approximation

$$\sum_{n=0}^{\infty} f^{(n)}(a)\frac{(x-a)^n}{n!} = f(a)+f'(a)(x-a)+\frac{f''(a)}{2!}(x-a)^2+\ldots+\frac{f^{(k)}(a)}{k!}(x-a)^n+\ldots \tag{8}$$

The Taylor series can be centered around any number a and is written as Eq. (8). The Maclaurin Series is a Taylor series centered about 0. A Maclaurin series is a power series that allows the calculation of an approximation of a function $f(x)$ for input values close to zero, given that values of the successive derivatives of the function at zero is known. A Maclaurin series can find the antiderivative of a complicated function, approximate a function, or compute an uncomputable sum. In addition, Maclaurin series partial sums provide polynomial approximations for the function. We propose to use the Taylor approximation of $\frac{1}{(1+e^{(-x)})}$ of degree 5 with center 0:

$$\sigma_5(x) = 0.5 + 0.25x - 0.020833x^3 + 0.002083x^5 \tag{9}$$

3 Proposed Solution

Our threat model consists of two entities - SAS (CSC) and SIEM (CSP). SAS wants to generate alerts from its' logs while preserving privacy. Therefore, SIEM should refrain from learning about the log information. On the other hand, SIEM wants to protect the weights and coefficients of the Machine Learning (ML) model used to detect anomalies and generate alerts. Therefore, SAS should not learn about the model information. This is achieved by SAS using FHE to encrypt the log inputs. Then, SIEM performs homomorphic calculations and passes the encrypted result(s) to SAS. SAS decrypts the result(s), learns whether there was an anomaly, and generates an alert accordingly (Table 1).

Table 1. Comparing "Ubiquitous" and "Aggregate" configurations. n is the number of logs, $T_E(p)$ is the time taken to encrypt a single message, $S_E(p)$ is bytes occupied by a single ciphertext, $T_D(c)$ is the time taken to decrypt a single ciphertext, and $S_D(c)$ is bytes occupied by a single (decrypted) message.

Configuration	Encryption Time	Encryption Size	Decryption Time	Decryption Size
Ubiquitous	$n \cdot T_E(p)$	$n \cdot S_E(p)$	$n \cdot T_D(c)$	$n \cdot S_D(c)$
Aggregate	$n \cdot T_E(p)$	$n \cdot S_E(p)$	$T_D(c)$	$S_D(c)$

We use FHE using the CKKS scheme, as CKKS is better suited for floating value computations.[3] We present two deployment configurations: (i) "Ubiquitous" and (ii) "Aggregate". While configuration (i) is similar to prevalent

[3] CKKS is more suited for arithmetic on real numbers, where we can have approximate but close results, while BFV is more suited for arithmetic on integers.

research works, (ii) reduces the computation and communication on the SAS. However, they differ in how the result is generated from SIEM and processed at SAS.

1. Ubiquitous - SIEM sends one encrypted result per encrypted user input.
2. Aggregate - Only one result is sent in the encrypted domain for all inputs. This technique helps reduce communication costs and uses much fewer resources on SAS to decrypt a single encrypted result than one encrypted result per encrypted input.

For both configurations, SAS collects logs, parses unstructured log to a structured form, and normalize the data. Data normalization helps to improve ML model accuracy and Sigmoid output. SAS is also responsible for generating keys for encryption (pk/sk), decryption (sk), and evaluation (evk) of the data using CKKS. SAS encrypts the parsed data and sends it to SIEM with an evaluation key. SIEM uses the evk for that particular SAS and performs homomorphic computation with the ML model's coefficients in plaintext and encrypted inputs. The results of these computations are achieved in the encrypted domain.

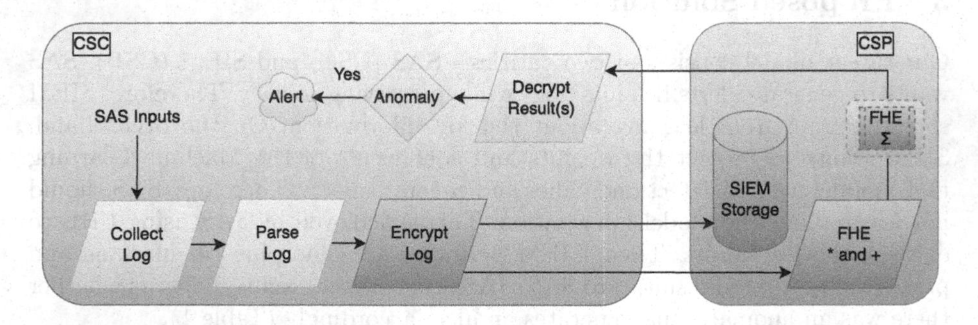

Fig. 2. Encrypted log anomaly detection in the "Ubiquitous" and "Aggregate" configurations. Yellow blocks represent plaintext and maroon blocks FHE operations. The dotted green block is an additional component in the "Aggregate" configuration. (Color figure online)

In the "Ubiquitous" configuration (Fig. 2), SAS sends encrypted inputs to SIEM for analysis, and SIEM performs homomorphic computations on encrypted data. SIEM sends one encrypted result per input to SAS. SAS decrypts all the results and evaluates the labels for all the inputs. In this configuration, the disadvantage is leaking the data used for training or the model weights, as a dishonest client can perform inference attacks.

In the "Aggregate" configuration (Fig. 2), SAS sends a block of encrypted inputs, and SIEM performs homomorphic calculation with plaintext ML weights and applies Sigmoid for each input. Sigmoid approximates the labels in the range [0, 1], where 0 corresponds to a "normal" class and 1 to "anomalous". Then, SIEM performs homomorphic additions of all these encrypted results. In this scenario, SAS receives only one result per block. It saves network bandwidth, and SAS needs to decrypt only one cipher per block instead of individual ciphers, saving computation and storage overhead. SAS decrypts the result and evaluates the sum for the block of messages. If there are no anomalies in the block, the sum should be 0. Otherwise, it should be the count of anomalous inputs.

Another benefit of this configuration is realized by employing an anomaly score per block. It can serve as a 'litmus test' for log anomalies. E.g., a security analyst may want to give a higher preference for examination of the block of logs with a higher anomaly score compared to a block with a much lower score. Also, if there are consecutive blocks with higher than usual anomaly scores, it may serve as an IoC. The disadvantage of this mode is that SAS can not pinpoint which of the message in the block is anomalous.

We present four models - one with unencrypted data and three with encrypted data and compare their performance.

1. We refer to the model with unencrypted data as a "Plain model". This model was used to create a baseline.
2. Encrypted model with a "1st-degree" Sigmoid polynomial approximation.
3. Encrypted model with a "3rd-degree" Sigmoid polynomial approximation.
4. Encrypted model with a "5th-degree" Sigmoid polynomial approximation.

Algorithm 1. σ_{LR} (Sigmoid-LR)

Input: $X, Y, r_{learn}, r_{iter}$
Output: W, b

```
1: m = size of X
2: W = b = 0
3: for l = 0 to r_iter do
4:     U = UpdateWeights(X, Y, m, r_learn, W, b)
5:     W, b = U
6: end for
```

We observed that the LR (from sklearn.linear_model) and SVC (from sklearn.svm) models from scikit-learn [21] do not perform well with Sigmoid activation for "Aggregate" configuration. Hence we designed Sigmoid-LR (σ_{LR}) model (Algorithm 1) to improve the outcomes. Essentially, we use a kernel $A = X \cdot W + b$ and reduce the errors of $Sigmoid(a)$ with a learning rate r_{learn} and r_{iter} iterations, where the inputs and labels are $X, Y \in [0, 1]$.

4 Related Work

To reduce latency in data processing and minimize the raw data revealed to CSP, [35] proposed to deploy services on CSC's devices at the Internet edge rather than putting everything on the cloud. They present the design of a novel

Algorithm 2. UpdateWeights

Input: X, Y, m, r_{learn}, W, b
Output: W, b
1: $A = X \cdot W + b$
2: $P = Q = \phi$
3: **for** every element a in A **do**
4: append $sigmoid(a)$ to P
5: **end for**
6: **for** $l = 0$ to size of Y **do**
7: append $P[l] - Y[l]$ to Q
8: **end for**
9: $\delta_W = X^T \cdot Q/m$
10: $\delta_b = \Sigma Q/m$
11: $W = W - r_{learn} \times \delta_W$
12: $b = b - r_{learn} \times \delta_b$

system, "Zoo", to support the construction, composition, and easy deployment of ML models on the edge and local devices. The Zoo system is implemented on Owl [31], an open-source numerical computing system in the OCaml language. However, their deployment is wider than edge devices. It can also be on cloud servers or a hybrid of both cases to minimize the data revealed to the CSP and the associated communication costs. However, moving ML-based data analytics from cloud to edge devices brings challenges like resource limitations, lack of usable models, and difficulties deploying user services. Furthermore, deploying services on a CSC's environment imposes challenges for CSP where the privacy of ML models (weights) should also be protected from the CSC.

[25] proposed anonymous upload, retrieve and delete protocols on log records in the cloud using the Tor network [30]. They proposed a comprehensive scheme that addresses security and integrity issues during the log generation phase and other stages in the log management process, including log collection, transmission, storage, and retrieval. However, their logging client depends on the chosen operating system and logs privacy is not addressed since the logs can be identified through their tag values.

[33,34] introduced Secure Logging as a Service (SecLaaS), which stores various logs generated for the activity of virtual machines running in clouds and provides access to forensic investigators ensuring the confidentiality and integrity of such logs of the CSC. After saving a log entry in the log database, the system will additionally store the proof of this entry in the proof database. To prove the logs as admissible evidence, the investigator can provide proof of the logs along with the logs. To ensure the confidentiality of CSC's logs, some Log Entry (LE) information can be encrypted using a shared public key of the security agencies. The private key to decrypt the log can be shared among security agencies. Additionally, an auditor can check the integrity of the logs using the Proof of Past Log (PPL) and the Log Chain (LC). However, they cannot encrypt all the fields

of the LE as the CSP needs to search the storage by some fields. Also, these schemes use a shared public key, violating a CSC's data privacy.

[24] proposed a forensically aware Blockchain-assisted Secure Logging-as-a-Service (BlockSLaaS) for a cloud environment to securely store and process logs by tackling multi-stakeholder collision problems and ensuring integrity and confidentiality. They aim to make the cloud more auditable and forensic-friendly. The integrity of logs is ensured using the immutable property of blockchain technology. Cloud Forensic Investigator (CFI) can only access the logs for forensic investigation by BlockSLaaS, which preserves the confidentiality of logs. To ensure the privacy of the CSC, the Node Controller (NC) encrypts each log entry using CFI's public key CFI_{PK}. CFI can use their secret key CFI_{SK} to decrypt the logs, so the confidentiality of CSC's logs gets preserved. This scheme uses a CFI_{PK}, which violates data privacy for a CSC.

[1] proposed a principled systems architecture - Encode, Shuffle, Analyze (ESA) for performing large-scale monitoring with high utility while also protecting user privacy. With ESA, the privacy of monitored users' data is guaranteed by its processing in a three-step pipeline. Their implementation, PROCHLO, develops new techniques to harden those steps, including the Stash Shuffle, a novel scalable and efficient oblivious-shuffling algorithm based on Intel's SGX, and new cryptographic secret sharing and blinding applications. However, using a Trusted Execution Environment (TEE) like Intel SGX may only be practical for some devices, considering legacy and low-resourced systems. Also, TEE limits the size of data that can be secured.

[20] designed a Collective Learning protocol, a secure protocol for sharing classified time-series data within entities to train the binary classifier model's parameters partially. They approximate the Sigmoid function (σ) to a polynomial of degree 7 and present a Collective Learning protocol to apply Homomorphic Encryption to fine-tune the last layer of a deep neural network securely. However, degree-7 approximation using an FHE scheme is inefficient with a resource-strained machine like a wireless sensor or an Internet-of-Things (IoT) device. In our work, we approximate σ to polynomials of degree-1 ($\sigma_1(x)$), degree-3 ($\sigma_3(x)$), and degree-5 ($\sigma_5(x)$) and compare their performances in terms of performance metrics and execution time.

The closest work on the confidentiality of log data, both during transmission and analysis, using Homomorphic Encryption techniques is proposed by [2]. In their scheme, the authors investigate the possible use of Homomorphic Encryption to provide a privacy-preserving log management architecture. They use SVM with a linear kernel to evaluate the homomorphic classification of IDS alerts from the NSL-KDD set. In their scheme, they encrypt the SAS input with the BFV scheme and perform homomorphic calculations with SIEM weights in plain. The encrypted result for each log is sent back to SAS for decryption. This can result in an inference attack by a malicious SAS.[4] We use the CKKS scheme and present two configurations in our work. (1) Ubiquitous configuration similar

[4] Cloud-based models are susceptible to training data inference attacks, e.g., attribute inference attacks, membership inference attacks, and model inversion attacks.

to their work and a (2) Aggregate configuration to reduce the communication and computation requirements. We also compare $\sigma_1(x), \sigma_3(x), \sigma_5(x)$ approximation for a Logistic Regression (LR) and a Support Vector Machine (SVM) model.

5 Experimental Analysis

The computations were performed on a MacBook Pro with a 2.4 GHz Quad-Core Intel Core i5 processor and 8 GB 2133 MHz LPDDR3 memory. We used the SEAL-Python [17] library for Python 3.10 [23] to provide CKKS encryption. In addition, we have used sklearn APIs [5] for binary classifiers.

5.1 Evaluation Criteria

We compare the performance of the four models based on the following measures. Precision, Recall, Accuracy, and F1-Score are helpful to evaluate the models in Ubiquitous configuration, while Σ-Ratio is used for Aggregate configuration. We repeat the experiments both for NSL-KDD and the balanced HDFS dataset.

Precision is the ratio of True Positives over the sum of False Positives and True Negatives. It is also known as positive predictive value.

$$Precision = \frac{TP}{TP + FP}$$

Recall is the ratio of correctly predicted outcomes to all predictions. It is also known as sensitivity or specificity.

$$Recall = \frac{TP}{TP + FN}$$

Accuracy is the ratio of correct predictions out of all predictions made. It can be calculated by dividing Precision by Recall or as $1 - \frac{FalseNegativeRate(FNR)}{FalsePositiveRate(FPR)}$.

$$Accuracy = \frac{TP + TN}{TP + FP + TN + FN}$$

F1-Score combines both *Precision* and *Recall* into a single metric that ranges from 0 to 1.

$$F1 - score = 2 * \frac{Precision * Recall}{Precision + Recall}$$

Σ-Ratio is the ratio of sums of all label values predicted by the model to the sum of actual labels. This measure is used for Sigmoid activation with binary outcomes. E.g., a balanced test set with 100 inputs has 50 inputs with label 0 and 50 with label 1. Then the sum of all actual labels is 50, and Σ-Ratio is the sum of all predicted labels divided by 50.

$$\Sigma - Ratio = \frac{\sum_{i=1}^{n} \text{Predicted } y_i}{\sum_{i=1}^{n} \text{Actual } y_i}, \text{ where } y_i \in \{0, 1\}$$

5.2 Dataset

Generally, log anomaly datasets are skewed and dominated by either 'normal' or 'anomalous' samples. Hence we use two balanced datasets to mitigate the "pseudo-high" Accuracy of ML models. To demonstrate the balance of classes in our dataset, we used a "Return-1 Model" to always classify the data as "anomalous". As a result, we observed 48.11% Accuracy and 2.07 Σ-ratio for NSL-KDD, and 49.99% Accuracy and 2.00 Σ-ratio for HDFS. For both datasets, we achieved a recall of 100%, as the model always returns 1 for an anomaly.

NSL-KDD. NSL-KDD [11] solves some of the inherent problems of the KDD'99 data set mentioned in [27]. This dataset contains 148,517 inputs with 41 features and two observations for label and score. We modify the labels to adhere to a binary classification problem (consolidating all attack categories in label 1) with 77,054 inputs with label 0 ("normal") and 71,463 inputs classified to label 1 ("anomalous"). The testing set contained 29,704 inputs with 15,351 inputs of label 0 and 14,353 inputs of label 1.

Loghub-HDFS. We have used Loghub [16] HDFS_1 labeled data from Logpai, which is a 1.47 GB of HDFS log data set generated through running Hadoop-based map-reduce jobs on more than 200 Amazon's EC2 nodes for 38.7 h and labeled by Hadoop domain experts. Among 11,175,629 log entries collected, 2.58% (288,250) data is anomalous. We have used Drain [15] log parser[5] to transform our unstructured log data to a structured format. We created a much smaller, balanced dataset of 576,500 inputs with seven observations equally distributed among "normal" and "anomaly" classes. We used 20% of the entire dataset as testing data having 115,300 inputs, out of which 57,462 inputs belonged to label 0 and 57,838 belonged to label 1.

5.3 Test Results

First, we created baselines with unencrypted data for all models. Results are shown in Table 2. For the NSL-KDD dataset, we achieved 93.52% Accuracy and 0.99 Σ-Ratio with LR, 93.30% Accuracy and 1.06 Σ-Ratio with SVM, and 90.92% Accuracy and 0.99 Σ-Ratio with σ_{LR}. Similarly, for the balanced HDFS dataset, we achieved 96.83% Accuracy and 1.00 Σ-Ratio with LR, 96.81% Accuracy and 0.86 Σ-Ratio with SVM, and 94.71% Accuracy and 1.00 Σ-Ratio with σ_{LR}.

Then, we measured the results for $\sigma_1(x)$. For the NSL-KDD dataset, we achieved 93.52% Accuracy and 1.39 Σ-Ratio with LR, 93.30% Accuracy and 1.22 Σ-Ratio with SVM, and 90.92% Accuracy and 1.16 Σ-Ratio with σ_{LR}. For the HDFS dataset, we achieved 96.83% Accuracy and −6.31 Σ-Ratio with LR, 96.81% Accuracy and −2.88 Σ-Ratio with SVM, and 94.71% Accuracy and 0.99 Σ-Ratio with σ_{LR}.

---

[5] We omit the details of textual log data parsing for brevity.

Table 2. $\sigma(x)$ (encrypted) comparison with Plain (unencrypted) models.

Dataset	Model	Type	Accuracy	Precision	Recall	F1-Score	Σ-Ratio
NSL-KDD	LR	Plain	0.9352	0.9502	0.9138	0.9317	0.9966
		$\sigma_1(x)$	0.9352	0.9502	0.9138	0.9317	1.3925
		$\sigma_3(x)$	0.7923	0.9272	0.6186	0.7421	0.6336
		$\sigma_5(x)$	0.7448	0.7014	0.8217	0.7568	23.4544
	SVM	Plain	0.9330	0.9550	0.9039	0.9287	1.0614
		$\sigma_1(x)$	0.9330	0.9550	0.9039	0.9287	1.2227
		$\sigma_3(x)$	0.9326	0.9550	0.9031	0.9283	1.0993
		$\sigma_5(x)$	0.9133	0.9223	0.8961	0.9090	1.5204
	σ_{LR}	Plain	0.9092	0.9217	0.8874	0.9042	0.9947
		$\sigma_1(x)$	0.9092	0.9217	0.8874	0.9042	1.1674
		$\sigma_3(x)$	0.9092	0.9217	0.8874	0.9042	0.9776
		$\sigma_5(x)$	0.9092	0.9216	0.8875	0.9042	2.9291
HDFS	LR	Plain	0.9683	0.9412	0.9992	0.9693	1.0001
		$\sigma_1(x)$	0.9683	0.9412	0.9992	0.9693	−6.3162
		$\sigma_3(x)$	0.5308	0.5167	0.9992	0.6812	292.6803
		$\sigma_5(x)$	0.7753	0.9204	0.6043	0.7295	−513337.2173
	SVM	Plain	0.9681	0.9402	1.0000	0.9692	0.8649
		$\sigma_1(x)$	0.9681	0.9402	1.0000	0.9692	−2.8863
		$\sigma_3(x)$	0.5605	0.5330	1.0000	0.6953	36.6039
		$\sigma_5(x)$	0.9681	0.9402	1.0000	0.9692	−14093.2109
	σ_{LR}	Plain	0.9471	0.9046	1.0000	0.9499	1.0076
		$\sigma_1(x)$	0.9471	0.9046	1.0000	0.9499	0.9973
		$\sigma_3(x)$	0.9471	0.9046	1.0000	0.9499	0.9996
		$\sigma_5(x)$	0.9363	0.8874	1.0000	0.9403	1.0071

Next, we measured the results for $\sigma_3(x)$. For the NSL-KDD dataset, we achieved 79.23% Accuracy and 0.63 Σ-Ratio with LR, 93.26% Accuracy and 1.09 Σ-Ratio with SVM, and 90.92% Accuracy and 0.97 Σ-Ratio with σ_{LR}. For the HDFS dataset, we achieved 53.08% Accuracy and 292.68 Σ-Ratio with LR, 56.05% Accuracy and 36.60 Σ-Ratio with SVM, and 94.71% Accuracy and 0.99 Σ-Ratio with σ_{LR}.

Last, we measured the results for $\sigma_5(x)$. For the NSL-KDD dataset, we achieved 74.48% Accuracy and 23.45 Σ-Ratio with LR, 91.33% Accuracy and 1.52 Σ-Ratio with SVM, and 90.92% Accuracy and 2.92 Σ-Ratio with σ_{LR}. For the HDFS dataset, we achieved 77.53% Accuracy and −513337.21 Σ-Ratio with LR, 96.81% Accuracy and −14093.21 Σ-Ratio with SVM, and 93.63% Accuracy and 1.00 Σ-Ratio with σ_{LR}.

Table 3. Average time comparisons in milliseconds (ms).

Dataset	Model	$\sigma_1(x)$			$\sigma_3(x)$			$\sigma_5(x)$		
		Enc	Dec	Sig	Enc	Dec	Sig	Enc	Dec	Sig
NSL-KDD	LR	15.1875	4.9902	10.1789	14.9943	3.9329	30.5417	15.0870	3.4859	61.4301
	SVM	15.3696	5.0939	10.3910	15.1596	3.9666	30.8182	15.6873	3.5897	64.0202
	σ_{LR}	19.8979	6.7749	13.7552	15.0424	3.9493	30.6180	15.0098	3.4396	61.0236
HDFS	LR	16.0792	5.2545	10.8421	15.2856	3.8630	30.9884	18.4485	4.1289	75.5258
	SVM	19.0548	6.1745	12.9145	15.5804	3.9436	31.5125	15.5052	3.4766	63.2896
	σ_{LR}	15.6823	5.0948	10.5092	15.1515	3.8340	30.5709	15.2478	3.4893	62.1698

Table 4. Average $\sigma_1(x)$ time in milliseconds (ms) with CKKS scale $\in \{2^{25}, 2^{35}, 2^{45}\}$.

Dataset	Model	scale-2^{25}			scale-2^{35}			scale-2^{45}		
		Enc	Dec	Sig	Enc	Dec	Sig	Enc	Dec	Sig
NSL-KDD	LR	4.0819	0.3863	2.2571	3.9922	0.3786	2.2025	4.0924	0.3893	2.2661
	SVM	5.3294	0.5014	2.9479	3.9758	0.3763	2.1841	4.2591	0.4027	2.3583
	σ_{LR}	6.2829	0.5884	3.4755	3.9980	0.3836	2.2172	6.0822	0.5747	3.3664
HDFS	LR	4.6180	0.3956	2.4876	6.4875	0.5553	3.5301	4.9126	0.4239	2.6454
	SVM	3.9620	0.3487	2.2317	4.2095	0.3660	2.3677	4.1502	0.3621	2.3338
	σ_{LR}	3.8840	0.3408	2.1693	3.9570	0.3482	2.2110	3.9257	0.3466	2.1948

Overall we observed that with more multiplications we lost more precision in performance metrics due to the approximate nature of CKKS. For some tests, e.g. LR and SVM for the HDFS dataset, we observed unusual Σ-Ratio due to scaling and relinearization requirements imposed by the SEAL library sanity checking for CKKS.

We also measured Encryption, Decryption, and Sigmoid-approximation time per input (Table 3). On average, it took about 15.32 ms for encryption and 3.94 ms for decryption. For Sigmoid-approximations, $\sigma_1(x)$ took about 10.67 ms, $\sigma_3(x)$ took about 30.71 ms, and $\sigma_5(x)$ took about 62.72 ms execution time. $\sigma_3(x)$ was about 3X slower and $\sigma_5(x)$ was 6X slower than $\sigma_1(x)$ approximation due to additional multiplications in the encrypted domain. The common CKKS parameters with a polynomial modulus degree of 2^{14} and scale of 2^{40} were used.

We further experimented with different scale sizes of $2^{25}, 2^{35}$, and 2^{45} with polynomial modulus of degree 2^{13} and 1st-degree Sigmoid approximation. We observed similar results for different scales. For NSL-KDD, the LR model achieved 93.52% Accuracy with 1.39 Σ-Ratio, and SVM achieved 93.30% Accuracy with 1.22 Σ-Ratio, σ_{LR} achieved 90.92% Accuracy with 1.16 Σ-Ratio. For HDFS, LR had 96.83% Accuracy with -6.31 Σ-Ratio, and SVM had 96.81% Accuracy with -2.88 Σ-Ratio, σ_{LR} had 94.71% Accuracy with 0.99 Σ-Ratio. We also measured average execution time per input as shown in Table 4, and we did not observe any significant changes concerning scales, models, or datasets.

6 Conclusion

We implemented a fully homomorphic encryption-based solution using the CKKS algorithm for supervised binary classification for log anomaly detection. We also approximated the Sigmoid function with 1st-order, 3-rd order, and 5th-order polynomials and proposed an "Aggregate" configuration to reduce communication and computation requirements for wireless sensors and IoT devices.

From our experiments, we observed that for "normalized" inputs, 1st-order Sigmoid approximation outperformed higher degree approximations in terms of accuracy metrics and computation time. We achieved the same accuracy metrics as the Plain-model baseline with 1st-order Sigmoid approximation for all three models and both datasets. We observed the worst Σ-Ratio for all cases with 5th-order approximation. For our σ_{LR} model, the first and third order performed equally well in accuracy and Σ-Ratio. For a polynomial modulus of degree 2^{14} and a scale of 2^{40}, $\sigma_1(x)$ approximation took more than 10 ms while $\sigma_5(x)$ approximation took more than 60 ms on average.

We also experimented with different scale parameters for CKKS of 2^{25}, 2^{35} and 2^{45} and did not see any impact on accuracy, Σ-Ratio, or computation times. For a polynomial modulus of degree 2^{13} and a scale of 2^{45}, $\sigma_1(x)$-approximation took more than 2 ms on average.

While we based our evaluation on linearly separable binary classification, future work includes implementing FHE with other models, like Random Forests and Recurrent Neural Networks. In addition, we plan on using Chimera and combining BFV/TFHE for evaluating the Sigmoid function by approximating it by the Sign (Signum) function provided by the TFHE bootstrapping.

References

1. Bittau, A., et al.: Prochlo: strong privacy for analytics in the crowd. In: Proceedings of the 26th Symposium on Operating Systems Principles, pp. 441–459 (2017)
2. Boudguiga, A., Stan, O., Sedjelmaci, H., Carpov, S.: Homomorphic encryption at work for private analysis of security logs. In: ICISSP, pp. 515–523 (2020)
3. Brakerski, Z.: Fully homomorphic encryption without modulus switching from classical GapSVP. In: Safavi-Naini, R., Canetti, R. (eds.) CRYPTO 2012. LNCS, vol. 7417, pp. 868–886. Springer, Heidelberg (2012). https://doi.org/10.1007/978-3-642-32009-5_50
4. Brakerski, Z., Gentry, C., Vaikuntanathan, V.: Fully homomorphic encryption without bootstrapping. Cryptology ePrint Archive, Paper 2011/277 (2011). https://eprint.iacr.org/2011/277, https://eprint.iacr.org/2011/277
5. Buitinck, L., et al.: API design for machine learning software: experiences from the scikit-learn project. In: ECML PKDD Workshop: Languages for Data Mining and Machine Learning, pp. 108–122 (2013)
6. TITLE 1.81.5. California Consumer Privacy Act of 2018 [1798.100 - 1798.199.100] (2018). https://leginfo.legislature.ca.gov/faces/codes_displayText.xhtml?division=3.&part=4.&lawCode=CIV&title=1.81.5
7. Chen, H., et al.: Logistic regression over encrypted data from fully homomorphic encryption. BMC Med. Genom. **11**(4), 3–12 (2018)

8. Cheon, J.H., Kim, A., Kim, M., Song, Y.: Homomorphic encryption for arithmetic of approximate numbers. Cryptology ePrint Archive, Report 2016/421 (2016). https://eprint.iacr.org/2016/421
9. Chillotti, I., Gama, N., Georgieva, M., Izabachène, M.: Faster fully homomorphic encryption: bootstrapping in less than 0.1 seconds. In: Cheon, J.H., Takagi, T. (eds.) ASIACRYPT 2016. LNCS, vol. 10031, pp. 3–33. Springer, Heidelberg (2016). https://doi.org/10.1007/978-3-662-53887-6_1
10. S.3195 - Consumer Online Privacy Rights Act (2021). https://www.congress.gov/bill/117th-congress/senate-bill/3195
11. for Cybersecurity, C.I.: Nsl-kdd—datasets—research—canadian institute for cybersecurity (2019). https://www.unb.ca/cic/datasets/nsl.html
12. Durumeric, Z., et al.: The security impact of https interception. In: NDSS (2017)
13. Fan, J., Vercauteren, F.: Somewhat practical fully homomorphic encryption. Cryptology ePrint Archive, Report 2012/144 (2012). https://eprint.iacr.org/2012/144
14. EUR-Lex - 02016R0679-20160504 - EN - EUR-Lex (2016). https://eur-lex.europa.eu/eli/reg/2016/679/2016-05-04
15. He, P., Zhu, J., Zheng, Z., Lyu, M.R.: Drain: an online log parsing approach with fixed depth tree. In: 2017 IEEE International Conference on Web Services (ICWS), pp. 33–40. IEEE (2017)
16. He, S., Zhu, J., He, P., Lyu, M.R.: Loghub: a large collection of system log datasets towards automated log analytics (2020). https://doi.org/10.48550/ARXIV.2008.06448, https://arxiv.org/abs/2008.06448
17. Huelse: Huelse/seal-python: Microsoft seal 4.x for python (2022). https://github.com/Huelse/SEAL-Python. Accessed 9 May 2022
18. Principles for the processing of user data by Kaspersky security solutions and technologies—Kaspersky. https://usa.kaspersky.com/about/data-protection
19. Nakashima, E.: Israel hacked Kaspersky, then tipped the NSA that its tools had been breached (2017). https://www.washingtonpost.com/world/national-security/israel-hacked-kaspersky-then-tipped-the-nsa-that-its-tools-had-been-breached/2017/10/10/d48ce774-aa95-11e7-850e-2bdd1236be5d_story.html
20. Paul, J., et al.: Privacy-preserving collective learning with homomorphic encryption. IEEE Access **9**, 132084–132096 (2021)
21. Pedregosa, F., et al.: Scikit-learn: machine learning in Python. J. Mach. Learn. Res. **12**, 2825–2830 (2011)
22. Perlroth, N., Shane, S.: How Israel caught Russian hackers scouring the world for U.S. secrets (2017). https://www.nytimes.com/2017/10/10/technology/kaspersky-lab-israel-russia-hacking.html
23. Python Core Team: Python: A dynamic, open source programming language. Python Software Foundation (2021). https://www.python.org/. Python version 3.10
24. Rane, S., Dixit, A.: BlockSLaaS: blockchain assisted secure logging-as-a-service for cloud forensics. In: Nandi, S., Jinwala, D., Singh, V., Laxmi, V., Gaur, M.S., Faruki, P. (eds.) ISEA-ISAP 2019. CCIS, vol. 939, pp. 77–88. Springer, Singapore (2019). https://doi.org/10.1007/978-981-13-7561-3_6
25. Ray, I., Belyaev, K., Strizhov, M., Mulamba, D., Rajaram, M.: Secure logging as a service—delegating log management to the cloud. IEEE Syst. J. **7**(2), 323–334 (2013)
26. Remez, E.Y.: Sur le calcul effectif des polynomes d'approximation de tschebyscheff. CR Acad. Sci. Paris **199**(2), 337–340 (1934)

27. Tavallaee, M., Bagheri, E., Lu, W., Ghorbani, A.A.: A detailed analysis of the KDD cup 99 data set. In: 2009 IEEE Symposium on Computational Intelligence for Security and Defense Applications, pp. 1–6. IEEE (2009)
28. Taylor, S.: Is your antivirus software spying on you?—Restore privacy (2021). https://restoreprivacy.com/antivirus-privacy/
29. Temperton, J.: AVG can sell your browsing and search history to advertisers (2015). https://www.wired.co.uk/article/avg-privacy-policy-browser-search-data
30. The Tor Project—Privacy & Freedom Online. https://www.torproject.org/
31. Wang, L.: Owl: A general-purpose numerical library in OCaml (2017)
32. Wang, Q., Feng, C., Xu, Y., Zhong, H., Sheng, V.S.: A novel privacy-preserving speech recognition framework using bidirectional LSTM. J. Cloud Comput. **9**(1), 1–13 (2020)
33. Zawoad, S., Dutta, A.K., Hasan, R.: SecLaaS: secure logging-as-a-service for cloud forensics. In: Proceedings of the 8th ACM SIGSAC Symposium on Information, Computer and Communications Security, pp. 219–230 (2013)
34. Zawoad, S., Dutta, A.K., Hasan, R.: Towards building forensics enabled cloud through secure logging-as-a-service. IEEE Trans. Dependable Secure Comput. **13**(2), 148–162 (2015)
35. Zhao, J., Mortier, R., Crowcroft, J., Wang, L.: Privacy-preserving machine learning based data analytics on edge devices. In: Proceedings of the 2018 AAAI/ACM Conference on AI, Ethics, and Society, pp. 341–346 (2018)

HBSS: (Simple) Hash-Based Stateless Signatures – Hash All the Way to the Rescue! (Preliminary Version)

Shlomi Dolev[1(✉)], Avraam Yagudaev[1], and Moti Yung[2]

[1] Department of Computer Science, Ben-Gurion University of the Negev, Beer Sheva, Israel
shlomidolev@gmail.com

[2] Google, Mountain View, USA

Abstract. One-time signatures (originated by Lamport) and their extensions to many signatures has gained recent momentum with the need for Post-Quantum secure signing since they are essentially based on standard hash (one-way) functions (rather than number theoretic assumptions). Such signatures, to date, have been based on an array of commitments that are de-committed (private key/preimage revealed) only based on bits of a representation of the hashed message signed, and their stateless versions have been based on pseudorandom functions. In this work, a new approach to the above is presented, based on probabilistic "set membership data structure," which in turn is based on hash functions. A signature based on hash access to a suitably long array, where k hash functions which are independent are used for each message to tag an index in the array. The above access is similar to the access performed in Bloom filters. The resulting signature scheme is stateless and can be tuned to support any given upper-bound number of signatures (by tuning the array's length). The central idea is that a de-committed array is only partially loaded with de-commitments to be valid, a fact that assures correctness (signed message is validated), soundness (unsigned message is not fully validated), and unforgeability of the signature (an attempt to forge a signature is reduced to decommitment without access to the private key/decommited preimage). Constructions that are based on enhancing a one-time or bounded-message construction (such as the Naor-Yung extension from bounded messages to regular signatures) are valid for the new Hash-Based Stateless Signature (HBSS).

1 Introduction

Binding a public key to an entity is the hardest chain in connecting an entity in the real world, be it a person, a company, or an organization, to its digital representation. It is a cumbersome process that may involve certificate authorities, courts, lawyers, offline and online documents, cameras, biometrics, etc., binding the entity (detailed description) with the public key.

Partially supported by the Rita Altura Trust Chair in Computer Science and the Israeli Science Foundation (Grant No. 465/22).

S. Dolev et al. (Eds.): CSCML 2023, LNCS 13914, pp. 389–397, 2023.
https://doi.org/10.1007/978-3-031-34671-2_27

One would like to minimize the number of binding a public key (that fits a public value that is carefully kept privately by the user) to the entity he/she represents (either themselves or another physical or organizational entity).

Post-quantum one-way functions and cryptographic schemes based on: Lattice-based cryptography, multivariate cryptography, hash-based cryptography, code-based cryptography, supersingular elliptic curve isogeny cryptography, see e.g., [1]. Out of the above, non-number-theory based hash primitives, such as the SHAs, are not based on long-standing unsolved (in polynomial time) mathematical problems for which a solution may be (or secretly already has been) found; SHA512 is commonly believed to be post-quantum.

Lamport's signature can be based on any one-way function as a primitive, and in particular, those that are PQ-secure and hence are not proven to be breakable by the Shor algorithm [16]. In particular, the Secure Hashing Algorithm (SHA) family is believed to be quantum-safe. The main possible cavity of Lamport's signature is the one-time usage, requiring a new binding process of another public key for each signature. Merkle trees allow performing many signatures (exponentially growing number with the tree depth) but requires tracking the state, namely the leaves that are already used [12,13].

A complicated scheme to yield a stateless signature is presented in [8] whereas lately, more efficient stateless schemes, SPHINCS, were presented, [2,10]. The SPHINCS schemes, in principle, are based on Merkle trees and the idea of trees of trees (where the capability of further signing is increased dynamically), signing the next signing object is a basic idea originating from [14], where this idea was proposed as a way to base signatures on one-way only functions (without the trapdoor property, which was needed till then for secure signatures), yet has now been incorporated into NIST proposals.

See also [2,3,5,15] for more recent (hyper) tree based stateless schemes. We believe the various SPHINCS schemes are more complicated to implement and maintain than HBSS, the scheme we suggest here.

2 Our HBSS Scheme

This scheme is based on several facets involving hash functions and it seems to be simpler (to implement and maintain) as a stateless bounded (multi-time) signature. This is so since it is inherently stateless. Further, HBSS has a new approach: it is inspired by Hash tables and Bloom filters (which is an array data structure that keeps membership of strings compactly), where a load factor α plays an essential performance/correctness factor.

A long array *Preimage* of dimensions $2 \times m$ is created (where m is a parameter), together with a *Commitment* array of the same dimensions. Then $2m$ random (or pseudorandom) numbers are created and assigned to the $2m$ entries of the *Preimage* array. Lastly, each *Commitment*$[i, j]$ entry is assigned by the SHA result of the *Preimage*$[i, j]$ entry. The value of the *Commitment* array is made public, binding with the entity description that holds the *Preimage* private.

A comment: usually, a certificate with both a public key (commitment array) and the entity description is signed by (the private key of) a certificate authority for which, in turn, the public key is known worldwide. Obviously, exposure of a private key of certificate authority can cause tremendous harm to internet security. This PKI infrastructure weakness is out of the scope of our paper (see, e.g., [6] for a relevant discussion).

Back to our technique: To sign a signature to a message, a message digest D of length k is computed, by using, for example, SHA512 (where the other particular hash functions can be chosen, e.g., SHA256) is the size of the hash function output which is tuned to the security level chosen.

For each message, it will be hashed by the collection of k independent random hash functions (assuming SHA is a random oracle) into indices in the array, and these indices will open (i.e., expose) their primages corresponding to the committed value at that index. In detail: k indices, $i_0, i_1, \ldots i_{k-1}$ are computed, by assigning the value of SHA(j, D) mod m, $0 \leq j \leq k-1$ to i_j. The signer then exposes *Preimage*$[D[j], i_j]$ for every $0 \leq j \leq k-1$.

Note that some of the *Preimage* entries can be already exposed; however, m is chosen to be large enough to yield a small ratio (say, less than 1/2) of exposed entries of *Preimage* with relation to m, taking in account all the signatures made in a lifetime.

There is a tradeoff between storing all the values of *Preimage* array versus computing them from s_0 and s_1 as needed. Note that it is possible to store several representative values of nested hashes for s_0 and s_1, spread in the $2m$ domain, and continue the nested hash from them. Thus, keeping a small memory and reducing the needed number of nested hashes when an entry of *Preimage* should be revealed.

Pseudocode for initializing the public and private arrays, *Commitment* and *Preimage*, respectively, appears in Algorithm 1. Algorithm 1 allocates two arrays, each of two dimensions. The first dimension corresponds to a digest bit being either 0 or 1, just as Lamport's signature does. The second dimension corresponds to the length of the digest. Random numbers, in the pseudocode, a particular concrete choice of 512 bits long is made, are assigned to the $2m$ entries of *Preimage* (lines 4-5). Then each of the entries of *Preimage* is hashed. Again a particular concrete choice of SHA512 is made to compute the corresponding entry of *Commitment* array (lines 6-7).

The pseudocode for signing appears in Algorithm 2. The pseudocode starts with the digest, D, computation, here too, SHA512 is the particular concrete hash function used (line 1). Then, an array, *Signature*, of $k = |D|$ entries, each of size *KeySize* bits, is allocated (line 2). Each *Signature* entry is assigned by an entry from the *Preimage* array. Where the j'th index of the *Signature* array is assigned by one of the two i_j entries of *Preimage*, where i_j is a result of a hash function[1] over the signed *Message* (line 4). The choice of which of the two entries of the i_j index is assigned to the j'th entry of *Signature* is made by the value of the j'th bit in the digest (lines 5 to 9).

[1] Here SHA512 over the index j together with the message, is used to make each index pseudo-random and pseudo-independent. Many other possibilities for using k results of hash functions as done in Bloom filter are possible as well.

The pseudocode for signature verification appears in Algorithm 3. First, the message digests D is calculated (line 1). Then the *Signature* entries are verified to correspond to the indexes i_j, as computed in the signature process, and the value of the j'th bit of the digest. A *True* value is returned only when all entries of *Signature* are correct, and *False* is returned otherwise.

Algorithm 1. HBSS – keygen()

```
1: Preimage ← array [2, m] of KeySize
2: Commitment ← array [2, m] of KeySize
3: for i = 0 to m − 1 do
4:     Preimage[0][i] ← Random(512bits)
5:     Preimage[1][i] ← Random(512bits)
6:     Commitment[0][i] ← SHA512(Preimage)[0][i]
7:     Commitment[1][i] ← SHA512(Preimage)[1][i]
8: end for
return key(Preimage, Commitment)
```

Algorithm 2. HBSS – sign(Message, Preimage)

```
1: D ← hash with SHA512(Message)
2: Signature ← array(k · KeySize)
3: for j = 0 to k − 1 do
4:     i_j ← SHA512(j, Message) mod m
5:     if D[j] = 0 then
6:         Signature[j] ← Preimage[0, i_j]
7:     else
8:         Signature[j] ← Preimage[1, i_j]
9:     end if
10: end for
return Signature
```

Algorithm 3. HBSS – Verify(Message,Signature,Commitment)

```
1: D ← hash with SHA512(Message)
2: for j = 0 to k − 1 do
3:     i_j ← SHA512(j, Message) mod m
4:     if D[j] = 0 then
5:         if (SHA512(Signature[j]) ≠ Commitment[0, i_j] ) then return False
6:         end if
7:     else
8:         if (SHA512(Signature[j]) ≠ Commitment[1, i_j] ) then return False
9:         end if
10:    end if
11: end for
return True
```

The size of m is a function of the upper bound on n, the number of signatures made during the lifetime of the system/entity, and k the number of entries of *Preimage* exposed (αm of which, for the first time) with each signature. If we chose $m \geq 2nk$, then even the last signature of the n exposes, for the first time, an expected number of $k/2$ entries of the *Preimage* array. Thus, if k is 512 bits (when using SHA512), then it is expected that 256 of them are verified to be originated by the signer, which in turn may imply a sufficient security level. Hence, the choice of k, the upper bound on α, and the size of each entry of the *Preimage* array can be tuned to imply the required security level.

Note that several one-way functions (OWF) can be chosen; (a) one that is used to define the entries of *Commitment* array as the OWF function of the entries of *Preimage* array, (b) one that is used to define D the digest of messages, and (c) one that defines the k indices in m for which a corresponding entry (of the two) of *Preimage* should be revealed.

Next, we outline the claims that ensure the post-quantum security of our scheme.

Theorem 1. *There exist parameter choices for which our HBSS scheme is post-quantum secure.*

Proof Outline. We set our signing scheme parameters according to the following policies: (a) entries of *Preimage* are random (or strong enough pseudorandom, possibly based on SHA512, as described in the sequel), each such entry consists of enough bits, say 512 bits, (b) entries of *Commitment* are a result of a strong enough one-way cryptographic hash, say SHA512, (c) α is kept small, say $\alpha < 1/2$, or kept small in terms of an upper bound on n, (m) k is large enough, say $k \geq 1024$ (one may concatenate the result of SHA512$(1, msg)$ with SHA512$(2, msg)$ to obtain 1024 bits digest from the message msg), and (e) the probing indices for revealing entries of the *Preimage* array are results of post-quantum hash function, say, SHA512.

Let M be the set of messages signed so far by exposing the entries of the *Preimage* array. Our policy choices imply that it is essentially impossible, even for a quantum computer (using Grover's algorithm [9]), to find a message $msg \notin M$ that uses only the already exposed entries in the *Primage* array. Note that even one not exposed entry implies the need for inverting SHA512. Mining a message (similarly to blockchain preimage search) that yields indexes of only already exposed entries in the *Primage* array implies the need to scan an expected number of potential messages in the order of 2^{512}, which in turn is essentially impossible, also when equipped with a quantum computer. □

- *Soundness: If the signer did not sign, there is a way to show it, given access to the signer's signing history.*

This is clear when at least one of the signature entries does not correspond, via hash, to the public key array entry. If all signature entries correspond to the public key array, finding a message with a signature that is only mapped to the revealed entries is essentially impossible.

- *Correctness: If the signer signed, there is no way to repudiate.*

  See the non-deniability arguments above.

- *Unforgeability: The signer or any poly-time adversary cannot forge a new message.*

  The choice of parameter choices implies that forging a new message is essentially impossible.

3 One Dimensional Bloom Filter

The new technique used here is the hash mapping of the indexes to the array. Apparently, a single-dimension Bloom filter will fully capture the power of our suggestion. Instead of mapping the digest value 0 digits to a single dimension array and the value one digit to a separate one, we can use k hash functions from the message itself or from the digest of the message.

The single-dimension solution is equivalent to the two dimensions solution. We choose one of the two single-dimensional arrays with equal probability and then choose indexes within the chosen array.

The security level, the resources, and the performance are functions of the particular choices made; for which the detailed discussion is omitted from this short version.

Bloom filter [4] common analysis exhibits a false positive probability for n items each using k hash mappings in an array of size m to be less than: $(1 - e^{-(k(n+0.5)/(m-1)})^k$.

The probability for an entry of *Preimage* to be not yet revealed is $p = (1 - 1/m)^{kn} \approx e^{-kn/m}$. Which yield $me^{-kn/m}$ expected number of unrevealed *Preimage* entries, or expected $\alpha = me^{-kn/m}/m$.

Choosing k to be optimal for n and m, one should choose $k = (m \ln 2)/n$, yielding $\alpha = 1/2$ after the last signature is issued. Thus, the expected number of entries of the *Preimage* revealed for the last signature, when $k = 1024$, is 512.

A malicious signer may act toward a deniability claim, choosing many messages that together reveal all the entries a message msg reveals. By doing so, the signer wishes to claim later that he/she did not sign msg. However, the malicious signer may need to "mine" (in the head) messages that (together) expose the k indexes that msg exposes. The probability of hitting one index of these k indexes is $1/m$. The expected number of messages that should be examined to have full coverage for the k entries of msg is m^k/k; for example, when $m = 2$ Billion and $k = 1024$, the adversary is doomed to surrender.

If $k = 1024$, then the number of signatures n, is approximate to be $n = m/1477$, say we use one TeraBytes of memory; in fact, for now, we use even two TeraBytes for *Preimage* and *Commitment*, in the sequel, we suggest ways to reduce the need for storage for the *Preimage* array or using Merkle tree (and re-computation of the tree) for the *Commitment* array. Divided the two TeraBytes by 512 bits for each entry for both *Preimage* and *Commitment*, we get $m = 2$ billion entries (as one TeraByte divided by half KiloByte is more than two billion), which implies n, to be more than 1.3 million signatures.

To avoid storing the *Preimage* array, one may use the technique suggested in [7] that uses two random seeds s_0 and s_1 that are used to produce the $2m$ entries of *Preimage*. Namely, to produce a pseudo-random sequence from a one-way function, such as SHA. To simplify the discussion, consider the entries of *Preimage* as a single sequence where *Preimage*$[1, i]$ immediately follows *Preimage*$[0, i]$. To produce the j'th entry in the above $2m$ (or m, in terms of the one-dimensional case) length sequence, s_0 is hashed j times. Namely, the result of $r_1 = \text{SHA}(s_0)$ is hashed again to obtain $r_2 = \text{SHA}(r_1)$ and so on, until r_j is reached. Then s_1 is hashed $2m - j$ times to obtain t_{2m-j}. At last the corresponding entry of *Preimage* is assigned by the xor of r_j and t_{2m-j}, namely, $r_j \oplus t_{2m-j}$. The xor operation serves as a "lock" for the possibility of predicting other entries of *Preimage* (by the use of SHA) once the value of *Preimage* in the j'th index is revealed.

4 HBSS with Committeed Merkle (Tree and) Root

When we are restricted in the commitment storage, we can employ the Merkle tree scheme over our *Commitment* array entries and publish only the tree's root. Computing and exposing in the signature the relevant paths (in fact, a subtree) to the root from the leaves commitments and the corresponding preimage of these leaves.

There is a tradeoff between the storage and processing used in different settings for the preimages (whether seed(s) for reproducible pseudorandom or actual random values), the storage used by the signing party for the commitment array (whether actual commitment array or Merkle tree for the commitment), the storage is publicly verified and maintained by authority/blockchain to be associated with the signer (whether the entire commitment array or only the Merkle root value). The particular setting implies different messages' signature lengths.

5 Extensions

The entity may sign a new public key (extension) once the ratio α of the number of exposed entries of *Preimage*, over m, is bigger than a threshold. Thus, the signer may upper bound the number of signatures made during a period, say a year, and sign and commit to a new additional *Commitment* array at the end of the period. Such a signature over the new *Commitment* implies the binding of the new *Commitment* to the entity. The verifier will not accept signatures from a *Commitment* array for which the threshold is violated and will verify the binding chain of the (entries indexes in the) *commitment* array used for the signature; a binding chain that leads to the first *Commitment* array. The signature will be regarded as valid only when the binding chain leads to the (first) trusted authentication process. Obviously, the signer will not reveal too many entries of any *Preimage*, avoiding the possibility of forging his/her signature by others.

Note that finding an input (collision) when several (say, half of the) bits of the inputs for the cryptographic hash function, e.g., SHA512, are known resembles the mining task in blockchain; mining for an input with so many zeroes. The level of security can be tuned to be harder when choosing SHA with more bits. To facilitate, for our purposes, an SHA result of 1024 bits (or more), we can define D to be a concatenation of i results of SHA512, where, say, the j'th, $1 \leq j \leq i$ concatenated result is SHA512(j, m).

All along, we suggest using a post-quantum hash function, such as SHA512, for which there exist fast and efficient implementations, see, e.g., [11].

References

1. Bernstein, D.J.: Post-quantum cryptography. In: van Tilborg, H.C.A., Jajodia, S. (eds.) Encyclopedia of Cryptography and Security, 2nd Ed, pp. 949–950. Springer (2011). https://doi.org/10.1007/978-1-4419-5906-5_386
2. Bernstein, D.J., et al.: SPHINCS: practical stateless hash-based signatures. In: Oswald, E., Fischlin, M. (eds.) EUROCRYPT 2015. LNCS, vol. 9056, pp. 368–397. Springer, Heidelberg (2015). https://doi.org/10.1007/978-3-662-46800-5_15
3. Bernstein, D.J., Hülsing, A., Kölbl, S., Niederhagen, R., Rijneveld, J., Schwabe, P.: The sphincs$^+$ signature framework. In: Cavallaro, L., Kinder, J., Wang, X., Katz, J. (eds.) Proceedings of the 2019 ACM SIGSAC Conference on Computer and Communications Security, CCS 2019, London, UK, 11–15 November 2019, pp. 2129–2146. ACM (2019). https://doi.org/10.1145/3319535.3363229
4. Bloom, B.H.: Space/time trade-offs in hash coding with allowable errors. Commun. ACM **13**(7), 422–426 (1970). https://doi.org/10.1145/362686.362692
5. Bos, J.N.E., Chaum, D.: Provably unforgeable signatures. In: Brickell, E.F. (ed.) CRYPTO 1992. LNCS, vol. 740, pp. 1–14. Springer, Heidelberg (1993). https://doi.org/10.1007/3-540-48071-4_1
6. Dolev, S.: Overlay security: quantum-safe communication over the internet infrastructure. IntechOpen (2019). https://doi.org/10.5772/intechopen.78088
7. Dolev, S.: System and method for Merkle puzzles symmetric key establishment and generation of Lamport Merkle signatures (2019). US Patent 0140819
8. Goldreich, O.: Two remarks concerning the goldwasser-micali-rivest signature scheme. In: Odlyzko, A.M. (ed.) CRYPTO 1986. LNCS, vol. 263, pp. 104–110. Springer, Heidelberg (1987). https://doi.org/10.1007/3-540-47721-7_8
9. Grover, L.K.: A fast quantum mechanical algorithm for database search. In: Miller, G.L. (ed.) Proceedings of the 28th Annual ACM Symposium on the Theory of Computing, Philadelphia, Pennsylvania, USA, 22–24 May 1996, pp. 212–219. ACM (1996). https://doi.org/10.1145/237814.237866
10. Hülsing, A., Kudinov, M.A.: Recovering the tight security proof of sphincs$^+$. In: Agrawal, S., Lin, D. (eds.) Advances in Cryptology - ASIACRYPT 2022–28th International Conference on the Theory and Application of Cryptology and Information Security, Taipei, Taiwan, 5–9 December 2022, Proceedings, Part IV. Lecture Notes in Computer Science, vol. 13794, pp. 3–33. Springer (2022). https://doi.org/10.1007/978-3-031-22972-5_1

11. Martino, R., Cilardo, A.: SHA-2 acceleration meeting the needs of emerging applications: a comparative survey. IEEE Access 8, 28415–28436 (2020). https://doi.org/10.1109/ACCESS.2020.2972265
12. Merkle, R.: Secrecy, authentication and public key systems, pp. 32–61 (1979)
13. Merkle, R.C.: A digital signature based on a conventional encryption function. In: Pomerance, C. (ed.) CRYPTO 1987. LNCS, vol. 293, pp. 369–378. Springer, Heidelberg (1988). https://doi.org/10.1007/3-540-48184-2_32
14. Naor, M., Yung, M.: Universal one-way hash functions and their cryptographic applications. In: Johnson, D.S. (ed.) Proceedings of the 21st Annual ACM Symposium on Theory of Computing, 14–17 May 1989, Seattle, Washington, USA, pp. 33–43. ACM (1989). https://doi.org/10.1145/73007.73011
15. Reyzin, L., Reyzin, N.: Better than BiBa: short one-time signatures with fast signing and verifying. In: Batten, L., Seberry, J. (eds.) ACISP 2002. LNCS, vol. 2384, pp. 144–153. Springer, Heidelberg (2002). https://doi.org/10.1007/3-540-45450-0_11
16. Shor, P.W.: Algorithms for quantum computation: discrete logarithms and factoring. In: 35th Annual Symposium on Foundations of Computer Science, Santa Fe, New Mexico, USA, 20–22 November 1994, pp. 124–134. IEEE Computer Society (1994). https://doi.org/10.1109/SFCS.1994.365700

Improving Performance in Space-Hard Algorithms

Hatice Kübra Güner(✉), Ceyda Mangır, and Oğuz Yayla

Institute of Applied Mathematics, Middle East Technical University, 06800 Çankaya, Ankara, Turkey
{kubra.guner,oguz}@metu.edu.tr, ceyda.mangir@alumni.metu.edu.tr

Abstract. Protecting secret keys from malicious observers is a major problem for cryptographic algorithms in untrusted environments. White-box cryptography suggests hiding the key in the cipher code with an appropriate method such that extraction of the key becomes impossible in the white-box settings. The key is generally embedded into the confusion layer with suitable methods. One of them is using encoding techniques. Nevertheless, many encoding methods are vulnerable to algebraic attacks and side-channel analysis. Another is the space hardness concept, which creates large lookup tables that cannot be easily extracted from the device. In (M,Z)-space hard algorithms, the secret key is embedded in large tables created as a substitution box with a suitable block cipher. So the key extraction problem in the white-box settings turns into a key recovery problem in the black-box case. One of the main issues in (M,Z)-space hard algorithms is accelerating the run-time of the black-box/white-box implementation. In this study, we aim to use the advantage of the efficiency of lightweight components to speed up the diffusion layer of white-box algorithms without decreasing the security size. Therefore, we compare the linear components of NIST Lightweight Standardization candidates for efficiency and suitability to white-box settings in existing space hard ciphers. The performance results of the algorithms are compared with WARX and SPNbox-32. According to the results, using the lightweight components in the diffusion layer accelerates the performance of white-box algorithms by at least nine times.

Keywords: White-box Cryptography · Space-hard ciphers · Lightweight components · Efficiency

1 Introduction

Products used in an untrusted environment are vulnerable to capturing encryption keys by a malicious observer, as the observer has the ability to gain access to the cryptographic algorithms and the encryption keys. From DRM similar products and cloud servers to endpoint users such as mobile phones, laptops, or lightweight devices require protection against third parties. White-box cryptography suggests software protection using an appropriate method to hide the key

S. Dolev et al. (Eds.): CSCML 2023, LNCS 13914, pp. 398–410, 2023.
https://doi.org/10.1007/978-3-031-34671-2_28

in the algorithm phases. The key is generally embedded into the confusion layer with suitable methods. According to Delerablée et al. [11], the security primitives of white-box implementations are unbreakability, onewayness, incompressibility, and traceability. Incompressibility is an important property for the white-box algorithms to prevent key extraction from the device.

The first SPN-based white-box algorithm, white-box AES [9] in 2003, was proposed by Chow et al. The secret key was embedded into the Sbox by transforming the algorithm layers into lookup tables with internal and external encodings. The suggested encoding methods to prevent key extraction were broken by the algebraic attacks [3,17]. Some other white-box AES variants were proposed [19,31], but they were also broken [10,23].

A dedicated white-box block cipher based on ASASA structure was proposed by Biryukov et al. [4]. The paper defines weak white-box security and proposes a memory-hard white-box block cipher against code lifting attacks. In code lifting attacks, an attacker uses the original implementation as a large secret key for encryption on a different device. Unfortunately, key recovery attacks were applied in [15,25] against the ASASA structure. Even if the structure was broken, the proposed methods inspired space-hard approaches for white-box algorithms.

After AES, and ASASA white-box block ciphers, a new dedicated algorithm called space-hard ciphers has been proposed by Bogdanov and Isobe [5,6]. In the new cipher, large lookup tables, constructed with a small block cipher, are used to embed secret keys. The constructed tables are used as a nonlinear layer in the algorithm. With this approach, key extraction in white-box settings becomes a key recovery problem in the black-box case. Against code lifting attacks, weak *(M,Z)-space hardness* is defined such that the algorithm provides Z-bit security until the size of the leakage from the code (table) is reached M bits.

Two main issues exist for (M,Z)-space hard white-box algorithms. The first one is updating the lookup tables after a particular leakage to provide security against code lifting attacks. Generally, the security size of the algorithms is limited to leaking $\frac{1}{4}$ of the lookup tables [5,6,8,20,24]. When the leak limit is reached, the tables must be updated, either on the device or by remotely loading [20] from the server.

The other issue is accelerating the black-box and the white-box implementations. The encryption of the space-hard ciphers is mainly based on Feistel [5,14,21,22] and SPN [6,20,24] structures. When examining SPN-based white-box implementations, the most time-consuming part is the linear layer with the MDS matrix, which consists of matrix multiplication and modular reduction. Hence, speeding up the white-box and even the black-box implementations depend on the linear layer. Among the (M,Z)-space hard ciphers, SPN-box aims to improve performance against Feistel based space-hard algorithm Space [5] by taking advantage of parallelism and single instruction multiple data (SIMD) instructions. Another space-hard cipher WARX [24] is proposed with the motivation that improving the performance by decreasing the round number of the algorithm through a random MDS matrix in the linear layer.

1.1 Our Contribution

Lightweight algorithms are resource constraint designs with heavy-weight security and fewer computational costs. Some of the lightweight algorithms offer new design ideas, including getting rid of the computational cost of the MDS matrix [12,27]. In this study, we considered using those lightweight components in the linear layer of space-hard white-box structures to speed up the encryption without reducing the security level. With this purpose, the linear components of NIST Lightweight competition candidates [16] were discussed to find alternative approaches to existing MDS matrices in the white-box settings.

We suggested fixing the security size to $M \cdot 2^{-keysize}$ bits to more precisely calculate the round numbers of the algorithms based on the recommended security level. The run-time of the white-box schemes was compared with the algorithms WARX and SPNbox-32. We observed that these lightweight components in white-box settings to be faster than WARX and SPNbox-32, with appropriate round numbers and security sizes.

The design approaches and algorithm specifications of white-box conversions are discussed in Sect. 2. Security notions of the white-box settings and details of round number computation are stated in Sect. 3. Performance comparisons of the algorithms are given in Sect. 4.

2 White-Box Implementations with the Lightweight Components

Ten finalists and 19 second-round candidates of the NIST Lightweight competition were examined by compatibility of the white-box settings. Algorithms with 16-bit/32-bit word sizes were chosen because the structure of lookup tables is based on these word sizes. The round constant and linear layers of the white-box implementations were adapted from the lightweight designs. The diffusion of the white-box conversions was calculated with the strict avalanche criteria (SAC) to check the reliability of the algorithms. According to SAC test results in Table 2, the white-box adaptations of Sparkle [1] from finalists, Spook-Shadow [2] and Saturnin [7] among the second-round candidates provide diffusion in desired white-box settings.

Saturnin [7] is an SPN-based lightweight algorithm that uses MDS matrix to provide diffusion property. Sparkle and Shadow algorithms are permutation constructs with no key scheduling or key addition layer. Sparkle [1] is based on ARX design consisting of addition, rotation, and xor operations [12]. The ARX design uses the long trail strategy (LTS) [12], based on the powerful built-in Sbox and the use of a linear layer with reduced computational cost. Shadow is a LS-design based algorithm with an additional diffusion layer.

The lookup table used as a substitution box in a white-box context can be created using small-block ciphers [6,24]. Round keys of the small-block cipher are generated using a trusted extendable output function (XOF) with a master key, and all values in the $2^{wordsize}$ space are encrypted with the cipher.

The table size is 128 KiB for the 16-bit word size and 16 GiB for the 32-bit case. The generated table is used as a substitution box to ensure nonlinearity in the white-box implementation. In any case, SHAKE [13] algorithm was used to generate the round keys.

Round numbers of the white-box algorithms were computed according to the code lifting security criteria in Sect. 3.2. The run-time performance of 16-bit designs was compared with WARX, and 32-bit designs were compared with SPNbox-32.

2.1 Saturnin

Specifications of White-Box Conversion. The linear layer and round constants of white-box Saturnin are adapted from lightweight Saturnin design [7]. In the linear layer, an MDS matrix is applied to columns in parallel and defined as

$$M : GF(2^4)^4 \rightarrow GF(2^4)^4$$

$$(u_0, u_1, u_2, u_3) \rightarrow \begin{bmatrix} \alpha^2(u_0) \oplus \alpha^2(u_1) \oplus \alpha(u_1) \oplus u_2 \oplus u_3 \\ u_0 \oplus \alpha(u_1) \oplus u_1 \oplus \alpha^2(u_2) \oplus u_2 \oplus \alpha^2(u_3) \oplus \alpha(u_3) \oplus u_3 \\ u_0 \oplus u_1 \oplus \alpha^2(u_2) \oplus \alpha^2(u_3) \oplus \alpha(u_3) \\ \alpha^2(u_0) \oplus u_0 \oplus \alpha^2(u_1) \oplus \alpha(u_1) \oplus u_1 \oplus u_2 \oplus \alpha(u_3) \oplus u_3 \end{bmatrix}$$

The α transformation used in the matrix is taken as

$$\alpha : GF(2)^4 \rightarrow GF(2)^4$$

$$(v_0, v_1, v_2, v_3) \rightarrow (v_1, v_2, v_3, v_0 \oplus v_1)$$

Linear transformations SR_slice and SR_sheet are used to provide diffusion into 16-bit state words. In SR_slice, rotation is applied inside each 4-bit part of the state-word, while 4-bit parts of the state-word are rotated in SR_sheet. The round constants RC0 and RC1 are generated using two different LFSR with feedback polynomials $X^{16} + X^5 + X^3 + X^2 + 1$ and $X^{16} + X^6 + X^4 + X + 1$.

Implementation Details. When Saturnin is converted to white-box settings, the key addition part is removed. The lookup table and nonlinear layer of Saturnin are replaced without changing the order of the layers of the lightweight design. Also, instead of applying layers SR_slice and SR_sheet only in even rounds, as in lightweight design, SR_slice is applied on odd rounds and SR_sheet on even rounds. The white-box Saturnin is implemented in 8 rounds.

The computational cost of the MDS matrix multiplication is $38 \cdot xor$ operations. SR_slice, and SR_slice_inv layers have $24 \cdot and + 12 \cdot or$ operations, respectively. Moreover, SR_sheet, and SR_sheet_inv layers have $12 \cdot or$ operation, respectively. There are $2 \cdot xor$ operation from adding round constants. Thus, computational cost of the white-box Saturnin is $320 \cdot xor + 192 \cdot or + 192 \cdot and$ operations.

Algorithm 1. WBI of Saturnin.

1: Input: $x_i \in \mathbb{F}_2^{16}$, $i \in \{0, \cdots, 15\}$, T-Table
2: Output: $x_i \in \mathbb{F}_2^{16}$, $i \in \{0, \cdots, 15\}$
3: **for** $i = 0$ to R-1 **do**
4: **for** $j = 0$ to 15 **do**
5: $x_j \leftarrow T(x_j)$
6: **end for**
7: **if** $((i \wedge 1) == 0)$ **then**
8: $(x_0, \cdots, x_{15}) \leftarrow SR\_slice((x_0, \cdots, x_{15})$
9: $(x_0, \cdots, x_{15}) \leftarrow MDS(x_0, \cdots, x_{15})$
10: $(x_0, \cdots, x_{15}) \leftarrow SR\_slice\_inv((x_0, \cdots, x_{15})$
11: **else**
12: $(x_0, \cdots, x_{15}) \leftarrow SR\_sheet((x_0, \cdots, x_{15})$
13: $(x_0, \cdots, x_{15}) \leftarrow MDS((x_0, \cdots, x_{15})$
14: $(x_0, \cdots, x_{15}) \leftarrow SR\_sheet\_inv((x_0, \cdots, x_{15})$
15: **end if**
16: $x_0 \leftarrow x_0 \oplus RC0[i]$
17: $x_8 \leftarrow x_8 \oplus RC1[i]$
18: **end for**

2.2 Sparkle

Specifications of White-Box Conversion. The linear transformation $\mathcal{L}_w$ [1] is defined as

$$\mathcal{L}_w : GF(2^w)^l \rightarrow GF(2^w)^l$$
$$((x_0, y_0), \cdots, (x_{l-1}, y_{l-1})) \rightarrow ((u_0, v_0), \cdots, (u_{l-1}, v_{l-1}))$$

where (u_i, v_i) pairs are computed from the equations

$$u_i \leftarrow x_i \oplus \alpha(\bigoplus_{j=0}^{l-1} y_j), \; v_i \leftarrow y_i \oplus \alpha(\bigoplus_{j=0}^{l-1} x_j), \; i \in 0, \cdots, l-1$$
$$\alpha(t) \leftarrow (t \lll w/2) \oplus (t \,\&\, 0 \times FFFF),$$

Round constants c_i are taken directly from the lightweight design.

Implementation Details. Although Sparkle is designed with a 32-bit word size, we also implement the 16-bit word size option when converting to the white-box settings. The Sparkle's white-box implementations use a lookup table for the nonlinear layer instead of the *Alzette* box. The round numbers of white-box Sparkle-16 and white-box Sparkle-32 are computed as 8 and 14, respectively, in the white-box settings.

One round of white-box Sparkle-16 consists of $2 \cdot xor$ operations from round constants, and $26 \cdot xor + 2 \cdot or$ operations from the linear layer. Hence, computational cost of the white-box Sparkle-16 is $224 \cdot xor + 16 \cdot or$ operations. Similarly, one round of the white-box Sparkle-32 has $2 \cdot xor$ operations from round constants, and $14 \cdot xor + 2 \cdot or$ from the linear layer, so total operation cost is $240 \cdot xor + 30 \cdot or$ for the white-box Sparkle-32.

Algorithm 2. WBI of Sparkle256.

1: Input: $(x_i, y_i) \in \mathbb{F}_2^w \times \mathbb{F}_2^w,\ i \in \{0, \cdots, l\},\ w = 16, 32,\ l = 256/w$, T-Table
2: Output: $(x_i, y_i) \in \mathbb{F}_2^w \times \mathbb{F}_2^w,\ i \in \{0, \cdots, l\}$
3: $(c_0, \cdots, c_7) \leftarrow algorithm\_constants$
4: **for** $i = 0$ to R-1 **do**
5: $\quad y_0 \leftarrow y_0 \oplus c_{(i \bmod 8)}$
6: $\quad y_1 \leftarrow y_1 \oplus i$
7: $\quad$**for** $i = 0$ to l **do**
8: $\quad\quad (x_i, y_i) \leftarrow (T(x_i), T(y_i))$
9: $\quad$**end for**
10: $\quad (x_0, y_0), ..., (x_l, y_l) \leftarrow \mathcal{L}_w((x_0, y_0), ..., (x_l, y_l))$
11: **end for**

2.3 Spook-Shadow

Specifications of White-Box Conversion. Round constants generated from a 4-bit LFSR, as well as linear components L-box and D-box, are taken from the lightweight design [2] to white-box conversion. L-box is a bit-slice implementation applied to two 32-bit state words to increase branch number. Since L-box is applied to 128-bit subblocks, an additional linear component D-box is used to diffuse 128-bit subblocks to each other.

Implementation Details. In both white-box conversions, the lookup table is added to the beginning of the Shadow permutation instead of the nonlinear layer. Additional linear layer *dbox_layer* for 384-bit case is taken from [2] as in Algorithm 3. However, for 256-bit case, we implemented the layer as in Algorithm 4 since the D-box in lightweight design cannot be directly adapted. Round numbers for 256-bit and 384-bit block sizes of white-box implementations are calculated as 14 and 15, respectively.

Algorithm 3. *dbox_layer* for 384-bit.

1: Input: $x_i \in (\mathbb{F}_2^{32}), i \in \{0, \cdots, 11\}$
2: Output: $x_i \in (\mathbb{F}_2^{32}), i \in \{0, \cdots, 11\}$
3: **for** $i = 0$ to 3 **do**
4: $\quad u \leftarrow x_{4 \cdot i}$
5: $\quad v \leftarrow x_{4 \cdot i+1}$
6: $\quad t \leftarrow x_{4 \cdot i+2}$
7: $\quad x_{4 \cdot i} \leftarrow u \oplus v \oplus t$
8: $\quad x_{4 \cdot i+1} \leftarrow u \oplus t$
9: $\quad x_{4 \cdot i+2} \leftarrow u \oplus v$
10: **end for**

Algorithm 4. *dbox_layer* for 256-bit.

1: Input: $x_i \in (\mathbb{F}_2^{32}),\ i \in \{0, \cdots, 7\}$
2: Output: $x_i \in (\mathbb{F}_2^{32}),\ i \in \{0, \cdots, 7\}$
3: **for** $i = 0$ to 3 **do**
4: $\quad x_i \leftarrow x_i \oplus (x_{i+4} \lll 15)$
5: $\quad x_{i+4} \leftarrow x_i \oplus (x_i \lll 19)$
6: **end for**

One round of the white-box Shadow256 has $48 \cdot xor + 48 \cdot or$ from *lbox_layer*, $8 \cdot xor$ and $8 \cdot or$ operations from *dbox_layer*, and $16 \cdot xor$ operations from *add_rc*. Therefore, total computational cost of the white-box Shadow256 is $1080 \cdot xor + 840 \cdot or$ operations. Similarly, one round of the white-box Shadow384 has

$72 \cdot xor + 72 \cdot or$ from *lbox_layer*, $16 \cdot xor$ operations from *dbox_layer*, and $24 \cdot xor$ operations from *add_rc*. Total computational cost of the white-box Shadow384 is $1680 \cdot xor + 1080 \cdot or$ operations.

Algorithm 5. WBI of Shadow.

1: Input: $x_i \in (\mathbb{F}_2^{32})$, $i \in \{0, \cdots, 4 \cdot n - 1\}$, T-Table
2: Output: $x_i \in (\mathbb{F}_2^{32})$, $i \in \{0, \cdots, 4 \cdot n - 1\}$
3: **for** $i = 0$ to R-1 **do**
4: **for** $j = 0$ to $4 \cdot n - 1$ **do**
5: $x_j \leftarrow T(x_j)$
6: **end for**
7: **for** $j = 0$ to n-1 **do**
8: $(x_{4 \cdot j}, \cdots, x_{4 \cdot j+3}) \leftarrow lbox\_layer(x_{4 \cdot j}, \cdots, x_{4 \cdot j+3})$
9: $(x_{4 \cdot j}, \cdots, x_{4 \cdot j+3}) \leftarrow add\_rc((x_{4 \cdot j}, \cdots, x_{4 \cdot j+3}), 2 \cdot i, j)$
10: **end for**
11: $(x_0 \cdots x_{4 \cdot n}) \leftarrow dbox\_layer(x_0 \cdots x_{4 \cdot n})$
12: **for** $j = 0$ to n-1 **do**
13: $(x_{4 \cdot j}, \cdots, x_{4 \cdot j+3}) \leftarrow add\_rc((x_{4 \cdot j}, \cdots, x_{4 \cdot j+3}), 2 \cdot i + 1, j)$
14: **end for**
15: **end for**

3 Security

The white-box security of an algorithm is evaluated by its resistance to key extraction from the lookup table and by its inability to use the table as a large key outside of the white-box environment called code lifting [5].

3.1 Key Extraction Security

In the space-hard ciphers, extracting the secret key from the white-box structure turns into recovering the key from the lookup table in black-box settings [5] since the secret key is embedded into the table. Therefore, the algorithm is as resistant to key extraction attacks as the reliability of the table-creation method. A malicious observer should not be able to obtain the secret key from the table values used in the internal steps of encryption or from the leaked portion of the table. The key extraction security of the our white-box conversions relies on the table creation methods of WARX and SPNbox-32. The WARX's method is inspired by the SPARX [12]. The details of optimal differential characteristic probability and optimal linear trail are given in the article [24]. The table generation method of SPNbox-32 [6] is used MDS matrix and S-box of AES in 16 rounds. The minimum number of active S-boxes was calculated using the method described by Mouha et al. in [26]. According to the result, the table creation method provides 128-bit security with 10 rounds against differential and linear cryptanalysis attacks.

3.2 Code Lifting Security

Resistance to code-lifting attacks is a crucial security measurement for space-hard ciphers since the tables are used as a large key in the design. Therefore, incompressible tables are needed to limit the leakage and prevent code lifting attacks [11]. Code lifting security is defined with weak (M,Z)-space hardness [5].

Definition 1 (Weak (*M,Z*)-space hardness [5]). *A white-box block cipher is called weak (M,Z)-space hard if it is not possible to encrypt/decrypt a randomly selected text with a probability greater than* 2^{-Z} *until the size of the leakage from the code (table) is reached to M bits.*

When computing the success probability in the encryption process in a white-box context, we also include correctly guessing the corresponding entry of the table if the entry is not located in the leaked part of the lookup table, as specified in [24]. Let T represent all table sizes in the memory, and M represents the size of the leaked part of the table. We can generalize the success probability as

$$p = (\frac{M}{T} + (\frac{1}{T-M}) \cdot (1 - \frac{M}{T}))^{r \cdot t} \tag{1}$$

where r represents the round number, and t represents the table lookup number for one round. If the corresponding entry is in the leaked part of the table, the probability of being correct is 1. The probability of encountering such entries is at most $\frac{M}{T}$, depending on the leak size. If the corresponding value is not in the leaked part, the probability of correctly guessing the value is $\frac{1}{T-M}$. The probability of encountering such an entry is $1 - \frac{M}{T}$.

The maximum achievable security (MAS) for a white-box algorithm is calculated by considering the leaked size of the table [8,14,24] and limited to *keysize* - $\log_2(T)$ bits. Since $\frac{M}{T}$ is considered a small rate, the level of security is generalized to the leak of the entire table. However, we think it is more convenient to take the security level as *keysize* - $\log_2(M)$ bits for more precise calculations on round numbers. The number may be smaller than the desired security level, assuming the entire table has been leaked. The round numbers of white-box implementations are given in Table 1.

Table 1. Round numbers of the white-box implementations

Algorithm	(keysize,wordsize)	$t = \frac{keysize}{wordsize}$	$r = \frac{\log_2(M) - keysize}{t \cdot \log_2(\frac{M+1}{T})}$
Saturnin	(256,16)	16	8
Sparkle-16	(256,16)	16	8
Sparkle-32	(256,32)	8	14
Shadow256	(256,32)	8	14
Shadow384	(384,32)	12	15

3.3 Diffusion Criteria

The strict avalanche criteria [30] is used to measure the diffusion property of the white-box conversions. The SAC test measures the effect on the output bits when one bit of the input is changed. The experiments were performed on 100000 random samples. For each sample, i-th bit of the input was complimented, and its effect on each output bit was examined, respectively. If the j output bit was changed, then the entry of $M_{i,j}$ was increased by one. According to the SAC test results in Table 2, the white-box Saturnin provides diffusion property after two rounds. The white-box Sparkle needs four rounds to diffuse every bit of the output, while the white-box Shadow supplies diffusion property after the second round.

Table 2. SAC test results with 100000 random samples.

Algorithm	(n, round)	Min. Value	Max. Value	$\mu = \frac{1}{n^2}\sum_{i=0}^{n-1}\sum_{j=0}^{n-1} M_{i,j}$
WARX	(128,2)	49401	50600	49999
Saturnin	(256,3)	49285	50792	50000
Sparkle-16	(256,4)	49310	50642	49998
SPNbox-32	(128,1)	49487	50563	50001
Sparkle-32	(256,4)	49264	50697	50000
Shadow256	(256,2)	49353	50703	50001
Shadow384	(384,2)	49306	50757	49999

4 Performance Results

The run-time of the white-box implementations was compared with the white-box algorithms WARX and SPNbox-32. Since the lookup tables differ according to word size, the white-box Saturnin and the white-box Sparkle-16 were compared with WARX and other algorithms with SPNbox-32. WARX code using Givaro library [28] for finite field calculations was taken from GitHub [18]. The SPNbox-32 was implemented without using a library by us. While using linear layers of lightweight designs in white-box settings, we referenced the NIST GitHub repository [29]. The algorithms run on randomly generated 3072 bytes messages with 100000 cycles and -O2 optimization on a laptop equipped with x86-64 architecture and a 2.80 GHz Intel Core i7-1165G7 CPU. The operating system is Ubuntu 20.04.5 LTS with Linux Kernel 5.15.0, and the compiler is gcc-9.4.0. The performance results for white-box and black-box implementations are given in Table 3 and Table 4.

According to performance results in Table 3, white-box Saturnin is almost nine times faster than the WARX, with a MAS level of 238-bit. The memory usage of WARX is almost twelve times higher than white-box Saturnin's. Based on performance comparisons for 32-bit word size algorithms, white-box Sparkle-32 and white-box Shadow are faster than SPNbox-32. White-box Sparkle-32 is eighteen times faster than SPNbox-32 with 221-bit security, while the white-box Shadow256 is fourteen times faster than SPNbox-32. The white-box Shadow384 is twelve times faster than SPNbox-32 with a 349-bit maximum achievable security level. Although the memory usage of the algorithms is the same, Sparkle-32's white box implementation is 21% faster than Shadow256.

Table 3. Performance results of the white-box implementation.

Algorithm	Word Size (bit)	Key Size (bit)	Round	MAS (bit)	WBI in Cycle (per byte)	Memory Usage
WARX	16	128	7	114	304	852.5 KiB
Saturnin	16	256	8	238	33	72 KiB
Sparkle-16	16	256	8	238	9	72 KiB
SPNbox-32	32	128	16	128	1555	16 GiB
Sparkle-32	32	256	14	221	84	16 GiB
Shadow256	32	256	14	221	106	16 GiB
Shadow384	32	384	15	349	120	16 GiB

The black-box Saturnin is 15% faster than the WARX, while the black-box Sparkle-16 is 2 times faster than WARX. Similar to the white-box implementation, memory usage black-box WARX is almost 12 times higher than the 16-bit implementations. Black-box Sparkle-32 is 12% faster than the remaining algorithms. Black-box Shadow256 and Shadow384 are 2% faster than SPNbox-32. Memory usage of the black-box implementations is the same.

Table 4. Performance results of the black-box implementation.

Algorithm	Key Size (bit)	Round	BBI in Cycle (per byte)	Memory Usage (KiB)
WARX	128	7	498	852.5
Saturnin	256	8	422	72
Sparkle-16	256	8	236	72
SPNbox-32	128	16	117848	72
Sparkle-32	256	14	103453	72
Shadow256	256	14	116262	72
Shadow384	384	15	116355	72

The reason white-box implementations are categorized by word size is that the table used in the nonlinear layer. Since white-box security is based on the leak limit, we think it's fair to compare leak sizes for the same security levels. According to Table 5, there is an inverse relationship between the block size of the algorithm and the table leak size at the same security level.

Table 5. Table leakage size for 2^{-114} and 2^{-98} success probability.

Algorithm	Security Size (bit)	Table Size (T)	Leakage Size (bit)
WARX	114	128 KiB	$T/2^2$
Saturnin	114	128 KiB	$T/2^{0.89}$
Sparkle-16	114	128 KiB	$T/2^{0.89}$
SPNbox-32	98	16 GiB	$T/2^{2.45}$
Sparkle-32	98	16 GiB	$T/2^{0.82}$
Shadow256	98	16 GiB	$T/2^{0.82}$
Shadow384	98	16 GiB	$T/2^{0.54}$

5 Conclusion

This study examined lightweight designs for efficiency and suitability to white-box settings. With this approach, linear layers of the appropriate algorithms from the NIST Lightweight Standardization candidates were adapted to the white-box settings to speed up the run-time of the white-box/black-box implementations according to WARX and SPNbox-32. The nonlinear layers, and if there was a round key addition part, of the lightweight designs were discarded in the white-box settings. The probability of correctly guessing the unknown part of the lookup table was included in the computing round number of the algorithm. In order to make more accurate calculations, the security size of the algorithms was taken as $keysize - \log_2(M)$ bits. According to the performance results, all white-box transformations are faster than (M, Z)-space hard algorithms WARX and SPNbox-32 without decreasing the white-box security level. The use of lightweight components in white-box settings enables reasonably fast algorithm designs.

References

1. Beierle, C., et al.: Schwaemm and esch: lightweight authenticated encryption and hashing using the sparkle permutation family (2021). https://csrc.nist.gov/CSRC/media/Projects/lightweight-cryptography/documents/finalist-round/updated-spec-doc/sparkle-spec-final.pdf
2. Bellizia, D., et al.: Spook: sponge-based leakage-resistant authenticated encryption with a masked tweakable block cipher (2019). https://csrc.nist.gov/CSRC/media/Projects/lightweight-cryptography/documents/round-2/spec-doc-rnd2/Spook-spec-round2.pdf
3. Billet, O., Gilbert, H., Ech-Chatbi, C.: Cryptanalysis of a white box AES implementation. In: Handschuh, H., Hasan, M.A. (eds.) SAC 2004. LNCS, vol. 3357, pp. 227–240. Springer, Heidelberg (2004). https://doi.org/10.1007/978-3-540-30564-4_16

4. Biryukov, A., Bouillaguet, C., Khovratovich, D.: Cryptographic schemes based on the ASASA structure: black-box, white-box, and public-key (extended abstract). In: Sarkar, P., Iwata, T. (eds.) ASIACRYPT 2014. LNCS, vol. 8873, pp. 63–84. Springer, Heidelberg (2014). https://doi.org/10.1007/978-3-662-45611-8_4
5. Bogdanov, A., Isobe, T.: White-box cryptography revisited: space-hard ciphers. In: Proceedings of the 22nd ACM SIGSAC Conference on Computer and Communications Security, CCS 2015, pp. 1058–1069. Association for Computing Machinery, New York (2015). https://doi.org/10.1145/2810103.2813699
6. Bogdanov, A., Isobe, T., Tischhauser, E.: Towards practical whitebox cryptography: optimizing efficiency and space hardness. In: Cheon, J.H., Takagi, T. (eds.) ASIACRYPT 2016. LNCS, vol. 10031, pp. 126–158. Springer, Heidelberg (2016). https://doi.org/10.1007/978-3-662-53887-6_5
7. Canteaut, A., et al.: Saturnin: a suite of lightweight symmetric algorithms for post-quantum security (2019). https://csrc.nist.gov/CSRC/media/Projects/lightweight-cryptography/documents/round-2/spec-doc-rnd2/saturnin-spec-round2.pdf
8. Cho, J., et al.: WEM: a new family of white-box block ciphers based on the even-mansour construction. In: Handschuh, H. (ed.) CT-RSA 2017. LNCS, vol. 10159, pp. 293–308. Springer, Cham (2017). https://doi.org/10.1007/978-3-319-52153-4_17
9. Chow, S., Eisen, P., Johnson, H., Van Oorschot, P.C.: White-box cryptography and an AES implementation. In: Nyberg, K., Heys, H. (eds.) SAC 2002. LNCS, vol. 2595, pp. 250–270. Springer, Heidelberg (2003). https://doi.org/10.1007/3-540-36492-7_17
10. De Mulder, Y., Roelse, P., Preneel, B.: Cryptanalysis of the Xiao–Lai white-box AES implementation. In: Knudsen, L.R., Wu, H. (eds.) SAC 2012. LNCS, vol. 7707, pp. 34–49. Springer, Heidelberg (2013). https://doi.org/10.1007/978-3-642-35999-6_3
11. Delerablée, C., Lepoint, T., Paillier, P., Rivain, M.: White-box security notions for symmetric encryption schemes. In: Lange, T., Lauter, K., Lisoněk, P. (eds.) SAC 2013. LNCS, vol. 8282, pp. 247–264. Springer, Heidelberg (2014). https://doi.org/10.1007/978-3-662-43414-7_13
12. Dinu, D., Perrin, L., Udovenko, A., Velichkov, V., Großschädl, J., Biryukov, A.: Design strategies for ARX with provable bounds: SPARX and LAX. In: Cheon, J.H., Takagi, T. (eds.) ASIACRYPT 2016. LNCS, vol. 10031, pp. 484–513. Springer, Heidelberg (2016). https://doi.org/10.1007/978-3-662-53887-6_18
13. Dworkin, M.J., et al.: SHA-3 standard: permutation-based hash and extendable-output functions (2015)
14. Fouque, P.-A., Karpman, P., Kirchner, P., Minaud, B.: Efficient and provable white-box primitives. In: Cheon, J.H., Takagi, T. (eds.) ASIACRYPT 2016. LNCS, vol. 10031, pp. 159–188. Springer, Heidelberg (2016). https://doi.org/10.1007/978-3-662-53887-6_6
15. Gilbert, H., Plût, J., Treger, J.: Key-recovery attack on the ASASA cryptosystem with expanding S-boxes. In: Gennaro, R., Robshaw, M. (eds.) CRYPTO 2015. LNCS, vol. 9215, pp. 475–490. Springer, Heidelberg (2015). https://doi.org/10.1007/978-3-662-47989-6_23
16. Information Technology Laboratory CSRC: Lightweight cryptography. https://csrc.nist.gov/Projects/lightweight-cryptography
17. Jacob, M., Boneh, D., Felten, E.: Attacking an obfuscated cipher by injecting faults. In: Feigenbaum, J. (ed.) DRM 2002. LNCS, vol. 2696, pp. 16–31. Springer, Heidelberg (2003). https://doi.org/10.1007/978-3-540-44993-5_2

18. JunLiu9102: Warx-project (2021). https://github.com/JunLiu9102/WARX-Project
19. Karroumi, M.: Protecting white-box AES with dual ciphers. In: Rhee, K.-H., Nyang, D.H. (eds.) ICISC 2010. LNCS, vol. 6829, pp. 278–291. Springer, Heidelberg (2011). https://doi.org/10.1007/978-3-642-24209-0_19
20. Koike, Y., Isobe, T.: Yoroi: updatable whitebox cryptography. IACR Trans. Cryptogr. Hardw. Embed. Syst. **2021**(4), 587–617 (2021). https://doi.org/10.46586/tches.v2021.i4.587-617. https://tches.iacr.org/index.php/TCHES/article/view/9076
21. Koike, Y., Sakamoto, K., Hayashi, T., Isobe, T.: Galaxy: a family of stream-cipher-based space-hard ciphers. In: Liu, J.K., Cui, H. (eds.) ACISP 2020. LNCS, vol. 12248, pp. 142–159. Springer, Cham (2020). https://doi.org/10.1007/978-3-030-55304-3_8
22. Kwon, J., Lee, B., Lee, J., Moon, D.: FPL: white-box secure block cipher using parallel table look-ups. In: Jarecki, S. (ed.) CT-RSA 2020. LNCS, vol. 12006, pp. 106–128. Springer, Cham (2020). https://doi.org/10.1007/978-3-030-40186-3_6
23. Lepoint, T., Rivain, M., De Mulder, Y., Roelse, P., Preneel, B.: Two attacks on a white-box AES implementation. In: Lange, T., Lauter, K., Lisoněk, P. (eds.) SAC 2013. LNCS, vol. 8282, pp. 265–285. Springer, Heidelberg (2014). https://doi.org/10.1007/978-3-662-43414-7_14
24. Liu, J., Rijmen, V., Hu, Y., Chen, J., Wang, B.: WARX: efficient white-box block cipher based on ARX primitives and random MDS matrix. Sci. China Inf. Sci. **65**(3), 1869–1919 (2021). https://doi.org/10.1007/s11432-020-3105-1
25. Minaud, B., Derbez, P., Fouque, P.-A., Karpman, P.: Key-recovery attacks on ASASA. J. Cryptol. **31**(3), 845–884 (2017). https://doi.org/10.1007/s00145-017-9272-x
26. Mouha, N., Wang, Q., Gu, D., Preneel, B.: Differential and linear cryptanalysis using mixed-integer linear programming. In: Wu, C.-K., Yung, M., Lin, D. (eds.) Inscrypt 2011. LNCS, vol. 7537, pp. 57–76. Springer, Heidelberg (2012). https://doi.org/10.1007/978-3-642-34704-7_5
27. Stoffelen, K., Daemen, J.: Column parity mixers. IACR Trans. Symmetric Cryptol. **2018**(1), 126–159 (2018). https://doi.org/10.13154/tosc.v2018.i1.126-159. https://tosc.iacr.org/index.php/ToSC/article/view/847
28. linbox team: Givaro (2021). https://github.com/linbox-team/givaro
29. usnistgov: Lightweight-cryptography-benchmarking (2021). https://github.com/usnistgov/Lightweight-Cryptography-Benchmarking
30. Webster, A.F., Tavares, S.E.: On the design of S-boxes. In: Williams, H.C. (ed.) CRYPTO 1985. LNCS, vol. 218, pp. 523–534. Springer, Heidelberg (1986). https://doi.org/10.1007/3-540-39799-X_41
31. Xiao, Y., Lai, X.: A secure implementation of white-box AES. In: 2009 2nd International Conference on Computer Science and its Applications, pp. 1–6. IEEE (2009)

A Survey of Security Challenges in Automatic Identification System (AIS) Protocol

Silvie Levy(✉), Ehud Gudes, and Danny Hendler

Ben-Gurion University of the Negev, Be'er Sheva, Israel
silvie.levy@gmail.com, gudes@bgu.ac.il, hendlerd@cs.bgu.ac.il

Abstract. The world of maritime transport is a significant part of the global economy. Traffic control relies, among other means, on the Automatic Identification System (AIS) device, which reports dynamic and fixed data. Vessels use advanced cyber capabilities to falsify data that the AIS transmits and impersonate an innocent ship while carrying out illegal activity, especially the vessel's location data, without control. A significant part of the work done to find false AIS reports looks for location reports. Each AIS device uses a transceiver based on SOTDMA (Self-Organized TDMA) and determines its transmission schedule (slot) based on data link traffic history and an awareness of other stations' possible actions. The SOTDMA protocol was developed in the late 1990s and does not have built-in security features, which leaves communication networks vulnerable to cyber threats such as eavesdropping, tampering with data, unauthorized access, and cyber-attacks. This Protocol is widely used in wireless communication systems where no central authority manages the communication between nodes, dynamically adjusts to changes in network topology, and nodes can come and go at any time. This article reviews the cybersecurity challenges in the AIS protocol used in vessels. Most of those challenges imply a variety of areas using SOTDMA protocols like Wireless Sensors (WSNs), Mobile (MANETs), Military, Disaster Relief Networks, Healthcare Monitoring Systems, Industrial Automation Systems, Vehicle-to-Vehicle (V2V) Communication Networks, Wireless Metropolitan Area Networks (WMANs), Internet of Things (IoT) Networks, and Machine-to-Machine (M2M) Communication Networks.

1 AIS OverView

AIS [1]
Automatic Identification System (AIS) is an automatic tracking system used in vessels and vessel traffic control services to identify and locate vessels through wireless information transmissions with other vessels and AIS base stations. The AIS information is the primary tool for preventing collisions at sea and is complementary to the vessel's radar information. The AIS system provides various

The paper is a regular submission to the 3rd International Symposium on Cyber Security Cryptology and Machine Learning (CSCML 2019).

S. Dolev et al. (Eds.): CSCML 2023, LNCS 13914, pp. 411–423, 2023.
https://doi.org/10.1007/978-3-031-34671-2_29

types of information, such as the unique identification signal of the vessel, its position, sailing direction, and speed. It can be displayed on a dedicated text screen or an electronic map. The AIS system is designed to aid in navigation for the officers of the ship's watch as well as the authorities to track and monitor marine movements. An AIS system contains a standard VHF transmitter and receiver, a positioning system, such as LORAN-C or GPS, and other electronic sensing devices, such as a compass. In 2010, the European Union required most commercial vessels in the inland waterways of the European continent to install an AIS system in classification A. All fishing vessels of the EU countries whose lengths exceeded 15 m were required to install the system until 2014. The estimated number of vessels equipped with an AIS system in classification A is more than 40,000.

Uses of AIS

Collision Avoidance. When a vessel navigates at sea, the information on the movements and identity of other vessels nearby is essential for deciding the sailing route that avoids the risk of collision with other vessels or stationary obstacles (such as reefs).

Vessel Traffic Service. A Vessel Traffic Service (VTS) may operate in busy waterways and ports designed to assist in the safe direction of vessel traffic. The AIS system provides additional information to the VTS service about the vessels' nature and movement.

Maritime Security. The AIS system allows the security authorities to identify vessels and their activities in and near the country's exclusive economic zone.

Navigation Aids. The AIS system can broadcast the location and name of stationary objects and the sea conditions near a utility, such as the speed of sea currents and weather conditions. These aids may be located on the shore, like a lighthouse, or in the middle of the sea, like a rig or a buoy.

Search and Rescue. The AIS system provides essential information to search and rescue operations, helps better utilize participating vessels, and even enables the placement of an AIS transmitter on aircraft participating in such procedures.

Accident Investigation. Information from AIS systems received at vessel traffic service stations is recorded at the vessel's "black box" and serves as an essential tool for the investigation of marine accidents by extracting from the AIS data the accurate data information on the time, identity, location, direction, speed, and turn rate of a vessel.

Binary Messages. Seven types of binary messages in the AIS, "Application Specific Messages" (ASM), allow the "competent authorities" to define additional messages in the AIS systems. These messages can be transmitted to a specific address, like informing vessels about water depth, order of entry to the vessel's cabins, and weather.

How AIS works
AIS devices automatically transmit information about their position, speed of movement, and other data, at pre-set intervals through a VHF transmitter that is part of the system. The AIS transceiver is installed on a vessel and operates using the CSTDMA or STDMA method. Based on SOTDMA (Self-Organized TDMA), each AIS station determines its transmission schedule (slot) based on data link traffic history and an awareness of other stations' possible actions. The transmission slot map is built in memory for each transceiver and is unavailable on a network. RF space is organized in frames. Each frame lasts precisely 1 min and starts on each minute boundary. Each frame is divided into 2250 slots. Transmission can happen on two channels. Thus, there are 4500 available slots per minute. AIS stations continuously synchronize themselves to each other to avoid overlap of slot transmissions. When a station changes its slot assignment, it announces the new location and the timeout for that location. Thus vessels will always receive new stations, including those that suddenly come within radio range close to other vessels. The transmission range of AIS transmitters on vessels is not uniform, but a typical horizontal AIS transmission range is up to 74 km. The vertical transmission range is more significant and reaches up to 400 km into space, which is the height of the orbit of the International Space Station.

It is important to note that these schemes can also have some limitations, such as **limited flexibility**, as each vessel is assigned a fixed time slot, which may not always be optimal for its specific situation, **increased complexity**, as the scheduling of time slots, must be coordinated and managed to ensure proper operation, and **limited scalability**, as the number of available time slots, limits the number of vessels sharing the same channel [12].

The SOTDMA protocol is widely used in different areas such as:

Military Communications: SOTDMA protocol provides a reliable and efficient communication system that can operate efficiently even in hostile environments.

Wireless Sensor Networks (WSNs): The SOTMDA protocol has low overhead and efficient resource utilization. Thus it is widely used in wireless sensor networks, where energy efficiency is crucial.

Mobile Networks (MANETs): MANETs are wireless networks formed spontaneously without central management; SOTMDA protocol manages the communication between nodes in a decentralized manner.

Disaster Relief Networks: SOTMDA protocol creates a decentralized communication network, allowing disaster relief teams to transmit data and coordinate their efforts effectively.

Internet of Things (IoT): SOTMDA provides a communication system between various devices and systems reliably and efficiently, allowing the devices and systems to interact and exchange data, such as smart home appliances and wearable devices.

Healthcare Monitoring Systems: Healthcare monitoring systems must have a reliable and efficient communication system that can transmit data from patients' devices to healthcare providers; SOTMDA protocol ensures that the data is transmitted quickly and efficiently.

Industrial Automation Systems: SOTMDA protocol enables the communication between various devices and systems reliably and efficiently, allowing the various devices and systems to interact and coordinate their actions.

Vehicle-to-Vehicle (V2V) Communication Networks: SOTMDA protocol manages the communication between vehicles in real-time, allowing vehicles to communicate with each other and coordinate their actions to improve road safety.

Wireless Metropolitan Area Networks (WMANs): SOTMDA protocol provides a communication system that can manage the communication between nodes over a large area, allowing the nodes to communicate with each other even when they are far apart.

Machine-to-Machine (M2M) Communication Networks: SOTMDA protocol provides a reliable and efficient communication system between various devices and systems, enabling them to interact and exchange data.

2 Related Work

Anomaly Detection in Maritime AIS Tracks: A Review of Recent Approaches [2]
Wolsing et al. surveyed 44 research articles on anomaly detection of maritime AIS tracks and identified different AIS anomalies, including temporal, spatial, and behavioral anomalies. They examined the use cases and landscape of recent AIS anomaly research and the limitations and approaches to deal with those anomalies. Pattern-based anomaly detection may indicate erroneous AIS data, such as automatic ship identification, ship tracking, and traffic analysis, to identify potentially suspicious or dangerous vessels. The authors reviewed a wide range of recent model-based techniques for anomaly detection in AIS tracks, including both unsupervised and supervised methods. The techniques include density-based clustering, Gaussian mixture models, support vector machines, and deep learning-based approaches. Wolsing et al. provided a detailed analysis of each technique, highlighting their strengths and weaknesses and discussing how they can be applied in practice, including the current challenges and future directions regarding anomaly detection in AIS tracks.

Cyber Security Attacks on Software Logic and Error Handling Within AIS Implementations: A Systematic Testing of Resilience [10]
Khandker et al. discuss the potential vulnerabilities of AIS and how cyber attackers can exploit them. The authors present a systematic testing method, called the "AIS Resilience Testing Framework", to evaluate the resilience of AIS against cyber attacks. The framework includes three main components:**the AIS model** represents the AIS system under test, **the attack model** represents the different cyber attacks that can be launched against the AIS, and **the resilience**

assessment evaluates the AIS's ability to withstand these attacks. Khandker et al. demonstrated and evaluated the impact of multiple cyberattacks on AIS via remote Radio Frequency (RF) links and provided an example of how the framework can test an AIS system's resilience using a power grid system case study. The testing approach combines automated and manual methods. Fuzz testing involves sending many random inputs to the system to identify any unexpected behavior or errors, and penetration testing simulates a real-world attack scenario to identify vulnerabilities in the system.

AIS Data Vulnerability Indicated by a Spoofing Case-Study [9]
Androjna et al. presented a case study of a vulnerability in the AIS System. The study is based on a real-life incident in the Adriatic Sea in 2016, where a ship was observed to be transmitting false AIS data. Androjna et al. used a software-defined radio (SDR) to easily change the AIS messages' information and send false messages from a ship. They presented a spoofing attack on a GNSS-based vehicle tracking system and successfully transmitted messages with altered time offsets, fooling the receivers into processing them as genuine satellite signals. The false AIS messages were received by other ships and were displayed on their navigation system.

Blind separation of complex-valued satellite-AIS data for marine surveillance: a spatial quadratic time-frequency domain approach [11]
Cherrak et al. discussed a new method for separating complex-valued satellite AIS data to improve marine surveillance. The authors presented a spatial quadratic time-frequency domain approach that can separate the data into multiple sources without prior knowledge of the source or its properties. They demonstrated their method's effectiveness by applying it to simulated AIS data and comparing the results to other commonly used methods. They presented a detailed analysis of the performance of their method, including a comparison of the number of sources that can be separated, the separation's accuracy, and the technique's computational complexity.

A Security Evaluation of AIS, Trend Micro Research Team [13]
Balduzzi et al. are part of the Trend Micro Forward-Looking Threat research team. They conducted a comprehensive security evaluation of AIS from both a software and a Radio Frequency (RF) perspective. The researchers identified threats that affected AIS implementation and protocol specifications, including disabling AIS communications causing Denial of Service (DoS), tampering with existing AIS data by modifying information of ships' broadcast, triggering Save and Rescue (SAR) alerts to lure ships into navigating to hostile, attacker-controlled sea space, or spoofing collisions to bring a ship off course. The authors divided the threats into three main categories: spoofing, hijacking, and availability disruption: software, RF-based, or both software and hardware threats. They proposed some possible mitigation strategies, such as **applying anomaly detection techniques** to the collected AIS data to detect suspicious activities, and **adopting a Public Key Infrastructure (PKI) schema** in the AIS protocol used in RF communications, such as X.509.

Peripheral authentication for autonomous vehicles [14]
Dolev et al. present a novel approach to secure autonomous vehicles through peripheral authentication, which involves authenticating the external devices and sensors used by the vehicle. The authors highlight the importance of securing autonomous vehicles due to their increasing adoption and the potential for them to be targeted by malicious attacks. The authors provide a detailed overview of the different types of peripheral authentication methods that can be used for autonomous vehicles, including physical authentication verifying the physical properties of external devices and sensors, such as their shape, weight, and texture, behavioral authentication verifying the behavior of external devices and sensors, such as their movement patterns and response to stimuli, and context-based authentication verifying the context in which the external devices and sensors are being used, such as the location and time of day. They describe the advantages and limitations of each method and provide examples of how they can be implemented in practice.

From Click to Sink: Utilizing AIS for Command and Control in Maritime Cyber Attacks [15]
Amro et al. analyze the current state of AIS security and discuss the various methods by which attackers can exploit AIS vulnerabilities, like using GPS spoofing techniques to manipulate AIS data, causing ships to misreport their positions, using jamming techniques to interfere with AIS transmissions, and effectively causing ships to "disappear" from the network, and the risks of insider attacks, where malicious actors with access to AIS systems can intentionally manipulate data or cause disruptions. They present potential countermeasures that can be used to mitigate these risks, like implementing encrypted and authenticated communication channels between AIS systems, improving network monitoring and intrusion detection systems, and using redundant communication channels, such as satellite-based systems, to ensure that vessels can continue communicating and discuss the potential consequences of a successful AIS cyber-attack on the maritime industry.

A novel technique to identify AIS transmissions from vessels which attempt to obscure their position by switching their AIS transponder from normal transmit power mode to low transmit power mode [16]
Kelly addresses a specific vulnerability of AIS, which is the ability of some vessels to obscure their position by switching their AIS transponder from normal to low. The author presents a novel technique for identifying these vessels, which involves analyzing the characteristics of the AIS transmissions, including the frequency and duration of the transmissions. The technique is based on the observation that vessels switching to low transmit power mode tend to have longer transmission duration and higher transmission frequencies than vessels that transmit continuously at normal power levels. Kelly describes the methodology for developing and testing the technique and the experimental analysis results and provides a step-by-step explanation of the process, including the selection of sample data sets, the criteria for selecting relevant vessels, and the statistical analysis of the transmission characteristics.

Triggering Mechanism for Cyber-Attacks in Naval Sensors and Systems [17]
Leite et al. propose a novel approach to identify potential cyber-attacks on naval sensors and systems by identifying a triggering mechanism for attackers to initiate a cyber-attack on naval sensors and systems; focusing on the triggering mechanism makes it possible to identify potential cyber-attacks before they occur. The authors introduce a methodology combining qualitative and quantitative methods to develop their approach, including a survey of naval personnel, a risk assessment, and a mathematical model providing insights into the potential vulnerabilities of the systems from the perspective of those who use them regularly.

Secure, Flexible, and Backward-Compatible Authentication of Vessels AIS Broadcasts [18]
Sciancalepore et al. propose a novel authentication protocol for vessels' AIS broadcasts, called 'Auth-AIS' protocol, which allows for flexible authentication of AIS broadcasts while maintaining backward compatibility with existing AIS systems and propose a modular approach that can accommodate different authentication methods, such as digital signatures, message authentication codes, or public key cryptography, making it possible to adapt to different security requirements while providing a detailed description of the Auth-AIS protocol and its implementation. The authors evaluated the Auth-AIS protocol's performance and security by conducting several experiments to assess the protocol's performance in terms of latency and energy consumption. They showed that the Auth-AIS protocol has low latency and energy consumption compared to existing AIS authentication methods.

Cyber Security in the Maritime Industry: A Systematic Survey of Recent Advances and Future Trends [19]
Farah et al. provide a comprehensive and systematic review approach review of the current state, recent advances, and future trends of cyber security in the maritime industry; They conducted a thorough literature search and identified relevant studies using strict inclusion and exclusion criteria, identified the key challenges facing the maritime industry regarding cyber security, highlighting the maritime ecosystem's complexity and the lack of standardization in security protocols as critical challenges, and identified emerging threats, such as ransomware attacks and using unmanned aerial vehicles (UAVs) for cyber-attacks, as potential future challenges. The authors suggested several potential areas for future research, including developing standardization frameworks for security protocols in the maritime industry and investigating the effectiveness of different security measures, such as intrusion detection systems and machine learning algorithms, in detecting and preventing cyber-attacks.

Secure AIS with Identity-Based Authentication and Encryption [20]
Goudosis et al. present a new approach to securing the AIS used in the maritime industry using a new authentication and encryption mechanism based on Identity-Based Cryptography (IBC), where the public key is generated using a

user's identity information, such as an email address or name, simplifying the key management process and proposing a novel scheme for generating and distributing public and private keys, including an online registration authority and a key escrow authority. Since the existing security mechanisms for communication systems, such as Transport Layer Security (TLS), are unsuitable for AIS due to the constraints of the maritime environment, the IBC provides a suitable solution, as it can authenticate and encrypt AIS messages without requiring a pre-established trust relationship between the sender and receiver, this approach differs from traditional encryption methods that require a certificate authority to issue and manage public keys. The system can be implemented on existing AIS infrastructure without significant modifications.

3 Challenges Using AIS Systems

The world of maritime transport is a significant part of the global economy. Traffic control relies, among other means, on the AIS device, which reports dynamic data (position, course, and speed) and fixed data (name, type, cargo, size, destination, etc.). Vessels use advanced cyber capabilities to falsify data that the AIS transmits and impersonate an innocent ship while carrying out illegal activity, especially the vessel's location data, without control. Most of the work to find false AIS reports looks for location reports. AIS devices use a transceiver based on SOTDMA (Self-Organized TDMA) protocol that was developed in the late 1990s; it was designed as an intentionally open standard and did not have any built-in message authentication mechanism, un-encrypted nature, nor built-in security features making it vulnerable to different threats like spoofing, hijacking, and availability disruption.

Types of Data Manipulations

AIS was always designed to be an open standard so it would completely function using different inter-operable products. Given that it relies on radio transmissions, all you need to disrupt the system is a high-power transmitter tuned to the AIS frequencies. This is not spoofing, but even anti-jamming technology has its limitations, and with the latest technologies, it becomes more accessible [10]. Someone could buy an AIS base station, develop their own AIS transmitter using an SDR platform, or use an existing exploitable interface that an AIS transponder manufacturer might have to transmit unauthorized messages.

In [8,10], Khandker et al. used transmission-enabled SDR to create a spoofing attack scenario, where the attacker sends false AIS messages from a ship, making it appear as if the ship is in a different location or under a different identity. The SDR allows the attacker to change the AIS messages' information easily. The attack succeeded by transmitting messages with altered time offsets, which fools the receivers into processing them as genuine satellite signals. Such attacks threaten the security of autonomous and semi-autonomous vehicles, as well as the safety of travelers.

Malicious vessels can use advanced cyber capabilities to falsify data that the AIS transmits, such as the following [13]:

Vessel Spoofing: Each vessel has a unique Maritime Mobile Service Identity (MMSI); this is static information, which is entered into the AIS on installation and needs only be changed if the ship changes its name, location, the electronic position fixing system (EPFS) antenna, or undergoes a significant conversion from one ship type to another. Ship spoofing refers to the process of crafting a valid but nonexistent ship. It involves assigning static information such as vessel name, identifiers (i.e., MMSI and call sign), flag, ship type (e.g., cargo), manufacturer, and dimensions, as well as dynamic information such as **ship status** (e.g., underway or anchored), **Position, Speed Over Ground (SOG), Course Over Ground (COG), Heading**, and **destination** to the fictitious ship. A malicious team can collect the MMSI of vessels at far routes and use it in their area to impersonate a different vessel or change the dynamic information to hide illegal activities, such as the close approach between vessels, forbidden zone entry, route deviations, AIS turn off-on, position deviations, and any other unexpected activity [3–7].

Faking Weather Forecasts. Attackers craft fake weather information to prevent target ships from entering their activity zone or getting into a bad-weather area.

Closest Point of Approach (CPA) SPOOFING. AIS allows automatic response when a collision is detected or expected. The CPA is an algorithm that computes the minimal distance between two ships and how much time and distance is left before they collide with another ship, at least one of which is in motion. CPA can be configured to trigger an alert, visually or acoustically, when a possible collision is detected so the ship can change course. CPA spoofing might fake a possible collision with a target ship that will trigger a CPA alert and could lead the target off course to hit a rock or run aground during low tide.

AIS-SART Spoofing. Attackers (pirates or unfriendly countries) can trigger SART alerts to lure victims into navigating hostile or territorial waters and attacker-controlled sea spaces.

Slot Starvation. Attackers impersonate maritime authorities to reserve the entire AIS transmission "address space" to prevent all stations within coverage from communicating with one another.

Frequency Hopping. Attackers impersonate maritime authorities to instruct one or more AIS transponders to change the frequencies on which they operate.

Timing Attacks. Attackers can instruct AIS transponders to delay transmission times by simply renewing commands, thus preventing further communications about vessels' positions. This enables vessels to "disappear" from AIS-enabled radars.

Aids to Navigation (AtoN) Spoofing. AtoNs are commonly used to assist vessel traffic management along channels or harbors or warn about hazards, low tides, rocky outcroppings, and shoals in the open sea. In AtoN, spoofing attackers craft fake information to lure target ships into making wrong maneuvers. For example, attackers can place one or more fake buoys at a harbor entrance to tamper with traffic or trick ships into navigating in low waters.

AIS Hijacking. AIS hijacking involves altering any information about existing AIS stations (e.g., cargo, speed, location, and country). Attackers can maliciously modify the information provided by AtoNs installed in ports by authorities for vessel assistance and monitoring.

Fuzz and Penetration Testing. Fuzz testing involves sending numerous random inputs to the system to identify unexpected behavior or errors. Penetration testing simulates a real-world attack scenario to identify vulnerabilities in the system.

Application Specific Messages (ASM). AIS messages where the application defines the data content are application-specific messages such as binary Messages. The data content does not affect the operation of the AIS, and the AIS transfers the data content between stations without format limitations; the functional message's data structure consists of an application identifier (AI) followed by the application data. The AIS device is connected to the vessel's systems via RS232 or Ethernet; sophisticated attackers can use binary messages to inject malware into the vessel's system.

4 Possible Solutions

The SOTDMA protocol was developed in the late 1990s as an intentionally open standard. It did not have built-in security features, a message authentication mechanism, or an unencrypted nature, making it vulnerable to threats and cyber attacks like spoofing, hijacking, ransomware, privacy breach, data manipulations, and availability disruption. These threats affect both the implementation in online providers and the protocol specification, making the problems relevant to many transponder installations and leaving communication networks vulnerable to cyber threats such as eavesdropping, tampering with data, unauthorized access, and cyber-attacks in many areas using the SOTDMA protocol. To overcome some of the threats, We suggest some possible countermeasures, like the following:

* Using security methods such as encryption, digital signatures, and hash to ensure the integrity and validity of messages; introduces the problem of key management and adopting a Public Key Infrastructure. Some of the papers described several approaches to handling authentication and encryption [2,20]
* Using anomaly detection methods and model-based techniques to investigate normal and abnormal patterns in message traffic [2,17]
* Integrating Endpoint Detection and Response (EDR) pattern-based and ML-based software to detect and alert spoofed messages using various data analytic techniques by recording and storing AIS behaviors to detect suspicious messages, providing contextual information and providing remediation suggestions to restore the affected system.

* Using pattern-based and ML-based software to monitor the vessels' devices and combine the data from various sources to detect and alert anomalous behavior.
* Using Sophisticated authentication methods like ASM messages combined with the MMSI and private data should be sent and validated with the device's physical structure in a similar method to the one described in [14]; a three-way handshake scheme for a vehicle to keyfob. A keyfob is a small security hardware device with built-in authentication to control and secure access to mobile devices, computer systems, network services, and data. The keyfob displays a randomly generated access code, which changes periodically, usually every 30 to 60 s. The protocol is based on generalized peripheral authentication, with an additional attribute verification of the keyfob holder. Conventionally, vehicle-to-keyfob authentication is realized through a challenge-response verification protocol. An authentic coupling between the vehicle identity, like MMSI, and the keyfob avoids illegal vehicle access.
* Using cooperation between several AIS stations or vessels to combine data, detect and fight malicious AIS messages, the way Byzantine agents work together.
* Using a standalone computations device that scans and analyzes the AIS data in the background and uses systematic testing on the data (Fuzz testing, Penetration testing, usage of exploitation, etc.) to detect breaches and anomalies in the received AIS data.
* Maintenance of continuous risk assessment model updated to new risks, IOCs, Known threats, and intelligence from various sources.
* Employee awareness of cybersecurity and team training to ensure that all relevant people know about cyber threats and handle the vessels' systems with precaution (suspicious e-mails, not using DOKs or personal devices, etc.).
* Cyber-security professionals training to overcome the shortage of Cybersecurity professionals.
* Cooperation with trusted entities, sharing data and knowledge.
* Regulation defines cyber security requirements and methods [21].

5 Conclusion

A significant part of the research done regarding AIS protocol looks for erroneous AIS data, such as location manipulation or GPS spoofing. One method of manipulation is to turn off the device while performing the inappropriate activity; a significant part of those algorithms looks for the time when the device was closed. For a vessel to be picked up by the control systems as an innocent ship while carrying out illegal activity elsewhere, it is to use other methods, like SDR, that include advanced cyber capabilities to falsify data that the AIS transmits, especially the vessel's location data. Some algorithms try to find anomalies in the AIS data using different pattern-based methods and techniques, while others use model-based methods. Not many studies combine several data sources and layers such as images, satellite information, and timing, nor use vessel cooperation like Byzantine agents.

Devices using SOTDMA protocols are becoming common and are widely used in different areas. Smart transportation systems use such protocols, and security vulnerabilities in automated vehicle information systems (AVISs) are a growing concern as the deployment of AVISs in mainstream transportation increases. However, the vulnerabilities of AVISs to spoofing and sophisticated attacks have received little attention. Most of the challenges using the AIS protocols apply to autonomous vehicles, medical devices, IoTs, and other devices using SOTDMA protocols. Cyber protection must be introduced when using protocols of this type. We must fill the gap in identifying security vulnerabilities and address the cyber protection challenges. In future work, we will develop the above protection methods in detail.

References

1. ITU-R Radiocommunication Sector of ITU, Technical characteristics for an automatic identification system using time division multiple access in the VHF maritime mobile frequency band, Recommendation ITU-R M.1371-5(02/2014)
2. Wolsing, K., Roepert, L., Bauer, J., Wehrle, K.: Anomaly detection in maritime AIS tracks: a review of recent approaches. J. Maritime Sci. Eng. **10**, 112 (2022)
3. Positions of Two NATO Ships Were Falsified Near Russian Black Sea Naval Base. USNI News, 21 June 2021. Accessed 23 June 2021
4. Bateman, T.: Fake ships, real conflict: How misinformation came to the high seas, 28 June 2021. Euronews. Accessed 29 June 2021
5. Harris, M.: Phantom warships are courting chaos in conflict zones - the latest weapons in the global information war are fake vessels behaving badly. Wired (magazine), 29 July 2021
6. UT Austin Cockrell School of Engineering, UT Austin Researchers Spoof Superyacht at Sea, 29 July 2013. www.engr.utexas.edu/features/superyacht-gpsspoofing. Accessed 8 Sept 2015
7. Cyberkeel, Maritime Cyber-Risks, 15 October 2014. www.cyberkeel.com/images/pdf-files/Whitepaper.pdf. Accessed 8 Sept 2015
8. Grant, A., Williams, P., Ward, N., Sally B.: GPS jamming and the impact on maritime navigation. The General Lighthouse Authorities of the United Kingdom and Ireland. www.navnin.nl/NIN/Downloads/GLAs%20-%20GPS%20Jamming%20and%20the%20Impact%20on%20Maritime%20Navigation.pdf. Accessed 10 Sept 2015
9. Androjna, J., Perkovic, A., Pavic, M., Miskovic, I.: AIS data vulnerability indicated by a spoofing case-study. Appl. Sci. J. **11**, 5015 (2021)
10. Khandker, A., Turtiainen, S., Costin, H., Hamalainen, T.: Cybersecurity attacks on software logic and error handling within AIS implementations: a systematic testing of resilience. IEEE Access **10**, 29493–29505 (2022)
11. Cherrak, O., Ghennioui, H., Moreau, N.T., Abarkan, E.: Blind separation of complex-valued satellite-AIS data for marine surveillance: a spatial quadratic time-frequency domain approach. Int. J. Electr. Comput. Eng. (IJECE) **9**(3), 1732–1741 (2019)
12. IS, All About. AIS TDMA access schemes (2012). http://www.allaboutais.com/

13. Balduzzi, M., Wilhoit, K., Pasta, A.: A security evaluation of AIS, trend micro forward-looking threat research team
14. Dolev, S., Panwar, N.: Peripheral authentication for autonomous vehicles. In: NCA, pp. 282–285 (2016)
15. Amro, A., Gkioulos, V.: From click to sink: utilizing AIS for command and control in maritime cyber attacks. In: Atluri, V., Di Pietro, R., Jensen, C.D., Meng, W. (eds.) ESORICS 2022. LNCS, vol. 13556, pp. 535–553. Springer, Cham (2022). https://doi.org/10.1007/978-3-031-17143-7_26
16. Kelly, P.: A novel technique to identify AIS transmissions from vessels which attempt to obscure their position by switching their AIS Transponder from normal transmit power mode to low transmit power mode. Expert Syst. Appl. (2022). https://doi.org/10.1016/j.eswa.2022.117205
17. Leite Junior, W.C., de Moraes, C.C., de Albuquerque, C.E.P., Machado, R.C.S., de Sá, A.O.: A triggering mechanism for cyber-attacks in naval sensors and systems. Sensors **21**, 3195 (2021). https://doi.org/10.3390/s21093195
18. Sciancalepore, S., Tedeschi, P., Aziz, A., Di Pietro, R.: Auth-AIS: secure, flexible, and backward-compatible authentication of vessels AIS broadcasts. IEEE Trans. Dependable Secure Comput. **19**(4), 2709–2726 (2022). https://doi.org/10.1109/TDSC.2021.3069428
19. Ben Farah, M.A., et al.: Cyber security in the maritime industry: a systematic survey of recent advances and future trends. Information **13**, 22 (2022). https://doi.org/10.3390/info13010022
20. Goudosis, A., Katsikas, S.: Secure AIS with identity-based authentication and encryption. TransNa Int. J. Marine Navig. Saf. Sea Transp. **14**(2), 287–298 (2020). https://doi.org/10.12716/1001.14.02.03
21. Tam, K., Jones, K.: Factors affecting cyber risk in maritime. In: 2019 International Conference on Cyber Situational Awareness, Data Analytics And Assessment (Cyber SA), Oxford, UK, pp. 1–8 (2019). https://doi.org/10.1109/CyberSA.2019.8899382

A New Interpretation for the GHASH Authenticator of AES-GCM

Shay Gueron[1,2(✉)]

[1] University of Haifa, Haifa, Israel
[2] Meta, Menlo Park, USA
sgueron@univ.haifa.ac.il

Abstract. AES-GCM authenticated encryption scheme has a significant role in modern secure communications. It combines AES CTR encryption with authentication that is based on a polynomial evaluation hash function (GHASH) computed in $\mathbb{F}_{2^{128}}[x] / \mathtt{P_{GCM}}(x)$, where $\mathtt{P_{GCM}}(x) = x^{128} + x^7 + x^2 + x + 1$. AES-GCM operates on 128-bit strings: it views them as AES inputs/outputs for the encryption, and as elements in $\mathbb{F}_{2^{128}}$ for the authentication. Unfortunately, the order of the bits, by which GHASH parses 128-bit strings as field elements is inconsistent with the way that AES uses 128-bit ciphertext/plaintext strings as arrays of 16 bytes. This leads to one of the following conclusions: a) GHASH does not operate directly on the ciphertext blocks. In this case, AES ciphertext blocks need to be bit-reflected before they are input to the GHASH computations; b) the field representation is *not* $\mathbb{F}_{2^{128}}[x] / \mathtt{P_{GCM}}(x)$. In this case, field multiplications are not directly expressed by polynomial arithmetic modulo $\mathtt{P_{GCM}}(x)$. The specification AES-GCM bypasses this discrepancy by describing the GHASH field operations as bit-level algorithms, rather than in terms of polynomial arithmetic, as expected. We resolve the inconsistency by introducing a description of GHASH that uses polynomial arithmetic in $\mathbb{G} = \mathbb{F}_{2^{128}}[x] / \left(x^{128} + x^{127} + x^{126} + x^{121} + 1\right)$. This formulation helps parsing 128-bit strings as AES inputs/outputs and as field elements, in a consistent manner. It also leads naturally to several recent AES-GCM software optimizations which are now already in use by leading open source cryptographic libraries.

Keywords: AES-GCM · finite field arithmetic · software optimization

1 Introduction

AES-GCM is currently one of the leading authenticated encryption schemes. It was defined in [13] (2005), later (2007) formalized as a NIST specification [2], and gained significant popularity after being included in the list of acceptable TLS 1.2 cipher suites.

The attractiveness of AES-GCM at the time it was proposed, was its security proof, and efficient software and hardware design. However, until 2009, optimized AES-GCM software had the same performance on high-end processors, as

S. Dolev et al. (Eds.): CSCML 2023, LNCS 13914, pp. 424–438, 2023.
https://doi.org/10.1007/978-3-031-34671-2_30

the leading TLS alternative AES-SHA1 (i.e., AES in CBC mode with HMAC-SHA1 authentication): both were performing at ~22 cycles per byte (C/B), being slower than the popular combination RC4-SHA1. At that time, the AES-GCM optimized software used byte-level lookup tables (see [3,14]). AES-GCM gained performance advantage on high-end processors with Intel's introduction of AES-NI (2010), instructions that are designed to speed up AES computations (see [4,5]). Due to the difference between instructions' latency and throughput, and adequate software pipelining (see [4,5]), parallelizable modes of operations have significantly higher performance than serial AES-CBC encryption. In particular, AES-GCM enjoyed significant speedup due to it use of CTR mode. Together with AES-NI, Intel introduced a carry-less multiplication instruction (PCLMULQDQ), designed for speeding up binary polynomial multiplications, and in particular offered advantage to the computations of the Galois Hash (GHASH) of AES-GCM (see [11,12]). AES-GCM software has the performance of ~0.64 C/B on the latest Intel Architecture Codename Skylake, practically achieving the full authenticated encryption at the same performance of the CTR encryption only (this processor has AES-NI ($AESENC$, $AESENCLAST$) and PCLMULQDQ instructions with throughput 1 and latency 4. Subsequent processor have even higher throughput for these instructions) The increasing AES-GCM performance from 2010, are due to the improved performance of AES-NI and PCLMULQDQ instructions across the CPU generations, bundled with improved algorithms for the $\mathbb{F}_{2^{128}}$ computations of GHASH. These computations are the focus of this paper. AES-GCM enjoyed optimized software implementations contributed to the open source cryptographic libraries (OpenSSL for server platforms, and NSS for client platforms), and the result is that practically, all the modern browsers (e.g., Chrome, Firefox, Internet Explorer)

We discuss an inherent inconsistency with the definition of the finite field representation used by AES-GCM. This difficulty is (implicitly) hidden by defining the field operations in [2,14] as bit-level algorithm, and similarly, in software implementations that use byte-level tables. However, it is unavoidable when GHASH computations use multi-bytes structures, as the implementations that use PCLMULQDQ, which are the source for the recent performance improvements.

The paper is organized as follows. In Sect. 2, we analyze the GHASH definitions and explain why the computations are actually *not* in $\mathbb{F}_{2^{128}}[x]/\left(x^{128}+x^7+x^2+x+1\right)$. In Sect. 3, we resolve the discrepancy by proposing an equivalent description for GHASH, using polynomial arithmetic in another representation of $\mathbb{F}_{2^{128}}$. In Sect. 4, we show how our formulation leads to the new efficient polynomial reduction algorithm that improves AES-GCM software performance.

2 The AES-GCM GHASH Definition Inconsistency

2.1 AES-GCM Description

We provide a brief description of AES-GCM (for full details, see [2]). The input is an encryption key $K \in \{0,1\}^{\kappa}$ ($\kappa = 128, 192, 256$), an Initialization Vector $IV \in \{0,1\}^{96}$ (for simplicity), a message M with $len(M)$ bits, and additional associated data D with $len(D)$ bits. The output is the ciphertext C with $len(C) = len(M)$ bits, and an authentication tag $Tag \in \{0,1\}^{128}$. For brevity, assume that $len(M)$ and $len(D)$ are divisible by 128 (the general case is defined in [2]), and parse M and D as the concatenation of 128-bit blocks $M = M_1, M_2, \ldots, M_{\ell_1}$, and $D = D_1, D_2, \ldots, D_{\ell_2}$, respectively. By requirement, $\ell_1 \leq 2^{32} - 2$.

For $0 \leq j \leq 2^w - 1$, let $\mathtt{encode}_w(j)$ denote the encoding of j as a w-bit string (e.g., $\mathtt{encode}_7$ (10) = 0001010; the string of 128 bits of zero is $\mathtt{encode}_{128}$ (0)). Define the following 128-bit blocks:

$$CTRBLK_j = IV \| \mathtt{encode}_{32}(j+1), \quad j = 0, 1, \ldots, \ell_1, \tag{2.1}$$

$$LENBLK = \mathtt{encode}_{64}\left(len(D)\right) \| \mathtt{encode}_{64}\left(len(M)\right), \tag{2.2}$$

$$H = AES_K\left(\mathtt{encode}_{128}\left(0\right)\right) \tag{2.3}$$

The encryption (in CTR mode) of $M_1, M_2, \ldots, M_{\ell_1}$ produces the ciphertext blocks

$$C_j = M_j \oplus AES_K(CTRBLK_j), \quad j = 1, \ldots \ell_1 \tag{2.4}$$

For the authentication, denote $\ell = \ell_1 + \ell_2 + 1$ and let $X_1, X_2, \ldots, X_\ell$ be the concatenation of the ℓ blocks blocks $D, C, LENBLK$. The authentication tag is

$$Tag = GHASH_H(X_1, X_2, \ldots, X_\ell) \oplus AES_K(CTRBLK_0) \tag{2.5}$$

where $GHASH$ (aka Galois Hash) is the polynomial evaluation

$$GHASH_H(X_1, X_2, \ldots, X_\ell) = X_1 \bullet H^\ell \oplus X_2 \bullet H^{\ell-1} \oplus, \ldots, \oplus X_{\ell-1} \bullet H^2 \oplus X_\ell \bullet H \tag{2.6}$$

The computations in (2.6) are carried out in $\mathbb{F}_{2^{128}}$, where "$\bullet$" denotes the field multiplication and $\oplus$ denotes field addition. The AES-GCM specification (in [2,13]) expresses GHASH by the equivalent iterations

$$Y_0 = 0, \quad Y_j = (Y_{j-1} \oplus X_j) \bullet H, \; j = 1, \ldots \ell, \quad GHASH_H(X_1, X_2, \ldots, X_\ell) = Y_\ell \tag{2.7}$$

that implements Horner's method for polynomial evaluation (the expression (2.6) is also mentioned in [2]).

Underlying the description of AES-GCM is the implicit assumption that per the context, a 128-bit string can be viewed, unambiguously and interchangeably as one of: a) an AES plaintext / ciphertext; b) an element in $\mathbb{F}_{2^{128}}$. In addition, it is implicitly assumed that the encoding $\mathtt{encode}_w(j)$ is used in an unambiguous way.

In the context of AES, the specification [1] defines the way by which a state of 128 bits is parsed as input, output, and intermediate value during the encryption/decryption flows. Specifically, the state is viewed as 16 bytes, and the operations are expressed in terms of byte-wise operations, by viewing bytes as elements of $\mathbb{F}_{2^8}$ given in a specific representation.

Defining GHASH requires a specific representation of $\mathbb{F}_{2^{128}}$. In our case, it is the polynomial representation with the irreducible reduction polynomial $x^{128} + x^7 + x^2 + x + 1$. Here, field elements are polynomials, and the definition [2] specifies the following shorthand notation: the 128-bit block $a_{127}a_{126}\dots a_0$ (the bits are a_i, $i = 0, \dots, 127$) is viewed as the polynomial $a_0x^{127} + a_1x^{126} + \dots + a_{126}x + a_{127}$ in GF.

This seems to be only a convention for labeling 128-bit strings, but we show here why it raises a difficulty in the context of AES-GCM.

A string of 128 bits is also a block of 16 bytes. Reading the same string of bits from different directions flips the order of the bytes *and* also reverses the order of the bits in each byte. All computer architectures define bytes with a standard order of bits, which is opposite to the order implied by AES-GCM specification, when a block is interpreted as an element in $GF(2^{128})$. Even if one is willing to overlook compatibility with storing/loading data on computer systems, the conflict with the AES definition remains: AES is defined in [1] over a state of 16 bytes, via byte-wise operations, and determines the order of the bits in the bytes of AES plaintexts/ciphertexts in the standard way. It follows that when F and H are ciphertexts, the operation $F \bullet H$ is, effectively, a three steps computational flow:

1. Bit-reflect F and H
2. Multiply the bit-reflected values in $GF(2^{128}) / \left(x^{128} + x^7 + x^2 + x + 1\right)$
3. Bit-reflect the resulting product[1]

This fact is not explicitly stated in [2] nor in [13]. Instead, steps 1 and 3 are absorbed into the definition of "$\bullet$" as a bit-level algorithm. This implies that the representation of $GF(2^{128})$ over which AES-GCM operates *is not* $GF(2^{128}) / \left(x^{128} + x^7 + x^2 + x + 1\right)$, as mistakenly perceived. In fact (by 1–3 above) the representation is obtained from permuting $GF(2^{128}) / \left(x^{128} + x^7 + x^2 + x + 1\right)$ with a bit-reflection permutation. This formulation has the drawback of introducing a non-algebraic operation into the description of the field's multiplication, which we wish to avoid.

Apparently, the AES-GCM was defined with a hardware implementation mindset, where input bits trickle, as input, bit-after-bit. Such environments do not (necessarily) work with the notion of bytes, and a bit-level algorithm is a natural description. Furthermore, the original optimized software implementations (e.g., recommendations in [13] and implementation in [3]) used byte-level lookup tables. In such cases, the bit reflection is embedded in these tables, so again, the discrepancy can be hidden. By contrast, modern software implementations use the PCLMULQDQ instruction, thus operate on chunks of multiple (8, 16) bytes, and also move 16-byte blocks to/from memory. In such cases, reversing the order of the bits within the bytes of the ciphertexts is required in order to implement AES-GCM. Ironically, the techniques that speed up AES-GCM and make it outperform alternative schemes, also manifest the discrepancy in its definition. Our goal here is to resolve the difficulty.

[1] Note that in AES-GCM, the hash key H is fixed (for a give key). It can be bit-reflected only once (at generation), so bit-reflection is necessary to only (but all) the ciphertext blocks.

3 Finite Binary Fields and Bit-Reflection

Let $n > 1$ be a positive integer, and denote $s = n-1$. Consider the finite field with 2^n elements, $G = GF(2^n)/P(x)$, given in polynomial representation with the irreducible reduction polynomial $P(x) = \sum_{j=0}^{n} p_j x^j$ ($p_j \in \{0,1\}, p_n = p_0 = 1$). A field element $A \in G$ is viewed as a binary polynomial of degree s, namely $A = A(x) = \sum_{j=0}^{s} a_j x^j$ where $a_j \in \{0,1\}$. Interchangeably, we also view A as the string of n bits, $a_s a_{s-1} \ldots a_0$. By convention, we write the bits of the string from left to right, where the leftmost bit is the coefficient of x^s in $A(x)$, and the rightmost bit is its independent term.

For every two elements in G, $A = A(x) = \sum_{j=0}^{s} a_j x^j$ and $B = B(x) = \sum_{j=0}^{s} b_j x^j$, the field operations are defined as follows. The sum $C = A + B$ is obtained by polynomial addition where the coefficients are added in $GF(2)$, namely $C = C(x) = \sum_{j=0}^{s} ((a_j + b_j) \mod 2)\, x^j$. The string representation of C is the bit-wise XOR $A \oplus B$ (where A and B are viewed as strings).

Denote the field multiplication by $\otimes$. $A(x) \otimes B(x)$ is defined by $A(x) \otimes B(x) = A(x) \times B(x) \pmod{P(x)}$, where "$\times$" denotes polynomial multiplication with coefficients being added and multiplied in $GF(2)$. If $T(x) = A(x) \times B(x)$ then $T(x)$ is a binary polynomial of degree $2s$ (the precise expressions for the coefficients are given in Appendix A). It follows that $A(x) \otimes B(x)$ can be computed in two steps: polynomial multiplication to compute $T(x)$, and then dividing $T(x)$ by $P(x)$, using polynomial division with $GF(2)$ arithmetic, where the remainder, a polynomial of degree s, is $A(x) \otimes B(x)$. This step is called reduction modulo $P(x)$.

We continue with a few definitions.

Definition 1 (Compact representation of element in $GF(2^n)$). *Suppose that $n = 8m$ for some m, and consider an element $F = F(x) = \sum_{j=0}^{s} f_j x^j \in G$. A compact representation of the string F groups chunks of 8 bits into a single symbol ("bytes"), to represent F as a sequence of m bytes $B_{m-1}B_{m-2} \ldots B_0$. The bits in the bytes are $B_k = f_{8k+7} \ldots f_{8k}$, $k = 0, \ldots, m-1$. By our convention, we write B_0 in the rightmost position. For convenience, we label bytes in hexadecimal notation (i.e., as 2-symbol numbers, written in in base 16, where the symbols are $0-9, A, B, C, D, E, F$).*

Example 1. Take $n = 16$, $m = 2$, and $F(x) = x^{15} + x^{13} + x^8 + x + 1$. The corresponding bit string representation of F is 1010000100000011, and in compact notation it is $A103$. The order of the bits in the bytes is the "standard" order.

Remark 1. *The writing convention adopted here, namely byte zero and bit zero in the rightmost position, is consistent with the way that the corresponding integers are written in base 2 or in base 256. For a string of n bits $F = f_s f_{s-1} \cdots f_0$, the corresponding integer is $\mathcal{F} = \sum_{j=0}^{s} 2^j f_j$. When $\mathcal{F}$ is written in base 2, its digits (bits) form the string $f_s f_{s-1} \cdots f_0$. When $\mathcal{F}$ is written in base 256, its digits form the string $B_{m-1}B_{m-2} \ldots B_0$.*

Definition 2 (Bit reflection). *Consider the bits string $F = f_s f_{s-1} \dots f_0$. The bit reflection of the string F is the string $f_0 f_1 \dots f_{s-1} f_s$, and is denoted $REF(F)$. When F is viewed as an element in G, $F = F(x) = \sum_{j=0}^{s} f_j x^j$, its bit reflected image is the field element $REF(F) = REF(F(x)) = \sum_{j=0}^{s} f_{s-j}\, x^j$.*

Remark 2 (Bit reflection and bytes). *Take $n = 8m$, and an m bytes string $F = B_{m-1} B_{m-2} \dots B_0$. The bit reflected string $REF(F)$ can also be grouped into m bytes, but they are* different *from the bytes of F: byte j ($j = 0, \dots, m-1$) of $REF(F)$ equals B_{m-j-1}, with the reversed order bits.*

Example 2 The 16-bit string $F = 1010000100000011$, written in compact notation is $A103$. Here, $REF(F)$ is 1100000010000101. In compact notation it is $C085$ (note that the bytes are different).

Remark 3 (Bit reflection is not a change of Endianness). *On modern computer architectures, bytes are the smallest unit of information that has an address (and can be written/read to memory/from memory). Suppose that the consecutive sequence of bytes $B_{m-1} B_{m-2} \dots B_0$ is stored in consecutive memory locations, starting from address ADDRS. In a "Little Endian" architecture B_0 resides in address ADDRS, and in a "Big Endian" architecture B_0 resides in address $ADDRS + m$ (x86 (IA) architectures use Little Endian convention). In both cases, the bits within the bytes are organized in the same order.*

3.1 Expressing Bit Reflection by $GF(2^n)$ Operations

Definition 3. *Let $\otimes$ be the multiplication operation in a given representation of $GF(2^n)/P(x)$, and let A, B be two elements in the field. Define the operation $\bullet$ by*

$$A \bullet B = A \otimes B \otimes x^{-s} = A \times \left(B \times x^{-s}\right) \pmod{P(x)}$$

Note that x^{-s} is some fixed element in the field.

The following proposition will be used to handle bit reflection via different field representations.

Proposition 1. *Let $P(x) = \sum_{j=0}^{n} p_j x^j$ and $Q(x) = \sum_{j=0}^{n} q_j x^j$ be irreducible binary polynomials of degree n, such that $p_j = q_{n-j}$, $j = 0, \dots, n$. Consider the two representations of $GF(2^{128})$: $G_1 = GF(2^{128})/P(x)$, and $G_2 = GF(2^{128})/Q(x)$ with field multiplication denoted $\otimes_1$ and $\otimes_2$, respectively. Consider two binary polynomials of degree s, $A = A(x) = \sum_{j=0}^{s} a_j x^j$ and $B = B(x) = \sum_{j=0}^{s} b_j x^j$. They can be viewed as field elements in both G_1 and G_2. Define*

$$U_1 = REF\left(REF(A) \otimes_1 REF(B)\right)$$

(operations in G_1) and

$$U_2 = A \bullet B = A \otimes_2 B \otimes_2 x^{-s}$$

(operations in G_2). Then $U_1 = U_2$.

Proof. Denote $\beta = x^{-1}$ ($\beta \in G_1$). We have $REF(A) = \sum_{j=0}^{s} a_{s-j}x^j = \sum_{j=0}^{s} a_j x^{s-j}$ and, similarly, $REF(B) = \sum_{j=0}^{s} b_j x^{s-j}$. We write

$$REF(A) = x^s \otimes_1 \sum_{j=0}^{s} a_j \otimes_1 \beta^j$$

$$REF(B) = x^s \otimes_1 \sum_{j=0}^{s} b_j \otimes_1 \beta^j \tag{3.1}$$

as an equation in G_1 Therefore (with computations in G_1),

$$L = REF(A) \otimes_1 REF(B) = x^{2s} \otimes_1 \sum_{j=0}^{s} a_j \otimes_1 \beta^j \ \otimes_1 \sum_{j=0}^{s} b_j \otimes_1 \beta^j \tag{3.2}$$

(we omit hereafter the $\otimes_1$ notation in expressions like $\sum_{j=0}^{s} a_j \otimes_1 \beta^j$).

Note that in G_2, $t^{-s} = t^{-s}$ (mod $Q(t)$) is an element in the field, say the polynomial $v(t)$ (of degree s). We write the *polynomial identity* for dividing $v(t) \times A(t) \times B(t)$ by $Q(t)$, with arithmetic in $GF(2)$, denoting the quotient $M(t)$ and the remainder $R(t) = \sum_{j=0}^{s} r_j t^j$.

$$v(t) \times \sum_{j=0}^{s} a_j t^j \ \times \sum_{j=0}^{s} b_j t^j = M(t) \times \sum_{j=0}^{n} q_j t^j + \sum_{j=0}^{s} r_j t^j \tag{3.3}$$

Substituting (3.3) into (3.2), and considering the equation in G_1, we get

$$L = \beta^{-s} \otimes_1 M(\beta) \otimes_1 \sum_{j=0}^{n} q_j \beta^j + \beta^{-s} \otimes_1 \sum_{j=0}^{s} r_j \beta^j \tag{3.4}$$

This simplifies to

$$L = x^s \otimes_1 M(\beta) \otimes_1 \sum_{j=0}^{n} q_j x^{-j} + x^s \otimes_1 \sum_{j=0}^{s} r_j x^{-j} =$$

$$= x^s \otimes_1 x^{-n} \otimes_1 M(\beta) \otimes_1 \sum_{j=0}^{n} q_j x^{n-j} + \sum_{j=0}^{s} r_j x^{s-j}$$

$$= x^s \otimes_1 x^{-n} \otimes_1 M(\beta) \otimes_1 \sum_{j=0}^{n} q_{n-j} x^j + \sum_{j=0}^{s} r_{s-j} x^j =$$

$$= x^s \otimes_1 x^{-n} \otimes_1 M(\beta) \otimes_1 \sum_{j=0}^{n} p_j x^j + \sum_{j=0}^{s} r_{s-j} x^j = \sum_{j=0}^{s} r_{s-j} x^j \tag{3.5}$$

It follows that $U_1 = REF(L) = R = \sum_{j=0}^{s} r_j x^j$. By the definition (of $v(t)$), we also have $U_2 = A(x) \times B(x) \times x^{-s}$ (mod $Q(x)$) and this completes the proof.

Example 3. Take $G_1 = GF(2^4)/\left(x^4+x+1\right)$, $G_2 = GF(2^4)/\left(x^4+x^3+1\right)$, $A = ED$, $B = 9F$.
$REF(A) = B7$, $REF(A) = F9 \Longrightarrow REF(A) \otimes_1 REF(B) = 8A \Longrightarrow$
$REF(REF(A) \otimes_1 REF(B)) = REF(8A) = 51$.
By comparison, $A \bullet B = A \otimes_2 B \otimes_2 x^{-3} = A \otimes_2 B \otimes_2 59 = 51$.

Proposition 2. *Let $\ell \geq 1$ be an integer, and let $X_1, X_2, \ldots, X_\ell$, and H be elements in $GF(2^n)/Q(x)$.*
Define the iterations

$$H_1 = H, \quad H_j = H_{j-1} \bullet H, \quad j = 2, \ldots, \ell \tag{3.6}$$

$$Y_0 = 0, \quad Y_j = (Y_{j-1} + X_j) \otimes H, \quad j = 1, \ldots \ell, \quad POLYVAL = Y_\ell \tag{3.7}$$

Then,

$$H_j = H^j \otimes x^{-(j-1)s}, \quad j = 1, 2, \ldots, \ell \tag{3.8}$$

and

$$POLYVAL = X_1 \bullet H_\ell + X_2 \bullet H_{\ell-1} + \ldots + X_{\ell-1} \bullet H_2 + X_\ell \bullet H \tag{3.9}$$

Proof. The proof follows from the definition of $\bullet$.

3.2 Application to AES-GCM GHASH

We can now use Propositions 2 and 1 with the special case $n = 128$ ($s = 127$) and the reduction polynomial $Q(x) = \left(x^{128} + x^{127} + x^{126} + x^{121} + 1\right)$, to establish an algebraic formulation of the GHASH operations.

GHASH Interpretation. Define $\overline{G} = \mathbb{F}_{2^{128}}[x]/\left(x^{128} + x^{127} + x^{126} + x^{121} + 1\right)$ and denote its multiplication operation by $\otimes$, namely $\Gamma_1 \otimes \Gamma_2 = \Gamma_1 \times \Gamma_2 \pmod{x^{128} + x^{127} + x^{126} + x^{121} + 1}$
Define the operation $\bullet$ as follows:

$$\Gamma_1 \bullet \Gamma_2 = \Gamma_1 \otimes \Gamma_2 \otimes x^{-127} \tag{3.10}$$

Then, the GHASH (the Galois Hash) of ℓ blocks $X_1, X_2, \ldots, X_\ell$, computed over the hash key H, as defined for AES-GCM in (2.7), can be expressed by

$$\begin{aligned} POLYVAL &= POLYVAL_H(X_1, X_2, \ldots, X_\ell) = \\ &= X_1 \bullet H_\ell \oplus X_2 \bullet H_{\ell-1} \oplus, \ldots, X_{\ell-1} \bullet H_2 \oplus X_\ell \bullet H_1 \end{aligned} \tag{3.11}$$

where $H_j = H^j \otimes x^{-127(j-1)}$, for $j = 1, 2, \ldots, \ell$. H_j and $GHASH$ can be computed by the iterations

$$H_1 = H, \quad H_j = H_{j-1} \bullet H, \quad j = 2, \dots, \ell \tag{3.12}$$

$$Y_0 = 0, \quad Y_j = (Y_{j-1} + X_j) \otimes H, \quad j = 1, \dots \ell, \quad GHASH = Y_\ell \tag{3.13}$$

In $\overline{G}$, we have $x^{-127} = x^{127} + x^{126} + x^{125} + x^{122} + x^{121} + x^{115} + x + 1$ (i.e., $E6080000000000000000000000000003$ in compact form).

Remark 4. *It seems like computing "•" requires two field multiplication before reduction can start. However, we show below that there is an efficient way to compute "•" in the context of GHASH.*

4 Computing $A \otimes B \otimes x^{-n}$ in $GF(2^n)$, for Special Form Polynomials

This section shows how, in some cases, it is easy to compute $A \otimes B \otimes x^{-n}$ in $GF(2^n)$.

Proposition 3. *Let $n = 2t$ and let $T(x)$ be a polynomial of degree $4t - 1$ ($= 2n - 1$). Let $Q(x)$ be a polynomial of degree $2t$ ($= n$), of the form $Q(x) = x^{2t} + W(x) \times x^t + 1$ where $W(x)$ is a polynomial of degree $t-1$. Then, it is possible to compute $T(x) \times x^{-2t} \pmod{Q(x)}$ with the computational cost of multiplying, twice, two polynomial of degree $t - 1$, and three field additions.*

Proof. Write $T(x) = D(x) \times x^{3t} + C(x) \times x^{2t} + B(x) \times x^t + A(x)$ where $A(x)$, $B(x)$, $C(x)$, $D(x)$ are polynomials of degree $t - 1$. Note that $T(x) + A(x) \times Q(x)$ is divisible by x^t, and write

$$T(x) + A(x) \times Q(x) = D(x)x^{3t} + U(x)x^{2t} + V(x)x^t \tag{4.1}$$

This determines $U(x)$ and $V(x)$ via the relation

$$U(x)x^t + V(x) = A(x) \times x^t + W(x) \times A(x) \tag{4.2}$$

Now compute

$$T_1(x) = (T(x) + A(x) \times Q(x)) / x^t = D(x)x^{2t} + U(x)x^t + V(x) \tag{4.3}$$

Note that $T_1(x) + V(x) \times Q(x)$ is divisible by x^t, and write

$$T_2(x) = (T_1(x) + V(x) \times Q(x)) / x^t = G(x)x^t + F(x) \tag{4.4}$$

This determines $G(x)$ and $F(x)$ via the relation

$$G(x)x^t + F(x) = (D(x) + V(x))\, x^t + U(x) + W(x) \times V(x) \tag{4.5}$$

Note that adding multiples of $Q(x)$ does not change the result modulo $Q(x)$. Since $T_2(x)$ is a polynomial of degree $2t-1$, $T_2(x) \pmod{Q(x)} = T_2(x)$. Therefore, we conclude that $T_2(x) = T(x) \times x^{-2t} \pmod{Q(x)}$.

Note that the computations, encapsulated in (4.2) and in (4.5), required a total of two polynomial multiplications of degree $t-1$ and three additions (the two multiplications by x^t are ignored).

Remark 5. *Proposition 3 offers an efficient way to compute $A \otimes B \otimes x^{-n}$. However, we are interested in computing "$\bullet$" ($A \otimes B \otimes x^{-s}$). To this end, we use the following identity for $\Gamma_1, \Gamma_2 \in GF(2^n)$:*

$$\Gamma_1 \bullet \Gamma_2 = \Gamma_1 \otimes \Gamma_2 \otimes x^{-s} = \Gamma_1 \otimes (x \otimes \Gamma_2) \otimes x^{-n} \tag{4.6}$$

In particular, if one operands, say Γ_2, is fixed, $x \otimes \Gamma_2$ can be pre-computed.

4.1 Aggregation Before Reduction

We note that (3.9) offers a way to save computations, at the cost of pre-computing a table. Suppose that H is the GHASH hash key, and denote $\bar{H} = x \times H \pmod{Q(x)}$. Suppose that the values

$$\bar{H}_j = \bar{H}^j \otimes x^{-n(j-1)}, \quad j = 1, 2, \ldots, \ell \tag{4.7}$$

are pre-computed. Then, POLYVAL can be computed by

$$T = X_1 \times \bar{H}_\ell + X_2 \times \bar{H}_{\ell-1} +, \ldots, X_{\ell-1} \times \bar{H}_2 + X_\ell \times \bar{H}_1 \tag{4.8}$$

(obtaining a polynomial of degree $2s$) and subsequently

$$GHASH = T \otimes x^{-n} = T \times x^{-n} \pmod{Q(x)} \tag{4.9}$$

In other words, the cost of reduction modulo $Q(x)$ can be alleviated by using a pre-computed table, computing ℓ polynomial multiplications, and then, reducing the result only once.

4.2 Putting It All Together: Application to AES-GCM

For AES-GCM, we use the special case of $n = 128$, $t = 64$ and the polynomial

$$Q(x) = x^{128} + x^{127} + x^{126} + x^{121} + 1 = x^{128} + x^{64}\left(x^{63} + x^{62} + x^{57}\right) + 1$$

Note that $Q(x)$ has the special form prescribed in Proposition 3 with $W(x) = x^{63} + x^{62} + x^{57}$ (which is $C200000000000000$ in compact form). In the field, $x^{-128} = x^{127} + x^{124} + x^{121} + x^{114} + 1$ (i.e., the block 92040000000000000000000000000001).

For GHASH, computing "$\bullet$" can be done by first multiplying $\bar{H} = x \times H \pmod{Q(x)}$, and then using the efficient procedures described above. A multiplication of the type $A(x) \times b(x) \times x^{-128} \pmod{Q}$ can be implemented by means of 6 carry-less multiplication instruction PCLMULQDQ that multiplies two polynomials of degree 63. A detailed example for such computation, and a code snippet that shows how they are implemented, are shown in Appendix B.

Efficient code that implements AES-GCM using our methods, has been contributed to the open source libraries NSS [9] and OpenSSL [10]. These implementations use the aggregation method. To further improve the performance from other aspects of the AES-GCM computations, they also interleave encryption and GHASH computations, and pipeline the CTR encryption ([4,5]). By now, this code has been integrated to the current versions of NSS (hence used by Chrome and Firefox browsers) and OpenSSL (version 1.0.2). For sufficiently large buffers (e.g., 8 Kilobytes), AES-GCM performs now at the rate of 0.76 C/B on the latest Intel architecture (Codename Broadwell).

5 Discussion

This paper resolved some discrepancy in the description of the field representation that AES-GCM (GHASH) operates in. This discrepancy leads to an awkward description in the AES-GCM NIST specification, where GHASH defined through a bit-level algorithm [2,14]. This discrepancy becomes more noticeable when a polynomial multiplication instruction (PCLMULQDQ) appears and GHASH computations can be carried out at a granularity larger than a byte (where lookup tables can be pre-calculated in any bit order inside the byte). Note that the definition of PCLMULQDQ *is consistent* with the natural bits order inside bytes (and a Little Endian memory settings).

We proposed an algebraic formulation for the field operations, to replace the bit-level algorithm that was used in [2,14]. In fact, the operation "$\bullet$" (of AES-GCM) is a "Montgomery Multiplication" in $\mathbb{F}_{2^{128}}[x]/\left(x^{128}+x^{127}+x^{126}+x^{121}+1\right)$. With this viewpoint, we note that the bit-level algorithm of [2,14] is actually a bit-by-bit implementation of a (special) Montgomery multiplication in that field representation (although without directly calling this out).

We showed how our interpretation leads to an efficient reduction algorithm in $\mathbb{F}_{2^{128}}[x]/\left(x^{128}+x^{127}+x^{126}+x^{121}+1\right)$. Optimized AES-GCM software we contributed to the leading open-source cryptographic libraries ([9,10]), and this algorithm is now used in OpenSSL and NSS[2].

We point out that our reduction algorithm does not depend on the particular choice of $P(x)$ ($Q(x)$) that is used by AES-GCM and can be applied with equal efficiency for any other representation with the prescribed form. Furthermore, a similar reduction algorithm, to compute $A(x) \times B(x) \pmod{P(x)}$ (i.e., not only to $\bullet$) with any $P(x)$ with the prescribed form. One example is the field $\mathbb{F}_{2^{128}}[x]/(x^{128}+x^{12}+x^{11}+x^{5}+1)$, for which $Q(x)=x^{128}+x^{123}+x^{117}+x^{116}+1$.

[2] The reduction algorithm is called there "magic reduction" in OpenSSL, because the contributed code did not include a proof. The reduction algorithm was stated by us ([6]), though not proven there, and we used it for the efficient AES-GCM implementation.

Finally, we point out to the Authenticated Encryption with nonce misuse resistance algorithm AES-GCM-SIV that is defined in RFC 8452 [7] from 2019. This algorithm also uses a polynomial evaluation hash function but it is already defined directly over $\mathbb{F}_{2^{128}}[x]/\left(x^{128}+x^{127}+x^{126}+x^{121}+1\right)$ with arithmetic in this field representation of $\mathbb{F}_{2^{128}}$ (see explanations in [8]).

Acknowledgments. This research was partly supported by: NSF-BSF Grant 2018640; The Israel Science Foundation (grant No. 3380/19); The Center for Cyber Law and Policy at the University of Haifa, in conjunction with the Israel National Cyber Bureau in the Prime Minister's Office.

Appendix A: Polynomial multiplication

Let $A = A(x) = \sum_{j=0}^{s} a_j x^j$ and $B = B(x) = \sum_{j=0}^{s} b_j x^j$ be two binary polynomials of degree s. Define $T(x) = A(x) \times B(x)$. Then, $T(x)$ is a binary polynomial of degree $2s$, with coefficients t_i satisfying

$$t_i = \sum_{j=0}^{i} a_j b_{i-j} \pmod 2 \quad 0 \leq i \leq s, \quad t_i = \sum_{j=i-s}^{s} a_j b_{i-j} \pmod 2 \quad s < i \leq 2s$$

Appendix B: AES-GCM examples

AES-GCM Example: Test Case 2 in [13] (p. 27)

Input parameters:
(note that [13] prints the inputs (and outputs) in a different order of bytes compared to the convention used here. Nevertheless, the quantities are, of course, identical).

$H = 66E94BD4EF8A2C3B884CFA59CA342B2E$
$C = 0388DACE60B6A392F328C2B971B2FE78$
$LEN = len(A)\|len(C) = 00000000000000000000000000000080$
$GHASH = F38CBB1AD69223DCC3457AE5B6B0F885$

Computations in $GF(2^{128})/\left(x^{128} + x^7 + x^2 + x + 1\right)$ ($\otimes_1$):

$REF(H) = 74D42C539A5F3211DC3451F72BD29766$
$REF(C) = 1E7F4D8E9D4314CF49C56D06735B11C0$
$U_1 = REF(H) \otimes_1 REF(C) = REF(H) \times REF(C) \pmod{x^{128} + x^7 + x^2 + x + 1} =$
$= ED7BCACA160DA13411460E8962E3747A$
$REF(U_1) = 5E2EC746917062882C85B0685353DEB7$
$U_2 = \left(U_1 + REF(LEN)\right) \times REF(H) \pmod{x^{128} + x^7 + x^2 + x + 1} =$
$= A11F0D6DA75EA2C33BC4496B58DD31CF$
$REF(U_2) = F38CBB1AD69223DCC3457AE5B6B0F885$
Finally, $GHASH = U_2$.

Computations using Proposition 3 and Remark 5:

(Computations ($\otimes_2$) modulo $x^{128} + x^{127} + x^{126} + x^{121} + 1$)
Pre-compute $\bar{H} = x \otimes_2 H = CDD297A9DF1458771099F4B39468565C$.
$TEMP_1 = C \otimes \bar{H} \otimes_2 x^{-128} = 5E2EC746917062882C85B0685353DEB7$
(note: $TEMP_1 = REF(U_1)$)
$TEMP_2 = (T1 + LEN) \otimes \bar{H} \otimes_2 x^{-128} = F38CBB1AD69223DCC3457AE5B6B0F885$
Finally, $GHASH = TEMP_2$.

A Code Example. The code listing in Fig. 1 shows the computation of a *single* GHASH update computation $A(x) \times B(x) \times x^{-128} \pmod{Q}$ for the special case where $Q = (x^{128} + W(x)x^{64} + 1)$. Lines 4–12 implement a Schoolbook multiplication using four invocations of PCLMULQDQ. Lines 13–19 implement Proposition 3 for the reduction step, using two invocations of PCLMULQDQ (lines 13–15 correspond to (4.2) and lines 16–19 correspond to (4.5)).

```
1  # Input: A (= RES); B
2  # Output: RES (overwriting A; B remains unchanged);
3  # poly 0xc200000000000000  ( = W(x))
4  # Schoolbook multiplication (via 4 vpclmulqdq)
5  vpclmulqdq  $0x00, H, RES, TMP1
6  vpclmulqdq  $0x11, H, RES, TMP4
7  vpclmulqdq  $0x10, H, RES, TMP2
8  vpclmulqdq  $0x01, H, RES, TMP3
9  # Organizing the 4 pieces
10 vpxor    TMP3, TMP2, TMP2
11 vpslldq $8, TMP2, TMP3
12 vpsrldq $8, TMP2, TMP2
13 vpxor    TMP3, TMP1, TMP1
14 vpxor    TMP2, TMP4, TMP4
```

Step 1: $A \times B$

```
15 ; T (256 bits) resides in T1:T7
16 ; W = 0xc200000000000000
17 vmovdqa          T3, [W]
18 vpclmulqdq       T2, T3, T7, 0x01
19 vpshufd          T4, T7, 78
20 vpxor            T4, T4, T2
21 vpclmulqdq       T2, T3, T4, 0x01
22 vpshufd          T4, T4, 78
23 vpxor            T4, T4, T2
24 vpxor            T1, T4, T1          ; the result is in T1
```

Step 2: Reducing $T \times x^{-128}$ (mod Q) $(x^{128} + W(x)x^{64} + 1)$

Fig. 1. GHASH code snippet: $A \times B \times x^{-128}$ (mod Q) $(x^{128} + W(x)x^{64} + 1)$. Step 1 shows the computation of $A \times B$ (Schoolbook multiplication) using 4 calls to PCLMULQDQ. Step 2 reduces the product modulo $Q(x)$ using 2 invocations of PCLMULQDQ (as in (4.2) and in (4.5)).

References

1. - National Institute of Standards and Technology (NIST), FIPS-197: Advanced Encryption Standard (2001). http://csrc.nist.gov/publications/fips/fips197/fips-197.pdf
2. Dworkin, M.: Recommendation for Block Cipher Modes of Operation: Galois/-Counter Mode (GCM) for Confidentiality and Authentication. Federal Information Processing Standard Publication FIPS 800–38D (2006). http://csrc.nist.gov/publications/nistpubs/800-38D/SP-800-38D.pdf
3. Gladman, B.: AES and Combined Encryption/Authentication Modes; Public Domain Source Code (2006). http://ccgi.gladman.plus.com/oldsite/cryptography_technology/index.php
4. Gueronm S.: Intel Advanced Encryption Standard (AES) Instructions Set, Rev 3.01. Intel Software Network (2012). https://software.intel.com/en-us/articles/intel-advanced-encryption-standard-aes-instructions-set
5. Gueron, S.: Intel's new AES instructions for enhanced performance and security. In: Dunkelman, O. (ed.) FSE 2009. LNCS, vol. 5665, pp. 51–66. Springer, Heidelberg (2009). https://doi.org/10.1007/978-3-642-03317-9_4
6. Gueron, S.: AES-GCM for efficient authenticated encryption - ending the reign of HMAC-SHA-1?. In: Workshop on Real-World Cryptography, Stanford University (2013). http://crypto.stanford.edu/RealWorldCrypto/program.php
7. Gueron, S., Langley, A., Lindell, Y.: AES-GCM-SIV: nonce misuse-resistant authenticated encryption (2019). https://www.rfc-editor.org/rfc/rfc8452.html

8. Gueron, S., Langley, A., Lindell, Y.: AES-GCM-SIV: Specification and Analysis (2017). https://eprint.iacr.org/2017/168
9. Gueron, S., Krasnov, V.: Efficient AES-GCM implementation that uses Intel's AES and PCLMULQDQ instructions (AES-NI), and the Advanced Vector Extension (AVX) architecture, for the NSS library (2012). https://bugzilla.mozilla.org/show_bug.cgi?id=805604
10. Gueron, S., Krasnov, V.: [openssl.org #2900] [PATCH] Efficient implementation of AES-GCM, using Intel's AES-NI, PCLMULQDQ instruction, and the Advanced Vector Extension (AVX) (2012). https://openssl-dev.openssl.narkive.com/zwHrxBBY/openssl-org-2900-patch-efficient-implementation-of-aes-gcm-using-intel-s-aes-ni-pclmulqdq
11. Gueron, S., Kounavis, M.E.: Intel Carry-Less Multiplication and Its Usage for Computing The GCM Mode, Rev. 2.01. Intel Software Network (2014). https://www.intel.com/content/dam/develop/external/us/en/documents/clmul-wp-rev-2-02-2014-04-20.pdf
12. Gueron, S., Kounavis, M.E.: Efficient implementation of the Galois counter mode using a carry-less multiplier and a fast reduction algorithm. Inf. Process. Lett. **110**, 549–553 (2010)
13. McGrew, D., Viega, J.: The Galois/Counter Mode of Operation (GCM) (2004). https://csrc.nist.rip/groups/ST/toolkit/BCM/documents/proposedmodes/gcm/gcm-spec.pdf
14. McGrew, D.A., Viega, J.: The security and performance of the Galois/Counter mode (GCM) of operation. In: Canteaut, A., Viswanathan, K. (eds.) INDOCRYPT 2004. LNCS, vol. 3348, pp. 343–355. Springer, Heidelberg (2004). https://doi.org/10.1007/978-3-540-30556-9_27

Fast ORAM with Server-Aided Preprocessing and Pragmatic Privacy-Efficiency Trade-Off

Vladimir Kolesnikov[1], Stanislav Peceny[1(✉)], Ni Trieu[2], and Xiao Wang[3]

[1] Georgia Tech, Atlanta, GA, USA
{kolesnikov,stan.peceny}@gatech.edu
[2] Arizona State University, Tempe, AZ, USA
[3] Northwestern University, Evanston, IL, USA

Abstract. Data-dependent accesses to memory are necessary for many real-world applications, but their cost remains prohibitive in secure computation. Prior work either focused on minimizing the need for data-dependent access in these applications, or reduced its cost by improving oblivious RAM for secure computation (SC-ORAM). Despite extensive efforts to improve SC-ORAM, the most concretely efficient solutions still require ≈0.7 s per access to arrays of 2^{30} entries. This plainly precludes using MPC in a number of settings.

In this work, we take a pragmatic approach, exploring how concretely cheap MPC RAM access could be made if we are willing to allow one of the participants to learn the access pattern. We design a highly efficient Shared-Output Client-Server ORAM (SOCS-ORAM) that has constant overhead, uses one round-trip of interaction per access, and whose access cost is independent of array size. SOCS-ORAM is useful in settings with hard performance constraints, where one party in the computation is more trust-worthy and is allowed to learn the RAM access pattern. Our SOCS-ORAM is assisted by a third helper party that helps initialize the protocol and is designed for the honest-majority semi-honest corruption model.

We implement our construction in C++ and report its performance. For an array of length 2^{30} with 4B entries, we communicate 13B per access and take essentially no overhead beyond network latency.

Keywords: Secure Computation · Oblivious RAM

1 Introduction

Real-world applications rely heavily on data-dependent accesses to memory. Despite many recent improvements, such accesses remain a bottleneck when evaluated in secure two-party and multiparty computation (2PC, MPC). While in plaintext execution such accesses are cheap constant-time operations, they are expensive in MPC, since access pattern must remain hidden. A naive secure

S. Dolev et al. (Eds.): CSCML 2023, LNCS 13914, pp. 439–457, 2023.
https://doi.org/10.1007/978-3-031-34671-2_31

solution to this problem is linear scan, which hides the access pattern by touching every element in memory and multiplexing out the result. This, of course, incurs overhead linear in memory size for each access. A much more scalable approach is to instead use more complex Oblivious RAM (ORAM) protocols [GO96], which achieve polylog complexity, while still hiding access patterns.

The first ORAM considered the client-server setting [GO96], where a client wishes to store and access her private array on an untrusted server. Soon after, initiated by [OS97,GKK+12], ORAM was shown applicable to RAM-based MPC: Secure RAM access was achieved for MPC simply by having the parties execute ORAM client inside secure computation, while both parties share the state of the server.

Despite extensive research focused on optimizing ORAM for secure computation (SC-ORAM) and ORAM in general, the overhead remains prohibitive for many applications. For example, a recent SC-ORAM Floram [Ds17] takes $\approx$2 s per access, communicates $\approx$5 MBs, and requires 3 communication rounds on arrays of size 2^{30} with 4-byte elements.

Such ORAM performance is unacceptable in settings where many accesses of large arrays are needed. Examples include network traffic or financial markets analyses, where data is continuously generated and frequently accessed.

3PC: 2PC with a Helper Server. Fortunately, many real-world applications can use a third party to help with computation. This third party may already be a participant of the computation (e.g. provide input and/or receive output) or can be brought as an (oblivious) helper server. As MPC of many functions is much faster in a 3-party honest-majority setting than in the two-party setting, [FJKW15] ask whether SC-ORAM can also be accelerated. [FJKW15] present a solution and report total wall-clock time of 1.62 s on a 2^{36}-element array. The rest of the measurements focus on the online costs; based on the discussion in the paper we estimate the total cost for 2^{30}-element array is $\approx$1.25 s. A follow-up work [JW18] then asymptotically reduced the bandwidth of [FJKW15], but still reports $\approx$0.7 s CPU time per access on a 2^{30}-element array.

Although 3-party SC-ORAM improves over 2-party SC-ORAMs, a 0.7 s RAM access time will still be considered prohibitive in many (most?) realistic use scenarios. Note, accessing smaller-size memories would be, of course, cheaper: [JW18] reports 0.1 s CPU time per access on a 2^{10}-element array. For context, note that garbled circuit linear scan of 2^{10}-element array would require about 2^{15} gates and would take less than 0.1 s on a 1 Gbps LAN.

Our Goals. In this work, we are interested in exploring what secure computation is possible in settings with hard performance constraints. We thus seek maximizing performance at the cost of relaxing the security guarantees.

We start in the easier 3-party setting, and ask whether we can get further significant improvement if one party in the computation is more trust-worthy and is allowed to *learn the access pattern.*

This trust model may naturally occur in real-world scenarios (see Sect. 1.1) e.g. if one of the parties is an established entity with trusted oversight, such as a government or a law enforcement agency.

Our Setting. We summarize our considered setting. Our Shared-Output Client-Server ORAM (SOCS-ORAM) protocol is run by three parties $\mathcal{A}$, $\mathcal{B}$, and $\mathcal{C}$. $\mathcal{B}$ holds an array $\mathbf{d}$ of n l-bit entries. $\mathcal{A}$ requests up to k read or write accesses to $\mathbf{d}$. For each access, $\mathcal{A}$ inputs index $i \in [n]$ and operation op (`read` or `write`). $\mathcal{A}$ and $\mathcal{B}$ hold a sharing of a value to write $[\![x]\!]$. $\mathcal{C}$ holds no input and *does not* participate in ORAM access; it is used to help initialize SOCS-ORAM.

All parties are semi-honest and do not collude with one another. We allow $\mathcal{A}$ to learn the access pattern – indeed Alice can be viewed as ORAM client; $\mathcal{B}$ and $\mathcal{C}$ learn nothing from the computation.

1.1 Motivation

Recall that our work explores a trade-off between maximizing performance at the cost of relaxing security guarantees. This is a natural and pragmatic research direction. For example, a similar trade-off was also considered in Blind Seer [PKV+14], a scalable privacy-preserving database management system that supports a rich query set for database search and addresses query privacy. [PKV+14] motivate the trade-off, warn of potential pitfalls, and convincingly argue its benefits. Our work is complementary. SOCS-ORAM can be used as a drop-in *no-cost* replacement to improve security of Blind Seer's unprotected RAM access. Indeed, Blind Seer similarly uses three parties but allows two parties (i.e. all parties other than helper server (Server in their notation)) to learn the access pattern, compared to only one party in our work. We believe this can be a crucial difference as trust is unbalanced in natural settings (e.g., bank may be trusted more than clients, wireless service provider – more than each individual customer, and government agency – more than private businesses).

We now briefly discuss several motivating applications spanning network security, financial markets, and review Blind Seer's air carrier's passenger manifest analysis.

Network Data Analysis. There is a significant benefit in operation of large-scale analysis centers, such as Symantec's DeepSight Intelligence Portal. These centers collect network traffic information from a diverse pool of sources such as intrusion detection systems, firewalls, honeypots, and network sensors, and can be used to build analysis functions to detect network threats [PS06].

Network data is highly sensitive; revealing network configuration and other details may significantly weaken its defences. Using MPC instead to enable expert network analysis and vulnerability reporting is a (costly) solution. Network analysis works with large volumes of data (e.g. Symantec's DeepSight has billions of events) and requires a large number of RAM accesses. Paying $\approx$1 s per RAM access is clearly not feasible for even trivial analyses.

Using SOCS-ORAM and placing, arguably, a reasonable trust in the analysis center (allowing it to learn RAM access pattern), may potentially enable this application.

Financial Markets Analysis. SOCS-ORAM can be used to identify fraudulent activity, such as insider trading in financial markets. In this use case, a regulatory agency such as SEC or FINRA investigates and analyzes data from brokerages. Typically SEC initiates its investigation based on suspicious activity in an individual security. SEC next makes a regulatory request. So-called *blue sheets data* brokerage response contains trading and account holder information. SEC's Market Abuse Unit (MAU) then runs complex analyses on the data, which may contain billions of rows. We note that there are privacy concerns for both parties. SEC does not want to reveal what they are investigating, while brokerages do not want to share their clients' data that is not essential for the investigation. This scenario is a fit for our SOCS-ORAM: The brokerage learns nothing about the investigation, while SEC learns only the output of the analysis functions, alongside the access pattern.

Passenger Manifests Analysis. Passenger manifests search and analysis is one of the motivating applications of the Blind Seer DBMS [PKV+14]. It considers a setting where a law enforcement agency wants to analyze or search air carrier's manifests for specific patterns or persons. The air carrier would like to protect its customers' data, and hence reveal only the data necessary for the investigation. The law enforcement agency would like to protect its query. Today's approach may be to simply provide the manifests to the agency. Using MPC (and keeping the private data private) would help allay the negative popular sentiment associated with large scale personal data collection by government.

1.2 Contributions

We present a highly efficient shared-output client-server ORAM (SOCS-ORAM) scheme. Here the client Alice knows the logical indices of the RAM queries, and the results are additively (XOR) secret-shared between her and the server Bob, allowing them, unlike the output of classical ORAM, to be directly used in MPC.

This construction is suitable for secure computation applications with hard performance constraints where one party is more trustworthy. While in MPC none of the parties learns the set of queried RAM locations, we reveal them to one of the parties. Further, our SOCS-ORAM uses a semi-honest third party who helps initialize our construction, but is not active when invoking `access`. In exchange, we achieve very high ORAM performance, whose *only* non-trivial cost is communication rounds. In particular, we present:

- **Efficient SOCS-ORAM Construction.** Our construction consists of an efficient third-party aided initialization protocol and an efficient 2-party access protocol.
  Our initialization protocol does not execute MPC; it runs PRG and generates a random permutation, all evaluated outside MPC. It requires 4 message

flows (the first 2 and the last 2 can be parallelized). To set up SOCS-ORAM for k accesses to an array of size n, we require sending $4n + 6k$ array entries, k bits, and a single permutation of size $n + k$, sent as a table. $2n$ entries are sent by $\mathcal{B}$, and the rest by $\mathcal{C}$.
Our access protocol communicates only 2 array entries, a single array index, and an additional bit, and requires a single round trip of interaction. No cryptography is involved in our access protocol: We only use the XOR operation and plaintext array access. The cost of our access protocol is independent of array size (but system level implementation costs manifest for larger array sizes).

- **Resulting efficient implementation.** We implement and experimentally evaluate our approach. Our experimental results indicate that on an array with 2^{30} entries each of 4B, we communicate 13B per access and run in 2.13 ms on a 2 ms latency network (as set by the Linux `tc` command; the actual latency, due to system calls overhead is closer to 2.13 ms).
Thus, our wall-clock time is extremely close to latency cost. While our setting is much simpler than that of SC-ORAM, state-of-the-art 3-party SC-ORAM [JW18] reports ≈0.7 s CPU time for arrays of the same size, while all our runs ran in less than 0.019 ms of computation. Similarly, our access communication is on the order of bytes instead of MBytes, and we use 1 round trip of interaction instead of $O(\log\, n)$. For a 2^{30} array of $4B$ entries (i.e. 4 GB size array) and 2^{20} accesses, the cost to initialize our SOCS-ORAM (preprocessing) is 4.8 min and 20 GB communication. In Sect. 6.2, we discuss a natural optimization that would reduce communication to 8 GB.

2 Notation

- Party $\mathcal{A}$ (Alice) inputs access indices i (i.e. client).
- Party $\mathcal{B}$ (Bob) inputs array $\mathbf{d}$ (i.e. server).
- Party $\mathcal{C}$ (Charlie) is a third party helper.
- κ denotes the computational security parameter (e.g. 128).
- $[n]$ denotes the sequence of natural numbers $0, \ldots, n-1$. $[n, n+k]$ denotes the sequence $n, \ldots, n+k-1$.
- We denote arrays in bold, index them with subscripts, and use 0-based indexing. E.g., $\mathbf{d}_0$ is the first element of array $\mathbf{d}$.
- We sometimes add subscript notation to arrays to indicate that for a bit array $\mathbf{f}$ and two arrays $\mathbf{s_0}, \mathbf{s_1}$, the array $\mathbf{s_f}$ holds entries from $\mathbf{s}_{\mathbf{f}_i}$ at index i. We index these arrays with a ',' (e.g. $\mathbf{s}_{\mathbf{0},i}$).
- We denote negation of a bit b as $\bar{b}$.
- We manipulate XOR secret shares.
  - We use the shorthand $[\![\mathbf{d}]\!]$ to denote a sharing of array $\mathbf{d}$.
  - Subscript notation associates shares with parties. E.g., $[\![\mathbf{d}]\!]_A$ is a share of $\mathbf{d}$ held by party A.

3 Oblivious RAM (ORAM) Review

Our notions of client-server oblivious RAM (ORAM) and secure-computation oblivious RAM (SC-ORAM) are standard.

Client-Server ORAM. A client-server ORAM [GO96] is a protocol that enables a *client* to outsource data to an untrusted *server* and perform arbitrary read and write operations on that outsourced data without leaking the data or access patterns to the server.

An ORAM specifies an initialization protocol that takes as input an array of entries and initializes an oblivious structure with those entries, as well as an access protocol that implements each *logical* (`read` and `write`) access on the oblivious structure with a sequence of polylog *physical* accesses.

We now present the ORAM functionality. Client inputs an array $\mathbf{d}$ of length n. For each access, client inputs operation op (`read` or `write`), index $i \in [n]$, and, if writing, the value x to write. Server inputs $\perp$. If $op = \texttt{read}$, client outputs $\mathbf{d}_i$ and server outputs $\perp$; if $op = \texttt{write}$, client and server set $\mathbf{d}_i = x$ and output $\perp$.

The ORAM's security guarantee is that the physical access patterns produced by the access protocol for any two sequences of logical accesses of the same length must be computationally indistinguishable. We take the security definition almost verbatim from [SSS12].

Definition 1. *Let* $y := ((op_0, i_0, x_0), (op_1, i_1, x_1), \ldots, (op_{m-1}, i_{m-1}, x_{m-1}))$ *denote a sequence of logical accesses of length* m*, where each* op *denotes* $\texttt{read}(i)$ *or* $\texttt{write}(i, x)$*. Specifically,* i *denotes the array index being read or written, and* x *denotes the data being written. Let* $A(\vec{y})$ *denote the (possibly randomized) sequence of physical accesses to the remote storage given the sequence of logical accesses* $\vec{y}$*. ORAM is said to be secure if for any two sequences of logical accesses* $\vec{y}$ *and* $\vec{z}$ *of the same length, their access patterns* $A(\vec{y})$ *and* $A(\vec{z})$ *are computationally indistinguishable by anyone but the client.*

RAM-Based Secure Computation. [OS97] noted the idea of using ORAM for secure multi-party computation (SC-ORAM). [GKK+12] proposed the first complete SC-ORAM construction. In SC-ORAM, the key idea is to have each party store a share of the server's ORAM state, and then execute the ORAM client access algorithms via a general-purpose secure computation protocol.

As the server's state is now secret-shared between both parties and the client is executed inside secure computation, we no longer refer to the physical parties as client and server but $\mathcal{A}$ and $\mathcal{B}$. In SC-ORAM, $\mathcal{A}$ and $\mathcal{B}$ input a sharing of an array $[\![\mathbf{d}]\!]$ of length n. For each access, they input a sharing of operation $[\![op]\!]$ (`read` or `write`), a sharing of index $[\![i]\!] \in [n]$, and a sharing of a value to write $[\![x]\!]$. If $op = \texttt{read}$, $\mathcal{A}$ and $\mathcal{B}$ output $[\![\mathbf{d}_i]\!]$; if $op = \texttt{write}$, set $[\![\mathbf{d}_i]\!] = x$ and output $\perp$.

There are a few key differences between client-server ORAM and SC-ORAM that [ZWR+16] explicates:

- In the client-server ORAM, the client owns the array and also accesses it. Hence, the privacy requirement is unilateral. In SC-ORAM, both the array and the access are distributed and neither party should learn anything about the array or the access pattern.
- In the client-server ORAM, the client's storage should be sublinear, whereas in SC-ORAM, linear storage is distributed across both parties.
- Client-server ORAMs have traditionally been measured by their bandwidth overhead and client storage. [WHC+14] observed that for SC-ORAMs the size of the client circuits is more relevant to performance.
- In SC-ORAM, the initialization protocol must be executed securely; in 2PC this cost is often prohibitive.

4 Related Work

We present a highly efficient 3-party SOCS-ORAM with applications in secure computation. We therefore review related work that improves (1) SC-ORAMs in the standard 2-party setting, (2) SC-ORAMs in the 3-party setting, and (3) Garbled RAM schemes that equip Garbled Circuit with a sublinear cost RAM without adding rounds of interaction. We also briefly discuss (4) differential obliviousness (DO) and (5) private information retrieval (PIR).

2-Party SC-ORAM. [OS97] proposed the basic idea of SC-ORAM, where the parties share the ORAM server role, while having the ORAM client algorithm executed via secure computation. [GKK+12] presented a specific SC-ORAM construction that started a long line of research to improve SC-ORAM. [WHC+14] observed that when using ORAMs for secure computation, the size of the circuits is more relevant to performance than the traditional metrics such as bandwidth overhead and client storage. Then they presented a heuristic SC-ORAM optimized for circuit complexity. [WCS15] followed up with Circuit ORAM, which further reduced circuit complexity. [ZWR+16] showed that by relaxing asymptotics, one can produce a scheme that outperforms Circuit ORAM for arrays of small to moderate sizes. We note that all [GKK+12, WHC+14, WCS15, ZWR+16] are recursively structured and as a result require $O(\log n)$ rounds of communication per access; they have expensive initialization algorithms, high memory overhead. E.g., [Ds17] observed they could not handle arrays of sizes larger than $\approx 2^{20}$ on standard hardware. With this in mind, [Ds17] introduced Floram that requires 3 rounds per access and significantly decreases memory overhead and initialization cost. Floram requires linear work per access. Crucially, this work is inexpensive since it is local and executed outside secure computation, unlike in the MPC-run linear scan. Still, despite a large concrete improvement, [Ds17] takes $\approx$2 s per access and communicates $\approx$5 MBs in communication on arrays of size 2^{30} with 4-byte elements.

3-Party SC-ORAM. [FJKW15] explore whether adding a third party to SC-ORAM can improve performance. They present a construction secure against semi-honest corruption of one party, which uses custom-made protocols to emulate the client algorithm of the binary tree client-server ORAM [SCSL11] in secure computation. For a 2^{36}-element array of 4-byte entries, their access runs in 1.62 s wall-clock time when executed on two co-located EC2 t2.micro machines. Their solution further requires $O(\log n)$ communication rounds for an array of size n. [JW18] followed up on their work and designed custom-made protocols to instead emulate the Circuit ORAM [WCS15] client. While their technique still requires $O(\log n)$ communication rounds per access, they asymptotically decrease the bandwidth of [FJKW15] by the statistical security parameter. Concretely, they report $\approx$0.7 s CPU time per access on a 2^{30}-element array of 4-byte entries, when run on co-located AWS EC2 c4.2xlarge instances. While we are not directly comparable, we execute one access in one communication round and all our runs took less than 0.019 ms on localhost on a same-size array.

[BKKO20] showed how to combine their 3-server distributed point function (DPF) with any 2-server PIR scheme to obtain a 3-server ORAM and then extended it to SC-ORAM. Their access protocol runs in constant rounds, requires sublinear communication and linear work, and makes only black-box use of cryptographic primitives. [FNO22] present 3-party SC-ORAM from oblivious set membership that aims to minimize communication complexity. These works do not offer implementation and evaluation, and we do not directly compare with their performance.

Garbled RAM (GRAM). GRAM is a powerful technique that adds RAM to GC while preserving GC's constant rounds of interaction. This technique originated in [LO13] but was not suitable for practice until [HKO22] introduced EpiGRAM. While EpiGRAM was not implemented, [HKO22] estimate that for an array of 2^{20} entries of 16B, the per-access communication amortized over 2^{20} accesses is $\approx$16 MB. In comparison, our work communicates $\approx$0.2 KB (initialization included) amortized over the same number of accesses.

Differential Obliviousness (DO). DO [CCMS19] is a relaxation of access pattern privacy. As opposed to simulation-based ORAM privacy guarantees, DO requires the program's access pattern to be differentially private. [CCMS19] showed that for some programs DO incurs only $O(\log\log n)$ overhead in contrast to ORAM's polylog complexity. We forfeit access pattern privacy against Alice.

Private Information Retrieval (PIR). PIR [CKGS95] enables a client to retrieve a selected entry from an array such that no information about the queried entry is revealed to the one (or multiple) server holding the array. Thus, PIR is concerned with the privacy of the client. There are many flavors of PIR, one of which is Symmetric PIR (SPIR) [GIKM98]. SPIR has an additional requirement that the client learns only about the elements she is querying, and nothing else. For our purposes, the main difference between PIR and ORAM is that PIR supports only read operations. While we do not further discuss PIR, we emphasize that PIR is sometimes used as a building block of ORAM constructions (e.g. in [Ds17, GKW18, JW18, KM19, BKKO20] discussed above).

5 Technical Overview

We introduce and construct, at the high-level, shared-output client-server oblivious RAM (SOCS-ORAM), a useful building block for efficient MPC. We present our construction by first simply achieving a basic limited functionality, and then securely building on that to achieve the goal. Full formal algorithms are in Sect. 6. For accompanying proofs of correctness and security see the full version.

Recall from Sect. 1, SOCS-ORAM is run by parties $\mathcal{A}$, $\mathcal{B}$, and $\mathcal{C}$, where $\mathcal{B}$ holds an array $\mathbf{d}$ of length n. On access, $\mathcal{A}$ inputs operation *op* (`read` or `write`) and an index $i \in [n]$. $\mathcal{A}$ and $\mathcal{B}$ also input a sharing of value $[\![x]\!]$ to write. $\mathcal{C}$ is a helper party that aids with SOCS-ORAM initialization and is not active during array access. Initialization provisions for k *dynamic* accesses. We consider honest majority with security against semi-honest corruption and allow $\mathcal{A}$ to learn (or know) the access pattern.

Goal. We aim to build a concretely efficient SOCS-ORAM using plaintext array lookup, masking, and PRGs, with constant access overhead and a single round trip of interaction, whose computational cost is close to plaintext array access. We design such SOCS-ORAM at the concession of allowing one party to learn the access pattern. We describe our construction next.

Basic Initialization for Our SOCS-ORAM. $\mathcal{A}$ and $\mathcal{B}$, with the help of $\mathcal{C}$, initialize $\mathbf{D}$ with $\mathbf{d}$ (cf Fig. 1; $\mathbf{d}$ is $\mathcal{B}$'s input array used to initialize the working array $\mathbf{D}$). $\mathcal{A}$ and $\mathcal{B}$ receive $[\![\mathbf{D}]\!]$, which is permuted according to a random permutation π unknown to $\mathcal{B}$ and secret-shared using randomness neither party knows. Uniform secret sharing ensures that upon access neither party learns anything about the value of the array entry they are retrieving; permuting ensures logical index is hidden from $\mathcal{B}$. Clearly, this *initially* (i.e. before any accesses) hides array entries and their positions. With $\mathcal{C}$'s help, this structure can be set up cheaply.

Handling Repeated Accesses. Following the above initialization, $\mathcal{A}$ will access $[\![\mathbf{D}]\!]$, possibly accessing the same logical index multiple times. Recall, only $\mathcal{A}$ is allowed to learn the access pattern. $\mathcal{C}$ is oblivious by not participating in the access protocol. The challenge is to preclude $\mathcal{B}$ from learning the access pattern.

As hinted above, if no logical index is accessed twice, $\mathcal{B}$ learns nothing, since each entry $[\![\mathbf{D}_i]\!]$ is placed in a random physical position $\pi(i)$. To access a logical index more than once, each time the physical location must be different: the value must be copied to a *new random* location.

We modify initialization to create the space for copied values. We *extend* the working array $\mathbf{D}$ with space for k entries (*shelter*), and secret-share and permute the *extended* $\mathbf{D}$ according to $\pi : [n+k] \mapsto [n+k]$. This is cheap with $\mathcal{C}$'s help.

We next show how to copy the read entry to a new index (corresponding to the next available shelter entry) in $[\![\mathbf{D}]\!]$, *obliviously* to $\mathcal{B}$. Then, at the next access to this element, $\mathcal{B}$ is accessing a random share at a random-looking index.

`read` *Access.* To clarify and extend the previous discussion, we allow for `read` in SOCS-ORAM as follows. Recall that $\mathcal{A}$ is allowed to learn the access pattern, and hence she can be given π. $\mathcal{A}$ can then track the position of each element in (extended) $[\![\mathbf{D}]\!]$ in a position map $\mathbf{pos}$, mapping logical indices $i \in [n]$ to physical indices $j \in [n+k]$. Initially $\mathbf{pos}_i := \pi_i \triangleq \pi(i)$ for all $i \in [n]$. $\mathcal{A}$ uses $\mathbf{pos}$ at each access to find her share of the sought entry i at position $\mathbf{pos}_i$ in $[\![\mathbf{D}]\!]_{\mathcal{A}}$ (i.e. $[\![\mathbf{D}_{\mathbf{pos}_i}]\!]_{\mathcal{A}}$). Since π is a random permutation, $\mathcal{A}$ simply gives $\mathcal{B}$ $\mathbf{pos}_i$, and $\mathcal{B}$ retrieves his share $[\![\mathbf{D}_{\mathbf{pos}_i}]\!]_{\mathcal{B}}$. $\mathcal{A}$ and $\mathcal{B}$ can now use $\mathbf{D}_{\mathbf{pos}_i}$ inside MPC.

We now explain how to arrange that the read entry at logical index i, stored at physical index $\mathbf{D}_{\mathbf{pos}_i}$, is prepared for a subsequent access. Intuitively, after the qth access (out of total k provisioned), entry's value is copied to position π_{n+q}. This is done as follows. $\mathcal{A}$ arranges that $\mathbf{D}_{\mathbf{pos}_{n+q}} = \mathbf{D}_{\mathbf{pos}_i}$ *solely* by updating her share $[\![\mathbf{D}_{\mathbf{pos}_{n+q}}]\!]_{\mathcal{A}}$. $\mathcal{A}$ can do this because at initialization $\mathcal{C}$ will perform an additional step: He generates a k-element random mask vector $\mathbf{m}$ and XORs it into the shelter positions of $[\![\mathbf{D}_{\mathbf{pos}_i}]\!]_{\mathcal{A}}$ (i.e. for $i \in [n, n+k]$). $\mathcal{C}$ sends $\mathbf{m}$ to $\mathcal{B}$. During the q-th access, where logical index i is read, $\mathcal{B}$ sends $[\![\mathbf{D}_{\mathbf{pos}_i}]\!]_{\mathcal{B}} \oplus \mathbf{m}_q$ to $\mathcal{A}$, who then XORs it with her share $[\![\mathbf{D}_{\mathbf{pos}_i}]\!]_{\mathcal{A}}$ to obtain $[\![\mathbf{D}_{\mathbf{pos}_{n+q}}]\!]_{\mathcal{A}}$. It is easy to see that this arranges for a correct sharing of $\mathbf{D}_{\mathbf{pos}_i}$ in physical position $n+q$.

Finally, $\mathcal{A}$ updates her map $\mathbf{pos}_i := \pi_{n+q}$. Next access to logical index i is set up to be read from $\mathbf{D}_{\mathbf{pos}_i}$, a new and random-looking location for $\mathcal{B}$.

General. `read`/`write` *access* is an easy extension of `read`. For access, in addition to opcode $op = (\texttt{read}, \texttt{write})$ known to $\mathcal{A}$, both parties also input $[\![x]\!]$, a sharing of the element to be written. `write` differs from `read` only in that $[\![x]\!]$, and not $[\![\mathbf{D}_{\mathbf{pos}_i}]\!]$, is used to arrange $[\![\mathbf{D}_{\mathbf{pos}_{n+q}}]\!]$. This extension is simple to achieve with an OT, which we implement efficiently with correlated randomness provided by $\mathcal{C}$ during initialization. One pedantic nuance we must address is that `write` must return a value. We set it to be the value previously stored in that location.

6 Our SOCS-ORAM

We now formally present our scheme. In Sect. 6.1, we define SOCS-ORAM's cleartext semantics. In Sect. 6.2, we specify Π - SOCS-ORAM, our protocol implementing SOCS-ORAM. We defer proofs of correctness and security to the full version of our paper.

6.1 Cleartext Semantics: SOCS-ORAM

Definition 2 *(Cleartext Semantics SOCS-ORAM). SOCS-ORAM$(\boldsymbol{d})_{n,k,l}$ is a 3-party stateful functionality executed between parties $\mathcal{A}$, $\mathcal{B}$, and $\mathcal{C}$ that receives a sequence of up to $k+1$ instructions. The first instruction is* `init`$(\boldsymbol{d})$, *where* $\mathbf{d}$ *is an array of n l-bit values and is input by $\mathcal{B}$.* `init` *associates the values in* $\mathbf{d}$ *with their corresponding indices in $[n]$ and sets $\boldsymbol{D} := \mathbf{d}$. The remaining up to k instructions are* `access`$_{\boldsymbol{D}}(op, i, [\![x]\!])$ *instructions.* `access` *is executed between $\mathcal{A}$ and $\mathcal{B}$ only; $\mathcal{A}$ inputs op, i and both input $[\![x]\!]$. Depending on op, they read the value at index i or write $[\![x]\!]$ to the value at index i in $\boldsymbol{D}$. See Fig. 1 for the* `init` *and* `access` *functionalities.*

init Functionality (3 Parties $\mathcal{A}$, $\mathcal{B}$, and $\mathcal{C}$)

init($\mathbf{d}$):

- INPUT: Party $\mathcal{B}$ inputs an array $\mathbf{d}$ of length n s.t. $\mathbf{d}_i \in \{0,1\}^l$. $\mathcal{A}$ and $\mathcal{C}$ input $\perp$
- Set $\mathbf{D} := \mathbf{d}$ // $\forall i \in [n]$, $\mathbf{D}_i := \mathbf{d}_i$

access Functionality (2 Parties $\mathcal{A}$ and $\mathcal{B}$)

read(i):

- INPUT: Party $\mathcal{A}$ inputs an index $i \in [n]$. $\mathcal{B}$ inputs $\perp$
- OUTPUT: $\mathcal{A}$ and $\mathcal{B}$ output $[\![\mathbf{D}_i]\!]$

write(i, $[\![x]\!]$):

- INPUT: Party $\mathcal{A}$ inputs an index $i \in [n]$. $\mathcal{A}$ and $\mathcal{B}$ input an element $[\![x]\!]$.
- Set $out := \mathbf{D}_i$
- Set $\mathbf{D}_i := x$
- OUTPUT: $\mathcal{A}$ and $\mathcal{B}$ output $[\![out]\!]$

access(op, i, $[\![x]\!]$):

- INPUT:
  - Party $\mathcal{A}$ inputs operation op (read or write) and an index $i \in [n]$
  - Parties $\mathcal{A}$ and $\mathcal{B}$ input additive sharing of an element $[\![x]\!]$
- OUTPUT:

$$\begin{cases} [\![\mathbf{D}_i]\!] \leftarrow \texttt{read}(i) & \text{if } op = \texttt{read} \\ [\![out]\!] \leftarrow \texttt{write}(i, [\![x]\!]) & \text{if } op = \texttt{write} \end{cases}$$

Fig. 1. The init and access functionalities for our SOCS-ORAM.

6.2 Protocol: Π-SOCS-ORAM

In this section, we formalize our protocol Π-SOCS-ORAM, which securely implements the semantics of SOCS-ORAM (Definition 2):

Construction 1 *(Protocol Π-SOCS-ORAM). Π-SOCS-ORAM$(\boldsymbol{d})_{n,k,l}$ is defined by first invoking* Π-init *in Fig. 2 and then up to k invocations to* Π-access *in Fig. 3.*

We include the proof of the following theorem in the full version of this paper:

Theorem 1. *Construction 1 implements the functionality SOCS-ORAM (Definition 2) and is secure in the honest-majority semi-honest setting.*

As Π-SOCS-ORAM consists of separate invocations to Π-init and Π-access (see Construction 1), we separate Π-SOCS-ORAM's description into Π-init (Fig. 2) and Π-access(Fig. 3), respectively.

Π-init. Π-init sets up data structures necessary to access $\mathbf{d}$ (see init in Fig. 1). It is a 3-party protocol, where $\mathcal{A}$ and $\mathcal{B}$ are aided by helper $\mathcal{C}$.

$\mathcal{B}$ inputs array $\mathbf{d}$ of n l-bit entries, sets $\mathbf{D} := \mathbf{d}$, and secret shares $\mathbf{D}$ between $\mathcal{A}$ and $\mathcal{C}$: $\mathcal{A}$ receives $[\![\mathbf{D}]\!]_\mathcal{A}$; $\mathcal{C}$ receives $[\![\mathbf{D}]\!]_\mathcal{B}$. $\mathcal{C}$ then helps to construct the data structures used in Π-access for $\mathcal{A}$ and $\mathcal{B}$.

$\mathcal{C}$ first generates the data structures for $\mathcal{B}$. He uniformly samples array $\mathbf{r}$ of same size as $\mathbf{D}$. $\mathcal{C}$ masks his share of $\mathbf{D}$ with $\mathbf{r}$, i.e. computes $[\![\mathbf{D}]\!]_\mathcal{B} := [\![\mathbf{D}]\!]_\mathcal{B} \oplus \mathbf{r}$. Simultaneously, $\mathcal{C}$ uniformly samples array $\mathbf{m}$, which will hold shelter values, where array entries will be written once they are accessed. $\mathbf{m}$ has k l-bit entries, where k determines the maximum number of array accesses. $\mathcal{C}$ secret-shares $\mathbf{m}$, and appends $[\![\mathbf{D}]\!]_\mathcal{B} := [\![\mathbf{D}]\!]_\mathcal{B} || [\![\mathbf{m}]\!]_\mathcal{B}$. Now $\mathcal{C}$ draws a random permutation $\pi : [n+k] \to [n+k]$ and permutes $[\![\mathbf{D}]\!]_\mathcal{B}$ according to π. $\mathcal{C}$ also uniformly samples two arrays $\mathbf{s}_0, \mathbf{s}_1$ of k l-bit entries, which will help $\mathcal{A}$ obliviously retrieve the message corresponding to either read or write operation during access. $\mathcal{C}$ then sends the masked and permuted $[\![\mathbf{D}]\!]_\mathcal{B}$ along with the masks $\mathbf{m}$, $\mathbf{s}_0$, and $\mathbf{s}_1$ to $\mathcal{B}$. $\mathcal{B}$ stores them for Π-access and additionally sets a counter $q := 0$ that counts the number of accesses.

$\mathcal{C}$ next generates and sends randomness to $\mathcal{A}$ that will enable it to construct its data structures. $\mathcal{C}$ already generated $[\![\mathbf{r}]\!]_\mathcal{A}$, $[\![\mathbf{m}]\!]_\mathcal{A}$, and π. He also uniformly samples k-bit $\mathbf{f}$ and constructs $\mathbf{s}_f$ such that for all $i \in [k]$ it contains $\mathbf{s}_{0,i}$ or $\mathbf{s}_{1,i}$ depending on $\mathbf{f}_i$. $\mathcal{C}$ then sends $[\![\mathbf{r}]\!]_\mathcal{A}, [\![\mathbf{m}]\!]_\mathcal{A}, \pi, \mathbf{f}, \mathbf{s}_f$ to $\mathcal{A}$.

$\mathcal{A}$ now constructs its data structures. First, she masks her share of $\mathbf{D}$ with $\mathbf{r}$, i.e. computes $[\![\mathbf{D}]\!]_\mathcal{A} := [\![\mathbf{D}]\!]_\mathcal{A} \oplus \mathbf{r}$, appends $[\![\mathbf{D}]\!]_\mathcal{A} := [\![\mathbf{D}]\!]_\mathcal{A} || [\![\mathbf{m}]\!]_\mathcal{A}$, and permutes $[\![\mathbf{D}]\!]_\mathcal{A}$ according to π. Then she computes a position map $\mathbf{pos}$ that tracks the position of the original n entries across accesses by setting $\mathbf{pos}_i := \pi_i$ for all $i \in [n]$. $\mathcal{A}$ stores $[\![\mathbf{D}]\!]_\mathcal{A}, [\![\mathbf{m}]\!]_\mathcal{A}, \pi, \mathbf{pos}, \mathbf{f}, \mathbf{s}_f$ along with a counter $q := 0$. As for $\mathcal{B}$, q tracks the number of accesses.

Optimizing Π-init *by Sending Randomness via Seeds.* For array $\mathbf{d}$ of n l-bit entries and k accesses, Π-init communicates $4n + 6k$ l-bit array entries, k bits, and a permutation (transferred as an array of length $n+k$). Communication can be reduced to $2n + 2k$ l-bit array entries and 7κ bits by sending randomness via short κ-bit pseudo-random seeds rather than large arrays, and locally expanding them with a pseudo-random generator[1]. In Π-init, this technique can be used when sending secret-shared arrays (i.e. send one of the secret shares as a seed), random arrays, and a permutation.

One must take care when using this optimization that the resulting protocol remains simulatable. A subtle technical issue here is that for modularity, we present and prove secure standalone Π-SOCS-ORAM, whose output is *shares* of the returned values. Because shares are explicit output of the parties, simulating above optimized protocol would require that the PRG output matches the fixed shares of the output. The solution is either to use programmable primitives (such as programmable random oracle), or to consider the complete MPC problem, where the wire shares are not part of the output.

[1] We did not implement this optimization as our focus was on efficient Π-access.

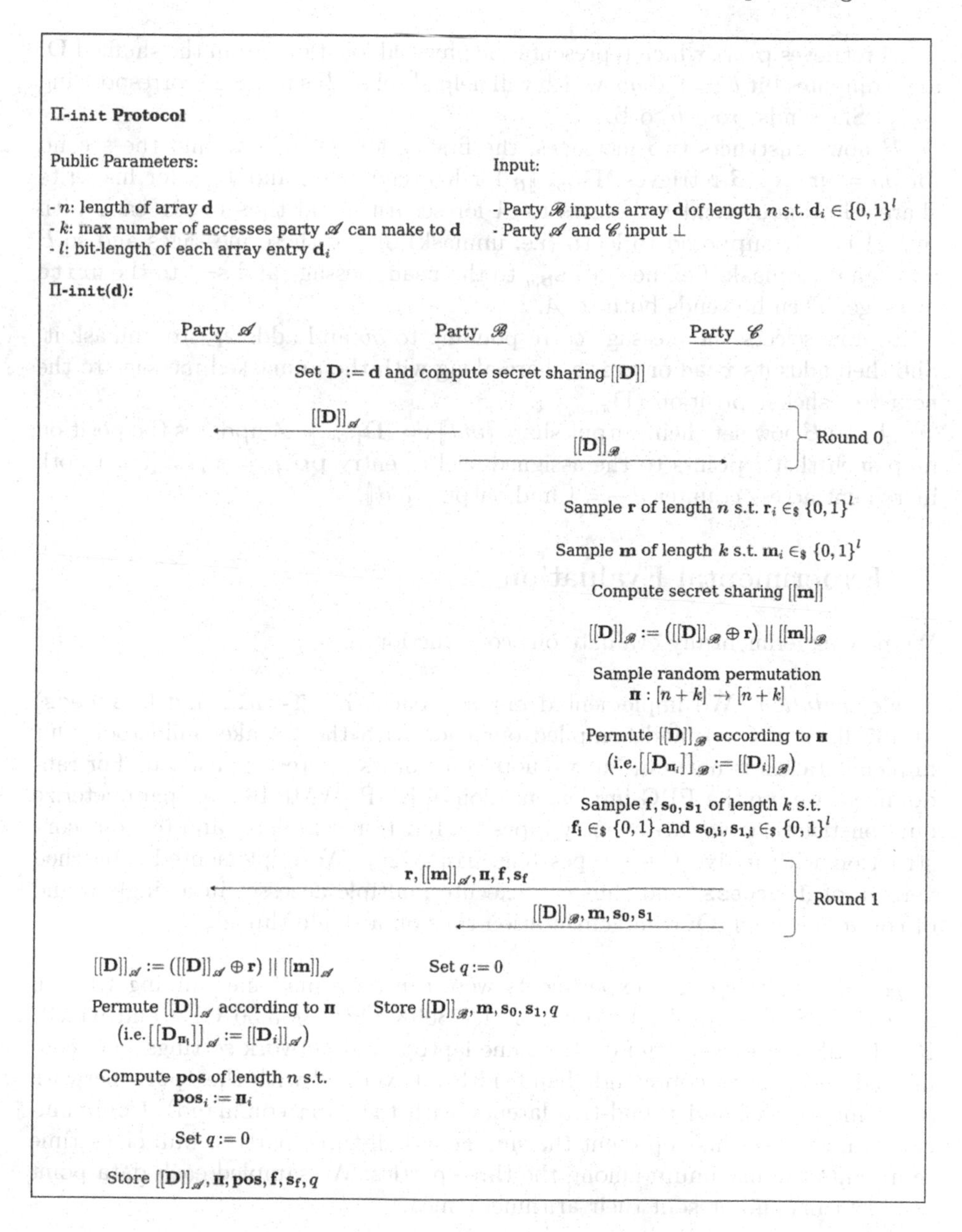

Fig. 2. `Π-init` is a subroutine of Π-SOCS-ORAM.

`Π-access`. `Π-access` accesses $\mathbf{D}$ (see `access` in Fig. 1). It is a 2-party protocol, run between $\mathcal{A}$ and $\mathcal{B}$, where $\mathcal{A}$ requests `read` or `write` to $\mathbf{D}_i$.

Recall that $\mathcal{A}$ inputs index i and operation op (`read` or `write`). Both input a sharing $[\![x]\!]$. $[\![x]\!]$ is input even if $op =$ `read` because $\mathcal{B}$ does not know op.

$\mathcal{A}$ retrieves $\mathbf{pos}_i$, which represents the physical location of i in the shuffled $\mathbf{D}$, and computes bit $b := \mathbf{f}_q \oplus op$, which will help $\mathcal{A}$ select $\mathcal{B}$'s message corresponding to op. She sends $\mathbf{pos}_i, b$ to $\mathcal{B}$.

$\mathcal{B}$ now constructs two messages: the first is for $op = \texttt{read}$ and the second for $op = \texttt{write}$. $\mathcal{B}$ retrieves $[\![\mathbf{D}_{\mathbf{pos}_i}]\!]_{\mathcal{B}}$ for his read share and $[\![x]\!]_{\mathcal{B}}$ for his write share. He cannot send his shares to $\mathcal{A}$ for security, and thus masks each with $\mathbf{m}_q$. $\mathcal{A}$ is only supposed to learn (i.e. unmask) one of these messages and so $\mathcal{B}$ adds another mask. I.e., he adds $\mathbf{s}_{\mathbf{b},q}$ to the `read` message and $\mathbf{s}_{\overline{\mathbf{b}},q}$ to the `write` message. Then he sends both to $\mathcal{A}$.

$\mathcal{A}$ now selects the message corresponding to op and adds $\mathbf{s}_{\mathbf{f},q}$ to unmask it. She then adds its `read` or `write` share along with the unmasked message to the next free shelter position $[\![\mathbf{D}_{\pi_{n+q}}]\!]_{\mathcal{A}}$.

$\mathcal{A}$ and $\mathcal{B}$ now set their output share $[\![out]\!] := [\![\mathbf{D}_{\mathbf{pos}_i}]\!]$. $\mathcal{A}$ updates the position map such that i points to the assigned shelter entry $\mathbf{pos}_i := \pi_{n+q}$. Then both increment access counter $q \mathrel{+}= 1$ and output $[\![out]\!]$.

7 Experimental Evaluation

We now experimentally evaluate our construction.

Implementation. We implemented our approach (i.e. `Π-init` and `Π-access`) in 437 lines of C++ and compiled our code with the CMake build tool. Our implementation is natural, but we note some of its interesting aspects. For randomness, we use the PRG implementation of EMP [WMK16]. We parameterize our construction over array entry types via function templates and test our construction with native C++ types (e.g. `uint32_t`). We implemented a batched version of `Π-access`, and thus can execute multiple accesses in a single round of communication. Our implementation runs on a single thread.

Experimental Setup. All experiments were run on a machine running Ubuntu 22.04.1 LTS with Intel(R) Core(TM) i7-7800X CPU @ 3.50 GHz and 64 GB RAM. All parties were run on the same laptop, and network settings were configured with the `tc` command (bandwidth was verified with the `iperf` network performance tool and round-trip latency with the `ping` command). Communication measurements represent the sum across all three parties; wall-clock time represents the maximum among the three parties. We sampled each data point over 10 runs and present their arithmetic mean.

Experiments. We performed and report on two experiments. The first evaluates our initialization protocol `Π-init` (see Sect. 7.1) while the second evaluates our access protocol `Π-access`(see Sect. 7.2). In both experiments, we measure communication and wall-clock time as a function of array size, which ranges from 2^{20} to 2^{30} with fixed $4B$ array entry size (i.e. `uint32_t`) and $4B$ position map entry size. We measure wall-clock time on 2 different simulated network settings:

Π-access Protocol

PARAMETERS (from Π-init):

- Parties $\mathcal{A}$ and $\mathcal{B}$ hold an array $[\![\mathbf{D}]\!]$ (processed in Π-init) of $(n+k)$ l-bit entries
- $\mathcal{A}$ and $\mathcal{B}$ access at most k elements; $q \in [k]$ is the current access number
- $\mathcal{A}$ holds position map $\mathbf{pos}$ of length n
- $\mathcal{A}$ holds a random permutation $\pi : [n+k] \to [n+k]$
- $\mathcal{B}$ holds two random arrays $\mathbf{s_0}, \mathbf{s_1}$ of k l-bit masks
- $\mathcal{A}$ holds random k-bit array $\mathbf{f}$ and array $\mathbf{s_f}$ of k l-bit masks
- $\mathcal{B}$ holds array $\mathbf{m}$ of k l-bit masks s.t. $\mathbf{m}_q := \mathbf{D}_{\pi_{n+q}}$

INPUT:

- $\mathcal{A}$ inputs op (read or write) and i s.t. $i \in [n]$; $\mathcal{A}$ and $\mathcal{B}$ input $[\![x]\!]$

Π-access$(op, i, [\![x]\!])$:

$\mathcal{A}$ sets $b := \mathbf{f}_q \oplus op$

$\mathcal{A}$ sends $\mathbf{pos}_i$, b to $\mathcal{B}$

$\mathcal{B}$ sets:

$$\begin{cases} md_0 := \mathbf{m}_q \oplus [\![\mathbf{D}_{\mathbf{pos}_i}]\!]_{\mathcal{B}} & \text{if } op = \mathtt{read} \quad // \ \mathbf{m}_q \text{ masks } [\![\mathbf{D}_{\mathbf{pos}_i}]\!]_{\mathcal{B}}. \ \mathcal{A} \text{ cannot learn } [\![\mathbf{D}_{\mathbf{pos}_i}]\!]_{\mathcal{B}} \\ md_1 := \mathbf{m}_q \oplus [\![x]\!]_{\mathcal{B}} & \text{if } op = \mathtt{write} \quad // \text{ Similarly, } \mathbf{m}_q \text{ masks } [\![x]\!]_{\mathcal{B}} \end{cases}$$

$\mathcal{B}$ sets: // This step ensures $\mathcal{A}$ learns only the message corresponding to op

$$\begin{cases} ms_0 := md_0 \oplus \mathbf{s}_{\mathbf{b},q} \\ ms_1 := md_1 \oplus \mathbf{s}_{\bar{\mathbf{b}},q} \end{cases}$$

$\mathcal{B}$ sends ms_0, ms_1 to $\mathcal{A}$

$\mathcal{A}$ unmasks exactly one of md_0 or md_1 depending on op:

$$md_{op} := \mathbf{s}_{\mathbf{f},q} \oplus \begin{cases} ms_0 & \text{if } op = \mathtt{read} \\ ms_1 & \text{if } op = \mathtt{write} \end{cases}$$

$\mathcal{A}$ sets:

$$tmp := md_{op} \oplus \begin{cases} [\![\mathbf{D}_{\mathbf{pos}_i}]\!]_{\mathcal{A}} & \text{if } op = \mathtt{read} \quad // \ tmp = \mathbf{D}_{\mathbf{pos}_i} \oplus \mathbf{m}_q \\ [\![x]\!]_{\mathcal{A}} & \text{if } op = \mathtt{write} \quad // \ tmp = x \oplus \mathbf{m}_q \end{cases}$$

$\mathcal{A}$ sets $[\![\mathbf{D}_{\pi_{n+q}}]\!]_{\mathcal{A}} := [\![\mathbf{D}_{\pi_{n+q}}]\!]_{\mathcal{A}} \oplus tmp$ // $\mathbf{D}_{\pi_{n+q}}$ now holds not permuted $\mathbf{D}_i$ (or x)

$\mathcal{A}$ and $\mathcal{B}$ set $[\![out]\!] := [\![\mathbf{D}_{\mathbf{pos}_i}]\!]$

$\mathcal{A}$ sets $\mathbf{pos}(i) := \pi_{n+q}$ // π_{n+q} is the new location of not permuted $\mathbf{D}_i$ (or x)

$\mathcal{A}$ and $\mathcal{B}$ increment $q \mathrel{+}= 1$

$\mathcal{A}$ and $\mathcal{B}$ return $[\![out]\!]$

Fig. 3. Π-access is a subroutine of Π-SOCS-ORAM.

1. **LAN 1:** A low latency 1 Gbps network with 2 ms round-trip latency.
2. **LAN 2:** An ultra low latency network also with 1 Gbps bandwidth but with 0.25 ms round-trip latency.

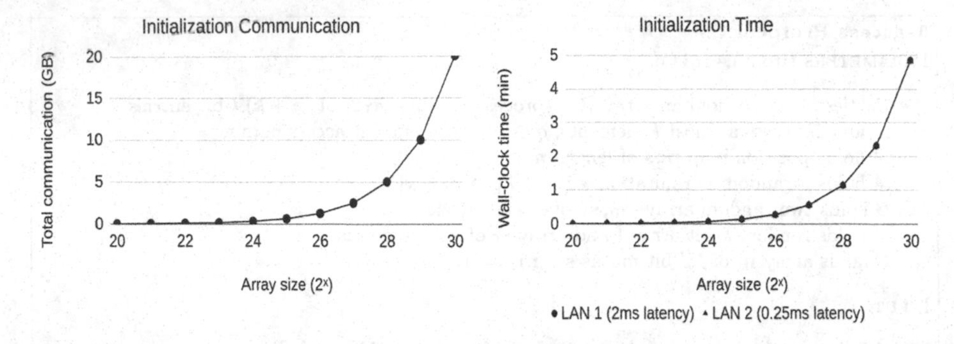

Fig. 4. Π-`init` performance. We fix the number of accesses to 2^{20} and plot the following metrics as functions of the *binary logarithm of the array size*: the overall communication (left) and the wall-clock time to complete the protocol on LAN 1 and LAN 2 (right). Note that the plots for LAN 1 and LAN 2 overlap.

7.1 Initialization Protocol

We first demonstrate that our Π-`init` is efficient for both small and large array sizes. In this experiment, we fix the number of array accesses to 2^{20}. Figure 4 plots the total communication and the wall-clock time in each network setting.

Discussion

- **Communication.** For an array of 2^{30} entries and for 2^{20} accesses, our implementation of Π-`init` communicates 20 GB (our plaintext array is 4 GB). For all runs of the initialization algorithm, our implementation matches exactly the number of bits incurred by our algorithm.
- **Wall-clock time.** For a large 2^{30}-entry array and for 2^{20} accesses, initialization runs for ≈4.8 min[2]. For a small 2^{20}-entry array with the same number of accesses, initialization takes ≈0.5 s. The wall-clock time is almost identical for both network settings as initialization consists of only 4 flows of communication (the first 2 and last 2 can be executed in parallel). Hence, initialization is not sensitive to latency.

7.2 Access Protocol

For our second experiment, we show that Π-`access` is fast and its performance is (almost) independent of array size. We show that wall-clock time is less than 0.019 ms per access on localhost for all runs and for all tested array sizes. Communication is 13B[3] per access.

[2] Sending 20 GB on 1 Gbps network takes ≈2.7 min. Remaining bottlenecks are generating permutation ≈71 s and permuting array according to a permutation ≈24 s.

[3] Note that this applies only to $4B$ array entries and $4B$ position map entries. The communication consists of sending two array entries ($8B$), a single entry in a position map ($4B$), and a single Boolean ($1B$).

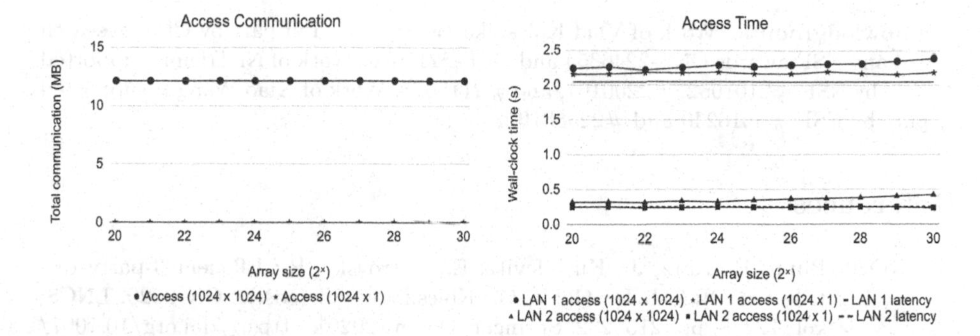

Fig. 5. Π-**access** performance. We consider two parameter regimes for the number of accesses: (1024 × 1024) and 1024 × 1. Then we plot the following metrics as functions of the *binary logarithm of the array size*: the overall communication (left) and the wall-clock time to complete the protocol on LAN 1 and LAN 2 (right). For the wall-clock time, we also plot cost because of latency on LAN 1 and LAN 2 to demonstrate our technique incurs almost no overhead over latency. Note that LAN 2 latency almost exactly overlaps with LAN 2 access (1024 × 1).

In this experiment, we consider 2 different parameter regimes for the number of accesses. The first (1024 × 1024) considers 1024 *sequential* accesses with each sequential access containing 1024 *batched* accesses. The second (1024 × 1) considers 1024 sequential accesses executed in batches of only 1 access. Figure 5 plots the total communication and the wall-clock time in each network setting.

Discussion.

- **Communication.** In `Π-access` communication is independent of array size[4]. In the (1024 × 1024) access number configuration, we use 12.125MB of communication. This matches exactly the theoretical communication in Fig. 3. In the (1024 × 1) setting, we communicate 13KB (i.e. 13B per access). Note that in this configuration we are losing 7 bits per access on the theoretical communication. This is because we send a single bit as one byte, which can be packaged with other bits in the batched setting.
- **Wall-clock time.** First note that in the (1024 × 1024) configuration and on a 2 ms round-trip latency network, `Π-access` takes ≈2.24 s on a 2^{20}-entry array (2.19 ms per 1024 parallel accesses) and ≈2.39 s on a 2^{30}-entry array (2.33 ms per 1024 parallel accesses). We believe the difference between the two experiments (and over the 2 ms latency baseline) is due to low-level costs such as effects of caching, system calls, interprocess communication, precision of `tc` timing, etc. From algorithmic perspective, the performed work is independent of array size.

[4] This is true as long as the array size stays small enough so that the entries in the position map need not increase (e.g. to 8B i.e. `uint64_t`).

Acknowledgments. Work of Vlad Kolesnikov is supported in part by Cisco research award and NSF awards CNS-2246354 and CCF-2217070. Work of Ni Trieu is supported in part by NSF #2101052, #2200161, and #2115075. Work of Xiao Wang is supported in part by NSF #2016240 and #2236819.

References

[BKKO20] Bunn, P., Katz, J., Kushilevitz, E., Ostrovsky, R.: Efficient 3-party distributed ORAM. In: Galdi, C., Kolesnikov, V. (eds.) SCN 2020. LNCS, vol. 12238, pp. 215–232. Springer, Cham (2020). https://doi.org/10.1007/978-3-030-57990-6_11

[CCMS19] Chan, T.-H.H., Chung, K.-M., Maggs, B.M., Shi, E.: Foundations of differentially oblivious algorithms. In: Chan, T.M. (ed.) 30th SODA, pp. 2448–2467. ACM-SIAM (2019)

[CKGS95] Chor, B., Kushilevitz, E., Goldreich, O., Sudan, M.: Private information retrieval. In: FOCS (1995)

[Ds17] Doerner, J., Shelat, A.: Scaling ORAM for secure computation. In: Thuraisingham, B.M., Evans, D., Malkin, T., Xu, D. (eds.) ACM CCS 2017, pp. 523–535. ACM Press (2017)

[FJKW15] Faber, S., Jarecki, S., Kentros, S., Wei, B.: Three-party ORAM for secure computation. In: Iwata, T., Cheon, J.H. (eds.) ASIACRYPT 2015. LNCS, vol. 9452, pp. 360–385. Springer, Heidelberg (2015). https://doi.org/10.1007/978-3-662-48797-6_16

[FNO22] Falk, B.H., Noble, D., Ostrovsky, R.: 3-party distributed ORAM from oblivious set membership. In: Galdi, C., Jarecki, S. (eds.) SCN 2022. LNCS, vol. 13409, pp. 437–461. Springer, Cham (2022). https://doi.org/10.1007/978-3-031-14791-3_19

[GIKM98] Gertner, Y., Ishai, Y., Kushilevitz, E., Malkin, T.: Protecting data privacy in private information retrieval schemes. In: 30th ACM STOC, pp. 151–160. ACM Press (1998)

[GKK+12] Gordon, S.D., et al.: Secure two-party computation in sublinear (amortized) time. In: Yu, T., Danezis, G., Gligor, V.D. (eds.) ACM CCS 2012, pp. 513–524. ACM Press (2012)

[GKW18] Gordon, S.D., Katz, J., Wang, X.: Simple and efficient two-server ORAM. In: Peyrin, T., Galbraith, S. (eds.) ASIACRYPT 2018, Part III. LNCS, vol. 11274, pp. 141–157. Springer, Cham (2018). https://doi.org/10.1007/978-3-030-03332-3_6

[GO96] Goldreich, O., Ostrovsky, R.: Software protection and simulation on oblivious RAMs. J. ACM **43**(3), 431–473 (1996)

[HKO22] Heath, D., Kolesnikov, V., Ostrovsky, R.: EpiGRAM: practical garbled RAM. In: Dunkelman, O., Dziembowski, S. (eds.) EUROCRYPT 2022, Part I. LNCS, vol. 13275, pp. 3–33. Springer, Cham (2022). https://doi.org/10.1007/978-3-031-06944-4_1

[JW18] Jarecki, S., Wei, B.: 3PC ORAM with low latency, low bandwidth, and fast batch retrieval. In: Preneel, B., Vercauteren, F. (eds.) ACNS 2018. LNCS, vol. 10892, pp. 360–378. Springer, Cham (2018). https://doi.org/10.1007/978-3-319-93387-0_19

[KM19] Kushilevitz, E., Mour, T.: Sub-logarithmic distributed oblivious RAM with small block size. In: Lin, D., Sako, K. (eds.) PKC 2019, Part I. LNCS, vol. 11442, pp. 3–33. Springer, Cham (2019). https://doi.org/10.1007/978-3-030-17253-4_1

[LO13] Lu, S., Ostrovsky, R.: How to garble RAM programs? In: Johansson, T., Nguyen, P.Q. (eds.) EUROCRYPT 2013. LNCS, vol. 7881, pp. 719–734. Springer, Heidelberg (2013). https://doi.org/10.1007/978-3-642-38348-9_42

[OS97] Ostrovsky, R., Shoup, V.: Private information storage (extended abstract). In: 29th ACM STOC, pp. 294–303. ACM Press (1997)

[PKV+14] Pappas, V., et al.: Blind seer: a scalable private DBMS. In: 2014 IEEE Symposium on Security and Privacy, pp. 359–374. IEEE Computer Society Press (2014)

[PS06] Porras, P., Shmatikov, V.: Large-scale collection and sanitization of network security data: risks and challenges. NSPW (2006)

[SCSL11] Shi, E., Chan, T.-H.H., Stefanov, E., Li, M.: Oblivious RAM with $O((\log N)^3)$ worst-case cost. In: Lee, D.H., Wang, X. (eds.) ASIACRYPT 2011. LNCS, vol. 7073, pp. 197–214. Springer, Heidelberg (2011). https://doi.org/10.1007/978-3-642-25385-0_11

[SSS12] Stefanov, E., Shi, E., Song, D.X.: Towards practical oblivious RAM. In: NDSS 2012. The Internet Society (2012)

[WCS15] Wang, X., Chan, T.-H.H., Shi, E.: Circuit ORAM: on tightness of the Goldreich-Ostrovsky lower bound. In: Ray, I., Li, N., Kruegel, C. (eds.) ACM CCS 2015, pp. 850–861. ACM Press (2015)

[WHC+14] Wang, X.S., Huang, Y., Chan, T.-H.H., Shelat, A., Shi, E.: SCORAM: oblivious RAM for secure computation. In: Ahn, G.J., Yung, M., Li, N. (eds.) ACM CCS 2014, pp. 191–202. ACM Press (2014)

[WMK16] Wang, X., Malozemoff, A.J., Katz, J.: EMP-toolkit: efficient MultiParty computation toolkit (2016). https://github.com/emp-toolkit

[ZWR+16] Zahur, S., et al.: Revisiting square-root ORAM: efficient random access in multi-party computation. In: 2016 IEEE Symposium on Security and Privacy, pp. 218–234. IEEE Computer Society Press (2016)

Improving Physical Layer Security of Ground Stations Against GEO Satellite Spoofing Attacks

Rajnish Kumar(✉) and Shlomi Arnon

Ben-Gurion University of the Negev, Be'er Sheva, Israel
rajnish@post.bgu.ac.il, shlomi@bgu.ac.il

Abstract. The integration of satellite and terrestrial communication systems has the potential to revolutionize worldwide services, enabling ubiquitous and dependable coverage. However, these systems' unique complexity and design also present opportunities for cyber attacks. The "new space revolution" makes it possible to launch many commercial satellites at a meager cost equipped with smart phased array beam having electronic steering capabilities. The wide spread use of space technologies creates opportunities for cyber attackers to hijack satellites and direct their beams toward specific geographic areas, posing a threat to ground stations and other terrestrial networks. To counter this problem, we propose a cutting-edge signal processing algorithm that can authenticate signals transmitted by satellites to ground stations. Our innovative algorithm can differentiate between signals originating from different satellites by leveraging the spatio-temporal imprint on the signal as it travels through different atmospheric paths in the channel. Simulation studies show that the algorithm achieves a remarkable authentication rate of more than 95% for signals received by a ground station located in Beer Sheva, Israel, based on a control experiment involving two different GEO satellites, AMOS-17 and AsiaSat-7. By enhancing physical layer security, our algorithm can protect satellite and terrestrial communication systems against cyber-attacks, ensuring reliable and secure global communication services.

Keywords: Satellite Communication · Terrestrial Network · Physical layer security · Deep learning

1 Introduction

The integration of terrestrial, satellite and aerial platforms network will create a vertical heterogeneous network (VHetNet) providing ubiquitous and ultra-reliable communication services globally [1,2]. The VHetNets are expected to play a critical role in future wireless communication systems, particularly in 5G and beyond, where the demand for high-speed data services and low-latency applications are expected to increase significantly [3]. Satellites in the VHetNet will provide extended coverage to remote or hard-to-reach areas, such as rural, hilly and mountainous regions, deserts, oceans and remote locations where traditional backhaul solutions may not be feasible [4]. This can help extend the coverage of the VHetNet to areas that would otherwise be under-served, improving connectivity and bridging the digital divide [5].

S. Dolev et al. (Eds.): CSCML 2023, LNCS 13914, pp. 458–470, 2023.
https://doi.org/10.1007/978-3-031-34671-2_32

The satellite networks comprises the low-earth orbit (LEO), Medium earth-orbit (MEO) and Geosynchronous orbit (GEO). The LEO satellite are placed between the altitude of 400 to 2000 km km above the earth, while MEO is placed between 8000 and 20000 km and GEO at 36000 km [6]. The GEO satellites have wider footprint that cover larger area on earth while having higher latency. GEO satellites play critical role in providing many government and private consumer services. They are essential to provide various communication services including television and broadcasting, broadband internet, and navigation [7]. They are used for military and defense purposes, providing critical information for reconnaissance and surveillance operations, providing real-time information about potential threats and enabling military forces to respond quickly and effectively [8]. They are also used for earth observation purposes, such as monitoring climate change, and natural disasters like hurricanes, floods, and wildfires [9,10]. Thus, their role in the society has huge impact in our day-to-day life and therefore, satellite networks including GEO are expected to play significant roles in the VHetNet.

The dynamic and complex VHetNet due to its unique characteristics would be more vulnerable to cyberattacks [11]. This complexity and heterogeneity will create new attack surfaces and vulnerabilities that may not exist in traditional wireless networks [12]. The real-time mitigation of cyberattacks in such networks require many cross-layer solutions. Such solutions may include intrusion detection and prevention systems, firewalls, access control, encryption, and other security measures including physical layer security (PLS) [13,14]. The PLS leverages the physical properties of the wireless channel and wave propagation to provide security at the lowest layer of the network stack. The wireless channel is subject to various physical phenomena, such as fading, interference, and noise, which can be exploited to achieve PLS [15,16]. Such cross-layer security design incorporating PLS and upper layer authentication will improve the robustness of wireless communication system against spoofing, eavesdropping and jamming [17].

Many recent events have brought forth the issue of satellite based cyber attacks [18–20]. In [21], the secrecy performance of hybrid satellite-terrestrial network has been studied using amplify-and-forward and decode-and-forward cooperative protocols. In [22], authors have provided a precoding algorithm that optimize the minimum secrecy capacity to secure a satellite downlink. In [14], relay selection and power allocation is employed to minimize the secrecy outage probability. In [23], cooperative jamming using relays are utilized to maximize the secrecy rate with power constraints on the relays for a hybrid satellite-terrestrial relay network. In [24], a robust beamforming is proposed to improve the physical layer security of the downlink of a multibeam satellite. In [25], artificial noise is used to improve the security of a multi-relay hybrid satellite-terrestrial systems.

With the modern technological advancements, satellites itself can used to launch a spoofing attack on the ground station networks [26]. The modern satellites are equipped with phased array antennas that are capable of electronic steering of the beams in a certain direction and thereby capable of focusing a particular geographical area [27]. These satellites having multi-beam antennas and adaptive power control mechanism can tune its Effective isotropic radiated power (EIRP) and thus can mimic as a legitimate satellite transmitter intended to provide the services. The signal-to-noise ratio (SNR) as well as frequency of the incoming signal from an illegitimate satellite can be tuned to match the SNR and frequency of the signal being received from a legitimate satellite. Thus, the ground station (GS) receiver can be spoofed to lock onto a signal being received from a illegitimate satellite mimicking the legitimate satellite.

To overcome this challenge, in this study, we propose a novel signal processing algorithm that will be able to differentiate between the two signals being received from an legitimate and illegitimate satellite. The algorithm is based on the taking advantage from the spatio-temporal characteristics of the signal being received at the GS. The signal from two different GEO satellites will traverse different spatial paths in the atmospheric channel and hence will have the imprint of the spatio-temporal characteristics of the channel. Such saptio-temporal variations of the channel cause rapid fluctuations in the received SNR at the GS antenna. We extract such rapid fluctuations from the SNR at the GS using a data-adaptive empirical mode decomposition (EMD). Furthermore, the output of EMD is then provided as input to a deep learning algorithm that will differentiate between the source of the signal i.e. legitimate or illegitimate satellite. A very high authentication rate is achieved using our proposed algorithm for a GS located at Beer Sheva, Israel receiving signals from two different GEO satellite namely AMOS-17 and AsiaSat-7. The scenario under consideration is depicted in Fig. 1.

2 SNR Modelling

In this section, we model the SNR of a signal at the GS being received from a GEO satellite at an elevation angle θ_0. The propagation of electromagnetic wave through the gaseous atmosphere causes the absorption loss in the satellite channel. The major gases responsible for such losses include water vapor and oxygen. The signal absorption loss can be found using the ITU recommendation that is valid from 1–1000 GHz. The loss at a given elevation angle θ_0 is expressed as [28]

$$L_A(\theta_0) = \sum_{i=1}^{N} \gamma_i d_i \tag{1}$$

where, N is the number of homogeneous layers in which the atmosphere is divided and γ_i and d_i denotes the specific attenuation (dB/km) and distance through each of these layers respectively. The specific attenuation is dependent on the gaseous constituents and atmospheric parameters i.e. pressure, temperature and

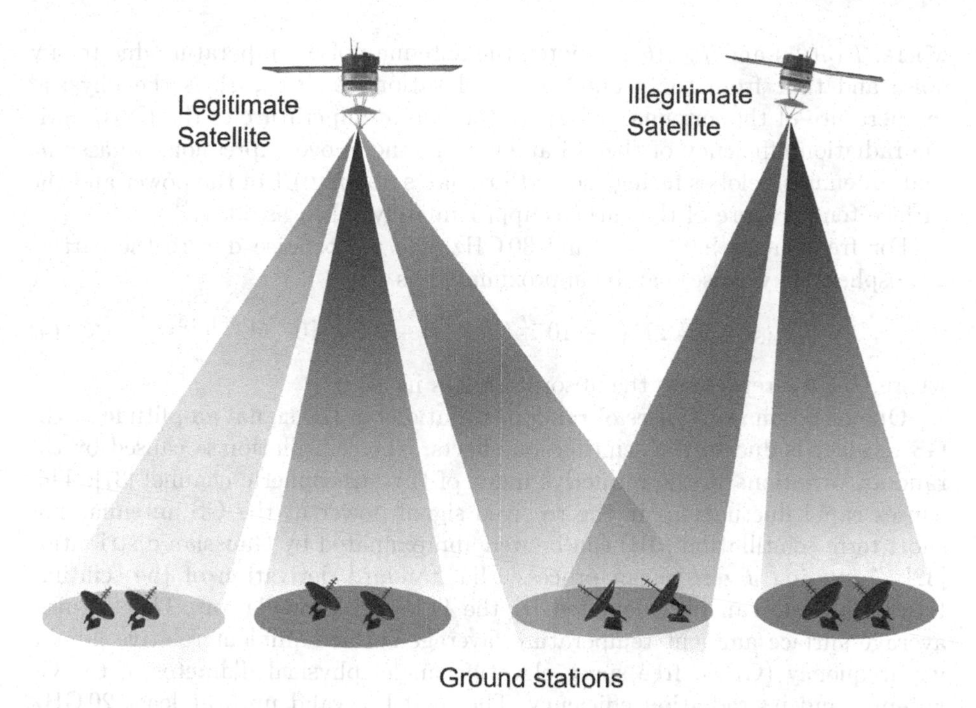

Fig. 1. Scenario showing a satellite based spoofing attack on the ground station by steering the beam towards the coverage area of a legitimate satellite

water vapor density. These atmospheric parameters are stochastic in nature that will cause a time-varying nature of absorption loss at the GS. The distance travelled through the atmosphere is also dependent on the refractive index of the medium that can be determined using the ITU recommendation. The refractive index too will show stochastic variations.

In addition, the free-space path loss will be given by [15]

$$L_{FS}(\theta_0) = \left[\frac{4\pi}{\lambda}(-R_e \sin\theta_0 + \sqrt{(R_e+h)^2 - R_e^2\cos^2\theta_0})\right]^2 \tag{2}$$

where, λ is the wavelength, h is the satellite altitude and R_e is the radius of the earth.

Besides signal losses, there will noise caused by the atmosphere as well as electronic circuit connected to the antenna. The sources of noise will include the sky noise, the noise from the ground surface, antenna physical temperature and noise due to low noise block (LNB) down-converter connected to the GS antenna terminal. The equivalent system noise temperature at the output terminal of the antenna will be expressed as [29]

$$T_S(\theta_0) = \eta_R[T_{AS}(\theta_0) + 0.1 \times T_{AG}(\theta_0)] + (1-\eta_R)T_p + T_{rx} \tag{3}$$

where, $T_{AS}(\theta_0)$ and $T_{AG}(\theta_0)$ denoted the antenna noise temperature due to sky noise and that from the ground at an elevation angle θ_0, T_p is the physical temperature of the antenna and T_{rx} is the noise temperature of the LNB, η_R is the radiation efficiency of the GS antenna. In the above expression, we assume that antenna sidelobes facing the earth radiates about 10% of the power and the surface temperature of the earth is approximately taken as 290 K.

For frequencies between 2 and 30 GHz, the radio noise due to the earth's atmosphere (sky noise) can be approximated as [30]

$$T_{AS}(\theta_0) = 275(1 - 10^{-L_A(\theta_0)/10}) + 2.7 \times 10^{-L_A(\theta_0)/10} \tag{4}$$

where, $L_A(\theta_0)$ represents the absorption loss in dB.

One of the major causes of random variations in the signal amplitude at the GS receiver is due to the scintillation effects. The scintillation is caused by the random variations in the refractive index of the atmospheric channel [31]. This causes rapid fluctuations in the received signal power at the GS antenna. The short term scintillation (dB) can be well approximated by Gaussian distribution [32]. Assuming a zero mean process, the standard derivation of the scintillation amplitude can be calculated by the ITU recommendation. This requires average surface ambient temperature, average surface ambient relative humidity, frequency (GHz), free space elevation angle, physical diameter of the GS antenna and its radiation efficiency. The model is valid upto at least 20 GHz. The standard deviation (dB) of the scintillation amplitude fluctuation is given by [33]

$$\sigma = \sigma_{ref} f^{7/12} \frac{g(x)}{(\sin\theta)^{1.2}} \tag{5}$$

where, σ_{ref} is the standard deviation of reference signal amplitude, f is frequency in GHz, $g(x)$ is antenna averaging factor. The computation of these parameters are provided in ITU P. 618-13.

Now, we can express the SNR at the output terminal of GS antenna as [34]

$$SNR(\theta_0) = \frac{\eta_T \eta_R G_T G_R \sigma_x^2}{kT_S(\theta_0) B L_{FS}(\theta_0) L_A(\theta_0)} \tag{6}$$

where, η_T and η_R are the radiation efficiency of the transmit and receive antennas respectively, G_T and G_R are the gain of transmit and receive antennas respectively, k is the Boltzmann constant, B is the bandwidth, σ_x^2 is the scintillation power. The antenna gains has been modelled as

$$G = (\frac{\pi D}{\lambda})^2 \tag{7}$$

where, D is the diameter of the parabolic reflector.

3 Proposed Algorithm

The algorithm is based on extracting the rapid fluctuations caused by the channel that is later processed by a deep learning (DL) classifier. As discussed earlier, the rapid fluctuations in the signal is caused by the stochastic atmosphere. The major contribution comes from the scintillation effects. Additionally, there will be change in antenna temperature due to the time varying gaseous attenuation [35]. Such variations will depend upon the elevation angle at which the signal arrives at the GS and the spatio-temporal properties of the channel. These rapid fluctuations in the signal is caused by the channel and hence beyond the control of the transmitter. Therefore, we extract such variations from the received SNR samples.

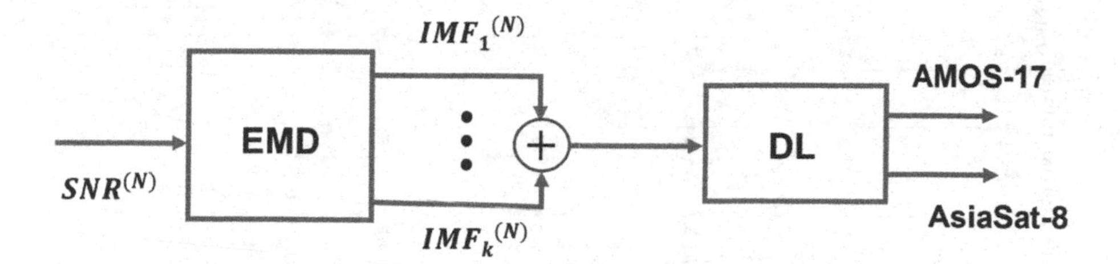

Fig. 2. Proposed signal processing algorithm to authenticate the signal being received at the ground station receiver antenna

The variations in the SNR (dB) will be caused by the signal power and noise power. While the variations in the signal power is very slowly varying, the noise power caused by the channel vary rapidly. So we can decompose the SNR (dB) in many high pass and low pass signals and filter out the slowly varying signal term. As the signal power can be tuned to match any given SNR by tuning the EIRP of a satellite, we will not consider the slowly varying signal term. The decomposition of the SNR time samples is done by using empirical mode decomposition (EMD). EMD is an adaptive data-driven multi-resolution method to separate the signal into various terms of different resolutions. EMD is widely used for the analysis of nonlinear and non-stationary data [36]. It considers the signal as fast oscillation superimposed on slow varying signal. After extracting the fast varying component, it again repeats the process until a certain criteria is reached e.g. maximum number of IMF.

The proposed signal processing algorithm is shown in Fig. 2. Initially, we process the time samples of the received signal-to-noise ratio (SNR) through EMD, which breaks it down into multiple intrinsic mode functions (IMFs) and a residual component. These IMFs capture the fast variations of different frequencies induced by the channel on the SNR. Next, we sum all IMFs to obtain the rapid-varying component of the SNR while discarding the residual component. The motivation to discard the residual component is the reason that a satellite can always tune its signal power to match that of the other satellite. The residual component will represent the slowly varying signal power. The IMFs collectively capture the spatio-temporal imprint of the atmospheric channel on the SNR. The signal obtained after summing all IMF components is then passed through a deep learning classifier which identifies the specific satellite that transmitted the signal.

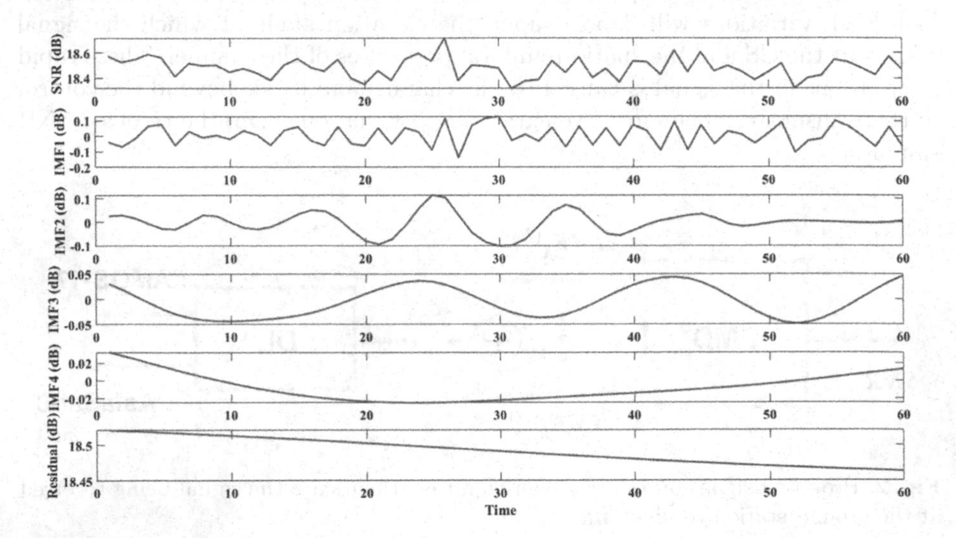

Fig. 3. EMD decomposition of the SNR (dB) time samples into four fast varying IMFs and a very slowly varying residual component

We show the EMD with an example of SNR samples received at the GS antenna in Fig. 3. The EMD decomposes SNR time series signal into a number of IMF and a residual function as shown in Fig. 3. After the decomposition, we sum all the IMFs and discard the residual component. The sum of all the IMF will be able to capture the random rapid variations caused in the SNR due to the stochastic atmospheric channel. As can be observed the residual component represents the high signal power term that is due to the very slow varying signal attenuation. The sum of the IMFs can be called as the noise spatio-temporal signature associated with the received SNR as it captures the rapid noise like variations in the signal caused by the stochastic atmospheric channel.

4 Numerical Results

The atmospheric model used in the simulation is mean annual global reference atmosphere provided by ITU [37]. The pressure, temperature and water vapor density as a function of height is calculated as given in the reference atmosphere. To generate the time samples of the SNR, we have assumed a Gaussian distribution of the atmospheric parameters in consideration. The average values of the pressure, temperature and water vapor density is taken as provided in the ITU Recommendation while a standard deviation of 10 hPa, 5 K and 0.5 g/m^3 are considered respectively. The refractive index is calculated as provided in the

ITU Recommendation [38]. A surface temperature of 30 °C and relative humidity of 63% is assumed for the calculation of scintillation effects. The short-term scintillation is assumed to be a zero mean Gaussian process with the standard deviation calculated as shown in Eq. (5) [39]. Other simulation parameters are mentioned in Table 1. The atmosphere is divided into $N = 922$ homogeneous layers and the specific attenuation and distance through each layer is calculated as provided in ITU Recommendation [28].

Table 1. Simulation Parameters

Definition	Symbol	Value
Satellite altitude	h	36000 km
Satellite antenna diameter		3 m
Satellite antenna efficiency	η_T	0.85
GS antenna diameter		1.2 m
GS antenna temperature	T_p	290 K
GS antenna efficiency	η_R	0.85
Atmospheric height	H	100 km
Ground surface temperature	T_G	290 K
LNB noise figure		0.8 dB
Frequency	f	12 GHz
System bandwidth	B	500 MHz
Earth radius	R_e	6371 km km

Our scenario considers AMOS-17 as the legitimate satellite and AsiaSat-7 attempting to spoof the GS antenna located at Beer Sheva, Israel. The GS antenna is located at latitude 31.2530 °C and longitude 34.7915 °C, resulting in an elevation angle of 48.8 °C for AMOS-17 and 7.8 °C for AsiaSat-7. We assume that AMOS-17 uses a fixed transmitter power of 10 W, while AsiaSat-7 uses varying power levels to match the SNR of the legitimate satellite at the GS. Specifically, we vary the power used by AsiaSat-7 at 5, 10, 15, 20, 25 and 30 W. For each power level, we create 1000 datasets for equal positive and negative examples, each having 60 discrete time samples of SNR.

The received signal-to-noise ratio (SNR) undergoes processing via empirical mode decomposition (EMD) before being fed into a deep learning (DL) algorithm for classification. The DL algorithm is designed as a feedforward classification network, with two fully connected layers, each consisting of 10 neurons. The output layer employs a softmax function and is fully connected with two outputs. The loss function for the DL algorithm is cross-entropy, and the activation function used is Rectified Linear Unit (ReLU). The input to the network has been standardized. The dataset is divided into 60: 40 ratio for training and test set.

The true positive (TP) denotes the correct prediction of the legitimate satellite and true negative (TN) denotes the correct rejection of illegitimate satellite. The false positive (FP) denotes incorrect rejection of legitimate satellite and false negative (FN) denotes incorrectly accepting the legitimate satellite. We can define the missed detection rate (MDR) and false alarm rate (FAR) as [40]

$$\text{FAR} = \frac{FN}{TP + FN} \tag{8}$$

$$\text{MDR} = \frac{FP}{TN + FP} \tag{9}$$

Using the FAR and MDR, we can find the authentication rate as [40]

$$\text{AR} = \frac{1}{2}[(1 - FAR) + (1 - MDR)] \tag{10}$$

The training loss of the DL classification network is shown in Fig. 4 for different power level used by the spoofing satellite. We observed that the training loss of the deep learning classifier quickly converged within only 16 iterations. Furthermore, the loss approached close to zero, indicating that the model had learned to accurately classify the incoming signals from the different satellites. This result suggests that the proposed signal processing algorithm, combined with the deep learning classifier, effectively handles the spatio-temporal imprint caused by the atmospheric channel on the SNR to accurately classify the legitimate source of received signal. Additionally, this suggests that the model has learnt meaningful representations of the signals, enabling it to make accurate predictions with a high degree of confidence.

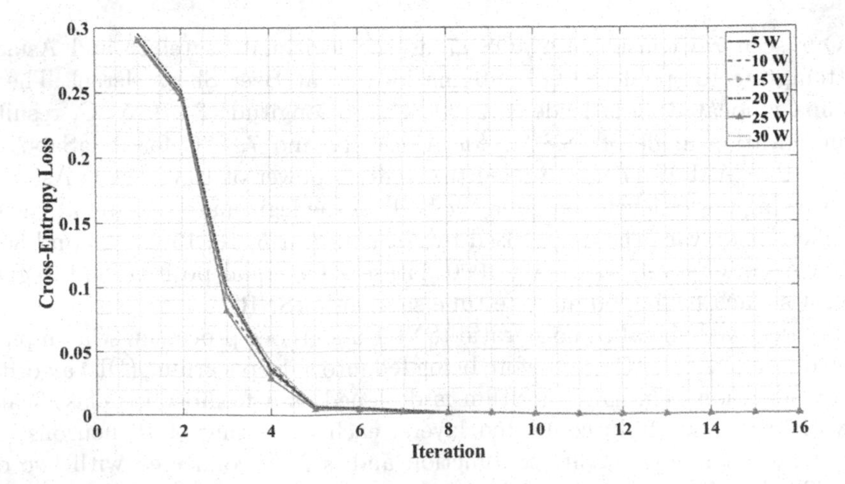

Fig. 4. Training loss of the DL classification network for different power levels of the spoofing satellite while the power of legitimate satellite is fixed at 10 W

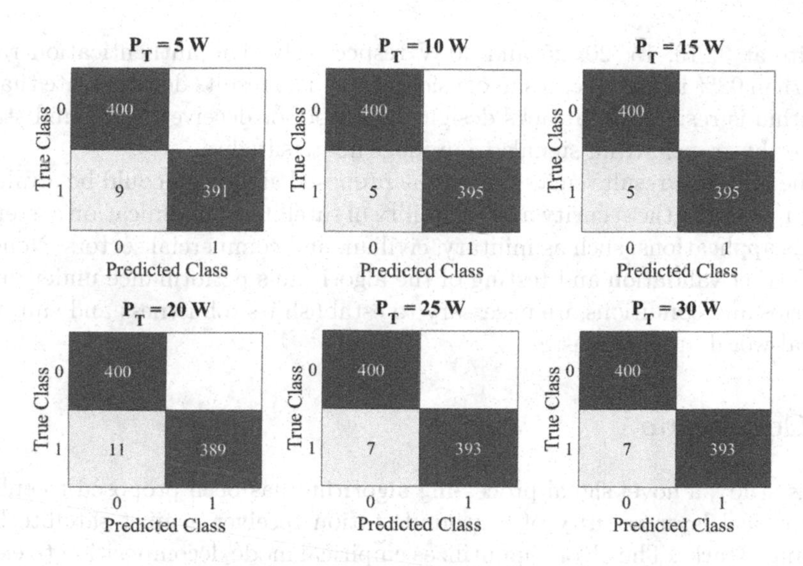

Fig. 5. Confusion matrix of the spoofing attack on GS as the illegitimate satellite varies different level of transmitted power while the power of legitimate satellite is fixed at 10 W

The performance of the DL over the test set is shown in Fig. 5. It is notable that despite using many power levels, it is very difficult for the illegitimate satellite to spoof the GS receiver. The confusion matrix plots for various power levels of the spoofing satellite show that the algorithm is robust against the GEO satellite based spoofing attacks at the terrestrial ground station. The false positive in all the cases is observed to be zero.

Table 2. Performance Metrics

P_T (W)	FAR	MDR	AR
5	0.0225	0	0.9888
10	0.0125	0	0.9938
15	0.0125	0	0.9938
20	0.0275	0	0.9863
25	0.0175	0	0.9913
30	0.0175	0	0.9913

In our case, for a transmit power of 10 W by AMOS-17 satellite, we show the FAR, MDR and AR in Table 2 as the transmit power of AsiaSat-7 varies. It can be seen that we obtain very high authentication rate using the proposed signal processing algorithm. The classification accuracy of 98.88%, 98.38%, 99.38%, 98.63%, 99.13%, and 99.13% is obtained for the transmit power of illegitimate satellite at 5, 10, 15, 20, 25 and 30 W respectively. The authentication rate is more than 98% in all the cases in consideration. The results demonstrate that the algorithm is resilient to attacks designed to spoof or deceive the ground station receiver by transmitting signals from illegitimate satellite.

The obtained results suggest that the proposed algorithm could be a valuable tool for ensuring the security and reliability of satellite communication systems in various applications, such as military, civilian, and commercial sectors. Nonetheless, further validation and testing of the algorithm's performance under diverse scenarios and conditions are necessary to establish its robustness and suitability for real-world applications.

5 Conclusion

In this study, a novel signal processing algorithm has been proposed to enhance the physical layer security of a ground station receiver against satellite based spoofing attacks. The algorithm utilizes empirical mode decomposition to extract fast varying component of the SNR time samples at the GS antenna that is then processed by a deep learning classifier to authenticate the legitimate and illegitimate satellite. Only the rapid fluctuations in the signal is considered while discarding the very slow varying signal power. It was shown that despite varying the transmitter power of the satellite transmitter, the illegitimate transmitter will not be able to spoof the GS antenna. As the authentication rate of more than 95% shows, the algorithm can provide a robust and efficient way to improve the physical layer security of the ground stations and terrestrial networks. In future, scenarios of many satellites as spoofing satellite can be considered in order to see the robustness and efficacy of the algorithm while maintaining a fixed power of the legitimate satellite. Furthermore, the testing and validation of the algorithm's performance with experimental data under various scenarios and conditions are necessary to establish its robustness and generalizability.

References

1. Alzenad, M., Yanikomeroglu, H.: Coverage and rate analysis for vertical heterogeneous networks (VHetNets). IEEE Trans. Wireless Commun. **18**(12), 5643–5657 (2019)
2. 3GPP. 3GPP TR 38.811 : Study on new radio (NR) to support non-terrestrial networks, September 2020
3. Ali, S., Abu-Samah, A., Abdullah, N.F., Kamal, N.L.M.: A review of 6g enabler: vertical heterogeneous network (v-HetNet). In: 2022 IEEE 20th Student Conference on Research and Development (SCOReD), pp. 180–183 (2022)
4. Darwish, T., Kurt, G.K., Yanikomeroglu, H., Senarath, G., Zhu, P.: A vision of self-evolving network management for future intelligent vertical HetNet. IEEE Wireless Commun. **28**(4), 96–105 (2021)
5. Dicandia, F.A., Fonseca, N.J.G., Bacco, M., Mugnaini, S., Genovesi, S.: Space-air-ground integrated 6g wireless communication networks: a review of antenna technologies and application scenarios. Sensors **22**(9), 3136 (2022)

6. Vemuri, S.S., Dappuri, B.: Walker-delta constellation design for LEO-based navigation using small satellites. In: 7th International Conference on Computing in Engineering & Technology (ICCET 2022), vol. 2022, pp. 250–253 (2022)
7. Abdu, T.S., Lagunas, E., Kisseleff, S., Chatzinotas, S.: Carrier and power assignment for flexible broadband geo satellite communications system. In: 2020 IEEE 31st Annual International Symposium on Personal, Indoor and Mobile Radio Communications, pp. 1–7 (2020)
8. Routray, S.K., Javali, A., Sahoo, A., Sharmila, K.P., Anand, S.: Military applications of satellite based IoT. In: 2020 Third International Conference on Smart Systems and Inventive Technology (ICSSIT), pp. 122–127 (2020)
9. Nemani, R., Lee, T., Kalluri, S., Ichii, K., Yeom, J.-M.: GeoNEX: earth observations from operational geostationary satellite systems. In: EGU General Assembly Conference Abstracts, p. 2463 (2020)
10. Kalluri, S., Zou, C.-Z., Flynn, L.E.: Applications of joint polar satellite system data and products for severe weather events and climate monitoring. In: 2021 IEEE International Geoscience and Remote Sensing Symposium IGARSS, pp. 695–698 (2021)
11. Guo, H., Li, J., Liu, J., Tian, N., Kato, N.: A survey on space-air-ground-sea integrated network security in 6G. IEEE Commun. Surv. Tutor. **24**(1), 53–87 (2022)
12. Yulei, W., Ma, Y., Dai, H.-N., Wang, H.: Deep learning for privacy preservation in autonomous moving platforms enhanced 5G heterogeneous networks. Comput. Netw. **185**, 107743 (2021)
13. Fratty, R., Saar, Y., Kumar, R., Arnon, S.: Random routing algorithm for enhancing the cybersecurity of LEO satellite networks. Electronics **12**(3), 518 (2023)
14. Li, J., Han, S., Tai, X., Gao, C., Zhang, Q.: Physical layer security enhancement for satellite communication among similar channels: relay selection and power allocation. IEEE Syst. J. **14**(1), 433–444 (2020)
15. Kumar, R., Arnon, S.: Enhancing cybersecurity of satellites at sub-THz bands. In: Dolev, S., Katz, J., Meisels, A. (eds.) CSCML 2022. LNCS, vol. 13301, pp. 356–365. Springer, Cham (2022). https://doi.org/10.1007/978-3-031-07689-3_26
16. Wei, Z., Masouros, C., Liu, F., Chatzinotas, S., Ottersten, B.: Energy- and cost-efficient physical layer security in the era of IoT: the role of interference. IEEE Commun. Mag. **58**(4), 81–87 (2020)
17. Mathur, S., et al.: Exploiting the physical layer for enhanced security [security and privacy in emerging wireless networks]. IEEE Wirel. Commun. **17**(5), 63–70 (2010)
18. Han, S., Li, J., Meng, W., Guizani, M., Sun, S.: Challenges of physical layer security in a satellite-terrestrial network. IEEE Netw. **36**(3), 98–104 (2022)
19. Falco, G.: When satellites attack: satellite-to-satellite cyber attack, defense and resilience (2020)
20. Van Camp, C., Peeters, W.: A world without satellite data as a result of a global cyber-attack. Space Policy **59**, 101458 (2022)
21. Bankey, V., Upadhyay, P.K.: Physical layer security of multiuser multirelay hybrid satellite-terrestrial relay networks. IEEE Trans. Veh. Technol. **68**(3), 2488–2501 (2019)
22. Schraml, M.G., Schwarz, R.T., Knopp, A.: Multiuser MIMO concept for physical layer security in multibeam satellite systems. IEEE Trans. Inf. Forensics Secur. **16**, 1670–1680 (2021)
23. Yan, S., Wang, X., Li, Z., Li, B., Fei, Z.: Cooperative jamming for physical layer security in hybrid satellite terrestrial relay networks. China Commun. **16**(12), 154–164 (2019)

24. Zhang, J., Lin, M., Ouyang, J., Zhu, W.-P., De Cola, T.: Robust beamforming for enhancing security in multibeam satellite systems. IEEE Commun. Lett. **25**(7), 2161–2165 (2021)
25. Huang, M., Gong, F., Zhang, N., Li, G., Qian, F.: Reliability and security performance analysis of hybrid satellite-terrestrial multi-relay systems with artificial noise. IEEE Access **9**, 34708–34721 (2021)
26. Jung, D.-H., Ryu, J.-G., Choi, J.: When satellites work as eavesdroppers. IEEE Trans. Inf. Forensics Secur. **17**, 2784–2799 (2022)
27. 3GPP. 3GPP TR 38.821 : Solutions for NR to support non-terrestrial networks (NTN), May 2021
28. ITU. ITU Recommendation: Attenuation by atmospheric gases and related effects, p. 676-12, August 2019
29. Bousquet, M., Bousquet, G., Sun, Z.: Satellite Communications Systems: Systems, Techniques and Technology, 5th edn. Wiley, Hoboken (2009)
30. ITU. ITU Recommendation: Radio noise, p. 372-14, August 2019
31. Kumar, R., Arnon, S.: Deep learning based scintillation prediction for satellite link using measured data. In: 2022 45th International Conference on Telecommunications and Signal Processing (TSP), pp. 246–249 (2022)
32. Ortgies, G.: Probability density function of amplitude scintillations. Electron. Lett. **21**(4), 141–142 (1985)
33. ITU. ITU Recommendation: Propagation data and prediction methods required for the design of earth-space telecommunication systems, p. 618-13, December 2017
34. Kumar, R., Arnon, S.: SNR optimization for LEO satellite at sub-THz frequencies. IEEE Trans. Antennas Propag. **70**(6), 4449–4458 (2022)
35. Mattioli, V., Marzano, F.S., Pierdicca, N., Capsoni, C., Martellucci, A.: Modeling and predicting sky-noise temperature of clear, cloudy, and rainy atmosphere from X- to W-band. IEEE Trans. Antennas Propag. **61**(7), 3859–3868 (2013)
36. Stallone, A., Cicone, A., Materassi, M.: New insights and best practices for the successful use of empirical mode decomposition, iterative filtering and derived algorithms. Sci. Rep. **10**(1), 15161 (2020)
37. ITU. ITU Recommendation: Reference standard atmospheres, p. 835-6, December 2017
38. ITU. ITU Recommendation: The radio refractive index: its formula and refractivity data, p. 453-14, August 2019
39. Garcia-del Pino, P., Riera, J.M., Benarroch, A.: Tropospheric scintillation with concurrent rain attenuation at 50 GHz in Madrid. IEEE Trans. Antennas Propag. **60**(3), 1578–1583 (2012)
40. Abdrabou, M., Gulliver, T.A.: Authentication for satellite communication systems using physical characteristics. IEEE Open J. Veh. Technol. **4**, 48–60 (2023)

Midgame Attacks and Defense Against Them

Donghoon Chang[1,2,3] and Moti Yung[4,5](✉)

[1] National Institute of Standards and Technology, Gaithersburg, MD, USA
donghoon.chang@nist.gov
[2] Strativia, Largo, MD, USA
[3] Department of Computer Science, Indraprastha Institute of Information Technology Delhi (IIIT-Delhi), Delhi, India
[4] Google Inc., Mountain View, USA
motiyung@gmail.com
[5] Department of Computer Science, Columbia University, New York, USA

Abstract. In this paper, we propose the *Midgame Security* attack model, where it is assumed that at some point in the middle of computation with a secret key, and after some secure work (typically but not necessarily initial one), the powerful adversary sees the entire internal state and attempts key recovery/forgery. This security model is motivated by a few trends: First and primarily, it may represent a model in which part of the computation is done in a possibly insecure environment (e.g., the emerging modes of cloud server delegation, hosting environment, general pc, the cloud, etc.), where the insecure environment performs the bulk of the work, after some initial or intermediate (relatively small amount of) work at a trusted location which holds the cryptographic keys (client, co-processor, trusted hardware with leakage countermeasures, an enclave in the cloud, etc.). Secondly, from a leakage perspective, the model represents a total leakage in the system at some point after some secured work has been done without leakage (perhaps at a different location). The model is novel (though, superficially, it has a flavor of forward security), and is most meaningful to demonstrate exposures of constructions where there is an obvious lengthy progress of computation (e.g., MACing or (Authenticated) Encrypting of long messages) which is done without the cryptographic keys present, and when we want short usage of keys (to minimize their exposure). In these cases, initially in secure periods the key may be blended into the state of the computation and an attacker task is to recover that key in spite of the blending from the leakage from the environment which never hold any key. We employ the new model to analyze numerous concrete cryptosystems and mainly find key recovery or forgery attacks. We first compare HMAC based on the SHA-3 finalists in this new midgame security model. One thing we show is that the domain extension of Keccak, called the sponge construction, is exposed in a HMAC-Keccak mode, and thus if there is an exposed state, the key is recoverable. Secondly, we analyze the midgame security of several popular message authentication codes, encryption, and authenticated-encryption (AE) schemes. We show that all known (authenticated) encryption schemes based on block ciphers,

S. Dolev et al. (Eds.): CSCML 2023, LNCS 13914, pp. 471–492, 2023.
https://doi.org/10.1007/978-3-031-34671-2_33

and that six ECRYPT stream ciphers out of the seven we examined are not secure against the midgame attacks. We note that from the point of view of risk analysis of overall systems, midgame attacks which may use a strong (but realizable) state exposure attack, may nevertheless open the door for new exposure deserving of considerations.

Keywords: Midgame Security · HMAC · Authenticated Encryption · Stream Cipher · Key Exposure · Leakage Attacks · Implementation Cost · Side Channel Countermeasures · Confidential Computing · Risk Management

"Hit 'em where they ain't" *William "Wee Willie" Keeler; MLB player in 1892–1910.*

1 Introduction

Our basic question in this work is the following: Assume a cryptographic computation has short periods protected from the adversary where secret keys are employed, and then it has another lengthy periods which do not hold or use the secret keys. In the latter periods the adversary may act in any given moment in the middle of the cryptographic operation (midgame). Can we protect the secret key (or currently encrypted/authenticated data under the key), when the entire current state information (which neve has the keys in it) is suddenly disclosed to the acting adversary while access to and use of the key are not available? Otherwise, can the attacker hit us with key recovery from the computation state in spite of the key not being there!?

Previously, in "forward security" works, the key is modelled as evolving and the goal is to evolve it in a one-way fashion so that the past is protected while moving forward (and forgetting past keys). Our game is somewhat different, where the secret key is fixed throughout, and after some short secured periods of the computation (say, in a secure co-processor) the computation is moved to a different (less secured but faster, say) environment (while the keys do not move to that environment), and then the entire state of the computation is leaked. Unlike forward security, if the scheme is secure the initial key is reusable for future needs. In addition, forward security research is about designing new secure schemes for key evolution, whereas our initial concentration is concrete existing symmetric key primitives which are examined under the new "midgame scenario."

Midgame security is related to delegation schemes where a cloud client may do some small work in a relatively secure environment/device (e.g., where security efforts and cost is invested and is costly in term of delay), but the bulk operation on long files can be then performed in the cloud servers which may not be as secure (or may not be known to be as secure as the initially acting device). The resulting cryptographic output is not sensitive to whether the environment is such that takes midgame attacks into account (and uses delegation) or whether the entire computation is done at one place (which is, btw, another difference from forward security where the key evolution is explicitly known).

From a practical point of view, it is important to minimize key exposure, e.g., the time of usage of the secret key, or to store the key in a secure place so that no adversary can see the key, or to delete the secret key information from memory or from any storage device right after the usage of the key is completed (as in, e.g., enclave/TEE based confidential computing). For example, if a secret key in a system remains exposed for a long time, the system may be insecure against memory-dump or cold-boot attacks [23]. Even if the secret key remains available for a short time period and also the key is encrypted or deleted, the system may still be insecure if the secret key can be computed from a leaked internal state. Note that maintaining the key in secure environments and operating in such an environment is typically more costly than operating in a general purpose computing environment (it requires countermeasures and better physical security, for example, such as TEE and leakage resistent operations, resp.). Hence, operating in the secured key environment is desired to be minimized.

Concretely, we introduce a security notion, called *Midgame Security*, in order to crystallize the security issues around the problem of the time of exposure of secret keys and the problem of making it difficult to obtain the key or key-related information from the total leakage of a typical internal state (especially after some secure processing, e.g. computing in a secure device has taken place).

We note that the midgame security is also somewhat different from leakage resilient security, because these works considered partial secret information or partial internal state which is disclosed to an adversary, or a computationally limited (e.g., log-space) adversary. In contrast, in the midgame security, we assume that at a point, a regular efficient computation (i.e., poly-time) adversary may get all the information of the internal state, called total leakage. It can be a penetration by a virus which leaks the internal state. We note that the notion of "remotely keyed encryption" by Blaze *et al.* [12,13] is also somewhat related to our notion, in the sense that it concerns a specific design allowing small work at a secure location and the rest in an insecure one (but it is a specific design, rather than being a security property of a general scheme). Further, the security notion of remotely keyed encryption is different from midgame security since the former considers security of future operation once the adversary is out of the system, whereas midgame security is about future (i.e., non-randomness-finding or distinguishing attack, key recovery and forgery attacks) and present security (i.e., backwardly-distinguishing attack or shortly backward attack, currently-used key recovery and forgery attacks).

The midgame security is further closely related to the secure implementation optimization in terms of time, power, and energy consumption (since secure implementations need sidechannel protection which is costly in terms of the above resources!). For example, if we base HMAC [3] on a hash function whose underlying compression function is one-way, then we need only protect some parts of and not the entire lengthy computation. Namely, only some actions need to be performed in a secure but slow device, which greatly improves the overall efficiency in terms of time, power, and energy consumption overhead. This basic idea was introduced by Hoerder *et al.* [25] (it is our opinion that the idea was long over due!). In [25], the flexibility of implementation of some of the SHA-3 candidates was also considered with some practical processors such as power-trust platform. In this paper, we will further investigate this with HMAC based on the SHA-3 finalists in detail. We will then show that many known (authenticated) encryption schemes including stream ciphers without a completely secure black-box implementation during their entire execution are not secure against midgame attacks, which means that secure implementation of encryption schemes against midgame attacks is very costly. Finally, we also show that some stream ciphers shown to be leakage resilient, are still secure against the midgame attack with their secure initial processing.

The midgame security notion gives rise to a new design criterion for cryptosystems: typically it is known that a primitive has to have overall, namely global, cryptographic hardness (i.e., one-wayness). Our conclusion will discuss how the new notion refines this criterion regarding steps of the primitive computation. This applies broadly, beyond the actual primitives we deal with.

As an important remark, let us note that the techniques we use are not the usual “complex distinguisher” type of cryptanalytic attacks (Differential, linear cryptanalysis, etc.) that the cryptographic community is used to and are traditionally involved techniques. But, we assert that a combination of state exposure (physical attack on the system level) and sometimes simple analysis which is *not scrutinized* by cryptanalytic papers may be the preferred way to attack cryptosystem deployments in the field (especially, perhaps, by powerful organizations), exploiting the lack of interest of the cryptanalytic community in attacks which do not involve overly sophisticated techniques. This justify paying attention to issues like the one presented here: While not claiming sophistication of the typical cryptanalysis level, from an overall risk analysis perspective our issues are worthy of attention!

Remark: when a system performs an operation which is one-way, for our initial scrutiny of systems in this work, we assume the one-way operation hides its entire preimage (i.e., we give designs the benefit of the doubt that their concrete one-way functions are perfect, while we are fully aware that further scrutiny of these assumptions is needed.)

Further note that the style of the presentation here is informal and intuitive, concentrating on attacks and lack of attacks on concrete systems. After the initial presentation in the 2014 real world cryptography symposium (https://realworldcrypto.wordpress.com/program/), some more formal framework oriented style [30] work on the subject has taken place [5,8].

1.1 Our Security Notions and Concrete Results

Let us next present the goal of security under the midgame attacks, and review our results in more details. The exposition is geared towards practical attacks, so we present the security specifications informally rather than as formal definitions with security parameter and probabilities (we believe the latter can be derivable from the former, but are not necessary for our treatment).

Important remark: We assume in the specification below that operations which involve the key directly (or some other well defined operation) at the start or at a few places in the computation are performed securely (i.e. by the secure element not accessible to the attacker). In case the adversary attacks during such secure operation it gets the prior state of the system (thus effectively it cannot attack during these periods). Below, an attack takes place at a natural period i which is always assumed to not to be a secured operation period.

Midgame Security in (Authenticated) Encryption Schemes. Let $M_1||....||M_t$ be a t-block message to be currently encrypted under a secret (session) key K in a system such as computer or a smartcard, etc. Suppose that at a point, $M_1||....||M_i||M_{i+1}||...||M_t$ would be securely replaced into $C_1||....||C_i||M_{i+1}||...||M_t$ using an (Authenticated) Encryption scheme, where C_i is the ciphertext corresponded to M_i. And finally, $M_1||....||M_t$ will be replaced with $C_1||....||C_t$. Suppose, during such encryption process, due to an unexpected incident, all the information stored in the system is disclosed to an adversary while M_i is being encrypted into C_i (period i of the computation). Informally, we say that the system is secure against *Midgame backward attack* on the (Authenticated) Encryption scheme if no adversary can decrypt or find non-randomness of any of $C_1||....||C_{i-1}$ when the adversary knows all the information stored in the system at the disclosure point. We say that the system is secure against *Midgame key recovery attack* on the (Authenticated) Encryption scheme if no adversary can obtain the secret key of the scheme when the adversary knows all the information stored in the system at the disclosure point during the encryption or decryption processes. We say that the system is secure against *Midgame forgery attack* on the (Authenticated) Encryption scheme if no adversary can forge (new challenge decryption/encryption) even when the adversary knows all the information stored in the system at the disclosure point during the encryption or decryption processes.

We will show that many known encryption schemes, including stream ciphers, as well as authenticated encryption schemes (including all the NIST-approved ones) fail to provide any of the above three midgame security requirements, unless the entire scheme is performed within a secure (black-box) implementation environment. For example, consider CTR mode of operation [16]. In order to encrypt a message under a key, the CTR encryption scheme should keep the secret key until it finishes completely encrypting the message. Without keeping the key to the end, there is no way to encrypt the message correctly. Therefore, if a system encrypts a message using the CTR mode, the key should be stored throughout somewhere, and if the entire state is handed over to the adversary, then it can easily decrypt all the encrypted information so far. A new authenticated-encryption scheme by Bertoni, Daemen, Peeters, and Assche [10], is design to operate without keeping its secret key to the end. But, still, the scheme is not secure in our attack model, because it is easy to compute the secret key from the internal state (unless the scheme is totally implemented within a black box, i.e., securely). There is one pseudo-random bit generator [4] which can be efficiently implemented and is secure against the midgame backward attack. Also, there are known leakage resilient stream ciphers proposed in [26,27] and we will show they are still secure against the midgame backward attacks.

Midgame Security in Message Authentication Code (MAC) Schemes. Let $M_1||....||M_t$ be a t-block message whose MAC value will be computed under a secret (session) key K in a system such as computer or a smartcard, etc. But, again, during the process of the MAC scheme, the entire information stored in the system is disclosed to an adversary while M_i is being processed. Informally, we say that the system is secure against *Midgame key-recovery attack* on the MAC scheme if no adversary can compute the key K when the adversary knows all the information stored in the system at a point. We say that the system is secure against *Midgame forgery attack* on the scheme if no adversary can forge a new MAC value, even when the adversary knows all the information stored in the system at the disclosure point.

We will show that many of known MAC schemes including hash-based MACs fail to provide the midgame key-recovery security. Unfortunately, an adversary can easily find the secret key K when several hash-based MACs are based on Keccak [11]. On the other hand, BLAKE [1], Grøstl [22], JH [32], and Skein [21] are still secure against the midgame key-recovery attack (recall that in our analysis here, being the first on the subject, we ideally assume that all one-way functions within the concrete designs of hash functions are ideal and hide their preimage completely). Also, we consider another hash-based MAC constructions and show that when based on Keccak it is not secure against the midgame forgery attack but when based on BLAKE, Grøstl, JH, and Skein, it is secure even against midgame forgery attack. Thus, our results are interesting since they present a new context (not considered by the SHA-3 competition) which separates the security of the SHA-3 finalists. (Note: since our setting was not considered earlier we do not claim anything w.r.t. the SHA-3 competition; our findings are to be considered from a pure scientific point of view). Also, MACs based on a block cipher such as CMAC [18] are not secure against the midgame attack because once the attacker knows any internal state of the MAC, he gets the fixed key of the underlying block cipher and generate all the subkeys from the key, and can, in turn, do whatever he wants.

2 Midgame Security Analysis of HMAC Based on the SHA-3 Finalists

In this section, we consider the midgame security of HMAC based on the SHA-3 finalists. In the midgame security model, the domain extension of Keccak, called the sponge construction, may provide poor efficiency for HMAC-Keccak with sidechannel countermeasures in terms of time, power, and energy consumption overhead for a long message, when compared with HMAC based on the domain extensions of the other four hash algorithms candidates (due to the need of securing large portion of the execution and hide it from our attacker). We first explain what HMAC is.

HMAC [3]. Let K be a n-bit key. We define $\overline{K} = K||0^{b-n}$ where b indicates the size of the message block of a hash algorithm H. For example, in cases of SHA-1 and SHA-256, the size of the message block is 512-bit. opad is formed by repeating the byte '0x36' as many times as needed to get a b-bit block, and ipad is defined similarly using the byte '0x5c'. Then, HMAC is defined as follows:

$$\mathrm{HMAC}_K(M) = H(\overline{K} \oplus \mathsf{opad}||H(\overline{K} \oplus \mathsf{ipad}||M)).$$

2.1 The Midgame Security of HMAC-Keccak

As shown in Fig. 1, if we know any internal state of the first hash call, we can compute the key K because f is efficiently invertible. More in detail, Keccak consists of five functions as follows.

The Permutation f of Keccak. Let $x, y \in \mathbb{Z}_5$ and $z \in \mathbb{Z}_{64}$. Let $a[x][y][z]$ represent each bit of 1600-bit by the values of x, y, z. The permutation f of Keccak, is an iterated permutation on 1600-bit, consisting of a sequence of 24 rounds $\mathbf{R}$. A round $\mathbf{R}=\iota \circ \chi \circ \pi \circ \rho \circ \theta$ consists of five steps:

$$\begin{aligned}
&\theta: a[x][y][z] \leftarrow a[x][y][z] \oplus \textstyle\bigoplus_{y'=0}^{4} a[x-1][y'][z] \oplus \bigoplus_{y'=0}^{4} a[x+1][y'][z-1]\\
&\rho: a[x][y][z] \leftarrow a[x][y][z-(t+1)(t+2)/2],\\
&\qquad \text{with } t \text{ satisfying } 0 \le t < 24 \text{ and } \begin{pmatrix} 0\,1 \\ 2\,3 \end{pmatrix}^t \begin{pmatrix} 1 \\ 0 \end{pmatrix} = \begin{pmatrix} x \\ y \end{pmatrix} \text{ in } \mathbf{GF}(5)^{2\times 2},\\
&\qquad \text{or } t=-1 \text{ if } x=y=0,\\
&\pi: \ a[x][y] \ \leftarrow a[x'][y'], \text{ with } \begin{pmatrix} x \\ y \end{pmatrix} = \begin{pmatrix} 0\,1 \\ 2\,3 \end{pmatrix} \begin{pmatrix} x' \\ y' \end{pmatrix},\\
&\chi: \ a[x] \ \leftarrow a[x] \oplus (a[x+1] \oplus 1)a[x+2],\\
&\iota: \ a \ \leftarrow a \oplus \mathbf{RC}[i_r],
\end{aligned}$$

where the round constants $\mathbf{RC}[i_r][0][0][2^j - 1] = \mathrm{rc}[j + 7i_r]$ for all $0 \le j \le \ell$ and $0 \le i_r \le 23$, and all other values of $\mathbf{RC}[i_r][x][y][z]$ are zero. The values $\mathrm{rc}[t] \in \mathbf{GF}(2)$ are defined as $\mathrm{rc}[t] = (x^t \bmod x^8 + x^6 + x^5 + x^4 + 1) \bmod x$ in $\mathbf{GF}(2)[x]$.

Among the five functions, ι, χ, π, ρ, and θ, it is clear that we can easily reverse the functions ρ, π, and ι, because the functions ρ and π just change each bit position and the function ι uses only one simple invertible operation $\oplus$. The remaining functions of θ and χ are invertible as well: In [9], the authors clearly showed how to construct their invertible functions which are efficient. Therefore, the Permutation f of Keccak is efficiently reversible. Based on this, for the secure sidechannel implementation of HMAC-Keccak, we argue that the entire computation should be protected as shown in Fig. 2, since otherwise an attacker can find the key K by knowing any state of the first call of Keccak during the HMAC-Keccak process.

2.2 The Midgame Security of HMAC-JH

Now, we consider the midgame security of HMAC-JH as shown in Fig. 3, where F_8 is defined in Fig. 4. Let us see Fig. 4. Once given an input message and

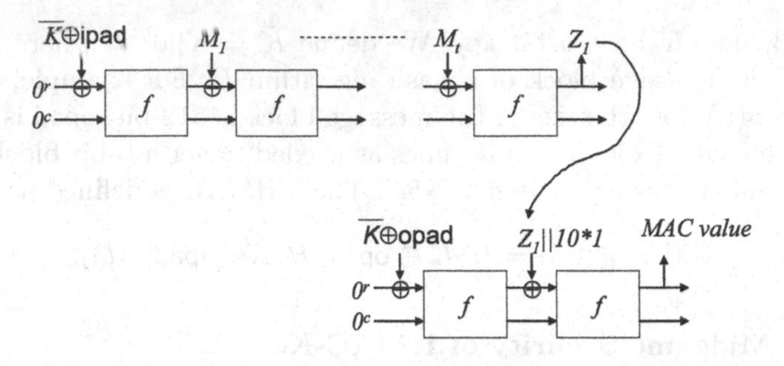

Fig. 1. HMAC-Keccak with one-block K

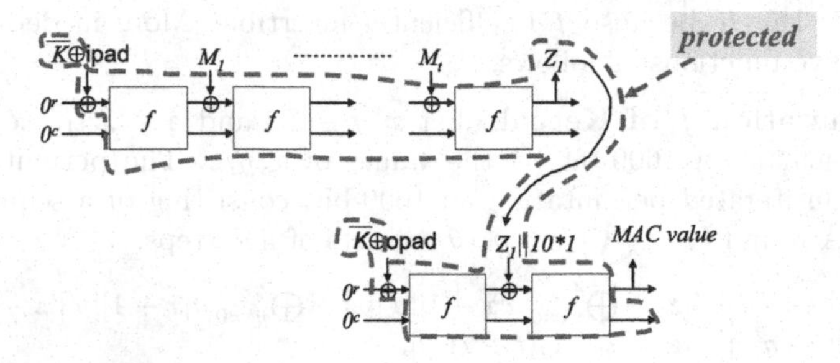

Fig. 2. Protection of HMAC-Keccak with one-block K

an output chaining values of F_8, we can compute easily the value of its input chaining value due to the invertibility of the underlying permutation E_8, because JH [32] uses simple invertible tables, S_0 and S_1 S-boxes, and invertible matrix L_8 and a position changing permutation P_8. In more detail, $L(A, B) = (5 \bullet A + 2 \bullet B, 2 \bullet A + B)$, where A and B are 4-bit words and the multiplication $\bullet$ is defined as the multiplication of binary polynomials modulo the irreducible polynomial $x^4 + x + 1$. So, we can easily define the inverse of L as $L^{-1}(C, D) = (1 \bullet C + 2 \bullet D, 2 \bullet C + 5 \bullet B)$. Thus, the Permutation E_8 of JH is efficiently reversible. Once an attacker knows the value of any internal state in the middle of the HMAC computation without directly knowing the key K, the attacker can backtrack up to the point of the output chaining value of the second F_8 computation, due to the invertibility of E_8. But, it is still difficult to compute the key K as shown in Fig. 5, because the key K is used as the input and output masking. This means that even if an attacker knows an internal state except the state of the second or the last two F_8 invocation of HMAC-JH, he cannot find the key and also cannot forge any message. Therefore, For the sidechannel-secure implementation of HMAC-JH, the protection of the second and the last two invocations of F_8 is enough for its security. Though it is not only a protection at the start of the computation, it is a much more local protection as in Fig. 6, when comparing to HMAC-Keccak (which requires to protect the entire computation). In this context, as shown in Fig. 7, for the sidechannel-secure implementation of HMAC-JH, once the countermeasures are designed, then we may optimize the *time*, *power*, and *energy* consumption overhead by turning ON or OFF the sidechannel countermeasures.

2.3 The Midgame Security of HMAC Based on the Other SHA-3 Finalists and Other Hash Algorithms

The compression functions of the other the SHA-3 finalists, BLAKE [1], Grøstl [22], and Skein [21], are one-way, because they use a simple forward operation. For SHA-1 and SHA-2, their compression functions also are one-way. (Recall we assumed these one-way function are ideal and conceal their pre-image for an initial design scrutiny purposes). Thus, we only need to protect the initial and the last computation of the HMAC, a fact which was already mentioned in [25].

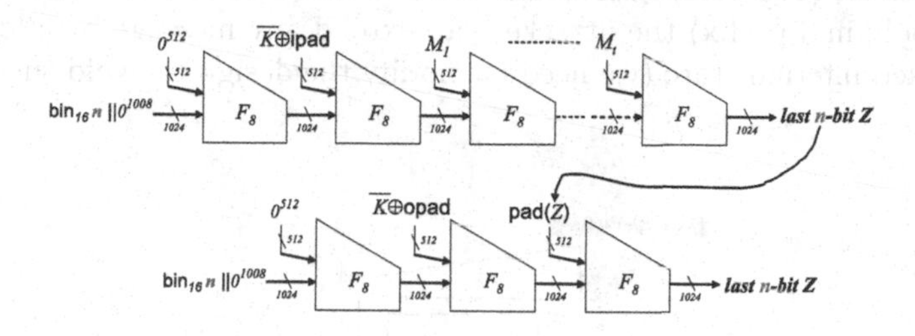

Fig. 3. HMAC-JH with one-block K

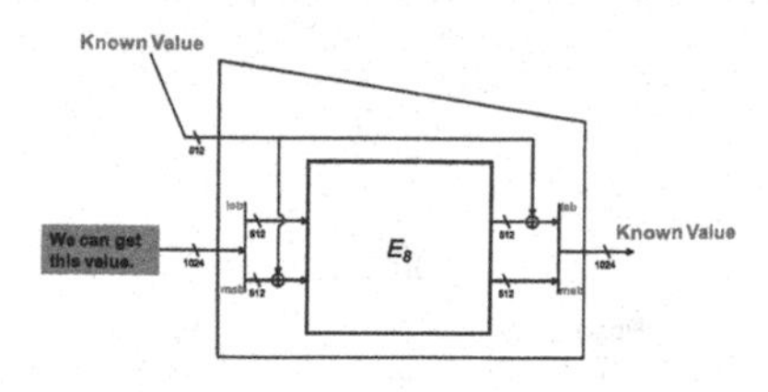

Fig. 4. Invertibility of F_8 of JH: E_8 is a permutation on 1024-bit strings.

3 Midgame Security of Other Hash Function-Based MAC Constructions

HMAC is one of MAC constructions based on a hash function. There are other simple MAC constructions such as a hash function with a prefix key or a hash function with a prefix and a suffix keys. In this section, we consider their security in the midgame security model.

3.1 Hash MAC with a Prefix Key

Let H be a hash function. Let K be a 128-bit secret key. Let $h = H(K||M)$ be the MAC value for a given message M, under the secret key K. All the final SHA-3 candidates, BLAKE, Grøstl, JH, Keccak, and Skein are designed to guarantee the security of such MAC construction, as long as the attack is allowed to know only the input message M, which is generated by the adversary, and its output h. However, in the midgame security model, an attacker can easily find a key from any internal state when H is Keccak. On the other hand, BLAKE, Grøstl, and Skein are still secure against the midgame "key recovery" attack if the computation of a function whose one of input is the key is protected (i.e., performed securely). But, we have a separation between the attacks and can easily observe that all of them are not secure against midgame "forgery" attack, because (due to the structure of the MAC and the fact that the key is involved only in a prefix) the attacker can extend any message by added blocks to the leaked internal state (we need to modify the design to avoid such message expansion).

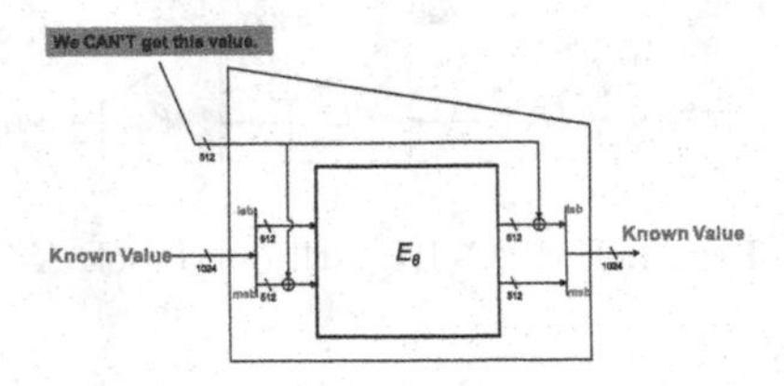

Fig. 5. Non-invertibility of F_8 of JH: E_8 is a permutation on 1024-bit strings.

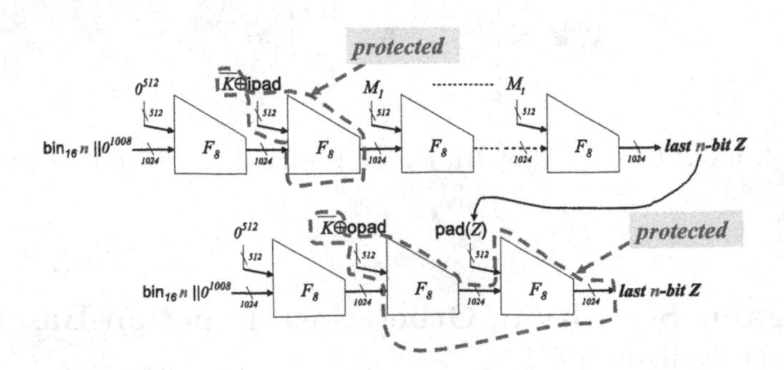

Fig. 6. Protection of HMAC-JH with one-block K

3.2 Hash MAC with a Prefix and Suffix Keys

Next we note that it is interesting to design a MAC construction based on a hash function in a way that the MAC construction based on BLAKE, Grøstl, JH, and Skein can provide a strong security against the midgame forgery attacks. For example, we can construct a MAC based on a hash function with a prefix and suffix keys like $MAC_K(M) = H(\mathsf{pad}(K||M||K))$, where H is a hash function and pad is an appropriate padding scheme (For detailed discussion on a padding scheme, refer to [28,33].). If H is one of BLAKE, Grøstl, JH, and Skein, then we can show their security as shown for the HMAC under the same condition of key operation being performed securely.

4 The Midgame Security Analysis of Stream Ciphers

In this section we examine the midgame security of some stream ciphers including ECRYPT stream ciphers.

4.1 The Midgame Security Analysis of ECRYPT Stream Ciphers

Here, we provide the security analysis for seven ECRYPT stream ciphers against the midgame backward attackers. Interestingly, six algorithms turned out not to be secure against the midgame backward attackers!

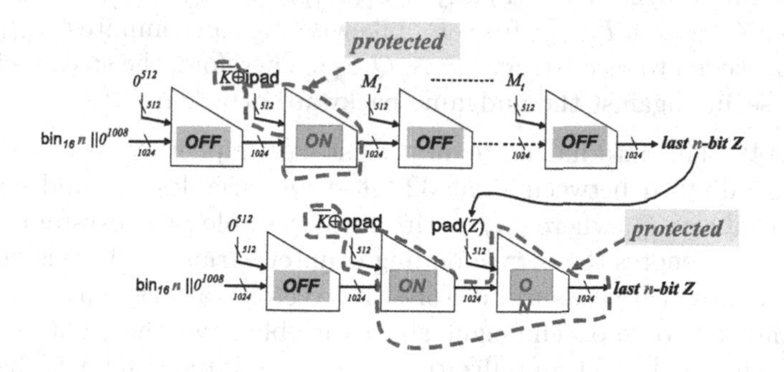

Fig. 7. Secure Implementation of HMAC-JH with one-block K

HC-128 [31]. The HC-128 algorithm is a software-efficient and synchronous stream cipher. The cipher makes use of a 128-bit key (K) and 128-bit initialization vector (IV); its secret state consists of two tables P and Q, each with 512 registers of 32 bits in length. Once P and Q are defined by K and IV, the key stream s_i's are defined as follows, where we don't need to explain what g_i's and h_i's are in below.

```
110     i = 0;
100     repeat until enough keystream bits are generated.
101     {
102         j = i mod 512;
103         if (i mod 1024) < 512
104         {
105             P[j] = P[j] + g1(P[j ⊟ 3], P[j ⊟ 10], P[j ⊟ 511]);
106             s_i = h1(P[j ⊟ 12]) ⊕ P[j];
107         }
108         else
109         {
110             Q[j] = Q[j] + g2(Q[j ⊟ 3], Q[j ⊟ 10], Q[j ⊟ 511]);
111             s_i = h2(Q[j ⊟ 12]) ⊕ Q[j];
112         }
113         end-if
114         i = i + 1;
115     }
116     end-repeat
```

Let the i-th internal state be (P_i, Q_i), where P_i and Q_i indicate P and Q at time i respectively. In the midgame backward security, we assume that an attacker know (P_i, Q_i). Then, he wants to reconstruct (P_{i-1}, Q_{i-1}). Each time, only one of $P[j]$ or $Q[j]$ (in lines 105 and 110) is updated. So, we only need to compute $P_{i-1}[j]$ or $Q_{i-1}[j]$ from P_i and Q_i. Firstly, let's see the line 105. We can rewrite the line 105 as $P_i[j] = P_{i-1}[j] + g_1(P_i[j \boxminus 3], P_i[j \boxminus 10], P_i[j \boxminus 511])$ from which we can compute $P_{i-1}[j]$ from P_i. Likewise, we can compute $Q_{i-1}[j]$. Therefore, we succeeded to reconstruct (P_{i-1}, Q_{i-1}). Therefore, the stream cipher HC-128 is not secure against the midgame backward attackers.

Rabbit [14]**.** The internal state of the stream cipher consists of 513 bits. 512 bits are divided between eight 32-bit state variables $x_{j,i}$ and eight 32-bit counter variables $c_{j,i}$, where $x_{j,i}$ is the state variable of subsystem j at iteration i, and $c_{j,i}$ denotes the corresponding counter variable. There is one counter carry bit, $\phi_{7,i}$, which needs to be stored between iterations. This counter carry bit is initialized to zero. The eight state variables and the eight counters are derived from the key at initialization. So, it is interesting whether we can compute $\{x_{j,i-1}, c_{j,i-1}\}_{0 \le j \le 7}$ and $\phi_{7,i-1}$ from $\{x_{j,i}, c_{j,i}\}_{0 \le j \le 7}$ and $\phi_{7,i}$. It is easy to efficiently compute $\{c_{j,i-1}\}_{0 \le j \le 7}$ and $\phi_{7,i-1}$ from $\{c_{j,i}\}_{0 \le j \le 7}$ according to the counter system defined in Sect. 2.5 of [14]. However, even if we know $\{c_{j,i-1}\}_{0 \le j \le 7}$ and $\phi_{7,i-1}$, it is still hard to compute $\{x_{j,i}\}_{0 \le j \le 7}$ because the Next-state Function of Rabbit is one-way (hard to invert, and here we assume that it is completely hard to invert), that is, each $x_{j,i}$ is defined by three elements of $\{x_{j,i-1}\}_{0 \le j \le 7}$. More in detail, its update is defined as shown in Fig. 11. Therefore, we can conclude that the stream cipher Rabbit provides strong security against the midgame backward attacks.

Salsa20/12 [7]**.** Salsa20/12 is the stream cipher based on a transformation Salsa20 with 64-byte input and 64-byte output. For an input 64-

byte x, the transformation Salsa20(x) $= x + doubleround^{12}(x)$, where the round function *doubleround* is repeated twelve times. The algorithm supports keys of 128 bits and 256 bits. During its operation, the key, a 64-bit nonce (unique message number), a 64-bit counter and four 32-bit constants are used to construct the 512-bit initial state. More in detail, let $\sigma_0 = (101, 120, 112, 97)$, $\sigma_1 = (110, 100, 32, 51)$, $\sigma_2 = (50, 45, 98, 121)$, and $\sigma_3 = (116, 101, 32, 107)$. If K_0, K_1, and n are 16-byte sequences then $\text{Salsa20}_{K_0||K_1}(n) = \text{Salsa20}(\sigma_0, K_0, \sigma_1, n, \sigma_2, K_1, \sigma_3)$. Let $\tau_0 = (101, 120, 112, 97)$, $\tau_1 = (110, 100, 32, 49)$, $\tau_2 = (54, 45, 98, 121)$, and $\tau_3 = (116, 101, 32, 107)$. If K and n are 16-byte sequences then $\text{Salsa20}_K(n) = \text{Salsa20}(\tau_0, K, \tau_1, n, \tau_2, K, \tau_3)$. Then, the key stream of Salsa20/12 is defined as follows: $\text{Salsa20/12}_K(v)$ is the 270-byte sequence

$$\text{Salsa20}_K(v, \underline{0}), \text{Salsa20}_K(v, \underline{1}), \text{Salsa20}_K(v, \underline{2}), \cdots, \text{Salsa20}_K(v, \underline{2^{64} - 1}),$$

where K is a 16-byte or 32-byte key, v is a 8-byte nonce, and $\underline{i}$ is the unique 8-byte sequence $(i_0, i_1, ..., i_7)$ such that $i = i_0 + 2^8 i_1 + 2^{16} i_2 + ... + 2^{56} i_7$.

For the mid-game backward security of Salsa20/12, we assume that an attacker knows an internal state, $(K, v, \underline{i})$. Then, the attacker knows the key K so he can compute the previous key stream by himself. Therefore, the stream cipher Salsa20/12 is not secure against the midgame backward attackers.

SOSEMANUK [6]. SOSEMANUK is a software-efficient and synchronous stream cipher. The cipher has a variable key length, ranging from 128 to 256 bits, and takes an initial value (IV) of 128 bits in length. A 384-bit internal state at time t consists of ten 32-bit $S_t = (s_t, s_{t+1}, ..., s_{t+9})$ for the FDR state, and two 32-bit registers $R1_t$ and $R2_t$. The update of the internal state at time $t \geq 1$ is defined as follows, where lbs(x) is the least significant bit of x, mix(c, x, y) is equal to x if $c = 0$, or to y if $c = 1$, $trans(z) = (\texttt{0x54655307} \times z \bmod 2^{32})_{\lll 7}$.

100 $s_{t+10} = s_{t+9} \oplus \alpha^{-1} s_{t+3} \oplus \alpha s_t$
101 $R1_t = (R2_{t-1} + \text{mix}(\text{lbs}(R1_{t-1}), s_{t+1}, s_{t+1} \oplus s_{t+8})) \bmod 2^{32}$
102 $R2_t = Trans(R1_{t-1})$

Once the attacker knows $S_t = (s_t, s_{t+1}, ..., s_{t+9})$, $R1_t$ and $R2_t$, (1) he can compute $R1_{t-1}$ from the line 102, because $Trans$ is efficiently invertible, (2) he can compute $S_{t-1} = (s_{t-1}, s_t, ..., s_{t+8})$ because the line 100 is invertible. And finally, he can compute $R2_{t-1}$ from $R1_t$, $R1_{t-1}$, and S_{t-1} in the line 101. Therefore, the stream cipher SOSEMANUK is not secure against the midgame backward attackers.

Grain v1 [24]. Grain v1 describes two stream ciphers: one for 80-bit (with 64-bit initialization vector) and another for 128-bit keys (with 80-bit initialization vector). As shown in Fig. 13, the key stream is generated by one NF SR with $g(x)$ and one FDR with $f(x)$ and a output function h. An internal state is defined with a 160-bit string, where the first 80-bit and the second 80-bit are defined by NF SR and FDR states respectively. Each state defines each bit of key stream. Since NF SR and FDR are invertible, we can compute $(i-1)$-th internal state from i-th state. Therefore, so we can easily reconstruct the previous key stream from the knowledge of the current leaked state, that is, the stream cipher Grain v1 is not secure against the midgame backward attackers.

MICKEY 2.0 [2]. MICKEY 2.0 is a hardware-efficient and synchronous stream cipher. The cipher makes use of a 80-bit key and an initialization vector with up to 80 bits in length. Each internal state is defined by two 100-bit registers, R and S, as shown in Fig. 14. It is easy to check that R and S are efficiently invertible, that is, we can compute R_{t-1} and R_{t-1} at time $t-1$ from R_t and R_t at time t. Therefore, the stream cipher MICKEY 2.0 is not secure against the midgame backward attackers.

Trivium [15]. The Trivium algorithm is a hardware-efficient and synchronous stream cipher. As shown in Fig. 15, the cipher makes use of a 80-bit key and 80-bit initialization vector (IV); its secret state has 288 bits, consisting of three interconnected non-linear feedback shift registers of length 93, 84 and 111 bits, respectively. More in detail, the update of the 288-bit internal state and the key stream z_i's are defined as follows.

101 for $i = 1$ to N do $(N \leq 2^{64})$
102 $\quad t_1 \leftarrow s_{66} + s_{93}$
103 $\quad t_2 \leftarrow s_{162} + s_{177}$
104 $\quad t_3 \leftarrow s_{243} + s_{288}$
105 $\quad z_i \leftarrow t_1 + t_2 + t_3$
106 $\quad t_1 \leftarrow t_1 + s_{91} \cdot s_{92} + s_{171}$
107 $\quad t_2 \leftarrow t_2 + s_{175} \cdot s_{176} + s_{264}$
108 $\quad t_3 \leftarrow t_3 + s_{286} \cdot s_{287} + s_{69}$
109 $\quad (s_1, s_2, ..., s_{93}) \leftarrow (t_3, s_1, ..., s_{92})$
110 $\quad (s_{94}, s_{95}, ..., s_{177}) \leftarrow (t_1, s_{94}, ..., s_{176})$
111 $\quad (s_{178}, s_{279}, ..., s_{288}) \leftarrow (t_2, s_{178}, ..., s_{287})$
112 end for

Based on the above key stream generation algorithm, each time the values of s_{93}, s_{177}, s_{288} are deleted from the next internal state. However, from lines 109,110, and 111, we can compute t_1, t_2, and t_3. Then, finally, from lines 102,103, and 104, we can reconstruct the values of s_{93}, s_{177}, s_{288}. So, we can compute $(i-1)$-th internal state from i-th state. Therefore, the stream cipher Trivium is not secure against the midgame backward attackers.

4.2 The Midgame Backward Security Analysis of Some Leakage-resilient Stream Ciphers

In this subsection, we consider the midgame backward security of two known leakage resilient stream ciphers. They are stream ciphers proposed by Petit *et al.* [26] (designed against a specific attack mode) and by Pietrzak [27] (designed against any bounded leakage attacker). As we will show these stream ciphers provide strong security against midgame backward attacks.

A Stream Cipher by Petit *et al.* [26]. Let E_1 and E_2 be two block ciphers in Fig. 8. The input of the first block cipher is initialized to a public IV, and each block cipher is initialized with its own master key, denoted k and k^* respectively, these keys playing the role of seed for the stream cipher. Then the ith block output of the stream cipher is computed as $y_i = E_2(k_i^*, m_i)$, the keys to be used by the block ciphers in the next round as $k_{i+1} = k_i \oplus m_i$ and $k_{i+1}^* = k_i^* \oplus m_i$, and the new input for the first block cipher as $x_{i+1} = IV \oplus y_i$. In the following, k and k^* are referred to as the master keys and to k_i, k_i^* as the running keys.

Let us assume that at some point $(k_{i+1}, k_{i+1}^*, y_i)$, which is an internal state, is revealed to an attacker. Then, in order to backtrack, the attacker has to compute m_i, k_i, and k_i^*, where $k_{i+1} = k_i \oplus m_i$, $k_{i+1}^* = k_i^* \oplus m_i$, and $y_i = E_2(k_i^*, m_i)$, without knowing k_i and k_i^*. As long as E_1 and E_2 are a secure block cipher (i.e., pseudorandom permutations), it is difficult to get m_i, k_i, and k_i^* as long as the attacker cannot have direct access to the initial keys.

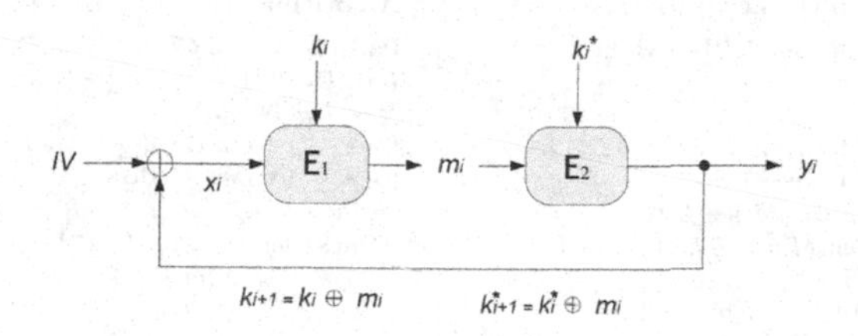

Fig. 8. A stream cipher by Petit *et al.* [26].

A Stream Cipher by Pietrzak [27]. Let $\mathsf{F} : \{0,1\}^k \times \{0,1\}^n \to \{0,1\}^{k+n}$ be a function. The stream cipher S^F is defined as follows:

- INITIALIZATION: The initial state is $S_0 = [K_0, K_1, X_0]$, where $K_0, K_1 \leftarrow_r \{0,1\}^k$ and $X_0 \leftarrow_r \{0,1\}^n$. Only K_0, K_1 must be secret, X_0 can be public.
- STATE: The state before the ith round is $S_{i-1} = [K_{i-1}, K_i, X_{i-1}]$.
- COMPUTATION: In the ith round, $\mathsf{S}^\mathsf{F}(S_{i-1})$ computes $(K_{i+1}, X_i) := \mathsf{F}(K_{i-1}, X_{i-1})$ and outputs X_i. Then the state $S_{i-1} = [K_{i-1}, K_i, X_{i-1}]$ is replaced with $S_i = [K_i, K_{i+1}, X_i]$.

Now, if we know any state $S_i = [K_i, K_{i+1}, X_i]$, we cannot backtrack to compute X_{i-1} and K_{i-1} because it is hard to compute K_{i-1}, X_{i-1} from (K_{i+1}, X_i) without knowing K_{i-1} when F is one-way. Even if the attacker knows the X_j's (for all j) and K_i and K_{i+1}, he cannot compute K_{i-1} when F is a weak pseudorandom function, where we say that F is weak pseudorandom if no attacker, who is allowed only to make random queries, can distinguish $F(K, \cdot)$ and a random function $\$(\cdot)$. So, S^F is secure against the midgame backward attackers as long as the attacker cannot have direct access to the initial key pair (K_0, K_1).

5 The Midgame Security Analysis of Known (Authenticated) Encryption Schemes

5.1 The Midgame Security Analysis of Known (Authenticated) Encryption Schemes Based on a Block Cipher

In case of encryption or authenticated encryption (AE) schemes with a block cipher, any internal state in the middle of the computation always has the key value, because the underlying block cipher needs a fixed key for the computation of every block. For example, we investigate OCB [29], which is a block-cipher-based authenticated-encryption scheme.

Algorithm $\mathrm{OCB.Enc}_K\,(N, M)$

Partition M into $M[1] \cdots M[m]$
$L \leftarrow E_K(0^n)$
$R \leftarrow E_K(N \oplus L)$
for $i \leftarrow 1$ **to** m **do** $Z[i] = \gamma_i \cdot L \oplus R$
for $i \leftarrow 1$ **to** $m-1$ **do**
  $C[i] \leftarrow E_K(M[i] \oplus Z[i]) \oplus Z[i]$
$X[m] \leftarrow \mathrm{len}(M[m]) \oplus L \cdot \mathbf{x}^{-1} \oplus Z[m]$
$Y[m] \leftarrow E_K(X[m])$
$C[m] \leftarrow Y[m] \oplus M[m]$
$C \leftarrow C[1] \cdots C[m]$
Checksum $\leftarrow$
  $M[1] \oplus \cdots \oplus M[m-1] \oplus C[m]0^* \oplus Y[m]$
$T \leftarrow E_K(\mathrm{Checksum} \oplus Z[m])$ [first τ bits]
return $\mathcal{C} \leftarrow C \parallel T$

Algorithm $\mathrm{OCB.Dec}_K\,(N, \mathcal{C})$

Partition $\mathcal{C}$ into $C[1] \cdots C[m]\,T$
$L \leftarrow E_K(0^n)$
$R \leftarrow E_K(N \oplus L)$
for $i \leftarrow 1$ **to** m **do** $Z[i] = \gamma_i \cdot L \oplus R$
for $i \leftarrow 1$ **to** $m-1$ **do**
  $M[i] \leftarrow E_K^{-1}(C[i] \oplus Z[i]) \oplus Z[i]$
$X[m] \leftarrow \mathrm{len}(C[m]) \oplus L \cdot \mathbf{x}^{-1} \oplus Z[m]$
$Y[m] \leftarrow E_K(X[m])$
$M[m] \leftarrow Y[m] \oplus C[m]$
$M \leftarrow M[1] \cdots M[m]$
Checksum $\leftarrow$
  $M[1] \oplus \cdots \oplus M[m-1] \oplus C[m]0^* \oplus Y[m]$
$T' \leftarrow E_K(\mathrm{Checksum} \oplus Z[m])$ [first τ bits]
if $T = T'$ **then return** M
  else return INVALID

Fig. 9. OCB encryption [29]. The message to encrypt is M and the key is K. Message M is written as $M = M[1]M[2]...M[m-1]M[m]$ where $m = \max\{1, \lceil |M|/n \rceil\}$ and $|M[1]| = |M[2]| = ... = |M[m-1]| = n$. Nonce N is a non-repeating value selected by the party that encrypts. It, along with ciphertext $\mathcal{C} = C[1]C[2]C[3]...C[m-1]C[m]T$, is needed to decrypt. The Checksum is $M[1] \oplus ... \oplus M[m-1] \oplus C[m]0^* \oplus Y[m]$. Offset $Z[1] = L \oplus R$ while, for $i \geq 2$, $Z[i] = Z[i-1] \oplus L(\mathsf{ntz}(i))$. String L is defined by applying E_K to the fixed string 0^n. For $Y[m] \oplus M[m]$ and $Y[m] \oplus C[m]$, truncate $Y[m]$ if it is longer than the other operand. By $C[m]0^*$ we mean $C[m]$ padded on the right with 0-bits to get to length n. Function len represents the length of its argument, mod 2^n, as an n-bit string.

As we can see from Fig. 9, all the secret information, K, L, R, and $Z[i]$ (for all i), are generated from the value of K. And the key K is used as the key of every block cipher which means that once an attacker knows any internal state during the entire OCB processing, then he can compute all the secret information and can do whatever he wants. Likewise, we cannot guarantee any midgame security for many (authenticated-) encryption schemes including NIST-approved ones such as ECB [16], CBC [16], CFB [16], OFB [16], CTR [16], CCM [17], GCM [19], AES Key-wrapping [20], unless the entire computation is securely protected.

5.2 The Midgame Security Analysis of SpongeWrap

SpongeWrap [10], as shown in Fig. 10, is an AE scheme based on a permutation f. Unlike other encryption and AE schemes based on the block cipher, except for the first several blocks related to the keys, any internal state of SpongeWrap has no key information. But, due to the invertibility of the permutation, we can backtrack and compute the key value once the attacker has a known plaintext-ciphertext pair. So, we cannot guarantee any security of SpongeWrap either, unless the entire computation is securely protected (Fig. 12).

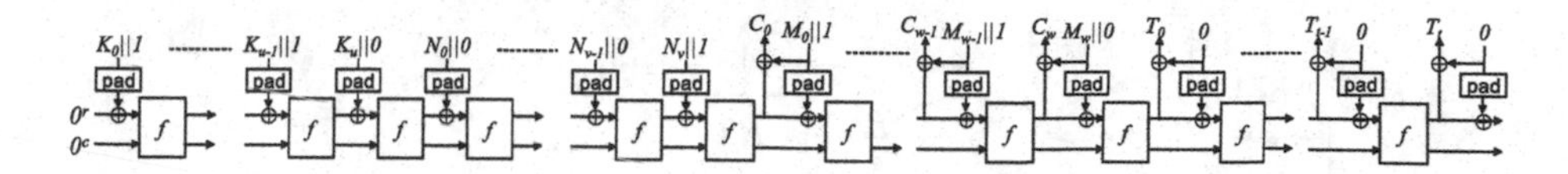

Fig. 10. SpongeWrap$[f, \text{pad}, r, \rho]$ [10], where $|K_i|, |A_i|, |M_i| \leq \rho$.

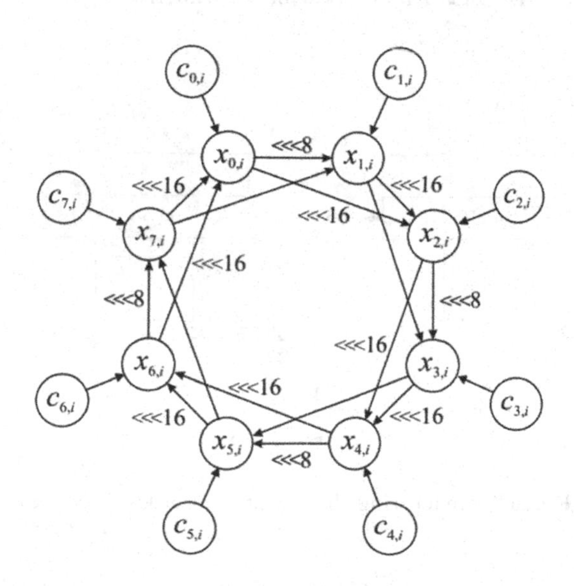

Fig. 11. Graphical Illustration of the Next-state Function of Rabbit [14].

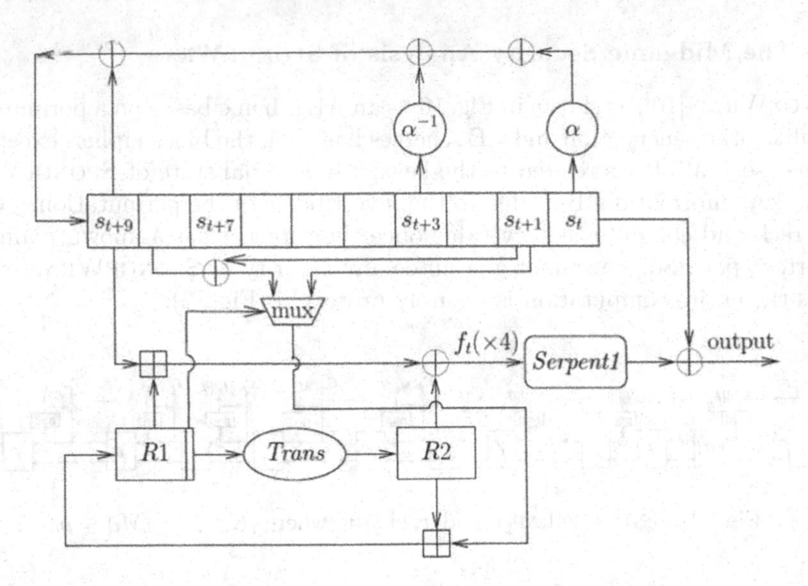

Fig. 12. An overview of SOSEMANUK [6].

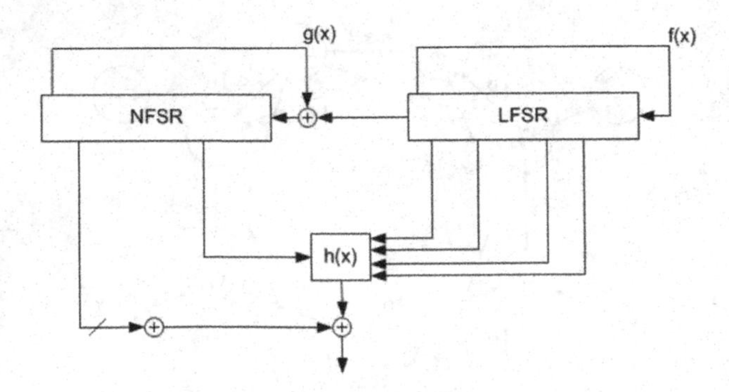

Fig. 13. Generating the key stream of Grain v1 [24].

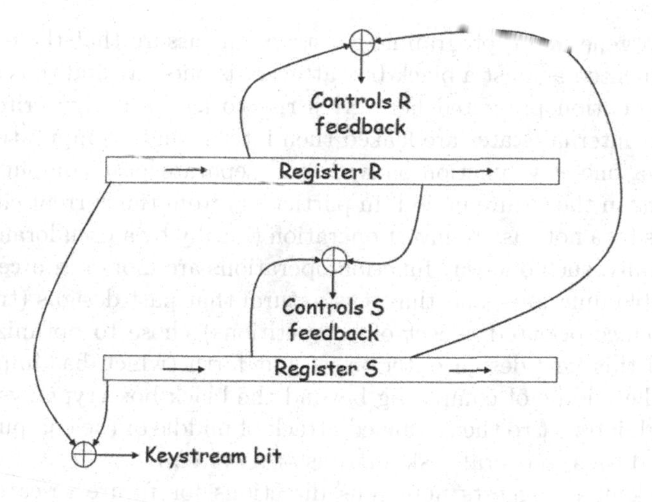

Fig. 14. Generating the key stream of MICKEY 2.0 [2].

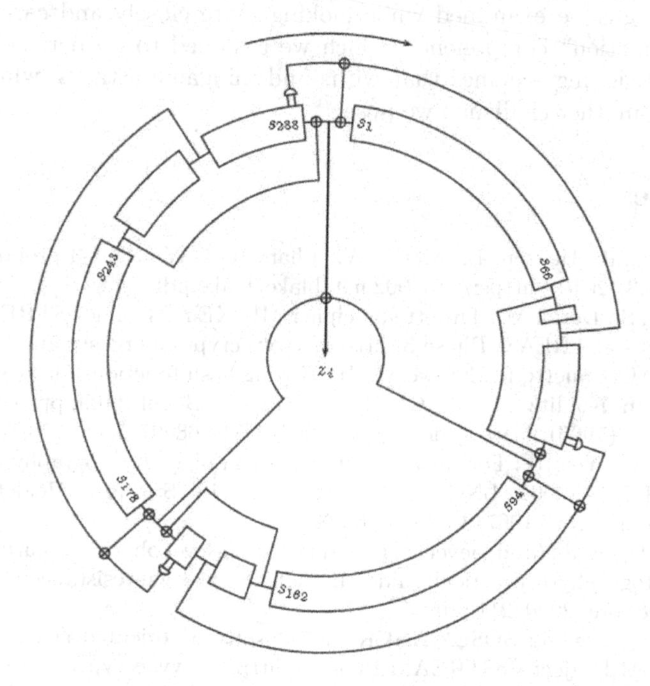

Fig. 15. Trivium [2].

6 Conclusion

In this paper, we proposed the midgame security and its application to the security analysis of MAC, AE, or encryption schemes, and its effect on the implementation optimization of MAC schemes in terms of time, power, and energy consumption. In general, cryptographic constructions assure that the entire design when implemented against a black-box attacker is one-way and thus secure. The new security notion presented here gives rise to a new design criterion which states that if internal states are leaked then intermediate computations need to behave like a one-way function as well, and separate past computations (and computations in the secure period, in particular) from the current computation; separation is by a not easy to invert operation (ideally by a pseudorandom operation). Naturally, such one-way function operations are more resource consuming than invertible functions, and thus it is natural that past designs (trying to win the performance oriented aspect of competitions) chose to optimize resources and ignored this new design criterion we put forth (which has implications to leakage in the middle of computing beyond the black box cryptosystem design, and is of high interest to the combined attack of middle of the computation state exposure and thus to overall risk analysis of a system.

The work here suggests numerous directions for future research. Refining the model to make it more theoretical is one direction; further scrutinizing the concrete designs we examined while looking more closely and carefully at the "one way function" components (which we assumed to be perfect) is another direction. Designing schemes that withstand midgame attacks, while retaining efficiency is another challenge we pose.

References

1. Aumasson, J., Henzen, L., Meier, W., Phan, R.C.-W.: SHA-3 proposal BLAKE-version 1.3 (2010). https://131002.net/blake/blake.pdf
2. Babbage, S., Dodd, M.: The stream cipher MICKEY 2.0. The eSTREAM Projosemanukect - eSTREAM Phase 3. http://www.ecrypt.eu.org/stream/
3. Bellare, M., Canetti, R., Krawczyk, H.: Keying hash functions for message authentication. In: Koblitz, N. (ed.) CRYPTO 1996. LNCS, vol. 1109, pp. 1–15. Springer, Heidelberg (1996). https://doi.org/10.1007/3-540-68697-5_1
4. Bellare, M., Yee, B.: Forward-security in private-key cryptography. In: Joye, M. (ed.) CT-RSA 2003. LNCS, vol. 2612, pp. 1–18. Springer, Heidelberg (2003). https://doi.org/10.1007/3-540-36563-X_1
5. Bellizia, D., et al.: Mode-level vs. implementation-level physical security in symmetric cryptography a practical guide through the leakage-resistance jungle. http://eprint.iacr.org/2020/211.pdf
6. Berbain, C., et al.: SOSEMANUK, a fast software-oriented stream cipher. The eSTREAM Project - eSTREAM Phase 3. http://www.ecrypt.eu.org/stream/
7. Bernstein, D.J.: Salsa20 specification. The eSTREAM Project - eSTREAM Phase 3. http://www.ecrypt.eu.org/stream/
8. Berti, F., Koeune, F., Pereira, O., Peters, T., Standaert, F.-X.: Ciphertext integrity with misuse and leakage: definition and efficient constructions with symmetric primitives. In: AsiaCCS 2018, pp. 37–50 (2018)
9. Bertoni, G., Daemen, J., Peeters, M., Assche, G.V.: Keccak sponge function family main document. Submission to NIST (Round 1) (2008). http://keccak.noekeon.org/Keccak-main-1.0.pdf

10. Bertoni, G., Daemen, J., Peeters, M., Assche, G.V.: Duplexing the sponge: single-pass authenticated encryption and other applications. Submission to the NIST second SHA-3 workshop (2010). http://csrc.nist.gov/groups/ST/hash/sha-3/Round2/Aug2010/documents/papers/DAEMEN_DuplexSponge.pdf
11. Bertoni, G., Daemen, J., Peeters, M., Assche, G.V.: The Keccak SHA-3 submission. Submission to NIST (Round 3) (2011). http://keccak.noekeon.org/Keccak-submission-3.pdf
12. Blaze, M.: High-bandwidth encryption with low-bandwidth smartcards. In: Gollmann, D. (ed.) FSE 1996. LNCS, vol. 1039, pp. 33–40. Springer, Heidelberg (1996). https://doi.org/10.1007/3-540-60865-6_40
13. Blaze, M., Feigenbaum, J., Naor, M.: A formal treatment of remotely keyed encryption. In: Nyberg, K. (ed.) EUROCRYPT 1998. LNCS, vol. 1403, pp. 251–265. Springer, Heidelberg (1998). https://doi.org/10.1007/BFb0054131
14. Boesgaard, M., Vesterager, M., Christensen, T., Zenner, E.: The Stream Cipher Rabbit. The eSTREAM Project - eSTREAM Phase 3. http://www.ecrypt.eu.org/stream/
15. Cannière, C.D., Preneel, B.: Trivium Specifications. The eSTREAM Project - eSTREAM Phase 3. http://www.ecrypt.eu.org/stream/
16. Dworkin, M.: Recommendation for Block Cipher Modes of Operation. NIST Special Publication 800-38A 2001 Edition. http://csrc.nist.gov/publications/nistpubs/800-38a/sp800-38a.pdf
17. Dworkin, M.: Recommendation for Block Cipher Modes of Operation: The CCM Mode for Authentication and Confidentiality. NIST Special Publication 800-38C 2004 Edition. http://csrc.nist.gov/publications/nistpubs/800-38C/SP800-38C_updated-July20_2007.pdf
18. Dworkin, M.: Recommendation for Block Cipher Modes of Operation: The CMAC Mode for Authentication. NIST Special Publication 800-38B 2005 Edition. http://csrc.nist.gov/publications/nistpubs/800-38B/SP_800-38B.pdf
19. Dworkin, M.: Recommendation for Block Cipher Modes of Operation: Galois/Counter Mode (GCM) and GMAC. NIST Special Publication 800-38D 2005 Edition. http://csrc.nist.gov/publications/nistpubs/800-38D/SP-800-38D.pdf
20. Dworkin, M.: Recommendation for Block Cipher Modes of Operation: Methods for Key Wrapping Morris Dworkin. NIST Special Publication 800-38F 2011 Edition. http://csrc.nist.gov/publications/drafts/800-38F/Draft-SP800-38F_Aug2011.pdf
21. Ferguson, N., et al.: The Skein Hash Function Family. Submission to NIST (Round 3) (2010). http://www.skein-hash.info/sites/default/files/skein1.3.pdf
22. Gauravaram, P., et al.: Grøstl - a SHA-3 candidate. Submission to NIST (Round 3) (2011). http://www.groestl.info/Groestl.pdf
23. Halderman, J.A., et al.: Lest we remember: cold boot attacks on encryption keys. In: USENIX Security Symposium, pp. 91–98 (2008)
24. Hell, M., Johansson, T., Meier, W.: A Stream Cipher Proposal: Grain-128. The eSTREAM Project - eSTREAM Phase 3. http://www.ecrypt.eu.org/stream/
25. Hoerder, S., Wójcik, M., Tillich, S., Page, D.: An evaluation of hash functions on a power analysis resistant processor architecture. In: Ardagna, C.A., Zhou, J. (eds.) WISTP 2011. LNCS, vol. 6633, pp. 160–174. Springer, Heidelberg (2011). https://doi.org/10.1007/978-3-642-21040-2_11
26. Petit, C., Standaert, F., Pereira, O., Malkin, T., Yung, M.: A block cipher based pseudo random number generator secure against side-channel key recovery. In: Proceedings of the 2008 ACM Symposium on Information, Computer and Communications Security, ASIACCS 2008, pp. 56–65 (2008)

27. Pietrzak, K.: A leakage-resilient mode of operation. In: Joux, A. (ed.) EUROCRYPT 2009. LNCS, vol. 5479, pp. 462–482. Springer, Heidelberg (2009). https://doi.org/10.1007/978-3-642-01001-9_27
28. Preneel, B., van Oorschot, P.C.: On the security of two MAC algorithms. In: Maurer, U. (ed.) EUROCRYPT 1996. LNCS, vol. 1070, pp. 19–32. Springer, Heidelberg (1996). https://doi.org/10.1007/3-540-68339-9_3
29. Rogaway, P., Bellare, M., Black, J.: OCB: a block-cipher mode of operation for efficient authenticated encryption. ACM Trans. Inf. Syst. Secur. (TISSEC) **6**(3), 365–403 (2003)
30. Standaert, F.-X., Malkin, T.G., Yung, M.: A unified framework for the analysis of side-channel key recovery attacks. In: Joux, A. (ed.) EUROCRYPT 2009. LNCS, vol. 5479, pp. 443–461. Springer, Heidelberg (2009). https://doi.org/10.1007/978-3-642-01001-9_26
31. Wu, H.: The Stream Cipher HC-128. The eSTREAM Project - eSTREAM Phase 3. https://www.ecrypt.eu.org/stream/
32. Wu, H.: The Hash Function JH. Submission to NIST (Round 3) (2011). https://www3.ntu.edu.sg/home/wuhj/research/jh/jh_round3.pdf
33. Yasuda, K.: "Sandwich" is indeed secure: how to authenticate a message with just one hashing. In: Pieprzyk, J., Ghodosi, H., Dawson, E. (eds.) ACISP 2007. LNCS, vol. 4586, pp. 355–369. Springer, Heidelberg (2007). https://doi.org/10.1007/978-3-540-73458-1_26

Deep Neural Networks for Encrypted Inference with TFHE

Andrei Stoian(✉), Jordan Frery, Roman Bredehoft, Luis Montero, Celia Kherfallah, and Benoit Chevallier-Mames

Zama, Paris, France
hello@zama.ai
http://zama.ai

Abstract. Fully homomorphic encryption (FHE) is an encryption method that allows to perform computation on encrypted data, without decryption. FHE preserves the privacy of the users of online services that handle sensitive data, such as health data, biometrics, credit scores and other personal information. A common way to provide a valuable service on such data is through machine learning and, at this time, neural networks are the dominant machine learning model for unstructured data.

In this work we show how to construct Deep Neural Networks (DNN) that are compatible with the constraints of TFHE, an FHE scheme that allows arbitrary depth computation circuits. We discuss the constraints and show the architecture of DNNs for two computer vision tasks. We benchmark the architectures using the Concrete stack (https://github.com/zama-ai/concrete-ml), an open-source implementation of TFHE.

CSCML 2023 Submission
This paper is submitted as a **short paper**.

1 Introduction

Neural Networks (NNs) are machine learning (ML) models that have driven the recent expansion of the field of Artificial Intelligence (AI). Their performance on unstructured data such as images, sound and text is unmatched by other ML techniques [19]. Though training deep NNs is more difficult, they obviate the need for complex feature engineering and process raw data directly, making them easier to deploy in production. Applications of NNs include image classification, face recognition, voice assistants, and search engines, tools which today are a staple of the user experience online. Deployment of such models in SaaS applications raises a security risk: they are a target of malevolent entities that seek to steal the sensitive user data these models process.

Privacy-preserving technologies, such as multi-party computing (MPC) and fully homomorphic encryption (FHE), provide a solution to the risk of data leaks, eliminating it by design. Notably, FHE encrypts user data and allows a

S. Dolev et al. (Eds.): CSCML 2023, LNCS 13914, pp. 493–500, 2023.
https://doi.org/10.1007/978-3-031-34671-2_34

third party to process the data in its encrypted form, without needing to decrypt it. Only the owner of the secret-key can decrypt the result of the computation. Thus, an attacker can only steal encrypted data they can not decrypt.

In this work we show how to build NNs that are FHE compatible, while minimizing the cryptography knowledge needed by the machine learning practitioner. We based our work on the Concrete Library [9] which uses TFHE [8], works over integers, provides a fast *programmable* bootstrapping (PBS) mechanism, and performs exact computation.

2 Related Work

Several alternative approaches exist for neural network inference over encrypted data. Many of these works rely on encryption schemes such as CKKS [7] and YASHE [4] which support both floating point and integer computation. Most works on NNs use quantized models where the weights and activations are fixed point or integers. With the exception of [15], these works avoid bootstrapping in order to reduce inference time. Avoiding boostrapping has the disadvantage of limiting the depth of the models that can be computed in FHE.

CryptoNets [12] uses YASHE which supports the computation of polynomials of encrypted values. CryptoNets are NNs quantized to integers (of 5–10 bits) with activation functions expressed as low-degree polynomials. CryptoNets achieve 99% accuracy on MNIST using a three layer network with an inference time of 570 s/image. To speed-up this technique, [6] optimize the ciphertext representation of clear values to leverage packing techniques specific to CKKS.

FHE-DiNN [5], a TFHE based approach, quantizes inputs, intermediate values and weights to binary values. During training the activation functions are set to `hardSigmoid`, while in inference they are swapped for the `sign` function. However, binary NNs are hard to train and do not perform well in many ML tasks such as object detection and speech processing.

Another TFHE approach, SHE [16], uses bit series representation of encrypted values and boolean gates. They run NNs that fit within a maximum multiplicative depth budget and, by avoiding expensive multi-bit PBSs, they achieve inference of a ShuffleNet on ImageNet with a latency of 18,000 s/image. They rely on logarithmic quantization of weights which allows to reduce multiplicative depth for the convolution layers by using bit-shifts. Sums, `relu` and `maxpool` are computed using boolean gates. Another TFHE-based approach, combining both boolean and integer ciphertext representations, [11], was shown to take advantage of the strengths of both types of representation.

Leveled approaches such as SHE and CryptoNets are limited by the maximum multiplicative depth budget, which, in turn, limits the supported network types and their depth. Moreover, to speed up the computations, many works that use CKKS set crypto-system parameters in a way that makes the computations approximate in the sense that noise corrupts some of the message bits.

In this work we propose an approach to train arbitrary NNs which can have any depth, number of neurons and activation functions. Furthermore, our approach performs exact computation in FHE: the noise of the encryption scheme does not corrupt the values that are processed. Thus results in FHE are the same as in the clear - there is no degradation of accuracy when moving to encrypted inference - which is a major advantage when putting models in production.

3 Neural Network Training for Encrypted Inference

Training NNs is usually done in floating point, but most FHE schemes, including TFHE, only support integers. Consequently, quantization must be used, and two main approaches exist:

1. Post-training quantization is commonly used [12,16], but, in this mode, NNs lose accuracy when the quantization bit-width is lower than 7–8 bits. With per-channel quantization, or logarithmic quantization, which are more complex to implement, as few as 4 bits were used for weights and activations without loss of accuracy [20].
2. Quantization-aware Training (QAT), used in this work and in [5], is an approach that adds quantizers to network activations and weights during training. QAT enables extreme quantization with less than 4 bit weights and activations.

To support arbitrarily deep NNs and any activation function, we make use of the PBS [10] mechanism of TFHE. PBS reduces the noise in accumulators of ciphertext leveled operations (addition, multiplication with clear constants) but also allows to apply a lookup-table (TLU) on its input ciphertext.

The TFHE PBS mechanism has a rather high computational cost, and this cost depends on the number of bits of the encrypted value to be boostrapped. It is convenient to keep the accumulator size low, in order to speed up the PBS computation. However, reducing accumulator bit-width has a negative impact on network prediction performance, so a compromise needs to be found.

We describe here a QAT strategy that can process all the intermediate encrypted values as integers. In this way, training an FHE compatible network becomes purely a machine learning problem and no cryptography knowledge is needed by the practitioner. To build a TFHE compatible NN, the constraints on the network architecture are the following:

- All layers that sum or multiply two encrypted values, such as convolution `conv` and fully-connected `fc`, must have quantized inputs. This is easily achieved using QAT frameworks.
- The bit-width of the accumulators of layers such as `conv`, `fc` must be bounded. To achieve this, similarly to [1], we use pruning of neuron synapses.

To control the accumulator bit-width while keeping the training dynamics stable, we use L^1-norm unstructured pruning. Figure 1 shows the impact of pruning on the accumulator size for two quantization modes: narrow and wide range.

While the inputs of `conv` and `fc` layers need to be quantized, it is possible to use floating point layers for all univariate operations such as batch normalization, quantization, and activations.

In our FHE compatible NNs the outputs of a `conv` or `fc` are processed by a sequence of univariate operations that ends with quantization. This sequence of functions takes integers and has integer outputs, but the intermediary computations in these operations can use float parameters. Thus, batch normalization, activation functions, neuron biases and any other univariate transformation of `conv` or `fc` outputs does not need quantization. Figure 1 shows the architecture of the network during training and inference.

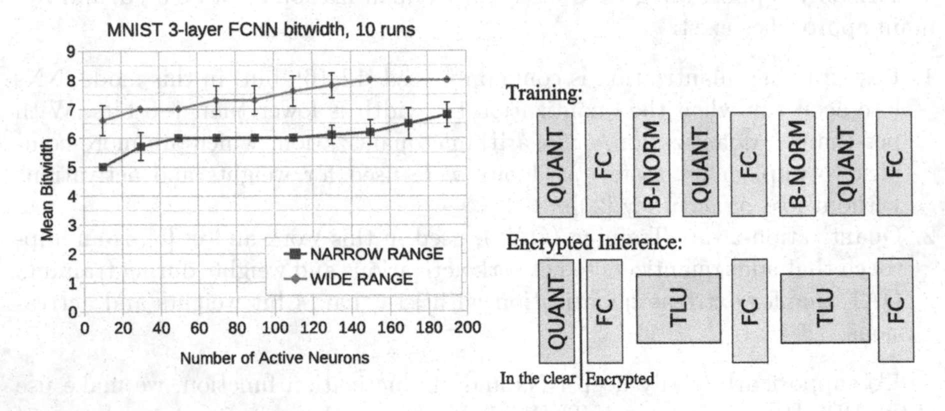

Fig. 1. Left: accumulator size while varying the number of active neurons during pruning for a 3-layer fully-connected network with 2 bit weights and activations. Two quantization modes are shown: Narrow range uses values $[-2^{b-1}+1, 2^{b-1}-1]$, while Wide range uses $[-2^{b-1}, 2^{b-1}-1]$. Right: the structure of a 2 layer convolutional network in training and during inference. Univariate layers are fused to table-lookups, implemented with PBS.

4 Neural Network Inference Using TFHE

Inference of our FHE NNs is based on quantized implementations of NN operators that add or multiply together encrypted values. Convolutional, fully connected and average pooling layers use the quantized formulation from [13]. Since uniform quantization is used, we can define a quantized value r as $r = S(q - Z)$ where S is the quantization scale, Z is the quantization zero-point and q is the integer representation of the value. Next, the fully connected layer, with inputs x, weights w and outputs o, with per-tensor quantization parameters (S_x, Z_x), (S_w, Z_w) can be written as:

$$S_o(q_o^k - Z_o) = \sum_{i=0}^{N} S_x(q_x^i - Z_x)S_w(q_w^{(i,k)} - Z_w) + b^k \quad (1)$$

where k is the index of a neuron in the layer and N is the number of connections of the neuron and b^k is the bias of the k-th neuron. A convolutional layer can be expressed by extending the sum to the height, width and channel dimensions. Equation 2 can be re-written to separate integer and floating point computations (note that zero-points Z_x, Z_o, Z_w are integers).

$$q_o^k = b^k + Z_o + \frac{S_x S_w}{S_o} \sum_{i=0}^{N} (q_x^i - Z_x)(q_w^{(i,k)} - Z_w) \tag{2}$$

Therefore we can separate the equation in a floating point univariate function f and a sum over products of encrypted inputs and clear weights:

$$q_o^k = f(\Sigma) \text{ where } f(q) = b^k + Z_o + \frac{S_x S_w}{S_o} q \text{ and } \Sigma = \sum_{i=0}^{N} (q_x^i - Z_x)(q_w^{(i,k)} - Z_w) \tag{3}$$

The univariate function f in Eq. 3 takes integer inputs. We compose this function with the batch-normalization, and, finally, with the quantization function $Q(x) = floor\left(\frac{x}{S_x}\right) + Z_x$. Thus f becomes a function defined on $\mathbb{Z}$, with values in $\mathbb{Z}$ and can be implemented as a lookup table with a PBS in FHE, without any loss of precision. Our approach - fusing all univariate functions - contrasts with other works [11,13] which, instead, prefer to express the batch normalization and quantization with integer operations. However, expressing all these univariate operations with integers only adds to the difficulty of the quantization process and decreases accuracy. Our PBS approach allows fusing all univariate floating point operations applied to an encrypted integer to a lookup-table, thus reducing quantization constraints.

The complete NN computation can now be expressed over integers using the following operations: multiplication of an encrypted value and a clear constant, sums of encrypted integer values, table lookup of encrypted integer values. In our implementation of TFHE, Concrete, we encode integers in two different ways: integers up to 8 bits are encoded into a single ciphertext, and integers between 9–16 bits are encoded with a CRT representation into several ciphertexts as described in [3]. This contrasts to previous works, such as [16], that encode each bit of an integer as an individual ciphertext and use boolean gates to build arithmetic circuits.

An automated optimization process [3] determines the cryptographic parameters of the circuit, based on several factors: (1) the *circuit bit-width*, defined as the minimum bit-width necessary to encode the largest integer value obtained anywhere in the NN's integer-based evaluation, (2) the maximum 2-norm of the integer weight tensors of the layers, and (3), the desired probability of error of the PBS. The optimization process determines the cryptosystem parameters (LWE dimension, polynomial size, GLWE dimension, etc. - see [14]) to ensure a fast execution, the target probability of failure and the security level (using the lattice-estimator [2]). We set the PBS error probability sufficiently low to ensure full correctness of the results, i.e. the results in the clear are always the same as those in FHE, up to a user-defined error-rate, e.g. 10^{-6}, for one full NN inference.

5 Experimental Results

The networks were implemented in PyTorch with Brevitas [18] and converted to FHE with Concrete-ML [17]. We ran experiments on two datasets with several neural network architectures, in two quantization modes (see Fig. 1, left). In Table 1, the latency results for encrypted inference are given. Unless otherwise indicated by a footnote, the test machine had an Intel i7-11800H CPU with 8 cores (16 threads were used). For one of the configurations, as indicated in the table, the first layer of the network was executed in the clear to keep the overall accumulator size low.

Table 1. Experimental results obtained with Brevitas and Concrete-ML. Unless otherwise stated, the timing is obtained on an 8-core machine.

Network	Quant. bits	Active Neurons	Narrow range	Data-set	Circuit bit-width	Accuracy	Inference time (s)
3-layer FCNN[a]	2/2	150	Yes	MNIST	6	92.2%	31
3-layer FCNN	2/2	90	No	MNIST	7	96.5%	77
3-layer FCNN	2/2	190	No	MNIST	8	97.1%	300
LeNet	2/2	190	No	MNIST	8	97.6%	2780
6-layer CNN[b]	2/2	190	No	MNIST	8	98.7%	5072
VGG-9[c]	2/2[d]	all	Yes	CIFAR10	8	62.3%	624[e]
VGG-9	6/6	110	No	CIFAR10	15	90.0%	25000[e]
VGG-9	6/6	110	No	CIFAR100	15	69.0%	25000[e]

[a] Three FC layers with 192, 192 and 10 neurons
[b] Four conv layers with 8,8,16 and 16 filters followed by a FC layer with 120 neurons and a final FC layer for classification
[c] Six conv layers with: 64, 64, 128, 128, 256, 256 filters, followed by two 512 neuron FC layers and a final FC layer for classification
[d] Inputs are quantized in 8 bits and the first layer is executed non-encrypted. All activations use 2 bits
[e] On a `c6i.metal` AWS instance with 128 vcpu.

6 Conclusion

Our approach to encrypted inference for neural networks shows several advantages over other methods. First, we believe our method is easier to use than other works, since the problem of making an FHE compatible network becomes strictly an ML problem and no cryptography knowledge is needed. Second, the computations in FHE are correct with respect to the computations in the clear and, using TFHE, noise does not corrupt the encrypted values. Thus, once a network is trained incorporating the quantization constraints, the accuracy that is measured on clear data will be the same as that on encrypted data. Finally, our approach, using PBS, shows competitive accuracies in FHE and allows to convert arbitrary-depth networks using any activation function to FHE. Networks up to 9 layers were shown, but deeper NNs can easily be implemented. Code for the MNIST, CIFAR10 and CIFAR100 classifiers is available[1].

Many possible strategies can be employed to improve upon this work, in order to support larger models, such as ResNet, on larger data-sets like ImageNet. For example, a better pruning strategy could decrease the PBS count, per-channel quantization can improve accuracy, and faster step functions in FHE could improve the overall speed.

References

1. Aharoni, E., et al.: HE-PEx: efficient machine learning under homomorphic encryption using pruning, permutation and expansion. CoRR abs/2207.03384 (2022). https://doi.org/10.48550/arXiv.2207.03384
2. Albrecht, M.R., Player, R., Scott, S.: On the concrete hardness of learning with errors. Cryptology ePrint Archive, Paper 2015/046 (2015). https://eprint.iacr.org/2015/046
3. Bergerat, L., et al.: Parameter optimization & larger precision for (T)FHE. Cryptology ePrint Archive, Paper 2022/704 (2022). https://eprint.iacr.org/2022/704
4. Bos, J.W., Lauter, K., Loftus, J., Naehrig, M.: Improved security for a ring-based fully homomorphic encryption scheme. In: Stam, M. (ed.) IMACC 2013. LNCS, vol. 8308, pp. 45–64. Springer, Heidelberg (2013). https://doi.org/10.1007/978-3-642-45239-0_4
5. Bourse, F., Minelli, M., Minihold, M., Paillier, P.: Fast homomorphic evaluation of deep discretized neural networks. In: Shacham, H., Boldyreva, A. (eds.) CRYPTO 2018. LNCS, vol. 10993, pp. 483–512. Springer, Cham (2018). https://doi.org/10.1007/978-3-319-96878-0_17
6. Brutzkus, A., Gilad-Bachrach, R., Elisha, O.: Low latency privacy preserving inference. In: Chaudhuri, K., Salakhutdinov, R. (eds.) Proceedings of the 36th International Conference on Machine Learning, ICML 2019, 9–15 June 2019, Long Beach, California, USA. Proceedings of Machine Learning Research, vol. 97, pp. 812–821. PMLR (2019). http://proceedings.mlr.press/v97/brutzkus19a.html
7. Cheon, J.H., Kim, A., Kim, M., Song, Y.: Homomorphic encryption for arithmetic of approximate numbers. In: Takagi, T., Peyrin, T. (eds.) ASIACRYPT 2017. LNCS, vol. 10624, pp. 409–437. Springer, Cham (2017). https://doi.org/10.1007/978-3-319-70694-8_15
8. Chillotti, I., Gama, N., Georgieva, M., Izabachène, M.: TFHE: fast fully homomorphic encryption over the torus. J. Cryptol. **33**, 34–91 (2019)
9. Chillotti, I., Joye, M., Ligier, D., Orfila, J.B., Tap, S.: CONCRETE: concrete operates on ciphertexts rapidly by extending TFHE. In: WAHC 2020–8th Workshop on Encrypted Computing & Applied Homomorphic Cryptography, vol. 15 (2020)
10. Chillotti, I., Joye, M., Paillier, P.: Programmable bootstrapping enables efficient homomorphic inference of deep neural networks. IACR Cryptol. ePrint Arch. **2021**, 91 (2021)
11. Folkerts, L., Gouert, C., Tsoutsos, N.G.: REDsec: running encrypted discretized neural networks in seconds. In: Proceedings 2023 Network and Distributed System Security Symposium (2023)

[1] https://github.com/zama-ai/concrete-ml/tree/release/0.6.x/use_case_examples.

12. Gilad-Bachrach, R., Dowlin, N., Laine, K., Lauter, K., Naehrig, M., Wernsing, J.: CryptoNets: applying neural networks to encrypted data with high throughput and accuracy. In: Balcan, M.F., Weinberger, K.Q. (eds.) Proceedings of The 33rd International Conference on Machine Learning. Proceedings of Machine Learning Research, vol. 48, pp. 201–210. PMLR, New York (2016). https://proceedings.mlr.press/v48/gilad-bachrach16.html
13. Jacob, B., et al.: Quantization and training of neural networks for efficient integer-arithmetic-only inference. In: 2018 IEEE/CVF Conference on Computer Vision and Pattern Recognition, pp. 2704–2713 (2017)
14. Joye, M.: Guide to fully homomorphic encryption over the [discretized] torus. Cryptology ePrint Archive, Paper 2021/1402 (2021). https://eprint.iacr.org/2021/1402, https://eprint.iacr.org/2021/1402
15. Lee, E., et al.: Low-complexity deep convolutional neural networks on fully homomorphic encryption using multiplexed parallel convolutions. In: Chaudhuri, K., Jegelka, S., Song, L., Szepesvari, C., Niu, G., Sabato, S. (eds.) Proceedings of the 39th International Conference on Machine Learning. Proceedings of Machine Learning Research, vol. 162, pp. 12403–12422. PMLR (2022). https://proceedings.mlr.press/v162/lee22e.html
16. Lou, Q., Jiang, L.: SHE: a fast and accurate deep neural network for encrypted data. In: Wallach, H., Larochelle, H., Beygelzimer, A., d'Alché-Buc, F., Fox, E., Garnett, R. (eds.) Advances in Neural Information Processing Systems, vol. 32. Curran Associates, Inc. (2019). https://proceedings.neurips.cc/paper/2019/file/56a3107cad6611c8337ee36d178ca129-Paper.pdf
17. Meyre, A., et al.: Concrete-ML (2022). https://github.com/zama-ai/concrete-ml
18. Pappalardo, A.: Xilinx/brevitas (2021). https://doi.org/10.5281/zenodo.3333552
19. Sejnowski, T.J.: The unreasonable effectiveness of deep learning in artificial intelligence. Proc. Natl. Acad. Sci. **117**(48), 30033–30038 (2020). https://doi.org/10.1073/pnas.1907373117, https://www.pnas.org/doi/abs/10.1073/pnas.1907373117
20. Yvinec, E., Dapogny, A., Cord, M., Bailly, K.: SPIQ: data-free per-channel static input quantization. CoRR abs/2203.14642 (2022). https://doi.org/10.48550/arXiv.2203.14642

On the Existence of Highly Organized Communities in Networks of Locally Interacting Agents

Vasiliki Liagkou[1,4], Panagiotis E. Nastou[5,6], Paul Spirakis[2], and Yannis C. Stamatiou[1,3](✉)

[1] Computer Technology Institute and Press - "Diophantus", University of Patras Campus, 26504 Patras, Greece
liagkou@cti.gr

[2] Department of Computer Science, University of Liverpool, Liverpool, UK
P.Spirakis@liverpool.ac.uk

[3] Department of Business Administration, University of Patras, 26504 Patras, Greece
stamatiu@ceid.upatras.gr

[4] Department of Informatics and Telecommunications, University of Ioannina, 47100 Kostakioi Arta, Greece

[5] Department of Mathematics, Applied Mathematics and Mathematical Modeling Laboratory, University of the Aegean, Samos, Greece
pnastou@aegean.gr

[6] Center for Applied Optimization, University of Florida, Gainesville, USA

Abstract. In this paper we investigate phenomena of emergence of highly organized structures in networks. Our approach is based on Kolmogorov complexity of networks viewed as finite size strings. We apply this approach to study the emergence of simple organized, hierarchical, structures based on Sierpinski Graphs and we prove a Ramsey type theorem that bounds the number of vertices in Kolmogorov random graphs that contain Sierpinski Graphs as subgraphs. Moreover, we show that Sierpinski Graphs encompass close-knit relationships among their vertices that facilitate fast spread and learning of information. In the context of our work we can investigate questions related to the emergence or formation of suitably large groups of malicious individuals in a networks of interacting agents, which are able to take control of the networks.

Keywords: Sierpinski Graphs · Kolmogorov Complexity · Ramsey Theory

1 Introduction

In this paper, we address the problem of emergence of certain configurations as *highly organized structures* in evolving networks of agents (e.g. social networks) based on combining the concepts of *Ramsey Numbers in graphs* and *Kolmogorov Complexity*.

Broadly speaking, *Ramsey theory* refers to mathematical statements that a specific structure (e.g. string, graph, number sequence etc.) is certain to contain a large, highly organised, substructure. In this paper, our focus is on the Ramsey

S. Dolev et al. (Eds.): CSCML 2023, LNCS 13914, pp. 501–510, 2023.
https://doi.org/10.1007/978-3-031-34671-2_35

Theory of *graphs.* This research area was inaugurated in 1930 by Ramsey in [9] in which he stated and proved the, so called, *Ramsey's Theorem*: for *any* graph H, there exists a natural number n such that for *any* colouring of the edges of K_n (i.e. the clique on n vertices) with two colours, K_n contains a *monochromatic* copy of H as a subgraph, not necessarily induced. The *least* such value n is called the *Ramsey number* of H and is denoted by $r(H)$ (if $H = K_t$, we simply write $r(t)$). For our purposes, we consider the Ramsey numbers for the class of *Sierpinski Graphs.* This class of graphs has *bounded* maximum degree for its vertices. This fact leads to *linear*, in the number of vertices of H, upper bounds for the Ramsey numbers of these graphs. This follows from a more general result of Chvátal, Rödl, Szemerédi and Trotter in [3] which states that if H is a graph on n vertices and maximum degree Δ, then the Ramsey number $r(H)$ is bounded by $c(\Delta)n$ for some constant $c(\Delta)$ depending only on Δ. That is, the Ramsey number of bounded-degree graphs grows linearly in the number of vertices. Moreover, a linear bound also holds for the *induced* Ramsey numbers which are defined as the Ramsey numbers only, now, the monochromatic copy of H needs to form an *induced* subgraph, which is a stricter form of subgraph that allows the existence of edges in the subgraph H if and only if they are edges of H. Other edges, which may exist among vertices of H as edges of the larger graph should not exist, as it may be the case in the general subgraph notion. The *Induced Ramsey Theorem* states that $r_{\text{ind}}(H)$ exists for every graph H (see, e.g., Chap. 9.3 in [5]).

On the other hand, the *Kolmogorov complexity* of a *finite object* (most often a bit sequence modelling a finite structure such as a graph) is defined as the *minimum* number of bits into which the object can be compressed without losing information, i.e. so as the compressed string is recoverable through some algorithm, running on a reference machine, which is most often the universal Turing machine (see [7]).

Based on Ramsey theory and Kolmogorov Complexity arguments, we investigate, in this paper, phenomena of existence or formation of highly organized structures, such as organizations, people's networks, and societal patterns, in evolving networks of agents. As a specific case, we focus on the class of *Sierpinski Graphs* which have, also, the important property of being *close-knit*, i.e. agents interconnected with connections based on these graphs are very "cohesive". In [10], Young defined a parameter of networks of agents, modelled as *graphs*, that describes their "coherence" as well as their vulnerability to outsiders' views against the innovation to be diffused. A group of agents S (i.e. nodes in a given graph with no isolated vertices) is *close-knit* if the following condition is true for every $S' \subseteq S, S' \neq \emptyset$, given appropriate values for r (see discussion below):

$$\min_{S' \subseteq S} \frac{d(S', S)}{\sum_{i \in S'} d_i} \geq r \tag{1}$$

where $d(S', S)$ is the number of links between agents in S' and S while d_i is the total number of links that agent i possesses (i.e. its degree in the graph). Intuitively, to have a large such ratio in a group of agents we need, relatively, many internal connections and few external connections. Close-knit graph families play an important role in diffusion processes in graphs (see [10]) in contexts such as the emergence of shared and agreed upon innovations or ideas over large population of agents.

We believe that in the context of our work further studies can be initiated to investigate, without relying on random graphs or asymptotic results, when and how organized structures appear in finite size networks of interacting agents and answer questions such as the following: (i) given a network of interacting agents of a specific size, how large can an organized subnetwork can be with a common goal to overtake the whole network? (ii) can a subgroup of a specific organization structure appear in networks of a given size? and (iii) how large a network of interacting agent should be before it becomes vulnerable to the possibility of formation of large subnetworks of malicious agents (i.e. agents in which a common goal emerges due to the close-knittedness property of interactions)?

2 Sierpinski Triangle Based Group Formations

In this section we describe a class of graphs based on the *Sierpinski Triangle* fractal and prove that it forms a *close-knit graph family.*

Definition 1 (Sierpinski Triangle gasket of level l). *Given an integer l, $l \geq 1$, we define the Sierpinski Triangle gasket of level l or, simply, Sierpinski Triangle of level l as follows: for $l = 1$ the Sierpinski Triangle is an equilateral triangle while for $l > 1$, the Sierpinski Triangle of level l is composed of three copies of a Sierpinski Triangle of level $l - 1$ connected at their corners.*

Based on the Sierpinski Triangles, we can define a corresponding family of graphs, called *Sierpinski Graphs.*

Definition 2 (Sierpinski graphs). *The Sierpinski Graph of level l, $l \geq 1$, denoted by S_l is formed as follows: if $l = 1$ then the Sierpinski Graph of level 1 is formed if we replace the three vertices and the edges of the Sierpinski Triangle of level 1 with graph vertices and edges otherwise, for $l > 1$, the Sierpinski Graph of level 1 is formed by three copies of the Sierpinkski Graph of level $l - 1$ by identifying their vertices corresponding to the corners of the corresponding Sierpinski Triangle of level $l - 1$.*

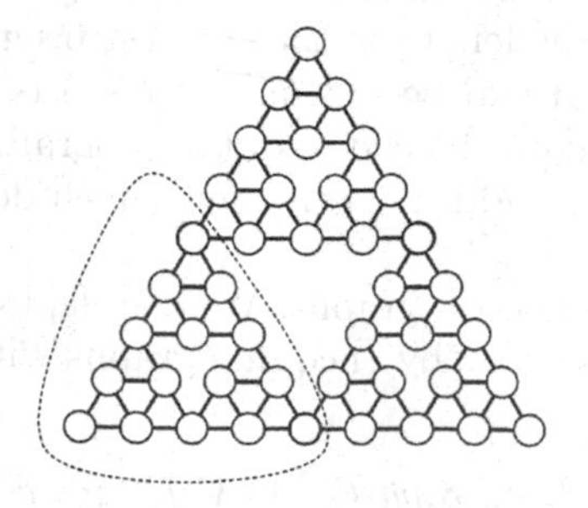

Fig. 1. The Sierpinski Graph of level 4

In particular, in Fig. 1 we see how the Sierpinski graph of level 4, S_4, is composed of three copies of the Sierpinski Graph of level 3, S_3, which is shown in a dashed enclosure. It is not hard to see that the following properties hold for S_l: (i) For every value of $l > 1$, all vertices of S_l have degree 4 except three vertices (the "corner" ones) which have degree 2. (ii) The number of vertices of S_l is $n_l = \frac{3}{2}(3^{l-1} + 1)$. (iii) The number of edges of S_l is $m_l = 3^l$.

The degree and connectivity properties helps satisfy Inequality (1). Intuitively, having too many nodes with large degrees, exposes a group of agents to much external interference. This affects negatively the close-knit property. We can prove the following (proof omitted due to lack of space):

Theorem 1. *The family of Sierpinski Graphs is close-knit.*

3 Existence of Highly Organized Structures in Networks of Interacting Agents

In this section we investigate conditions upon which highly organized structures can appear in networks of interacting agents. These structures may represent any organized community of individuals with a common aim or shared beliefs.

More specifically, we will study the problem of the existence of Sierpinski Graphs, as subgraphs, in sufficiently large graphs, which can model evolving interacting agents, social networks and societies. In this section we deploy techniques from Kolmogorov Complexity and Ramsey Theory. In what follows, we briefly state the main definitions and some useful results from *Kolmogorov Complexity* and *Ramsey Theory*. We, then, apply both theories in order to investigate conditions that enable or hinder the emergence of organized substructures in evolving structures which are modelled as graphs.

Kolmogorov Complexity. In informal terms (see [7]) the *Kolmogorov Complexity* of a (binary) string x is the *length* of the*shortest algorithmic description* of x. In other words, the *Kolmogorov Complexity*, denoted by $C(x)$ of a finite string x is the length of the *shortest program* (or Turing machine in general) encoding as a binary sequence of bits, which produced x as output, without taking any input. Similarly, the *conditional* Kolmogorov Complexity of x given y, denoted by $C(x|y)$, is the length of the shortest program which produces x as output given y as input. It can be shown that $C(x)$ is, in some sense, *universal* in that it does not depend on the choice of the programming language or Turing machine model, up to fixed additive constant, which depends on this choice but not on x.

In this paper, our focus is on graphs. We can deploy the notion and properties of Kolmogorov Complexity by encoding graphs with strings as follows (see, e.g., [7]):

Definition 3. *Each labelled graph $G = (V, E)$ on n nodes $V = \{1, 2, \ldots, n\}$ can be represented (up to automorphism) by a binary string $E(G)$ of length $\binom{n}{2}$.*

We simply assume a fixed ordering of the $\binom{n}{2}$ possible edges in an n-node graph, e.g. lexicographically, and let the ith bit in the string indicate presence (1) or absence (0) of the ith edge. Conversely, each binary string of length $\binom{n}{2}$ encodes an n-node graph. Hence we can identify each such graph with its binary string representation.

Definition 4. *A labelled graph G on n nodes has* randomness deficiency *at most $\delta(n)$, and is called $\delta(n)$-*random, *if it satisfies*

$$C(E(G)|n) \geq \binom{n}{2} - \delta(n). \tag{2}$$

Also, the following holds (see, e.g., [7]):

Lemma 1. *A fraction of at least $1 - 1/2^{\delta(n)}$ of all labelled graphs G on n nodes is $\delta(n)$-random. In particular, for $\delta(n) = \log n$, a fraction of $(1 - \frac{1}{n})$ of all graphs on n vertices is $\log n$-random.*

Definition 5. *Let $G = (V, E)$ be a labelled graph on n nodes. Consider a labelled graph H on k nodes $\{1, 2, \ldots, k\}$. Each subset of k nodes of G induces a subgraph G_k of G. The subgraph G_k is an ordered labelled* occurrence *of H when we obtain H by relabelling the nodes $i_1 < i_2 < \cdots < i_k$ of G_k as $1, 2, \ldots, k$.*

Ramsey Theory. As discussed in the introduction, Ramsey Theory is concerned with questions involving the appearance of certain patterns in sufficiently large graphs. For instance, Ramsey Theory started with the following question: given a graph H, determine the Ramsey number $r(H)$, which is defined as the *smallest* natural number n such that *any* two-colouring of the edges of K_n contains a monochromatic copy of H.

In this paper, we are interested in the *Induced Ramsey Number* of a given graph H, which is defined as follows:

Definition 6. *(Induced Ramsey numbers) A graph H is an induced subgraph of a graph H if $V(H) \subset V(G)$ and two vertices of H are adjacent if and only if they are adjacent in G. The induced Ramsey number $r_{ind}(H)$ is defined as the* minimum *integer for which there is a graph G on $r_{ind}(H)$ vertices such that every two-colouring of the edges of G contains an induced monochromatic copy of H.*

Note that an induced monochromatic copy of H is, also, an *induced* copy of H as an induced subgraph, in the ordinary graph-theoretical sense, regardless of the colour of the edges of G.

In [4] the following is proved:

Theorem 2. *There exists a constant c such that any graph H on k vertices with maximum degree Δ satisfies $r(H) \leq 2^{c\Delta \log \Delta} k$.*

There is a number of results that provide upper bounds on induced Ramsey numbers for sparse graphs. For instance, Beck in [1] focused on the case when H is a tree. Also, Haxell, Kohayakawa, and Łuczak [6] showed that the cycle of length k has induced Ramsey number linear in k. Moreover, Łuczak and Rödl [8] proved that the induced Ramsey number of a graph with bounded maximum degree is at most polynomial in the number of its vertices, settling a conjecture of Trotter. More precisely, they proved the following:

Theorem 3. *For every integer d, there is a constant c_d so that every graph H on k vertices and max-degree at most d satisfies $r_{ind}(H) \leq k^{c_d}$.*

The proof provides an upper bound on c_d that is a tower of 2's of height proportional to d^2. Since a Sierpinski Graph has maximum degree equal to 4, then an immediate corollary from Theorem 3, applied for $d = 4$, is the following:

Corollary 1. *Let $H = S_l$ be a labelled Sierpinski Graph of level l with n_l vertices. Then $r_{ind}(S_l) \leq n_l^{c_d}$, where c_d is a positive constant, independent from n_l and, thus, from l.*

Although random graphs (see e.g. [2]) are a powerful tool for proving limit properties its main limitation is that it does not say anything about *specific*, finite, graph instances, for a fixed size n. Kolmogorov complexity theory, on the other hand, focuses on the study of specific, finite, objects (graphs in our case). Based on this theory, we can prove the following (see [7] for a similar result on cliques on which we were based - the proof here is omitted due to lack of space):

Theorem 4. *Let S_l be a labelled Sierpinski Graph on n_l vertices. Let G be a labelled incompressible graph on n vertices that contains S_l as an induced subgraph. Then $n \geq 2^{\frac{n_l-1}{2}}$.*

Theorem 4 states that no incompressible graph with *fewer* than $2^{\frac{n_l-1}{2}}$ vertices can contain S_l as an *induced* subgraph. Consequently, an incompressible graph cannot contain S_l as a *monochromatic* induced subgraph either, in *any* two colouring of its edges. Thus, from Corollary 1 and Theorem 4, we have the following:

Theorem 5. *No incompressible graph on $r_{ind}(S_l)$ vertices can contain S_l as an induced subgraph except, possibly, for a finite set of values for l.*

Proof. Let G be an incompressible graph on n vertices that contains S_l as an induced subgraph. Since $r_{\text{ind}}(S_l) \leq n_l^{c_d}$ from Corollary 1, it follows that $n \leq n_l^{c_d}$. Also, the bound $n \geq 2^{\frac{n_l-1}{2}}$ holds from Theorem 4. Thus, it follows that $n \leq n_l^{c_d}$ can only hold for a finite set of values for n_l, whose cardinality depends on the constant c_d, since the growth rate of n is exponential in n_l while the growth rate of the bound $n_l^{c_d}$ for $r_{\text{ind}}(S_l)$ is polynomial in n_l. □

Moreover, based on Lemma 1, we obtain the following *stronger*, than Theorem 5, result:

Theorem 6. *Almost all graphs on $r_{ind}(S_l)$ vertices (a fraction of $(1 - \frac{1}{r_{ind}(S_l)})$ of them) are such that no colouring of their edges contains an induced monochromatic copy of S_l, as l grows.*

Proof. Following the same line of proof as in Theorem 4, we now start with a labelled graph G such that

$$C(E(G)|n, P) \geq \frac{n(n-1)}{2} - \log n. \tag{3}$$

These graphs form a fraction of at least $(1 - \frac{1}{n})$ of all labelled graphs on n vertices, according to Lemma 1. For $n = r_{\text{ind}}(S_l)$ these graphs form a ratio of at least $(1 - \frac{1}{r_{\text{ind}}(S_l)})$ of all graphs with $r_{\text{ind}}(S_l)$ vertices. What is stated below, applies to *all* of these graphs which, as l tends to infinity, including *almost all* possible graphs on $r_{\text{ind}}(S_l)$ vertices.

The rest of the proof follows closely the proof of Theorem 4, setting $n = r_{\text{ind}}(S_l)$, but now the following inequality must be satisfied:

$$n_l \log n + \log n \geq \frac{n_l(n_l-1)}{2} \Leftrightarrow n \geq 2^{\frac{n_l(n_l-1)}{2(n_l+1)}} \Leftrightarrow r_{\text{ind}}(S_l) \geq 2^{\frac{n_l(n_l-1)}{2(n_l+1)}}. \tag{4}$$

However, since $r_{\text{ind}}(S_l) \leq n_l^{c_d}$ for some constant depending only on k, according to Theorem 3, Inequality (4) would require $n_l^{c_d} \geq 2^{\frac{n_l(n_l-1)}{2(n_l+1)}}$, which cannot hold from some value of l onwards. □

Definition 7 (Size Constructible Graphs). *Let $\mathcal{F}$ be a family of graphs. We call $\mathcal{F}$ size constructible if each of the graphs in $\mathcal{F}$ can be uniquely constructed by an algorithm $P_{\mathcal{F}}$ which takes as inputs the graph's size, i.e. the number of vertices of the graph, and, possibly, a permutation that gives some ordering information about the vertices.*

Thus, Sierpinski Graphs are size constructible. A Sierpinski Graph S_l with n_l vertices can be described with only information n_l and the permutation that denotes the ordering of its vertices which the reconstruction algorithm P that we described in the proof of Theorem 4 uses in order to reconstruct the graph.

Then, Theorem 4 can be generalized as follows (the proof is omitted due to lack of space but follows a similar approach):

Theorem 7. *Let $\mathcal{F}$ be a family of size constructible graph. Let G be a labelled incompressible graph on n vertices that contains as an induced subgraph a graph in $\mathcal{F}$ with k vertices. Then $n \geq 2^{\frac{k-1}{2}}$ if $P_{\mathcal{F}}$ requires information about the ordering of the vertices and $n \geq k2^{\frac{k}{2}}\left(\frac{1}{e\sqrt{2}} - o(1)\right)$ if $P_{\mathcal{F}}$ does not require this information.*

Theorems 4 and 7 show that the existence of organized structures, such as the Sierpinski Graphs, in *incompressible* networks requires the networks to be *exponentially large* with respect to the size of the organized structure. Smaller interconnection networks almost certainly do not contain such organized structures.

4 Induced Ramsey Numbers for Incompressible Graphs

In this section we investigate Ramsey Numbers in the context of Kolmogorov random graphs. More specifically, we prove bounds on the size of Kolmogorov Random graphs so as to contain a Sierpinski Graph (or other size constructible graph) as a subgraph.

Definition 8. *The induced Ramsey number for incompressible graphs, $r_{ind}^{INC,\delta(n)}(H)$, is defined as the* minimum *integer for which there is a Kolmogorov Random graph G with randomness deficiency at least $\delta(n)$ on $r_{ind}^{INC,\delta(n)}(H)$ vertices such that every two-colouring of the edges of G contains an induced monochromatic copy of H.*

Theorem 8. *Let H be a size constructible graph on k vertices and of maximum degree at most d, whose description needs vertex ordering information (a similar result holds for size constructible graphs that do not require such information). Then for $\delta(n) \geq n_1 n_2 + \binom{n_2}{2}$, it holds that (i) $r_{ind}^{INC,\delta(n)}(H)$ exists, and (ii) $r_{ind}^{INC,\delta(n)}(H) < 2^{\frac{k-1}{2}} + r_{ind}(H)$.*

Proof. Let G_1 be any *incompressible* graph on $n_1 = 2^{\frac{k-1}{2}} - 1$ vertices, so that (according to Theorem 7) it does not contain H as an induced subgraph. Since G_1 is incompressible, it holds $C(E(G_1)) \geq \binom{n_1}{2}$.

Let, also, G_2 be any graph on $n_2 = r_{\text{ind}}(H) \leq k^{c_d}$ vertices such that for every colouring of its edges with two colours it contains an induced monochromatic copy of H (the existence of such a graph is guaranteed by Theorem 3). Let $n = n_1 + n_2$. We focus on the graphs G with n vertices which, simply, consist of on copy of G_1 and one copy of G_2, that is, two subgraphs isomorphic to G_1 and G_2. There are no additional edges except those in these two subgraphs. Obviously, every colouring of G contains an induced monochromatic copy of H, since G_2 does. This proves that $r_{\text{ind}}^{\text{INC},\delta(n)}(H)$ exists for Kolmogorov random, i.e. incompressible, graphs (at least for the deficiency function $\delta(n)$ which will be defined below).

Also, G has $n = n_1 + n_2 = 2^{\frac{k-1}{2}} - 1 + r_{\text{ind}}(H)$ vertices. We will show that for appropriate randomness deficiency function $\delta(n) = \delta(n_1, n_2)$ the following holds, i.e. G is $\delta(n_1, n_2)$-incompressible:

$$C(E(G)) \geq \binom{n_1 + n_2}{2} - \delta(n_1, n_2), \delta(n_1, n_2) \geq \binom{n_2}{2} + n_1 n_2. \tag{5}$$

Assume, towards a contradiction, that $C(E(G)) < \binom{n_1+n_2}{2} - \delta(n_1, n_2)$ for a randomness deficiency function $\delta(n_1, n_2)$ that obeys the inequality in (5). Then the following holds:

$$C(E(G)) < \binom{n_1 + n_2}{2} - \delta(n_1, n_2) \leq \binom{n_1 + n_2}{2} - \left[\binom{n_2}{2} + n_1 n_2\right] = \binom{n_1}{2}.$$

We will describe an algorithm that can reconstruct G_1 from a description of G of $C(E(G)) < \binom{n_1}{2}$ bits contradicting our assumption that $C(E(G_1)) \geq \binom{n_1}{2}$, i.e. that G_1 is incompressible.

Let us assume that we have a description of G of $C(E(G)) < \binom{n_1}{2}$ bits. Then we can reconstruct G_1 as follows. We first reconstruct G using its description of $C(E(G))$ bits. We, then, compute its *connected components* using, e.g., a *Depth First Search (DFS)* algorithm (sec [11]). Let these components be $G_{i_1}, G_{i_2}, \ldots, G_{i_s}$. One of these components must be the graph G_2. The union of the rest of the components should be the graph G_1. However, it is possible that only *one* component exists, beyond G_2, if G_1 is a *connected* graph. Our next step is to identify G_2. If we succeed in this task, then we have identified G_1: it is the graph that is composed of vertices and edges of G which do not belong in G_2.

In order to identify G_2 we work on the components $G_{i_1}, G_{i_2}, \ldots, G_{i_s}$, one at a time. We should be cautious here, since G_2 may not be a *connected* graph and, thus, we need to locate *all* its components. The crucial observation is that G_2 cannot contain a component such that no colouring of its edges contains an induced monochromatic copy of H. Otherwise, we could dispense with this component and have a smaller graph satisfy the definition of $r_{\text{ind}}(H)$, contradicting its *minimality* requirement.

Let G_{i_l} be the currently examined component. We produce all possible edge colourings of this component and for *each* of them we check whether G_2 contains a monochromatic copy of H. Observe that this can be true only for G_2 components since all other components belong to G_1, which is an incompressible graph with less than $2^{\frac{k-1}{2}}$ vertices and thus, according to Theorem 4, its edge colourings (and, consequently, its components' colourings) cannot contain a monochromatic copy of H. Thus, having identified G_2 we can reconstruct G_1 as the subgraph of G containing the rest of the components. In this way, we have managed to reconstruct G_1 using less than $\binom{n_1}{2}$ bits, which is a contradiction. Consequently, $C(E(G)) \geq \binom{n_1+n_2}{2} - \delta(n_1, n_2)$, that is G is $\delta(n_1, n_2)$-incompressible.

In conclusion, G is a graph of $n = n_1 + n_2$ which, for $\delta(n) \geq \delta(n_1, n_2)$, is $\delta(n)$-incompressible and all the colourings of its edges contain an induced monochromatic copy of H. Thus, it follows that $r_{\text{ind}}^{\text{INC},\delta(n)}(H) \leq n = 2^{\frac{k-1}{2}} - 1 + r_{\text{ind}}(H)$ or, in simpler form, $r_{\text{ind}}^{\text{INC},\delta(n)}(H) < 2^{\frac{k-1}{2}} + r_{\text{ind}}(H)$. □

5 Conclusions and Directions for Further Research

In this paper we considered two well established mathematical theoretical frameworks targeting the concept of complexity of finite objects, the *society and its interacting agents* in our case: *Kolmogorov Complexity* and *Ramsey Theory*. Both of these theories, each from another perspective, study the conditions upon which certain substructures appear (or do not appear) in large, evolving, network structures also deriving estimates on how large the structures should become in order to contain such regularities. In this context, questions related to the formation of suitably large groups of common goal, e.g. malicious, agents can be investigated related to network security. We hope that our work will contribute to the further exploitation of the rich mathematical theories of complex structures and their long-term evolutionary properties in the context of cybersecurity related properties of large networks.

References

1. Beck, J.: On size Ramsey number of paths, trees and circuits. II. In: Nešetřil, J., Rödl, V. (eds.) Mathematics of Ramsey Theory. Algorithms and Combinatorics, vol. 5, pp. 34–45. Springer, Heidelberg (1990). https://doi.org/10.1007/978-3-642-72905-8_4
2. Bollobás, B.: Random Graphs, 2nd edn. Cambridge University Press, Cambridge (2001)
3. Chvátal, V., Rödl, V., Szemerédi, E., Trotter, W.T., Jr.: The Ramsey number of a graph with bounded maximum degree. J. Combin. Theory Ser. B **34**, 239–243 (1983)
4. Conlon, D., Fox, J., Sudakov, B.: On two problems in graph Ramsey theory. Combinatorica **32**, 513–535 (2012)
5. Diestel, R.: Graph Theory, 2nd edn. Springer, Heidelberg (1997)
6. Haxell, P.E., Kohayakawa, Y., Łuczak, T.: The induced size-Ramsey number of cycles. Combin. Probab. Comput. **4**, 217–240 (1995)
7. Li, M., Vitányi, P.M.B.: An Introduction to Kolmogorov Complexity and its Applications, 4th edn. Springer, New York (2019). https://doi.org/10.1007/978-0-387-49820-1
8. Łuczak, T., Rödl, V.: On induced Ramsey numbers for graphs with bounded maximum degree. J. Combin. Theory Ser. B **66**, 324–333 (1996)
9. Ramsey, F.P.: On a problem of formal logic. Proc. London Math. Soc. **30**, 264–286 (1930)
10. Peyton Young, H.: The diffusion of innovations in social networks. In: Blume, L.E., Durlauf, S.N. (eds.) Proceedings of the Economy as a Complex Evolving System, vol. III. Oxford University Press (2003)
11. Sedgewick, R., Wayne, K.: Algorithms, 4th edn. Addison-Wesley Professional (2011)

Patch or Exploit? NVD Assisted Classification of Vulnerability-Related GitHub Pages

Lucas Miranda[1], Cainã Figueiredo[1], Daniel Sadoc Menasché[1(✉)], and Anton Kocheturov[2]

[1] Federal University of Rio de Janeiro (UFRJ), Rio de Janeiro, Brazil
cainafpereira@cos.ufrj.br, sadoc@dcc.ufrj.br
[2] Siemens Corporation, Princeton, USA
anton.kocheturov@siemens.com

Abstract. This paper presents a semi-automated approach to distinguish between patches and exploits published on GitHub, by using the National Vulnerability Database (NVD) as a reference. For this purpose, we leverage two interpretable algorithms, FP-Growth rule mining and decision trees, to extract patterns from the data and provide insights into the relationships between variables. To mitigate the risk of overfitting, we use more than 30,000 GitHub pages labeled by NVD as ground truth and focus on simple models. Among our findings, we discover that it is feasible to semi-automatically identify GitHub webpages containing patches and exploits. In particular, after pre-filtering webpages of interest, we discovered that most commits refer to patches, whereas URLs containing screenshots correspond to exploits showcasing how to exploit a given target system. Our results suggest that NVD is valuable to bootstrap machine learning algorithms for assisting in the analysis of increasingly larger amounts of cybersecurity data shared over the Internet.

Keywords: Cybersecurity · National Vulnerability Database · GitHub

1 Introduction

GitHub is widely considered the most significant platform for sharing source code on the Internet, counting with more than 200 million repositories as of 2023. Given the vast diversity of loosely structured repositories, their manual analysis is challenging, motivating semi-automated approaches to label webpages based on their content. In particular, for the cybersecurity community, it is key to rapidly characterize new repositories containing patches or exploits.

Over the past few years, a number of different websites and tools targeted the problem of organizing source codes of patches and exploits, including PatchDB [20], ExploitDB [19] and Metasploit [9]. However, given the speed at which new repositories are added to GitHub, alternative sources have not been covering all new additions to GitHub, complementing rather than replacing each other. In addition, the problem of determining the rules that distinguish

S. Dolev et al. (Eds.): CSCML 2023, LNCS 13914, pp. 511–522, 2023.
https://doi.org/10.1007/978-3-031-34671-2_36

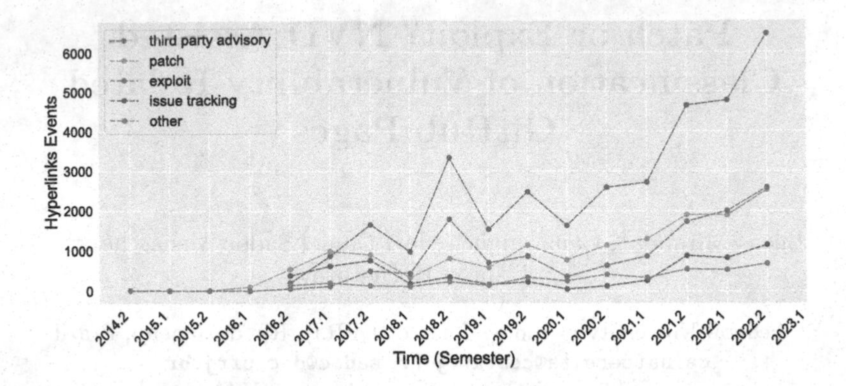

Fig. 1. Number of NVD tags in GitHub hyperlinks, over time (semesters)

patches from exploits, by itself, sheds insights into the cybersecurity ecosystem. In particular, patches and exploits are artifacts that play a key role in the cybersecurity landscape, given that these artifacts directly impact the risk of vulnerabilities. To further illustrate that, the CVSS temporal score of a vulnerability is affected by the existence and the maturity of patches and exploits – they are captured by remediation level and exploitability dimensions of CVSS, respectively. In addition to CVSS, other scoring systems (e.g., EPSS [8] and Expected Exploitability [18]) rely on these artifacts to assess the risk of vulnerabilities being exploited.

Motivated by the intuition that these systems could benefit from the automatic detection of patches and exploits, we pose the following question: how to distinguish between patches and exploits published on GitHub, in a semi-automated fashion? Semi-automated labeling is helpful to cope with intrinsic delays associated to manual labeling [17], to assist in the manual labeling process [2] and to provide additional artifacts for vulnerability risk assessment [8,12,18]. Our key insight consists in using the GitHub webpages cited by the National Vulnerability Database (NVD) as our reference. NVD curates the links associated with vulnerabilities by adding manual labels to each of its external resources. Despite the incompleteness of such database of curated links, it is a valuable resource to bootstrap machine learning algorithms. Figure 1 illustrates the evolution of the number of labeled GitHub hyperlinks cited by NVD, indicating that the number of labels set by the NVD team has been steadily increasing over the years.

Leveraging the labels of GitHub webpages cited by NVD, we consider two interpretable algorithms: FP-Growth rule mining and decision trees. Both algorithms extract interpretable patterns from data. The rules generated by FP-Growth are human-readable and can provide insights into the relationships between variables. Similarly, decision trees are intuitive and easy to interpret, as they represent the data as a tree structure with clear branching conditions, e.g., indicating that if a given webpage contains screenshots it is likely describing an exploit as opposed to a patch.

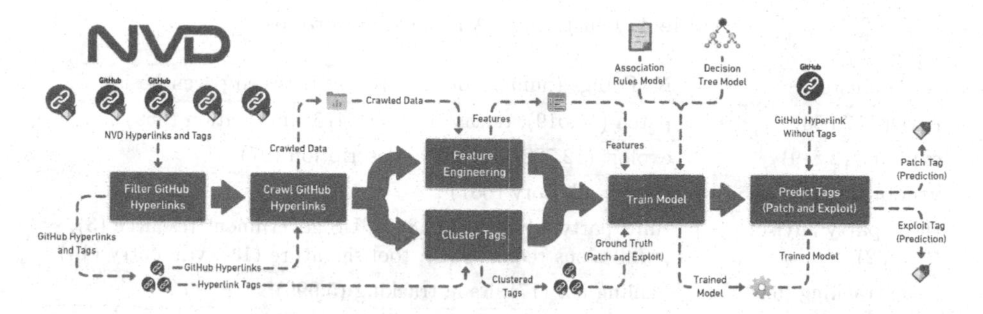

Fig. 2. Proposed framework for detecting patches and exploit on GitHub pages.

In terms of accuracy, the FP-Growth algorithm may not accurately predict the outcome of a target variable. In particular, the FP-Growth algorithm does not differentiate between features and target variables. Decision trees, in contrast, can be designed with the goal of predicting a given target variable, but may also suffer from overfitting. Indeed, trees with too many branches, or rules that are too specific to the training data, can lead to overfitting.

To avoid overfitting, in this paper we count with NVD tags to serve as ground truth. The large dataset of labeled NVD links helps to mitigate overfitting by providing more than 30,000 labeled examples for our models to learn from. However, even with a large labeled dataset, a complex model may still overfit the data. Therefore, in this paper we focus on simple models, namely simple rules derived from rule mining and decision trees with small number of branches, and leave the extension of those models as a subject for future work. In addition to these models, we also leverage ChatGPT to classify some GitHub pages and extract additional insights about the key features for detecting the existence of patches and exploits. Figure 2 summarizes our methodology.

Related Work. There is a vast literature on the analysis of GitHub data, for security purposes [3,4,15,16]. However, to the best of our knowledge, our work is the first to bridge NVD and GitHub as a way to bootstrap machine learning algorithms to analyze GitHub pages. Previous work [3,21] indicates that some users maliciously add malware to source code deployed at GitHub. In our work, we focus on malware that is explicitly published as such on GitHub, for the purposes of offensive security. Detecting covert malware is out of our scope.

Rokon et al. [15,16] proposed Repo2Vec and SourceFinder as tools to address problems that are related to the ones considered in this paper, namely extracting meaningful embeddings from GitHub repositories and finding malware at GitHub pages. However, they have not leveraged publicly known databases, such as NVD, for that purpose. As we indicate in this paper, NVD provides valuable labels to be coupled with non-supervised methods.

Outline. The upcoming sections of this paper are structured as follows. In Sect. 2 we briefly discuss our data gathering and processing pipeline, including a manual inspection of the ground truth provided by NVD. Then, Sect. 3 discusses the experiments and reports the results using rule mining and decision trees

Table 1. Clustering NVD tags into groups

NVD cluster	NVD tags (number of occurrences between parenthesis)
patch (17,853)	patch (15,019), release notes (2,475), mitigation (359)
exploit (12,819)	exploit (12,722), technical description (97)
vendor advisory (657)	vendor advisory (657)
third party advisory (33,922)	third party advisory (33,889), US government resource (3), permissions required (3), tool signature (13), vdb entry (14)
issue tracking (6,012)	mailing list (17), issue tracking (5,995)

to classify GitHub vulnerability-related pages. We also comment on the use of ChatGPT for the purpose of classifying GitHub pages in that section. Section 4 concludes.

2 Producing the Dataset

NVD is one of the most important vulnerability databases. Among the information provided for each CVE, NVD discloses references to webpages relating to CVEs along with tags. In order to construct our dataset, we filter the references and crawl each of the GitHub hyperlinks cited by NVD (see Fig. 2). We also extract their tags according to NVD. The crawling leads to the download of content from 33,702 webpages, including HTMLs and source codes. For each of the webpages, a total of 81 features were extracted. The features are divided into 5 categories, as further described below.

Keywords: There are 17 features that characterize the presence of keywords. The keywords are extracted from GitHub webpages, and were manually selected based on expert knowledge. It includes the keywords *poc*, *payload*, *inject* and *fix*.

URL Type: There are 6 URL types: *pull*, *release*, *issue*, *commit*, *wiki* and *other*. They are extracted directly from GitHub URLs. *Pulls*, for instance, correspond to pull requests, serving to request the merge of branches. Note that we classify a URL as *other* when its content does not fit into any of the other 5 classes. This is the case, for instance, when the URL contains a single image.

URL File Extension: There are a total of 51 possible URL extensions, directly identified from GitHub URLs, including "*.md*", "*.c*", "*.csv*" and "*.jpg*".

Screenshot: This feature indicates whether there is at least one image in the webpage. In most cases, images correspond to screenshots.

NVD Cluster: There are 5 variables in this category, and each of them indicates whether a GitHub page belongs to a given cluster or not. The clusters are defined from NVD tags as specified in Table 1. Note that some webpages may not belong to any cluster, whereas others may belong to multiple clusters. NVD clusters are the only class of variables that rely on information from outside of GitHub, namely from NVD.

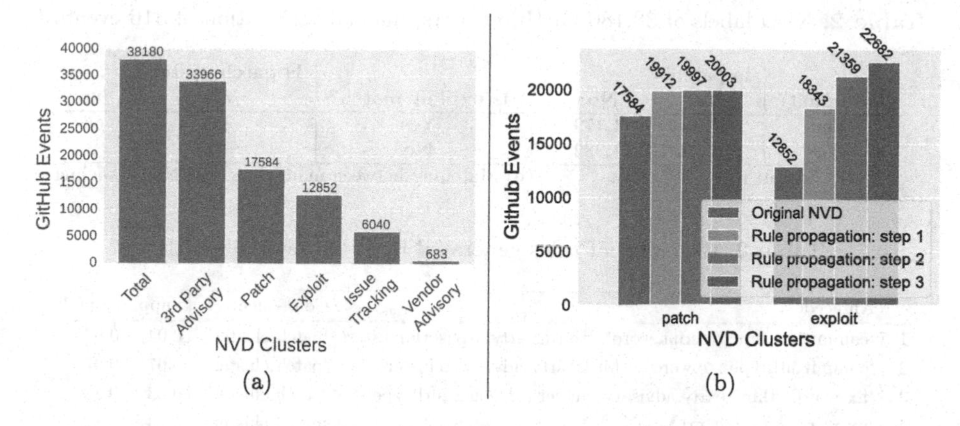

Fig. 3. (a) Initial dataset comprising 38,180 events (33,702 unique URLs) cited by NVD; (b) NVD patch and exploit tags after each rule propagation step.

From the standpoint of rule mining, all variables are treated similarly. From the standpoint of decision trees, the variables in the NVD cluster class are our **target variables**, noting that for decision trees we particularly focus on the distinction between "patch" and "exploit" (more details in Sect. 3.2).

2.1 Ground Truth Verification by Manual Labeling

In this section, our goal is to investigate to what extent the tags defined by NVD are suitable to be used as a ground truth to our model. More specifically, we 1) verify the accuracy of the labels set by NVD and 2) determine the incompleteness of the labels.

We refer to each NVD citation to a GitHub URL as an **event**. Figure 3(a) shows the distribution of tags for these events (see also Table 1), noting that each event typically receives multiple tags, as illustrated in Table 2(a) for the tags patch and exploit. There are 71,263 tags (sum of numbers between parenthesis in Table 1) towards 38,180 GitHub events (sum of numbers in Table 2(a)), that occurred between November 17, 2014 and January 18, 2023. Those events correspond to 33,702 unique GitHub URLs, i.e., 4,478 (38,180-33,702) citations occur to URLs that appear at least twice in the dataset. Among all events, 12,819 correspond to exploits[1] and 17,853 to patches.

In our investigation, we manually labeled a sample of 316 GitHub events. These manually labeled events are compared against NVD in order to evaluate the quality of its labels. Tables 2(a) and 2(b) report the distribution of labels in NVD and the distribution of labels from our manual inspection compared against NVD, respectively. The findings from our manual inspection follow below.

Finding 1 (NVD is accurate). *NVD's labeling of GitHub links as patch or exploit proves consistently accurate upon manual inspection.*

[1] To be contrasted against 5,323 exploits considered by Cyentia as of February 2023: https://www.cyentia.com/services/exploit-intelligence-service/.

Table 2. NVD labels of 38,180 GitHub events; manual verification of 316 events

	Is patch?	
Is exploit?	**Yes**	**No**
Yes	1,543	11,179
No	13,476	11,982

(a) Sample from NVD data

	Is patch match?	
Is exploit match?	**Yes**	**No**
Yes	233	66
No	15	2

(b) Matching between manual against NVD labeling

Table 3. Examples of rules generated by FP-Growth algorithm

	Antecedent	Consequent	Supp.	Confid.
1	'commit_url_type', 'script_word', 'third_party_advisory_cluster'	'patch_cluster'	0.04	0.97
2	'commit_url_type', 'c_word', 'third_party_advisory_cluster'	'patch_cluster'	0.07	0.97
3	'fix_word', 'third_party_advisory_cluster', 'releases_url_type'	'patch_cluster'	0.03	0.88
4	'fix_word', 'commit_url_type'	'patch_cluster'	0.14	0.87
5	'third_party_advisory_cluster', 'releases_url_type'	'patch_cluster'	0.05	0.87
6	'screenshot', 'payload_word'	'exploit_cluster'	0.03	0.95
7	'other_url_type', 'inject_word', 'third_party_advisory_cluster'	'exploit_cluster'	0.05	0.82
8	'md_url_file_ext', 'third_party_advisory_cluster'	'exploit_cluster'	0.07	0.82
9	'issues_url_type', 'third_party_advisory_cluster'	'exploit_cluster'	0.15	0.71
10	'c_word', 'issues_url_type'	'exploit_cluster'	0.05	0.71

When NVD labels a GitHub link as a patch or exploit, manual inspection indicates that the labeling of that link by NVD is typically correct and complete.

Out of the 316 manually labeled events, 214 were tagged either as patches or exploits by NVD, or both (this number is not shown in the table). Out of those, 208 fully match with our tagging (in agreement with Finding 1). In terms of accuracy, it indicates that NVD tags are a reliable source of information for the purposes of this paper and that we can use GitHub links labeled either as patch or exploit as our ground truth. In particular, we use such ground truth to compute the accuracy of the considered classification algorithms. The remaining 6 events were incompletely tagged by NVD, i.e., one of the tags was missing. This finding along with Finding 2 indicate that NVD is incomplete, further motivating our approach to detect patches and exploits in GitHub pages.

Finding 2 (NVD has missing data). *When NVD does not label a GitHub link, manual inspection indicates that this typically corresponds to an opportunity for semi-automated labeling.*

Out of the 316 manually labeled events, 102 were not tagged neither as patches nor as exploits by NVD (this number is not shown in the table). However, 63 were identified by our manual labeling as patches, and 14 as exploits (those mismatches correspond to most of the elements in the off-diagonal in Table 2(b)). The remainder $25(102 - (63 + 14))$ were not manually labeled neither as patches nor as exploits. This is in agreement with Finding 2, as 75.5% (77/102) of the events without tags in NVD received at least one tag in our manual labeling.

3 Experimental Results

3.1 Rule Mining with FP-Growth Algorithm

Next, we discuss the use of FP-Growth algorithm for rule mining. FP-Growth has been successfully compared against another prominent rule mining algorithms [6, 14]. We begin by considering the classical FP-Growth algorithm, and then extend it through an iterative process wherein additional rules are added every iteration, progressively increasing the confidence threshold required to add a new rule to the ruleset.

Classical FP-Growth Algorithm. Applying the FP-Growth algorithm [5] we filtered rules with a minimum support of 0.02, a minimum confidence of 0.7 and a minimum of 2 antecedents (definitions of these classical metrics can be found in [1]). Then, 1,069 rules were selected and applied. Ten of those rules are illustrated in Table 3. It is worth noting that in Table 3 all rules have either patch or exploit as consequent, but this is not necessarily the case for all the rules produced by the algorithm.

In Table 3, the first five rules serve to tag pages as patches. The first two rules, for instance, indicate that *commits* typically correspond to patches. The following two rules indicate that the presence of the word *fix* in the page also leads to patches. Finally, the third and fifth rules indicate that the same holds for *releases* previously classified by NVD as third party advisories. The last five rules serve to tag pages as exploits, e.g., when the keywords *payload* or *inject* appear in the page, as indicated by the sixth and seventh rules, respectively.

Figure 3(b) illustrates the behavior of FP-Growth algorithm. The blue bars correspond to the numbers shown in Fig. 3(a). The orange bars correspond to the results produced by the first application of the FP-Growth algorithm. It shows that, after the algorithm was applied, there was a significant increase in the number of elements tagged as exploits. In addition, we also observe that the number of patches and exploits becomes more balanced as compared to the statistics before the application of the algorithm.

Iterative FP-Growth. Next, we apply the FP-Growth algorithm iteratively for two additional iterations. We follow the same criteria as described in the beginning of Sect. 3.1 to select the rules, except for the confidence threshold. In the second and third executions of the FP-Growth algorithm, we set the confidence threshold at 0.8 and 0.9, respectively. The progressive threshold increase serves to avoid propagating errors between iterations.

Finding 3 (Iterative rule mining can fill missing data). *Iterative application of FP-Growth significantly increases the number of labeled pages.*

As shown in Fig. 3(b), after two additional executions of the FP-Growth algorithm, we were able to tag additional 4,339 GitHub events as exploits and 91 as patches. Note that there is a tradeoff in the choice of confidence thresholds: lower thresholds could lead to additional tags, at the cost of lower accuracy (not shown in the figure). Indeed, we experimentally found that the proposed

Table 4. Evaluation of predictions under FP-Growth algorithm and decision trees.

	FP-Growth algorithm				Decision tree			
	Accuracy	Precision	Recall	F1	Accuracy	Precision	Recall	F1
Patch	0.86	0.89	0.86	0.86	0.89	0.94	0.87	0.90
Exploit	0.74	0.79	0.74	0.73	0.91	0.87	0.95	0.91

increase of confidence threshold from 0.7 to 0.9 leads to an overall increase in accuracy. To discriminate pages between exploits and non-exploits, for instance, we observed that (precision, recall and F1-score) increased from (0.73, 0.71, 0.70) to (0.79, 0.74, 0.73) between the second and third applications of the FP-Growth algorithm (see Table 4).

FP-Growth Against Apriori. Interestingly, when we compared FP-Growth against Apriori for the task at hand, the learned rule set was exactly the same. However, FP-Growth took 20x less time to execute, accounting for all the iterations of the proposed iterative FP-Growth approach when compared against the corresponding iterative Apriori approach.

3.2 Decision Trees

Next, we consider the use of decision trees for classifying GitHub links (Fig. 2). The supervised nature of decision trees complements the rule mining methodology (Sect. 3.1), noting that both produce interpretable results.

Figure 4 shows the tree learned for predicting the presence of a patch in a GitHub link. Due to space restrictions, the corresponding tree to predict the presence of an exploit is reported in [13]. In those trees, each leaf that is colored in green corresponds to a node that is reached mostly by examples corresponding to patches, whereas red corresponds to exploits. Additionally, each leaf also reports the fraction of examples correctly classified at that node.

To generate the trees, we split the dataset into two disjoint sets, one for training (80%) and the other for testing (20%), and use the C4.5 algorithm for training. We performed a hyperparameter tuning by training multiple trees with the train set under 5-fold cross-validation. Then, the best hyperparameters are used to train a new model using the whole training set. The same approach is taken for patch and exploit prediction models. The evaluation of the final models on the test set is summarized in Table 4.

As with the rule mining approach, we extract a number of interesting insights. First, we observed that there is complementarity between the patch and exploit prediction trees (the patch tree is in Fig. 4 and the exploit tree is reported in [13]). For instance, the roots of the trees share a common condition, i.e., it checks if the page corresponds to a *commit*. However, as we move further down the nodes, variations between the trees begin to emerge. This suggests that many features that are key for classifying GitHub pages as patches are also relevant when classifying them as exploits. In addition, the leaves of the

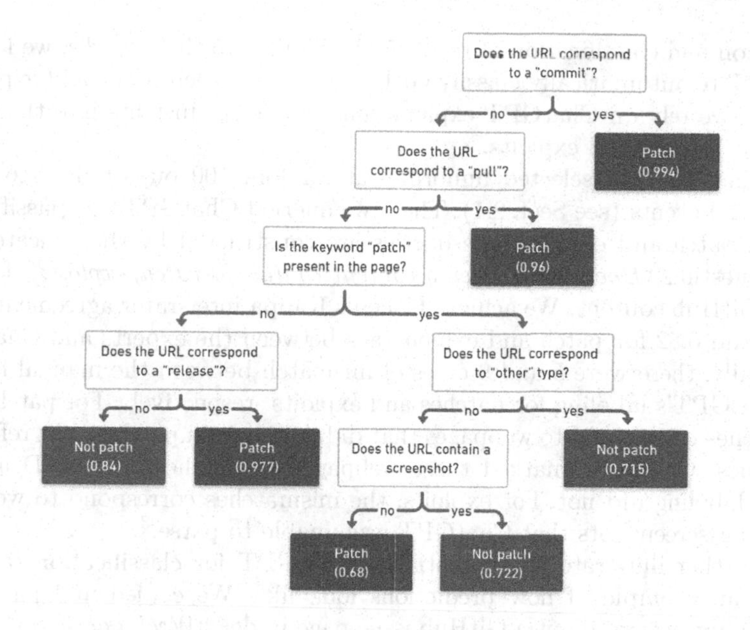

Fig. 4. Decision tree trained in the task of predicting if a link contains a patch.

trees indicate that pages publishing patches and exploits are typically different and non-overlapping. In fact, the statistics in Table 2(a) indicate that webpages simultaneously containing patches and exploits are not frequent.

Finding 4 (URL type, e.g., commit versus issue, is key). *URL type is the most important class of features to classify GitHub links as patches or exploits.*

Our findings show that URLs referring to GitHub commit, pull, or release events have a high likelihood of containing patches. Similarly, exploits are commonly published as GitHub issues (not shown in the figure).

Finding 5 (Screenshots suggest exploits). *The presence of certain keywords or screenshots proved to be effective to find exploits on GitHub.*

As previously described, we observed that many exploits published on GitHub relied on visual descriptions (screenshots) to describe attack steps. We also noted that a number of rules in the trees refer to the presence of keywords. Motivated by the intuition that the textual content of the pages plays an important role for predicting the nature of the links, in the following section we further investigate the feasibility of classifying links as patches or exploits using ChatGPT.

3.3 ChatGPT for GitHub Page Classification

ChatGPT is a large language model based on transformers. Among the extensive variety of tasks it can accomplish, it has been effectively used for the automatic annotation and classification of texts [7,11]. Similar to these works, we leverage ChatGPT to automatically classify GitHub pages between patch and exploit. In addition, we rely on ChatGPT explanations to extract insights into the task of detecting patches and exploits.

To that aim, we selected uniformly at random 100 out of the 316 manually labeled events (see Sect. 2.1). Then, we queried ChatGPT[2] to classify them between patch and exploit. The queries are constructed by the concatenation of the question "*Does the following content contain a patch/exploit?*" followed by the GitHub content. We achieve Cohen's Kappa inter-rater agreement scores of 0.90 and 0.82 for patch and exploit tags between the experts and ChatGPT. Specifically, there were 5 and 9 cases of mismatch between the manual labeling and ChatGPT's labeling for patches and exploits, respectively. For patches, the mismatches correspond to webpages that did not contain patches, but references to patches. ChatGPT marked those webpages as patches, but NVD and our manual labeling did not. For exploits, the mismatches correspond to webpages containing screenshots that ChatGPT was unable to parse.[3].

To further illustrate the potential of ChatGPT for classification tasks, we provide an example of how predictions look like. We copied and pasted the textual information about a GitHub issue page under "*Weak encryption (MD5) leads to backstage violent cracking #188*"[4]. Then, we queried ChatGPT whether it contains an exploit and its answer was: *"Yes, the content contains a proof of concept (POC) exploit (...) The POC shows how an attacker can grab data packets to get fields, such as the MD5 hash of the 'admin' account's password, and then load a dictionary for violent cracking to gain access to the website's backstage."* In addition, it did not find any evidence of patches in that page.

As another example, we queried ChatGPT about a webpage containing a commit.[5] Its answer was: *"Yes, the content contains a patch. The commit message mentions two modifications made to the zblogphp project, specifically adding the hash_equals function and making corrections to the CloseTags function."*

Insights. We noticed that ChatGPT typically focuses on code snippets and steps describing how to reproduce a vulnerability in order to find evidence of exploits. To illustrate that, ChatGPT automatically detected that the code "<img src=1 onerror=alert(1)>" represents a PoC to an XSS vulnerability.

More broadly, we noticed that ChatGPT leverages the context around the source code to make its judgement about the nature of the webpages. ChatGPT can leverage interactions between users and other events related to user activity to find evidence about whether a page is a patch or exploit. Weak evidence of a patch includes events closing issues and contributor assignments to issues, and strong evidence includes version update instructions and explicit mentions of fixes in users comments or in the descriptions of the commits.

[2] ChatGPT version released on February 13, 2023.

[3] All datasets and material to reproduce our results is made available at https://tinyurl.com/githublabelingexpl.

[4] https://github.com/zblogcn/zblogphp/issues/188.

[5] https://github.com/zblogcn/zblogphp/commit/a67607fc984f976d6b36b8870dffaabd9d6c9d5e

.

Remarks on the Use of ChatGPT. ChatGPT is instrumental to detect patches and exploits. However, further work is needed to choose the best ChatGPT prompt. Recently, prompt engineering has been advocated as a way to improve performance in classification tasks, and we leave its analysis in our domain of study as a subject for future work.

4 Conclusion

We presented a semi-automated approach to distinguish between patches and exploits on GitHub pages cited by NVD. Our results show that simple algorithms such as FP-Growth rule mining and decision trees are effective in extracting interpretable patterns and relationships from the data, e.g., commits typically refer to patches and URLs containing screenshots correspond to exploits.

Our work paves the way towards the hybrid use of supervised and non-supervised methods for mining cybersecurity data available at GitHub. In particular, we focus on the usefulness of NVD as a tool to support machine learning algorithms for analyzing the growing amount of cybersecurity data shared on the Internet, and we envision a number of different directions for future work, including a broader analysis of GitHub pages beyond those cited by NVD and the automatic classification of the maturity level of exploits. We also envision the application of alternative rule mining techniques, to increase scalability or to leverage temporal patterns in our data [10].

References

1. Agarwal, R., Srikant, R., et al.: Fast algorithms for mining association rules. In: Proceedings of the 20th VLDB Conference, vol. 487, p. 499 (1994)
2. Anwar, A., Chen, S., et al.: Cleaning the NVD: comprehensive quality assessment, improvements, and analyses. IEEE TDSC **19**(6), 4255–4269 (2021)
3. Cao, A., Dolan-Gavitt, B.: What the fork? Finding and analyzing malware in GitHub forks. In: Proceedings of the NDSS, vol. 22 (2022)
4. Di Rocco, J., Di Ruscio, D., Di Sipio, C., et al.: Hybridrec: a recommender system for tagging GitHub repositories. Appl. Intell. 1–23 (2022)
5. Han, J., Pei, J., Yin, Y.: Mining frequent patterns without candidate generation. ACM SIGMOD Rec. **29**(2), 1–12 (2000)
6. Heaton, J.: Comparing dataset characteristics that favor the Apriori, Eclat or FP-Growth frequent itemset mining algorithms. In: SoutheastCon 2016, pp. 1–7 (2016)
7. Huang, F., Kwak, H., An, J.: Is ChatGPT better than human annotators? Potential and limitations of ChatGPT in explaining implicit hate speech. arXiv:2302.07736 (2023)
8. Jacobs, J., Romanosky, S., Edwards, B., Adjerid, I., Roytman, M.: Exploit prediction scoring system. Digit. Threats Res. Pract. **2**(3), 1–17 (2021)
9. Kennedy, D., O'gorman, J., Kearns, D., Aharoni, M.: Metasploit: the penetration tester's guide. No Starch Press (2011)
10. Kocheturov, A., Momcilovic, P., Bihorac, A., Pardalos, P.M.: Extended vertical lists for temporal pattern mining from multivariate time series. Expert Syst. **36**(5), e12448 (2019)

11. Kuzman, T., Ljubešić, N., Mozetič, I.: Chatgpt: beginning of an end of manual annotation? Use case of automatic genre identification. arXiv:2303.03953 (2023)
12. Miranda, L., et al.: On the flow of software security advisories. IEEE Trans. Netw. Serv. Manag. **18**(2), 1305–1320 (2021)
13. Miranda, L., et al.: Patch or exploit? NVD assisted classification of vulnerability-related GitHub Pages (2023). https://tinyurl.com/githublabelingexpl
14. Mythili, M., Shanavas, A.M.: Performance evaluation of Apriori and FP-Growth algorithms. Int. J. Comput. Appl. **79**(10) (2013)
15. Rokon, M.O.F., Islam, R., et al.: SourceFinder: finding malware source-code from publicly available repositories in GitHub. In: RAID, pp. 149–163 (2020)
16. Rokon, M.O.F., Yan, P., et al.: Repo2vec: a comprehensive embedding approach for determining repository similarity. In: ICSME, pp. 355–365. IEEE (2021)
17. Ruohonen, J.: A look at the time delays in CVSS vulnerability scoring. Appl. Comput. Inform. **15**(2), 129–135 (2019)
18. Suciu, O., Nelson, C., et al.: Expected exploitability: predicting the development of functional vulnerability exploits. In: USENIX Security, pp. 377–394 (2022)
19. Sun, J., Xing, Z., et al.: Generating informative CVE description from ExploitDB posts by extractive summarization. arXiv preprint arXiv:2101.01431 (2021)
20. Wang, X., Wang, S., Feng, P., Sun, K., Jajodia, S.: Patchdb: a large-scale security patch dataset. In: 2021 51st Annual IEEE/IFIP DSN, pp. 149–160. IEEE (2021)
21. Yadmani, S.E., The, R., Gadyatskaya, O.: How security professionals are being attacked: a study of malicious CVE proof of concept exploits in GitHub. arXiv preprint arXiv:2210.08374 (2022)

Author Index

A
Afek, Yehuda 1
Aharoni, Ehud 96
Arnon, Shlomi 458

B
Bhasin, Shivam 296
Blumberg, Dan G. 85
Boudguiga, Aymen 117, 372
Bredehoft, Roman 493
Breier, Jakub 296
Bremler-Barr, Anat 1

C
Cakir, Ceyhun 170
Carminati, Michele 135, 234, 337
Chang, Donghoon 471
Chevallier-Mames, Benoit 493
Choukroun, Daniel 85
Clet, Pierre-Emmanuel 117
Cohen, Asaf 74, 200
Cyprys, Paweł 74

D
D'Onghia, Mario 234
Ding, Jintai 273
Dolev, Shlomi 74, 389
Dror, Itai 85
Drucker, Nir 65, 96

F
Figueiredo, Cainã 511
Frery, Jordan 493

G
Gudes, Ehud 411
Gueron, Shay 424
Güner, Hatice Kübra 398
Gurewitz, Omer 200

H
Hadar, Ofer 85
Hendler, Danny 411
Hou, Xiaolu 296

I
Iancu, Tiberiu 170
Israeli, Dor 1

J
Jap, Dirmanto 296
Jena, Riyanka 188
Jurcut, Anca Delia 285

K
Kherfallah, Celia 493
Kocheturov, Anton 511
Kolesnikov, Vladimir 439
Komargodski, Ilan 49
Korzilius, Stan 307
Kumar, Rajnish 458
Kushnir, Eyal 96

L
Last, Mark 151
Levy, Silvie 411
Liagkou, Vasiliki 501
Longari, Stefano 135, 337

M
Maman, Shimrit 85
Mangır, Ceyda 398
Masalha, Ramy 96
Meinke, Klaas 170
Menasché, Daniel Sadoc 511
Mendelson, Avi 216
Miranda, Lucas 511
Mishra, Divya 85
Mohanty, Manoranjan 188
Montero, Luis 493

S. Dolev et al. (Eds.): CSCML 2023, LNCS 13914, pp. 523–524, 2023.
https://doi.org/10.1007/978-3-031-34671-2

Mukhopadhyay, Debdeep 320
Mukhopadhyay, Madhurima 18

N
Nair, Madhav 320
Nastou, Panagiotis E. 501
Nichelini, Alessandro 135
Noseda, Francesco 337
Novick, Deborah 151
Noy, Alon 1

O
Onwuegbuche, Faithful Chiagoziem 285

P
Paladini, Tommaso 234
Pasquale, Liliana 285
Peceny, Stanislav 439
Pozzoli, Carlo Alberto 135

S
Sadhukhan, Rajat 320
Samajder, Subhabrata 181
Sarkar, Palash 18, 181
Sarmah, Dipti Kapoor 251
Schoenmakers, Berry 32, 307
Segers, Toon 32
Segev, Ron 216
Shaul, Hayim 96
Singh, Priyanka 188
Sirdey, Renaud 117
Spirakis, Paul 501
Stamatiou, Yannis C. 501
Stoian, Andrei 493

T
Tamir, Yoav 49
Tan, Jiayi 251
Tatulli, Maria Paola 234
Trama, Daphné 117
Triandopoulos, Nikos 372
Trieu, Ni 439
Trivedi, Devharsh 372

V
van der Laan, Tamis Achilles 170
Vershinin, George 200

W
Wang, Xiao 439

Y
Yagudaev, Avraam 389
Yayla, Oğuz 398
Yung, Moti 389, 471

Z
Zanero, Stefano 135, 234, 337
Zhao, Ziyu 273
Zimerman, Itamar 65
Zirdelis, Giorgos 353